Flexible Automation
and
Information Management
1992

Flexible Automation
and
Information Management
1992

Proceedings of the Second International FAIM Conference
Falls Church, Virginia, U.S.A.
June 30 - July 2, 1992

Edited by

Osama K. Eyada, Ph.D., P.E.
Department of Industrial and Systems Engineering
Virginia Polytechnic Institute and State University
Blacksburg, Virginia

M. Munir Ahmad, Ph.D., CEng, EurIng, P.E.
Department of Mechanical and Production Engineering
University of Limerick
Limerick, Ireland

CRC Press
Boca Raton   Ann Arbor   London   Tokyo

Library of Congress Cataloging-in-Publication Data

Catalog record is available from the Library of Congress.

Direct all inquiries to CRC Press, Inc., 2000 Corporate Blvd., N.W., Boca Raton, Florida, 33431.

© 1992 by CRC Press, Inc.

International Standard Book Number 0-8493-0138-6

Printed in the United States of America   1 2 3 4 5 6 7 8 9 0

Printed on acid-free paper

PREFACE

This book features the proceedings of the Second International Flexible Automation and Information Management Conference (FAIM '92). FAIM is an annual, international conference jointly organized by Virginia Polytechnic Institute and State University (Virginia Tech) and the University of Limerick. The first conference was hosted in 1991 by the University of Limerick in Limerick, Ireland. FAIM '92 is being hosted by Virginia Tech and is endorsed by more than 20 societies, associations, university affiliates, and international organizations.

The last decade has been characterized by global competitiveness and rapidly advancing technology in flexible automation and information management. FAIM conferences provide a forum to address problems relevant to these areas. FAIM '92 further focuses on state-of-the-art and future trends within the general area of flexible automation and information management. FAIM differs from many other conferences in that it presents an integrated picture of these related technologies rather than focusing on a single technology.

Authors from 20 countries have contributed more than 80 papers, and each paper has been thoroughly refereed by at least two reviewers. The papers are organized into six tracks which spanned the three day conference:

1. Managerial Aspects of World Class Manufacturers,
2. Concurrent Engineering Techniques,
3. Computer Integrated Manufacturing,
4. CAD/CAM Databases and Applications,
5. Flexible Manufacturing Systems, and
6. Increasing Competitiveness Through Technology.

FAIM aims at providing high quality refereed papers for upper and middle level managers, industrial and manufacturing engineers, and practitioners/researchers in the field of computer assisted manufacturing.

Osama Eyada and Munir Ahmad

June 1992

CONFERENCE ORGANIZATION

Conference Co-sponsors

Society of Manufacturing Engineers
Institute of Industrial Engineers
Computer Aided Manufacturing - International
The Association for Manufacturing Technology
Institute of Electrical and Electronics Engineers - IAS
National Center for Manufacturing Sciences
Virginia's Center for Innovative Technology
Boart Hardmetals (Europe) Ltd.
Institute of Engineers of Ireland
EOLAS - The Irish Science and Technology Agency

University Affiliates

Lehigh University - The Department of Industrial Engineering and
      The Center for Manufacturing Systems Engineering
Ohio State University
Oregon State University
Texas A&M University
Trinity College - Dublin
University of Texas-Arlington - Automation and Robotics Research Institute
University of South Florida - The Department of Industrial Engineering and
      The Center of Computer Integrated Engineering & Manufacturing

Conference Chairpersons

Professor William G. Sullivan
Virginia Polytechnic Institute and State University, Blacksburg, VA, USA

Professor M. Munir Ahmad
University of Limerick, Limerick, Ireland

Organizing Committee

Virginia Polytechnic Institute and State University

Dr. Michael P. Deisenroth
Dr. Robert D. Dryden
Dr. Osama K. Eyada

Dr.   Wolter J. Fabrycky
Dr.   Roderick J. Reasor
Dr.   C. Patrick Kolleing
Dr.   Subhash C. Sarin
Mr.   Ben E. Selvey

University of Limerick

Mr.   Leo Coglin
Dr.   Huw J. Lewis
Ms.   Gay P. Moynihan
Dr.   Seamus O'Shea
Mr.   Donal Walshe
Dr.   H. Kenneth Wylie

Conference Secretary

Ms. Dot Cupp
Virginia Polytechnic Institute and State University

FAIM '92 Reviewers

Osama M. Ettouney
Miami University

Osama K. Eyada
Virginia Tech

Wolter J. Fabrycky
Virginia Tech

Jeffrey E. Fernandez
The Wichita State University

Donald R. Flugard
Iowa State University

Keith M. Gardiner
Lehigh University

Brian J. Gilmore
Penn State

B. Gopalakrishnan
West Virginia University

Timothy J. Greene
Oklahoma State University

Mikell P. Groover
Lehigh University

Inyong Ham
Penn State

Andrew K. Hansborough
Virginia Tech

Robert A. Hanson
Virginia Tech

Yasser A. Hosni
University of Central Florida

Faizul Huq
The University of Texas - Arlington

Sankar Jayaram
Virginia Tech

Albert Jones
*National Institute of Standards
and Technology*

S. Joshi
Penn State

Krishna K. Krishnan
Virginia Tech

Way Kuo
Iowa State University

Amine Lehtihet
Penn State

Steven Y. Liang
Georgia Tech

T. Warren Liao
Louisiana State University

Richard R. Lindeke
University of Minnesota, Duluth

Ming C. Liu
The Wichita State University

James T. Luxhoj
Rutgers University

Elin L. MacStravic
Virginia Tech

Don E. Malzahn
The Wichita State University

S. Manivannan
Georgia Tech

Abu S.M. Masud
The Wichita State University

Thomas McLean
The University of Texas - El Paso

O.O. Mejabi
Wayne State University

Lamine M. Mili
Virginia Tech

Hal L. Moses
Virginia Tech

Saied Motavalli
The Wichita State University

Taysir Nayfeh
Virginia Tech

Bartholomew O. Nnaji
University of Massachusetts - Amherst

J. Pablo Nuno
Arizona State University

Nicholas G. Odrey
Lehigh University

Jin B. Ong
Virginia Tech

Hamid R. Parsaei
University of Louisville

Luis C. Rabelo
Ohio University - Athens

William Ranson
University of South Carolina

Roderick J. Reasor
Virginia Tech

Fred J. Robinson
University of Minnesota, Duluth

Kevin Rong
Southern Illinois University - Carbondale

Paul Rossler
Virginia Tech

Jai Saboo
Virginia Tech

Subhash C. Sarin
Virginia Tech

Joseph Sarkis
The University of Texas - Arlington

J. William Schmidt
Virginia Tech

Greg Shekita
IBM Entry Systems Division, NC

Gene R. Simons
Rensselaer Polytechnic Institute

William G. Sullivan
Virginia Tech

Robert T. Sumichrast
Virginia Tech

Jay M. Teets
Virginia Tech

Janis P. Terpenny
Virginia Tech

Jeffrey D. Tew
Virginia Tech

Paul E. Torgersen
Virginia Tech

Phillip Wadsworth
Edison Welding Institute

George R. Wilson
Lehigh University

Jeffrey C. Woldstad
Virginia Tech

Yaoen Zhang
Virginia Tech

Chen Zhou
Georgia Tech

CONTRIBUTORS

M. Munir Ahmad
Department of Mechanical and
Production Engineering
University of Limerick
Limerick, Ireland

Melvin T. Alexander
Westinghouse
P.O. Box 1693, MS 1297
Baltimore, Maryland 21203, U.S.A.

Leo Alting
Institute of Manufacturing Engineering
Tech. University of Denmark
DK-2800 Lyngby, Denmark

Douglas N. Anderson
3M Center/Building 525-5E-02
3M Corporate Quality Services
St. Paul, Minnesota 55144, U.S.A.

Antonio Armillotta
Istituto di Tecnologie Industriali e
Automazione
CNR
Via Ampere, 56
I-20131 Milano, Italy

Girish Bakshi
Industrial and Systems Engineering
Department
Ohio University
Athens, Ohio 45701, U.S.A

Heinz W. Bargmann
Swiss Federal Institute of Technology
CH-1015 Lausanne, Switzerland

Arne Bilberg
Institute of Manufacturing Engineering
Tech. University of Denmark
DK-2800 Lyngby, Denmark

Samir B. Billatos
Department of Mechanical Engineering
University of Connecticut
191 Auditorium Road
Storrs, Connecticut 06269-3139, U.S.A.

Benjamin S. Blanchard
333C Norris Hall
College of Engineering
Virginia Polytechnic Institute and State
University
Blacksburg, Virginia 24061, U.S.A.

Thomas A. Bock
Institut fur Maschinenwesen im
Baubetrieb
Universitat Karlsruhe
P.O. Box 6980
W-7500 Karlsruhe 1, Germany

Marcos G.D. Bortoli
Embraco, S.A.
Joinville, S.C., Brazil 89200

Carlos F. Bremer
LAMAFE-EESC-USP
P.O. Box 359-13560
Sao Carlos, Brazil

John K. Bryant
Dunlop Cox, Ltd.
Glarsdale Parkway
Nottingham NG8 4GP, UK

Christian P. Burger
Institute for Innovation and Design in
Engineering
Texas A&M University
College Station, Texas 77843, U.S.A.

Patrice Chollet
Laboratoire de Mecatronique
Institut Superieur des Materiaux et de la
Construction Mecanique
Saint Ouen, 93407, France

Andre Clement
Laboratoire de Mecatronique
Institut Superieur des Materiaux et de la
Construction Mecanique
Saint Ouen, 93407, France

Paul H. Cohen
Industrial and Management Systems
Engineering
207 Hammond Building
Pennsylvania State University
University Park, Pennsylvania 16802,
U.S.A.

Noel Conaty
Apple Computer
Holly Hill
Cork, Ireland

Sridhar S. Condoor
Institute for Innovation and Design in
Engineering
Texas A&M University
College Station, Texas 77843, U.S.A.

Paul C. Conway
Department of Manufacturing
Engineering
Loughborough University
Loughborough LE11 3TU, UK

Edison Da Rosa
Mechanical Engineering Department
Universidade Federal
De Santa Catarina 88049, S.C., Brazil

Neil H. Darracott
Department of Manufacturing
Engineering
University of Nottingham
Nottingham NG7 2RD, UK

Kenneth Dawson
Department of Computer Science
Trinity College
Dublin 2, Ireland

Michael P. Deisenroth
Industrial and Systems Engineering
Department
Virginia Polytechnic Institute and State
University
Blacksburg, Virginia 24061, U.S.A.

Edward C. De Meter
Industrial and Management Systems
Engineering
151-A Hammond Building
Pennsylvania State University
State College, Pennsylvania 16803,
U.S.A.

Paolo Denti
Istituto di Tecnologie Industriali e
Atomazione
CNR
Via Ampere 56
I-20131 Milano, Italy

Bernie Dillon
Amdahl Ireland Limited
Balheary, Swords, Co.
Dublin, Ireland

Jian Dong
Department of Industrial Engineering
University of Louisville
Louisville, Kentucky 40292, U.S.A.

Elizabeth A. Downey
Downey & Small Associates, Inc.
10103 East Bexhill Drive
Kensington, Maryland 20895, U.S.A.

Alfredo Drago
Institute of Technology and Mechanical
Plants
University of Genoa
Via Allopera Pia 15
I-16145 Genoa, Italy

John Driscoll
Department of Mechanical Engineering
University of Surrey
Guildford, Surrey GU2 5XH, England

Jacques-Olivier Dufraigne
Laboratoire de Mecatronique
Institut Superieur des Materiaux et de la
Construction Mecanique
F-93407 Saint Ouen, France

Dieter Eden
Deutsche Aerospace, MBB
Unternehmensbereich Flugzeuge
Postfach 1149
8072 Manching, Germany

Osama K. Eyada
Department of Industrial and Systems
Engineering
Virginia Polytechnic Institute and State
University
Blacksburg, Virginia 24061, U.S.A.

Jeffrey E. Fernandez
Wichita State University
Box 35
Wichita, Kansas 67208, U.S.A.

Labiche Ferreira
IEMS Department
University of Central Florida
P.O. Box 25000
Orlando, Florida 32816, U.S.A.

Mark A. Fugelso
Department of Industrial Engineering
University of Minnesota
Duluth, Minnesota 55812, U.S.A.

Gerard J.C. Gaalman
Faculty of Economics, Business
Administration
and Management Science
University of Groningen
P.O. Box 800
NL-9700 AV Groningen,
The Netherlands

Keith M. Gardiner
200 Mohler Laboratory
Lehigh University
Bethlehem, Pennsylvania 18015, U.S.A.

Friedrich S. Gebhart
Institut fur Maschinenwesen im
Baubetrieb
Universitat Karlsruhe
P.O. Box 6980
W-7500 Karlsruhe 1, Germany

Paul J. Gilders
Department of Manufacturing
Engineering
Loughborough University of Technology
Loughborough, LE11 3TU, England

Pietro Giribone
Institute of Technology and Mechanical
Plants
University of Genoa
Via Allopera Pia 15
I-16145 Genoa, Italy

Vivek Goel
Louisiana State University
Baton Rouge, Louisiana 70803, U.S.A.

B. Gopalakrishnan
Department of Industrial Engineering
West Virginia University
P.O. Box 6101
Morgantown, West Virginia, 26506,
U.S.A.

Thomas D. Halpin
Computer Integrated Manufacturing and
Engineering Research Centre
University of Limerick
Limerick, Ireland

Robert A. Hanson
Industrial and Systems Engineering
Department
Virginia Polytechnic Institute and State
University
Blacksburg, Virginia 24061, U.S.A.

Howard H. Harary
Building 220, Room A107
National Institute of Standards and
Technology
Gaithersburg, Maryland 20899, U.S.A.

Alexander Hars
University of Saarbrucken
Im Stadtwald, Geb. 14.1
W-6600 Saarbrucken 11, Germany

Paul Healy
Department of Computer Science
Trinity College
Dublin 2, Ireland

Ralf Heib
University of Saarbrucken
Im Stadtwald, Geb. 14.1
W-6600 Saarbrucken 11, Germany

Marto J. Hoary
Digital Equipment Corporation
60 Codman Hill Road
Boxborough, Massachusetts 01719,
U.S.A.

Brian J. Hogan
Westinghouse
P.O. Box 746
Baltimore, Maryland 21203, U.S.A.

Anthony D. Hope
Engineering Division
Southampton Institute
East Park Terrace
Southampton S094WW, UK

Yasser Hosni
IEMS Department
University of Central Florida
P.O. Box 25000
Orlando, Florida 32816, U.S.A.

Chien-Nan Huang
Department of Industrial Engineering
Wichita State University
Wichita, Kansas 67208-1593, U.S.A.

Faizul Huq
Information Systems and Management
Sciences
College of Business Administration
University of Texas
Arlington, Texas 76019-0437, U.S.A.

Simon F. Hurley
Industrial and Systems Engineering
Virginia Polytechnic Institute and State
University
Blacksburg, Virginia 24061, U.S.A.

Richard H.F. Jackson
National Institute of Standards and
Technology
Gaithersburg, Maryland 20899, U.S.A.

Hyeon H. Jo
Department of Industrial Engineering
University of Louisville
Louisville, Kentucky 40292, U.S.A.

Albert T. Jones
Building 220, Room B124
National Institute of Standards and
Technology
Gaithersburg, Maryland 20899, U.S.A.

Sagar V. Kamarthi
Industrial and Management Systems
Engineering
207 Hammond Building
Pennsylvania State University
University Park, Pennsylvania 16802,
U.S.A.

Ali K. Kamrani
215 Engineering Management Building
University of Missouri
Rolla, Missouri 65401, U.S.A.

Christos Kassiouras
Zenon SA
Egialias 48
Athens 151 25, Greece

Frances Kerrigan
University of Limerick
Limerick, Ireland

John F.C. Khaw
Gintic Institute of CIM
Nanyang Avenue
Singapore

Donald J. Knapp
StepsAhead Support Systems
3715 Oakes Drive
Hayward, California 94542-1718, U.S.A.

Estomih M. Kombe
CIM Systems Research Center
Arizona State University
Tempe, Arizona 85287-5106, U.S.A.

Thomas R. Kramer
Building 220, Room B124
National Institute of Standards and
Technology
Gaithersburg, Maryland 20899, U.S.A.

Jochen Kreutzfeldt
Institute for Production Engineering
and Machine Tools
3000 Hannover 1, Germany

Christian Kruse
University of Saarbrucken
Im Stadtwald, Geb. 14.1
W-6600 Saarbrucken 11, Germany

Rishi Kumar
Department of Mechanical Engineering
University of Connecticut
191 Auditorium Road
Storrs, Connecticut 06269-3139, U.S.A.

Tsuang Y. Kuo
Department of Industrial Engineering
University of Cincinnati
Cincinnati, Ohio 45221-0116, U.S.A.

Eberhard Kurz
Fraunhofer-Institut
fur Arbeitswirtschaft
und Organisation
Holzgartenstr. 17
7000 Stuttgart 1, Germany

Apostolos Kyritsis
Zenon, SA
Egialias 48
Athens 151 25, Greece

Brian K. Lambert
Wichita State University
Box 35
Wichita, Kansas 67208, U.S.A.

Kwan S. Lee
Louisiana State University
Baton Rouge, Louisiana 70803, U.S.A.

El-Amine Lehtihet
Department of Industrial Engineering
207 Hammond Building
Pennsylvania State University
University Park, Pennsylvania 16802,
U.S.A.

Kunibert H. Lennerts
Institut fur Maschinenwesen im
Baubetrieb
Universitat Karlsruhe
P.O. Box 6980
W-7500 Karlsruhe 1, Germany

Huw J. Lewis
University of Limerick
Plassey Technological Park
Limerick, Ireland

T. Warren Liao
Louisiana State University
Baton Rouge, Louisiana 70803, U.S.A.

Hampton R. Liggett
Department of Industrial Engineering
North Carolina State University
Box 7906
Raleigh, North Carolina 27695-7906,
U.S.A.

Beng S. Lim
Gintic Institute of CIM
Nanyang Avenue
Singapore

Lennie E.N. Lim
School of Mechanical and Production
Engineering
Nanyang Technological University
Nanyang Avenue
Singapore

Rasaratnam Logendran
Department of Industrial and
Manufacturing
Engineering
118 Covell Hall
Oregon State University
Corvallis, Oregon 97331, U.S.A.

Magdi S. Mahmoud
Department of Electrical Engineering
Kuwait University
P.O. Box 5969
Safat 13060, Kuwait

James A. Mahon
Department of Computer Science
Vision and Sensor Research Unit
School of Engineering
Trinity College
Dublin, Ireland

Anil Malkani
Electrical and Computer Engineering
Department
Ohio University
Athens, Ohio 45701, U.S.A.

Don E. Malzahn
Wichita State University
Box 35
Wichita, Kansas 67208, U.S.A.

Robert J. Marley
Industrial and Management Engineering
Department
University Technical Assistance Program
Montana State University
Bozeman, Montana 59717, U.S.A.

James O. McEnery
Boart Europe
Shannon Industrial Estate
Clare, Ireland

Thomas J. McLean
Mechanical and Industrial Engineering
University of Texas
500 West University Avenue
El Paso, Texas 79968-0521, U.S.A.

Declan M. Melody
Department of Mechanical and
Production
Engineering
University of Limerick
Limerick, Ireland

John L. Michaloski
Building 220, Room B124
National Institute of Standards and
Technology
Gaithersburg, Maryland 20899, U.S.A.

Atony R. Mileham
School of Mechanical Engineering
University of Bath
Claverton Down
Bath BA2 7AY, England

Moonkee Min
Building 220, Room A127
National Institute of Standards and
Technology
Gaithersburg, Maryland 20899, U.S.A.

Anil Mital
Industrial Engineering Department
University of Cincinnati
Cincinnati, Ohio 45221-0116, U.S.A.

Joe H. Mize
Industrial Engineering Department
Oklahoma State University
Stillwater, Oklahoma 74078-0540, U.S.A.

Roberto Mosca
Institute of Technology and Mechanical
Plants
University of Genoa
Via Allopera Pia 15
I-16145 Genoa, Italy

Saeid Motavalli
Department of Industrial Engineering
Wichita State University
Wichita, Kansas 67208, U.S.A.

Bartholomew O. Nnaji
Department of Industrial Engineering
and Operations Research
Automation and Robotics Laboratory
University of Massachusetts
Amherst, Massachusetts 01003, U.S.A.

James S. Noble
Industrial Engineering Program
University of Washington
Seattle, Washington 98195, U.S.A.

Jose Pablo Nuno
CIM Systems Research Center
Arizona State University
Tempe, Arizona 85287-5106, U.S.A.

Chris O'Brien
Department of Manufacturing
Engineering
University of Nottingham
Nottingham NG7 2RD, UK

Edward M. O'Mahony
Department of Mechanical and
Production
Engineering
University of Limerick
Limerick, Ireland

Sean M. O'Neill
Department of Computer Science
Vision and Sensor Research Unit
School of Engineering
Trinity College
Dublin, Ireland

Seamus O'Shea
Department of Computer Science
and Information Systems
University of Limerick
Limerick, Ireland

Bechir N. Ouederni
Department of Industrial and Systems
Engineering
Virginia Polytechnic Institute and
State University
Blacksburg, Virginia 24061, U.S.A.

Olagunju Oyeleye
Department of Industrial Engineering
207 Hammond Building
Pennsylvania State University
University Park, Pennsylvania 16802,
U.S.A.

Paul R. Painter
Engineering Division
Southampton Institute
East Park Terrace
Southampton S094WW, UK

Vijayakumar Pandiarajan
Industrial Engineering Department
West Virginia University
Morgantown, West Virginia 26506,
U.S.A.

Hamid R. Parsaei
Industrial Engineering Department
University of Louisville
Louisville, Kentucky 40292, U.S.A.

Anita M. Pietrofitta
Intel Corporation
941 E. Ivyglen Street
Mesa, Arizona 85203, U.S.A.

Frederick M. Proctor
Building 220, Room B124
National Institute of Standards and
Technology
Gaithersburg, Maryland 20899, U.S.A.

Luis C. Rabelo
274 Stocker Engineering and Technology
Center
Ohio University
Athens, Ohio 45701, U.S.A.

Abdur Rahman
University of Limerick
Limerick, Ireland

Parthasarathi Ramakrishna
Department of Industrial and
Manufacturing
Engineering
Oregon State University
Corvallis, Oregon 97331, U.S.A.

Michael D. Reimann
Information Systems and Management
Sciences
College of Business Administration
University of Texas
Arlington, Texas 76019-0437, U.S.A.

William G. Rippey
Building 222, Room B212
National Institute of Standards and
Technology
Gaithersburg, Maryland 20899, U.S.A.

Frederick J. Robinson
Department of Industrial Engineering
University of Minnesota
Duluth, Minnesota 55212, U.S.A.

Terry L. Ross
Pacific Northwest Laboratories
Richland, Washington 99352, U.S.A.

Henrique H. Rozenfeld
LAMAFE-EESC-USP
P.O. 359
Sao Carlos, SP, Brazil

Elfatih A. Rustom
School of Mechanical Engineering
University of Bath
Claverton Down
Bath BA2 7AY, England

Michael Ryan
School of Computer Applications
Dublin City University
Dublin 9, Ireland

Medhat M. Sadek
Department of Mechanical Engineering
Kuwait University
Safat, Kuwait

Joseph Sarkis
Information Systems and Management
Sciences
College of Business Administration
University of Texas
Arlington, Texas 76019-0437, U.S.A.

Heiko Schafer
Department of Manufacturing
Engineering
University of Nottingham
Nottingham NG7 2RD, UK

August-Wilhelm Scheer
University of Saarbrucken
Im Stadtwald, Geb. 14.1
W-6600 Saarbrucken 11, Germany

Bernd C. Schmidt
CIM Fabrik Hannover GmbH
Hollerithaller 6
3000 Hannover 21, Germany

Thomas J. Schriber
University of Michigan
Ann Arbor, Michigan 48109, U.S.A.

Quirico Semeraro
Dipartimento di Meccanica
Politecnico di Milano
Via Bonardi 9
I-20133 Milano, Italy

Srinivasa R. Shankar
Institute for Innovation and Design in
Engineering
Texas A&M University
College Station, Texas 77843, U.S.A.

Kim Siggaard
Institute of Manufacturing Engineering
Tech. University of Denmark
DK-2800 Lyngby, Denmark

Gene R. Simons
Rensselaer Polytechnic Institute
Troy, New York 12180-3590, U.S.A.

John A. Simpson
National Institute of Standards and
Technology
Gaithersburg, Maryland 20899, U.S.A.

Jannes Slomp
Faculty of Management and
Organization
University of Groningen
P.O. Box 800
NL-9700 AV Groningen,
The Netherlands

Albert W. Small
Downey and Small Associates, Inc.
10103 East Bexhill Drive
Kensington, Maryland 20895, U.S.A.

Graham T. Smith
Engineering Division
Southampton Institute
East Park Terrace
Southampton S094WW, UK

Jason G. Song
Mechanical and Industrial Engineering
Department
University of Texas
500 West University Avenue
El Paso, Texas 79968-0521,
U.S.A.

Jianfeng Song
Department of Industrial Engineering
West Virginia University
Morgantown, West Virginia 26506,
U.S.A.

Chelliah Sriskandarajah
Department of Industrial Engineering
University of Toronto
Toronto, Ontario, Canada M5S 1A4

Frank A. Stam
Digital Equipment International B.V.
Ballybrit Industrial Estate
Galway, Ireland

Kathryn E. Stecke
University of Michigan
Ann Arbor, Michigan 48109, U.S.A.

William G. Sullivan
Department of Industrial and Systems
Engineering
Virginia Polytechnic Institute and State
University
Blacksburg, Virginia 24061, U.S.A.

Edward T. Sweeney
Department of Engineering
University of Warwick
Coventry CV4 7AL, UK

J.M.A. Tanchoco
School of Industrial Engineering
Purdue University
West Lafayette, Indiana 47907, U.S.A.

Janis P. Terpenny
Department of Industrial and Systems
Engineering
Virginia Polytechnic Institute and State
University
Blacksburg, Virginia 24061, U.S.A.

Marlin U. Thomas
200 Mohler Laboratory
Lehigh University
Bethlehem, Pennsylvania 18015, U.S.A.

Anand V. Thummalapalli
Department of Mechanical Engineering
University of Connecticut
191 Auditorium Road
Storrs, Connecticut 06269-3139, U.S.A.

Jaime Trevino
Department of Industrial Engineering
North Carolina State University
Box 7906
Raleigh, North Carolina 27695-7906,
U.S.A.

Ingrid J. Vaughan
Westinghouse
P.O. Box 746
Baltimore, Maryland 21203, U.S.A.

Patrick C. Walsh
University of Limerick
Plassey Technological Park
Limerick, Ireland

Donald Walshe
Symnet International
Shannon Industrial Estate
Shannon, Co. Clare
Ireland

Robert R. Wehrman
Industrial and Management Engineering
Department
University Technical Assistance Program
Montana State University
Bozeman, Montana 59717, U.S.A.

Liang-Sung Wei
Chinese Naval Academy
Tsoying, Taiwan 813, R.O.C.

David C. Whalley
Department of Manufacturing
Engineering
Loughborough University
Loughborough LE11 3TU, UK

David J. Williams
Department of Manufacturing
Engineering
Loughborough University
Loughborough LE11 3TU, UK

George R. Wilson
200 Mohler Laboratory
Lehigh University
Bethelem, Pennsylvania 18015, U.S.A.

John R. Wilson
Department of Manufacturing
Engineering
University of Nottingham
Nottingham NG7 2RD, UK

Kung C. Wu
Department of Mechanical and Industrial
Engineering
University of Texas
500 W. University Avenue
El Paso, Texas 79968-0521, U.S.A.

Jun Yan
School of Computer Applications
Dublin City University
Dublin 9, Ireland

Mahmoud A. Younis
Engineering Department
American University
113 Sharia Kasr el Aini
P.O. Box 2511
Cairo, Egypt

Yaoen Zhang
Industrial and Systems Engineering
Department
Virginia Polytechnic Institute and State
University
Blacksburg, Virginia 24061, U.S.A.

ACKNOWLEDGMENTS

The organizing committee of FAIM '92 Conference is indebted to many individuals and organizations for their contribution, support and endorsement. We wish to thank all the keynote speakers who shared their views and visions of flexible automation and information management. We also wish to acknowledge, with many thanks, the contributions of all the authors who presented their work at the Conference and submitted the manuscript for publication. Our special thanks are extended to all FAIM '92 reviewers and session moderators who helped insure the high quality of the conference.

We wish to acknowledge, with gratitude, the support and endorsement received from the Conference cosponsors and university affiliates. We also wish to acknowledge, with great gratitude, the help and support received from Dr. James D. McComas, President at Virginia Tech, Dr. Edward Walsh, President at the University of Limerick, Dr. Wayne Clough, Dean of the College of Engineering and the Lewis A. Hester Professor of Civil Engineering at Virginia Tech, and Dr. Eamonn McQuade, Dean of Engineering and Science at the University of Limerick.

Although there are many who deserve credit for their various contributions to the success of FAIM '92, there are some who stand far above all others in deserving to be honored here. Our sincere thanks to Ms. Dot Cupp, FAIM '92 Conference Secretary, Pui-Mun Lee, and other staff members in the Industrial and Systems Engineering Department at Virginia Tech for their patience and hard work. We also wish to express our sincere thanks to all colleagues and student volunteers, especially Elin MacStravic, Jin Ong, and Krishna Krishnan, for their help, ideas, and suggestions. Finally, we wish to acknowledge the excellent work done by CRC Press in publishing the FAIM '92 proceedings.

TABLE OF CONTENTS

Keynote papers

INNOVATION, QUALITY, TIME — THE PARTNERSHIP FOR CUSTOMER SATISFACTION

Douglas N. Anderson

3M Corporate Quality Services, St. Paul, Minnesota, U.S.A.

ABSTRACT

During the past few years, changes in technology, competition and economic conditions have happened more rapidly than ever before. Global competition has increased significantly. Changes in technology are forcing shorter product life; i.e., obsolescence. In turn, the time to develop new products is being forced to be reduced. Forecasters predict that these changes will continue even faster through the rest of the century. Clearly, the challenge for maintaining growth and success in such an environment is to plan to manage change. The key ingredients require leveraging, innovation, quality and time in concert. Translating these three ingredients into understanding, commitment and action throughout the organization is paramount. Key aspects include establishing the foundation, mobilizing commitment and ownership, early cross functional teams, and strategic integration.

INNOVATION, QUALITY, TIME - THE PARTNERSHIP FOR CUSTOMER SATISFACTION

3M is a thirteen billion dollar company marketing more than 50,000 products worldwide. Since its founding in 1902, 3M has enjoyed continued success. Our philosophy of product diversification, innovation, and most importantly, quality products and services has enabled 3M to grow into new markets and expand existing ones.

The company has modified its paths and posture frequently over the years to adapt to changing markets, economic situations, and competitive pressures. The common denominator in our 89 year history can be described in a single word, change.

During the past few years, changes in technology, competition and economic conditions have happened more rapidly than ever before. Global competition has increased significantly. Changes in technology are forcing shorter product life; i.e., obsolescence. Also, the time to develop new products is being forced to be reduced. Forecasters predict that these changes will continue even faster through the rest of the century. Clearly, the challenge for maintaining growth and success in such an environment is to plan to manage change.

Within all of this flurry of competitive changes, customers have an increasingly growing opportunity to be more selective in their buying decisions. Customers are demanding and expecting products and services that meet their exact wants and needs. Further, they expect that these products and services perform 100% of the time and that these wants and needs be delivered exactly when requested.

Fundamental in 3M's strategies for the 90's is our focus on innovation, quality and time. Innovation encourages us to expand our business opportunities of existing technologies and in the development of new technologies. Quality allows us to manage innovation to achieve worldwide competitive success. Our quality improvement process is a crucial link toward our directions in the expanded applications of existing technologies and in our development of new technologies. Time sharpens our focus toward the right products and services at the right time.

Collectively, innovation, quality and time are the foundation for first to market with products and services that meet and exceed customer expectations 100% of the time.

Innovation is highly ingrained in 3M's culture. It represents 3M's spirit toward leveraging existing and new technologies toward exciting new business opportunities. So important is this desire, employees are empowered to spend 15% of their time on any product idea they choose. This time allocation is not directly measured and enforced. Rather, it is nurtured, protected and constantly communicated. Further, the corporation provides funds and resources, in the form of internal grants, to assist employees in turning their ideas into marketable opportunities. Also, the corporation invests heavily toward communication and recognition of emerging and new opportunities.

Time represents a new dimension to competitive business. Concepts such as Just In Time (JIT) have traditionally been viewed as applicable to only manufacturing. However, merging technologies, increasingly global competition, and increasing customer demands have forced another paradyme change. JIT is moving from the factory floor to all areas of business. In order to remain competitive, we must do things better and faster. We learned in manufacturing that JIT and quality are not conflicting approaches. Instead, they are fully compatible and supportive of each other. Now, we are learning that these two concepts are paramount in our goal of 100% product and service performance and first to market.

Quality improvement is the glue that binds innovation and time into a successful partnership. It leverages all aspects of the organization into a complimentary chain focused on continuous improvement of meeting and exceeding customer expectations. Positioned as a global management system, it represents the road map for doing business now and in the future. It insures that the right decisions are made by the right individuals at the right time.

3M'S BEGINNING WITH TOTAL QUALITY

3M first committed to the total quality process in 1979 under the leadership of Lew Lehr, then CEO. In 1980, we established a Corporate Quality Department responsible for defining quality objectives and designing a strategy to implement continuous quality improvement throughout the corporation.

We began by defining Five Essentials of Quality as the basis for 3M's new quality philosophy.

1. QUALITY IS CONSISTENTLY MEETING CUSTOMERS' EXPECTATIONS

Quality has been defined by quality management experts in terms such as "fitness for use,
"meeting specifications" and "meeting customers' requirements." Our experience has been that these definitions fall quite short in meeting the true wants and needs of the marketplace.

The 3M definition, quality is consistently meeting customers' expectations, represents three key components of quality: consistency, expectations and the customer.

Consistency implies doing it right every time.

Expectations move the process from a static focus on requirements to a dynamic focus on continual improvement to meet the changing needs and desires of customers. We are also led to the understanding that customer satisfaction is strongly influenced by innovations and non-product services. When we spoke of "fitness for use" or "meeting requirements," we found most people wanted to simply document the specifications and resisted changing them. Using "expectations" as part of the definition involves people in sales, marketing, billing, shipping, accounts payable, accounts receivable, purchasing, and other staff support functions in the quality improvement process.

3M is often admired for developing new and innovative products. Products, for example, as different as coated abrasives, video cassettes, printing plates and surgical drapes share common origins in the technology of precision coating and bonding. Being a worldwide company with many diverse products and services reflects our management's encouragement toward innovation. However, our experience has shown that innovation toward technology is not enough. To make the system work, we have found that getting close to the customer belongs at the center of our drive for innovation. This allows us to channel our encouragement of innovation toward products and services that meet wants and needs in the worldwide marketplace.

2. MEASUREMENTS OF QUALITY ARE THROUGH INDICATORS OF CUSTOMER SATISFACTION, RATHER THAN INDICATORS OF SELF-GRATIFICATION

Companies often get caught up in measurements of certain activities which can make them look good on paper yet show little significant information on how well they are competing on the basis of quality. These are indicators of self-gratification, measurements that management, owners and stockholders like to see which seem to say that their investment is being productively employed. Examples of these measurements are profit, annual sales growth, number of acquisitions, number of sales calls per day, number of daily/weekly/monthly reports turned out by a company department.

While such indices will always be useful for measuring the owners' interest, 3M has adopted a set of quality measurements that directly relate to our effectiveness in satisfying our customers' expectations.

These are such measurements as returns, complaints, sales adjustments, on-time deliveries, missed deadlines (external and internal), exception reports, customer surveys and competitive analysis.

Our laboratory research personnel have focused their attention on "technology transfer" as a significant contribution toward internal and external customer satisfaction. Effective technology transfer allows our manufacturing units to make products according to specification with fewer experimental runs. It also increases the effectiveness of our marketing personnel, as they can feel more confident that our products will conform to the specifications published. From this position, marketing and manufacturing are viewed as internal customers of our laboratory. Effective technology transfer also allows us to insure 100% product and service performance with respect to the users (external customer) of our products and services.

Consequently, the effectiveness of technology transfer has become an important laboratory quality measurement. Our definition of technology transfer are those activities which converts the scientific and engineering efforts of our laboratories into new manufacturing processes or into marketable new products which provide new sales and profit dollars. It also refers to the process of transferring new scientific information from universities. Successful technology transfer is a key factor in our ability to compete on a worldwide basis.

By focusing on these kinds of measurements, we get a more accurate picture of our customers' level of satisfaction with our products and services. We see how well we are satisfying their expectations.

3. THE OBJECTIVE IS CONSISTENT CONFORMANCE TO EXPECTATIONS 100% OF THE TIME

This concept is perhaps the most difficult to understand and accept. Traditionally, we have set performance standards with the attitude that errors are inevitable. For example, in manufacturing, we have accepted, as fact, that new product start up problems are the norm. However, when we view manufacturing as an internal customer of our laboratory, our emphasis is immediately focused on understanding and meeting their wants and needs. This leads us to conclude that the product of a laboratory is information. A laboratory when viewed as part of a larger business unit together with all other parts of that organization creates products. However, viewed as a separate entity with manufacturing as its customer, the product of the laboratory is information. Consequently, attention is focused on effective technology transfer from laboratory to manufacturing leading to conforming to manufacturing expectations 100% of the time.

4. QUALITY IS ATTAINED THROUGH PREVENTION ORIENTED IMPROVEMENT PROJECTS

The fourth key concept in 3M's strategy is "prevention." We adopted a new quality system that involves identifying the causes of errors, and then implementing specific projects to change the process, procedure or materials to prevent errors from re-occurring. The system of prevention replaces the system of inspection which inspects and sorts mistakes after they have occurred.

5. MANAGEMENT COMMITMENT LEADS THE QUALITY PROCESS

This is the final and most important quality concept 3M has adopted. Commitment must start at the top of an organization. Management must recognize that quality improvement doesn't just happen. It has to be planned and actively managed, like every other aspect of the business, if it is to become a way of life.

At 3M, quality improvement is a people process. Management takes a leadership role in demonstrating commitment and cultivating that same attitude in every employee. Management's role is to provide training, to properly define expectations and to prevent problems. The critical challenge is to create an environment that encourages individual involvement in decision making and personal "ownership" of the improvement process.

Our overall corporate improvement process is guided by a corporate Quality Steering Team. This team is comprised of 3M's senior management: the Chief

Executive Officer and Chairman of the Board, the Executive Vice Presidents of our business sectors, the Vice President of finance and our Senior Staff Vice Presidents. Together, the members of our Quality Steering Team have defined corporate values and overall business objectives, and each member works closely with individual organizations to integrate quality measurements into their business evaluation systems.

QUALITY IMPLEMENTATION STRATEGY

Translating our corporate values and objectives into operational activities is a major undertaking, complicated by 3M's size and diversity. Our individual business units and staff departments are each unique in function and they operate in diverse parts of the world. It became obvious that the best way to implement the improvement process was to provide guidelines and methods that could be adapted and modified to suit the cultural needs and personality of each operating area.

Formal training for all supervisory and management personnel worldwide became our first step towards institutionalizing the quality process. Initial training involved a "Quality Workshop" covering the Five Essentials of Quality and Eight Elements of implementation.

Implementation begins with management establishing a vision for the organization. The process for creating the vision includes and "awareness stage" to generate an understanding of quality and how it applies to the individual environment of each manager. This vision includes a plan for taking advantage of opportunities; it is an involved approach to establishing the organization's goals three to five years into the future.

Next, management develops the measurements used as evidence of success. These indicators help us quantify our progress and nearness to our vision. They measure how well we are meeting customer, employee and owner expectations.

After defining our vision and measurement system, we begin the journey of achievement by creating a quality action plan, to be reviewed and updated annually. The Quality Steering Team of each business unit is responsible for identifying major improvement opportunities and developing a specific action plan to achieve them. Improvement teams are also created in each area of operation to work on the improvement projects and to act on specific problems related to their own area. It is important that each part of the organization, from top to bottom, has a voice in updating the plan. This is typically achieved with a team approach.

The cycle is then ongoing, and quality improvement becomes a continuous process of identifying opportunities, establishing measurements, and setting goals for improvement.

QUALITY OUTSIDE THE FACTORY

Just as most of the successes by the Japanese have been achieved and documented within factory or production environments, 3M also experienced many of our early improvements in this area. Historically, a factory uses a wide range of standards, measurements methods, and well defined processes, more so than any other business operation. Initiating programs for improvement in the factory was much easier than in other areas; the opportunities are more obvious and the indices for measurement are frequently already in place.

Implementing the process is more difficult in less rigid areas, such as marketing, R&D, administration and staff support groups. Often, the "product" produced in these environments is so interrelated with other functions or so broadly defined that it is extremely difficult to break it into manageable segments for improvement.

It quickly became obvious that to extend quality improvement beyond the factory we needed additional techniques and methods. We developed a training package that includes a wide range of statistical and analytical techniques to promote brainstorming, data gathering, data analysis, presentation of information, and prioritizing. We also developed an extensive selection of internal measurement tools for improving our services and determining customer satisfaction.

Rather than train every person in all techniques, our strategy is to tailor the training to suit the requirements of each individual area.

TEAMS

Analyzing "products and customers" emphasized the importance of team efforts as never before. It became obvious that quality objectives could not be achieved without, first, having personal commitments to quality and, second, having a group's commitment to work together for improvement.

Teamwork, however, is not a working method that comes naturally in our society. It must be learned. 3M, like most American companies, has systems in place to stress and reward individual efforts and achievements more than group accomplishments. Undertaking quality improvement projects on a team basis needed new insights into group behavior, from top management on down.

Using our own Human Resources Department and outside services, we provided a multitude of courses and techniques on group processes, dynamics, and motivation to increase understanding of group activity. We then tied group training and management techniques, such as performance management, to tangible programs that are quality related. These strategies have accelerated the integration of teamwork within the company.

SUCCESSES

3M has found that if we focus on customer expectations, ultimately customer satisfaction will increase, costs will go down, and market share will increase. The objective is to create profit through customer's willingness to buy, and buy again.

Our customers buy when they are satisfied by quality and cost. We raise quality and create savings with our improvement process. We look at effectiveness (how well we interpret what the customer wants) and efficiency (how well and consistently we meet that expectation).

Our divisions, staff organizations, and overseas subsidiaries are each independently implementing the quality process, and enjoying a very broad range of success. In the few years since we began the total quality improvement process, achievements have ranged from exciting new successful products, to dramatic hard dollar savings.

WHAT IS NEXT?

Building on our quality essentials, 3M has created a new vision "Quality Vision 2000" as a means to bridge the gap between our current profile and our vision for the future. This allows us to establish an overall goal to focus on, a goal which is part of the quality process model. This translates into quality improvement plans at all levels, plans that address the major issues defined in our vision.

Issue: Global competition will significantly increase, and the total quality process will be an increasingly important strategy for achieving success in the global marketplace. As such, quality will not only be part of the overall strategic business plan, it will also be strongly positioned as a sales strategy. We will become even more dynamic in our efforts to understand our customers, developing aggressive strategies that address customer expectations for all our services.

Issue: Technology will force shorter product and service life because of obsolescence, and the demand will be for shorter new product and service development time. Quality and innovation will become even more important, along with the strategy of "doing the right things right."

Issue: Customers will increasingly make their purchasing decisions based on perceived value. Such concerns as service quality, product quality, life cycle cost, delivery quality - all key ingredients of perceived value - will become more important. We will continue to develop programs to quantify customers' perceptions of key buying criteria for both products and services. We will also use systems to benchmark how our customers perceive us and our competitors. Continued measurement of customer perceptions, at all levels, helps us to stay on top of their changing needs.

Issue: Partnerships between buyers and sellers will eventually become a requirement for doing business. Industrial manufacturers will be positioning certification programs for their suppliers as well as seeking to be certified and preferred suppliers to their customers. We are working to develop partnerships with our customers, beginning with the product development phase to better meet their expectations with innovative products and services in exchange for long-term business relationships.

Issue: Customers will expect 100 percent conformance in product and service performance as well as a responsive and friendly transaction. We must strategically translate customer wants and needs into design, manufacturing and service specifications and procedures. We will continue to use and develop methodologies to establish service improvements as well as product improvements. Developing partnerships with our customers will foster this goal.

All of these issues can be summarized into a customer orientation strategy leveraged toward responsiveness, flexibility, and time compression. In other words, "Do it right and do it fast." This becomes a large challenge considering that skilled resources are becoming a limiting factor.

3M is addressing this challenge by positioning our process for customer focused continuous improvement as our global management system. It represents our roadmap for the way we do business. We call this approach Q90s which stands for, "Quest For Business Excellence Into The 90s." Key aspects of this approach include: 1) Self-assessment of our current business practices with an improvement plan toward excellence; 2) the integration of the wants and needs of our customers, both current and potential, as a priority of continuous improvement; 3) the utilization of specialized tools and best practices such as hi-performance teams, quality function deployment, customer partners, benchmarking, etc.

Q90s as a global management system postures the entire organization, from laboratory to the factory floor, as an unbroken process of teamwork all focused toward continuous improvement. As such, innovation is enhanced in that it is directly leveraged toward customer wants and needs. Time is significantly reduced in that all areas are working together with a common focus and a common goal. That is, the pursuit of customer focused continuous improvement. Most of all, the expectations of our customers are met and exceeded with a diversification of innovative products and services that perform 100% of the time. This allows us to grow into new markets and expand our existing ones.

MANUFACTURING FOR THE GLOBAL MARKETPLACE

Richard H. F. Jackson and John A. Simpson

National Institute of Standards and Technology, Gaithersburg, Maryland, U.S.A.

ABSTRACT

This paper explores the current status of manufacturing technology in the United States, describes a vision of next-generation manufacturing systems, and discusses ongoing research at the National Institute of Standards and Technology (NIST) in support of U.S. industry's efforts to achieve that vision.

Keywords: Twenty-First Century Manufacturing, Advanced Manufacturing, Deterministic Metrology, Quality in Advanced Manufacturing.

INTRODUCTION

The United States is at a critical juncture in its manufacturing history. Since the first industrial revolution, the U.S. manufacturing sector has maintained a position of strength in competition for world market share. The strength of this industrial base has provided incredible growth in the U.S. gross national product and contributed immensely to the material well-being of the citizenry. Unfortunately, this position and its beneficial effects on the standard of living can no longer be taken for granted. U.S. industries are being threatened from all sides: market share has been slipping, capital equipment is becoming outdated, and the basic structure of the once mighty U.S. corporation is being questioned.

A growing national debate has focused on the decline of U.S. industry's competitiveness and the resultant loss of market share in the global marketplace. This rapid loss of competitiveness of American industry in international markets is an extremely serious problem with wide-ranging consequences for the United States' material well-being, political influence, and security. The national debate on this subject has identified many possible culprits, ranging from trade deficits to short-term, bottom-line thinking on the part of U.S. management to inappropriate and outdated management and engineering curricula at U.S. universities. Nevertheless, among the reasons for this loss of competitiveness certainly are the slow rate at which new technology is incorporated in commercial products and processes, and the lack of attention paid to manufacturing. There is a clear need to compete in world markets with high-value-added products, incorporating the latest innovations, manufactured in short runs with flexible manufacturing methods. Research, management, and manufacturing methods that support change and innovation are key ingredients needed to enhance our nation's competitive position.

In fact, efforts in these areas seem to be paying off already. In his upbeat message on technology opportunities for America [1], John Lyons, the NIST Director, reports impressive gains recently in the cost of labor, productivity and the balance of trade. As Lyons notes, one key area in which we must focus continued effort is in commercializing new technologies. As a nation, we have been slow to capitalize on new technology developed from America's own intellectual capability. Many ideas originating in the American scientific and technical community are being commercially exploited in other parts of the world. The nation must now find ways to help such companies meet the demands of global competition, when the speed with which firms are able to translate research results into quality commercial products and processes is of utmost importance.

Continued success in this effort will only come from full cooperation among government, industry, and academe. This maxim was clearly stated by the President's Commission on Industrial Competitiveness [2]:

"Government must take the lead in those areas where its resources and responsibilities can be best applied. Our public leaders and policy must encourage dialogue and consensus-building among leaders in industry, labor, Government, and academia whose expertise and cooperation are needed to improve our competitive position.... Universities, industry, and Government must work together to improve both the quality and quantity of manufacturing related education."

Many agendas have been written for how such cooperation should proceed and what issues must be addressed by each of these sectors. (See, for example, [3], [4], [5], [6], [7], [8], [9], [10], [11], [12], [13], [14], [15], [16], [17], and [18]). This paper describes some efforts underway at NIST to aid U.S. manufacturers in their own efforts to compete in the global marketplace, and thrive in the next century.

At the center of the NIST efforts are the programs and projects of the Manufacturing Engineering Laboratory (MEL) at NIST. NIST is the only federal research laboratory with a specific mission to support U.S. industry by providing the measurements, calibrations, quality assurance techniques, and generic technology required by U.S. industry to support commerce and technological progress. The Manufacturing Engineering Laboratory is tasked with bringing the resources of NIST to bear in support specifically of the mechanical manufacturing industries. Our programs are organized around several thrust areas: Precision Engineering, Robotics, Manufacturing Data Interface Standards, and Automated Manufacturing. These thrust areas are aimed at supporting the advanced manufacturing needs of U.S. industry both today and in the future. To develop a program that supports future needs, we have developed a vision of the next-generation manufacturing system, and this is discussed next.

A VISION OF MANUFACTURING IN THE TWENTY-FIRST CENTURY

To thrive in the twenty-first century, a manufacturing enterprise must be globally competitive, produce the highest quality product, be a low cost, efficient producer, and be responsive to the market and to the customer. In short, the next century's successful manufacturing firms will be "World Class Manufacturers" who make world class products for world class customers; i.e, customers who know precisely what they want and are satisfied only with world-class products.

There are many perspectives from which one can view a world class manufacturing firm. It can be viewed from the perspective of the shop floor and its interplay of the hardware and software of production. It can be viewed from the perspective of the legal environment in which it operates, both nationally and internationally. It can be viewed from the standpoint of the business environment, with its complex of tax and capital formation policies that affect long- and short-term planning. It can be viewed from the perspective of its corporate structure and embedded management techniques, which may facilitate or impede progress toward a manufacturing system of the twenty-first century. It may be viewed through the eyes of its employees, and the way it incorporates their ideas for improving the manufacturing process, the way it seeks to train them and upgrade their skills, and the way it strives to integrate them with the intelligent machines of production. It may be viewed from the perspective of its policies for performing, understanding, incorporating and transferring state-of-the-art research in advanced manufacturing technology.

A world class manufacturer may be viewed from these and many other perspectives, but, as depicted in Figure 1, the essential issue of importance for a

21ST CENTURY MANUFACTURING

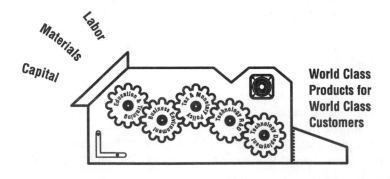

FIG. 1 Twenty-First Century Manufacturing

successful twenty-first century manufacturing enterprise is to learn how to operate smoothly and efficiently within each of the regimes discussed above, and, on the basis of this knowledge, create a smoothly functioning, well-oiled engine of production. Such an engine takes as input the classical categories of labor, capital, and material, and produces world class products for world class customers. The engine works best when each of the internal gears is precisely machined and carefully maintained. Each of the cogs is important and none can be ignored in the twenty-first century manufacturing enterprise. It is a part of the NIST mission to concentrate in two of these areas: manufacturing technology research and development, and technology deployment. The next section summarizes our work in these areas.

THE THREE PILLARS OF MANUFACTURING TECHNOLOGY

We have organized our programs at NIST around the three basic components of any successful mechanical manufacturing system: machine tools and basic precision metrology, intelligent machines and robotics, and computing and information technology, all overlaid on a background of standards and measurement technology, as shown in Figure 2. Since the mid-twentieth century and the onslaught of the second

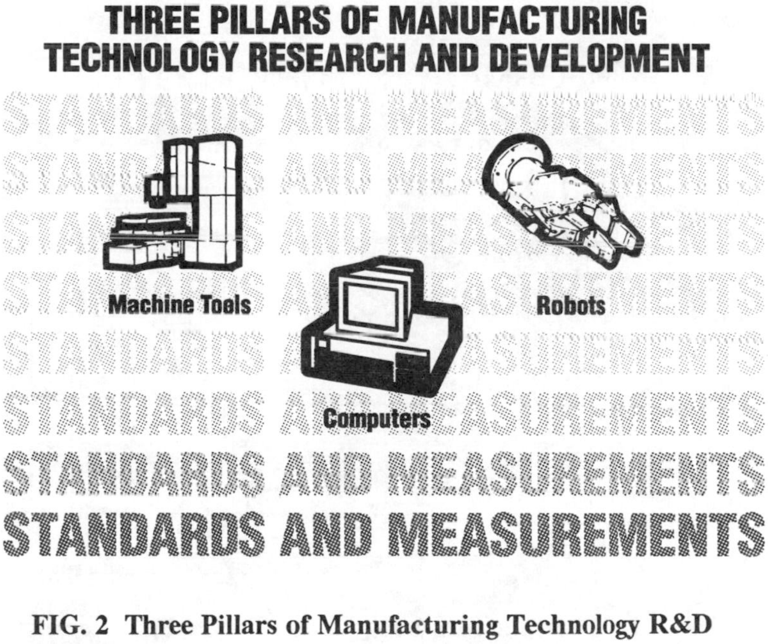

THREE PILLARS OF MANUFACTURING TECHNOLOGY RESEARCH AND DEVELOPMENT

Machine Tools   **Robots**

Computers

STANDARDS AND MEASUREMENTS

FIG. 2 Three Pillars of Manufacturing Technology R&D

industrial revolution, these three pillars have formed the foundation upon which all new techniques and advances in the technology of factory floor systems have been built. This is of course not to deny the importance of advances in non-technology areas such as lean production, quality management, continuous improvement, and workforce training. On the contrary, improvements in these areas can provide significant gains in industrial productivity and competitiveness. Nevertheless, in the area of manufacturing systems the three pillars are just that: the foundation. These three pillars of manufacturing technology research and development have been, are, and will be for some time to come, the quintessential components of factory floor systems from the very small to the very large. That is, in one fashion or another they have been addressed on factory floors since the establishment of the second manufacturing paradigm [19]. Further, in one fashion or another, they will continue to be addressed by those who seek to refine this paradigm over the next twenty to fifty years. Measurement is important because, simply put, if you cannot measure, you cannot manufacture to specifications. Standards are important to this foundation because, among other things, they help define the interfaces that allow interconnections among the elements of manufacturing enterprises, and the subsequent interchange of process and product information.

These three pillars form the foundation of the advanced manufacturing research program of the Manufacturing Engineering Laboratory at NIST, and since that work is at the center of the U.S. government's programs in advanced manufacturing research and development, it is featured here. It is not the intent of this paper to dwell in detail on these three pillars: that is accomplished elsewhere [20]. Nevertheless, it is important to provide some additional detail here.

In the area of computing and information technology, a twenty-first century manufacturer can be depicted as shown in Figure 3. The figure depicts the enormous

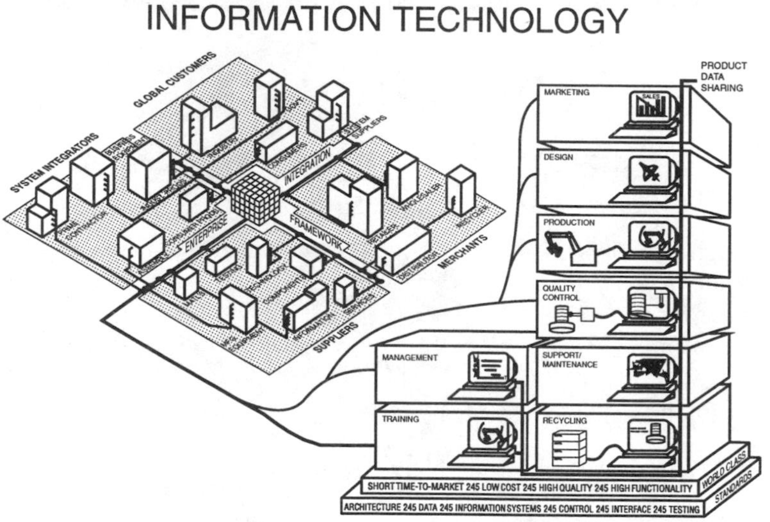

INFORMATION TECHNOLOGY

FIG. 3 Information Technology in the Twenty-First Century

amount of information that will be available in such an enterprise, and the importance of providing a smoothly functioning, seamless mechanism for making all the information available to whomever needs it, whenever and wherever it may be needed. These enterprises may be collocated as indicated on the left or they may be distributed throughout the country, or indeed the world. In any case, these information-laden enterprises must find ways to collect, store, visualize and access the information that is required for process-monitoring and decision-making. The exploded view at the right shows how one portion of such an enterprise would be integrated through the sharing of process and product data. The development of product data exchange standards is a critical component in the development of such an integrated enterprise.

Indeed, the development of product data exchange technology and international agreement on standards for the exchange of such data may just form the last real opportunity for U.S. industry to catch up to, and perhaps even leapfrog, the Japanese manufacturing juggernaut. Some in this country believe that there is very little about Japanese manufacturing techniques that we do not know. The challenge for us is to find a way to apply those techniques within our distinctly U.S. culture of creativity and entrepreneurial drive of the individual and the independent enterprise, and to apply it both in the advancement of technology and in the conduct of business. Product data exchange standards and the technologies like concurrent engineering and enterprise integration that are subsequently enabled by them may just be the answer to this challenge. For example, students of manufacturing technology have long held that one of the keystones of Japanese manufacturing success is the "Keiretsu," a large, vertically integrated group of companies with complete control of the manufacturing process from raw material to distribution, achieved through stable corporate structures and person-to-person interactions. Complete integration of all aspects of a manufacturing enterprise from design to manufacturing to marketing and sales, in a homogeneous, collocated or widely diverse and distributed enterprise is, in a sense, an "electronic Keiretsu." Such electronic Keiretsu could provide for U.S. manufacturers the "level playing field" required to compete successfully against the vertically integrated traditional Keiretsu in Japan. Product data exchange standards are the heart and soul of concurrent engineering and enterprise integration, and are critical to the ability of U.S. Manufacturers to thrive in the next century.

In the area of machine tools and basic metrology, the next pillar, impressive gains have been made in recent years, and will continue to be made in the future. The machine tools of the next century will exploit these gains and will be capable of heretofore unheard-of precision, down to the nanometer level. These gains will be possible through the combination of new techniques in machine tool quality control and in our ability to see, understand, and measure at the atomic level. After all, the truism that in order to build structures, one must be able to measure those structures, applies at the atomic level also. Even today, there are U.S. automobile engines manufactured with tolerances of 1.5 micrometers, pushing the limits of current machine tool technology. Next generation machine tool technology is depicted in Figure 4. It is noteworthy that both machine tools and coordinate measuring machines will make gains in precision based on the same kind of technologies, and thus it will be possible to have both functions available in one collection of cast iron.

Intelligent machines today and tomorrow go far beyond the simple industrial robots and automated transporters of yesterday. These new machines are a finely tuned combination of hardware components driven by a software controller with enough built-in intelligence as to make them almost autonomous. These software controllers will be capable of accumulating information from an array of advanced sensors and probing devices, matching that information against their own world-model of the environment in which they operate, and computing optimum strategies for accomplishing their tasks. The strategies must be determined in real time and take into account the dynamic environment in which these machines will operate. The sensory information obtained must of course conform to existing standards of manufacturing data exchange. This conformance can only be accomplished through a full understanding of the nature of intelligence and the development of software tools and structures which facilitate their efficient implementation and realization in a software controller. The architecture for such a controller is illustrated in Figure 5, which indicates the hierarchical nature of the controller and the processes for interaction with sensors and world models.

MANUFACTURING AND MEASURING TO SUPER-PRECISION

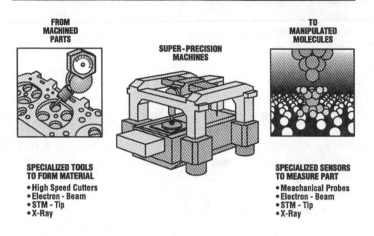

FROM
MACHINED
PARTS

SUPER-PRECISION
MACHINES

TO
MANIPULATED
MOLECULES

SPECIALIZED TOOLS
TO FORM MATERIAL
- High Speed Cutters
- Electron - Beam
- STM - Tip
- X-Ray

SPECIALIZED SENSORS
TO MEASURE PART
- Meachanical Probes
- Electron - Beam
- STM - Tip
- X-Ray

FIG. 4 Machine Tools in the Twenty-First Century

Intelligent Machines and Processes

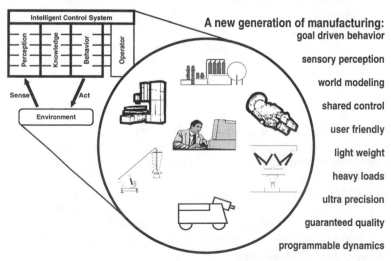

Intelligent Control System

Perception | Knowledge | Behavior | Operator

Sense   Act

Environment

A new generation of manufacturing:
goal driven behavior

sensory perception

world modeling

shared control

user friendly

light weight

heavy loads

ultra precision

guaranteed quality

programmable dynamics

FIG. 5 Intelligent Machines in the Twenty-First Century

THE AUTOMATED MANUFACTURING RESEARCH FACILITY AT NIST

The core of the NIST effort in advanced manufacturing is the Automated Manufacturing Research Facility (AMRF), and no discussion of NIST work in advanced manufacturing would be complete without some discussion of it. The AMRF has played a significant role in the identification and development of new and emerging technologies, measurement techniques, and standards in manufacturing. In fact, it was a catalyst in the legislative process that resulted in the Technology Competitiveness Act of 1988, which changed the name of the National Bureau of Standards to NIST and enhanced our mission. Further, the successful work summarized above would not have been possible without the development of this research facility. The AMRF is a unique engineering laboratory [21]. The facility provides a basic array of manufacturing equipment and systems, a "test bed," that researchers from NIST, industrial firms, universities, and other government agencies use to experiment with new standards and to study new methods of measurement and quality control for automated factories. In this sense, it serves as kind of "melting pot" of ideas from researchers across a broad spectrum of disciplines, talents, and skills. NIST is fundamentally committed to promoting the development of standards for automated manufacturing systems and to transferring technology to American industry. Credit should be given here to the U.S. Navy's Manufacturing Technology Program for helping to support the AMRF financially as well as in identifying fruitful opportunities for research.

The AMRF includes several types of modern automated machine tools, such as numerically controlled milling machines and lathes, automated materials-handling equipment (to move parts, tools, and raw materials from one "workstation" to another), and a variety of industrial robots to tend the machine tools. The entire facility operates under computer control using an advanced control approach, based on the ideas in Figure 5, and pioneered at NIST. The AMRF incorporates some of the most advanced, most flexible automated manufacturing techniques in the world. The AMRF includes donations and purchases from four different robot manufacturers, three machine tool vendors, and every major computer company. Basic principles from physics, computer science, behavioral sciences, control theory, operations research, and engineering disciplines have been used to transform these individual components into a fully integrated, flexible, small batch manufacturing system. To achieve maximum flexibility and modularity, the AMRF control system was:

o   partitioned into a five-level functional hierarchy in which the control processes are completely data driven and communicate via NIST-developed hardware and software interfaces,

o   designed to respond in real-time to performance data obtained from machines equipped with sophisticated sensors, and

o   implemented in a distributed computer environment using state-of-the-art techniques in software engineering and artificial intelligence.

The hierarchical command/feedback control structure ensures that the size, functionality, and complexity of individual control modules is limited. Although the flow of control in this hierarchy is strictly vertical and between adjacent neighbors, it is necessary and even desirable to share certain classes of data across one or more levels. In addition, each control level is completely data-driven. That is, the data required to perform its functions is separated from the actual control code. All data is managed by a distributed data administration system [22] and transmitted to and from control processes via a communication network.

Several hierarchical models for controlling shop floor activities have been proposed. The models typically decompose manufacturing activities into five hierarchical levels: facility, shop, cell, workstation, and equipment. Activity at each of these levels is data-driven, and each can be expanded to fill out the tree. This structure provides a convenient mechanism for describing the functions of an automated facility and the databases needed to meet manufacturing requirements.

When originally conceived in the late 1970's, the AMRF was unique. Since that time, and since having solved some of the hardware and communications problems in the AMRF, more automated facilities have begun to appear. During this time, most of the resources expended on automated manufacturing have concentrated on demonstrating feasibility of the idea of fully flexible, integrated, automated manufacturing facilities. It is now time to turn our attention to the difficult tasks confronting us in the areas of software engineering, data handling, data analysis, process modeling, and optimization. That is, having demonstrated that the idea works, we must now make it work smoothly and efficiently.

QUALITY IN ADVANCED MANUFACTURING: DOING IT RIGHT

It has been two decades since the Manufacturing Engineering Laboratory turned its attention to the challenge of Advanced Manufacturing. In the decade of the seventies, the target of our research was the industrial metrology laboratories of advanced manufacturing enterprises. The flexibility of production methods was rapidly increasing as was the number of products. The metrology laboratories were faced rapidly shrinking tolerances and with an ever increasing number of artifact standards to maintain. To aid U.S. industry in coping with this first wave of what is often called the New Manufacturing Paradigm or the Second Industrial Revolution, MEL developed the Accuracy-Enhanced Computer-Controlled Coordinate Measuring Machine. This machine was as flexible as the production process, supported the tighter tolerances and, in many cases, virtually eliminated the need for artifact standards. Where this technology was not applicable, MEL showed how to use the computer to apply the methods of Statistical Process Control to the process of measurement, greatly reducing the number of required measurements, while at the same time increasing their reliability. This latter technology, known as the Measurement Assurance Program (MAP), has become the world standard operating procedure wherever metrology is still considered an independent manufacturing function.

The decade of the eighties was devoted to demonstrating that, by proper choice of control architecture, it was possible to retro-fit existing production facilities and turn them into Flexible Manufacturing Systems. This entailed developing interface standards between manufacturing sub-systems. One of the first and most widely used of these interface standards is the Initial Graphic Exchange Specification (IGES) which provides for communication between Computer Aided Design systems through neutral file formats. It was also shown that in an advanced manufacturing environment, it was possible to construct software systems to provide complete integration from design to fabrication, using only available software tools. As metrology for Quality Control moved from the metrology laboratory to the factory floor, MEL followed it, and developed the concept of "Deterministic Metrology" in which quality control is applied, not to final inspection of the product, but to the real-time control of the **process**. Drill-up [23], a system for detecting drill wear, is perhaps the best known Deterministic Metrology. In the late eighties this project was formalized into the Quality in Automation Program.

In both these decades, the emphasis was on proving that automation of the production process was **possible** at all levels of complexity, from Total Enterprise Integration of a Fortune 500 company to PC-based CAD/CAM for a 50-person shop. This task is essentially complete at the research level, and examples of successful application of advanced manufacturing technology, influenced by these efforts, exist at all levels of complexity. The task now before a research effort in advanced manufacturing is to ensure that these automated production processes operate as smoothly and efficiently as possible; i.e., now that we know how to do it, we must turn our attention to ensuring that we do it right. The tools for "doing it right" are sought in our program on **Quality in Advanced Manufacturing**.

The Manufacturing Engineering Laboratory's earlier efforts were restricted to the control of unit-processes on the factory floor. Somewhat in the spirit of Total Quality Management (TQM), this new effort targets the total enterprise. Hence, the field of action covers design, process planning, resource management, all unit processes, factory management, and enterprise integration. The concept of quality is also expanded from the earlier definition: conformity to specification of the product. For example, it is generally agreed that the measures of performance of an enterprise are:

o the degree to which products meet or exceed customer expectations, (quality in its narrowest sense);

o the degree to which an enterprise meets or beats a competitors price for comparable product; and

o the speed with which new products are brought to market and existing products are delivered.

The first is a measure of product quality and is, to a large extent, a measure of the quality of one's design capability. The second two are a measures of the quality of production capability. The quotient of the first two (quality per unit cost) is often called product value. Thus the expanded view of quality incorporates all these parameters as measures of Quality in Advanced Manufacturing.

As MEL enters its third decade of contributing to the manufacturing productivity of the Nation, the nexus of the activity is increasingly Quality in Advanced Manufacturing. Our efforts on PDES/STEP are aimed at developing the tools for concurrent engineering, which we recognize as doing-it right in design. Our work in intelligent machines and robotics is aimed at reducing cost by doing-it-right on the factory floor. The efforts in precision engineering are aimed at developing products whose performance exceeds anything in the customer's experience. Much of the metrology developed in the 70s has been rendered obsolete, with the introduction of new processes and the refinement of old ones. Our program to revisit basic metrology and develop new techniques based on these concepts is doing-it-right in metrology.

Although the work of the Laboratory over the years has varied from development of Measurement Assurance Programs for gage blocks in a completely manual calibration system, to development of standards for product data, to real time control for robots, there is a strong unifying theme: control using feedback. In every case there is a **loop,** comparing the consequence of an action with its intent, and modifying the action accordingly. This loop contains actuators, sensors, expressions of intent, and comparators. For example, a coordinate measuring machine is an actuator, a touch probe is a sensor, and an IGES/PDES/STEP file is an expression of intent. This loop may be closed in hours or days when the sensor is manual inspection of the end product, or in microseconds when the loop is under robot control. Nonetheless, all quality, and its control, is the consequence of such loops, and the degree of quality depends on the "tightness" of a particular loop.

CONCLUSION

We believe that in order to be successful in solving the kinds of problems confronting this nation's manufacturing industry, an interdisciplinary approach to problem-solving is needed. This interdisciplinary approach must concentrate on obtaining the most efficient, cost-effective solutions to problems in advanced manufacturing as is possible. Too often system designers are content to settle for feasible solutions to a problem with little or no effort expended to find optimal or even improved solutions. This cannot continue if the U.S. manufacturing industry is to survive in the global marketplace of the 21st Century. The projects and programs of the Manufacturing Engineering Laboratory have been, and continue to be, aimed at providing U.S. industry with the tools it needs to measure, monitor, and improve manufacturing processes, improve productivity, and achieve more market share. A continued focus on these topics through the Quality in Advanced Manufacturing Program is, quite simply, doing-it right.

REFERENCES

1.   Lyons, John W., "America's Technology Opportunities," The Bridge, National Academy of Engineering, 21(2), summer 1991.

2.   Presidential Commission on Industrial Competitiveness, January, 1985.

3.   Executive Office of the President, U.S. Technology Policy, Office of Science and Technology Policy, Washington, DC 20506, September 26, 1990.

4.   National Academy of Engineering, The Challenge to Manufacturing: A Proposal for a National Forum, National Academy Press, 2101 Constitution Avenue, NW, Washington, DC, (1988).

5.   U.S. Congress, Office of Technology Assessment, Making Things Better: Competing in Manufacturing, OTA-ITE-443, U.S. Government Printing Office, Washington, DC 20402-9325, February (1990).

6.   National Academy of Engineering, Manufacturing Studies Board, The Role of Manufacturing Technology in Trade Adjustment Strategies, National Academy Press, 2101 Constitution Avenue, NW, Washington, DC 20418, (1986).

7. National Academy of Engineering, Manufacturing Studies Board, <u>Toward a New Era in U.S. Manufacturing - The Need for a National Vision</u>, National Academy Press, 2101 Constitution Avenue, NW, Washington, DC 20418, (1986).

8. <u>Bolstering Defense Industrial Competitiveness - Preserving Our Heritage the Industrial Base Securing Our Future</u>, Report to the Secretary of Defense by the Under Secretary of Defense (Acquisition), July 1988.

9. U.S. Congress, Office of Technology Assessment, <u>Paying the Bill: Manufacturing and America's Trade Deficit</u>, OTA-ITE-390, U.S. Government Printing Office, Washington, DC 20402, June, (1988).

10. Kelley, Maryellen R., and Brooks, Harvey, <u>The State of Computerized Automation in U.S. Manufacturing</u>, Center for Business and Government, Weil Hall, John F. Kennedy School of Government, Harvard University, 79 John F. Kennedy Street, Cambridge, MA 02138, October, (1988).

11. National Academy of Engineering, Manufacturing Studies Board, <u>A Research Agenda for CIM: Information Technology</u>, National Academy Press, 2101 Constitution Avenue, NW, Washington, DC 20418, (1988).

12. Magaziner, Ira C., and Patinkin, Mark, <u>The Silent War - Inside the Global Business Battles Shaping America's Future</u>, Random House, Inc., New York, (1989).

13. Grayson, C. Jackson, and O'Dell, Carla, <u>American Business - A Two-Minute Warning - Ten Changes managers must make to survive into the 21st Century</u>, The Free Press, A Division of Macmillan, Inc., 866 Third Avenue, New York, NY 10022, (1988).

14. Cohen, Stephen S., and Zysman, John, <u>Manufacturing Matters: The Myth of the Post-Industrial Economy</u>, Basic Books, Inc., New York, (1987).

15. Dertouzos, Michael L., Lester, Richard K., Solow, Robert M., and the MIT Commission on Industrial Productivity, <u>Made in America - Regaining the Productive Edge</u>, The MIT Press, Cambridge, MA, (1989).

16. National Academy of Engineering, Manufacturing Studies Board, <u>The Competitive Edge: research priorities for U.S. manufacturing</u>, National Academy Press, Washington, DC, (1991).

17. Womack, James P., Jones, Daniel T., and Roos, Daniel, <u>The Machine that Changed the World</u>, Rawson Associates, New York, (1990).

18. <u>21st Century Manufacturing Enterprise Strategy: an industry- led view</u>, Iacocca Institute, Lehigh University, Bethlehem PA, 1991.

19. Simpson, John A., "Mechanical Measurement and Manufacturing," in <u>Three Pillars of Manufacturing Technology</u>, R.H.F. Jackson, ed., Academic Press, New York, (1991).

20. Jackson, Richard H.F., <u>Three Pillars of Manufacturing Technology</u>, Academic Press, New York, 1991.

21. Simpson, John A., Hocken, Robert J., and Albus, James S., "The automated manufacturing research facility at the National Bureau of Standards," <u>Journal of Manufacturing Systems</u>, 1(1)17-31, (1982).

22. Barkmeyer, Edward M., Mitchell, Mary J., Mikkilineni, K.P., Su, S.Y.W, and Lam, H., "An Architecture for Distributed Data Management in Computer Integrated Manufacturing," NIST Technical Report NBSIR 86-3312, National Institute of Standards and Technology, Gaithersburg, MD, January 1986.

23. Yee, K.W., and Blomquist, D.S.,"Checking tool wear by time-domain analysis," <u>Manufacturing Engineering</u>, 88(5)74-76 (1982).

DEVELOPMENT OF AN OBJECT ORIENTED MODELING AND SIMULATION ENVIRONMENT FOR ADVANCED MANUFACTURING SYSTEMS

Joe H. Mize

Oklahoma State University, Industrial Engineering Department,
Stillwater, Oklahoma, U.S.A.

ABSTRACT

A research team at Oklahoma State University has been exploring alternative approaches to the modeling and simulation of complex manufacturing systems since 1985. This work has led to an expanded conceptual framework for the development and use of analytical and simulation modeling within a manufacturing enterprise. The software paradigm called "object-oriented programming" has emerged as a powerful software environment in which to develop our concepts of manufacturing system modeling. Several conceptual breakthroughs have been achieved which yield a modeling and simulation capability that is superior to those modeling approaches based upon traditional programming concepts. This paper describes the current state of development of the modeling environment.

INTRODUCTION

Modeling, analysis, and optimization are critical activities in the planning, design, implementation, reconfiguration, and day-to-day operations of a complex manufacturing system. Discrete-event simulation and queueing networks are tools that are typically used for dynamic systems analysis. Additional tools such as mathematical programming techniques, statistical experimental design techniques, expert systems, etc., come into play during the optimization phase. Due to the lack of a common modeling framework, the current practice is to create models tailored to specific techniques. The resulting models and solutions tend to be single purpose, throw-away efforts. The availability of a common modeling framework would enable the creation of multipurpose, reusable models. In addition to greatly improving the efficiency of the modeling activity, this would also lead to better utilization and maintenance of models within an operating environment.

A research team at Oklahoma State University in the Center for Computer Integrated Manufacturing began examining these issues in 1985. In 1986, after several months of research that explored alternate approaches to the modeling and simulation of complex manufacturing systems, the CIM research team decided to pursue an approach based upon a new programming paradigm, object-oriented programming. The focus originally was on simulation model construction only. Initially, the Smalltalk V object-oriented environment was utilized due to hardware restrictions in our laboratory. Several manufacturing simulation models were developed, including a model of a department in an electronics assembly factory (AT&T plant, Oklahoma City), to demonstrate the superiority of the object-oriented approach. The development later shifted to Smalltalk-80, which was chosen as our "permanent" environment.

OBJECT ORIENTED PARADIGM

The Object Oriented Programming (OOP) paradigm was considered initially due to the claims of its proponents that it facilitates software reuse. Solutions obtained via traditional modeling approaches are characterized by their single use nature. That is, once a particular model is used for its original purpose, it is rarely used again. When a new problem arises, a new model is developed. This approach leads many in manufacturing to question the viability of computer based modeling from a benefit/cost standpoint. This concern is particularly true in today's world characterized by constant change. What is needed is a modeling environment that facilitates rapid and flexible development of models which can be continuously calibrated and updated and are easily re-usable. Object oriented programming (OOP), a paradigm in which all programming elements are represented as independent "objects", appears to offer significant advantages in this regard.

To appreciate the distinctions offered by OOP, the nature of OOP objects and three key OOP concepts must be understood. As stated earlier, all programming elements within OOP are "objects". An object consists of some private memory and a set of operations. The nature of its memory and operations depend upon what type of object it represents. For instance, a 'number' object stores a value and can carry out basic arithmetic operations. Similarly a 'dictionary' object stores a set of character strings and understands an operation of the type 'retrieve by key-word'.

Three key OOP concepts are *message passing*, *encapsulation*, and *inheritance*. Objects communicate with one another through *message passing*. A message is a request for an object to carry out one of its operations. The message specifies the operation which is desired but not how the operation is to be accomplished. The job of implementing (or choosing not to implement) the operation is solely the responsibility of the receiving object. Thus, the only way an object's private memory can be modified is if the object executes one of its own operations to change its own private memory. This concept is known as *encapsulation*. The benefits derived from the combination of message passing and encapsulation include modularity, reusability, modifiability, and understandability. Figure 1 illustrates these two important OOP concepts.

The objects within an OOP environment are defined in a hierarchical structure of classes. A class describes the implementation of a set of objects which all represent the same type of system component. The individual members of a class are called instances of the class. A class describes the form of its instances' private memory and the set of operations which its instances understand. The instances of a class actually allocate private memory space through storage locations known as instance variables. An instance variable can be a particular value, like '5', or can be an instance of another class. For example, an instance of class 'Machine' may have instance variables 'inputQueue' and 'outputQueue' each of which is an instance of class 'Queue'.

The hierarchical structure of the classes within OOP facilitates the concept of *inheritance*. Through inheritance, a class need not redefine any operation which is defined within one of its super-classes (i.e. a class higher up in its hierarchy). On the other hand, if a class needs a special message it can be implemented to augment the inherited messages. This capability greatly enhances the reusability of modeling objects and procedures (i.e. messages). Figure 2 illustrates this concept for a hypothetical class 'Machine' and its subclasses, 'Robot', and 'Lathe'. Class 'Lathe' further extends the hierarchy with its own subclasses 'Automatic Tool Change Lathe' and 'Manual Lathe'. Note that each subclass of 'Machine' inherits the data element 'status' and that each subclass of 'Lathe' inherits the messages 'Load' and 'Unload' from 'Lathe'. As an example of a special message required by a particular class, note the message 'change tool' for the class 'Automatic Tool Change Lathe' which is not found in the 'Manual Lathe' class. Similarly, the data element 'worker' is found in the class 'Lathe' but not in 'Automatic Tool Change Lathe'.

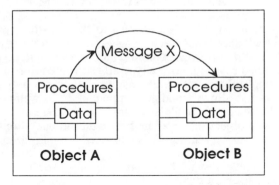

Figure 1
OOP Message Passing and Encapsulation

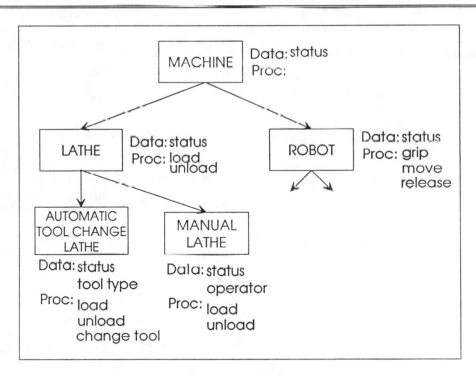

Figure 2
OOP Inheritance

OBJECT ORIENTED MODELING OF
MANUFACTURING SYSTEMS

With this as background, one can turn to the question, what makes the OOP approach to modeling advanced manufacturing systems superior to traditional approaches? The differences manifest themselves in three major ways. First, objects within an object oriented modeling (OOM) environment are typically structured to closely parallel real world objects, as in the example above with class 'Machine'. This not only facilitates more rapid "natural" modeling but also allows manufacturing engineers as well as simulation experts to build and run models. Second, a modeler's ability to understand, modify, and maintain a model is improved since objects incorporate both data and operations on the data within a single coherent entity. Finally, reusability is enhanced through the use of inheritance. For example, a new model of a machine can be quickly defined by making it a subclass of the previous machine in the class hierarchy and defining (or modifying) operations to incorporate its new characteristics.

Traditional and object-oriented programming approaches for simulation modeling are contrasted in Figure 3, while Figure 4 illustrates the "natural" manner in which real world elements can be abstracted with an object-oriented model. Close examination of these two figures reveals the significant advantages of OOP over modeling approaches based in traditional programming languages.

Implementing these OOP concepts within a manufacturing modeling framework results in a set of objects which are relatively independent of the environment within which they are used. The objects can be designed, tested, and documented as stand-alone units and stored for future use and reuse. This results in objects which have value outside of a specific system, because they can be reused in many systems.

THE CONCEPT OF MODELING PRIMITIVES

One approach to implementing an OOM environment which specifically targets model reusability is through a bottom-up modeling approach. The bottom up approach starts with the development of a modeling object database. One component of this database is a library of modeling "primitives". A modeling primitive is a physical, information, or control/decision object whose representation portrays the structure and behavior of a component of the real world system. An object is considered a primitive only if further decomposition into simpler modeling objects is not required for any potential simulation experiment objective.

An important point implied by the above definition for primitives needs to be made explicitly. Unlike current modeling approaches, the proposed OOM environment will separately model physical, information, and control/decision elements of a manufacturing system. This approach has profound ramifications. The most significant ramification is the capability to implement logical/decision modularity. Logical/decision modularity is equivalent to "plug compatible" decision elements, which can be replaced as desired in any location in the model. For the first time within a modeling environment the decision/control structure of a model is not imbedded within other parts of the system, it does not reside within the other modeling elements, it is separate and distinct and can be easily modified. The primitives needed to model a generic workstation are shown in Figure 5. Notice that the physical, information, and control elements are modeled separately and distinctly.

Factors	Traditional Modeling Paradigm	Object-Oriented Modeling Paradigm
Model Construction		
Software	Simulation languages based on procedural programming style	Simulation environments based on object oriented programming style
Translation into code	Process is abstract	Process is natural and intuitive
Interface	Usually textual	Usually graphical with icons and dialog boxes
Level of detail	Usually not much detail due to programming complexity	At user's discretion, but requires detailed object library
Treatment of distinct system elements	Different element types are not distinctly modeled; Aggregation to reduce program complexity	Physical , information, and decision / control elements are modeled distinctly and independently
Effort/time/cost	Moderate costs of model development, but a "throw away" type	Initial cost of establishing detailed model is very high, but cost of subsequent reuse is relatively low
Model Attributes		
Purpose	Usually a unique model is created for a specific purpose	More general models possible for multiple purposes
Usage	Single usage, throw away models	Repeated usage and continuous refinement
Flexibility	Highly inflexible; changes almost always result in a complete rewrite of program	Highly flexible, due to the ability to modify fundamental building blocks; quick reconfiguration is possible.
Accuracy	Useful for measuring relative differences in alternative configurations	With greater degree of detail and realism, can also estimate absolute performance with greater accuracy
System/Model Relationship	Not connected via data links	Detailed model can be imbedded in control structure of the firm, with linkages to databases; continuous model calibration and parameter updating

Figure 3
Contrasting Traditional and Object -Oriented Paradigms for Simulation Modeling

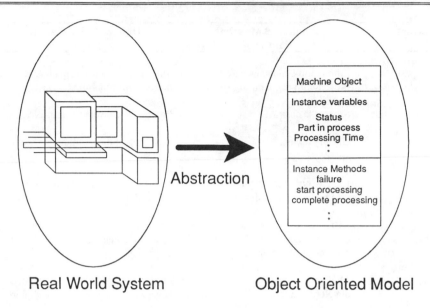

Real World System   Object Oriented Model

Figure 4
Modeling Abstraction

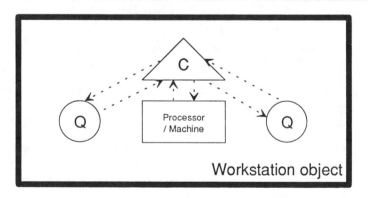

Figure 5
Primitives Required to Model a Workstation

INTEGRATED ENVIRONMENT FOR
MODELING AND ANALYSIS

A set theoretic formalism has been developed that represents a descriptive model of the system of interest in a context that is independent of any particular modeling methodology. As such, within the OOM environment, it provides a descriptive model which can be used either for creating analytical models such as queueing network models or for creating simulation models, depending upon the particular interest and purpose at the time. This sort of dual purpose environment can be easily facilitated within the object oriented paradigm. Figure 6 illustrates this structure.

The ability to perform quick "rough cut" analysis using analytical techniques and detailed analysis using simulation analysis via the same descriptive model dramatically reduces the time needed for planning, designing, and reconfiguring complex manufacturing systems. Additionally, it provides a cohesion among the various methodologies so that solutions obtained using a particular model for one part of the system are consistent and feasible in the context of a different solution obtained using a different model for a related part of the system. It also provides a mechanism which allows a model to be exercised at varying degrees of detail at different times, yet maintain a reasonable degree of consistency in the model results.

RESULTS TO DATE

Significant progress has been made in several key areas. The current status of the research team's efforts can be characterized as follows:

o   We have acquired an in-depth understanding of the fundamental concepts underlying various approaches to system modeling (i.e. queueing, scheduling, mathematical programming, networks, simulation, control theory, etc.).

o   We have carried out conceptual and software development to achieve logical decision modularity among the decision elements of a modeled system, and we have an initial concept of a means to express these elements such that they can be "plugged" into appropriate points in a model.

o   We have acquired a strong capability in our knowledge of the object-oriented programming paradigm in general, and of the Smalltalk-80 environment in particular.

o   A set-theoretic based formalism was developed for expressing embedded decision elements in a consistent, abstract manner.

o   A new random number generator and several random variate generators were added to the Smalltalk-80 core simulation module.

o   The current modeling environment permits the separate and distinct specification of physical elements (machines, material handlers, etc.) and data/information elements (e.g., part routing).

o   We have successfully constructed and implemented a modeling and simulation environment for discrete part manufacturing systems in Smalltalk-80. We have:
-implemented a menu-driven user interface that permits quick and easy specification of the plant structure in a hierarchical manner (plant, workcenter, and machine), product structure via bill of materials, part routing, etc.
-created reusable classes and methods in Smalltalk-80 to support simulation of systems operating under either a "push" or "pull" philosophy and systems with alternate routings.
-successfully constructed a model of a hypothetical manufacturing system in which the physical elements, the data/information elements, and the decision elements are modeled separately and distinctly.

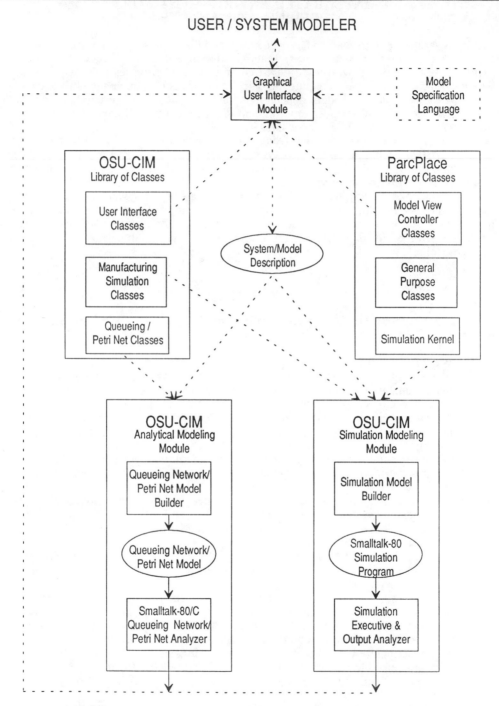

Figure 6
Software Architecture for Modeling and Analysis Environment

o We have developed a preliminary understanding of the "design specification requirements" for an object oriented modeling and simulation environment.

o We have completed literature search and preliminary work to develop a formal, structured model specification language that permits the separate specification of physical, data/information, and decision elements.

o Four completed Ph.D. dissertations have demonstrated the usefulness of the object-oriented modeling environment as a test-bed for theoretical research investigations pertaining to advanced concepts in operations planning and control. One additional dissertation is in progress.

o We have developed improved analytical methods based on two-moment queueing approximations for the analysis of manufacturing systems; new analytical approximations were developed and validated for modeling material handling systems, operator availability, blocking due to limited storage areas, etc.

o We have successfully constructed and implemented "metamodels" which permit the replacement of certain segments of a detailed model with an equivalent single step model; this feature shows promise of permitting mixed level modeling, a goal of model builders for many years.

SUMMARY

A major paradigm shift is beginning to occur relative to the modeling, analysis, and design of complex manufacturing systems. Recent developments in several areas, including object-oriented programing, modeling formalisms, and logical/decision modularity, are now to the point that a meaningful convergence can be crafted to yield a modeling environment far superior to any we have known in the past. This environment currently under development at OSU's Center for Computer Integrated Manufacturing offers continuing opportunities for major advancements in systems modeling from both the conceptual and methodological viewpoints. The ultimate goal is a modeling workstation that will be useful for various levels of system designers and managers within a manufacturing environment.

The approach to modeling and simulation described in this paper is clearly revolutionary, not evolutionary. Thus, it represents a fundamental paradigm shift, not only in the construction of models, but in their utilization and maintenance within an operating environment. One of the significant features of this approach is the integration of analytical and simulation methodologies within a single software environment. The ability to perform quick "rough cut" analysis using analytical techniques and detailed analysis using simulation within the same descriptive model will dramatically reduce the time needed for planning and designing or reconfiguring complex manufacturing systems.

ACKNOWLEDGMENTS

The author gratefully acknowledges all the faculty and graduate students at Oklahoma State University and the University of Oklahoma who have made contributions to the modeling environment described in this paper. We also acknowledge the financial support received from the AT&T Foundation and the Oklahoma Center for Integrated Design and Manufacturing.

REFERENCES

1. Adiga, S. (1989), "Software Modelling of Manufacturing Systems: A Case for an Object-Oriented Programming Approach," In *Analysis, Modelling and Design of Modern Production Systems*, A. Kusiak and W. E. Wilhelm, Eds. J. C. Baltzer AG, Basel Switzerland.

2. Beaumariage, T.G. (1990), "Investigation of an Object Oriented Modeling Environment for the Generation of Simulation Models," Ph.D. Dissertation, School of Industrial Engineering and Management, Oklahoma State University, Stillwater, OK.

3. Cox, B. (1986), *Object Oriented Programming: An Evolutionary Approach*, Addison-Wesley, Reading, MA.

4. Endesfelder, T. and H. Tempelmeier (1987), "The SIMAN Module Processor - A Flexible Software Tool for the Generation of SIMAN Simulation Models," In *Simulation in Computer Integrated Manufacturing and Artificial Intelligence Techniques*, J. Retti, Ed. SCS, San Diego, CA 38-43.

5. Goldberg, A. and D. Robson (1989), *Smalltalk-80 The Language*, Addison-Wesley, Reading, MA.

6. Karacal, S.C. (1990), "The Development of an Integrative Structure for Discrete Event Simulation, Object Oriented Programming, and Imbedded Decision Processes," Ph.D. Dissertation, School of Industrial Engineering and Management, Oklahoma State University, Stillwater, OK.

7. Khoshnevis, B. and A. Chen (1987), "An Automated Simulation Modeling System Based on AI Techniques," In *Simulation and AI*, P.A. Luker and G. Birtwistle, Eds. The Society of Computer Simulation, San Diego, CA, 87-91.

8. King, C.U. and E.L. Fisher (1986), "Object-Oriented Shop-Floor Design, Simulation, and Evaluation," In *Proceedings of the 1986 Fall Industrial Engineering Conference*, Institute of Industrial Engineers, Norcross, GA, 131-137.

9. Kreutzer, W. (1986), *System Simulation Programming Styles and Languages*, Addison-Wesley, Reading, MA.

10. Law, A.M. and S. W. Haider (1989), "Selecting Simulation Software for Manufacturing Applications: Practical Guidelines and Software Survey," *Industrial Engineering*, 31, 5, 33-46.

11. Meyer, B. (1987), "Reusability: The Case for Object-Oriented Design," *IEEE Software*, 4, 3, 50-64.

12. Meyer, B. (1988), *Object-Oriented Software Construction*, Prentice Hall International (UK) Ltd., Hertordshire, Great Britain.

13. Mize, J.H. and T.G. Beaumariage, and S.C. Karacal (1989), "Systems Modelling Using Object-Oriented Programming," In *Proceedings of the 1989 Spring Conference*, Institute of Industrial Engineers, Norcross, GA, 13-18.

14. Mize, J.H. and D.B. Pratt, "A Comprehensive Object Oriented Modeling Environment for Manufacturing Systems," *Factory Automation and Information Management*, M.M. Ahmad and W.G. Sullivan, Eds., CRC Press, Boca Raton, FL, 1991, pp. 250-257.

15. Nance, R.E. (1984), "Model Development Revisited," In *Proceedings of the 1984 Winter Simulation Conference*, S. Sheppard, U.W. Pooch, and C.D. Pegden, Eds., IEEE, Piscataway, NJ, 75-80.

16. Nilsson, N.J. (1980), *Principles of Artificial Intelligence*, Tioga Publishing Co., Palo Alto, CA.

17. Oren, T. and K. Aytac (1985), "Architecture of MAGEST: A Knowledge-based Modeling and Simulation System," In *Simulation in Research and Development*, Elsevier Science Pub., North-Holland, Amsterdam, The Netherlands, 99-109.

18. Pritsker, A.A.B. (1986), *Introduction to Simulation and SLAM II*, Third Edition, Halsted Press, New York, NY.

19. Thomasma, T. and O.M. Ulgen (1988), "Hierarchical, Modular Simulation Modeling in Icon-based Simulation Program Generators for Manufacturing," In *Proceedings of the 1988 Winter Simulation Conference*, M. Abrams, P. Haigh, and J. Comfort, Eds. IEEE, Piscataway, NJ, 254-262.

20. Ulgen, O.M., T. Thomasma, and Y. Mao (1989), "Object Oriented Toolkits for Simulation Program Generators," In *Proceedings of the 1989 Winter Simulation Conference*, E.A. MacNair, K.J. Musselman, and P. Heidelberger, Eds. IEEE, Piscataway, NJ, 593-600.

21. Zeigler, B.P. (1976), *Theory of Modeling and Simulation*, Robert E. Kreiger Pub. Co., Malabar, FL.

A EUROPEAN'S VIEW OF TQM AND ITS SUCCESSFUL IMPLEMENTATION

Bernie Dillon

Amdahl Ireland Limited, Balheary, Swords, Co., Dublin, Ireland

ABSTRACT

Today, Total Quality Management (TQM) has become a strategic priority for business firms around the world, because of its proven significance for acquiring and maintaining competitive advantage. This reality of business life is now well understood by a growing number of top European companies and has led to the founding of the European Foundation for Quality Management in October 1988. However, when it comes to implementing a real TQM process many organisations fall far short of expectations, and, indeed, many fail outright. A quality approach to establishing a continuous improvement process is the best guarantee of success. Without the discipline and structure to make it happen, best efforts and aspirations, however well intentioned, will lack credibility. It is the demonstrated commitment of an organisation in general and its leadership in particular, that underwrites its success. This paper outlines:

- **The successful implementation and evolution of TQM within Amdahl Ireland since adopting the Crosby approach in 1985**
- **The need for a structured approach and the extensive use of teamwork to ensure success**
- **How it has worked, the results and the benefits**
- **Significant improvement has been achieved in operational performance as well as employee satisfaction**
- **Key lessons learnt by Amdahl in bringing about a culture of continuous improvement. The most important of these is that implementation takes a lot longer than expected, significant education/training is required, management must demonstrate their commitment and finally - keeping the process moving is hard work.**

A EUROPEAN'S VIEW OF TQM AND ITS SUCCESSFUL IMPLEMENTATION

European companies have been relatively slow to adopt and embrace the concepts of **Total Quality Management** (TQM). The leading TQM companies have typically 5 to 10 years' experience in TQM as compared with close to 15 years for U.S. companies and in excess of 25 years for the leading Japanese TQM companies. Europe has fallen behind, ironically, because it traditionally put a lot of stock in quality and thus was less vulnerable to attacks from Japan than the U.S. back in the 1970s. The problem now is that Europe's methods of assuring quality, (high in line inspection to find and fix defects), have become expensive, inefficient and outdated. The first Europeans to jump on the TQM bandwagon in the early 1980s were, not surprisingly, those who first ran up against Japanese competition: Philips in consumer electronics, Olivetti & ICL in computers, Volvo & Jaguar in automobiles.

However as many U.S. converts to TQM found, implementing quality circles, work teams and other faddish techniques that merely copied the Japanese style - while missing its substance - fell short of the mark and rarely translated into ongoing quality programs or results. Competition, (especially with 1993 single market status), and high quality costs have galvanised Europe into action. In September 1988 presidents of 14 top companies took the initiative to establish the **European Foundation for Quality Management** (EFQM). One year later they were joined by 53 more presidents, committed to the total quality management movement in Europe. Their objective was clear: to enhance the competitiveness of Western European business in the world markets. Research shows that customer-perceived total quality is the key to success in the marketplace, with the lowest consumption of resources. This reality of business life is now understood by a growing number of top European managers.

Customer-perceived total quality exceeds quality system certification, still a high priority objective in Europe. More and more companies and supporting organisations understand the essential but relative significance of quality system certification. In the past, ISO9000 and EN29000 certification have been final objectives, but they are now correctly being seen as the starting point to achieving total quality.

In the coming years EFQM will focus its activities around the development of recognition and education, training and research programs. The top priority in the recognition program is the implementation of the **European Quality Company** award. To win this award, a company must demonstrate that its approach to total quality management has contributed significantly to achieving benefits of value to its customers, its employees, other key stakeholders in the company and society in general.

The first award is to be presented during the European Quality Management Forum in Madrid, Spain in October 1992.

In a Business Needs Survey carried out in 1990 and commissioned by the Commission of the European Communities and supported by EFQM, respondents were asked to rank a list of 15 factors so as to determine what TQM means to an organisation. The top five factors were:

- **Satisfying external customers**
- **Reducing cost**
- **Partnership with customers**
- **Each person satisfying their internal customers**
- **Employee involvement and development**

The issue facing companies is how to effectively focus on these priorities and make the changes necessary to become successful. Many manufacturing entities are mediocre manufacturing performers. They are not responding well to changes in technology, materials, manufacturing processes and to the skills and career expectations of their employees. They are not responding to opportunities in global business. In many instances they seem paralysed, either unable or unwilling to act or make changes and many of their senior managers are not knowledgeable of the science of manufacturing. In general there are three elements missing in many of them:

1. *Vision* - a vision of how business in general and manufacturing will exist in the world, and in a particular industry, in the next 20 years.
2. *A willingness to take action* - senior managers seem unwilling to lead people toward greater competitive success in the future and
3. *A Process for taking action* - in order to implement the vision to support their company's manufacturing and business strategies and to improve the competitive advantage of their companies. The problem is further compounded by an ever increasing list of three letter acronym s **(TLAs)**, all designed to be quick fixes to the latest dilemma facing manufacturing.

The best approach for a company to address these issues is to embark on a "programme' of **Total Quality Management** (TQM). This enables the organisation to get the key issues of customers, quality, employees and cost effectiveness into a coherent approach throughout the organisation.TQM provides a means of focusing the attention and energies of everybody in a company towards looking after customers,institutionalising quality and thereby improving profitability.Amdahl started down such a road seven years ago (1985) and I would like to share some of our experiences/observations with you. As a result we have changed significantly the way the company is being managed on a day-to-day basis. We have endeavoured to develop Teamwork to a very high extent and empower our employees by affording them more participation and involvement in the running of the business. We used an evolutionary approach and worked our way forward from one level to the next e.g. employees first participate in problem solving, progress to participation in day-to-day decisions and then to participation in organisation change. The management or managers had to make the biggest change, and in general this is taking place successfully. As a result of these changes significant benefits are accruing to both the company and its employees. I will trace briefly some of the highlights of our progress, the good and the bad, over this period and on how the manager's job has changed and what the requirements are for new and successful managers in this changed environment. First I would like to tell you a little about Amdahl so that you can understand where I am coming from and our starting point which would be different for each organisation.

Amdahl Ireland Limited (AIL) is a wholly owned subsidiary of Amdahl Corporation who designs, develops, manufactures, markets and services large-scale, high performance data processing systems. The company's product line includes large general purpose mainframe computers, data storage subsystems, data communications products and software. The company concentrates primarily on one segment of the data processing market - the users of large systems who have standardised their operating systems environment on System/370 software and its extensions. This makes Amdahl a manufacturer of IBM compatible systems dedicated to providing choice to the marketplace. Our customers include a broad range of private and public corporations, financial institutions, government bodies, universities etc. around the world. Amdahl currently employs over 9,000 people and has had an impressive growth record. Since shipping our first product in 1975, sales have grown to over $2 billion with an unbroken record of profitability.

Amdahl set up its manufacturing facility in Ireland in 1978. We are located just north of Dublin in a building of over 260,000 sq. ft. We currently have over 600 employees and annual sales of $500 million. We have a relatively young, well-educated workforce with in excess of 30% having third-level qualifications. Since 1985 our management methods have changed from what could be termed conventional to a system based almost entirely on teamwork. In 1985 we started a formal Quality Improvement Process as we were concerned about future product quality levels and future competitiveness. The process was initiated using the concepts and philosophies of Philip Crosby & Associates. Crosby's core definition of Quality is Conformance to Requirements, i.e. you need to meet or exceed your customers' expectations. To ensure success and maintain the momentum in our Total Quality Improvement effort, known in Amdahl as World Class Quality Improvement (WCQI), we focused on ten key activities:

1. *Management must be committed and involved:*

Senior management must accept responsibility for implementing the process, for Quality improvement, for providing education, for making resources available, for making Quality the top priority etc. In order for the improvement process to be successful and continue to evolve and grow, the management of the company must be fully committed and involved. In addition, all management, but especially top management, must continually and consistently demonstrate its commitment and be involved. Management must convince all employees that the Continuous Improvement Process and Teamwork is an integral way of how the company is now being managed.

It cannot be overemphasised that management must **"walk the talk"**. This means demonstrating commitment to the principles and practice of Total Quality by actually doing (and being seen to do) at least what is expected from other employees. This is one of the most powerful ways in which management can show the rest of the company that Total Quality matters, and that management is committed to employee involvement and participation through teamwork.

2. *Extensive Training & Education must be provided for everybody:* Initially, we used the **Crosby Quality Education System** (QES). The program was aimed at Management and Professional Staff. However we chose to use it for all employees, leaving out a few modules (such as Role of Manager and External Vendor Relations) for non-management employees. For managers the program consisted of 8 x 4 hour modules and for non-managers 5 x 4 hour modules. The QES program ensured that all employees were given the same substantial education. This firmly established the common language of quality and emphasised the "equality" of responsibility for quality. The QES packaged training program (contains videos, overheads, workshop books, participants workbooks etc.) focuses on developing a new quality culture (including language) and on implementing the quality improvement process. Our education and training has continued to evolve and our approach is as follows:

- Company-wide education on major topics - all employees get same level of education.
- Specific individual/groups then get further education on special topics

so as to increase their level of expertise.

- Education given is coordinated so as to continually add to each individual's knowledge and provide understanding of how the jig-saw fits together. We follow a philosophy of gradually building on the foundations firmly established. Training/education is essential in any commitment to the participative (teamwork) process. Hence when the basic Quality education was complete we immediately focused on providing employees with education/training on how to be an effective team, how to run meetings and to improve their problem solving skills. Effective training programs help teams/groups reach objectives, such as enhanced teamwork, improved communications, creative thinking, quality/service attitudes, and more effective manager/employee relationships. Most of our education is provided by managers on a part-time basis rather than specialist training staff.

The education must be relevant to employee's day to day activities and they must then be given the opportunity to put it into practice through being involved in teams. How the education is delivered is obviously extremely important and must be done in such a way to maintain people's interest with plenty of opportunity for participation.

3. *Focus on prevention of errors* - not inspection or correction: We endeavour to be prevention oriented not on detection. Zero errors must be the ultimate goal, with focus on doing it right first time. The use of the word 'errors' is important so as to include all company activities, not just a zero 'defect' products focus.

4. *Continuous Improvement:* We have a philosophy of continuous improvement coupled to the attitude that (ultimately) no level of error is acceptable. The focus for improvement must be on job processes and cross-functional processes.

5. *Customer Orientation:*

Awareness that everyone has a customer (internal or external) that must be satisfied. The focus on internal customer/supplier relations has been a major factor in being focused on addressing cross-functional issues and helps significantly in breaking down barriers across departmentals and functions.

We have progressed from a complete lack of appreciation of internal customer/supplier relations, to suppliers now being acutely aware of their internal customers and agreeing requirements with them. To this end measurements were put in place to monitor compliance or to promote customer awareness. Customers and suppliers are working together as a team and there is keen awareness of satisfying customer requirements.

6. *Everyone Involved through Teamwork:* Everybody in the company must be involved and be responsible for doing quality work. To achieve this we use Teamwork extensively. People are the common denominator for continuous improvement and success depends on people involvement. All employees must accept responsibility for doing quality work. People are quite prepared to do quality work if given the opportunity, management support and resources.

The main strength of Crosby's programme is the attention it gives to transforming quality culture and to teamwork. Crosby involves everyone in the organisation in the improvement process by stressing individual conformance to requirements.

The main thrust of our Improvement Process has been and still is continuous improvement while achieving employee participation through teamwork. All departments within the company operate with a Workgroup Team improvement approach called a **Quality Work Team (QWT)**.

Initial efforts by teams were focused on the elimination of hassle/problems in the workplace. The focus then gradually changed to defect reduction/elimination and then to the making of improvements. The results from QWT activity has been very impressive and all areas of the company have become very aware of satisfying Customer requirements. The focus is now being changed to Defect Prevention and the continuous improvement of all activities with special attention being given to waste elimination (non value adding activities) and simplification.

Many of the workgroups are moving towards being self-directed with members chairing meetings in rotation, and the manager acting as another member or not present a lot of the time. The managers are gradually moving towards the role of coach/facilitator and managing of the boundary activities only.

Today we have over 80% of employees involved on approximately 90 formal teams and a very strong demand exists for expansion of team participation by those not on teams.

7. *Measurements / Benchmarking:* Measurement is at or near the heart of any continuous improvement activity. It is essential to know what your starting point is so that improvement can be tracked. The adage of what gets measured gets done is a very valid one, as the simple activity of measuring something ensures its improvement.

Benchmarking is a process of measuring your products services and/or processes against your competitors or against leaders in other industries. Benchmarking can be used to determine competitive requirements and to determine ways to improve products, services and processes.

8. *Recognition / Sharing in success:* Recognition has proven to be very significant in maintaining continuing interest and awareness. In Amdahl, the formal forms of recognition have been specifically oriented towards being non-monetary and for group achievement.

These are the major activities that we at Amdahl focus on, however to ensure that the vision and the improvement process are kept in sync with each other, and to get everybody aboard the same quality bandwagon, you need to have a structure in place for facilitating the process and also for good employee communications. The following is a brief discussion of how these operate at Amdahl:

9. *Facilitation Structure*

The structure helps to provide overall direction and co-ordination for all the teams.

Why do you need a structure and what is it?

You need a structure to communicate the vision, to coordinate the implementation and to maintain the effort once started. Maintaining the effort is the 99% portion of the whole endeavour.

Our structure and its main elements are as follows:

MQIT (Management Quality Improvement Team) is the overall steering committee and is composed of the Chief Executive, 8 Function Directors, 8 Senior Managers, one of whom is a fulltime administrator for the overall coordination. The team serves as the focal point for all improvement activity and meets every week for 2 hours. The purpose of MQIT is to:

- **Develop a companywide strategy to implement the continuous improvement process**
- **Be the driving force for improvement, based on overall strategic company initiatives based on competitive manufacturing strategies**

MQIT coordinates the activities of two types of teams:

Facilitation Teams: These teams are chaired by a Function Director and include 5 to 10 employees (predominantly managers). The purpose of these teams is the maintenance and improvement of the Improvement Process and ensure that the Improvement Process gets institutionalised into the fabric of the company. Examples of these teams are; Teamwork, Education, Measurements, Awareness, Recognition, Prevention.

Improvement Teams: These teams are similar in make-up to the Facilitation Teams but are focused on specific improvement activities which are driven by the company's competitive manufacturing strategies. The number of these teams can vary based on the major issues being addressed at any point in time.   This structure operates alongside the functional structure. The function structure is responsible for the implementation of improvements while the WCQI structure is responsible for overall coordination of improvement activities and for providing focus. Examples of these teams are: Product Reliability, Responsiveness (JIT), New Product Introduction.

The activity of MQIT and its sub-teams must be communicated throughout the company to provide regular status, to reinforce the vision, direction, focus etc.

The facilitation teams regularly use themes and awareness activities to provide focus and emphasis. Examples are Poster competitions, competitions, Quizs, Theme awareness weeks, videos etc.

10. Communications Structure

Good two-way communications is a key factor in creating a committed workforce.AIL has a very formal communications structure which is designed to keep every employee informed of how the company is doing against its objectives/strategies and also allow employees to   raise issues/questions that concern them and get answers quickly and promptly.   Management needs to have a formal system in place for listening to employees and being seen to be responsive to what they hear. The more important elements of AIL's system are as follows:

a) **Monthly Management Meetings (Update)**
b) **Monthly Communications Meetings (all employees by work group)**

These two types of meetings form the core of our formal communications cycle. The cycle starts with the Management Update Meeting (2 hours) which reviews status and company performance over a wide range of issues, discusses employee interest issues, communicates proposed policy changes and seeks input on proposed changes in the way that the company does things.

The purpose of the monthly workgroup communications meetings is to convey to all workgroups details of all news items for the month, which would include responses to previous employee queries. In addition, adequate time is allowed for employees' questions or even complaints which are recorded and responded to without fail.

c) Coffee Sessions:
The Chief Executive holds a weekly meeting with 12 employees (usually 1 1/2 to 2 hours) with an open agenda to discuss and answer queries from employees on any issues of interest to them. All Function Directors also hold Coffee Sessions such that they meet all their employees at least twice a year.

d) Company Performance Status Meetings: for all employees given by the Chief Executive at end of each quarter.

e) Daily" Work Group Meetings: These short meetings (15/20 minutes) take place at least 3 times/week in the work area and address any issues.

All of these activities are ongoing, never-ending and we are constantly seeking ways to improve the way that we do them.

The Manager's Role - The manager's role has obviously changed significantly with continuing emphasis on Teamwork and workgroups becoming more autonomous. The manager is also the key individual in the facilitation of these changes and in they being successful. Hence it is essential that all managers have a thorough understanding of how their role is changing and also that they be given enough education, training and time to become comfortable with their new role.

First line managers will need to become much better at being coaches, trainers, facilitators, educators, team leaders, skilled change agents and communicators, while some of the more conventional practices must be dropped. Managers themselves will spend an increasing percentage of their time on teamwork throughout the organisation as they participate in companywide cross-functional improvement activity and in managing the boundaries for their own workgroups.

HOW HAS WCQI WORKED - RESULTS/BENEFITS

The benefits have been significant and far-reaching, but first a few of the issues that sometimes concern employees, especially managers:

It takes considerable time, effort and commitment. Managers were not always convinced that time spent away from the job was beneficial.

You do end up having a lot of meetings/discussions both for information dissemination and for gaining consensus. This can slow up the decision making process and gives rise to lots of meetings.

A concern that team decisions are not always the best ones and the risk of team domination by one or two individuals.

For us at Amdahl these are no longer concerns as education/training of both employees and managers resolved these issues/concerns. The benefits on the other hand have been very significant and the following list is just a broad overview.

Employee Benefits:
- Improved job satisfaction
- Employee potential more fully utilized
- Employee involvement & participation
- Participate in & influence change
- Employee's pride in their work/company

- Appreciation of other employee's contribution
- High level of commitment

Specific feedback on these areas is obtained annually through a 130 question climate survey of all employees. Levels of satisfaction are high and continually improving on a year to year basis.

Company Benefits:

- Produces quality decisions
- Significant improvement in Quality - reliability of products has increased by a factor 15 to 50
- Increased responsiveness - time to build most complex product decreased by 60%
- Ongoing simplification - e.g. work orders decreased by 95% to build a more complex product
- Continuous improvement in all activities
- High level of employee satisfaction
- High level of customer satisfaction

Customer satisfaction is obviously the ultimate payoff. Amdahl has in recent years consistently outperformed our competitors in achieving best results in independent customer satisfaction surveys. One such survey is Datapro and Amdahl has over the last 3 years come out tops in at least 75% of the categories. The most important of these have been:

Amdahl was **Number 1** in 'Overall Customer Satisfaction' in each of the last three years. We have also been **Number 1** in Reliability, Maintenance, Service and Technical Support.

Amdahl has also come out on top when customers were asked the questions:

Did the product meet all of their expectations?

Would they recommend the product to a fellow user?

In addition a new 'Best Price/Performance' rating was introduced in 1990 in which Amdahl has rated **Number 1**.

KEY LESSONS

In summary, Amdahl has learnt a number of lessons in its efforts to bring about a culture of continuous improvement:

- to establish TQM takes longer than first thought. Culture and attitudes change slowly and in small steps, not giant leaps. Therefore you have to be constantly looking for progress and recognising achievement. don't be cautious about recognition. emphasise the long, slow journey of TQM to avoid it being consigned to the scrap heap of discarded magic management fads. You must be patient.
- Avoid overstating the benefits of TQM (disillusionment is often caused by unrealistic expectations), emphasise that quality improvement requires ongoing neverending commitment to reap the lasting benefits.
- TQM is a continuously learning experience, you must figure it out for yourself. You will make mistakes. Search for the simple measurements. Discuss and elaborate on concepts - try to stay away from 'exhortations' but don't completely overlook the value of slogans either.
- Don't underestimate people's ability and interest to become involved. Get started with teams early, empower people to solve problems. Listen to your

employees and react/respond/resolve.

- Planning is important - it just doesn't happen. Develop a realistic and integrated plan. Follow some system religiously but customise to your organisation-what will it allow you to do? don't copy anyone's specific formula/approach but learn from everyone. Each company or operation should have a unique approach/method. Organise to provide the right direction and support.
- Constancy of purpose (Deming) - consistency is a must.
- Good, open communications up/down and across the organisation drives the improvement process and challenges existing processes. Share business results and competitive data.
- Education/training must be a priority and on-going in order to achieve all of the above. Tie education to action i.e. provide when ready to implement.
- You need a workable day to day approach to raising the performance expectations of your organisation. focus on customer requirements and competitive benchmarking is an effective way of doing this.
- Teamwork and employee involvement are critical to implementing effective changes.
- Develop your vision early. Keep focused on the end state-don't compromise on Zero Defects (errors) as the ultimate goal in satisfying customers' requirements. give people a sense of mission e.g. being the best, survival, patriotism etc.
- Don't start without top management commitment and until totally committed as it will fail and get more difficult to get restarted - you get one chance to do it right (first time).
- Focus on continuous improvement - incremental steps by each employee (don't let 'innovation' concepts become dominant). there is no such thing as a tiny, insignificant improvement/idea. Challenge the process all the time.
- Finally, TQM is hard work. The toughest job is to keep the process moving and for this management must lead and actively demonstrate their commitment, Avoid understating the commitment required when trying to gain acceptance for the quality strategy. Avoid creating the impression that quality is a finite task, that once installed will last forever, with only minimum maintenance.

CONCLUSION

Competitive pressure is forcing industry to look at a broad range of success factors in its search for trading and business advantages. Time and service related capabilities will become more important. These increased competitive pressures are forcing changes on all players in the supply chain - manufacturers, raw material suppliers, distributors and retailers.

The most important issues for manufacturing management in the 1990s will be:

- **Coping with rapid change**
- **Being customer driven**
- **Dealing with structure and organisation changes**
- **Continuous improvement in all activities**
- **Releasing and directing the energy of all employees**

The key to success will be in being totally committed to achieve Continuous

Improvement in Manufacturing.

Implementing the TQM philosophy, which is based upon continuous improvement, is much more than implementing a set of manufacturing techniques. It is a philosophy of excellence for companies that requires a new way of thinking. In short, it is a cultural change. Cultural changes are very difficult and require an enormous degree of will, energy and leadership. Culture change must be led by management. The senior manager's behaviour is the most critical element, his behaviour must be perceived to be consistent with the new philosophy. Even then it must be sustained for a long time for the changes to become permanent.

CHALLENGES IN THE MANUFACTURE OF ELECTRONIC PRODUCTS

David J. Williams, Paul C. Conway, and David C. Whalley

*Department of Manufacturing Engineering, Loughborough University,
Loughborough, England*

This paper is intended to review some of the problems facing manufacturers of current and future generation electronic products. It will also show how academic research can support manufacturers, particularly in the definition of the limitations of the manufacturing process. The paper closes by identifying research areas that can be addressed by the academic community.

INTRODUCTION

In 1989 the electrical/electronics manufacturing sector overtook the transport sector as the largest sector of manufacturing industry. This was a result of the increasing influence of information technology in all our lives, for example in the ubiquity of computers that we use at home, work and in the cars we drive [1]. A recent survey by the Channel Marketing Company of Texas suggests that there will be 2.2 personal computers in each US household by the end of this decade. This paper will present some of the challenges that engineers face in the manufacture of such products. It will examine these in the context of some of the changes that have taken place in the manufacturing of the basic technologies, concentrating on the printed circuit board and second level packaging. It will identify the form of contribution that academic research can make in these areas and identify areas worthy of further work. The paper does not address the issues encountered in the manufacture of semiconductor devices.

CHANGES IN THE PATTERN OF ELECTRONICS PRODUCTION

Traditionally the USA has led the world in the manufacture of products containing a significant amount of electronics. This is perhaps demonstrated best by the history of the manufacture of the mainframe computer and the domination of world markets by IBM [2]. This dominance has now been challenged by the Japanese giants of Fujitsu and Hitachi across all areas of the product range. The manufacturing outputs of the Japanese giants are now being challenged in turn by the emerging manufacturing economics of the Newly Industrialised Countries (NIC's) as they serve the growing market of the Pacific Rim. The impacts of these global changes is particularly apparent in the changes in the data processing and consumer electronics sectors. Figure 1 shows the Relative Competitive Advantage (RCA) (the Ratio of Exports to Imports: a value of > 1 shows advantage) of regions manufacturing data processing equipment (after Gomes-Casseres [2]). Figure 2 shows similar data gathered by the CEC [3] for the production of consumer electronics.

Implicit in these data is a significant challenge to the competitive position of older economies. Responses to this challenge are now being significantly affected by the influence of multi- or trans-national enterprises on manufacturing wealth creation and the formation of alliances between multinationals from different continents. The issues that manufacturing economies face have been summarised in Porter [4]. The challenge that we have to address as production engineers is how can we create wealth in our own country or continent and have an enlightened approach to working in the global economy.

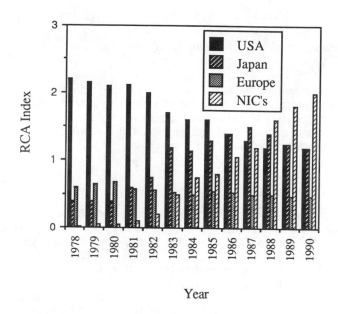

Figure 1. Relative Competitive Advantage in data processing equipment.

	Production	Imports	Exports	Balance	Market	Imp/ Market Ratio
EC	10.7	9.3	1.2	-8.1	18.8	49.5%
Japan	32.2	0.7	16.8	+16.1	16.1	4.3%
USA	5.4	11.2	0.9	-10.3	15.7	71.3%
Korea	7.6	0.5	5.2	+4.7	3	16.6%
ROW	11.1				12.2	

Figure 2. World-wide production of consumer electronics (units $Billion)

CHANGES IN THE METHODS USED
IN THE MANUFACTURE OF ELECTRONIC PRODUCTS

Electronics product production was, in the past, a comparatively unsophisticated manufacturing activity. The basic product, the populated (printed) circuit board, being assembled profitably due to the well understood nature of the production processes, for example, wave soldering. Recent years have seen changes in the techniques used both for the organisation of manufacture and the processes used in manufacture - these two taken together have resulted in a step change in the practice of electronics product manufacture.

Traditionally quality has been inspected into electronic assemblies - notionally because of the low costs of partially completed assemblies - to give a process "yield". Since the 80's and the increased understanding in the West of Just-In-Time (JIT) manufacturing environments, reduced inventory and "continuous improvement", it has been understood that sub-assemblies should be produced "right-first-time" to reduce overall costs encountered in the manufacturing system.

Also in the 80's there has been increased penetration of surface mount technology (SMT) from hybrid manufacture into the more traditional through hole printed board products. This trend is indicated in figure 3 which shows the penetration over time of SMT and Chip On Board (COB) packaging in one sector [5]. This change has been further aggravated by an increase in the complexity of the product produced due to the increased use of microprocessor technology, both in computers themselves and with the introduction of electronic and mechatronic replacements for purely mechanical devices. Furthermore, the reducing scale and increasing complexity of the devices and products has meant that increasing amounts of automation is used in what was traditionally a manual industry.

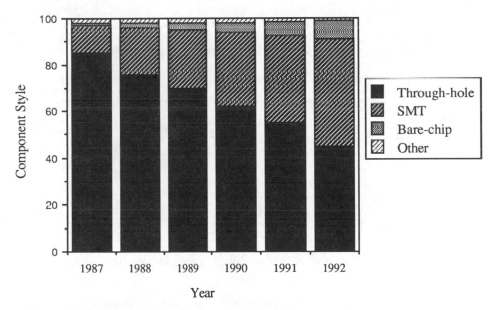

Figure 3. Changing packaging styles with time [5].

This has led to a change in both the organisational methods and the processes employed in many businesses. These changes mean that the industry now needs to understand its processes, particularly in volume production applications (for example personal computers, consumer and automotive applications) to ensure its output quality and to generate efficient production. Many of the processes, such as injection moulding, are as those used in other manufacturing sectors, the major difference being in the volume processes associated with the assembly of components onto printed circuit boards, sometimes known as second level packaging. This paper concentrates on trends in this area.

WORK TO DATE AT LOUGHBOROUGH

The understanding needed to allow improvements in the manufacturing process requires skills at the intersections of materials science, manufacturing technology and design. Work to assist this understanding needs to be carried out in the spirit of design for manufacture. It requires traditional engineering science and mechanical manufacturing skills to be applied to the embodiment of the electrical engineers' functional design. Design for manufacture requires management of the trade-offs between the product and process requirements. Such work must be carried out with an understanding of the coupling between the mechanical arrangement of the product and its long term reliability. Figure 4 attempts to show the relationships between these disciplines.

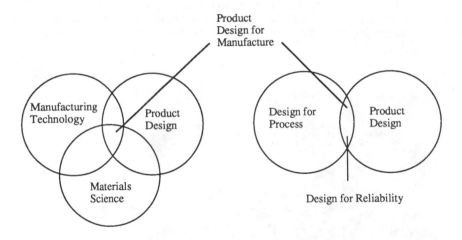

Figure 4. The disciplines that must be addressed in packaging.

Our own work at Loughborough concentrates on tackling quality bottle-necks in the manufacturing process. Increased understanding of the process here is likely to give significant quality and yield improvements by allowing product redesign for manufacture or improved process design or control.

The pre-reflow processes of screen printing and component placement have seen challenges posed by ever decreasing feature sizes and increasing pin counts for surface mount devices. Work at Loughborough has progressively addressed such impacts on these processes with attention to consistent volume deposition of solder paste across the PCB and around single high pin count devices, the influence of the directionality of printing with respect to the orientation of screen apertures and the effects of component placement forces on pre-reflowed solder paste geometry and component quality [6,7,8]. Work has also been directed at the understanding of heat transfer mechanisms in reflow soldering in order to minimise the manufacturability problems associated with this key process. An element of this work was reported in the first FAIM conference [9]. A process modelling tool is emerging which will allow off-line process design before the manufacture of production prototypes.

This work on reflow followed on from a research programme into the control of adhesive dispensing for surface mount [10] which brought a very unstable manufacturing process under control using knowledge based techniques. More recent work focuses on the understanding of the manufacturing process, joint reliability and the relationship between the two for anisotropic adhesives used in fine pitch interconnection and high temperature applications [11]. Such scientifically driven efforts dramatically improve our design for manufacture/design for process knowledge. Complimentary business led work concentrates on understanding how design for manufacture knowledge is held in multinational enterprises [12].

SECOND LEVEL PACKAGING - THE TRENDS

It is now appropriate to begin to discuss some of the challenges that will arise in the manufacture of the next generation of products containing significant amounts of electronics. We will concentrate here on second level packaging, the integration of devices into printed circuit sub-assemblies, the "guts" of the product. Many of the manufacturing challenges in second level packaging will be driven by the increasing number and complexity of microprocessors in all levels of products, computers, automotive components and instrumentation, and domestic appliances including HDTV.

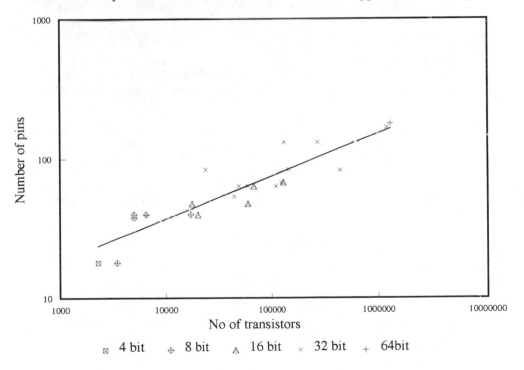

Figure 5. Increase in pin out for increasing microprocessor complexity.

This trend has two impacts, increasing pin out of devices, as is shown in figure 7, with its consequent increases in complexity and reduction in scale, and the potential use of Multi-Chip-Module (MCM) technologies as conventional single chip processors with increased functionality become more difficult to manufacture with acceptable yields. These advances show the necessity to work on technology improvements in conventional packaging and to understand the problems associated with the MCM technology scale up from mainframe production volumes to the volumes consequent on the production of automotive components and computing workstations.

TARGET PROCESS IMPROVEMENTS IN CONVENTIONAL PACKAGING

Electronic packaging is now becoming a major area of innovation world-wide. Effort is also increasingly addressing its scientific basis after pioneering work by Klein Wassink [13] and Lea [14]. The academic community has a unique opportunity to support industry by assisting it to develop a scientific basis on which it can make more informed decisions on manufacturing issues. Understanding and development is required in the following areas additional to those already discussed. Such a brief review can never be complete and reflect the most up to date issues (as many of these will be company confidential!). Other reviews of packaging issues can be found in [16,17,18,19].

Issues in the construction of the printed board itself include the design decision and associated trade-offs between fine line pitch and multilayer substrate's [20]. The generation of fine line (100um) tracks for high density with closely controlled geometry for low impedance (geometric tolerances < 5%) [21] gives many process problems. Manufacturers of surface mounted assemblies are increasingly concerned with the flatness being achieved in PCB' s [22]. The problems associated with this, particularly in "hot and wet" processes are likely to be exacerbated by the tendency to produce thinner boards and larger components. Users are also not satisfied that they understand the opportunities offered by novel polymeric substrate materials [20,p32]. The drilling of fine holes in multilayer boards remains an issue, it has even been suggested that the drilling process can be used to provide the direct metallisation of through holes [23].

Work on screen printing and soldering also proceeds world-wide. Ekere in a recent paper [24] reviews the mechanics of screen printing and indicates some of the active research issues. Understanding soldering and solder alternatives, including low temperature joining techniques [25], remains important. Understanding the behaviour of solder pastes [26] is key especially in the context of the rapid evolution of the products for fine pitch applications and non-CFC post-reflow PCB cleaning including the use of water based pastes.

Even with well characterised processes, because of the high complexity of present generation assemblies it is still necessary to put major efforts into inspection and test. A major US programme addresses design for test, O'Grady [27] and Kim [28]. Machine vision is increasingly used to inspect assemblies, some recent work uses high speed transputer based vision systems to inspect solder joints [29]. In more conventional test domains novel non-contact probes based on laser plasmas have been developed to overcome the disadvantages of the 'bed of nails' testers [30]. Work is still needed in defect identification support, and in the inexpensive, fast, mass non-destructive testing of solder joints [25].

One of the major problems that lead to failures in both production and service is the cleanliness of the manufacturing operation. There is little understanding of the cleanliness required for second level packaging manufacture, both in terms of gross particulate contamination and surface contamination from reactive gases in thermal processes [31]. "No clean" processing and the associated ecological issues are addressed in the book recently published by Dr. Colin Lea. Once faults have been identified the further development of rework techniques for interconnection, especially for fine pitch and adhesives, remains a major issue [25]. This has been compounded by the recent recognition by the industry that abandoned electronic products can give rise to significant ecological damage. Groups are now beginning to examine techniques in design for disassembly [20] and methods of disassembly that will allow recycling of the materials in the product.

Pecht [19] identifies that the complex interactions between all of the issues above require the production of tools to assist engineers in the resolution of the trade offs.

MULTICHIP MODULE CHALLENGES

MCM's are in effect "supercomponent's" consisting of more than one bare semiconductor device mounted directly onto a stable substrate characterised by high interconnection density. The MCM industry has classified the range of MCM's available into three types, namely:

(i) MCM - L : high density laminated printed circuit boards

(ii) MCM - C : ceramic substrate's, either co-fired or low dielectric constant

(iii) MCM - D : covers modules with deposited wiring on silicon or ceramic and metal substrate's.

Multi Chip Modules (MCM's) fill the technology gap between wafer scale integration and printed circuit boards laden with individual plastic packaged devices. They offer a solution to the yield problems currently being experienced in the manufacture of single die packaged microprocessors with the increasing functionality associated with, for example, the manufacture of personal computing workstations with increased functionality and performance [32]. MCMs offer the solution to the problems of VLSI by reducing the area of silicon made in a single device step. In addition they offer the advantages of reduced delays between chips, simplified power distribution, improved density, speed and enhanced dissipation capabilities. These issues are reviewed in Eakin [33].

The percentage yield, Y, in the manufacture of integrated circuits can be expressed as:

$$Y = k e^{-AD}$$

where D is the defect density per unit area and A the area of the silicon die [34]. This shows an exponential reduction in yield with increasing area of silicon and additionally, the defect density also increases with reducing feature size. The paper by Kumar [35] discusses the manufacturing cost implications of reduced feature size. Such solutions will not give the ideal increase in yield because of the yield and process problems associated with the manufacture of the substrate for MCM's and the interconnection needed between the devices.

The engineering workstation technology is driven by market demands for ever increasing computational power in desk top and desk side systems. The technology of traditionally packaged single processors is already at the limits of silicon die production capabilities. In the next two years the market shall see the response to this limitation with vendors venturing into the realm of multichip module technology for single processor desktop machines and multiprocessor technology, in the interim, based on traditional workstation packaging (for example Sun MicroSystems MP600 series), for deskside and data centre towers. The increase in processor speeds and the incorporation of multiprocessor technology will also demand developments in bus architectures and effective shielding of buses from the rest of the system. Despite the maturity of MCM technology resident in many companies in high end systems such as mainframes, the production volumes required for the manufacture of personal and engineering workstations are, however, two to three orders of magnitude larger than those associated with mainframes. The semiconductor manufacturers problems of today will therefore be passed on in the future to those needing to make the MCM's in volume. It is considered that there will be few further problems associated with the final assembly of such super-components into board level assemblies other than those that are exacerbated by their thermal mass and pin count. However, until demand becomes stronger, product costs will remain fairly high, typically \$4 - \$10 per cm^2 of an MCM substrate. This would equate at current prices to \$100 - \$250 per typical workstation processor, currently packaged in the form of 4 large quad flat packs (QFPs) on a printed circuit board. It is likely that MCM's packages aimed at such applications will be, like those aimed at telecommunication and automotive applications, based on ceramic and laminate substrate's rather than silicon [36].

Evolution in electronic packaging may eventually be limited by such fundamental constraints as quantum tunnelling or even quantum uncertainty, but for the foreseeable future the trade off between cooling limits and the requirement of small size to reduce time of flight delays is likely to remain a limiting factor. Time of flight is fundamentally limited by the speed of light (30cm/ns), but electrical signals typically propagate at one quarter to a half of this speed in typical interconnect structures [37].

MANUFACTURING SYSTEM LEVEL ISSUES

Many of the systems level issues in electronics manufacture are similar to those encountered in other industry sectors. As in all sectors overall manufacturing complexity is increased by poor process yields, attention to the quality problems at the process level will simplify many of the   systems that it has been considered necessary to tackle with Computer Integrated Manufacturing (CIM) based solutions. Such attention at the process level actually tackles the addition of value to the product rather than "fixing" manufacturing problems that arise from inadequacies in manufacturing organisations and philosophy.

Much progress has been made in the resolution of systems level issues in electronic manufacturing situations, for example in production activity control and scheduling [38,39,40] and in the informational issues surrounding databases for electronic production including the handling of engineering changes [41]. Significantly approaches developed for scheduling and planning at the system level have become important at the machine level when tackling the efficient sequencing of component placement machines as is shown in the work of Metalla [42]. Chiu [43] shows planning motion within such machines for placement sequencing taking into account the geometric factors of component size and potential collision between components and machine tooling.

CONCLUSIONS

This paper has presented an number of technical areas in which the scientific community can assist manufacturing companies to face the challenges of modern electronics manufacturing. One area in particular that can be identified as significant in the short term future are the problems associated with the scale up of the production of MCM's when applied in personal computers and workstations. In non-technical areas there is a need to understand the international implications of the regional growth of the electronics industry. Increased understanding in this area would both promote world harmony and allow businesses to make informed decisions on long term manufacturing strategies that take advantage of international opportunities. It is necessary for the economies of the world to take an enlightened attitude to each other as both market places and as manufacturing nations.

ACKNOWLEDGEMENTS

We would like to acknowledge the support of our colleagues, Dr Owen Boyle, Robert Jahn, Oluyinka Ogunjimi and Dr Andrew West, for their contributions to this paper both visible and invisible. We would like also to thank Professors David Upton and Marco Iansiti of the Science and Technology Interest Group of the Harvard Business School for helping to keep technologists operating within the constraints of doing business. We would also like to thank the ACME Directorate of the SERC for their continued funding of work in this area (particularly under contracts GR/F 55324 and GR/H12898). We would also like to thank the industrial collaborators of all our research grants in this area - there is nothing like input from those who have to do this for a living for keeping you on the right track.

REFERENCES

1. Jurgen, R. K., Putting Electronics to work in the 1991 car models, IEEE Spectrum, December 1990, 72-75.

2. B. Gomes-Casseres, International Trade, Competition and Alliances in the Computer Industry, Harvard Business School Working Paper, November 1991.

3. The European Electronics and Information Technology Industry: State of Play, Issues at Stake and Proposals for Action, DGXIII, Commission of the European Communities, 1991.

4. Porter, M. E., The Competitive Advantage of Nations, Fress Press, 1990.

5. Miyairi, K., Electronic Packaging, Handbook of Industrial Engineering, 1992, 489-504.

6. Conway, P. P., Williams, D. J., Whalley, D. C., Sargent P. M., Tang A. C. T., Process variables in the reflow soldering of surface mount. Proc. IEEE CHMT/ IEMTS Symposium, Stresa, Italy, 1990, 385-394.

7. Donoghue, J. G.: Final Year Project Report, Department of Manufacturing Engineering Loughborough University of Technology.

8. Donoghue, J. G. and Williams, D. J., Solder paste displacement on component placement, to appear in Proc. IEMT Symposium, Mainz, Germany, 1992.

9. Conway, P., Williams, D. J. and Whalley, D. C., Joint Level Modelling Of Infra Red Reflow For Different Termination Geometries, *FAIM Conference*, Limerick, March 1991, 219-228.

10. Chandraker, R., West, A. A. and Williams, D. J., Intelligent Control of Adhesive Dispensing, IJCIM, Vol 3, No 1, 24-34.

11. Ogunjimi, A. O., and Boyle, O. A., Review of Conductive Adhesive Technology, Processes Group Technical Report, Processes Group Technical Report, 92/2.1, Department of Manufacturing Engineering, Loughborough University of Technology, 1992.

12. Williams, D. J. and Conway, P. P., Technology Migration in Multinational Enterprises, Processes Group Technical Report 92/3, Department of Manufacturing Engineering, Loughborough University of Technology, January 1992.

13. Klein-Wassink, R. J. Soldering in Electronics, Second Edition, Electrochemical Publications, 1989.

14. Lea, C., A Scientific Guide to Surface Mount Technology, Electrochemical Publications, 1984.

16. Ohsaki, T., Electronic Packaging in the 1990's - A Perspective from Asia, IEEE Transactions on CHMT, Vol 14, No 2, June 1991, 254-261.

17. Tummala, R., Electronic Packaging in the 1990's - A Perspective from America, IEEE Transactions on CHMT, Vol 14, No 2, June 1991, 261-271.

18. Wessely, H., Fritz, O., Horn, M., Klimke, Koschink P. and Schmidt, K. H., Electronic Packaging in the 1990's - A Perspective from Europe, IEEE Transactions on CHMT, Vol 14, No 2, June 1991, 272-284.

19. Pecht, M. G., Watts, J. D. and Balakrishnan, S., Placement Design for Producibility, Journal of Electronics Manufacturing, Volume 1, No 2, 1991, 61-76.

20. A manufacturing dilemma, Electronics Manufacture and Test, October 1991, 31-36.

21. Low-cost design for manufacture - the role of statistical methods, Materials and Design, Vol 12, No 4, August 1991, 217-219.

22. Willis, R., How Flat is Flat, Electronics Manufacture and Test, October 1991, 13-14.

23. Lomerson, R., Direct metallizing Technology, IPC Technical Review, September-October 1991, 18-24.

24. Ekere, N.N. and Lo, E., New Challenges in Solder Paste Printing, Journal of Electronics Manufacturing, Volume 1, No 1, 1991, 29-40.

25. Clouder Richards, F. Research Priorities in Joining, Proceedings of a Community Meeting, The SERC ACME Directorate, 13 November 1989.

26. Hwang, J. S., Solder Paste for Electronics Packaging, Van Nostrand Reinhold, 1989.

27. O'Grady, P., Kim, C and Young, R. E, Issues in the Testability of Printed Wiring Boards, Journal of Electronics Manufacture, Vol 2, No 2, 1992.

28. Kim, C., O'Grady, P., and Young, R. E, Test: A Design for Testability System for Printed Wiring Boards, Journal of Electronics Manufacture, Vol 2, No 2, 1992.

29. Netherwood, P., Barnwell, P. and Forte, P., A Transputer Based System for the Visual Inspection of Surface Mount Solder Joints, SERC/DTI Transputer Initiative Mailshot, 1991.

30. Millard, D., Block, R. and Umstader, K., Laser-induced, Plamsa-based, Non-contact Electrical Testing of Functional Hardware, IEEE CHMT Symposium, 1991, 215-217.

31. Williams, D. J., Manufacturing Processes Study Tour Report, Europe, Japan and USA 1989-90, Processes Group Technical Report 90/4, Department of Manufacturing Engineering, Loughborough University of Technology, 1990.

32. Issac, J., Designing the next generation SPARC processor using Multichip-Modules, to be presented at the First International Conference on Multichip Modules, Denver, April 1992.

33. Eakin, W., Gardiner, K and Nayak, J, Cost and performance criteria for selection of matreials and processes in microelectronic packaging, Journal of Electronics Manufacturing, Volume 1, No 1, 1991, 13-23.

34. Ruska, W., Microelectronics Processing, An Introduction to the Manufacture of Integrated Circuits, McGraw-Hill, 1987, pp15-16

35. Kumar, S. and Maslin, G., Off and on-line techniques to optimse processes, Journal of Electronics Manufacturing, Volume 1, No 1, 1991, 23-28.

36. European collaboration on MCM development, Electronic Production, January 1992, 5.

37. Seraphim, R.L., and Lee, C., Principles of Electronic Packaging, McGraw Hill, 1989.

38. Lyons, G. J., Duggan, J. and Bowden, R., Project 477: Pilot implementation of a production activity controller in an electronics assembly environment, IJCIM, Vol 3, No 3/4, 1990, 196-205.

39. Salonna, R., Scheduling and Planning of electronics manufacturing operations, a case study of the MADEMA approach, IJCIM, Vol 3, No 6, 1990, 343-353.

40. Meyer, W. and Isenburg, R., Knowledge based factory supervision: EP 932 results, IJCIM, Vol 3, N0 3/4, 1990, 206-233.

41. Hidde, A. R., Management of information in a system of heterogeneous distributed databases using the example of a PCB assemblage, IJCIM, Vol 4, No 6, 1991, 323-330.

42. Mettala, E. G. and Egbelu, P. J., Alternative Approaches to Sequencing Robot Moves for PCB assembly, IJCIM, Vol 2, No 5, 1989, 243-256.

43. Chiu, C., Yih, Y. and Chang, T. C., A Collision Free sequencing algorithm for PWB Assembly, Journal of Electronics Manufacturing, Volume 1, No 1, 1991, 1-12.

Part I
Managerial Aspects of
World Class Manufacture

THE COMMON STRATEGIC THEMES OF WORLD-CLASS MANUFACTURERS

[1]Anita M. Pietrofitta and [2]Jose Pablo Nuno

[1]*Intel Corporation, Mesa Arizona, U.S.A.*
[2]*Arizona State University, Tempe, Arizona, U.S.A.*

ABSTRACT

The past several years' decline in U.S. manufacturing firms' competitiveness has prompted U.S. manufacturing management to ask two basic questions: (1) How can the current downward trends be reversed so that their companies can survive through the 1990s?, and (2) How can their companies regain leadership in the future world market? Past excuses for poor U.S. manufacturing performance have centered on fiscal, monetary, and trade policies at the national level. But there is a growing awareness that the real problem lies in American managers' strategies and attitudes, and, that these weaknesses are especially evident in the areas of manufacturing and technology development. The conclusions drawn by many researchers is that drastic changes in the management philosophies, perspectives, and approaches of most U.S. manufacturing firms are required before a competitive edge can be regained in the dynamic, global environment of the future.

Firms that are considered world-class manufacturing organizations possess the philosophies, perspectives, and approaches which result in superior manufacturing capabilities and provide a competitive advantage in the marketplace. Therefore, the pursuit of world-class manufacturing status must be the goal of any manufacturing organization that desires to be competitive in the future marketplace. This paper presents a framework which highlights four common strategic themes of world-class manufacturing organizations and explores their interrelationships as they pertain to the manufacturing strategic planning process. The themes are based on a global view of the strategic concepts of core competencies, horizontal strategies, technology orientations, generic strategies, time-based management, Just-In-Time, Total Quality Control, and Computer Integrated Manufacturing. Each section of this paper discusses one of the four themes and the strategic concepts on which they are based.

INTRODUCTION

Strategic literature emphasizes that a company's success is measured by its ability to sustain a competitive advantage in the marketplace. In the past, literature on how to achieve a competitive advantage has been based primarily on Porter's generic strategies: cost leadership or differentiation. However, in today's environment, the competitive edge is shifting to those companies which exhibit innovation, flexibility, and reduced time-to-market. This drastic shift in the factors that provide a competitive advantage requires an equally drastic shift in approaches to manufacturing strategic planning.

The loss of world-wide competitiveness that many U.S. manufacturing organizations' are currently experiencing indicates that these firms have not yet adapted to this new competitive environment. In the past, excuses for poor U.S. manufacturing performance have centered on fiscal, monetary, and trade policies at the national level. But there is a growing awareness that the real problem lies in American managers' strategies and attitudes, and, that these weaknesses are especially evident in the areas of manufacturing and technology development. For example, Wheelwright, 1985, highlights management's view of the manufacturing function, its role, and how that ought to be carried out as the single most important explanation for the worldwide decline in U.S. manufacturing competitiveness. Another study, performed by industrialists under the auspices of the National Research Council, identified manufacturing and technology weakness as the major reason for the U.S. decline in manufacturing competitiveness (1).

The competitive edge in the future marketplace will only belong to firms that exhibit excellence and continuous improvement in all aspects of their organization. Manufacturing firms that have achieved the required shift in management philosophies, perspectives, and strategic approaches to exhibit these traits are described as world-class manufacturing organizations. These firms possess superior manufacturing capabilities which provide their organizations with a competitive advantage in the marketplace. This paper presents a framework which highlights four strategic themes common among world-class manufacturing organizations and explores their interrelationships as they pertain to the manufacturing strategic planning process. Figure 1 illustrates these themes and interrelationships. Each of the following sections of this paper will address one of the four themes. The first section discusses the importance of Focused Goals and the relationships between core competencies, horizontal strategies, technology orientations, and generic strategies. The next section discusses the Focus on the Customer theme which explores the relationship between the core competency and

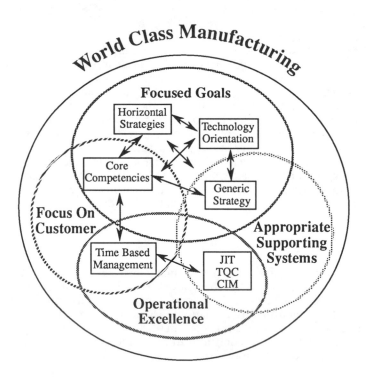

FIG 1. Common Strategic Themes of a World-Class Manufacturing Organization

time-based management approach (TBM). The third theme of Operational Excellence relates time-based management to the Just-In-Time (JIT), Computer-Integrated Manufacturing (CIM), and Total Quality Control (TQC) philosophies. The last section incorporates each of the previously discussed themes into the manufacturing strategy formulation process to aid in the selection of the Appropriate Supporting Systems. The discussion in this chapter is directed towards a multi-divisional firm in a medium- to high-technology industry which manufactures a related diversity of product lines.

FOCUSED GOALS

The first common theme of world-class manufacturing organizations is the proliferation of focused strategic goals throughout all levels of the organization. These goals provide an organization-wide awareness of the firm's relationship to its customers and competitors. As a result, organizational synergy and effectiveness is significantly increased since all internal activities and decisions are directed with a common purpose and towards the same objectives.

A review of manufacturing and business strategic literature identified the following four strategic concepts which provide the foundation for a multi-divisional manufacturing firm to develop a comprehensive set of corporate-level strategic goals: (1) core competencies, (2) horizontal strategies, (3) technology orientations, and (4) generic strategies. These four strategic concepts are illustrated in Figure 2.

Core Competencies

The core competency strategic approach emphasizes a strategic match of the firm's core skills with the marketplace needs and a philosophical separation of technological skills from the components and end-products that a firm manufactures. Identifying a firms core competencies should be the first step in a manufacturing organization's strategy formulation process. With this approach, the strategic vision of a corporation defaults to leveraging its core competencies in a manner which provides

Focused Goals

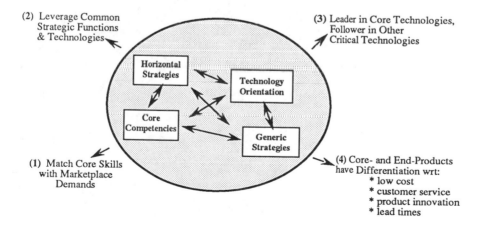

FIG 2. Focused Goals

the corporation with a competitive advantage. Subsequently, its strategic thrusts become developing core technologies and applying them into core products. With this top-down approach, the guidelines for the other major corporate-level strategic decisions are also determined, ie.:

* what businesses to pursue,
* how products and markets should be assigned among divisions or units, and
* the allocation of resources to each business.

As illustrated by Figures 3 and 4, this approach requires a shift in the the typical conceptual view of a corporation from a portfolio of end-products to a portfolio of core-competencies. In essence, as the phrase "structure follows strategy" implies, the structure of the corporation should be built to support its core technical competencies and core-products rather than focusing solely on end-products. For example, Intel Corporation is structured around its core technology, semi-conductor design, and core product, silicon wafers, with separate silicon fabrication plants that specialize in specific technologies. In a further extension of the core product concept, they have distinct semi-conductor assembly and test sites. And, illustrating the end-product production distinction, have distinct end-product production sites which assemble computer systems. This approach allows for high specialization of technological expertise while maximizing flexibility and responsiveness to marketplace demands.

Horizontal Strategies

Once core competencies, marketplace strategies, and structural requirements are defined at the corporate-level, common functional activities must be identified, coordinated, and standardized horizontally across the entire organization where ever it is advantageous to do so. This requires a horizontal strategy, as described by Porter,

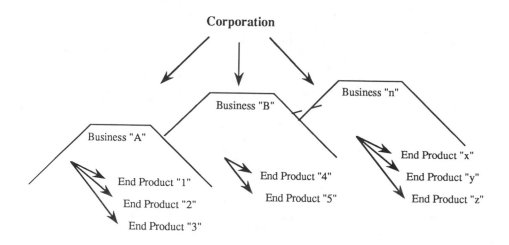

FIG 3. A Multi-Divisional Organization with an End-Product Orientation

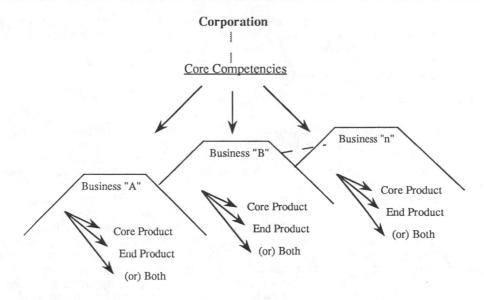

<u>FIG 4.</u> A Multi-Divisional Organization with a Core Competency Orientation

1985, to identify the functional and technological areas whose interrelationships or consolidation will benefit the firm. This process allows an organization to eliminate redundancy and leverage off of its commonalities. Functions may be consolidated at the corporate level or just coordinated at the business level through procedural guidelines and common data systems. Again, this requires an enterprise-wide view of a multi-divisional corporation as one entity which designs, produces, sells, and distributes core- and end-products.

Technology Orientation

In addition to identifying core technologies of the organization, other critical supporting technologies must also be identified. The distinction of these two categories of technologies is important since they require unique strategic approaches, management skills, and support mechanisms. Technological leadership implies the active pursuit of technological research and development and a goal to generate improved, substitute technologies ahead of the competition. These objectives are associated with a firm's core technologies. Therefore the optimal culture, organizational infrastructure, and managerial skills said to support this type of environment, namely those which support and encourage risk-taking and innovation, should surround these technological areas in the organization. The technology follower orientation applies to the non-core but critical technologies used in the design or manufacture of core-products and end-products. Advances in the development of these technologies outside of the organization must be constantly reviewed to ensure that they are rapidly adopted if it is advantageous to do so. Critical technological breakthroughs in these areas can be used to enhance core-product features or production capabilities and/or be incorporated to develop new end-products. The support systems and skills in areas which interface with these technologies must be designed with flexibility and responsiveness in mind.

It is important to explicitly define these core and critical technologies since the process itself forces management to understand them and subsequently develop explicit statements regarding their appropriate strategic approaches. This in turns aids in identifying the areas of emphasis and skills required to support each business division's technological requirements. As a result, the explicitness and focus of the strategic goals and objectives are further enhanced.

Generic Business Strategies

As discussed above, the firm's overall means for seeking a competitive advantage is addressed at the corporate-level by its core-competency strategy. In addition to this macro view of strategy, generic business-/product line-level strategies must be identified to further define common and unique strategic objectives and support requirements. Generic business strategies can be defined as a generalization of different strategic orientations that is used to categorize how a firm plans to achieve a competitive advantage in the marketplace. The categories are defined with respect to primary measures of performance (MOP) that describe the overall objectives of the firm. One such framework is Porter's generic business strategies which categorize business strategy into two dimensions: (1) market scope and (2) strategy, ie. the categories of cost leadership or product differentiation. Using Porter's framework, core product lines should follow a differentiation strategy since application of core technologies will provide product differentiation with respect to product features and/or the advantages provided through superior processing capabilities such as cycle time reduction. Once each product line's strategic orientation is understood, the corporation's organizational structure can be further streamlined to group those with common orientations together.

To summarize, one essential activity in world-class manufacturing organizations is the stringent practice of strategic management. This requires a comprehensive and global strategic analysis at the corporate level of an organization. The resulting strategic goals identify appropriate trade-off variables, common and unique support requirements for each business and product-line, and the information necessary to design the firm's organizational structure. This, in turn, maximizes the effectiveness and efficiency of the firm by providing a common focus for decisions and activities throughout a firm and ensuring the synergy of strategic initiatives in each business area. Each of the above four strategic analyses should be addressed in a relatively hierarchical manner in the corporate strategic planning process, ie. analyze core competencies, then horizontal strategies, followed by either the technology orientation or generic strategy analyses.

FOCUS ON THE CUSTOMER

As mentioned previously, the competitive edge is shifting to those companies which exhibit innovation, flexibility and reduced time-to-market in addition to low cost and high quality. The ability to achieve these competitive capabilities are directly related to the methods by which a company manages their product development, production, sales and distribution activities. The common thread among world-class manufacturing organizations in their management of these activities is their constant focus on customer requirements and emphasis on the continuous reduction of cycle times.

As illustrated in Figure 5, both the core competency and time-based management (TBM) strategic concepts consider the customer as the most important factor in effectively managing cycle times. In addition to supporting the development and application of the firm's core technologies, implementation of the core competency approach requires an organizational structure that is streamlined around activities which add value for the customer. Similarly, the concept behind time-based management is to develop superior insight into what customers value and then build the systems to support those needs. Both concepts emphasize that a continuous workflow of all value-added

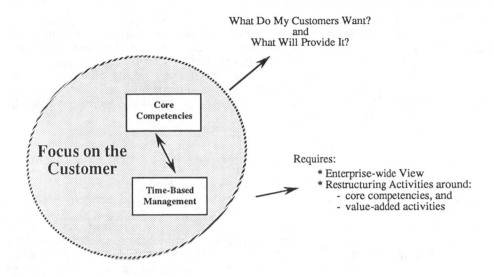

FIG 5. Focus on the Customer

activities in the firm must be created and managed to constantly reduce its cycle time. As a result of simplifying, streamlining and leveraging existing in-house knowledge, a company can consistently reduce product development and production cycle times which in turn reduce the costs and risks of new product development and increase the firm's responsiveness to the marketplace. Kenworth Mexicana is a good example of a successful implementation of these concepts. Since the commencement of their simplification program two years ago, they have increases their production capabilities by 250% while reducing indirect costs by 30%.

Initiation of the streamlining process requires asking two primary questions:

(1) What deliverables do my customers want?, and

(2) What organizations and processes inside my company will most directly provide those deliverables?

Once these questions are answered with regards to the customers external to the organization, the questions should be asked again at the functional level with regards to internal customers. The results of these questions should be provided as input to corporate restructuring requirements.

Effective implementation of the core competency and time-based management concepts requires that three fundamental changes occur within the corporation:

(1) a shift in operational goals to emphasize reducing cycle times as a means to reducing costs,

(2) a shift in the rules on how the organization should work to emphasize streamlining around product lines and value-added activities, and

(3) the development of new measures of performance which are time/cost-based rather than just cost-based.

The most important aspect of these changes is the focus on customer requirements throughout the organization. Ideally these changes should be coordinated across all functional areas from the corporate level and then implemented throughout the

organization. If the above analysis is not solicited by corporate-level management, the manufacturing organization or any department can take the initiative to strategically manage themselves using these principles of core competencies and time-based management. Since manufacturing interfaces with almost all functional areas, once it has its own internal operations streamlined and focused on its customers, it can encourage other areas to join in on the transition and implement
change in a bottom-up fashion. This is a proactive approach that is indicative of world-class manufacturing organizations.

OPERATIONAL EXCELLENCE

The core competency and time-based management strategic concepts address the need for streamlining around customer requirements and reducing cycle times across the organization. But in order to support these goals, becoming a world class manufacturer also requires an emphasis on continuous improvement of factory operations and the pursuit of operational excellence. If production operations are not continuously improved, a firm could easily lose a competitive edge to other firms which are emphasizing continuous improvement in their manufacturing operations. Therefore, the third theme of world-class manufacturing is the integration of principles of operational excellence into factory management. These principles emphasize simplification, continuous elimination of waste, first time quality, improved control over production processes, discipline, and adherence to commitments.

TBM, Just-In-Time (JIT), Total Quality Control (TQC), and Computer-Integrated Manufacturing (CIM), are unique manufacturing philosophies which emphasize different aspects of manufacturing operations management. In general, TBM emphasizes

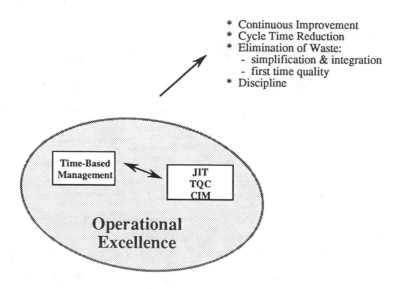

FIG 6. Operational Excellence

reduction of cycle times and a shift to time/cost-based performance measures; JIT emphasizes smooth production planning and elimination of variability in production processes; TQC emphasizes first-time quality; and CIM emphasizes integration of data and shared knowledge. The appropriate implementation of the different aspects of these philosophies is essential for a firm's competitiveness. But it is important to note that all four of these philosophies share the principles of operational excellence as a common foundation. Therefore, implementation of the operational aspects of TBM, JIT, TQC, or CIM will not be successful without management's commitment to these principles.

APPROPRIATE SUPPORTING SYSTEMS

The fourth common denominator of world-class firms is an emphasis on the implementation of the appropriate infrastructural systems to support their strategic goals and objectives. These systems provide a firm with the means to achieve operational excellence, reduce costs, and reduce cycle times. Since different areas of the corporation have both common and unique strategic goals and customer requirements, the assessment of current systems and implementation of new systems is most effective if each of these goals and requirements are explicitly understood and approached from a functional activity standpoint. The four corporate-level strategic analyses discussed previously define the corporate and business goals required to develop strategies for each functional area and subsequently the guidelines required to develop and implement the appropriate supporting systems for each area in the organization.

To accomplish this effort effectively, functional strategies should be initially developed at the corporate level and then at subsequent hierarchical levels of the firm as illustrated in Figure 7. This strategic approach ensures that an enterprise-wide view of all corporate resources is held to optimize the consolidation and sharing of critical

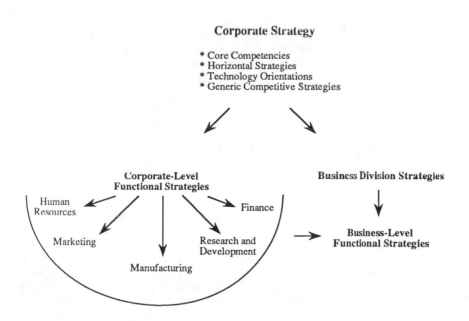

FIG 7. The Hierarchical Flow of Strategic Goals with a Corporate-Level Emphasis on Functional Strategies

manufacturing resources and skills. The main purpose of the manufacturing strategy at the corporate-level should be to determine how to gain the competitive edge in the production of core products and acquire the skills, knowledge, and behaviors necessary to meet individual divisions' competitive objectives. Achieving this objective will require: (1) exploiting the appropriate horizontal relationships, (2) coordinating with other functional strategies, (3) defining the appropriate organizational structure, and (4) coordinating the development and proliferation of core manufacturing process technologies. The resulting corporate-level manufacturing strategy should provide focused goals and operational standards from which lower hierarchical levels of the manufacturing organization can develop individual goals and operating procedures.

Leong, et. al's, 1990, summary of strategic manufacturing literature describes the contents of manufacturing strategy as consisting of two key elements: (1) competitive priorities, and (2) strategic decision category policies. Their study identified five distinct categories of strategic manufacturing priorities: Quality, Delivery Performance, Cost, Flexibility, and Innovativeness. The goal of a world-class manufacturing organization should be to continuously strive for improvements in each of these areas in a manner that is appropriate for their environment. Priorities should be dictated by their relative importance to the customer and internal benefits they provide to the firm. These priorities will in turn define the appropriateness of different supporting systems. To achieve synergy throughout the corporation, each production area's competitive priorities must be understood and then the organizational structure should be further redefined to combine production areas with common requirements.

As stated above, the second element in the contents of a manufacturing strategy is a comprehensive set of strategic decision category policies for each production area. Leong, et. al, 1990, compiled a list of strategic manufacturing decision categories such as production planning, quality, and technology. Policies to govern decisions in each of these areas should be developed for each category at each hierarchical level of the firm and be customized for individual business divisions, production facilities, and production areas.

The underlying philosophies of operational excellence and the relevant operational aspects of TBM, JIT, TQC, and CIM should be incorporated into each decision policy to ensure that the appropriate culture and systems are emphasized for each area. As illustrated in Figure 8, TQC can be incorporated with the Quality decision policy category, JIT with the Production Planning/Material Control category, and CIM with the Technology decision category. This example illustrates that the strategic use of manufacturing decision policies can be an effective method to guide every-day and long-term decisions which will impact manufacturing's strategic capabilities. Communicating and enforcing these policies throughout the organization is essential to ensure the desired organizational effectiveness is achieved.

SUMMARY

Achieving world-class manufacturing status requires stringent practice of strategic management. This paper presents a framework which identifies strategic philosophies, concepts, and processes required to develop a comprehensive manufacturing strategic plan. A corporate-level approach was emphasized due to the need for a much broader definition of manufacturing's role than what currently exists in most U.S. manufacturing organizations. New skills, new knowledge, and new methods for performing activities are required at all organizational levels within manufacturing and between manufacturing and other functions to make implementation of a corporate-wide plan successful. This requires significantly lowering interfunctional and interorganizational boundaries as well as managing projects differently to effectively combine components of the entire organization and subsequently accomplish the transition to and sustainment of world-class manufacturing status.

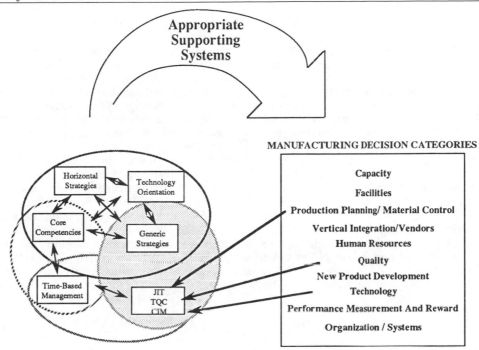

FIG 8. The Appropriate Supporting Systems

In summary, achieving world-class manufacturing status requires a global perspective in the strategic management of a corporation. World-class manufacturing status and a competitive advantage in the marketplace are achieved if each of the four described themes and related strategic concepts are viewed with a global perspective and made an integral part of the organizational culture and manufacturing strategy formulation process. The first theme, Focused Goals, emphasizes that the first step towards world-class manufacturing is the development of strategic goals through the analyses of the organization's core competencies, horizontal interrelationships, technological orientations, and generic strategies. The second common theme of world-class manufacturers is a constant Focus on the Customer with an emphasis on thoroughly understanding customer requirements and streamlining functional activities around activities that add value for the customer. The third theme is the relentless pursuit of Operational Excellence which requires continuous improvement, elimination of waste through simplification, insistence on first-time quality, and cycle time reduction. Finally, the fourth common theme is implementation of the Appropriate Supporting Systems required to achieve the strategic objectives. To select the appropriate systems for each production area, a comprehensive corporate-level manufacturing strategy must be developed. The manufacturing strategy formulation process should identify common and unique strategic requirements across the organization, define the appropriate strategic decision policies for each level of the organization, and determine proper resource allocations.

FURTHER READING

For additional information on the core competency strategic approach, reference an article by Prahalad & Hamel, "The Core Competence of the Corporation" in the May-June 1990 issue of the Harvard Business Review. Implementation of the time-based management approach is discussed in detail in Stalk, Jr. & Hout's article in the January-February, 1990, issue of Planning Review entitled "Redesign Your Organization for Time-Based Management." Eschenbach & Geistauts's article in the May, 1987 edition of IEEE Transactions on Engineering Management entitled "Strategically Focused Engineering: Design and Management" addresses the support requirements for differing technology orientations. Hayes, Wheelwright, & Clark's book called "Dynamic Manufacturing: Creating the Learning Organization" provides detailed information on different aspects of world-class manufacturing including strategic decision categories, competitive priorities, and the road to achieving operational excellence. Porter's book entitled "Competitive Advantage" provides detailed information on generic business and horizontal strategies. And, an article by Nuno, Shunk, & Cruz y Celis entitled "A CIM Architechture and Implementation in a Truck Company to Become a World-Class Manufacturer: Fostering the North American Economic Community", in March 1990's CRC Press, provides detailed information on Kenworth Mexicana's approach to simplification which resulted in the improvements discussed in this article.

ADDITIONAL REFERENCES

1. Badiru, AB: A Management Guide to Automation Cost Justification. *Industrial Engineering* 22: 27-30, 1990.

2. Bower JL, Hout TM: Fast-cycle Capability for Competitive Power. *Harvard Business Review* November-December: 27-30, 1988.

3. DeMeyer A, Nakane J, Ferdows, K: Flexibility: The Next Competitive Battle. *Strategic Management Journal* 10: 135-144, 1989.

4. Drucker PE: The Emerging Theory of Manufacturing. *Harvard Business Review* May-June: 94-102, 1990.

5. Kanter RM: From the Editor - How to Compete. *Harvard Business Review* July-August: 7-8, 1990.

6. Kotha S, Orne D: Generic Manufacturing Strategies. *Strategic Management Journal*: 211-231,1989.

7. Leong, GK, Snyder, DL, Ward, PT: Research in the Process and Content of Manufacturing Strategy. *Omega International Journal of Management Sciences.* 18:109-122, 1990.

8. Marucheck, A, Pannesi, R, Anderson, C: An Exploratory Study of the Manufacturing Strategy Process in Practice. *Journal of Operations Management* 9:101-123, 1990.

9. Stalk, G Jr.: Time - The Next Source of Competitive Advantage. *Harvard Business Review* November-December:41-51, 1988.

10. Wheelwright, SC: Restoring the Competitive Edge in U.S. Manufacturing. *California Management Review* Summer: 26-42, 1985.

GLOBAL LOGISTICS SUPPORT PLANNING FOR MANUFACTURING AND DISTRIBUTION

Keith M. Gardiner, Marlin U. Thomas, and George R. Wilson

Mohler Laboratory #200, Lehigh University, Bethelem, Pennsylvania, U.S.A.

ABSTRACT

Global issues for strategic manufacturing efforts are presented that require a "rapid response" capability for manufacturing and distribution. Included are high priority logistics issues and discussion on management structures for dynamic responsiveness. The features of a dynamic location policy are discussed with regard to the company, competitors, and economic factors.

1. INTRODUCTION

The marketplace of today is highly segmented and hungry for only the best quality. As a result, product life cycles have been reduced as new technologies and products are introduced as early and as often as possible. A good global logistics plan of today may be outdated by the realities of the marketplace of tomorrow; thus such issues as location of production and distribution resources need to be responsive to market stimuli and to the actions of competitors over time.

The purpose of this paper is to provide a background and framework for designing and developing logistical systems for global environments. We start with a discussion in Section 2 of the management structures for dealing dynamically with response to the needs of global markets followed in Section 3 with the logistics issues and manufacturing strategies that have emerged with the transition to global competition. In Section 4, we discuss planning for logistics support. The flexibility of highly competitive firms in the design and manufacturing areas must be duplicated in the logistics support area, as well. Historically, logistical decisions involving location and distribution have been based on some assumed steady state and stable conditions for supply and demand. A location analysis is presented that recognizes market disequilibria may be exploited for competitive advantage. Some concluding remarks are provided in Section 5.

2. MANAGEMENT STRUCTURES FOR DYNAMIC RESPONSIVENESS

To be successful in the future our organizations and manufacturing strategies must be structured with a facility to accept and adapt to change as being endemic. In general, change is anathema. Any enterprise creates an organization to prosecute its objectives and advance its interests. This organization, if it embodies anything more than very few people is compelled to develop bureaucratic structures to handle routine matters uniformly and expeditiously. Organizations of their nature try to establish "surprise-free" environments for their customers and employees. Thus we see that both people and the organizations in which they operate are change resistant, and yet our manufacturing strategies must be about producing change continually.

In any new manufacturing enterprise, or location, the organization and administration must grow to handle the challenges presented by the dynamic environment. The management of the manufacturing and distribution must be tightly coupled to the product development sequence and to marketing. All impedances must be minimized by maintaining "flat" organizations with authority matched to operational activities and responsibilities at the lowest feasible levels. Of course, in a dynamic environment with low inventories and fast cycle times, risks of large volumes of scrap and waste are lowered, and the rewards of employee empowerment in time-to-market, customer satisfaction, market share and overall cost reductions can be immense. It is vital though that all "engaged" employees have an informed appreciation of the whole product cycle, and of their customers.

The nodes of our future production and distribution systems will be small and highly focused with flexible 'empowered' workforces. They will work as teams, and be measured and rewarded according the value each team adds towards the goals for satisfying the customers of the enterprise. The whole product generation cycle from concept to attainment of customer satisfaction must be handled comprehensively, in the shortest possible time and with minimum trauma (or entropy losses). There will be a new compliant management structure and manufacturing strategy which will ensure that the teams possess the necessary expertise, facilities and resources to achieve their goals. It must be recognized that employee satisfaction and customer satisfaction go together. Hierarchical, organizational, and psychological barriers that customarily inhibit communications and delay the implementation of new ideas and sound engineering practice must be eliminated.

3. GLOBAL ISSUES AND MANUFACTURING STRATEGIES

While the U.S. still has the largest economy, we continue to experience a declining share of the global GNP and high technology trade. Today's global economy is changing drastically through new emerging economic powers, and increasing and expanding international integration.[1] In order to be competitive there must be systems and strategies that allow for quick and highly efficient response to changes. As we have already learned, doing nothing is not going to work for the U.S. or any other industrial country.

A business strategy must include a global vision and strategy for dealing with the respective global competition.[2] In any highly competitive environment, firms are constantly seeking ways to gain an advantage in the market place. A number of factors effect the competitiveness of a company. DeMeyer et al [3] surveyed a number of U. S. manufacturers to elicit a rank-ordered list of strategic capabilities a company must have to be competitive: conformance quality, on-time delivery, performance quality, delivery speed, product flexibility, after-sale service, price, broad line, broad distribution, volume flexibility, and promotion. In a comparison to the same study conducted four years previously, it was found that time-based capabilities such as delivery speed, after-sale service, product flexibility, volume flexibility, delivery increased in importance while other factors were the same or decreased in importance. This message is reinforced by Stalk and Hout [4] who point out that decreased lot sizes, increased quality, increased variety, and decreased response time ultimately push down costs rather than cause them to rise. Also, customer demand is extremely sensitive to these time-based factors, causing companies that stress these factors to quickly dominate the market.

Examples of the need for dynamic flexibility in the allocation of manufacturing resources can be found in every hemisphere. Many of the sadder episodes relating to the perceived decline of US industry are the result of a lack of responsiveness to opportunities occasioned by changes which were difficult to forecast either for reasons of their rapid and almost random occurrence, or because of chaos conditions in related international markets and technology.

There were significant concerns about the health and viability of US semiconductor manufacturers in the eighties as the density of memory chips reached 256 kilobit capacity. Heavy pressures were brought to bear upon Japanese industry to inhibit supplies to the US. As this happened the burgeoning US personal computer industry experienced memory shortages; prices rocketed with substantial negative impact upon the US consumer. A similar phenomena has occurred with respect to flat panel displays: In order to protect incipient US industrial developments in this field a very heavy tariff has been levied on imported unassembled panels. There is little levy on assembled products using flat panels, laptop PC's etc., thus firms with ability to relocate their laptop manufacture and assembly operations off-shore could gain substantial cost advantage over those with lesser choice of site. Apple relocated their laptop assembly activities from the US to Ireland, resulting in some loss to the US economy, but an overall gain as result of the increased competitiveness of Apple laptops. Improvements in the profitability and output of Apple are almost certainly better for the US economy in the long haul.

Clothing and textiles offer a further example of the need for fast response to changing market, taxes and tariff conditions: The US textile industry has marvelous capability for production of excellent quality and at a good price high volume standard pattern fabrics. However, when it comes to custom fabric designs, smaller less automated off-shore operations have an edge; but, the import of off-shore fabrics occasions high duties. The importation of finished clothing carries less burdensome tariffs, thus much high fashion seasonal clothing can be produced more economically off-shore, but this creates longer lead times. The fortunes of retail operations like The Gap, KMart, Sears and Bloomingdales depends upon how and where they place their large volume seasonal orders, and the pricing and manufacturing regimes of their suppliers. Apart from the vagaries of any fashion business there is a high sensitivity to political and tariff trends.

In the automotive arena which now comprises a major twelve(?) rather than the original "big three", flexibility in plant loading and assignment of primary manufacturing sites is a major factor in ensuring corporate profitability. Chrysler admits to allocating assembly of larger models to Canadian plants to give them greater flexibility in meeting federal "Corporate Average Fuel Efficiency" (CAFE) targets for their US domestic output. Honda are extremely active, adjusting fabrication venues and ratios of interplant intercontinental shipments to meet varying content requirements imposed by politicians in different marketplaces. It is clear that all multinationals must engage in such endeavors if they are to achieve their business objectives while promoting the greatest degree of customer and stakeholder satisfaction in all marketplaces.

4. AN APPROACH TO DYNAMIC LOGISTICS SUPPORT

Logistics includes all material flow, distribution, transportation, storage and inventory control. No matter how marvelous the product design, without effective logistics support of development and manufacturing, a company cannot compete in the global markets of today. The point of view taken in modeling the competition among N firms for market share with respect to a single product is that in which the firms are playing an N-person, noncooperative game. The actions that may be taken by a firm include changing the price, improving the technology of the production process, and improving the quality of the product. Some actions will immediately effect market share; e.g., a reduction in price or special promotion. Other actions may have a delayed effect on market share, such as developing production flexibility to decrease lot sizes and lead times. Geographic distribution of production and distribution facilities also effect market share, not only by reducing delivery times, but also for social and geo-political reasons.

Economic Foundation for a Dynamic Logistical Analysis

The usual long term goal is to establish an equilibrium condition among all the players, or exploit to advantage a disequilibrium caused by exogenous variables or the action of one or more of the companies. An equilibrium condition would appear elusive except for the case where the players adjust prices to which the market adjusts quickly. Such an equilibrium would be short-lived in the competitive environment just described. Samuelson [5] recognized many years ago that studying a static market equilibrium is of general interest only if it is stable, which is not likely in a market-place with highly competitive participants. Even earlier, Schumpeter [6] suggested that static equilibrium is more likely to occur if there is little competition. But with increasing competition, economies are constantly evolving precisely because they are "out of equilibrium". This theory of "evolutionary economics" is characterized by long term and progressive change with intervals of rapid change. During intervals not subject to rapid changes, conditions may exist for "temporary equilibrium". These conditions are made more precise in a survey article on temporary equilibrium by Grandmont [7].

Evolutionary economics is the driving theoretical force behind the Self-Organizing Systems concept of Nicolis and Prigogine [8]. Here, industries go through a process of industrial mutation, or "creative destruction", that revolutionizes the economic structures of the affected industries from within. The various features of self-organization are:

1. Knowledge creation and diffusion - presenting a catalyst for industrial change.
2. Generation of new alternatives - in production processes and/or products through mutation or innovation.
3. Exit and entry triggers - existing products exit and new ones enter as stimulated

by profit opportunities.

4. Selection mechanisms - evolution by competitive substitution; superior techniques are initiated and inferior ones are phased out or updated, new products with better attributes to price combination, replace old ones through customer preference, and output is increased through investment in production capacity.

5. Time-paths - every successful new product follows a logistic "time-path" creating successive waves of innovation and imitation each time superior production techniques or products are introduced.

6. Structural patterns - at any point in time, products and techniques of different "vintage" will coexist and form a distribution of profit, productivity, and capacity over firms that represents historical evolution and conditions for future change in each market.

The foregoing discussion gives a great deal of economic justification for the employment of a dynamic spatial location analysis. However, the literature is dominated by static spatial (or competitive) location models. The Dobson and Karmarkar [9] model is representative in that it establishes locations for production, service provision, warehousing, and retail sales. Customer demands are located at the nodes of the network, a subset of which provides the locations for facilities. Competitors jointly maximize their profit using a fixed price at all locations. Harker [10] on the other hand includes competitive pricing and establishes spatial price equilibrium in a spatial competition model. He overcomes the unrealistic assumption of perfect competition (i.e., price=marginal cost) by using a Cournot-Nash equilibrium framework. Lederer and Thisse [11] use a static, two-stage approach of first choosing locations, production techniques, and input mixes (which are now known) and then choosing "delivered prices"; i.e., equilibrium prices including transportation costs. For a summary of theoretical and algorithmic results for static games in location analysis, see Michandani and Francis [12].

A Sequential Competitive Location Model

The methodology we employ to capture the dynamic aspect of an evolutionary competitive location problem is the sequential game (see Heyman and Sobel [13], Fudenberg and Triole [14]). A sequential game is an N-person game where each person makes a sequence of decisions that influence the evolution of the process and affect the rewards garnered by each participant over time. A sequential game is actually a generalization of Markov decision processes and static game theory. Let

$$a_t^q \equiv \text{action taken by player q in period t}$$
$$a_t \equiv \{a_t^q : q \in Q, \text{ set of players}\}, \ |Q| = N$$
$$S_t \equiv \text{state at begining of period t}$$
$$A_s^q \equiv \{\text{actions for player q} | \text{state s}\}$$
$$A_s \equiv \underset{q \in Q}{\times} A_s^q \equiv \{\text{actions in period t} | S_t = s\}.$$

The state of the competitive location sequential game includes the location of each facility defined on nodes of a network of centroids representing countries or regions of countries. In addition, the strategic capabilities listed earlier (or some similar list of competitiveness attributes) are included with a level of achievement recorded for each using a suitable metric. Other states of the system would not be associated with product and process technology; rather, they would be the level of regulation or tariffs existing at a particular location. The state transition function satisfies

$$p_{sj}(a) = P\{S_{t+1} = j \mid S_t = s, a_t = a\}$$

which exhibits a Markov property. We assume that the competitiveness attributes are changed in discrete increments at a cost that is a function of the original level and the incremental jump. A transition from one state vector, s, to another state vector, j, once an action is taken to change one or more state variables in s, has a probability associated with the final state j because previous actions taken by all competitors may cause a change in state before the latest action by a specific competitor has a chance to be fully implemented. Thus, the state space is expanded for those states that have a "delayed" transition; in addition to the starting value, supplementary variables are appended giving incremental change and the number of periods until the change is implemented.

Let X_{tq} equal the reward received by player q in period t. The single-stage reward function is then

$$r_q(s,a) = E[X_{tq} \mid S_t = s, a_t = a] .$$

In the context of the dynamic competitive location problem, the reward at each stage can be, for example, the profit earned by each company for their product at the price set at each location and at a volume dictated by their market share. The market share is a function of the location of the companies facilities and the level of the competitiveness attributes as perceived by customers at various locations. A "learning curve" effect delays the full realization of an enhanced profit level. Taking full advantage of a new technology may take some time and the marketplace may not "learn" of the enhanced desirability of a product from a particular company right away.

Let P_{iqk} equal the proportion of the population at location i that prefers the product produced or made available at a distribution point by company q at location k. Further stratification of the population at location i into subgroups is possible. The multinomial logit model is used to characterize the market-share function since there is evidence to suggest the model is consistent with empirical evidence when used in conjunction with game-theoretic analysis, as discussed in Karnani [15].
Thus, we have

$$P_{iqk} = \frac{\exp(\overline{s}^T_{qk} \overline{\beta}_{iqk})}{\sum_k \sum_q \exp(\overline{s}^T_{qk} \overline{\beta}_{iqk})} ,$$

where

$s_{qk} \equiv$ vector of competitive attribute values for facility k of company q.

$\beta_{ijk} \equiv$ vector of preference parameters associated with the attribute set of the facility k of company q as perceived by customers at location i.

Notice, it is not the absolute value of product and production process attributes that lead to increased market share, but the values of attributes relative to competitors that makes a difference. It is not that a company is constantly evolving its technology but the rate of that evolution relative to the competition that is a critical consideration in this model. If we let D_{it} equal the forecasted demand for the product by population i in period t, then the reward function may be written

$$r_{qt}(s,a) = \sum_{k} \sum_{i} (v_{qkt} - c_{qk}(s)) p_{iqk} D_{it}$$

where

$v_{qkt} \equiv$ unit price for the product by company q at facility k in period t.[1]

$c_{qk}(s) \equiv$ unit cost to produce (or store) the product at location k given the attribute settings in that part of the state vector, s, associated with facility k of company q.

Although the duration of the game could be considered infinite, it proceeds for a finite interval in its present form. Ultimately, companies are added or are dropped, customer tastes change necessitating an updating of the model, either structurally or parametrically.

The model will be very useful in assessing whether it is more beneficial to be a "leader" or a "follower" with respect to major technological advances and relocation of facilties as a function of such things as the lag time before a follower can respond to a leader's initiative and the degree of imperfection in the information the competitors have about one another.

5. CONCLUDING REMARKS

Any manufacturing enterprise which desires to prosper and succeed must have capability to develop and deliver new products rapidly. There is an absolute need to be responsive to both the perceived needs of customers, and to the fast changing and often mandatory requirements of our global society. Economic, environmental, industrial, social and technological conditions are in a state of continual and dynamic change. The assumptions of static or quasi-static conditions which are implicit in much planning and theoretical development are inappropriate in today's environment.

Impedances within manufacturing and distribution systems must be reduced by

---

[1] v_{qk} is one of the attributes in s and an action in period t can set it to its value, v_{qjt}.

increasing spans of control, and other worker empowerment techniques to enable faster decision making. Likewise it is increasingly important to reduce inventories, and cycle times with corresponding improvements in responsiveness to special customer needs, costs, and quality.

In the future it is likely that the most successful enterprises will develop highly flexible capabilities with ability for "mass customization" and highest value-add in close proximity to the final customer. Manufacturing systems are likely to become globally integrated, but will comprise a series of smaller and more flexible focused "cellular" factories acting in concert to satisfy customer demands in fast changing markets. The output of these factories will be planned and allocated in response to globally changing regulations, tariff structures, materials supplies, workforce skills, other resources and variables to maximize enterprise and customer benefits. The work presented herein describes an approach which is worthy of consideration to facilitate this logistics support planning. The methodology discussed should be useful in developing a strategic policy for product and process improvement coupled with production and distribution placement over time.

REFERENCES

1.   R. Aggarwal, "The Strategic Challenge of the Evolving Global Economy," *Business Horizons*, July-August, 1987, pp. 38-44.
2.   W.C. Kim and R.A. Mauborgne, "Becoming an Effective Global Competitor," *Journal of Business Strategy*, January/February 1988, pp. 33-37.
3.   A. De Meyer, J. Nakane, J.G. Miller, and K. Ferdows, "Flexibility: The Next Competitive Battle the Manufacturing Futures Survey," *Strategic Management Journal*, Vol. 10, 1989, pp. 135-144.
4.   Stalk, G. and Hout, T. M., *Competing Against Time: How Time-Based Competition is Reshaping Global Markets*, The Free Press, New York, 1990.
5.   Samuleson, P. A. *Foundations of Economic Analysis*, Harvard University Press, Cambridge, Mass., 1946.
6.   Schumpeter, J. A., *Theory of Economic Development*, Oxford University Press, New York, 1934.
7.   Grandmont, J. M., Temporary General Equilibrium Theory, *Econometrica*, 45, pp. 535-572, 1977.
8.   Nicolis, G. and Prigogine, I., *Self-organization in Nonequilibrium Systems: From Dissipative Structures to Order through Fluctuations*, John Wiley, NY, 1977.
9.   Dobson, G. and Karmarkar, U. S., Competitive Location on a Network, *Operations Research*, 35 (4) pp. 565-574, 1987.
10   Harker, P. T., Alternative Models of Spatial Competition, *Operations Research*, 34 (3) pp. 410-425, 1986.
11.   Lederer, P. J., Competitive Location on Networks under Delivered Pricing, *Operations Research Letters*, 9, pp. 147-153, 1990.
12.   Mirchandani, P. B. and R. L. Francis, eds., Chapters 10 and 11, *Discrete Location Theory*, NY, John Wiley, pp. 439-501, 1990.
13.   Heyman, D. P. and Sobel, M. J., Chapter 9, Stochastic Models in Operations Research, *Stockastic Optimization,* Vol. II, NY, McGraw-Hill, 1984.

14. Fudenberg, D. and Tirole, J., *Game Theory*, The MIT Press, Cambridge, Mass., 1991.

15. Karnani, A., Strategic Implications of Market Share Attraction Models, *Management Science*, 31, (5), pp. 536-547, 1985.

THE FEDERAL MANUFACTURING TECHNOLOGY CENTER PROGRAM: WHAT WORKS AND WHAT DOESN'T

Gene R. Simons

Rensselaer Polytechnic Institute, Troy, New York, U.S.A.

ABSTRACT

The Manufacturing Technology Center Program was established in January, 1989 by the National Bureau of Standards and Technology to transfer advanced manufacturing technology to small/medium sized manufacturers to improve their competitive position. The Northeast Manufacturing Technology Center (NEMTC) at Rensselaer Polytechnic Institute was one of the first three MTC's established to serve the 350,000 foundation firms in the United States. NEMTC had focused on shop floor level assistance to fabricators who are or wish to be suppliers to major manufacturers. In serving over 400 companies in the past three years, NEMTC has gained valuable insight into both cost effective and successful approaches to technology transfer. NEMTC developed a wide range of services and capabilities centered around four functional areas: field services, education and demonstration, training, and technical projects. This paper examines the experience of NEMTC and describes the application of best technical practices, the effectiveness of supplier development programs, and how to overcome the technical roadblocks in Total Quality Management.

THE EMERGENCE OF THE SMALL MANUFACTURER

During the 1980's, manufacturing employment in the U.S. dropped by nearly 12% (U.S. Department of Commerce County Business Patterns). This loss, however, was not evenly distributed with larger firms (more than 500 employees) dropping over 19% while employment in smaller firms was steady. General business conditions and imports were the major factors in the loss of employment in the larger firms while the source of strength in the small firms came from the growing tendency of the larger firms (OEM's) to outsource fabrication of components and retain the design, assembly, and distribution functions.

The role of the smaller firm as a supplier to an OEM is further complicated by the transition of vender certification programs into supplier development programs, the lack of technical support from the OEM, the shift from outsourcing discrete components to functional subassemblies, the need to pass on technical requirements to their (second tier) suppliers, and the required participation in the product design process through concurrent engineering. This role has placed a substantial burden on the small manufacturer to become "world class" and embrace the principles of Total Quality Management to satisfy and hold on to their customers by turning out the highest quality product at the lowest possible cost in small lots while looking at continuous redesign of the product. A small manufacturer may find itself as a sole source supplier linked electronically to the OEM customer who demands perfect quality produced by a manufacturing system of advanced process technology.

The 360,000 small manufacturers in the U.S., or "foundation firms" as they have been described by Congress, employ half the people in manufacturing and produce 60% of the domestic components used in U.S. products. The Technology Administration of the U.S. Department of Commerce has defined success factors for these firms as quality, speed to market, and the ability to adapt rapidly to change [1]. To assess the needs of the small manufacturers to achieve these success factors, we must examine their capabilities in five areas:

> Technical readiness, i.e. the ability of the employees to adapt to new technologies in manufacturing.
>
> Degree of market and strategic planning.
>
> The financial capability of the firm.
>
> The quality control program in place.
>
> The current manufacturing facilities and processing equipment.

PUBLIC POLICY ISSUES

In analyzing the needs of the small manufacturers, we have to look beyond technical solutions and examine the public policy changes that would enable them to improve their competitive position. In February, 1990, at the New York State Governor's Conference on Manufacturing, policy workshops addressed the following issues [2]:

> Human resources
>
> > How to make manufacturing attractive as a career.

    Retraining of existing workforce.

    Improvement of basic reading and math skills.

  Quality and Productivity

    Education and training.

    Sources of capital for investment.

    Vender certification assistance.

    Who will provide technical assistance.

  Regulation

    The role of manufacturing in establishing regulation.

    The impact of regulation of manufacturing.

  Taxation

    Preservation of personal tax cuts.

    Reduction of government spending.

  Technology Research and Development

    Protection of intellectual property.

    Facilitation of technology transfer.

Therefore, the application of advanced technology is only one of a number of government actions that may be taken to to improve the competitiveness of U.S. foundation firms.

  In a recent policy statement [3], the Technology Administration listed five goals that will be used to guide the development and implementation of federal programs to support manufacturing. These were:

    The translation of technology into timely, cost competitive, high quality manufactured products.

    A quality workforce that is educated, trained, and flexible in adapting to technological and competitive change.

    A financial environment that is conducive to longer–term investment in technology.

    An efficient technological infrastructure, especially in the transfer of information.

    A legal and regulatory environment that provides stability for innovation and does not contain unnecessary barriers to private investments in R & D and domestic production.

However, at the current time, the only programs other than the Small Business Administration that have a direct impact on the technology used by small manufacturing firms are the Manufacturing Technology Center Program (MTC) and the State Technology Extension Program (STEP). The MTC program will be discussed in the next section. STEP provides small grants to help state governments improve their technology assistance efforts. NIST funding for these two programs will total $16.3 million in FY92. Compare that to the Japanese funding of their technology assistance centers which totals $500 million per year. It should be noted that, based on a recent study by the National Governors' Association [4], the state governments spend $83 million per year for technology assistance to manufacturers.

THE MANUFACTURING TECHNOLOGY CENTER PROGRAM

In January, 1989, the National Institute of Standards and Technology (NIST) of the U.S. Department of Commerce established three Manufacturing Technology Technology Centers with the mission of transferring advanced manufacturing technology to small manufacturers to improve their competitive position. The first three Centers were the Great Lakes MTC at the Cleveland Advanced Manufacturing Program in Ohio, the Southeast MTC at the University of South Carolina, and the Northeast MTC at Rensselaer Polytechnic Institute in Troy, New York. In late 1991, two more Centers were established: The Midwest MTC at the Industrial Technology Institute in Ann Arbor, Michigan and the MidAmerica MTC at the Kansas Technology Centers. In 1992, two more Centers will be established bringing the total to seven.

The concept of the MTC was first put forward by Senator Hollings in 1988 as a part of the Trade Bill vetoed by President Reagan. Senator Hollings felt that in addition to protecting American industry from foreign competition through tariffs and quotas, the long term solution was to improve the productivity and quality of U.S. products. The MTC concept not supported by the administrations of Presidents Reagan and Bush because it represented a step toward manufacturing policy which was strongly opposed by these two administrations as creating winners and losers. The concept, however, was strongly supported in Congress and additional legislation to expand the program was introduced at the last session by both Senators Hollings and Bingaman. There is some sign that administration opposition is softening with the expansion of the Technology Administration in the U.S. Department of Commerce run by Undersecretary for Technology, Robert White. The mission of the Technology Administration is to support and strengthen the competitiveness of U.S. industry through the effective use of technology.

The objectives of the MTC Program are to enhance productivity and technical performance of U.S. manufacturers through the transfer of manufacturing technologies and techniques developed at federal laboratories (especially the Advanced Manufacturing Research Facility at NIST) through the MTC's to manufacturer's, make new manufacturing technology and processes usable by the smaller firms, and actively disseminate manufacturing information [5]. During the first year of operation, the MTC's concentrated on identifying the technology needs of smaller manufacturers, identifying the sources of technology to meet these needs, and developing cost effective mechanisms to transfer these technologies in the smaller manufacturers.

NORTHEAST MANUFACTURING TECHNOLOGY CENTER

The Northeast Manufacturing Center (NEMTC) was established under the Center for Manufacturing Productivity and Technology Transfer (a manufacturing research facility established in 1979) at Rensselaer Polytechnic Institute in January, 1989. In 1991, NEMTC has served over 700 manufacturing firms with field project, training, education/demonstration, and technical project services with a staff of 10 full–time and 32 part–time professionals (including 18 students). Most of the outreach to make contact with client firms is through state industrial extension services, especially the Industrial Technology Extension Services in New York State, and includes Massachusetts, Maine, and Pennsylvania operations. The 1991 budget was $6 million including $3 million from NIST, $1.7 million for State (New York, Massachusetts, and Maine) sources, $.8 million from fees, and $.5 million from OEM support.

NEMTC has focused on the upper end (technologically) of the smaller manufacturers who fabricate components and who either have their own product lines or who are or desire to be suppliers to OEM's. The primary technology area provided is CAD/CAM/CNC integration using off–the–shelf technologies that are considered best business practices by trade associations and journals. We estimate that there are over 40,000 small/medium manufacturers in the Northeast who perform fabrication services (SIC's 34–39). Of these, we estimate that Of these, about 30% have the resources (people, skills, money) to be assisted but only 1/3 of these firms are interested in assistance thus providing NEMTC with a target group of about 4,000 firms.

During 1991, NEMTC and its agents have reviewed 1200 manufacturers, visited 358 plants, initiated 97 field and technical projects, and completed 57 of these projects with an estimated savings to the firms of $10.6 million. NEMTC has conducted 12 workshops, seminars, and forums for 898 participants from 330 companies and 53 training programs for 558 participants from 247 companies.

In January, 1992, control of NEMTC was shifted to the New York State Science and Technology Foundation to formalize the relationship between the Rensselaer operation and the Industrial Technology Extension Service and to insure 1992–4 funding. The change should be transparent to users of NEMTC services because the services at Rensselaer will remain in place as will the services at the satellite operations in Western New York, Hudson Valley Community College, Massachusetts, and Maine.

NEMTC SERVICES

The NEMTC has developed service capabilities in five areas:

1. Education/Demonstration

Technology transfer to the smaller manufacturer is usually done by a salesperson for a software/hardware vender. NEMTC's approach involves educational seminars ("What is CAD?", "What is CAM?", "What is EDI?", "What is ISO 9000?", etc) put on through community colleges and satellite broadcast. NEMTC develops the courseware and train instructors from community colleges and vocational high schools on the use of the courseware. This "Train the Trainers" approach has resulted in 32 presentations in the past year alone.

The second approach is the demonstration facility – Manufacturing Technology Resource Facility (MTRF) located in a separate building at Rensselaer. This facility offers 54 software packages on 17 different computers that may be reviewed by a company with the assistance of an engineer or a technician. 47 firms visited the facility in 1991 to obtain assistance in their decisions to acquire CAD, CAM, CAE, or other types of software. The cooperation of vendors has been excellent with over $2 million in equipment and software provided, along with training and technical support, at little or no cost.

At the end of 1991, a mobile facility was acquired through a STEP grant to New York State that will be used as part of an outreach activity for the MTRF. In addition, New York State has provided funding for satellite operations to be located at Community Colleges in Troy, Jamestown and New York City.

2. Field Projects

Individual consultancies are conducted on a 50–50 cost sharing basis with firms that have defined their problems and need solution implementation. These field projects are conducted by full–time staff, students, community college faculty, and private consultants using off–the–shelf (best business practice) solutions.

3. Training

A network of 17 community colleges and vocational high schools was established in three states to provide training services to firm in the local community. NEMTC subsidizes these training programs and provides "Train the Trainer" services described above.

4. Technical Projects

When an off–the–shelf solution cannot be found or is too expensive for the smaller manufacturer, a joint development project is set up on a cost–sharing basis to seek that solution. The service provider tends to be a university research center and the project usually takes a year. Another source of technical projects is the transfer of technology from a federal laboratory to a small manufacturer. This involves downsizing, documentation, and training.

5. Assessment Tools

A primary difficulty during the intake process is to identify not only the problems that needs solving in the smaller manufacturer but also to prioritize the needs. If the firm's primary problem is lack of a marketing plan, it is senseless to help them install a CAD system. During 1991, an assessment procedure – NEMTC Assessment Instrument Development Project (NAID) was developed that is currently being field tested [6]. It consists of a questionnaire that is filled out by the firm and then analyzed by an expert system using industry benchmarks to evaluate the capability of the firm and point out deficiencies that need correction. The firm can also be benchmarked against Malcolm Baldrige, Deming, or ISO 9000 criteria. The decision on the appropriate type of assistance – technical versus business – then can be made in conjunction with a state extension agent.

WHAT WORKS

The experience of NEMTC has provided empirical evidence on some cost–effective approaches to providing technology transfer services to small manufacturers. The program of services has evolved during the three years of operation from a heavy emphasis on technical projects to an emphasis on group projects through either clusters or networks of companies. The first question, however, is what are the best practices in general. Souder's study [7] of technology transfer in industry and government sited 37 practices that were of different interest depending on the stage of technology transfer. He identified four stages: prospecting, developing, trial, and adoption. Best practices were divided into analytical, facilities, pro–actions, people–roles, conditions, technological quality, organization.

Our own experience indicates that the following conditions are necessary for an effective transfer program:

1. The program should be geographically close to users in order to provide continual support.
2. The program's staff should have real world (practical experience), preferably in a small firm.
3. The program should focus on one or two areas and try to be all things to all people.
4. The program should have access to applied research facilities for assistance in solving more complex problems and for transfer of new technologies to the program.
5. The program should have the ability to educate, i.e. explain technologies to potential users.
6. The program must be able to demonstrate the technologies within its focus.
7. A commercial incentive must be established for manufacturers to use the program.
8. The program should be linked to existing providers, eg. state extension services, and augment rather than duplicate their capabilities.

This is based on the characteristics and needs of the small manufacturer which include:

1. Short runs with limited output.
2. Lack of skilled personnel.
3. Limited financial and technical resources
4. Lack of strategic planning and market/competitor identification.
5. Pressure from customer for quality, low prices, and JIT service.
6. Lack of quality improvement programs.

and is based on the following premises:

1. If there is no learning on the part of the firm, there is no technology transfer [8].
2. The small firms usually does not have the skilled personnel needed to take advantage of advanced technology.
3. Consultants generally do not do technology transfer.
4. The smaller firm needs neutral party to aid them in evaluating alternatives.
5. The smaller firm needs a series of projects timed on the ability of the firm to absorb each change. The overall impact may take over a year [9].

Therefore, NEMTC has concentrated its efforts on building a program that is focused and offers the ability for a smaller firm to establish a continuous relationship with the Center. In addition to the services described in the previous section (which are in line with the principles listed above), NEMTC has developed a series of approaches to work with groups or networks of companies that allows maximum leverage to the limited public funding. Examples of these networks are:

The Western Quality Network in Western New York consists of three groups of 6–8 companies that assist each other in the implementation of Total Quality Management programs through monthly meetings in each others plants to review operations under the tutelage of a state extension agent.

TECnet(Technologies for Effective Cooperation) in central Massachusetts consists of 44 small tool and die shops on a PC–based network that acts as a bulletin board and a resource directory.

Supplier development programs from OEM's offer the opportunity for a joint project that will improve the capability of their suppliers and will include partial financing by the OEM. An example is an Electronic Data Interchange project that will link 20 suppliers to an OEM on Long Island [10].

Another network being supported by NEMTC is the ISO 9000 user group which was initiated at the end of 1991 and currently has 7 members. These are firms working jointly on meeting first the registration and then the certification requirements of ISO 9000 with the technical assistance of trained NEMTC staff.

WHAT DOESN'T

We should also point out some of the failures during our three years of operation, the reasons for these failures and the corrective actions we have taken. In some cases, it is the opposite to the "what works" list, eg avoid being all things to all people and focus on one or two technical areas. However, the problem of implementing this is to determine what are the "right" technical areas to focus on. This should be a function of the needs of the industries to be served but who decides which industries should be served? The primary impediment to an effective program is lack of strategic direction – strategic industries and technical focus must be defined to avoid dilution of the services. NEMTC pre–empted this problem by choosing a strategic direction unilaterally which is a privilege that will not be available to future Centers.

The list of don'ts however should include the following:

1. Avoid centralization. Services should be available within a reasonable driving distance (one hour) of the target firm. Implementation of this strategy requires the establishing of satellite operations.
2. Although assessment is necessary, avoid assessment approaches that require substantial time from both the firm and the MTC.
3. Avoid being a beta–site for technology from federal laboratories. A better approach is to work with a firm that wishes to commercialize that technology.
4. Avoid technician–level training. There are many sources of this in most states.
5. Avoid becoming a consultant concentrating on individual projects to individual firms.
6. Require the firm to cost share, otherwise you become a grant–in–aid program.
7. Do not compete with existing state services. Cooperate with them.

8. Use students only on well–defined projects which are fully supported by the client firm.

CONCLUSIONS

The experience of NEMTC has provided an interesting laboratory for advancing the knowledge of technology transfer to smaller manufacturers. There are, however, some unanswered questions that should be studied during the next year. The first is how to structure the satellites needed by the MTC's to meet their service mission. Should the MTC set up its own branch office? Should it use a community college or an existing state extension program? How will the services be functionally defined?

Second, as we have already stated, who will set the strategic priorities for the MTC's – what industries should they serve and what technologies should they transfer?

Finally, how should we distinguish between the role of the MTC as a federal program versus the state extension services? Is there a need for a federal service or should the federal government simply subsidize existing state services? The future of the MTC program requires an answer to these questions.

REFERENCES

1. Technology Administration: *Strategic View*. U.S. Department of Commerce, November, 1991, p. 1.

2. Proceedings: *New York State Governor's Conference on Manufacturing*. NYS Department of Economic Development, Albany, NY, Feb 12–13, 1990.

3. Ibid., Technology Administration (1991), pp. 5–9.

4. Clark MK and Dobson EN: Increasing the Competitiveness of America's Manufacturers: A Review of State Industrial Extension Programs. National Governors' Association, Washington DC, 1991, p. vii.

5. National Institute of Standards and Technology: *Manufacturing Technology Centers*, Federal Register, Vol. 57, No 10, January 15, 1992, p. 1725.

6. Simons GR and Raghvachari M: Total Quality Control in Small Manufacturing Firms. *Transactions of the 45th Annual Quality Congress*, Milwaukee, Wisconsin, May, 1991.

7. Souder WE, Nashar AS, Padmanabhan V: A Guide to the Best Technology– Transfer Processes. *Technology Transfer*, Vol 15, Nos 1 & 2, 1990, pp 5–16.

8. Baughn CC, Osborne RN: Strategies for Successful Technological Development. *Technology Transfer*, Vol 14, Nos 3 & 4, 1989, pp 5–13.

9. Simons GR: The Experience of the Northeast Manufacturing Technology Center. *Proceedings of the Second Meeting on Modernizing America's Industrial Base,* Detroit, Michigan, May, 1991

10. Simons GR, Brandow R, Chank M: Technology Transfer in Supplier Development Programs. *Proceedings of the Technology Transfer Society Annual Meeting,* Denver, June, 1991

MANUFACTURING STRATEGIC PLANNING: TYPICAL ISSUES IN A "THIRD WORLD" ENVIRONMENT

[1]Jose Pablo Nuno and [2]Estomih M. Kombe

[1]*CIM Systems Research Center, Arizona State University, Tempe, Arizona, U.S.A.*
[2]*Arizona State University, Tempe, Arizona, U.S.A.*

ABSTRACT

Over the past several decades we have witnessed the further penetration of the international market place by scores of multi-national companies (MNC's), mostly from the countries of the west and Japan. This penetration has not spared the countries of the less industrialized third world. In fact, in view of the inability of these countries to produce the industrial goods needed for their livelihood, theirs is presently a small but certain market. The potential implications of an increased purchasing power for this huge population is very tempting to any investor. However, because of the reality of the operating environment in these countries, the performance of business units in these areas have in many instances raised serious questions with investors. Some have abandoned their operations, selling off to other multi-nationals or merely making 'offers' to host governments and partners. This kind of frustration is in no way restricted to foreign investors. Local businessmen and industrialists as well as local governments and their agencies are faced with the same hurdles as they try to succeed in such a situation. The major difference lies in the non-availability (or impracticality) of the option to sell and go elsewhere. This paper is an attempt to explore available literature on the strategic options/elements in such an environment and make some suggestions based on current readings and experiences.

INTRODUCTION

It appears appropriate to first mention the nature of the 'adverse environment' in question, as well as to take a brief look at a typical set of the generic strategies as found in literature.

The Adverse Environment

The third world, as a grouping of nations, is probably too broad to provide for a definition, whether it is used to infer political or economic position. The original post - World War II use of the grouping made reference to all those countries that were neither in the industrialized western block (the first world), nor in the eastern - 'red' block (the second world). Things have changed considerably since then. Early on in his publication, Richard Steade segments the international scene into five worlds as follows [11];

1). **The first world** - the advanced industrialized countries of Europe, North America and Asia.
2). **The second world** - The countries of the communist world with centrally planned economies. Today this situation is considerably changed.
3). **The third world** - countries that were perceived to have a need for time and technology (not massive finances) to build modern developed economies. These include nations that have key revenue generating resources or else which are developed enough to attract foreign investment or borrow on commercial terms. Steade included in this group OPEC nations, Morocco, Malaysia, Taiwan, Singapore, Mexico, Brazil, South Korea and Zambia amongst others.
4). **The fourth world** - those countries that have some promise in terms of technology, raw materials, trained personnel or economic infrastructure, but which need significant financial help and special treatment to export their goods and acquire technology. Examples in this group were Peru, Egypt, Liberia and Thailand.
5). **The fifth world** - made up of those countries that have few known resources, where most of the population is engaged in subsistence farming, life expectancy is normally below fifty years, and generally living in hard core poverty. Examples of the many that fall in this category are such countries as Chad, Mali. Ethiopia, Somalia, Rwanda and Bangladesh.

While the broader <u>third world</u> still has an important role as a group, particularly in major international negotiations, most reference to the 'third world' today is directed at the last group, as seen from the descriptions normally given. For purposes of this paper, this latter reference will be assumed.

The plight of the industrial performance of the underdeveloped countries can be viewed by looking, for instance, at their total exports in industrial products (machinery and transport equipment) relative to total global exports. Table 1 shows the exports of these products for five selected years in the last two decades, according to United Nations statistics [14]. Compared to their population, the share of the least developed nations of the world in these exports is negligibly small.

	1970	1980	1985	1986	1987
World Exports *(Total $mill.)*	89769	513081	600638	721059	857861
Developing Countries (Excluding South America and South/South East Asia) *$ mill.*	540	2097	6564	6355	7880
(Includes the whole of Sub-Saharan Africa and the Arab World) - % Global Total	0.6%	0.41%	1.1%	0.88%	0.92%

Table 1. Third World Share of Global Exports in Machinery and Transport Equipment (f.o.b. $mil, 1989).

We do not intend in this paper to dwell much further on the manifestations of the poor industrial and economic performance of the least developed countries except in related discussions in later sections.

The Generic Strategies

A look at the generic strategies, for example as presented by Porter, is a good starting point. This is not to suggest in any way that one is usually confronted with a situation as simple as picking one of these strategies. That would be a serious over-simplification   of the problem. However, these strategies do appear frequently in discussion, and they are definitely worth early mention.

In their most basic form, Porter's generic strategies fall into four categories, with *Focus* split into its two components. These are, Overall Cost Leadership, Differentiation, Cost Focus and Differentiation Focus [10]. One question that may be posed with respect to business unit performance in these locations is; *can the formulation of corporate/ business strategies that primarily address economic and technical considerations in these countries have a positive impact on performance?* Or more broadly, *do identifiable strategies exist that offer strategic competitive advantage under these conditions?* And will the successful implementation of these strategies transcend the political-legal and socio-cultural incompatibilities?

In his article on the performance of Multinational company units in the third world, Wright [18] suggests that "the creation of small scale production facilities in the third world can be thriving, competitive units, if their products/market scopes lend themselves to that of differentiation and/or focus general strategies". This observation is made after noting that a cost leadership strategy is associated with high production volumes (economies of scale). Such volumes, it is pointed out, are not sustainable in these countries either because of the low purchasing power of most of the customer base, or in view of the poor shape of the existing infrastructure. The latter seriously impairs the distribution of the products. By adopting a differentiation or focus strategy it is supposed that specific 'local' needs will be met. This, it is further argued, will provide an edge over imported 'standard' items, and would therefore permit charging higher prices.

The above conclusions and the arguments from which they are drawn are subject to discussion. For instance, what are the volumes associated with the 'small scale facilities'? Can the buyers afford the premium product?, and what price differential will make the units thrive? Whether or not the example given (a venture of that type in Kuwait) is typical of the situation in question, is subject to debate.

Although the infrastructure in most of the Least Developed Countries (LDC) is un-supportive of high volume- fast distribution, it is also true that fast moving items do reach the different corners of the countries one way or the other. The problem is that market areas are so dispersed, that a producer trying to have a good grip over the delivery of goods to customers will be frustrated, and in the process may find it quite costly. However, products having a long shelve life, if made available to regional centers, would be delivered to customers by local businessmen. Multilateral agreements between MNC's and government agencies in a number of countries, or joint ventures with local industrialists in these countries, can provide for an increased access to customers. At the same time such arrangements would minimize market entry bottlenecks.

SOME SOURCES OF A FIRM'S MANUFACTURING COMPETITIVENESS

In getting its product onto the market, a manufacturing firm must go through several stages.

    1). Procurement of raw materials and supplies.

    2). Processing of the raw materials into finished products.

    3). Distribution of the product to the customer.

Each one of these levels has a significant competitive edge on the performance of the firm. In the LDC's each one of these stages takes a somewhat extended dimension. As can be seen from Figure 1, at each of the three stages, there are numerous factors that frustrate the smooth operation of the system. The figure only shows the more obvious problems.

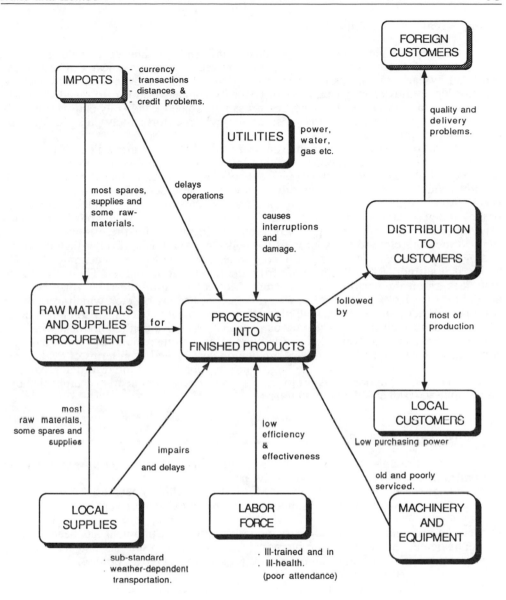

FIGURE 1. MAJOR FACTORS IN MANUFACTURING COMPETITIVE POSITION IN
THE EXISTING ENVIRONMENT.

Plant Location Considerations:

Figure 2 summarizes the major factors influencing plant location decisions. The performance of a manufacturing facility, and therefore its competitiveness, can be seriously hindered by early oversights in location decisions. Such disadvantages in competitive position once allowed to set-in, can not be easily compensated for. In the absence of a means to correct such a disadvantage, compensation would mean highly outperforming the competition by focussing on other cost drivers, which unfortunately puts the firm on a defensive position.

Location questions in the environment of the LDC's are thrown out of proportion by the seriously deficient communication infrastructure, erratic or non-availability of utilities, and not least important, government policies. In a country like Tanzania, many of the industries are directly or indirectly involved in the processing of some agricultural goods. Amongst these are textile industries, leather industries, rope products (mainly sisal based) and general food processing. A major proportion of these would qualify as typical raw material based industries [3]. Location considerations for these industries would suggest locating close to the source of the bulky raw materials, if other factors are not strategically affected. However, these companies are to be found concentrated in a few 'cities', sometimes 500 miles or more from the raw material sources. These aren't short distances given the existing infrastructure. Most of the time the reason has been the availability (or lack) of electricity, water, industrial heating oil etc., which in the inner country is either completely lacking or much more unreliable in comparison to the cities. The result of this situation is a heavily burdensome transportation requirement which can not be supported by the available infrastructure. This in turn precipitates a marked drop in competitive position, both in terms of production cost as well as in terms of quality, and therefore customer satisfaction. The huge transportation bill and the resulting production delays when materials don't make it to the plant on time, are frequently accompanied by a loss of material properties/ quality in transit.

Machinery and Equipment

The following can be cited as important factors with respect to machinery and equipment that have a bearing on competitive position.
1). Source of machinery and equipment: Mostly (if not wholly) purchased from overseas suppliers for which after-sale support is either non-existent or very expensive to the buyer.
2). Service and repair: Over the years, as machinery and equipment has become more sophisticated, particularly with increased electronic controls, it has also become more and more difficult to get local technicians for service and repair, or to use locally made replacement parts and supplies.
3). Foreign exchange constraints: Most of the time hard currency is either very difficult to obtain or very expensive, making the purchase of essential parts and supplies more complicated.
4). Training shortfalls: Technical Skill is scarce, particularly middle level technicians who are crucial for the smooth operation of machinery and equipment. In Tanzania for instance, while there has been a lot of effort to train engineers during the 70's and 80's, little corresponding effort has been directed toward creating the necessary middle level technical cadre.

THE CASE OF THE SOUTHERN PAPER MILLS [SPM] IN TANZANIA

Tanzania had earlier-on earmarked the pulp and paper industry as an essential basic industry along with iron and steel (currently shelved), metal fabrication, glass, cement, plastic products, leather, wood products, printing chemicals, fertilizers, ceramics and

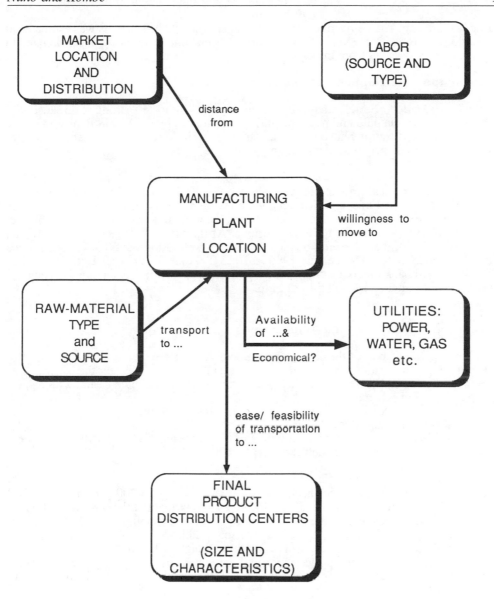

FIGURE 2. COMPETITIVE POSITION AND MANUFACTURING PLANT LOCATION
DECISIONS.

petroleum refining. The demand for paper and paper products in Tanzania was estimated at around 45,000 tons for the year 1980 (an early 70's estimate). So far all paper demand in the country was satisfied from imports [15].

Issues in the Development of the SPM's

Economies of Scale. Earlier studies by different parties indicate a strong scale sensitivity of cost components in pulp and paper manufacture. Studies by the World Bank and by the FAO of the United Nations have shown a marked drop in investment and labor costs per unit of capacity with increased plant capacity [15].

Tanzania paper consumption and projections for the period 1968 to 1995. Paper consumption figures and projections were studied by different consulting groups between 1969 and 1978. A total of ten projections, which were made by BIS of Britain, FAO, the World Bank, Poyry of Finland and local firms, were available at the time of project planning[15]. Figure 3 shows these projections. At the time of project appraisal by the world bank (1978), the following demand estimates were adopted: 58,800 tons for the year 1985, 82,500 tons for the year 1990 and 115,000 tons for the year 1995. As project leader, the world bank finally recommended a pulp and paper mill with an initial capacity of 60,000 tons of paper and 1,400 tons of pulp.

Expectations, Problems and Implications

The following were amongst the noted expectations;
1). Paper supply will satisfy national demand at least to the year 1990 (for the paper grades to be produced by the mill). This is based on the assumptions cited earlier with respect to the projected demand.
2). Some grades of paper will still need to be imported since the mill will only produce some of the paper grades on demand.
3). Extra supply will find export markets. For some period the mill will have excess capacity with respect to local demand.

The following problems are noted;
1). Initial Investment cost for the project was estimated at US $ 4200 per ton of installed capacity, later adjusted to some US $ 4867. This was thought at the time to be about 2-1/2 times the highest investment cost per ton of a similar facility in the industrialized countries. High cost was partly due to small scale, addition of infrastructure as part of the project, technical assistance, cost over-runs and high inflation.
2). The implementation of the project required agreement by the many parties that were involved in its financing. As will be noted later, this was not always easy to accomplish.

The implications would appear to be the following;
1). The operation can hardly be competitive in terms of cost, quality and service against external markets (eg. in the neighboring countries) given the initial cost of the project.
2). With the poor performance of the Tanzanian economy (demand projections had anticipated 'reasonable' growth in the economy), local paper consumption could prove that projections were overly optimistic. This will cause the project to rely heavily on success on external markets.
3). The expected high cost of the product from the mill if too high will inhibit local consumption and encourage competition from outside suppliers.

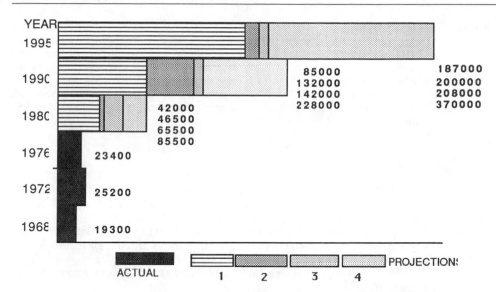

FIGURE 3.   PAPER DEMAND - PROJECTIONS FOR   1968 / 1995.

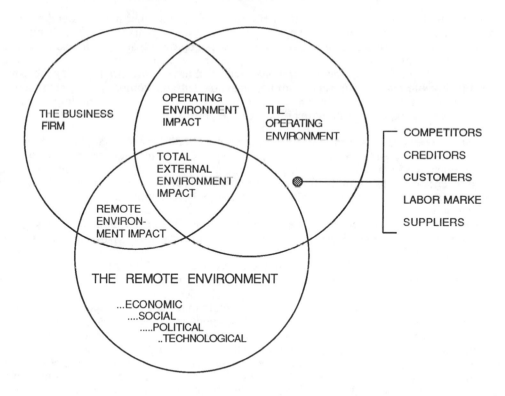

FIGURE 4. - THE FIRM'S EXTERNAL ENVIRONMENT.

Financing of the Project

One of the things that make this project a special one was the number of financiers for the project. Funds came from seven different sources including the World Bank and the Swedish SIDA, apart from the government of Tanzania itself. With such a collection of parties to finance the project, one is inclined to have the highest expectations in view of all the associated influences and expertise! The initial cost of the project was put at US $ 252 million at that time [15].

Project Implementation

The following issues can be raised in regard to the project implementation;
1). Coordination of the many sources of funds. Delays in the release of funds, disagreements in the award of tenders etc. meant delays to the project
2). The size and complexity of the project was beyond the existing technical and managerial capabilities in the country, thus necessitating a substantial amount of technical and managerial assistance.
3). Infrastructure problems. This was both a financial burden as well as a real bottleneck in the project execution.
4). Completion of the project came about four years behind the initial schedule!

Operation. The high import content of chemicals, fuel and other materials and supplies may end up as a serious bottleneck unless there are exports to guarantee a quick availability of hard currency by the company. Some estimates put upto 40% of the manufacturing cost of the firm in foreign inputs. Although it may be argued that supplying the local market makes available foreign currency that would otherwise have gone into importing the same, it is a fact of life that such currency will not be available to the company <u>when</u> it is needed. A delay, for say, one month in the release of needed funds during operation can cause tremendous consequences, including prolonged stoppages.

Recent Experiences. In respect of recent experiences, the following is worth mention.
1). In its initial years (1986, 1987), paper prices went so high that newspaper publishers in the country sought to double their prices. The government was forced to intervene and to offer some subsidy in the form of special sales tax reductions to these 'customers' in order to keep newsprint affordable. One can only speculate on how long this can continue.
2). The Quality of paper from the plant raised quite a number of questions with consumers. A number of user requirements called for serious attention by the paper producers.

Longer Term Questions. The following questions arise in the longer term.
1). Repayment of the multi-million dollar loans by the company (or by the country). Will the plant pay for itself or will it be a tax payer's burden ?
2). Liberalized markets in the country would dictate that the firm compete with imports. Will the operation survive the onslaught of competitors.
3). Imports of spares, replacements and supplies. Can the system provide for the necessary amount of hard currency to keep the plant running all the time ?
4). Environmental pollution. Paper industries are known for their notoriety in polluting the environment when operating conditions are allowed to deteriorate. Will the condition of the plant be kept high enough to spare the sorrounding environment in the years to come?

Case Remarks

Despite the many questions raised in respect to this project, there is no doubt as to the usefulness of the project to Tanzania in terms of serving the country's long term aspirations. The case is presented here so that it can highlight the issues underlying such an undertaking. The efforts of all the parties involved in the project deserve our appreciation, particularly noting that some of the questions they have had to address do not have clear cut answers.

LOOKING AT THE EXTERNAL ENVIRONMENT OF A FIRM

In earlier sections we have looked at factors relating to competitive position. The case of the Southern Paper Mills in Tanzania was presented to show the complex nature of typical investment undertakings of a sizeable project. As suggested by Pearce and Robinson [9], it is the external environment impact that is of most concern here as shown in Figure 4. The firm's external environment can thus be divided into two major categories. The first is the operating environment and the second is the remote environment.

Looking at the two major categories presented in Figure 4 in relation to the preceding discussions, the significance of the remote environment impact becomes apparent. First, factors in this category dictate upon the available support structure for project implementation. Where this structure is weak, projects can not be developed so as to enjoy maximum benefits from available 'theoretical' performance efficiency and effectiveness. This means that operating characteristics under these conditions are much worse than the ideal feasible conditions that are a target point of reference for such operations.

Second, the remote environment factors have some direct and indirect effects on the factors in the second category, the operating environment. The nature and strength of customers and the labor supply for instance, are not independent of the economic, political and technological position of the society in question. The remote environment therefore impacts very heavily on performance under the conditions of the poor, non-industrialized third world countries. In view of this reality, it is crucial that available tools of analysis be employed to make a thorough investigation of the implications of these factors in the implementation and operation of a project. One such tool is to be found in Scenario Analysis. Most researchers discussing scenario analysis concentrate on the use of scenarios as an industry analysis tool [10]. They focus mostly on the competition and on market analysis. We must direct our analysis more on the remote environment.

Uncertainties that have a low chance of occurrence but which may be associated with potentially serious consequences must receive adequate consideration. The tendency is for such cases to be overlooked for the more likely happenings, which on the other hand may only have minor consequences. Drawing partly from the experiences of a number of manufacturing projects in Tanzania, some suggestions may be put forward in respect of issues pertaining to the project planning stage and the operation stage.

Issues in the Initial Project Planning Stage

1). Avoid (overly) optimistic projections on the growth of the economy and associated commodity demand figures.
2). Work out details with all parties concerned in the project to avoid (or at least minimize) struggles at the time of implementation.
3). Do not fall into the trap of basing performance on "local consumption" alone. If the project is only feasible for local consumption, the chance is there is a problem.

4). Do away with any avoidable fixed costs (keep these to a bare minimum). When things do go wrong and the 'investment' stands idle for sometime, the less the fixed value the less severe is the impact.

5). Put some thought to such questions as power failures or fluctuations, water supply problems etc. and see what kind of precautions can be taken to minimize their effects.

6). Draw-out the project implementation schedule clearly and let every party to the different inputs know what is expected of them and when.

7). Engaging in government controlled industry sectors has its good and bad sides. Be sure you know how the industry operates and how much government control is exercised.

8). Important roles by third parties must be clearly documented and seen to be binding.

Issues In the Operational Stage

1). Follow up things constantly to see evidence of something been done. Some people will not necessarily stick to their promises (they may be tied up with other problems).

2). Schedule deliveries to arrive ahead of actual demand to allow for some uncertainties in the system.

3). Make all effort to develop good trading terms (credit facilities etc.) with your local as well as external suppliers. This is sometimes difficult in view of the rigidity of existing procedures, particularly the settling of external bills through the central banking systems.

4). Establish contacts in areas where procedures are unreasonably lengthy. If necessary get some insider to assist in some of the paper-pushing.

CONCLUDING REMARKS

Amongst the disadvantages posed by the adverse environment in many of the "third world" countries in respect to manufacturing operations are the following;

1). Very low purchasing power concentrations due to meagre real incomes.

2). Poor and unreliable infrastructure.

3). Scarcity of a specialized technical labor supply.

4). An unstable and therefore un-supportive political and social setting.

5). Lack of a supporting industry network (for reliable vendors etc.).

This means that the operation and success of a facility hinges on many more factors than would otherwise be the case in an 'industrialized' society. Surviving therefore means been able to adjust to a wider range of unexpected turns in the *operating* and in the *remote* environment in view of the many uncertainties in the system.

There isn't sufficient ground to justify a particular generic strategy as generally superior under these circumstances against others. One thing that stands out clearly is that planning must be thorough and be based on realistic assumptions. As seen in many of the failed or not so successful ventures in these areas, minor omissions and/or false assumptions can (and normally do) have devastating consequences. At the same time strategic options need to be re-evaluated continuously against new developments and/or possible scenarios. Markets can be expanded by establishing links across national boundaries. There are different approaches to this end, including the selling of shares to individuals and establishing agencies in different countries to break down entry barriers.

Given the low purchasing power of most local buyers (for ordinary items), product and service design and development should seek to first satisfy the basic need(s) at the lowest cost to the buyer. Differentiation should sometimes be interpreted to mean *not paying a premium* but paying so much less at the expense of secondary values (functions) in the product or service. Emphasis on concurrent engineering concepts can be of valuable use in this respect. Cost leadership can be viewed at two different levels. Cost leadership within the *local market zone* (against other local manufacturers and imports) and cost leadership at the *world market place* (against "local" goods in the foreign markets and other exports into the same market). The difference is to be found in the additional cost and complexity associated with exports in terms of freight, applicable duties and other selling expenses.

This paper addresses a few of the many issues relating to competitive advantage for manufacturing operations in the poor non-industrialized countries. The topic is no doubt a very wide one. However, we do believe that the issues which have been cited and discussed here have shed some light on the question. Strategic planning does definitely have a very central role in the industrial development of countries like Tanzania, which are struggling to convert their industrial priorities into reality.

REFERENCES

1. Ballance, R. H.; *"International Industry and Business: Structural Change, Industrial Policy and Industry Strategies"*. Allen & Unwin, 1987.
2. Barker, C. E., Bhagavan, M. R., Mitschke-Collande, P. V. and Wield, D. V.; *"African Industrialization: Technology and Change in Tanzania"*. Gower Publishing Company, Vermont USA, 1986.
3. Berry, B. J. L., Conkling, E. G. and Ray, D. M.; *"Economic Geography"*. Prentice Hall, Inc., New Jersey, 1987.
4. Crowe, T.J. and Nuño, José Pablo; *"Flexibility, Cost, Quality and Service in the Product Process Matrix"*. Long Range Planning, December 1991.
5. Hofer, C. W. and Schendel, D. S.; *"Strategy Formulation: Analytical Concepts"*. West Publishing Company, 1978.
6. McCormick, B. J.; :*The World Economy - Patterns of Growth and Change"*. Phillip Allen/ Barnes & Noble Books, 1988.
7. Mushi, J. J. and Kjekshus, H. (editors); *"Aid and Development: Some Tanzanian Experiences"*. Norwegian Institute of International Affairs, 1982.
8. Nuño, José Pablo, Shunk, D. L. & Cruz-y-Celis M.; *"A CIM Architecture and Implementation in a Truck Company To Become a World Class Manufacturer: Fostering the North American Economic Community"*. Factory Automation and Information Management, CRC Press, 1991.
9. Pearce II, J. A. and Robinson, R. B. Jr.; *"Strategic Management: Strategy Formulation and Implementation"*. Richard D. Irwin, Inc., 1988.
10. Porter, M. E.; *"Competitive Advantage: Creating and Sustaining Superior Performance "*. Free Press, NY., 1985
11. Steade, R. D., *"Multinationals and the Changing World Economic Order "*. Califonia Management Review, Vol. XXI, No. 2, 1978.
12. Steel, W. F. and Evans, J. W.; *"Industrialization in Sub-Saharan Africa: Strategies and Performance"*. World Bank Technical Paper Number 25. The World Bank, Washington DC, 1984.
13. Tostensen, A.; Research Report No. 62: *"Dependence and Collective Self-Reliance in Southern Africa"*. Scandinavian Institute of African Studies, 1982.
14. United Nations: *1989 "Handbook of International Trade and Development Statistics"*. United Nations, New York, 1990.
15. Wangwe, S. M. and Scarstein, R.; *"Industrial Development in Tanzania: Some Critical Issues"*. Scandinavian Institute of African Studies and the Tanzania Publishing House, 1986.
16. World Bank *Annual Report* -1990. Washington D. C. 20433,
17. World Economic Survey: *"Current Trends and Policies in the World Economy"*. United Nations, N.Y., 1990.
18. Wright, P.; *"MNC-Third World Business Unit Performance: Application of Strategic Elements"*. Strategic Management Journal, Volume 5, pp 231 - 240, 1984.

THE ROLE OF ABANDONMENT ANALYSIS IN ADVANCED MANUFACTURING ENVIRONMENTS

Bechir N. Ouederni and William G. Sullivan

Department of Industrial and Systems Engineering, Virginia Polytechnic Institute and State University, Blacksburg, Virginia, U.S.A.

ABSTRACT

In this paper, results from an extensive taxonomic study of the product abandonment analysis topic are briefly presented and discussed. The product abandonment problem is then reformulated in view of both the strengths and shortcomings of older models and the requirements of the new manufacturing environment. Finally, a global decision model which incorporates the product abandonment option into the company's overall strategic planning and control system is proposed and illustrated through a simplified case-study.

INTRODUCTION

The worldwide technological explosion has dramatically changed the basis of international competition. The accelerated rate of change in product engineering and process technology has led to decreasing product life cycles and made equipment obsolescence a primary concern to U.S. manufacturers. Researchers in academia, industry, and the government have unanimously agreed on the primary role that the investment in advanced manufacturing technologies (AMT, eg., Flexible Manufacturing Systems) must play in meeting the challenges of the new global business environment. However, U.S. manufacturing is still lagging far behind its technological innovation , and many manufacturers are practically unable to justify the needed modernization. Lately, many authors have written about the necessity to account for strategic, long-term benefits associated with acquiring new AMT's in order for manufacturers to justify more easily, and more realistically, such decisions. However, most authors have overlooked the fact that the decision to acquire a new AMT is most likely to displace old processes, and that unless manufacturers are offered a tool to evaluate the impact of abandoning obsolete or less-than-profitable products and/or processes and justify such decision, the needed modernization process will be hindered.

The objective of this paper is to reformulate the product/process abandonment problem from a new perspective which is consistent with the requirements of the new global business environment. Strategic considerations such as financial, social, and legal among others will be accounted for in the proposed model. In the following sections of this paper the terminology "product abandonment" will be used interchangeably to refer to product, product line, or process abandonment unless a distinction is explicitly made. It is our belief that by proactively incorporating the abandonment option into the company's strategic planning and control system, a better and more realistic evaluation of the merits of the new AMT as well as the opportunities from abandoning the old technology and/or product line can be achieved.

HISTORICAL PREVIEW: PRESENTATION AND DISCUSSION

In a literature survey recently completed by the authors, it has been shown that the product abandonment problem has received relatively little attention from researchers in both academia and industry [1]. The three major abandonment schools identified by the survey are briefly described below.

The first school included product abandonment research as treated in the "engineering economy, financial management, and management accounting" literature. In 1967 Professors Robicheck and Van Horne published the first article to systematically treat the product abandonment option as a critical dimension of the capital budgeting analysis process [2]. Two years later, they published a revised version of their original algorithm [3] as a reply to comments by Dyl and Long [4]. Subsequent efforts, although very few, have typically been directed towards improving the findings of their antecedents either by including one or more new aspects of the product abandonment option and its impact on the firm's financial profitability or simply by trying to develop a more generic abandonment algorithm [5]. A more generalized abandonment model was offered by Bonini in a 1976-article [6], where he proposed a dynamic programming algorithm based on Hillier's analytical model of capital budgeting under conditions of uncertainty [7]. Later attempts to model the product abandonment problem have been typically very limited in adding any intellectual substance to Bonini's model. While a few authors have tried to compare the product abandonment decision with other product decisions familiar to management, such as the asset replacement decision [8], other authors have either applied previous algorithms to specific case problems [9] or tried to suggest some alternative solution techniques such as Bayesian analysis [10] or graphical project evaluation methods [11].

Research in this first school has typically focused on solving the abandonment optimal timing problem. The abandonment decision rule had the general format of "hold on to the project as long as the net present value (NPV) of the expected future cash flows associated with its continued operations exceeds its current abandonment value." The product abandonment option was perceived as a strategic alternative in the capital budgeting problem. However, different authors have generally emphasized different dimensions of this option; and operationally, no one has offered a model which incorporates a company's overall strategic goals into the analysis. Instead, they were using, at best, some empirical models based on product profit/cost/volume data as key decision criteria. Project NPV was practically the decision criterion used by the majority of the companies to evaluate product profitability. The resulting profitability and cost reports were therefore distorted due to the shortcomings of traditional accounting practices [12]. Furthermore, basing the abandonment decision on specific product volume-related data usually results in an optimal abandonment time that occurs right after the time of maximum market penetration, or at the beginning of the product sales decline phase, as indicated by point T_A^* in FIG.1 below. In fact, T_A^* often corresponds to the maximum NPV generated by operating the product line over T periods, $T_A^{*'}$ being its estimated physical life. However, this decision process does not account for the possibility of newly available, better investment opportunities. For instance, an earlier abandonment of product A, at $T_B < T_A^*$ (see FIG.1), makes extra funds available to start producing product B. Investing in B at time T_B will ensure more sales volume and favor a technology leadership position. Moreover, the decision process does not account for the impact of terminating product A, or continuing its operation, on other operations in the system. It also fails to evaluate the merit of product A in terms of investment portfolios (i.e., various feasible combinations of A, B, and C). Thus, failing to account for non-financial factors influencing the strategic planning process, in general, and the abandonment decision, in particular, coupled with an apparent distortion and lack of relevance in the available financial accounting information, made it very difficult for practitioners to justify their abandonment decisions by simply using such models.

A second abandonment school, the "marketing school", has considered product abandonment as a pure marketing decision [13, 14]. The emphasis in this school was put on the design issue of a systematic product abandonment strategy which seeks to identify weak products, evaluate their profitability based on both financial and marketing criteria, and implement the abandonment decision. Researchers from the marketing school have typically viewed the abandonment option as an undesirable, but inevitable, decision that the firm has to make sooner or later. Accordingly, their proposed solutions have generally been

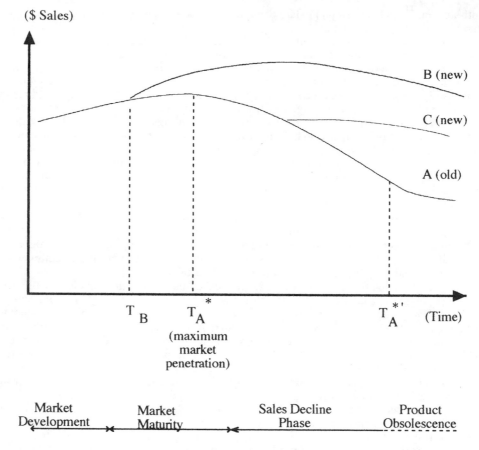

($ Sales)

B (new)

C (new)

A (old)

T_B T_A^* $T_A^{*'}$ (Time)

(maximum market penetration)

Market Development   Market Maturity   Sales Decline Phase   Product Obsolescence

FIG.1 Product Sales Profile

in the form of a reactive response to market decline rather than a proactive, planned capital reallocation strategy which takes advantage of the potential opportunities that a dynamic business environment may present. While such models have generally included non-financial criteria of relevance to the firm's competitive position in the final product abandonment decision, identification of weak products has often been based on product sales volume and/or profitability reports. The accuracy of such reports is obviously questionable because of the inadequacy of traditional accounting practices. Moreover, once a firm makes its decision to abandon a given product, there is not enough guidance offered by these models with respect to the abandonment strategy implementation issue. In particular, these models have at best referred to the importance of choosing an optimal timing for abandonment, and sometimes referred to the factors to be considered in making such choice, but they never went further to calculate this optimal timing or its financial impact on the firm's future profitability.

The third school included research on product abandonment as treated in the literature of "strategic management and corporate organizational and behavioral sciences." According to this school, product abandonment was viewed as a viable strategy among several alternatives to deal with a declining business environment [15]. The proposed abandonment strategy was, in general, similar to that of the marketing school. However, more emphasis was put on the firm's corporate strategy and its competitive position [15], factors influencing the divestment decision [15, 16, 17], behavioral dimensions of the abandonment decision [18], and its legal and social implications [19, 20, 21]. Although this school has offered a broader strategic perspective of the product abandonment option than the two previous ones, it has failed to develop a ready-to-use, step-by-step methodology which guides managers through the whole product review process from identification of weak products through implementation of the deletion decision. In particular, it has emphasized the importance of such social and behavioral factors as employee security and morale, decision-makers' personal attachment to the project, and impact of the abandonment decision on the community as a whole; but it did not offer any tool to realistically measure and account for these factors. Moreover, this school has limited its research to the problem of abandoning whole business units or divisions and assumed that single-asset abandonment is a short-term, tactical capital budgeting problem [16].

Although very little empirical research has been done, so far, with respect to the product abandonment problem, field evidence from the few completed attempts [22, 23, 24] reveals the following major observations common to the three abandonment schools discussed above :

(1) A general observation is that relatively little attention has been given by both academicians and practitioners to the subject of product abandonment.

(2) Another observation is the apparent disjoint efforts of the three abandonment schools working in parallel. Practically, no integrative efforts have been reported yet which could put together, in a single model, the concepts developed separately in the three schools. In particular, only the "engineering economy, financial management, and management accounting" school has gotten into the detailed, quantitative calculation of the financial repercussions on the firm of the abandonment decision and of its optimal timing.

(3) A third major observation has to do with the lack of relevant and reliable financial data customarily needed by researchers and practitioners from the three schools to evaluate product profitability and identify abandonment candidates. Also, the impact of the abandonment decision on the financial profitability of other products in the firm's portfolio has not been well documented. Although new concepts from the modern cost management movement have revolutionized product costing and managerial accounting theory with the advent of activity-based costing (ABC) systems [25], not a single abandonment model has been developed, so far, which incorporates these concepts.

(4) The results from the empirical studies made it clear how practitioners, whenever they had a formal product abandonment policy, had recourse primarily to much of their own intuitive judgement, heuristic approaches, and other empirical models believed to be well suited to their specific situation. Even though a few concepts and/or techniques might have eventually been borrowed from one of the three major abandonment schools, no one company has fully adopted one particular normative abandonment model.

REFORMULATING THE PRODUCT ABANDONMENT PROBLEM FOR THE NEW MANUFACTURING ENVIRONMENT

The development of a sophisticated, yet easy-to-implement, strategic decision model which incorporates the product abandonment option is then needed in order to meet the requirements of the new global business environment. In particular, manufacturers keep viewing their investments in a given venture as a commitment to such venture over its entire physical life. To compete effectively, these manufacturers must first rid themselves of antiquated management philosophies and proactively integrate the product abandonment option into their companies' global decision models. Positive features of, and concepts from, the three conventional abandonment schools should be pooled into a single product abandonment model. Accordingly, various strategic factors, financial and non-financial, of relevance to the company's corporate goals should be incorporated into the product abandonment decision rule. The model should also be dynamic in nature so that it allows for accurate forecasting of future states of the market. Thus, the role that a reliable and accurate cost accounting system can play in evaluating the merits of a given product or process is very important.

Redefining the Nature of the Product Abandonment Decision

Manufacturers are likely to view the product abandonment option from a totally different perspective because the decision to abandon is no longer perceived as a reactive response to conditions of declining demand. To the contrary, the abandonment option could be proactively recommended by management either to deal with some forecasted declining demand trends or simply to make funds available for an even more profitable investment alternative. This option must, therefore, be integrated into the company's overall planning and control system. It must be considered as a strategic component of the company's global capital expenditure decision model (GDM). The GDM should be based on various state-of-the-art managerial concepts and evaluation techniques. In particular, strategic performance measures, or critical success factors, should be defined to translate the company's corporate objectives into quantifiable terms. Production simulation techniques and activity-based costing systems (ABC) can be used to evaluate the performance of the existing system, highlight non-value added activities, and identify where the need for system improvement is. Accordingly, improvement alternatives are generated in terms of portfolios of combined acquisition, replacement, and abandonment projects. Then, each improvement portfolio is analyzed using again simulation techniques and ABC systems. Forecasts and computations should encompass all aspects of costs and benefits incurred by the continued operations of the proposed system as well as those incurred by its premature abandonment or partial alteration at any decision point during the project life cycle. In this phase of the analysis, the product abandonment algorithm is interfaced with the GDM in a highly interactive mode to determine, at each decision point in time, if it is more profitable for the company to abandon the project or hold on to it for one more decision period. FIG.2, below, shows the interface between the GDM and the abandonment algorithm. Once cost, performance, and risk profiles are established for each candidate improvement portfolio, rational preference ordering principles should be used to make trade-offs among these measures [26]. The multi-attribute decision module (MADM) of the GDM is then used to identify the course of action to be selected. In this module the actual improvement brought by each alternative to the existing system is calculated as a percentage of the total targeted improvement for each performance measure. Then, a weighted overall improvement measure is calculated for each alternative. The improvement portfolio which scores the highest is the one to be selected for implementation. After implementation of the selected improvement portfolio, its performance should be continuously tracked and controlled in view the company's corporate objectives.

Formulation of the Product Abandonment Algorithm

A recursive dynamic programming algorithm is developed which evaluates, at each decision point in the product life cycle, the financial merits of both the option to continue the project for one more period and the option to abandon it at that time, assuming all posterior decisions are optimal. Optimality here refers to decisions being based on the recommendations of the multi-attribute evaluation module of the GDM rather than solely on cost information as is the case with practically all conventional abandonment algorithms. Thus, the decision rule used by the proposed model extends the one suggested by Bonini [6] to include non-financial strategic factors. The proposed algorithm assumes that time-dependency of future cash flow streams can be described by a first-order Markov chain. At each stage (or decision point in time), state variables are the Markov-chain probabilistic cash flow states, and decision variables are either to abandon the project at that point in time or to keep it for one more period.

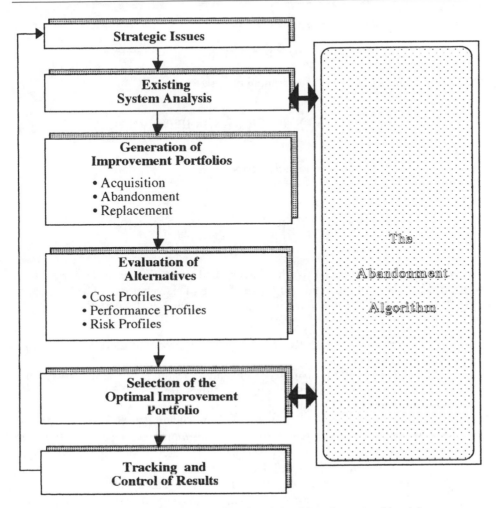

FIG.2 Interfacing the GDM and the Abandonment Algorithm

The framework of the abandonment algorithm is described below.

For each candidate alternative:

1) Calculate

$$IPFW_n^*(S_i) = A V_n(S_i)$$

for all states S_i, i=1 through d.

2) For states i=1 through d, and periods t=1 through n-1, solve backwards starting with period t=n-1 and using the most pertinent forecasts :

   (i) Call the GDM to compare the two alternatives of continued operations of the ongoing concern during period t and abandoning it at the beginning of this period.

(ii) Calculate $IPFW^*_t(Si)$ as:

$$IPFW^*_t(S_i) = A\,V_t(S_i)$$

if the decision is to abandon at the beginning of period t,
or

$$IPFW^*_t(S_i) = \frac{1}{1+k} \sum_{j=1}^{d} P_{ij}(CF_{t+1}(S_j) + IPFW^*_{t+1}(S_j))$$

if the decision is to continue for one more period.

3) for period 0 (the current time), calculate:

$$IPFW^*_0 = \frac{1}{1+k} \sum_{j=1}^{d} P_{0j}(CF_1(S_j) + IPFW^*_1(S_j)) - CF_0$$

4) $IPFW^*_0$ represents the financial worth of the studied alternative, as of the current time projections, if it is to be exploited optimally in the future. This value should be fed back to the GDM for each alternative as one input to the multi-attribute selection process.

where

S_i	is probabilistic cash flow state i. We assume that there are d possible states for each period t.
P_{ij}	is the transition probability from state i to state j. We assume that cross temporal correlations among cash flows are described by a first order Markov chain.
P_{0j}	is the probability for the initial cash flow state in period t=1.
$IPFW^*_t(S_i)$	is the optimal expected present value, or improvement portfolio financial worth of future cash flows at time t associated with being in state S_i at time t, and making optimal decisions from period t on.
$IPFW^*_0$	is the expected net present value of the improvement portfolio.
$CF_t(S_j)$	is the cash flow in period t associated with being in state S_j in that period.
$AV_t(S_i)$	is the abandonment value at the end of period t, given cash flow state S_i during period t.
CF_0	is the initial capital investment.
k	is the cost of capital rate from period t to t+1.

MODEL ILLUSTRATION

A simplified illustration of the proposed solution methodology and a summary of the analysis results are presented in this section.

XYZ, Inc. is a division of a typical medium-size, multi-product, U.S. manufacturing company of consumer products. In the early seventies, corporate management adopted a strategy of product diversification which has enabled their business to successfully expand into new market segments. Unfortunately, adverse performance trends began to show up by the end of the same decade. Not only did the company start losing market share in a number of its business segments to more aggressive competitors, but it also started finding it more and more difficult to manufacture its products at a competitive cost by simply relying on its existing production technologies. The general manager of XYZ realized that the performance of his division has been deteriorating because of the declining profitability trend exhibited during the last three years by products B and C in XYZ's portfolio. However, because of his strong attachment to these specific product lines, the general manager decided to try harder to make these product lines profitable again rather than to consider eliminating them and focusing only on product line A which offers some competitive advantage to the company. Later on, the company's annual report referred to a continued declining profitability trend for products B and C and a decision by HQ to assign a new general manager to the XYZ division. The new manager felt no personal attachment to these "sick" products and made a commitment to reposition XYZ as a more focused company. To carry out this new strategy, she formed a Strategic Planning and Control Committee (SPCC) whose task was to analyze the existing system, identify potential improvement alternatives, and select and implement an optimal course of action. The decision model proposed in this paper was, therefore, recommended as an appropriate framework to perform the required task. Input data and analysis results are summarized below.

Strategic input data

The SPCC initiated its analysis by collecting all relevant strategic inputs from HQ and translated them into quantifiable measures.

Strategic Performance Measures   TABLE 1, below, shows the company's strategic performance measures to be used by the model's MADM to evaluate various alternatives.

TABLE 1
Strategic Performance Measures

Name	Weight	Target
financial worth (M$)(*)	0.4	500.00
quality (% non-defective)	0.3	98
lead-time (min)	0.3	1

(*) M$=Millions of $

Strategic Exit Barriers   TABLE 2, below, shows strategic exit barriers which would eliminate the abandonment option whenever the studied alternative's score with respect to at least one of these barriers is higher than its corresponding prespecified "eliminatory height." Such eliminatory and actual heights are subjectively determined by the SPCC to reflect the impact of each exit barrier on the acceptance or rejection of the product abandonment decision and the the actual score of each improvement portfolio with respect to this barrier.

TABLE 2
Strategic Exit Barriers

Name	Eliminatory Height (%)
core competence	50
legal	70
social	80

Management's Risk Attitude   To simplify the computational burden, assume management is neutral to project riskiness. This means that the risk-aversion coefficient in the corporate utility function is equal to zero. Therefore, maximizing project certainty equivalence is simply equivalent to maximizing its expected monetary value.

Existing System Analysis

Input Data   System physical life: N= 3 years;
Using forecasting techniques, ABC analysis, and production simulation results, the following three probabilistic cash flow states, CF(Si, i=1,2,3), were identified for the system at any period t, 1<= t <=3:   CF(S1)=M\$1,750; CF(S2)=M\$2,000; and CF(S3)=M\$2,250.
The initial investment cost was estimated at :   CF0=\$M6,000.
The transition probability matrix, [Pij], governing this assumed Markov-process was determined using subjective probability assessment techniques and is shown in TABLE 3 below.
Abandonment values were assumed to be independent of the cash flow state at each period: AV(t=0)=CF0=M\$6,000;   AV(t=1)=M\$3,500;   AV(t=2)=M\$2,000;   and AV(t=3)=\$M1,000.

TABLE 3
Transition Probability Matrix--Existing System

From State i (in column 1) to State j (in row 1)	1	2	3
0	0.25	0.5	0.25
1	0.5	0.25	0.25
2	0.25	0.5	0.25
3	0.25	0.25	0.5

Evaluation results   By applying the proposed dynamic programming abandonment algorithm to the existing system, the calculated optimal decision policy showed that the financial worth of the existing system cannot exceed a net of (M\$5,724.99-M\$6,000), or negative M\$275.01 based on the data above if its operations are to be continued. The ABC analysis has particularly pointed to products B and C as being responsible for most of the poor performance of the existing system. Not only do they account for most of the current non-value-added activities within the production system, but they are also diverting management's attention from concentrating on product A which is obviously much more profitable than B and C.

Generation of Candidate Improvement Portfolios

Based on the results from the "existing system analysis" phase and other external market considerations, the SPCC has identified the following two feasible improvement portfolios: 1- Delete product line B and keep A and C as such;
2- Delete product lines B and C and expand the investment in product A.

Evaluation of Alternatives

Improvement portfolio-1   The physical life of the system was   N=3 years;
Four probabilistic cash flow states, CF(Si, i=1,2,3,4), were identified for the system at any period t, 1<= t <=3:   CF(S1)=M\$1,250; CF(S2)=M\$1,350; CF(S3)=M\$1,450; and CF(S4)=M\$1,550;
The initial investment cost (or residual value) was estimated at: CF0=M\$3,600;

The transition probability matrix, [Pij], governing the assumed Markov-process is shown in TABLE 4 below.

TABLE 4
Transition Probability Matrix--Improvement Portfolio-1

From state i (in column 1) to state j (in row 1)	1	2	3	4
0	0.2	0.3	0.3	0.2
1	0.4	0.3	0.2	0.1
2	0.1	0.4	0.3	0.2
3	0.1	0.2	0.4	0.3
4	0.1	0.2	0.3	0.4

Abandonment values were assumed to be independent of the cash flow state at each period: AV(t=0)=CF0=M\$3,600; AV(t=1)=M\$2,500; AV(t=2)=M\$1,200; and AV(t=3)=M\$500;

Improvement portfolio-2 The physical life of the system was: N= 3 years; Three probabilistic cash flow states, CF(Si, i=1,2,3), were identified for the system at any period t, $1 <= t <= 3$: CF(S1)=M\$1,200; CF(S2)=M\$1,300; and CF(S3)=M\$1,400; The initial (residual) investment cost was estimated at : CF0=M\$3,500; The transition probability matrix, [Pij], governing the assumed Markov-process is shown in TABLE 5 below.

TABLE 5
Transition Probability Matrix--Improvement Portfolio-2

From state i (in column 1) to state j (in row 1)	1	2	3
0	0.3	0.4	0.3
1	0.4	0.3	0.3
2	0.3	0.4	0.3
3	0.3	0.3	0.4

Abandonment values were assumed to be independent of the cash flow state at each period: AV(t=0)=CF0=\$3500, AV(t=1)=\$2700, AV(t=2)=\$1800, AV(t=3)=\$500;

Evaluation results Using the proposed decision model, the SPCC found that improvement portfolio-1 can generate a maximum financial value of (M\$3,903.64-M\$3,600), or positive M\$303.64, if kept until the end of its physical life. Whereas, improvement portfolio-2 can generate a maximal financial value of (M\$3,743.80-M\$3,500), or positive M\$243.80, if abandoned at the end of its second year of operation, based on the currently available information. In determining such optimal policy, the decision to abandon improvement portfolio-2 at the end of the second year had first to pass the test of "strategic exit barriers" as specified by the SPCC. In fact, this portfolio scored at exit barrier levels of 40, 50, and 30 respectively, which are all lower than their corresponding maximum allowable (eliminatory) heights shown in TABLE 2 above.

Selection of the "optimal" improvement portfolio

Based on the financial criterion alone, improvement portfolio-1 is obviously the "optimal" choice since it maximizes the expected certainty equivalence, or expected monetary value in this case, of the system. However, to account for the company's corporate objectives, the SPCC included various strategic performance measures shown in the table above as indicated by the MADM. TABLE 6 and TABLE 7 below summarize the results of this analysis.

TABLE 6
Actual Performance of Improvement Alternatives

System	Financial Worth($)	Quality (% non-defective)	Lead-Time (min)
Existing	-275.00	88	3.5
Improv. Portf.-1	303.64	92	3.2
Improv. Portf.-2	243.80	96	3.9

TABLE 7
Ratio of Actual vs. Targeted Improvement Figures

System	Financial Worth (%)	Quality (%)	Lead-Time (%)	Overall Weighted Score (%)
Existing	0.0	0.0	0.0	0.0
Impro. Portf.-1	74.8(*)	40.0	12.0	45.51(**)
Impro. Portf.-2	66.9	80.0	64.0	69.98

(*) [303.64-(-275.01)]/[500.00-(-275.010]=0.7477=74.8%
(**) (0.4*74.8)+(0.3*40.0)+(0.3*12.0)=45.51%

Thus, the weighted overall improvement figures are equal to 45.51% and 69.98% for improvement portfolios 1 and 2, respectively. Contrary to the decision reached based solely on financial criteria, the GDM recommends the implementation of improvement portfolio-2 since it contributes more to the achievement of the company's overall strategic objectives.

CONCLUDING REMARKS

In this paper, the three major schools of thought regarding the problem of product/process abandonment analysis have been briefly described. Pertinent observations about the economic validity and operational applicability of conventional abandonment models were also made. Several apparent limitations in these models were identified as being a major impediment to the modernization process in manufacturing. The lack and inadequacy of traditional cost accounting information customarily used by companies in their product profitability reports was a major origin of non-competitive manufacturing decisions. The paper has also described, in broad terms, the new business environment and the required managerial philosophy to meet the challenges of such environment . Finally, major findings from the two previous sections were integrated to synthesize a global decision model (GDM) which proactively incorporates the product abandonment option into the company's strategic planning and control system. The proposed model was then illustrated through a simplified case-study. Not only were the benefits from including the product abandonment option into the analysis shown, but in addition, the need to use a strategy-driven abandonment decision rule was clearly demonstrated.

REFERENCES

1. Ouederni, Bechir N.,"A Taxonomy of the Product Abandonment Analysis Topic," in an unpublished Ph.D. dissertation, VPI&SU, Spring 1992.
2. Robicheck, Alexander, and James C. VanHorne,"Abandonment Value and capital Budgeting," *Journal of Finance*, Dec. 1967, pp.577-589.
3. Robicheck, Alexander, and James C. VanHorne,"Abandonment Value and capital Budgeting," *Journal of Finance*, March 1969, pp.96-97.
4. Dyl, Edward A., and Hugh W. Long,"Abandonment Value and Capital Budgeting," *Journal of of Finance*, March 1969, pp.88-95.
5. Schwab, B., and P. Lusztig,"A Note on Abandonment Value and Capital Budgeting," *Journal of Financial and Quantitative Analysis*, Sept. 1970, pp.377-379.
6. Bonini, Charles P.,"Capital Investment Under Uncertainty With Abandonment Options," *Journal of Financial and Quantitative Analysis*, March 1977, pp.39-54.
7. Hillier, Frederick S.,"The Derivation of Probabilistic Information for Evaluation of Risky Investments," *Management Science*, Vol.9, No.3, April 1963, pp.443-457.
8. Howe, Keith M., and G. M. McCabe,"On Optimal Asset Abandonment and Replacement," *Journal of Financial and Quantitative Analysis*, Nov. 1983, pp.295-305.
9. Sibley, A. M.,"The Abandonment Decision and Capital Rationing," *Akron Business and Economic Review*, Vol.19, No.3, Fall 1988, pp.95-108.
10. Chen, Son-Nan, and William T. Moore,"Project Abandonment Under Uncertainty: A Bayesian approach," *Financial Review*, Vol.18, No.4, Nov. 1983, pp.306-313.
11. Aggrawal, Raj, and Luc A. Soenen,"Project Exit Value as a Measure of flexibility and Risk Exposure," *The Engineering Economist*, Vol.35, No.1, Fall 1989, pp.39-54.
12. Cooper, Robin, and R. S. Kaplan,"Measure Costs Right: Make the Right Decisions," *Harvard Business Review*, Sept.-Oct. 1988, pp.96-103.
13. Berenson, C.,"Pruning the Product Line," *Business Horizons*, Summer 1963, pp.63.
14. Kotler, Philip,"Phasing Out Weak Products," *HBR.*, March-April 1965, pp.107-118.
15. Harrigan, Kathryn, R., Strategies for Declining Businesses, D.C. Heath and Company, Lexington Books, Lexington, MA., 1980
16. Duhaime, Irene M., and John H. Grant,"Factors Influencing Divestment Decision-Making: Evidence from a Field Study," *Strategic Management Journal* (U.K.), Vol.5, No.4, Oct.-Dec. 1984, pp.301-318.
17. Porter, Michael E., "Please Note Location of Nearest Exit: Exit Barriers and Strategic and Organization Planning," *Calif. Mgmt. Rev.*, Vol.19, No.2, Winter 1976, pp.21-33.
18. Duhaime, Irene M., and Charles R. Schwenk,"Conjectures on Cognitive Simplification in Acquisition and Divestment Decision-Making," *Academy of Management Review*, Vol.10, No.2, April 1985, pp.287-295.
19. Bullock, Kathy,"Downsizing: Opportunities and dilemmas," *Cost and Management* (Canada), Vol.59, No.4, Jul.-Aug. 1985, pp.60-63.
20. DeWitt, Rocki-Lee,"Strategies for Downsizing," Unpublished Paper, Department of Management and Organization, The Pennsylvania State University, 1990.
21. Heenan, David A.,"The downside of Downsizing," *Across the Board*, Vol.27, No.5, May 1990, pp.17-19.
22. Hise, Richard T., and Michael A. McGinnis,"Product Elimination: Practices, Policies, and Ethics," *Business Horizons*, June 1975, pp.29-42.
23. Lambert, Douglas M., The Product Abandonment Decision, A Study by the NAA and the Society of Management Accountants of Canada, Montvale, N.J. 1985.
24. Rothe, J. T.,"The Product Elimination Decision," *MSU Bus. Topics*, Autumn 1970.
25. Brimson, James A., Activity-Based Investment Management, AMA, N.Y., 1989.
26. Canada, J. R., and W. G. Sullivan, Economic and Multiattribute Evaluation of Advanced Manufacturing Systems, Prentice Hall, Englewood Cliffs, N.J., 1989.

UNCERTAINTY MANAGEMENT SYSTEMS IN MANUFACTURING INVESTMENT APPRAISAL RISK ANALYSIS

Edward T. Sweeney

Warwick Manufacturing Group, Department of Engineering,
University of Warwick, Coventry, U.K.

ABSTRACT

The use of quantitative risk analysis techniques is becoming increasingly widespread in industry. The use of expert system techniques has been explored by many researchers in an attempt to improve upon the approaches currently adopted by managers responsible for the allocation of capital resources in manufacturing companies. An uncertainty management system (UMS) is an example of how the use of an approach associated with expert systems offers potential for improvement in industrial investment appraisal practice.

This paper reviews traditional approaches to risk analysis and describes how the four uncertainty management systems in most common use can be of use in analysing the risk associated with different aspects of the investment appraisal decision making process. A UMS selection methodology is also proposed as an important element of the process.

INTRODUCTION

Advances in manufacturing technology which have taken place in recent years have sharpened the focus on the need for effective investment appraisal practices in industry. It has long been accepted that the most appropriate method of investment appraisal is one which uses discounted cash flow (DCF) [1].

Empirical evidence points to increasing use of DCF techniques by industrial managers responsible for the allocation of capital resources particularly in firms in the U.K. and the U.S.A. [2]. This shift from the use of simpler and technically inferior methods, such as payback period and accounting rate of return, appears to be part of a general trend towards the use of more sophisticated techniques within the investment appraisal decision making process.

Recent advances in hardware and software have facilitated the use of expert systems in industry. Many systems have been developed across a range of manufacturing domains. A number of authors (e.g. [3]) have proposed the use of expert system techniques as a means of overcoming some of the weaknesses currently evident in industrial investment appraisal practice. An uncertainty management system (UMS) is an example of how an approach associated with expert systems offers potential for improvement.

The objectives of this paper are to review some of the traditional approaches to risk analysis and to demonstrate the use of the four UMSs in analysing different aspects of the risk element involved in appraising proposed investments in manufacturing technology.

However, each of these four approaches was conceived to deal with particular types of uncertainty which occur in practice. Selection of the appropriate approach can be a difficult task for the expert system developer. UMS selection methodology is, therefore, proposed as an important element of the process.

SOME TRADITIONAL APPROACHES TO RISK ANALYSIS

Introduction

In the appraisal of investment proposals one is attempting to predict future outcomes which, by their very nature, can never be known with absolute certainty. Many different approaches are used throughout industry in an attempt to account for this uncertainty. These range from the very simple (and often quite trivial) to the relatively sophisticated. Each approach has its own particular strengths and weaknesses in attempting to understand how uncertainty in the input factors to the investment appraisal decision analysis process (e.g. cashflows and discount rates) combine to produce uncertainty in the output of the process (e.g. project net present value).

"Crude" Methods

Such methods include reducing the project cashflows and increasing the criterion rate of return. In essence, the yardstick against which the proposal is being judged is changed or the input factors are being more conservatively estimated than would have been the case in a project perceived as being less risky. Empirical evidence suggests that such "crude" approaches are widely used in practice [4]. Increasing the so-called hurdle rate of return (i.e. the rate at or above which a project is deemed to be acceptable) appears to be particulary popular. Indeed, a number of authors have suggested that this practice is one of the most serious ways in which DCF techniques have been misapplied in industry (e.g.[5]).

Use of Payback Period

It may be perceived as being important with a risky investment proposal to at least recoup the initial capital outlay. Hence, the payback period or a form of the payback period which uses discounted cash flows is often used [6]. There is some evidence to suggest that many Japanese firms use an approach based on cash flow projections that include imputed interest charges on their investment in a project [7]. This is equivalent to using payback period with discounted cash flows. There are, however, several difficulties with the use of any form of payback period [1].

Use of DCF Methods

DCF methods place less reliance on future cash flows by discounting them back to present values. A *risk premium* can be built into the discount rate and this use of risk-adjusted discounted rates is widespread [8]. An obvious problem with the use of this approach to risk analysis is that the risk adjustment is compounded as the discount rate is compounded for future cash flows.

Sensitivity Analysis

Sensitivity analysis considers the effect on a project of changes in any of the parameters considered sensitive to change. Sensitivity graphs can be prepared which indicate the consequences of such changes on the overall profitability of an investment.

Variation analysis is a form of sensitivity analysis which involves projecting what are considered to be pessemistic and optimistic cash flows (in addition to the expected "most likely" cash flows) and looking at the resulting returns. A disadvantage of this approach is its inability to deal with situations where input parameters are interdependent [9].

Application of Probability Theory

Some of the concepts of probability theory can be applied in risk analysis. The main technique in use is known as the Monte Carlo method[10]. This technique simulates an investment proposal by taking random samples from probability distributions of all factors identified as affecting the profitability of a project. The probability of a range of possible outcomes is computed. A number of software packages exist which implement this technique.

Risk Analysis in Practice

Whilst the preceeding sections are by no means exhaustive it does give an indication as to the range of techniques available to the investment planner. Several independent surveys, conducted over the past thirty years, have provided some insight into industrial practice in this area. An interesting trend is the increasingly widespread use of the more sophisticated techniques within the investment appraisal toolkit. This trend is pointed out by Rosenblatt and Jucker [11] who analysed a number of surveys carried out in the U.S.A. between 1955 and 1977. A U.K. survey [12] points to a similar trend. It reports increased use of sensitivity analysis, probability analysis and computer simulation in the period 1975 - 1980. More recent surveys (e.g. [2] and [13]) provide further evidence of the trend towards increased use of quantitative techniques in investment appraisal and, in particular, in risk analysis.

EXPERT SYSTEMS IN INVESTMENT APPRAISAL

A number of authors ([14],[15],[16]) have proposed the use of expert systems as a way of overcoming some of the difficulties currently evident in investment appraisal practice. A number of distinct types of possible application within capital budgeting can be identified and include the following :

1. The embodiment within a computer of knowledge which is essential if an investment appraisal is to be as complete and comprehensive as possible - Relevant details about the corporate taxation system and the costs of various sources of funding are examples of such information.

2. The provision of checklists - An expert system, in reliably posing all relevant questions, can act as a checklist, reminding the user of all the factors to be taken into account.

3. The provision of heuristic rules - This involves giving advice based on procedural tips or incomplete methods of performing certain tasks.

In addition to the above, expert systems provide a mechanism for coping with uncertainty, i.e. an uncertainty management system (UMS). This would appear to add to their attractiveness as a tool in investment appraisal.

UNCERTAINTY MANAGEMENT SYSTEMS IN INVESTMENT APPRAISAL

Recent years have seen a growing necessity for including uncertain, imprecise and incomplete data and relations between them in a broad range of domains. Uncertainty is incorporated into all knowledge based system components. Most complex application domains involve multiple types of uncertainty and the interaction between them is often of critical importance [17]. Several approaches for treating uncertainty have been developed.

What is of interest from the point of view of the work described in this paper is their applicability to the field of investment appraisal and, in particular, to the analysis of the risk associated with the appraisal of investment proposals in manufacturing companies. The degree to which the application of the techniques can lead to improvements over the more traditional techniques (described above) needs to be explored. The following sections describe the application of each of the four most commonly used UMS to different aspects of the investment appraisal decision making process.

Bayes Theory in the Revision of Subjective Probabilities

Consider the example shown in the decision tree in Figure 1 (below).

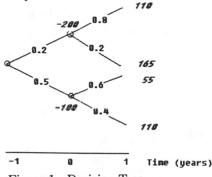

Figure 1 - Decision Tree

A number of hypotheses ($H_1..H_3$) concerning the cashflow for year 1 (C_1) can be put forward as follows.

$$H_1 : C_1 = 110$$
$$H_2 : C_1 = 165$$
$$H_3 : C_1 = 55$$

The probabilities associated with each of these possible outcomes is :

$$p(H_1) = (0.5 \times 0.8) + (0.5 \times 0.4) = 0.6$$
$$p(H_2) = 0.5 \times 0.2 = 0.1$$
$$p(H_3) = 0.5 \times 0.6 = 0.3$$

If it turns out that the cash flow for year 0 (C_0) is, in fact, -200, then this provides us with a piece of evidence (E) with which to revise the probability of occurence of the possible outcomes for C_1, i.e.

$$E : C_0 = -200.$$

The conditional probabilities in this case are :

$$p(E|H_1) = 0.5$$
$$p(E|H_2) = 1$$
$$p(E|H_3) = 0$$

Then by Bayes rule the posterior probabilities become :
$$p(H_1|E) = 0.75$$
$$p(H_2|E) = 0.25$$
$$p(H_3|E) = 0.$$

What results essentially represents revised probabilities for the three original hypotheses. For longer investment projects this revision of probability could be carried out dynamically so that the effect of new pieces of evidence on the subjective probabilities could be explored on an ongoing basis. The effects of these revisions on the overall project profitability (as measured by a NPV, for example) could then also be explored.

Other uses of Bayes theory in investment appraisal has been proposed by many authors (e.g.[18]).

Fuzzy Investment Decision Analysis

Fuzzy Sets in Investment Appraisal   Since the introduction of fuzzy logic in 1965 [19] several attempts have been made to relate the theory of fuzzy sets to practical decision making. The following sections outline how a number of these fuzzy decision models might be applied to investment appraisal.

Decision Making in a Fuzzy Environment   Bellman and Zadeh [20]suggested a fuzzy decision making model which they applied to the problem of determining the optimal dividend to be paid to a company's shareholders. Its basis is as follows :
Let $X = \{x\}$ be a given set of alternatives, i.e. the decision space. Then a fuzzy goal will be identified with a fuzzy set G in X. Similarly, a fuzzy constraint, C is defined to be a fuzzy set in X. Then G and C combine to form a decision D, which is a fuzzy set resulting from the intersection of G and C. The fuzzy set "decision", using fuzzy set theory is then characterized by its membership function :
$$\mu_D(x) = \min(\mu_G(x), \mu_C(x))^1.$$
However, many problems of interest to investment analysts are not covered by this model. For example, a decision as to which of several alternative strategies to adopt

1.   The detailed terminlogy, properties and calculus of fuzzy set can be found in [21].

based on maximum utility can not be addressed. For this type of problem, Watson et. al. proposed an algorithm [22].

Fuzzy Decision Analysis [22]   This approach is based upon the analysis of a simple decision tree such as that shown in Figure 2 (below). This is quite similar in form to the decision tree representation of a set of cash flows (Figure 1) discussed previously.

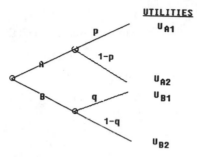

Figure 2 - Simple Decision Tree

The choice represented here is between two possibilities A and B, dependent on different uncertain events.   In A there is the probability p of gaining an amount expressed as utility U_{A1} and a probability $1 - p$ of gaining an amount expressed as utility U_{A2}. A similar statement can be made concerning B.

However, the quantities $p, q, U_{A1}, U_{A2}, U_{B1}$ and U_{B2} are only imprecisely known and, therefore, can be expressed as fuzzy sets. What is of interest is, as Watson et. al. point out, is the "fuzz" on the output given the "fuzz" on the inputs. Fuzzy calculus, is then used in the following way :

$$\mu_A(a) = \max_{p u_{A1} + (1-p) u_{A2} = a} [\min(\mu_P(p), \mu_{A1}(\mu_{A1}), \mu_{A2}(\mu_{A2}))]$$

$$\mu_B(b) = \max_{q u_{B1} + (1-q) u_{B2} = b} [\min(\mu_Q(q), \mu_{B1}(\mu_{B1}), \mu_{B2}(\mu_{B2}))]$$

where $\mu_A(a)$ and $\mu_B(b)$ are the degrees to which a and b belong to the possible sets of expected utilities for possibilities A and B, respectively, $\mu_P(p)$ is the degree to which p belongs to the set of possible values for this probability and the other functions are similarly defined.

Several difficulties were encountered when attempting to apply the approach outlined above to a simple investment analysis described using a decision tree. Foremost among these was the fact that in practical applications the probability variables p and q are not in the same universe of discourse as the utility variables (U_{A1} etc.). This fundamental problem is ignored by Watson et. al. who appear to have performed some mathematical transformation to enable the utility variables to be expressed as values in the range [0..1].   Freeling [23] discusses several other weaknesses of the approach.

A More Generalised Approach   In many investment appraisal situations one is concerned with making a decision in an environment in which both objectives and constraints exist. Often (but not necessarily always) these objectives and constraints will be fuzzy in nature. An example is now presented to illustrate the proposed approach.

Consider the case where one needs to decide in which of five possible proposals to invest. The expected cash flows for each of the five projects is shown in Figure 3 (below).

Project	1	2	3	4	5
Year					
0	-80000	-90000	-100000	-110000	-120000
1	40000	10000	20000	50000	10000
2	60000	20000	30000	50000	10000
3	40000	30000	50000	50000	20000
4		60000	30000	50000	60000
5		60000	20000		10000
6		40000			80000
7		20000			50000
8					70000
9					40000
10					30000

Figure 3 - Expected Cash Flows (for five projects)

The net present values, using a discount rate of 10%, of each of the five projects are :

Project 1 : 35964

Project 2 : 69185

Project 3 : 13402

Project 4 : 48455

Project 5 :124595

A number of objectives and constraints exist as follows :

Objective 1 - the NPV should be as great as possible.

Objective 2 - the variation in cash flows over a project's life should be as small as possible.

Constraint 1 - the life of the project should be "about five years".

Constraint 2 - the cash flow for year 0 should not be more than 100000

If the projects are re-arranged in descending order of NPV these objectives and constraints can be expresses as fuzzy (or, indeed, non-fuzzy sets).

Objective 1

If $\mu_{O1}(d)$ is the membership function for project d (1< d < 5, in this case) in relation to O1 (objective 1) then it might be expressed as :

$$\mu_{O1}(d) = 1 - (d - 1/5)$$

This clearly indicates the projects with relatively small values of d are favoured. This gives the fuzzy set

$$O1 = \{(1,1),(2,0.8),(3,0.6),(4,0.4),(5,0.2)\}$$

Objective 2

The fuzzy set representing this objective (O2) might be subjectively evaluated as follows:

$$O2 = \{(1,0),(2,0.4),(3,1),(4,0.7),(5,0.5)\}$$

Project number 1 with its very uneven cash flow profile is assigned a membership function 0, whereas project 3 with equal cash flows arising each year of the project is assigned a value 1.

<u>Constraint 1</u>

Again, a fuzzy set (C1) is subjectively evaluated as follows :

$$C1 = \{(1,0.2),(2,0.6).(3,0.9),(4,1),(5,0.6)\}$$

In this case project 4, with an expected life of exactly five years is assigned a membership function 1.

<u>Constraint 2</u>

This is a straightforward non-fuzzy set (C2).

$$C2 = \{(1,0),(2,1),(3,0),(4,1),(5,1)\}$$

These functions, representing the objectives and constraints in this particular situation, are shown graphically in Figure 3 (below). The fuzzy "decision function" (F(d)) is also shown. From fuzzy calculus (section 4.3 above) this is given by :

$$F(d) = \min \{O1(d),O2(d),C1(d),C2(d)\}$$

To obtain a "crisp" (i.e. non-fuzzy) solution to this problem, one would simply choose the maximum value in this decision function (project 2 or project 4 in this case).

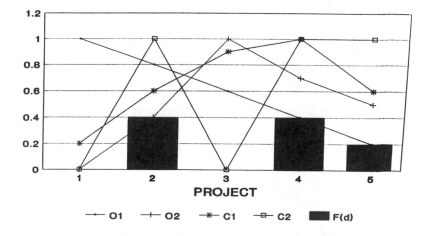

Figure 3 - Functions for objectives, constraints and decision

This approach to using fuzzy logic allows the effect a number of objectives and contraints of an investment proposal to be explored. It can, therefore, be used as an evaluation tool in conjunction with the conventional techniques of discounted cash flow, particularly in a situation such as the one described above when a number of alternative ways of proceeding are available to an investor.

Certainty Factors in Risk Analysis

Introduction　　The form of the knowledge representation used in MYCIN [24] is based on production (i.e. IF..THEN) rules. These rules are referred to as premise-action pairs where a premise consists of one or more conditions and an action is a conclusion or an instruction. The certainty factors are associated with the data elements that match conditions, indicating a degree of belief or disbelief in the information represented.　　　The certainty factor for a conclusion is derived using the simple formula :

$$CF<action> = CF<premise> \times CF<rule>$$

This form of uncertainty management could be used in any investment appraisal sub-domain where it is possible to represent the knowledge in the form of production rules.　　An example of such a sub-domain is that of forecasting.

Certainty Factors in FOREX - An Expert System for Demand Forecasting [25]
FOREX is an expert system which recommends appropriate forecasting techniques. It was initially designed as a decision support tool for investment analysts attempting to forecast market demand as part of the capital budgeting process. It is a production rule based system and hence the application of certainty factors is a possibility.

One example of a rule which forms part of FOREX is outlined in [25] and is expressed as follows :

> IF accuracy is not of importance
> > AND turning point identification is not of importance
> > AND ease of understanding is not of importance
> > AND time span is short term
> > AND time horizon is long term

AND available resources are limited

THEN the use of a visionary forecast appears to be appropriate.

The way in which the system is currently implemented indicates that this rule is believed with absolute certainty. In practice it might well be the case that the balance between the measures of belief and disbelief leads to the conclusion that the level of certainty is, say, 0.8, i.e.

CF<rule> = 0.8.

FOREX, in its consultation with the user, asks for information concerning the factors considered by the rule (above). If the user responds by telling the system that, for example, accuracy is not of importance, the implicit assumption therein is that this is <u>definitely</u> the case. Again, in practice it could be the case, for example, that in relation to a particular forecast that accuracy (denoted acc) is *not of importance* with a certainty of 0.8 and *of importance* with a certainty of 0.2. This can be expressed as :

acc = (unimportant 0.80) (important 0.2).

Similar statements might be made concerning the other factors which comprise the rule, e.g :

tpi = (unimportant 1.0)

eou = (unimportant 0.9) (important 0.1)

ts = (short term 1.0)

th = (short term 0.1) (medium term 0.1) (long term 0.8)

res = (very limited 0.1) (limited 0.8) (unlimited 0.1).

These statements relate to turning point identification (tpi), ease of understanding (eou), time span (ts), time horizon (th) and available resources (res).

Returning to the rule, the condition "accuracy is not of impotance" evaluates to 0.8, indicating that this condition is met with a certainty of 0.8. The other five conditions evaluate to 1.0,0.9,1.0,0.8 and 0.8 respectively. Because the recommendation is only made if all of the conditions are satisfied (i.e. the use of ANDs in the rule) the value for the conjuction becomes the minimum of these six values, i.e.

CF<premise> = 0.8.

The certainty factor associated with the recommendation is computed using the simple formula :

$$CF<action> = CF<premise> \times CF<rule>$$

so we conclude that "a visionary forecast appears to be appropriate" with a certainty of 0.64.

The Potential of Dempster Shafer Theory in Risk Analysis

The potential of Dempster Shafer theory [26, 27] as a tool in the management of uncertainty has been explored by the author in a number of investment appraisal sub-domains. The basis of the approach which was adopted was as follows :

1. Identification of a "frame of discernment", θ.
2. Identification of subsets of hypotheses in θ giving rise to new hypotheses.
3. Consideration of all possible subsets of θ (denoted 2^θ).
4. Assignment of bpa (basic probability assignment) to subsets of θ based on available evidence.
5. Computation of belief functions (Bel).
6. Computation of new belief functions representing impact of combined evidence.

Two specific types of potential application area were identified.

1. Choices between alternatives

Many such choices present themselves in investment appraisal. Indeed, the whole investment appraisal decision making process could be regarded as being about choosing between alternatives. The type of problem focussed upon was similar to the one described above to which fuzzy theory was applied (i.e. the problem of choosing between a number of alternative investment proposals).

2. Rule based systems

Any area of investment appraisal amenable to solution using a rule based approach may also benefit from the application of Dempster-Shafer theory. As Gordon and Shortcliffe [27] point out "the original MYCIN CF combining function is a special case of the Demster function". The FOREX system was again used as the basis of the work.

To date, little progress has been made in successfully applying this theory to either of these broad caterories in any meaningful or practical way. Among the obstacles encountered has been the mutual exclusivity requirement of the hypotheses. This issue has been referred to by other authors in relation to the possible application of the theory in different domains (e.g. Barnett [28] in relation to its application to MYCIN). Other difficulties relate to the sheer amount of input data required (i.e. a bpa must be assigned to every subset of possible events) and to the computational complexity involved.

UMS SELECTION METHODOLOGY

Complex real world domains, such as investment appraisal, usually involve several different types of uncertainty. A problem often faced by the expert system developer is the choice between the various alternative uncertainty management systems which exist. A methodology for selection of an appropriate system has been proposed [17] and implemented. The methodology's basis lies in the identification of eight seperate sources of uncertainty and assessing the power of each of the four UMS in relation to these. It is not proposed to discuss it in detail here. However, it is of importance in the context of the work described in this paper as potential novel applications of the four most commonly used UMS to particular sub-domains of the investment appraisal field have been explored and presented. The issue of which UMS is appropriate to which sub-domain is critical if their application is to be broadened beyond that outlined above.

DISCUSSION

The work outlined above indicates that at least three of the four UMS discussed is potentially of value in understanding the risk associated with specific aspects of the investment appraisal decision making process. The existence of the selection methodology should allow further areas where UMS application is appropriate to be identified and thus broaden the scope and usefulness of the approach.

It is envisaged that the use of UMS be complementary to (and **not** act as a replacement of) the traditional risk analysis techniques described above. Both approaches have a role to play and can easily be used in conjunction with each other.

The provision of a particular UMS with commercially available expert system shells is likely to add to the attractiveness of expert systems as a feasible tool in investment appraisal. If the expert system approach is adopted a development tool can be selected which incorporates an appropriate UMS.

The thrust of current and future work is centred around the identification of further investment appraisal sub-domains to which workable application of UMS can be made. The existence of the selection methodology facilitates this. Work on the application of Dempster-Shafer theory, in particular, is considered to be of importance. As mentioned above, a realistic application has yet to be identified.

The use of approaches such as those outlined in this paper enhances the capability of expert systems in handling investment appraisal problems. Such an enhancement makes the use of the expert system approach a more feasible proposition.

REFERENCES

1. Lumby, S : *Investment Appraisal* , 2nd Ed., Wokingham, Van Nostrand Reinhold, 1984.
2. Sweeney, ET : Investment appraisal in Irish manufacturing systems, in *Advances in Manufacturing Technology.* (O'Kelly ed.) Ireland, University College Galway, September 1986, pp. 593 - 617.
3. Sweeney, ET : An integrated approach to the appraisal of investment opportunities in advanced manufacturing technology, in *Proceedings of the Fifth International Manufacturing Conference in China,* Guangdong , April 1991, pp. B236 - B242.
4.Mills, RW : Capital budgeting techniques used in the U.K. and the U.S.A., in *Management Accounting,* pp. 26 - 27, September 1988.
5. Hayes, RH and Garvin, DA : Managing as if tomorrow mattered, in *Harvard Business Review,* pp. 71 - 79, May - June 1982.
6. McIntyre, A and Coulthurst, N : Planning and control of capital investment in medium sized U.K. companies, in *Management Accounting,* pp. 39 - 40, March 1987.
7. Hodder, JE and Riggs, HE : Pitfalls in evaluating risky projects, in *Harvard Business Review,* pp. 128 - 135, January - February 1985.
8. Ashford, RW, Dyson, RG and Hodges SD : The capital-investment appraisal of new technology : problems, misconceptions and research directions, in *J. Opl. Res. Soc.* Vol. 39, No. 7, pp. 637 - 642, 1988.
9. Townsend, EC : *Investment and Uncertainty - A Practical Guide.* Edinburgh, Oliver and Boyd, 1969, pp. 122 - 123.

10. Hertz, DB : Risk analysis in capital investment, in *Harvard Business Review*, pp. 169 - 181, September - October 1979.

11.Rosenblatt, MJ and Jucker JV : Capital investment decision making : some tools and trends, in *Interfaces*, Vol. 9, No. 2, pp. 146 - 149, 1979.

12. Pike, RH : A review of recent trends in formal capital budgeting processes, in *Accounting and Business Research*, pp. 201 - 208, Summer 1983.

13. Weiss, GE : Investment appraisal practice in U.K. firms, unpublished dissertation, University of Warwick, 1991.

14. Myers, SC : Notes on an expert system for capital budgeting, in *Journal of Financial Management*, Vol. 17, Iss. 3, pp. 23 - 31, 1988.

15. Sweeney, ET : Expert systems in manufacturing investment appraisal, in *Proceedings of the fourth International Conference on Computer-aided Production Engineering*, University of Edinburgh, pp. 433 - 442, November 1988.

16. Ash, N : How Cash Value appraises capital projects, in *The Accountant*, pp. 18 - 19, 2 October 1985.

17. Petrovic, D. and Sweeney, ET : An approach for selecting appropriate methods for treating uncertainty in knowledge based systems, in *Yugoslav Journal of Operations Research*, Vol. 1, No. 1, pp. 79 - 89, 1991.

18. Hull, JC : *The Evaluation of Risk in Business Investment*, Oxford, Pergamon, 1980.

19. Zadeh, LA : Fuzzy sets, in *Information and Control*, Vol. 8, pp. 338 - 353, 1965.

20. Bellman, RE and Zadeh, LA : Decision making in a fuzzy environment, in *Management Science*, Vol. 17, No. 4, pp. B141 - B164, 1970.

21. Dubois, D. and Prade, H : *Fuzzy Set and System - theory and application*, Academic Press, 1980.

22. Watson, SR, Weiss, JJ and Donnell, ML : Fuzzy decision analysis, in *IEEE Transactions on Systems, Man and Cybernetics*, SMC-9 (1), pp. 1 - 9, 1979.

23. Freeling, ANS : Fuzzy sets and decision analysis, in *IEEE Transactions on Systems, Man and Cybernetics*, SMC-10 (7), pp. 341 - 354, 1980.

24. Shortliffe, E H and Buchanan, B G : A Model of Inexact Reasoning in *Medicine, in Role-Based Expert Systems* (Buchanan and Shortliffe, E H eds.) Addison Welsey, 1985, pp.233 - 262

25. Sweeney, ET : The use of expert systems in demand forecasting, in *Integration and Management of Technology for Manufacturing* (Robson, EH, Ryan, HM and Wilcock, D. eds.), London, IEE, 1991, pp. 47 - 54.

26. Shafer, G : *A Mathematical Theory of Evidence*, Princeton, Princeton University Press, 1976.

27. Gordon, J and Shortliffe, EH : The Dempster Shafer theory of evidence, in *Rule-Based Expert Systems* (Buchanan, BG and Shortliffe, EH. eds.), Addison Wesley, 1985, pp. 272 - 292.

28. Barnett, JA : Computational methods for a mathematical theory of evidence, *in Proceedings of the 7th International Conference on Artificial Intelligence* (Vancouver B.C.), pp. 868 - 875, 1981.

MAINTAINING THE AUTOMATED SYSTEMS OF THE 21ST CENTURY: ISSUES, REQUIREMENTS, APPROACHES, AND IMPLEMENTATION

Donald J. Knapp

StepsAhead Support Systems, Hayward, California, U.S.A.

ABSTRACT

As advances in technology pertaining to factory automation and information systems move forward, there follows increasing dependence on these capabilities for delivering competitive, cost-effective results. However, largely overlooked at present is a companion requirement to keep the automated systems themselves maintained so that they perform reliably and continuously at peak level. Technical support (maintenance) of such technology is being stretched to the limits under current levels of work responses and conventional management approaches. An automated factory with manufacturing equipment or information systems on the blink, or operating at sub-optimum levels, surely will fall short of delivering intended results. This paper examines the growing requirements for maintenance support, the issues that must be addressed and resolved, some approaches that could be considered for meeting the needs, and the technical and management challenges of implementing the levels of professional maintenance that will be needed as the 21st century draws near.

ISSUES

As new technology continues to enter the factory, the proportion of effort being shifted to technical support increases each day. Indeed, the more complex devices and systems seem to require much greater levels of technical support than previously. Currently, more failures than successes appear to be counted in this arena, because there has been too little planning and mobilization for this function. Gearing up for "winning maintenance" for the next century will require innovative, bold, and informed concepts of what sort of technical support efforts are needed; what assets and resources it will take to meet those needs; and what plans for expending efforts and money should be implemented for getting to the desired levels of capability and competence.

At issue here are a number of important things happening in almost every industrial situation.

- Global competition in almost all products
- Increasingly expensive equipment and systems
- Increasingly complex equipment and systems
- More integration of equipment and systems
- More reliance on automated control designs
- Needs for increased product quality
- Cycle time reduction
- Productivity increases
- Increased yield / less scrap
- Competitive product costs
- Cost reduction programs in all functions
- Capacity increases / flexibility
- Capacity assurance
- Equipment life cycle attainment / extension
- Flexible manufacturing schemes
- Small lot manufacturing / more customizing
- More frequent set-ups in manufacturing
- Just-in-time, and other manufacturing concepts
- Shorter development and product life cycles
- More design changes with shorter lead times
- Safety considerations
- Environmental concerns / restraints
- Energy reduction demands
- Alternative energy possibilities
- Employee involvement / development
- Intervention by regulatory agencies
- More involvement of customer / user in the process
- Emphasis on production operations achieving total
  customer satisfaction and follow-on support

These, and perhaps many more issues, all have an underlying maintenance component of one sort or another. Clearly, how the factory performs impacts some or all of the items listed; and the maintenance of equipment and systems has a direct bearing on how individual pieces of the factory perform.

As we move along in this last decade of the 20th century, new technology devices continue to enter the workplace at an alarming rate. Efforts to cope with these increased needs are characterized by increased manpower, more support contracts, more overtime and emergencies, and interruption of operations at inopportune times. Those maintenance organizations that think they are behind the response curve at the present time will experience even more pressure for providing technical support as we enter the era of the 21st century.

Automation, in its various forms, is the driving influence for the situation described. However, this is but a reflection of a greater consideration having to do with competitiveness in the world market place: the need to increase productivity and quality, while reducing cost, scrap, inventory and development time.

Traditional maintenance functions have been structured and manned in the past to deal with individual bits and pieces of equipment, as opposed to integrated computer-based systems of immense complexity and cost. Redundant equipment has been the norm, and a reactive response to technical support usually has been good enough. Moving to a proactive stance will require much re-thinking of what technical support is all about, as well as raising the levels of competence of the maintenance function itself, and finding ways of greater participation in the production and/or operations management process.

A key to optimizing the technical support of new equipment and systems that are coming into the work environment is gaining some measure of integration and control of the individual devices that they contain. This implies gathering operating and performance data from the devices which can then be processed to enable decisions concerning technical support requirements. Sadly, these data are not always accurate or relevant because the individual devices being monitored may not be performing as they should, or even close to it. To compound the problem, the devices might also be maintained or operated incorrectly. In both situations productivity, costs, and quality control are adversely impacted.

A comprehensive program of technical support (referred to by many people as maintenance) is vital for ensuring all automated equipment and subsystems are performing as they should, so that data gathering for the control system's data base is the best it can be. Moreover, a consistent technical support effort itself can become a source of unique data that can be used as added inputs to the control system.

This technical support is often forgotten in the design and implementation of automated systems. Recent reports suggest technical support costs can add up to another 35% on top of the direct production costs; and most probably this percentage will grow as the number of devices and their complexity increase in the workplace. An automated factory cell or department that is out of action or degraded, due to lack of technical support, is in great peril. The same is true for a modern office building's environmental control system, or a public utility's distribution system that might go down: the consequences on people, profits, and performance are enormous.

TECHNICAL SUPPORT REQUIREMENTS

As equipment and system automation increases, both in terms of vertical and horizontal applications, it becomes increasingly difficult for decision makers to allocate funds and resources for technical support of individual devices and subsystems. Unlimited technical support costs are not acceptable, just as poorly performing automated systems are also not acceptable.

And yet, surely the requirements for technical support are there, because the newly-added units, by definition, are critical to the success of the organization's operations. Maintenance costs tend to rise with each new device added, with seemingly little prospect of economy-of-scale efficiencies, since technology varies so greatly across the units involved.

If an automated factory or other facility is to perform in an optimal fashion, the reliable and correct performance of each and every device contained in an automated system cannot be overemphasized. Redundant or parallel equipment is rarely cost-justified, and seldom incorporated in such expensive, high technology systems. The design philosophy of such systems usually includes expectations that the devices will perform within an acceptable range as long as the system is "on line", and that accurate and reliable data on such performance can be collected for processing by the system. If either expectation fails or falls short, the automated system faces a troubled future. The mission of the technical support function is to focus on the maintenance needs of the various devices and interfaces contained in the automated system, and see that they are kept in an operable state. A second (but equally important) mission is to ensure that data collection from all designated points in the system is relevant, timely, accurate, and trouble-free.

The fact to ponder here is that computer-controlled production systems are hungry for information which is gathered throughout the electro-mechanical and electronic environments in which they operate. This information is needed on a broad, detailed and continuous basis, so that control of automated operations may be achieved. If this information is not relevant and accurate, or is masked by poor performance of individual system devices themselves, a built-in bias is possible. It may take months or years to track down these device/data problems and eliminate them; meanwhile, productivity, quality, and production costs suffer. On the other hand, if the collected information correctly reflects poor performance (due to poor maintenance or technical support) measured on the part of a device or subsystem, the resulting process or control commands can only be the best response possible under the circumstances. Either way, the overall performance of the automated system will be something less than the combined potential of all of its elements. This is why we see so many modern factories, filled with the very latest in automated equipment and advanced control systems, that fall miserably short of the production and quality levels that their designers predicted. Continuous improvement in device performance as well as data collection is essential, and this makes a strong case for implementing total productive maintenance (TPM) which will be discussed later.

While a well-designed automation system has some room for making up slack caused by any substandard components it knows about and controls, the best survival philosophy is one of ensuring that all components in the system function as they should on a long-term basis. This philosophy is achieved by good design and selection of equipment for the automated system, and responsive technical support that begins well before the implementation and operation of the system.

THE CHANGING SCOPE OF SUPPORT REQUIREMENTS

With top management decisions for more numerous and more sophisticated applications of automated systems, increasingly complex requirements also emerge for technical support. Often, these requirements are not recognized when automated systems are planned and funded. All too frequently system problems due to insufficient technical support are encountered well down the road when serious or disabling situations are experienced in the workplace. Eventual recognition of these support requirements may come as an unwelcome revelation if not a surprise to some managers, and perhaps even to the designers and implementers of the systems.

It is certainly not such to the personnel who must actually operate the automated systems over the long-term; each day they see the impact of inadequate maintenance on their ability to get their own jobs done.

After all, one of the main reasons for automating a factory, or any other enterprise for that matter, is to increase productivity. And technical support is vital to keeping the automated systems functioning as they should. To the extent that this does not happen, much valuable potential can be lost, and making it up later is difficult and usually very costly.

THE CHANGING ROLE OF OPERATORS

The term "operator" usually has referred to someone who runs production equipment in a factory. However, in terms of maintenance thinking, it is more appropriate to think of an operator in many other environments, such as computer or laboratory equipment operators, airline pilots, public utility operators, and hundreds of other examples that might be listed. An operator, simply defined, is on one side of whatever man/machine interface is under consideration.

However, using operators of factory equipment as an example illustrates recent changes quite well. A look at what has happened in recent years as factory automation has increased shows how the role of operators has changed through evolution.

Industries in the U.S. and other industrialized countries have spent large capital sums to establish or expand automated factories. Not only are assembly and fabrication equipments being automated, but also material handling, inspection, factory and business information, and environmental control functions. As more organizations rush to automate, the effect on personnel who formerly operated the equipment will be profound: most of them will simply disappear from the production workplace. Some will be retained in operator/ monitor roles, and others in engineering roles, and still others in technical support roles. In all such cases, significant retraining of the operators who remain appears to be a large and on-going necessity. Where existing personnel cannot fill the new requirements for technical support, new technicians with required skills and specific training will enter the scene to work behind the equipment and maintain it, but most probably not operate it. Also, some operators will have a varying degree of equipment maintenance responsibilities (anywhere from limited to substantial, depending on the operation), as envisioned in the "total productive maintenance" concept that has gained many followers in automated factory organizations.

But, the previous utility and convenience of having numerous operators on the production line will be lost to a great degree. Their adaptability, loyalty, creativity, innovation and multiple skills won't be readily available. Therefore, production in the automated factory will be very sensitive to downtime of any sort. Even short episodes of downtime will have pervasive effects on productivity, Flexible Manufacturing , MRP/CIM and similar production control concepts, quality control, incentive plans, just-in-time manufacturing, and eventual profits. The irony is that tomorrow's automated factories will be in great trouble if management attempts to support them by today's concepts of maintenance techniques and methods of maintenance management. Hence, a whole new look at the concepts of staffing, organization and management of technical support in the automated factory is strongly indicated, if not mandatory.

As noted earlier, similar illustrations can be made for operators of any sort, as increased sophistication appears in the equipment and systems they will be operating in the 21st century. Of special note should be requirements in the high technology fields, such as communications, transportation, medicine, R & D, banking and finance, space programs, military operations, and environmental systems.

ARRANGING TECHNICAL SUPPORT

Some organizations purchase technical support from equipment manufacturers or suppliers, by way of warranties, service contracts, or paid field service calls for a period of time. A variation of this approach is a "buy-install-operate-turnover" concept that seemingly offers the purchaser assurance that the new device or system is up and running correctly before custody is taken. These plans may work for a while, but grow more risky as time passes and latent problems or the effects of usage begin to surface. Moreover, this approach fails to support a major aim of technical support in modern enterprises, namely, finding ways to achieve the complete avoidance of breakdowns and sub-standard performance. The supplier's technicians cannot always be on hand at all the moments critical to an organization's operations. And, because they are outsiders to the organization, they generally are unable to take pre-emptive actions to prevent or avoid breakdowns. Delays, reduced quality and productivity losses are the harvest of sole reliance on technical support from suppliers of equipment. However, such a resource can be quite useful if it is part of a broader, more comprehensive approach to managing technical support requirements.

When they consider in-house resources, the automated system managers find a bit more encouragement. There are the possibilities of achieving technical support by way of own-staff maintenance technicians, operators of the automated equipment, and contract maintenance firms for those needs requiring special expertise or equipment. Also, the information management function (in-house or outside contract) which is likewise concerned with the automated systems can serve as an another resource. Added to this list of resources is the focused support that might be selectively available through the various equipment suppliers, as discussed earlier.

All of these organizations working as a combined team have the potential of dealing quite effectively with the technical support requirements. However, this approach places a heavy burden on senior management for developing a vision for employing and directing all these resources in an effective way. But that is the crux of the matter. Technical support is a major top management concern, and will be increasingly so in the future, regardless of the industry or market sector that the organization might be following.

For the technical support efforts to succeed, major commitment by top management is imperative, as well as enlightened direction at the operating and working levels. And, as more activities are automated, the burden on management for ensuring good technical support increases. This is at once both a top management issue and a pocketbook issue. It could well be that responsibility for accomplishing technical support should reside with one of the most senior operations managers in the organization, even as the responsibility for production (or operations) does now. The two functions, once mutually exclusive, have become one as a result of increasing automation; they must be regarded in an integrated way if the full impact of equipment and/or system automation is to be gained.

Why? Because excellent technical support will be a major ingredient of production or operations targets being met, productivity achievement, quality attainment, and profits. Increasingly, top managers will learn that maintenance must receive the same level of attention as do design, fabrication and assembly, materials handling, purchasing, and logistics. Assignment of top managers to maintenance will be seen as a way to focus more closely on the technical support sector in the production or operations arena, and also as another way to reward exceptional performance with promotion, bonus and career prospects.

TYPES OF TECHNICAL SUPPORT

A fully or partially automated enterprise requires technical support responses on a number of levels. These include support for individual devices and components, subsystems, systems (sometimes referred to as cells), and integrated facility-wide systems.

There are many ways to accomplish technical support. Some formal strategies currently in vogue include the following:

> Preventive maintenance
> Routine maintenance
> Periodic maintenance
> Corrective (or breakdown) maintenance
> Opportunistic maintenance
> Operator attention and/or maintenance
> Emergency maintenance
> Project (or upgrading) work
> Predictive maintenance
> Reliability-centered maintenance
> Total productive maintenance(TPM)
> Cooperative maintenance

These strategies, and others that are constantly emerging, are available to the maintenance function to assist in addressing the problem of supporting the increasingly complex automated facility. The problem is that far too few organizations adopt or follow any of these strategies, and the poor performance of their factories offer mute testimony to this lack. In essence, these strategies help the right maintenance technicians do the right thing to the right equipment or system at the right time, and using the right techniques, tools, parts, and methods. While most of the listed strategies are well known as to the concept they represent, brief explanations of some of them follow in order to clarify their major features.

Preventive maintenance (PM) is planned, scheduled inspection, assessment, adjustment and minor repair of equipment. It is most often characterized by a prescribed interval (frequency) of activity. PM is also called routine maintenance. However, routine maintenance may also be characterized by an on-going or repetitious nature, such as daily lubrication, monitoring and/or testing, adjustment, cleaning, and supply of consumables to equipment. Preventive and routine maintenance often include proactive repairs to things noticed during the maintenance inspections which, if not done at once, could lead to breakdowns or inadequate performance. Preventive maintenance is generally accomplished by specialists from outside the manufacturing or operations department; routine maintenance is generally accomplished by personnel who operate the equipment. However, broad exceptions occur in both cases.

Periodic maintenance overlaps preventive maintenance, in that it is scheduled. Generally, bigger tasks are undertaken, with the equipment sometimes being removed from service for a time. This work might be accomplished either by contractor organizations, own staff, or equipment suppliers.

Corrective maintenance is also referred to as breakdown maintenance, but does not necessarily imply emergency conditions (although just about any breakdown in the automated factory or other types of facilities is a potential genuine emergency). This type of maintenance can range from adjusting or calibrating designated equipment/systems in order to improve quality or performance, to actual hard repairs requiring quick response and considerable expertise. Corrective maintenance usually results from the spontaneous appearance of a problem; it can also result from predictive and/or preventive maintenance activities, statistical process control analyses, or advice from the control subsystem of an automated system.

Opportunistic maintenance can be performed when almost any other unplanned event interrupts the operation of a device, subsystem, or system. For example, if a department experiences a parts shortage, or another machine goes down and thus stops activity at other machines around it, or a power outage occurs, there is an opportunity to use available time thus created as a maintenance opportunity. Also, if a specific machine or device is down due to some unrelated problem, opportunistic maintenance can be accomplished within the time envelope needed for the primary repair. Realizing good value from opportunistic maintenance requires innovative advanced planning by maintenance planner/schedulers, as well as preparation on the part of technicians and operators to avoid missing the opportunity or spending more time and effort on the situation than justified.

Operator attention/maintenance has gained more popularity since automated systems began to multiply. This concept is also central to the Total Productive Maintenance (TPM) system that will be described later. Operator attention/maintenance requires very detailed training of operator personnel, close supervision of what work is done, and coordination with others who might work on the equipment as well (such as technicians or contractors). Operator attention is an important contributor to equipment uptime, quality, and reduced costs. But it is rarely "free" maintenance because of the time it takes for completion and preparation of essential documentation. Also, some interruption of operations may be involved when this form of maintenance is performed.

Emergency maintenance is self-explanatory. As stated earlier, any failure that produces downtime in an automated system is most probably an emergency. However, failures of other types may also be classed as an emergency (such as health, fire and accident situations). All resources at every level should be used to respond to a genuine emergency situation. Avoidance of emergency situations is best addressed through design reviews, equipment redundancy, modular devices, preventive and predictive maintenance systems, expert and/or vision systems, and continuous training of personnel and supervisors.

Project (or upgrading) tasks are perhaps overlooked in the continuing quest for maintenance improvement. During the course of the life cycle of a factory automation system, a number of inventions, evolutionary technology improvements, new materials, and process improvements can enter the picture. Some of these items, if incorporated into the automated system, might be capable of improving reliability, cost, quality, or life extension of the system itself. Far from being a "rainy day project", this effort can pay handsomely if it is made a regular part of maintenance planning, and organized and pursued with a well-defined purpose.

Predictive maintenance has been described in detail in the literature associated with factory automation. It is an organized effort to predict potential problems by analyzing or monitoring the operations of a machine or system. A computer program analyzes equipment history and current operations, senses trends, diagnoses undesirable conditions, and makes this information available to appropriate people or control devices so that maintenance action may be taken before the situation gets beyond desirable limits. Contributors to systems of this type range from the human operator on the line, to computerized sensing and monitoring points in the automated system, and even expert and/or intelligent control systems. Yet, with all the potential inherent in predictive maintenance systems, it appears evident that they are little used by many of the organizations that invest so heavily to implement automated factories. Almost any investment in predictive maintenance efforts offers great returns.

Total Productive Maintenance (TPM) is a concept that can be described as improving the organization by improving the personnel and the equipment and systems of the plant - that is, as one of the founding developers of TPM put it, "changing the basic culture of the organization". Started in Japan in 1971, the TPM concept has spread throughout the industrialized world, and currently enjoys much interest and acceptance due to the revolutionary results in improved productivity and quality attributed to it. TPM has five features: it (1) maximizes unit equipment effectiveness, thus improving overall system effectiveness; (2) establishes a total system of preventive/ predictive maintenance covering the entire life of the equipment; (3) covers all departments, such as equipment users, planners, maintainers, and all related operations or administrative activities; (4) requires participation by all staff members from top management to workers at the machines; and (5) promotes productive maintenance through small group activities that are more responsive to motivation. There is no question as to the success of a carefully planned TPM system implementation. Systems similar to TPM, such as Cooperative Maintenance, embody some of the concepts defined by TPM. In the final analysis, all TPM-oriented systems encourage each employee at every level to take responsibility for the equipment, process, quality, productivity, and for all information involved with what is going on at his or her location.

Reliability-centered maintenance (RCM) systems are specialized systems that optimize preventive and predictive maintenance efforts to achieve very high levels of sustained operations and reliability, such as might be found in activities like nuclear power plants, public utilities, commercial aircraft, space craft and their launch facilities, and hazardous chemical processing plants. RCM depends heavily on analytical methods and structured decision logic to determine the maintenance tasks and schedules necessary to maintain equipment at its designed level. These systems are also very dependent on good, reliable and contemporaneous information concerning each component in the system, and focus greatly on the timely, thorough completion and documentation of each element of prescribed maintenance work.

ANALYTICAL SYSTEMS AND TECHNIQUES

A dizzying array of analytical systems and techniques are currently available to the technical support function. Some of the systems include: Decision Support systems; Machine Vision systems; Expert systems; Hybrid Knowledge Based systems; Intelligent systems; Simulation systems; Asset Evaluation & Management systems; and Group Decision Technology systems. These are but a few of the many analytical systems that can be used to meet the challenges of technical support and maintenance. They will not be discussed further here, except for some brief comments about Expert systems.

Expert systems, a branch of a field known as Artificial Intelligence, allow human expertise to be captured by computer programs. This knowledge is comprised of facts such as events, objects, techniques, and history related to the equipment or system involved, and what to do with it in terms of current activities. Expert systems show great potential for helping maintenance technicians troubleshoot automated equipment problems, especially if the equipment is linked to other automated equipment elsewhere in the factory or other facility. The use of Expert systems in diagnosis and correction work also provides some positive side benefits. They assist in collecting data on maintenance problems, which itself is a valuable input to the factory automation system; and they can be helpful in training maintenance personnel, promoting safe working procedures, and providing detailed working procedures for accomplishing repairs and adjustments to components of the automated factory system.

Some of the analytical techniques currently available to the technical support staff are: signal strength and waveform analysis; electronic signal analysis; vibration analysis; infrared analysis; nondestructive testing, such as analysis of magnetic particles, eddy currents, radiographic and liquid dye penetrant applications; oil (and other fluids) analysis; acoustic analysis; calibration procedures; cell status displays; shock pulse, wave form, and motor current (megger) analysis; shaft alignment data; device polling; capacity and parameter controller sensors; performance trending; thickness monitoring; machine condition monitoring; steam trap programs; alert data from energy control and monitoring systems; and information received from statistical process control activities. These are but a small number of the analytical tools available for use in maintenance. Many are oriented to the automated systems themselves, while others are generally thought to be useful only to operators of those systems. However, technical support interests in the automated environment, as earlier explained, are integral with those of the operators.

For example, a status display for a manufacturing cell indicates the status of the equipment in the cell in a readily understandable form. The display is continuously updated by the information management computer, and immediately indicates some of the problems that occur at the devices in use in that cell. This can help in quick diagnosis and response by either technicians or operators of the equipment. The key consideration here is the training that must be given all parties so that rapid use may be made of what is shown on the displays. Good exchange of information between the technical support function, equipment operators, and the cell designers early in the design process can help make the cell status display tremendously effective for all who might use it.

In a similar vein, statistical process control has been regarded as the province of the production managers and the quality control function. However, the maintenance engineers and technicians can also make great use of such information. In setting up the statistical process design, maintenance inputs can help set parameters which can then be monitored for both maintenance and quality data. The advisory information flowing from the statistical process monitoring is useful for: planning support for gaining machine access for scheduled maintenance projects; predictive and corrective maintenance; and indications for machine overhaul or replacement of components at forecasted points in time.

DEVICE INTERFACING AND DATA COLLECTION

Automated systems (or in the case of manufacturing, cells and/or departments) are usually made up of individual devices or subsystems connected in some fashion to form an integrated system. Integrating the range of the functions of these devices or subsystems is the objective of automated systems development. The functions can be quite broad and varied, spanning activities such as processing and assembly, CAD, production control, scheduling, numerical control, environmental control, piping and pumping management, processing, materials handling, and maintenance. Achieving total integration of all devices throughout the facility is almost impossible, so what is most often seen are islands of automation dedicated to specific functions, products or processes. The practical challenge is to get these islands of automation to play together, so that optimal control can be exercised across the many functions within the facility. Therefore, the technical support function should be involved and considered from the very beginning when planning an automated system.

Technical support focused in two critical areas is key to assisting in meeting this challenge. First, every effort should be exhausted to assure that clean and reliable data is gathered throughout the device/system interface. This will result in a form of standardized information flow between various parts of the factory and/or facility. Unfortunately, there are no recognized industry-wide standards for information flow between software applications between various automated systems that might be present in a particular facility. So, close liaison with the information management department is very important if optimal results in this area are to be had.

And second, as stated earlier, major contributions to good data collection can be made by maintaining the devices or subsystems in the condition and performance levels to which they were designed. (Preventive/predictive maintenance, TPM, etc.)

In this way, information management systems such as automated cell controllers, equipment area controller/managers, and the plant host system can operate to their best capacity.

ROLE DEFINITION AND TRAINING AS AN INGREDIENT OF SUCCESS

Technical support personnel in the automated system arena have a major responsibility they perhaps did not previously possess in a traditional facility. In their new role, they must have a thorough understanding of the functional design of the computerized automated equipment system (be it manufacturing or other), the hierarchy of functions that are included, and how information flows in and out. Particular emphasis is needed on how device data acquisition is handled (i.e., by event, or by preset frequency), and how support technicians may enter the system to obtain maintenance information or make electronic adjustments to devices. Major training in these new responsibilities, both on a current and an on-going basis, is an absolute necessity.

A comprehensive, on-going training program for support technicians and engineers is a basic ingredient for success in supporting system automation. The technology of new and existing devices in the system is always changing or becoming more advanced. On top of this, the design and implementation of the computerized automated system itself grows more complex as time passes and new devices and subsystems are added. The technical support personnel require an almost continuous hardware and automated systems education program if they are to be accountable for staying ahead of breakdowns and failures in the factory or facility environment. Lack of such training will throw more burden on the equipment suppliers and the information system specialists, with the likely result that unscheduled delays in production will take place as the systems increase in size and complexity.

Hardware training concerning the devices, components and other equipment in the automated factory or facility should be detailed enough to make the technical support person competent to test, maintain, evaluate, and take corrective actions necessary to keep assigned equipment operating. Substitution of devices is often the best way to proceed, when they are too complex or too costly, in terms of working time, to repair in place.

However, when it comes to training associated with the computerized automated system, it is not to necessary that technical support technicians or engineers should be information management specialists. But they should have a good general knowledge about the architecture and information flow in the system. This will help focus on where technical support efforts can be most important, and aid in faster response to problems that occur with the hardware in the operating environment.

Another issue develops as a result of the emergence of technical support engineers and technicians. The term "empowerment" is entering the literature on a frequent basis. Indeed, many organizations are intensely courting the concept of empowerment, not only for technical support (maintenance) personnel, but in a more limited way for some equipment operators. Potentially, personnel or social problems can develop wherein elitist groups (technical support) might exercise great imputed authority to interrupt and/or delay activities in automated system environments. Moreover, a wall of assumed superiority can develop to portray that technical support personnel are of a higher station than operators.

Recognition early-on by managers and supervisors is the best way to avoid or correct this issue, coupled with proper training, supervision, and motivation of both technical support and operator personnel. A good rule of thumb seems to be that technical support personnel (and operators, for that matter) should be empowered only up to the limits of their capability as defined by management, and vetted by their documented training. It is essential that the opportunity for these undesired side-effects of empowerment to occur be avoided.

APPROACHES

It is evident that a fresh management approach is needed to develop and implement the type of maintenance capability that will be needed in the factory of the future. Such an organization must be capable of offering responsive, sophisticated technical support in a timely, cost-effective manner. Ideally, it will be based on a team concept that works integrally with the functions it serves.... such as production, R & D, services or overall facilities. Technical expertise equal to that of the equipment and systems of the client will be a must. The maintenance managers will have to be capable of earning the trust of their clients, such that they are admitted to the highest levels and circles of the client's management environment to provide technical support advice and participate in business/production/business planning activities.

This is a tall order, in terms of most maintenance functions seen today. A definite image problem exists, and this must be overcome with improved technical and management expertise, and education efforts with top management. It is essential that top management's attitude about maintenance be changed from " a necessary evil " to a " a necessary good ". This is quite evident when maintenance is pictured and proven as a business unit that provides a product: ASSURED CAPACITY.

The devastating effect of not having the right plant capacity is reflected in missed schedules, poor quality, inability to accept orders or change order mixes quickly, and (most important) reduced profits. One of the chief, if not primary, contributors to the capacity equation is maintenance. Simply put, effective maintenance/technical support creates a situation wherein the clients being served can go about doing their business without interruption or restraint.

In high technology operations, maintenance already comprises about 6 to 8% (and much higher in certain types of organizations) of the product costs, and most likely will increase in the future. Unfortunately, most accounting systems mask this information by the traditional way costs are accumulated and allocated. So some fresh accounting practices, such as activity based accounting, are indicated so that the true impact of what effective maintenance can contribute to the overall operation can be shown.

A proactive approach to maintenance must be conceived and put into practice. This approach emphasizes reduction of the total maintenance required, and extending the life of equipment and systems through systematic removal of the sources of observed and potential failures.

Reduced maintenance, and elimination of sources of failures? Did we not say earlier that more and better maintenance is needed? These statements are not necessarily in opposition. The fact is that "new" management utilizes such techniques as predictive maintenance and TPM to get closer to the equipment and analyze the problems in a sharper way. The result is usually less but more focused maintenance activities, coupled with constant improvement efforts that eliminate or diminish the occurrence of failures over the long-term.

As an example, the National Research Council (an arm of the National Academy of Sciences), in 1991, published a report entitled, " The Competitive Edge: Research Priorities For U>S> Manufacturing ". The report urged manufacturers to shift from breakdown and preventive maintenance to predictive maintenance to keep up with worldwide competitors. It looked at research needs in five areas, one of which was equipment reliability and maintenance. The matter of maintenance was referred to as being in the backwater of manufacturing research, and needed to be brought to the forefront. The report said: "Predictive maintenance, in use in U.S. industry for only a few short years, is usually understood to involve the use of computer software to detect conditions that might eventually lead to equipment failure. Predictive maintenance is a little-used approach that has great potential".

So, maintenance management personnel with greater technical qualifications, and advanced systems and concepts of performing the maintenance work will necessarily be the hallmarks of the technical support organization of the future.

IMPLEMENTATION

The cornerstone of the implementation phase is creating a vision of what the new maintenance function should accomplish. Next the organization structure, management tools and assets, and personnel development objectives need to be conceived and agreed to, along with the relationship to, and commitment by, top management.

Various authors have written prescriptions for success. In sum, they seem to agree on the following:

- Develop a comprehensive plan that shows maintenance
    doing things that support what the rest of the
    operation is really doing

- Make a firm, long-term commitment to carry out
    the plan; get top management to buy in on
    this plan and commitment

- Publicize the plan to other functions in the
    organization; make sure that you have the
    involvement and support of your people and
    (where applicable) the union

- Establish the technical support personnel with
    firm job descriptions, appropriate on-going
    training, and empower them to do their jobs

- Keep abreast of the needs for maintenance by use
    of systems of PM, predictive maintenance, TPM
    or whatever system(s) best meet the needs of
    the enterprise

- Be prepared to make mid-course changes, as often
    as it takes to match what else is going around
    in the client organizations being served by
    maintenance

In short, develop a program that works for <u>you</u>, in <u>your</u> environment, with <u>your</u> people, tailor made to <u>your</u> equipment's needs, and to satisfy <u>your</u> goals and objectives.

SUMMARY

Technical support is a major factor in the success of automated facilities, factories, and other possible applications. Technical support will become more critical as more automation and equipment sophistication enters the workplace. Some considerations concerning device interfacing and data collection have been mentioned; actually, these are the basics of good operation of automated systems anywhere. Since automation is intended to increase productivity and reduce costs, downtime of any sort is a threat to automation; technical support must be aimed at reducing or eliminating these downtime conditions.

Technical support requirements should be considered in the early stages of the design and development of automated systems. The costs of technical support should be estimated, and their impact considered in judging the economic viability of the proposed system (and/or improvements) during its projected period of use. Management concerns about escalating costs associated with technical support should be addressed by tasking project managers who are developing systems to give close scrutiny to this area. Support costs should be analyzed and managed like any other major cost element of an automated system.

Technical support personnel must be competent in the hardware involved with the automated systems, as well as possess an understanding of the associated computerized control systems. And, they should work early-on and continuously with systems developers to ensure features and information vital to maintenance are provided via the systems on a useful, timely, on-going basis.

Major thrusts by management are needed to effect good technical support. Chief among these is recognition of the role that maintenance must play in the success of the high technology or automated systems being considered for the enterprise. Next, decisions concerning assignment of adequate numbers of qualified managers, engineers and technicians, sufficient budget, training, and other resources must be made and sustained. Finally, firm commitments must be made by senior managers to underscore the expectation that the technical support function will carry out its duties in such ways that contribute fully to the success of events happening on the factory floor, or within the total span of whatever the purpose of the enterprise might be.

What all this amounts to is a reinvention of the maintenance function. It will be necessary to do this if the needs of future manufacturing demands are to be met. The present format for getting technical support done simply won't do. Dramatic changes in the concepts and vision of maintenance will be necessary, along with some hard scoping-out of what sort of technicians and operators will be required to accomplish the new requirements. Management, professionals, technicians, operators, trainers, Unions, and a host of others will have to address a new agenda that calls for close interweaving of efforts once thought to be mutually exclusive. The organizations that confront the issues and challenges directly, and early, will find the most possibilities of success as they enter the 21st century.

TOTAL PRODUCTIVE MAINTENANCE: A LIFE-CYCLE APPROACH TO FACTORY MAINTENANCE AND SUPPORT

Benjamin S. Blanchard

Virginia Polytechnic Institute and State University, College of Engineering, Blacksburg, Virginia, U.S.A.

ABSTRACT

Many systems in use today are not performing as intended, nor are they cost-effective in terms of their operation and support. Manufacturing systems, in particular, often operate at less than full capacity, productivity is low, and the costs of producing products are high. In dealing with the aspect of cost, experience has indicated that a large percentage of the total cost of doing business is due to maintenance-related activities in the factory; i.e., the costs associated with maintenance labor and materials and the cost due to production losses. Further, these costs are likely to increase even more in the future with the added complexities of factory equipment through the introduction of new technologies, automation, the use of robotics, and so on. In response to these maintenance and support problems in the factory, the Japanese have introduced the concept of "Total Productive Maintenance (TPM)". TPM, which is an integrated life-cycle approach to factory maintenance and support, is being implemented in many factories throughout Japan today and is currently being introduced in a number of United States firms. The objective is to improve overall equipment effectiveness and increase productivity. This paper briefly describes the TPM concept, discusses current status relative to TPM implementation (primarily in the United States), and offers some recommendations for further improvement.

INTRODUCTION

Recent trends indicate that, in general, systems are (1) increasing in complexity with the introduction of new technologies, (2) are not meeting customer expectations in terms of performance and effectiveness, and (3) are becoming more costly relative to their operation and support. In the production of goods, manufacturing systems are often operating at less than full capacity, productivity is low, and the costs of factory operations are high. This is happening at a time when international competition is increasing worldwide.

In addressing the issues of cost-effectiveness as related to the commercial factory, the trends illustrated in Figure 1 are apparent and are of concern. In many companies, productivity is low and the costs of operating and maintaining equipment in the factory have become a significant factor in the production of goods. According to a recent study reported by R. K. Mobley, from 15% to 40% (with an average of 28%) of the total cost of finished goods can be attributed to maintenance activities in the factory [1]. Given the on-going addition of new technologies, the introduction of more automation, robotics etc., maintenance costs are likely to be even <u>higher</u> in the future with the continuation of existing practices. T. Wireman reports from a study conducted in 1989 that the estimated cost of maintenance for a selected group of companies increased from $200 Billion in 1979 to $600 Billion in 1989 [2].

With regard to the issue of costs due to maintenance activities, a large portion of this cost can be categorized under "production losses", which often range from 2 to 15 times the direct costs for the maintenance and repair of equipment (i.e., maintenance labor and materials). Further, these costs are often hidden as reflected in the "iceberg" illustrated in Figure 2. While the initial acquisition costs associated with factory design and construction have in most instances been quite visible, the costs of sustaining maintenance and support have not! As a result, there have been some surprises, with the cost of many products being higher than initially anticipated. This, of course, has had an impact on profits and a company's competitive position in the marketplace.

In evaluating "cause-and-effect" relationships, experience has indicated that a significant portion of the total projected life-cycle cost for a given system (i.e., those costs represented by the lower portion of the iceberg in (Figure 2) is the direct result of decisions made during the initial phases of advance planning and design. Those early decisions pertaining to the utilization of new technologies in design, the selection of components and materials, the selection of a manufacturing process, the identification of equipment packaging schemes, maintenance support policies, diagnostic routines, etc., have a great impact on life-cycle cost. In other words, there is a large "commitment" to life-cycle cost in the early phases of system development, and the maintenance and support costs for a given system (which often constitute a large percentage of the total) can be highly impacted by early design decisions [3].

Relative to past practices, a short-term rather limited view of system requirements has been predominant, as compared to a total integrated life-cycle approach in the system design, development, and evaluation process. While the technical characteristics of system performance have received emphasis in system design and construction, very little attention has been directed toward such design characteristics as reliability, maintainability, human factors, supportability, quality, and the like. The lack of consideration of reliability and maintainability in design, in particular, has resulted in high maintenance and support costs downstream. Additionally, these extensive maintenance and support requirements have had a definite degrading impact on overall system effectiveness, or productivity as it applies to a commercial factory operation.

With these past experiences in mind, the design and development process for <u>future</u> manufacturing systems (i.e., selection of new technologies and components, procurement practices, construction) should (1) consider <u>all</u> elements of the system

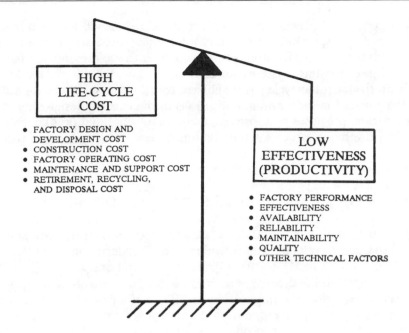

FIG. 1. The cost-effectiveness balance

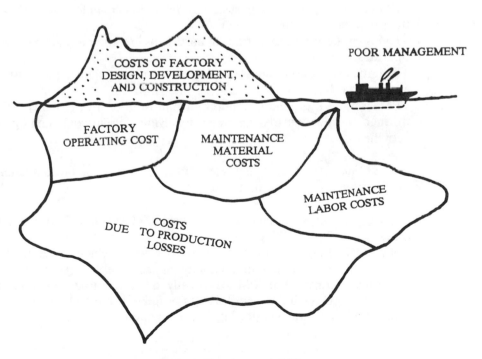

FIG. 2. Total cost visibility

on a total integrated and concurrent basis, and (2) should view the system from a long-term life-cycle perspective. The factory maintenance and support capability needs to be addressed, along with the prime equipment requirements for factory operation. Those existing systems already in being should be evaluated continuously on the basis life-cycle cost; high-cost contributors should be identified along with the causes for such; and modifications to the existing design should be incorporated for the purposes of improving factory efficiency and productivity. A "continuous improvement" process should be implemented using life-cycle criteria for evaluation.

TOTAL PRODUCTIVE MAINTENANCE (TPM) -- A PROPOSED SOLUTION

In response to maintenance and the support problems in the commercial factory, the Japanese developed and introduced the modern concept of "Total Productive Maintenance (TPM)" in 1971 [4]. From your author's perspective, TPM constitutes "an integrated life-cycle approach to factory maintenance and support."

Since this early period, the application of TPM methods and techniques has increased significantly in Japan and, during the past several years, TPM concepts have been adopted by companies in other countries as well.

According to S. Nakajima (Vice Chairman, Japan Institute of Plant Maintenance), the concept of TPM includes the following five elements [4]:

1.   TPM aims to maximize equipment effectiveness (overall efficiency).
2.   TPM establishes a thorough system of preventive maintenance (PM) for the equipment's entire life span.
3.   TPM is implemented by various departments in a company (engineering, operations, maintenance).
4.   TPM involves every single employee, from top management to the workers on the floor.
5.   TPM is based on the promotion of preventive maintenance (PM) through "motivation management" involving autonomous small-group activities.

To quote S. Nakajima further, the word "Total" in TPM has three meanings that describe the principle feature of concept:

1.   Total effectiveness (Item 1 above) indicates TPM's pursuit of economic efficiency and profitability.
2.   Total maintenance system (Item 2 above) includes maintenance prevention (MP) and maintainability improvement (MI), as well as preventive maintenance (PM). Basically, this refers to a "maintenance-free" design through the incorporation of reliability, maintainability, and supportability characteristics into the equipment.

3. Total participation of all employees (Items 3, 4, and 5 above) includes autonomous maintenance by operators through small group activities. Essentially, maintenance is accomplished through a "team" effort, with the operator being responsible for the ultimate care of his/her equipment.

To achieve overall equipment effectiveness, TPM addresses what are considered to be the "six big losses" (i.e., the major obstacles) identified in Table 1. The goal is to cause the elimination of these losses, improve productivity in the factory and, thus, increase the profitability of a company.

TABLE 1
The Six Big Losses Addressed By TPM

Losses due to downtime:

1. Equipment failures resulting in unscheduled maintenance actions.
2. Setup and adjustment requirements.

Losses due to slowdowns or reduced speeds:

3. Idling and minor stoppages due to the abnormal operation of sensors, blockages of work-flow processes, etc.
4. Reduced speed due to discrepancies between the designed speed and the actual speed of equipment.

Losses due to defective products:

5. Process defects leading to product rework (i.e., defects in quality).
6. Reduced yield from machine startup to stable production.

(Reference: Nakajima, S., Introduction To Total Productive Maintenance, Cambridge, Mass., Productivity Press, Inc., 1988.

In order to evaluate how a given company is progressing relative to meeting TPM objectives, one needs to establish a measure for Overall Equipment Effectiveness (OEE). OEE is a function of Availability (A), Performance Efficiency (PE), and Quality Rate (Q):

$$Availability\ (A) = \frac{Loading\ Time - Downtime}{Loading\ Time}$$

(1)

where <u>loading</u> <u>time</u> is the planned time available per day (or month) for production operations, and <u>downtime</u> is the total time that the system is not operating because of equipment failures, setup/adjustment requirements, exchange of dies and other fixtures, etc. Availability can also be expressed as the ratio of actual operating time to loading time. This factor primarily addresses the first two losses identified in Table 1.

$$\textit{Performance Efficiency (PE)} = \textit{(Operating Speed Rate)} \; \textit{(Net Operating Rate)}$$

(2)

and

$$\textit{Operating Speed Rate} = \frac{\textit{Theoretical Cycle Time}}{\textit{Actual Cycle Time}}$$

(3)

where the cycle times represent the <u>theoretical</u> time that it takes to process an item, as compared with the <u>actual</u> time. This factor measures the difference between the ideal speed (based on the equipment capacity as initially designed) and the actual operating speed of the equipment.

$$\textit{Net Operating Rate} = \frac{\textit{(Processed Amount) (Actual Cycle Time)}}{\textit{Operating Time}}$$

(4)

where <u>processed</u> <u>amount</u> refers to the number of items processed per day (or month), and <u>operating</u> <u>time</u> is the difference between loading time and downtime. Performance Efficiency (PE) is directed toward reducing the losses due to slowdowns or reduced speed; i.e., items 3 and 4 in Table 1.

The third factor in the OEE equation is

$$\textit{Quality Rate (Q)} = \frac{\textit{Processed Amount} - \textit{Defect Amount}}{\textit{Processed Amount}}$$

(5)

where the <u>defect</u> <u>amount</u> represents the number of items rejected due to quality defects of one type or another, and require rework or are scrapped. This factor addresses items 5 and 6 in Table 1.

When combining Equations (1), (2) and (5), the Overall Equipment Effectiveness (OEE) for a given company operation is determined from:

$$OEE - (Availability)\ (Peformance\ Efficiency)\ (Quality\ Rate)$$

$$(6)$$

This measure of effectiveness has been applied to numerous factory operations in Japan through the years. Based on experience in this area, an <u>ideal</u> OEE should be 85% or greater; where Availability (A) should be greater than 90%, Performance Efficiency (PE) should be greater than 95%, and Quality Rate (Q) should be greater than 99%.

TOTAL PRODUCTIVE MAINTENANCE (TPM) -- CURRENT STATUS

The number of TPM programs being implemented in Japan has increased significantly, particularly during the past ten years. The Japan Institute of Plant Maintenance (JIPM) has accomplished a great deal in promoting the concept of TPM, not only within Japan but internationally as well! Promotional materials have been prepared (brochures, newsletters, training manuals, video tapes), a number of TPM books have been published, prestigious awards have been established recognizing those companies who have successfully implemented TPM, and a first International TPM Conference was held in Tokyo in the fall of 1991.

In the United States, the popularity of TPM has grown, primarily as a result of a higher-level promotional campaign commencing in 1988 when Productivity, Inc., translated and published its first in a series of books on the subject. Additionally, two well-attended national conferences have been held in Chicago, Illinois (October 1990 and 1991); a "TPM Newsletter" is being published by Productivity, Inc.; an American Institute For Total Productive Maintenance (AITPM) was established in 1991; and a series of plant tours organized by the AITPM involving Hughes Aircraft Co., Eastman Kodak Co., Ford Motor Co., and DuPont - Fibers Division were scheduled early in 1992 to show examples of TPM program implementation.

With the objective of assessing the extent of TPM activity in the United States, C. Dyer (Director of Research, AITPM) initiated a research project in June 1991, where a questionnaire was distributed to many different companies representing process industries, producers of high-tech products, and so on. While the overall response rate was low, over 60 companies were identified as being active implementers of TPM in one form or another [5].

In attempting to predict the future, it is your author's opinion that the implementation of TPM programs (or some equivalent level of activity) will continue to grow in the United States. As international competition expands further, more and more companies will be forced to develop a more efficient means for manufacturing products. Any company that is going to be competitive in a world wide market is going to have to implement a management approach that (1) addresses the factory as a total "system", to include maintenance and support; (2) views the factory from an overall "life-cycle" perspective, versus assuming the usual short-term approach; and (3) involves all factory personnel from the top-down, committed to reduce cost and improve productivity. While these objectives appear to be relatively obvious, they will not be easily attained as there is a significant educational process ahead! A major challenge is to convince managers, engineers, operators, and supporting personnel that significant productivity gains can be realized by paying attention to the issues of maintenance and support.

SOME OBSERVATIONS AND RECOMMENDATIONS FOR THE FUTURE

From your author's perspective, the concept of Total Productive Maintenance (TPM) is an excellent one, and the complete implementation of such should lead to increased efficiency and productivity for any given factory operation. It is felt that many of the specific methods/techniques required in implementing TPM are not new. For instance, the concept of maintenance prevention through the introduction of reliability, maintainability, and supportability characteristics in equipment design has been applied for many defense systems; the development of preventive maintenance requirements based on reliability and life-cycle cost data has been accomplished in the airline industry and for many other systems; and the assignment of equipment operators to perform maintenance on their own equipment has also been accomplished. However, these methods/techniques have not been very well integrated, nor have they been applied to the commercial factory environment. The TPM approach, as described herein, does present a total integrated life-cycle approach to factory maintenance and support. Further, the experiences of those companies who have attempted to implement TPM have been very positive.

Based on a cursory observation from a review of the literature and through several company plant visits, it is believed that the majority of the emphasis thus far has been placed on the downstream "after-the-fact" activities associated with factory maintenance and support; i.e., the development of a good preventive maintenance program given an existing factory equipment configuration, the development of good diagnostic and test routines, and the structuring of an organization to effectively respond to maintenance actions. On the other hand, very little progress has been made in the areas of maintenance prevention (MP) and maintainability improvement (MI). This fact is also supported by C. Dyer's 1991 research and survey [5]. Yet it is believed that improving the equipment design through reliability and maintainability offers the greatest potential for meeting the overall objectives of TPM. Thus, it is felt that the complete implementation of TPM has not as yet been experienced by many companies, at least those in the United States.

With regard to the future, more attention needs to be given to the evaluation of a company's manufacturing capability in terms of life-cycle cost. High-cost contributors can be identified, cause-and-effect relationships should be established, and recommendations for maintainability improvement in equipment design need to be incorporated. This life-cycle-oriented evaluation process should be implemented on a continuing basis; i.e., for continuous improvement. This can be accomplished through the appropriate utilization of selected design tools such as reliability and maintainability prediction models; the failure mode, effects, and criticality analysis (FMECA); component failure analysis methods; the level of repair analysis (LORA); the maintenance task analysis (MTA); and others [6]. With the appropriate emphasis in this area, it is believed that the full benefits of implementing a TPM program (or an equivalent level of activity) can be achieved. This, in turn, should improve a company's overall productivity and competitive position worldwide.

REFERENCES

1. Mobley, R.K.: <u>An Introduction to Predictive Maintenance</u>. New York, Van Norstrand Reinhold, 1990.

2. Wireman, T.: <u>World Class Maintenance Management</u>. New York, Industrial Press, Inc., 1990, pp

3. Blanchard, B.S.: <u>System Engineering Management</u>. New York, John Wiley & Sons, Inc., 1991, pp 5-8.

4. Nakajima, S.: <u>Introduction To Total Productive Maintenance (TPM)</u>. Cambridge, Mass., Productivity Press, Inc., 1988 (translated into English from the original text published by the Japan Institute for Plant Maintenance, Tokyo, Japan, 1984.

5. Dyer, C., "Implementing TPM in America", Proceedings, <u>Second Annual Total Productive Maintenance Conference And Exhibition</u>, Co-sponsored by Productivity, Inc., JIPM, Maintenance Technology Magazine, and AITPM, Chicago, Illinois, October 9-11, 1991.

6. Blanchard, B.S.: <u>Logistics Engineering And Management</u>, 4th Edition. New Jersey, Prentice-Hall, Inc., 1992.

CIM-MAINTENANCE — INFORMATION AND DOCUMENT MANAGEMENT FOR THE MAINTENANCE OF AIRCRAFTS

[1]Eberhard Kurz and [2]Dieter Eden

[1]*Fraunhofer-Institut für Arbeitswirtschaft und Organisation, Stuttgart, Germany*
[2]*Deutsche Aerospace, MBB, Unternehmensbereich Flugzeuge, Manching, Germany*

ABSTRACT

For the maintenance of aircrafts still most information and documentation is paper-based. Existing systems are isolated solutions and do not find high levels of acceptance by the worker.

In this paper a system for the information and document management in the maintenance of aircrafts will be presented.

For the case of the "TORNADO" fighter aircraft of the German Airforce which is being maintained at the maintenance plants of Messerschmitt-Bölkow-Blohm (MBB), a prototype for the electronic handling of information and documents has been designed, developed and installed.

The core of the system is a NCR 3125 Notepad with handwriting recognition for pen-based input. Each worker will be equipped with this notepad. All technical information such as forms, drawings, service plans etc. is available on the notepad. A knowledge-based diagnosis system is assisting the worker in fault detection and repair.

The notepad is integrated via wireless transmission to the production planning and control system, to the material handling system, and to the technical information systems like technical databases (failure statistics etc).

The system received a high level of acceptance, because the transfer from a paper-based form to the notepad is easy. The envisaged reduction of failure rates in the execution of the maintenance work is 50%. The time for information retrieval is reduced to 30% of the former value.

The system can be used for all types of complex technical items such as chemical and power plants, complex machinery.

INTRODUCTION

The TORNADO aircraft is the backbone of the German Airforce. Three nations (Great Britain, Italy and Germany) have cooperated in achieving the design, manufacturing and assembly of the TORNADO. Roughly 800 aircrafts have been produced between 1980 and 1990. Currently the aircraft is under normal operation. After a fixed number of operating hours certain maintenance tasks and overhaul are necessary. Each user country has its own facilities for this maintenance. The airforces and the the industrial partners are working together in keeping the aircrafts ready for operation. In Germany a plant of the Deutsche Aerospace (DASA), Messerschmitt-Bölkow-Blohm (MBB) is responsible for the major maintenance tasks. There are approximately 350 aircrafts in use. Three versions have been produced: interdictor striker, trainer, and versions for reconnaissance.

This paper presents the results of a project which has been done in cooperation with MBB from the Fraunhofer-Institute for Industrial Engineering. In the second chapter the results of an information and document analysis are shown. The third chapter contains the elaborated concept reagarding hardware, software and communication aspects. A cost-benefit analysis and the summary are concluding the paper.

INFORMATION AND DOCUMENT ANALYSIS

The first step in the maintenance of the TORNADO is the inspection (by vision as well as by tools). Here, failures and damages are detected. In the following stations, these faults will be repaired as well as the "normal" maintenance tasks will be done. These maintenance procedures have been analyzed by the project team. The results are shown below.

"Paper flood" at the Maintenance Dock

In total 40 different documents (forms, Electronic Data Processing (EDP) printouts, drawings, circuit diagrams, handwriting notes) have been counted. Most of the information is paper-based. The same information is in multiple representations in different forms. Some documents are useless, because the information can be retrieved from other forms. Very often the forms are puffed-up and do contain more information to be filled in than necessary. Sometimes old forms are still being used. If the information could be retrieved by EDP solutions, software was missing which eases the way of use or which links existing software packages. In Germany, the electronic signature is still not valid. Thus, especially in such sensitive areas like aircraft maintenance, all documents have to be signed manually on paper after finishing the work. These paperwork has also to be stored.

Maintenance Data Acquisition Method

For each failure and repair, numerical codes have to be looked up in a thick handbook. Each retrieval takes, dependent on the failure, between 2 and 10 minutes. The coding takes place manually without EDP support. The numbers are being written down and typed into a midrange-sized computer. With an additional statistics program these data could be evaluated regarding frequency and time, costs and material of faults.

Technical Documentation

The aircraft documentation is very extensive. The workers have the technical documentation as drawings, circuit diagrams, lists of parts and repair instructions paper-based; the part catalogues are on microfiches.

The main problem with paper-based documentation is the configuration and version difficulty. Not all working documents have the latest version number. Thus it might happen, that repair is being done with an older instruction. In addition, there are differences between the versions on different media like microfiche and paper-drawings. The microfiche part catalogue is very rigid and cannot be handled the way the worker would like to. It does not present the information in a way which would fit to the user's needs. Functionality which is known from new hypertext/hypermedia systems is not available. The inflexible, hierarchical structure of information handling complicates the user's work flow.

The qualification of the workers is very high. They have a lot of experience beginning with the Lockheed Starfighter aircraft. The workers act very independently with a high degree of flexibility and quality of work. There is a strong wish of support for administrative details (material order, fault coding) and support for supply of technical information and documentation.

CONCEPT OF THE CIM-MAINTENANCE SYSTEM

Computer-Integrated Manufacturing in general describes the integrated use of electronic data processing in all internal and external areas, relevant to manufacturing. Integration covers not only the technical functions but also the organisational tasks necessary for planning and manufacturing products. The basic concept is to use information company-wide via suitable interfaces, data bases, and communication through networks.

Manufacturing in the sense discussed here is focussing on maintenance. Thus the abbreviation CIM-M for CIM-Maintenance has been coined.

The CIM-M concept does not begin from scratch. In addition, as mentioned above, CIM is besides functionality also integration. The single functions of CIM-M, especially at the shop-floor level, will be discussed in detail in the remainder of this paper. At the integration level, CIM-M focusses on process- and data integration.

The analysis of today's situation had as a result the following requirements for the support at the shop-floor level:

- the system has to be accepted by the worker as a suitable tool and it should be not be regarded as an instrument of control
- it must be easy to use the system. The worker has to use it like his individual handbook, adaptable to his own needs. It has to contain information in such a way that all drawings, texts, symbols, tables, and circuit diagrams are available on the spot.
- redundant information has to be avoided. The amount of paper at the maintenance plant has to be reduced to the minimum necessary.
- the information has to be presented fast, safe, and complete.
- the system has to be modular and extensible. It should be adaptable to the user's individual working style, qualification (expert or beginner modus), age, and to the difficulties and specialities of the maintenance task.
- interfaces to existing application programs should be available
- the concept should center on a hand-held computer
- the system should work with batteries or an accumulator.

The requirements led to the following solution (see figure 1)

Hardware:

The basis for the CIM-M concept is the use of notepad computer for the personal use of each worker. At the final step of implementation, each worker will be equipped with one notepad. Thus, he can accept it as his personal too. The notepad has a 80386 SL processor and can be used with hand-writing recognition. The weight of the system is about 2 kg. The accumulators allow operation of 4 hours; future developments aim at the extension of the capacity to a shift (8 hours).

Software:

Some software modules need to be implemented. Their prototype status can already be shown. Their hardcopies are included below.

Fault result form:

The fault result form has been a paper-based solution. This form has been implemented on the notepad basically the same way it appeared on paper.

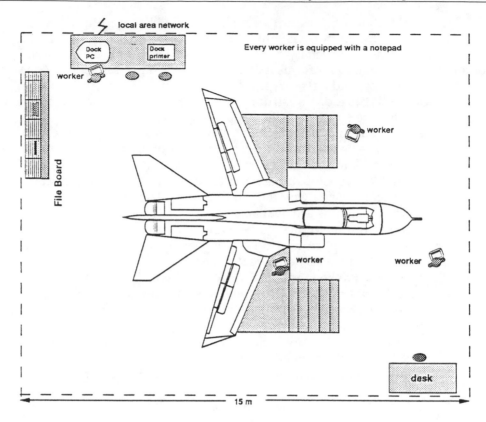

Fig. 1: Requirements for the CIM-M system at the maintenance dock

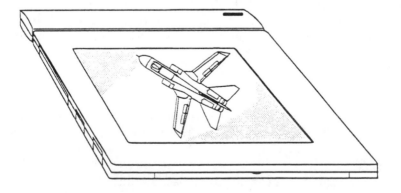

Fig. 2: Notepad Computer

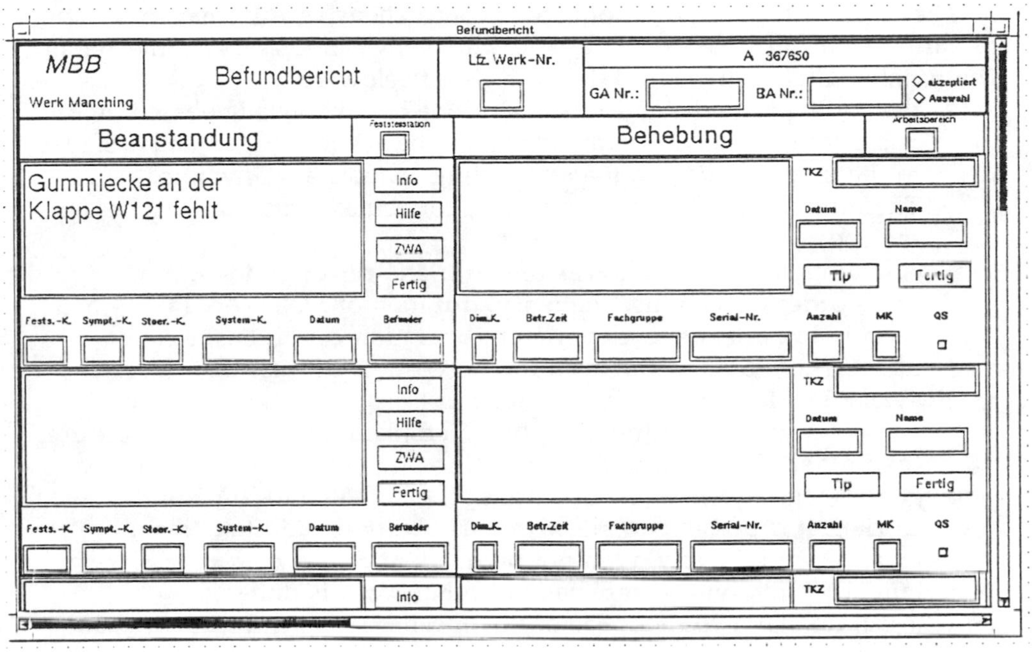

Fig. 3: hardcopy of the fault result form (english terms see Appendix)

The main principle can be described by the statement: "Information at your fingertips". The form is as easy to use as a sheet of paper. Today, the hand-writing recognition allows just a single letter recognition. Therefore, the input is for the most part given by dragged windows. It can be foreseen, that the continuous hand-writing recognition will be available in 1 to 2 years. During the analysis it was ascertained that the hand-writing quality of the worker is very good.

The form is the starting point to a comprehensive information system, containing all information, data, and knowledge, necessary to the repair. The functionality is given in datail and listed below:

- the coding of faults and repairs is done by a knowledge-based modul. Some variants for the same input can be acknowledged and assigned to the correct code. No manual coding is necessary anymore.
  The module checks for consistency, correctness, and completeness of the coding.
  A statistics module analyzes the fault and repair codes and interprets these codes concerning frequency, duration of repair, occurrence at certain aircrafts, dependency from operation conditions, etc.

- the working plan is given on the notepad. All repair instruction is available. Moreover, a certain stock of tasks can be given to the worker at the beginning of a shift. The planning of single procedures can be done by the worker, depending on the provision of spare parts, gauges, jigs and fictures. Because of the high qualification of the worker, it is possible to allow this flexibility. In addition, main tenance work is very often not predictable. Some work takes longer or involves complication ; other work is done in less time than expected. Factory data collection and worker or order related time could be stored.

- in addition, aircraft reports and logs can be retrieved at the notepad. For example, images with special information like corrosion data and location.

- a linkage to material management and logistics is provided. The demand for parts is determined at the shop-floor level. The input is done on the notepad. The serial numbers of the parts can be specified through the electronic part catalogue, explained below. Via the communication module, this need is transferred to the material management system.

- For the archiving and documentation of large volumes of data like texts, data, pictures, images, drawings, language etc., optical storage media will be the choice of future systems. For the CIM-M system the MBB product "CIROS" (Computer Aided Individual Part Catalogue on Optical Storage Media) will be used. Using CIROS, existing documentation (paper-based, microfiche or already as binary data) can be accessed on standard hardware. Thus all documentation needed for diagnosis, maintenance, and repair is available on the computer. For the CIM-M system a personal computer (80386 or 80486 processor, DOS operating system, compact disc drive) has been chosen. Each aircraft dock is equipped with one CIROS system. The essentials (part catalogue, list of parts, graphics etc.) are also available on the notepads (see fig. 4). These data for the notepads are configured in dependence on the jobs to be done by the worker.

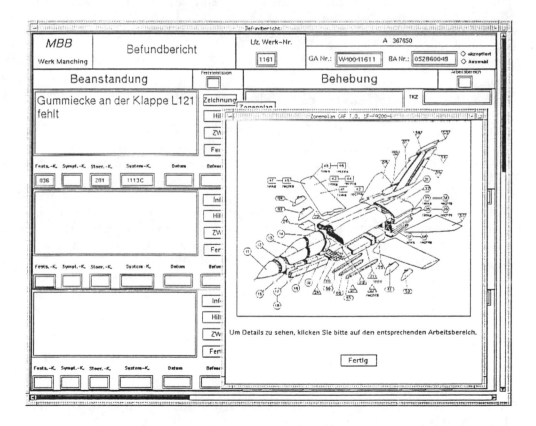

Fig. 4: hardcopy of the part catalogue available on the CIM-M system
(english terms see Appendix)

- as mentioned above, the German law still does not allow "an elec tronic signature". Yet, it is to expect that with on ongoing develop ment in office automation, concerning optical storage devices and new hardware concepts like notepads in field service and sales, the law will change. Furthermore, techniques for a safe coding and re cognition of an individual signature are already on the market or will soon be launched as products.
- the next step for user input will be language driven. The goal is to recognize speaker independent a limited number of single words.

Communication

The notepad will be operated wireless. Data transfer will be partially done via infra-red wireless communication with a built-in infra-red module, integrated in the notepad. Due to the low communication rate (nowadays about 40 Kbit/second) it is not advisable to transfer graphical data like drawings or images. On the other hand, alphanumerical data can be easiliy transferred between dock-PC and notepad. The dock-PC communicates with the other system architectures of the plant using Ethernet network and protocol.

COST-BENEFIT ANALYSIS

Cost

Up to now, just a prototype is being tested, having implemented the features listed above. At the final stage each worker will be equipped with his personal notepad.

The investment costs per work place are calculated to about $ 30,000 (software (user interface, pen input software, diagnosis support system, failure database, interfaces to existing programs, hardware (notepad, dock-PC, printer), communication (wireless communication)).

The running costs per work place amount to $ 3,000 per year. This sum includes all costs for update and software-maintenance as well as personnel costs for these tasks.

Benefit

The analysis phase led to the result, that about 10% of the total working time are used for information retrieval. After having implemented the CIM-M system, it is envisaged to reduce this time to 3 - 4%.

An economics calculation has been carried out. The amortization of the project investment costs is less than 4 years.

Besides the quantitative approach, there are qualitative arguments for introducing the CIM-M system. These arguments could be quantified for a cost-benefit analysis through the possible approaches listed in the table below:

Qualitative Argument	Approach for Quantification
Information/Documentation	
correct information	cost for personnel
more information	cost for material/personnel
processed information	training effort, cost for rework
higher quality of documentation	lower failure rate
less paperwork and records	cost of material and appliances
less multiple input of the same data	cost for personnel
less search time	cost for personnel
Manufacturing Facilities	
flexibility	time to retrofit to other types of aircraft
utilization ratio, productivity	costs for maintenance, competition
shorter times of delivery	revenue, profit
Personnel	
better motivation, lower work load	lower time of absence (e.g. illness)
new technologies in maintenance	qualification of the personnel

Table 1 Qualitative arguments and their possible quantification.

CONCLUSION

In this article a new technique for maintenance of aircrafts has been presented. The system has been designed and a software prototype is already implemented for the TORNADO aircraft. The core of the system is a notepad computer with handwriting recognition. It is designed as a personal tool and contains all information and data needed for a high qualitative maintenance work at the aircraft. The maintenance, production planning, job scheduling, and material management will be optimized.

Technical documentation, such as handbooks, drawings etc. are available on the spot. The input of failures, repair actions, and notes can easily be done by the worker with a pen similar to the conventional paper-based solution.

During the development of the CIM-M concept human factors have been taken into consideration in order to assure the acceptance of the worker. In addition, the work organisation of the job shop has been considered.

With future technical developments additional techniques such as speech recognition will be included in the system.

The CIM-M concept can be used not only for military but also for civilian aircraft. It is a modular concept and can be introduced step by step. The more information is available digitally, the more facile is the development for a certain product to maintain. Moreover, the maintenance philosophy can also be used for complex technical items such as field service of machine tools, servicing of trains and wagons like the high-speed train Inter-City Express or ships and trucks. In addition, maintenance teams of large facilities can be equipped with the system in order to repair their chemical, steel making or automotive plant. Also in civil engineering, it can used for the check of buildings, bridges etc.

In a nutshell, with a CIM-M system the first time in aircraft maintenance, electronic archiving of technical documentation on the spot is realized, using a new technology of personal computers. The notepad can be used portable, individual, and, most importantly, it is integrated in the process chain of aircraft maintenance.

APPENDIX

Befundbericht	fault result form
Lfz Werk Nr	aircraft serial no
GA Nr, BA Nr	production planning order numbers
akzeptiert	accepted
Auswahl	choice
Beanstandung	failure
Behebung	repair
Arbeitsbereich	working area
Gummiecke L121 defekt	rubber part L121 bad
Zeichnung	drawing
Hilfe	help
Zonenplan	zone, area drawing
ZWA (Zonenwerkstattauftrag)	work plan
Fertig	ready, done
TKZ (Teilekennzeichen)	part no.
Datum	date
Name	name
Tip	hint
Feststell-Kode	code for time of the failure
Symptom-Kode	code for the symptom for the failure

Stoerungs-Kode	characterization of the failure
System-Kode	code for the zone where the failure happened
Befunder	worker
Dim.-Kode	entity
Betr.Zeit	elapsed time of the aircraft
Fachgruppe	group of the worker
Serial Nr.	serial no. of the part
Anzahl	number
QS	quality assurance, control
Zonenplan, GAF T.O.	zone plan, German Airforce, Technical Order

Um Details zu sehen, klicken Sie bitte auf den entsprechenden Arbeitsbereich

For details, please click on the according working area

RATIONAL QUALITY ASSURANCE: THE THEORY OF QUALITY ENGINEERING AND FAILURE PREVENTION

Heinz W. Bargmann

Swiss Federal Institute of Technology, Lausanne, Switzerland

ABSTRACT

A general theory of quality assurance for complex, homogeneous, systems is presented. Quality is defined as system success with respect to a functional mission. It is measured in terms of the probability with which a success criterion is fulfilled. Such a theory must describe nonlinear as well as random systems under all types of interactions and redundancies. The theory requires the solution of two distinct problems : one of system decomposition, and one of physics of components. It contains safety and reliability theories as special cases. In particular, it removes the basic limitation of classical reliability theory, as the failure rates of components are no longer considered to be given *a priori* but may be predicted on the basis of physical failure mechanisms.

INTRODUCTION

There is an increasing awareness of the importance of quality and safety assessments for complex technical systems. The industrialized countries realize that advanced methods for quality assurance of products and processes are decisive for the economic survival of whole segments of national economies. Governments and public offices advocate a coherent approach to the safety assurance of structures, installations, and processes, as a prerequisite for the rational assessment of entire technologies. Increased public concern with the potential risk of large technological systems demands a coherent, integrated, and overall approach towards safety assurance. There is, finally, an increasing concern among scientists and engineers with quality and safety measures currently taken : frequently they come too late and are too much of an *ad hoc* nature, often they are incomplete and defective. These dispositions are not sufficiently integrated into the entire system, hence, not sufficiently "rational" in order to really guarantee what they claim to assure.

It appears more and more that what is really lacking is a clear and definite mathematical *theory of quality assurance* which allows a rational approach towards quality and safety of systems. The successful development of high quality components and systems during the last twenty-five years, in particular in aeronautical and nuclear engineering, makes it worthwhile to attempt a general formulation. In these areas the most difficult problems of deformable structures subject to non-uniform thermomechanical fields are treated, but we wish to present the basic ideas of the theory of quality assurance without these complications. The range of intended applications is just that of classical reliability theory : uniform components subject to homogeneous processes. However, the theory we shall present here, as simple as it is, is both far more embracing and far more explicit than classical reliability theory, as it allows to *predict* component quality, *a priori* knowledge of which is implicitly assumed in the classical theory.

We shall present the thory of quality assurance of complex, homogeneous, systems in two parts : first, the *system decomposition*, and second, the *physics of components*.

SYSTEM REDUNDANCY AND CONSTITUTION

We know from experience that failure of a component does not always give rise to failure of the system. In general, it may be sufficient for the success of the system that there is success of one or more subsystems, *i.e.* subsets of components. This fact suggests that, in order to describe the *redundant character* of a general system, alongside the event {success of the system}, denoted by x, the notion of the event {success of the system along a shortest path}, denoted by T, should be introduced, and the existence of *several* "shortest paths" (none of which containing subpaths along which succes of the system were possible) may be assumed. If, at any time t, a system S of a number of n components, Fig. 1, exhibits K paths, *there will be success of the system if there is success along at least one out of K paths*, and the probability of success of the system $P(x;t) \equiv P(x)$ is given by the probability of the *union* of the successes along the K paths, according to the **axiom of redundancy** :

$$P(x) = P(T_1 + T_2 + ... + T_K), \quad 1 \le K \le n \tag{1}$$

which is the *first part of the principle of decomposition*. In general, the number K of shortest paths will depend on time, $K = K(t)$.

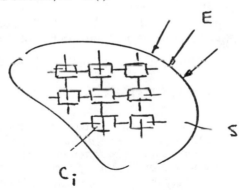

Fig. 1. System S of a number n of components C_i, i = 1,2,...,n, interacting with environment E, at any time t

While this first axiom is universal, valid for all systems, at any time t, the shortest paths themselves do not only change with time but are characteristic for a particular system. Such a path is defined by *constitutive relations* for the event T, *i.e.* {success along a shortest path}. The constitutive relations embody the **constitutive axiom** or the *second part of the principle of decomposition,* according to which *there is success along a shortest path* k *if and only if there is success of all components* k_j *of which this path is constructed*. The success of path k, $T_k(t) \equiv T_k$, is thus given by the *intersection* of the successes of its components k_j, successes denoted by x_{k_j}, at any time t,

$$T_k = x_{k_1}x_{k_2}...x_{k_{J_k}}, \quad 1 \le k_j \le n, \quad 1 \le j \le J_k \le n \tag{2}$$

where $1 \le k \le K \le n$. Again, the *grouping* of components may change with time, $k_j = k_j(t)$.

The two axioms (1) and (2) are thus equivalent to the **principle of decomposition** :

$$P(x) = P\left(\sum_{k=1}^{K} \prod_{j=1}^{J_k} x_{k_j}\right), \quad 1 \le k_j \le n, \quad 1 \le J_k, K \le n \tag{3}$$

Hence, the probability of success of a general system, under any type of interaction and redundancy of its n components, may always be represented by the probability of a logical form of a sum-of-products. The specific grouping of intersections and unions depends, in general, on time t. We note that, at this point, the success criterion which defines the functional mission of the system is only implicitly contained in (3).

Consider, as a particular case, the classical *series structure*. The logical form is the most simple one : there is *success* of the system if there is success of *all* components,

$$P(x) = P(x_1 x_2 ... x_n) = P(x_1)P(x_2|x_1)P(x_3|x_1 x_2)...P(x_n|x_1 x_2...x_{n-1}) \tag{4}$$

Equation (4) contains conditional probabilities which must be evaluated with care. For n *independent* components, (4) simplifies to

$$P(x) = \prod_{i=1}^{n} P(x_i) \tag{5}$$

The probability of success of a series system is thus always lower than the probability of success of its weakest component. If the probability of success is the same for all components, $P(x_i) \equiv p$, corresponding to a probability of failure of the component $P(\overline{x}_i) \equiv 1 - p \equiv q$, the probability of failure of the system becomes

$$1 - P(x) \equiv P(\overline{x}) = 1 - (1 - q)^n \sim nq \tag{6}$$

where (6)$_3$ is valid for sufficiently small values of the failure probability, nq « 1.

We may now formulate the **first problem of quality engineering** : *for a class of systems specified by both the physical configuration and a success criterion defining the functional mission, to formulate the probability of success of the system, at time t, as a function of the probabilities of success of the components and to discover the properties of this function.*

EXAMPLE 1 (The classical "reliability theory"). Classical reliability theory intends to predict system performance as a function of so-called *transition rates* between the two complementary events for the component C_i : x_i = {success at time t}, and $\overline{x}_i$ = {failure at time t), i = 1,2,...,n. To this end, we consider a component C with time-varying probability of success, where the time rates of change of the respective probabilities of success $P(x;t) \equiv P_0$ and of failure $P(\overline{x};t) \equiv P_1 \equiv 1 - P_0$, at any time t, are given by

$$\dot{P}_o = -\lambda P_o + \mu P_1$$
$$\dot{P}_1 = \lambda P_o - \mu P_1 \tag{7}$$

and where the transition rates, *i.e.* the *failure rate* λ and the *repair rate* μ, are supposed to be known. The corresponding initial conditions $P_0(0)$, hence $P_1(0)$, are assumed to be given.

The probability of the event {success at time t} is called *availability* $A(t) \equiv P_0 \equiv P(x;t)$ of the component. Substitution of $P_1 \equiv 1 - P_0$ into (7)1 yields

$$\dot{A} + (1 + \mu) A - \mu = 0 \tag{8}$$

where the general solution could easily be written down.

For the special and important case of *no repair*, $\mu \equiv 0$ for $t \geq 0$, the availability is called *reliability* $R(t) \equiv A(t) \equiv P(x;t)$. It decreases according to

$$R(t) = R(0) \exp\left(-\int_0^t \lambda \, d\xi\right) = P\{\underline{\tau} > t\} \equiv 1 - F(t) \tag{9}$$

from its initial value $R(0)$. The event {success at time t} is here equivalent to the event {success during the interval [0,t]} or $\{\underline{\tau} > t\}$, where the *lifetime* $\underline{\tau}$ of the component is a random variable (indicated by the lower bar). In (9)3, F denotes the distribution function of the lifetime $\underline{\tau}$. The *mean lifetime* is given, with $f \equiv \dot{F}$, upon partial integration, by

$$<\underline{\tau}> = \int_0^\infty t \, f(t) \, dt = \int_0^\infty R \, dt \tag{10}$$

which for *constant* failure rate becomes $<\underline{\tau}> = R(0)/\lambda$.

In complete analogy, for the special case of *no failure*, $\lambda = 0$ for $t \geq 0$, the availability is called *maintainability* $M(t) \equiv A(t) \equiv P(x;t)$ of the component. It increases according to

$$M(t) \equiv 1 - P_1(0) \exp\left(-\int_0^\infty \mu \, d\xi\right) = P\{\underline{\theta} \leq t\} \equiv G(t) \tag{11}$$

from its initial value $M(0)$. The event {success at time t} is now equivalent to the event {success during the interval $[t,\infty)$} or $\{\underline{\theta} \leq t\}$, where the *repair time* $\underline{\theta}$ of the component is a random variable, under the assumption that there was initial failure with probability $P_1(0) \equiv 1 - M(0)$. In (11)3, G denotes the distribution function of the repair time $\underline{\theta}$. The *mean repair time* is given, with $g \equiv \dot{G}$, by

$$<\underline{\theta}> = \int_0^\infty t \, g(t) \, dt = \int_0^\infty (1 - M) \, dt \tag{12}$$

which for *constant* repair rate becomes $<\underline{\theta}> = [1 - M(0)]/\mu$.

We note from (8) that, for *constant* transition rates, a stationary solution for the availability exists, with $\dot{A} \equiv 0$ for $t \to \infty$, given by the relative repair rate :

$$A(\infty) = \mu \, (\lambda + \mu)^{-1} = <\underline{\tau}> \, [<\underline{\tau}> + <\underline{\theta}>]^{-1} \tag{13}$$

where (10) and (12) with $R(0) = 1$ and $M(0) = 0$ have been substituted. We have thus proved the following **theorem** : *For a repairable component, if failure rate λ and repair rate μ are constants, the stationary value of the availability $A(\infty)$ is equal to the portion of the mean lifetime $1/\lambda$ (of the non-repairable component with full initial reliability) in the* *sum of this mean lifetime and the mean repair-time* $1/\mu$ *(of the non-failing component of zero initial maintainability)*, $1/\lambda + 1/\mu$.

This is one of the celebrated results of reliability theory which can be found in most textbooks. The classical work on reliability consists mainly of plugging numbers of mean lifetimes of components, *via* (9), into simplified versions of (3). Classical reliability theory permits to calculate the chances for two particular system qualities, for the events {success at time t} and {success up to time t}, under the supposition that the chances for the corresponding component qualities are delivered by "transition rates" which are assumed to be known *a priori*. This latter assumption is the main and serious limitation of classical reliability theory.

Classical reliability theory thus essentially solves the first problem of quality engineering. It does not provide any means to assess the performance of components. In order to predict the chance for *component quality*, *e.g.* to *derive* the transition rates, suitable assumptions must be made regarding the physical behavior of the components. We thus need to solve the **second problem of quality engineering** : *for an assigned class of systems, to determine the probability of success of the components.*

For this deeper aspect of quality assessment we have to invoke the physics of components.

THE PHYSICAL STATE OF A COMPONENT

The intended application of elementary system theory is to "components" subject to *homogeneous processes*: all physical quantities occurring are functions of time only; any space-dependence is excluded.

Just as in classical mechanics we consider the primitive concepts of place and force, in system analysis we consider two categories of "terminal variables" of a component C, Fig. 2 : a list of (input) *excitations* $u_1, u_2, ..., u_m$, collectively denoted by the symbol **u**, and a list of (output) *responses* **y**. In a particular application, u_1 might be an electrical tension, u_2 a temperature or the amount of a resultant force, but at this level of generality no specific interpretation is necessary.

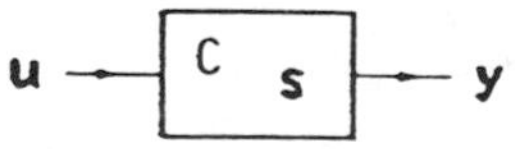

$$\mathbf{u} \longrightarrow \boxed{\begin{array}{cc} C & \\ & s \end{array}} \longrightarrow \mathbf{y}$$

Fig. 2. Component C responding, at any time t, by y(t) and state s(t) to the history of the excitation $\mathbf{u}_{(t_0, t]}$

In addition, a list of real parameters $s_1, s_2, ..., s_\ell$ is introduced which are called the *state variables* of C, collectively denoted by **s**, which assign the state **s** to the component C, at any time t,

$$\mathbf{s} = \mathbf{s}(t) \tag{14}$$

by which the component is *completely* characterized, at time t.

If $\varphi(t)$ is a function of time, its *history* $\varphi_{(t_0, t]}(\xi)$ from t_0 on and up to time t is defined by $\varphi_{(t_0, t]}(\xi) \equiv \varphi(t-\xi)$, $0 \leq \xi < t-t_0$. It is then asserted by the **axiom of coherence** (or *principle of physically compatible determinism*) that for a component a functional Y| exists such that a *coherent relation* can be established :

$$\mathbf{y}(t) = \mathbf{Y}| \langle \mathbf{s}(t_0); \mathbf{u}_{(t_0, t]} \rangle \tag{15}$$

which satisfies the principles of physics, *for any excitation process*. We also require that the state s(t) of C at time t be *uniquely* determined by $s(t_o)$, its value at time t_o, together with the history $u_{(t_o,t]}$; and that the functional dependence between s(t) and $s(t_o)$, $u_{(t_o,t]}$, the *state equation* for C be delivered, at least in principle, by the coherent relation (15),

$$s(t) = S| \langle s(t_o); u_{(t_o,t]} \rangle \tag{16}$$

These few formulae provide a foundation on which a theory of quality assurance can be built.

SUCCESS CRITERION

The fundamental problem of component design is to assess the component performance, *i.e.* the degree to which it fulfills a mission, defined by some criterion of success.

For a component completely characterized by its state s(t), at any time t, this implies the knowledge of the distance to a non-admissible or *critical state* $s_c(t)$. In order to remain sufficiently general, we define a function $\phi(s)$ of the actual state together with its critical value ϕ_c, corresponding to a particular failure mechanism of the component stipulated for a certain case. We then postulate as **success criterion** a list of finite differences **D**(t), at any time t :

$$D \equiv \phi_c - \phi > 0 \tag{17}$$

such that a critical state not be reached. The values ϕ_c, at any time t, are, in general, defined by scientific, technical, or economic considerations.

PROBABILITY OF SUCCESS

The number of l state variables, denoted by s = s(t), *cf.* (14), together with the list of r critical values of the state functions $\phi_c = \phi_c(t)$, *cf.* (17), are, in general, functions of time. We shall assume that they are *random processes*, $\underline{s}(t)$, $\underline{\phi}_c(t)$. For a specific t, they represent a set of $l + r$ *random variables* : $\underline{s}_1, \underline{s}_2, ..., \underline{s}_l$, and $\underline{\phi}_{c1}, \underline{\phi}_{c2}, ...,\underline{\phi}_{cr}$. We suppose that their joint distribution function is given, at any time t, by

$$\Psi(\phi_c, s; t) = P\{\underline{\phi}_c < \phi_c, \underline{s} < s; t\} \tag{18}$$

The joint probability density $p(\phi_{c1}, \phi_{c2}, ...,\phi_{cr}, s_1, s_2, ..., s_l; t)$ is obtained upon differentiation with respect to each one of the variables $\phi_{c1}, ..., s_l$. The probability that the success criterion (17) be fulfilled, *i.e.* the *probability of success* of a component C_i, is then given, at any time t, by

$$P(x_i;t) = P\{\underline{D}(t) > 0\} = \int_{D > 0} p(\phi_c,s; t) \, d\phi_c \, ds \tag{19}$$

The **second problem of quality engineering,** announced at the end of Example 1, may now be formulated in a more specific way : *Within a given class of components, defined by a functional* Y| ; *for an initial state* s(t_o) *given together with the history of an* *excitation process* $\mathbf{u}_{(t_o, t]}$; *and for a given joint probability density* p *of the state variables* **s** *and the critical values* ϕ_c *of given state functions* ϕ; *to determine the probabilities of success of the components.* The problem has now also a second, deeper part: *to determine, for the components of a given class and for a class of prescribed excitations, the success criterion.*

EXAMPLE 2 (Scalar state component). Consider a component which can be characterized by a single random state variable, $\underline{s}(t)$. Let us assume $\underline{\phi} \equiv \underline{s}$, $\underline{\phi}_c \equiv \underline{s}_c$. If $\underline{s}$ and $\underline{s}_c$ are random variables with given joint probability density $p(s_c, s; t)$, we have for the probability of success of component C_i, at any time t,

$$P(x_i; t) = \int_{D \equiv s_c - s > 0} p(s_c, s; t) \, ds_c \, ds \tag{20}$$

For *independent* variables, $p(s_c, s; t) \equiv g(s_c; t) h(s; t)$, we have

$$P(x_i; t) = \int_0^\infty g(s_c; t) \, ds_c \int_0^{s_c} h(s; t) \, ds \tag{21}$$

If, moreover, the critical value is *deterministic*, $s_c \equiv s_{co}$, *i.e.* $g(s_c) \equiv \delta(s_c - s_{co})$, and if the state variable has an *exponential* density, $h(s) = \alpha \exp[-\alpha(s - s_o)]$, $s \geq s_o$, we obtain the simple expression

$$P(x_i) = 1 - \exp[-\alpha(s_{co} - s_o)] \tag{22}$$

for the probability of component success. There, α is a measure for the concentration of the distribution; in fact, α^{-1} represents both the standard deviation and the difference between mean and lower threshold value s_o. In general, α, s_o, and s_{co} may depend on time t. Then, again, $P(x_i) \equiv P(x_i; t)$.

EXAMPLE 3 (The prediction of the failure rate). The advanced theory of reliability removes the basic limitation of classical reliability theory, insofar as the failure rates of components are no longer considered to be given *a priori* but may be predicted on the basis of both a particular physical failure mechanism and the solution of a so-called "first-passage problem". In these applications, one is interested in the statistical characterization of threshold-crossings of the excitation process (state function) $\underline{\phi}(\underline{s})$, that is of the points t_i such that $\underline{\phi}[\underline{s}(t_i)] = \underline{\phi}_c(t_i)$ Thus, the reliability and, hence, the failure rate are obtained, at least in principle, *via* ,

$$R(t|s_0; 0) = P\{\mathbf{D}(\xi) > 0, \ 0 \leq \xi \leq t; t \mid s_0; 0\}, \quad s_0 \equiv s(0) \tag{23}$$

where $R(t|s_0; 0) \to 1$ for $t \to 0$, and $R(t|s_0; 0) \to 0$ for $D = 0$, $\dot{D} < 0$. This is in general a difficult problem. For certain processes, simple conclusions about the statistics of these points can be drawn.

In some situations, however, an assessment of the failure rate may directly be carried out. Consider, for example, the probabilistic 'experiment' of repetitive failures in a set of identical components, where the failures are caused by the repetitive arrival of a particular loading condition which, for the given components, is defined by the probability density p in (19). Let the loading condition be characterized by the *conditional arrival rate* $\beta(t)$, where $\beta(t)$ dt equals the probability of the event p_{dt} : {the particular loading condition, hence the particular probability density p, arrives in the interval (t, t+dt), assuming that it did not arrive up to time t}. Since, in practice, every probability can be interpreted as a conditional probability relative to some hypothesis implied by the experiment under consideration, we identify the probability of component failure, (19), by

$1 - P(x_i) = P(\overline{x_i}) \equiv P(\overline{x_i}|p_{dt})$. With the frequency interpretation of probabilities, we thus arrive at the *conditional failure rate* $\lambda(t)$:

$$\lambda \sim P(\overline{x_i}|p_{dt}) \beta \equiv P(\overline{x_i}) \beta \tag{24}$$

where $\lambda(t)$ dt equals the probability of component failure occurring in the interval (t, t+dt),

assuming that it did not occur up to time t. $P(\overline{x_i}) \beta(t)$ may, hence, be used for the assessment of component failure rates in global system reliability analyses, *cf.* (9), (3).

EXAMPLE 4 ("Safety factor" *vs* probability of success). In classical design, use is frequently made of the so-called "safety factor" ν defined by the ratio of a representative value of the (random) "resistance" ϕ_c to a representative value of the (random) "loading" ϕ. For example,

$$\nu = <\phi_c>/<\phi> \tag{25}$$

where the *mean value* $< \cdot >$ of a random variable is taken as representative. In general, such a safety concept is rather narrow. In particular for complex systems, it does not reflect what is usually associated with the notion of "safety".

There, the concept of probability of success is appreciably wider. Consider, as an *example*, a system of n identical components in series with identical safety factors ν. As a consequence, the entire system represents this safety factor ν. The probability of success $P(x)$ of the system, however, is *reduced* to the n-th power of the probability of component success p, $P(x) = p^n$. In particular, as compared to the probability of component failure $q \equiv$

$1-p$, for sufficiently small values, the probability of system failure $P(\overline{x}) \equiv 1 - P(x)$ is n times larger,

$$P(\overline{x}) \sim nq \tag{6}_3$$

More generally, as the probability of success $P(x)$ reflects all essential physical parameters of a particular problem, knowledge of the mathematical structure of $P(x)$ provides the necessary and sufficient insight in order to arrive at the rational solution and optimization of the physical problem at hand.

REMARK. There are certain success criteria for which system quality is traditionally called "safety", either to stress the gravity of consequences of potential failure or to emphasize the extreme and/or unexpected nature of external excitations or of component behavior. It is in this case that the problem of quality ("safety problem") is characterized by a particular importance *(i)* to systematically decompose the complex system taking into account the *interdependencies* between the components as well as between the system and the external agencies (its environment), *cf.* (3), (4); *(ii)* to properly take into account the *nonlinear* physical component response to any excitation process, *cf.* (15); *(iii)* to systematically examine the potential *component failure mechanisms* in order to establish the relevant success criterion, *cf.* (17); *(iv)* to realistically take into account the *random* nature of both the system and the external agencies, *cf.* (18).

CONCLUDING REMARKS

System quality engineering requires the solution of two distinct problems : one of system decomposition, and one of physics of components. The theory is based on clear, elementary mathematics. By adopting it, the designer follows the rational approach towards quality assurance. The engineer who examines particular failure mechanisms sees how his results will fit when finally integrated into the overall quality assessment of the system. The scientist interested in attacking new problems has at his disposal a general methodology based on continuum physics and at the same time naturally integrated into the formalism of classical system theory. He will realize that the present quality theory of homogeneous systems can be generalized, in a straightforward manner, to a theory of systems exposed to non-uniform fields. Most important here is the distinction between the formal principle of system decomposition and the coherent relation of physics. In the presence of non-uniform fields, the latter will have to be separated again into the general principles of mechanics and physics and the constitutive relations describing the local material behavior. If, finally, he wishes to interpret gross component behavior on a microphysical scale, he can see at once what is to be explained. A great new area of interrelations between material science, mechanics, and global system analysis invites us to join our efforts for the rational quality, safety, and reliability assurance of complex systems.

REFERENCES

Some further motivation for the theory presented here may be found in the following memoirs:

1. A. M. FREUDENTHAL, "The Safety of Structures", *Transactions of the American Society of Civil Engineers*, Paper No. 2296, New York, 1947.

2. C. TRUESDELL & R. A. TOUPIN, "The Classical Field Theories", in FLÜGGE's *Handbuch der Physik*, Volume III/1, Berlin *etc.*, Springer-Verlag, 1960.

3. A. PAPOULIS, *Probability, Random Variables, and Stochastic Processes*, New York *etc.*, McGraw-Hill, 1965.

4. L. A. ZADEH & C. A. DESOER, *Linear System Theory*, Huntington *etc.*, R. E. Krieger, 1979.

5. C. TRUESDELL, *Rational Thermodynamics*, New York *etc.*, Springer-Verlag, 1984.

THE DEVELOPMENT OF QUALITY CONTROL EXPERT SYSTEMS: PAST, PRESENT, AND FUTURE TRENDS

Tsuang Y. Kuo and Anil Mital

Industrial Engineering, University of Cincinnati, Cincinnati, Ohio, U.S.A.

ABSTRACT

Recent development of computer applications for quality control includes automated test facilities, data collection, storage, and analysis. These lead to many implementations of statistical techniques which eliminate several routine works from the daily practice of quality control activities. Latest expert system technology enhances the capabilities of computers even further. It provides more user friendly, more automated and integrated systems which help not only data analysis but decision making for quality management.

Expert systems have attracted many quality control researchers and practioners as they provide the ability to preserve heuristic knowledge in computers. Most quality control expert systems available today have a major limitation: either they are unique to a specific process or do not have sufficient knowledge for practical use. To benefit from expert systems, a general purpose quality control expert system must be developed. The objective of this paper is to present a review of state-of-the-art expert systems in quality control and conclude a design framework for a general purpose quality control expert system.

Keywords: Statistical quality control, Statistical process control, Knowledge-based expert systems

1. INTRODUCTION

In statistical quality control applications, the computation of sample statistics and development of the control chart are routine exercises. However, the interpretation of chart patterns and trends and the resulting diagnosis of assignable causes require expert knowledge. The solutions to many manufacturing problems are not deterministic and many alternative courses of action may exist. All of these aspects require expert knowledge which is heuristic in nature. This heuristic knowledge is very difficult to implement in the traditional procedural programming. The expert system provides support to the process, or quality engineer, in the selection of proper type of control charts to use in tracking the state of the process and provides feedback.

Expert systems are programs that solve substantial problems employing facts and heuristics used by experts. Expert systems will become more important tools as the U.S. industry moves to total quality control, a process which involves many people who are not in quality control department to work on Q.C. teams.

Structurally, expert system programs consist of a knowledge base, a working memory and an inference mechanism. The knowledge base contains facts and heuristics associated with the problem. The working memory is a contextual database used for keeping track of the state of the "world." This includes input data for the particular problem, the problem status, and relevant history of what has been done. The inference mechanism is the control structure that defines the way the knowledge is used to reach a solution. The common inference paradigms are forward chaining and backward chaining. Forward chaining is the process of working from known facts towards a goal state (i.e., the process is data driven). Backward chaining is the process of selecting a hypothesis and determining if the hypothesis is supported by the context (i.e., goal driven).

2. COMPUTER-AIDED QUALITY CONTROL SYSTEMS: THE CONVENTIONAL WAY

Computer-aided quality (CAQ) represents the totality of the application of computers to quality control to increase the likelihood that quality problems could be detected early in the manufacturing process.

2.1 General Purpose CAQ Systems

There has been an explosion of quality control software and devices for data collection. Many new methods for measurement provide input directly to the computers. The March 1988 issue of quality progress contains a directory of software from vendors. The 1988 directory lists 415 software applications and packages in the following categories: (Note: some software is entered in more than one category)

Calibration	Capability studies
Design of experiments	Inspection
Management	Measurement
Process control	Quality costs
Reliability	Sampling
Software quality assurance	Statistics
Special (Nonconformity analyzers, process simulators, etc.)	
Supplier quality	Taguchi methods
Training	

2.2 Special Purpose CAQ Systems

Special purpose system use traditional procedural programming techniques to detect quality problems and provide limited advisory. CAQ application in fibre-cement pipe production (Elimam, 1989) is a typical example. The system emphasizes process quality control which includes the identification of significant process parameters, the use of economically designed control charts and the introduction of efficient feedback mechanisms.

This CAQ system was designed to provide production and quality control staff with the capabilities to:

(1)   generate control charts for monitoring and analyzing finished product quality characteristics and process parameters;

(2)   compute major statistics, including variabilities in quality parameters and process capability by shift;

(3)   sample by variable for accepting lots with a given level of defective products;

(4)   estimate trends in the process parameters for early detection of quality problems;

(5)   diagnose manufacturing process conditions to trace the causes of producing substandard products;

(6)   generate periodic and regular quality reports for feedback to production staff and for evaluation by plant management.

3. EXPERT SYSTEMS: THE FUTURE NECESSITY

One thing essential to quality control is expertise: the type of knowledge that is based on past experience and that can not easily be cast into mathematical formulae, or, with all its details, into conventional algorithmic programs.

3.1 Application for Product Design
Quality function deployment (QFD) is a good representative of expert systems application for product design (Braun, 1990). The aim of QFD is to address quality concerns as early as possible, starting with defining customer requirements early in the design phase, when corrective actions are easier to take and are much more effective than those taken further downstream.

The basic QFD tool is the product-planning matrix or means/ends matrix, which fits nicely with the equivalent premise/conclusion matrix frequently used in preparing data for expert systems. The matrix format, in turn, leads directly into the if-then rule structure of an expert or knowledge based system. In addition, a network of related systems can be used to tie the various parts of a total QFD plan together and provide the disciplined environment of a formal, yet very flexible, framework.

3.2 Application for Control Charts' Selection, Construction and Interpretation
Control charts, the essence of statistical quality control, are effective tools that aid the manufacture of quality products and provision of quality service. To be effective, care has to be taken in selecting an appropriate chart, in constructing proper parameters, such as control limits, sample size, and sample frequency, as well as consistently monitoring the control chart for out-of-control signals. Existing quality-oriented expert systems are primarily focused on either control charts selection or interpretation. Functionally, this kind of expert systems use interactive computer programs to communicate with end users. Expert system ask for necessary quantitative or qualitative information which is needed to determine most suitable control charts, set up control chart parameter, and perform non-random patterns checking on control charts to signal out-of-control situations.

The performance of different control charts can be varied in several ways, such as: type I and type II errors, sensitivity to detect changes in the process mean and/or variance, and the average run length. These factors have to be considered thoroughly in order to select appropriate control chart(s) for each potential application. Alexander and Jagannathan (1986) developed an expert advisory system to illustrate the framework for appropriate control chart selection.

Rules for control chart selection are based on the nature of quality characteristic, such as whether measurements are possible or not (visual inspection), or measurements are not practical (too expensive), or 100% check required. Other factors include type of defects, sensitivity requirement of shift, and type I error. The rules are divided into two categories: rules for attribute control charts and rules for variables charts.

Guildlines for construction and interpretation of control charts can be obtained through database query which is not performed by the expert system. Optimization of control charts' parameters is mentioned as a possible future enhancement.

Scott and Elgomayel (1987) developed a knowledge based diagnosis system for identification and interpretation of random and assignable patterns on x-bar and R/s charts. The knowledge base consists of five basic components:

- x-bar and R/S control limits determination:
    Control limits are obtained by either previously established or calculation based sample data statistics.
- Basic control charts testing:
    Control charts are first tested using 3 sigma limits rule which locates out-of-control sample(s). Further investigation of runs are implemented for process instability.
- Basic control charts interpretation:
    The presence of out-of-limit points, unnatural patterns, or trends on R/s chart are evidence of increased process variability which are usually caused by inaccurate inspection procedures or the level of automation of process: automated or manual operation. For an automated operation, process variability is a function of machine capability. An increased process variability indicates a fundamental breakdown in the machine. For a manual operation, it usually means an inconsistency in work method, or a lack of operator care or concentration.

    Frequent shift indicates tool wear problem which is unavoidable part of the process. In such a case, the process should be allowed to operate until an maximum degree is reached.

    The interpretation system provides two measurements which further evaluate the state of control. With processes in statistical control, histograms are produced by a FORTRAN program in order to understand the underlying data distributions. Process performance is compared against engineering specifications to estimate the percentage of units which exceed specification.

- Advanced control chart testing:
    The process is now tested against patterns, such as, cycles, freaks, gradual change in level, grouping or bunching, instability, interaction, mixtures, natural pattern, stable forms of mixture, stratification, sudden shift in level, systematic variables, tendency of one chart to follow another, trends, and unstable forms of mixture.

- Comparison of process capability against engineering specifications:
    The process is determined to be capable of meeting specifications if the following two conditions hold true:

    (1) $2 * (M1 + 1)\sigma \leq$ upper specification - lower specification
           where M1 is x-bar control limits multiplier; usually M1 is 3.
    (2)   the process is centered within $1/3$ σ of the nominal specification values.

Dagli and Stacey (1988) developed an expert system to assist the process or quality engineer in the selection of the best control chart for a given application. During the consultation, the knowledge based system asks questions to determine the type of attribute to be tracked, the resources available for chart calculation as well as the information needed from the control chart. The expert system then determines the best control chart for the situation along with recommendations to the suggested chart.

The reasoning process of control chart selection depends on chart type (variable or attribute). For variable chart, the expert system determines appropriate sample size depending on the nature of quality control test, destructive or non-destructive, and homogeneity of sample population. It then recommends a control chart with a confidence factor based on the resources available for calculating various values used in control charts. Attribute charts are determined by the type of data to be tracked, either defects per unit or number of proportion defective items, and sample variability. Duncan's formula (1956) is used to optimize control chart parameters: sample size, sample interval, and upper and lower control limits values. The expert system is currently structured to select at least one of the following control charts: P-chart, NP-chart, x-bar and sigma charts, x-bar and R charts, x-bar and Moving R charts, C-charts, U-chart, and Median chart.

Evans and Lindsay (1988) designed an expert system for interpretation of x-bar and R charts. The knowledge base is partitioned into three sets: analysis rules, interpretive rules, and diagnostic rules. The analysis rule base consists of supplemental runs rules for automatic determination of out-of-control conditions. The conclusion reached from the analysis rule base is one of the following: in control, out of control or suspected to be out-of-control. If the process is lacking control, the interpretative rule base uses these conclusions along with a decision tree to determine the type of pattern in the chart. The diagnostic rule base uses the pattern found with specific process information to conclude assignable cause(s). Assignable causes included: points at or near control limits, patterns and trends, and cyclic variation. For x-bar charts, diagnosis could be one of these: change in process setting, change in material, minor part failure, tool wear, environmental factors, and rotation of operators. For R chart, diagnosis includes operator error, poor material, operator fatigue, operator skill improvement, and maintenance cycles.

Hosni and Elshennawy (1988) used decision trees and decision matrices to develop a knowledge based system which selects and interprets control charts and assesses product capability. The decision of selecting a chart is made by the following factors: qualitative vs. quantitative, underlying distribution, type of inspection: destructive vs. non-destructive, cost of inspection, inspection time, available of sampling during production, bulk vs. discrete production, in-process vs. pre-process inspection, as well as the lot size. Control charts are then selected from individual chart, x-bar & s charts, x-bar & σ charts, x-bar & R charts, p chart, np chart, u chart, and c chart. Another decision matrix is used to relate out-of-control signals and possible assignable causes. Out-of-control signals are points beyond control limits and several run rules. Possible assignable causes are shifts in the process parameter, changes in the process variability, accuracy, and precision, and human errors. Process capability indices Cp and Cpk are used to interpret the process performance. Sample size is determined from MIL-STD 414, Normal Inspection, Level IV.

Love and Simaan (1988, 1989) developed a knowledge based system for the detection and diagnosis of out-of-control events of an aluminum strip rolling mill operation. Automatic detection and diagnosis is a combined application of statistical process control principles and knowledge of process. Data from the manufacturing process are collected into a signal database. Signal of interest is chosen from the database and the system is initiated to produce reports of diagnosed problems in the process. Based on nonlinear filtering techniques, a two-level procedure for automatic process diagnosis is provided. In the first level, interpretation is done by the following steps. Using numerical signal processing techniques, the raw signals are preprocessed to eliminate noise. A combination of nonlinear filters are then used to isolate particular primitive variations in the signal. These nonlinear filters are a median filter, a slope filter, a horizontal threshold filter, an integrator and an amplitude thresholder. Finally, input data are classified into three features: peaks (impulses), steps (mean-shift), ramps (linear trend). The second level is a rule based program which diagnoses special cause(s) of variation. The signal interpreter process is written in LISP. The filtered signals are segmented into pieces. A combination of the three features forms a data structure to represent the primitive variations. Combinations (events) are used to identify particular signal objects and transform the signals into their symbolic descriptions. The symbolic information are then used for interpretation. Each event has an associated ruleset which contains special cause of variation of that event. Separate rulesets help both conceptual and software modularity which makes it easier to add knowledge when it becomes available. Diagnosis is reached by backward chaining on the appropriate ruleset to a given event. Undefined events are automatically reported and new rules may be developed and added to rulesets later.

Harris Semiconductor developed a real-time expert system (P-R-EXPERT) for monitoring, characterizing, and controlling front-end photolithographic process for optimal product throughput, quality, and yield (Brillhart and Wible, 1989). PREXPERT is not only capable of detecting out-of-control products, it can also continuously improve the process aim and tighten the process variation. This application combines the concepts of real time data acquisition, recency-weighted process characterization, and automate system tuning. Lot history and product information are stored in COMETS (Consilium's On-Line Manufacturing and Engineering Tracking System) data base. Knowledge base of PREXPERT comprises of the statistical process control techniques, customer requirements and the engineering expertise of photolithographic process. Process capabilities indices (Cp and Cpk) are utilized to aim the process mean and to reduce the variability. With the linear regression from exposure energy to the resist measurement and the regression model from the resist measurement to the final etched measurement, PREXPERT is able to predict the final etched dimension with a 95% confidence interval which eliminate pilot run of production. The adjusted R-square of regression model is 0.95 indicating 95% of the variability has been accounted for by this model. This plant is capable of producing over 220 different product types and over 2600 masks. At any given time, there are more than 50 active products and over 700 masks in the line. Most of the products have their own unique process flow. To combine these products into a single control chart, the normalization z control chart is employed. Currently, run tests about the center line is the only non-random pattern detection; such as, eight points that fall on the same side of the center line on the x-bar chart or the range median of the R chart. Further analysis caused by the run detection is based on two parameters: slope and RMS. A line is fit through the subgroup points and the slope of this line is obtained. Then, the root mean square value of the detected run is calculated. A decision is made whether or not to notify the engineer and to its prioritization (process shutdown, immediate notification, or daily/weekly report).

Pandelidis and Kao (1990) developed a knowledge-based system (DETECTOR) for injection model diagnostics. It has been developed and implemented in the field. Based on the fuzzy set theory, the authors designed a general knowledge representation scheme to represent inexact and incomplete information. The first step in the diagnostic process is to accept the set of observed defects from the user, and associate appropriate weights for their respective severity. The second step is to determine the associated set of possible causes to the observed defects and obtain their associated weight in a matrix form. Minimum cover criterion is used to determine the smallest possible set of causes explaining the defects. The most obvious advantage is to narrow the search space in the specified domain and to produce the sequential problem-solving paradigm to seek further information for the human diagnostician. The Knowledge base of DETECTOR consists 5 modules: trouble shooting guide, inference rules, construction of plastic variables and machine variables, material data base, and machine data base.

3.3 Application for Attributes Control Charts

If the quality-related characteristics can not be represented in numeric form, such as characteristics for appearance, softness, and color, the control charts for attributes are used. Product units are either classified as 'conforming' or 'non-conforming', depending upon whether or not they meet specifications, or the number of non-conformities per product unit is counted.

The binary classification into conforming or non-conforming might not be appropriate in many situations where product quality dose not change abruptly from satisfactory to worthless, and there might be a number of intermediate levels. These intermediate levels may be expressed in the form of linguistic terms (Wang and Raz, 1990). For example, the quality of a product can be classified by one of the following terms, 'perfect', 'good', 'medium', 'poor', and 'bad', depending on its deviation from specifications. Appropriately selected continuous functions can then be used to construct control charts for the quality characteristic associated with each linguistic term.

The vagueness and ambiguity inherent in linguistic variables may be treated mathematically with the help of fuzzy set theory. This theory provides the tools for performing mathematical operations on linguistic descriptors by representing them as fuzzy sets of a numerical domain.

3.4 Application for Continuous Process Control

When "all" the data are available in a automatic control mode, time-series analysis may be more useful than Shewhart control charts because data might not be independent and normally distributed. Automated inspection and data collection at the point of manufacture may eliminate the sampling environment. The data acquisition system (DAS) acquires and conditions the data from the product sensors. Time series techniques and expert systems can be used to replace the Shewhart control charts and sampling techniques since "total Knowledge" about product quality variables and attributes are available (Hubele and Keats, 1987).

In time series (ARIMA) models, the next observation is predicted by a weighted-moving average forecast, often exponentially smoothed. Spectral density analysis can be used to gain information about periodicities. A control system approach to process control involves the use of transfer functions of closed-loop systems. An understanding of the feedback control theory and "expert systems" enables a process to be designed that never needs human intervention. The expert system is the automatic controller which process adjustments to make.

Kuo (1989) developed a prototype expert system for diagnosing continuous processes if the assumption of normality for data does not exist. Time series analysis gives precise information on the length of periodicity and slope of the trend. This feature provides a quantitative information which leads to more accurate interpretation and diagnosis. A two-step algorithm was developed to interpret the control charts. Supplemental runs rules (Nelson, 1984) are first employed to locate out of control conditions. If any out-of-control occurs, time series analysis is then used to obtain quantitative information about the out-of-control process. Parameters for time series analysis models are estimated by the FORTRAN program developed by Pandit and Wu (1983). Estimated parameters are then entered manually for expert system analysis. Assignable causes are acquired by utilizing information from time series analysis and generic information provided by Nelson (1985). The advantage of this two-step algorithm is that substantial computational efforts required by time series analysis are avoided while quantitative information about the process is secured.

Du Pont has discovered that process control systems are good for on-line real-time expert systems application in several ways: sensor validation, data reconciliation, process and equipment diagnosis, and model-based expert systems (Rowan, 1989). Process control systems receive data primarily from on-line process sensors and respond in real time to process problem. Individual measurements from sensors are critical because they supply feedback and closed-loop control. The expert system is segmented into a monitoring component and a rule-base component. The monitoring segment continuously scans on-line data for suspicious behavior, including the signal stability of the measurement and reference signal, and characteristics of internal signals, such as the automatic gain control. With a short term history of process data, suspicious sensor behavior is detected by specially developed algorithms which compute the rate of change, moving average and standard deviation, and other variables. If a fault has occurred, the sensor validation expert system alerts the operator. The diagnostic expert system reads a snapshot of information from the on-line scanner, processes the data, and collects the additional information necessary to diagnose the problem. The objectives of using expert systems for sensor validation is to improve the measurement system's overall reliability and to improve the overall availability of the process measurement. By providing trouble shooting assistance to the maintenance technician, the mean time to repair reduced. Drifting of individual sensors may occur over a long period. Although it may not be detectable by sensor-validation techniques, it can often be detected by data-reconciliation technique. Model-based expert systems involve deep knowledge about the process. Models are derived from first-principal relationships of physics and chemical engineering and from empirical models based on statistical regression of process data. Incorporating this type of deep knowledge into an expert system allows for a very compact and precise system design.

4. CONCLUSIONS

The advantages of expert systems, compared to human experts, include their availability, consistency, and unlimited testability ahead of deployment. Expert systems differ from conventional computer programs in that the domain knowledge can easily be updated without having to modify other portions of the system. Their major disadvantage is that building and maintaining useful expert systems is a long and tedious job.

Expert systems should be aimed at quality control more directly, either to help managing the QC process in manufacturing (e.g., an integrated diagnostic expert system) or designing a product (i.e., off-line quality control) to enhance its inherent reliability. To be useful, an expert system should at least have the following characteristics:

(1) It should be able to help quality engineer to identify quality characteristic and select suitable control chart(s) for the process. The parameters of selected control chart should be economically designed. It should be also possible to use the QCES to verify existing process parameters.

(2) The QCES should act as an experienced quality control engineer to monitor the process and perform the diagnostics.

(3) Training in quality control techniques should be possible if requested by the user.

(4) The efforts to fine tune an existing QCES should be minimized by providing an effective utility tool for the user.

REFERENCES

[1] Alexander, S. M. and Jagannathan, V., 1986, Advisory System for Control Chart Selection, **Computers and Industrial Engineering**, Vol. 10, No. 3, pp. 171-177.

[2] Braun, R. J., 1990, Turning Computer into Experts, **Quality Progress**, Feb. pp. 71-75.

[3] DaGli, C. and Stacey, R., 1988, A Prototype Expert System for Selecting Control Charts, **International Journal of Production Research**, Vol. 26, No. 5, pp. 987-996.

[4] Elimam, A. A., 1989, Computer-aided Quality in Fibre-cement Pipe Production, Int. J. of Production Research, Vol. 27, No. 10, pp. 1743-1756.

[5] Evans, J. R. and Lindsay, W. M., 1988, A Framework for Expert System Development in Statistical Quality Control, **Computers and Industrial Engineering**, Vol. 14, No. 3, pp. 335-343.

[6] Hosni, Y. A. and Elshennawy, A. K., 1988, Quality Control and Inspection: Knowledge-based Quality Control System, **Computers and Industrial Engineering**, Vol. 15, Nos. 1-4, pp. 331-337.

[7] Hubele, N.F. and Keats, J.B., 1987, Automation: The Challenge for SPC, **Quality**, Vol 26, No. 3, pp. 14-22.

[8] Kuo, T.Y., 1989, Control Charts Interpretation System - A Prototype Expert System for Patterns Recognition on Control Charts, Master Thesis, Ohio University, Athens, Ohio.

[9] Love, P.L., and Simaan, M., 1988, Automatic Recognition of Primitive Changes in Manufacturing Process Signals", **Pattern recognition**, Vol. 21, No. 4, pp. 333-342.

[10] Love, P.L., and Simaan, M., 1989, A Knowledge-based System for the Detection and Diagnosis of Out-Of-Control Events in manufacturing Processes, **Proceeding of 1989 American Control Conference** Vol. 3, Published by IEEE, pp. 2394-2399.

[11] Nelson, L.S., 1984, The Shewhart Control Chart - Test for Special Causes, **Journal of Quality Technology**, Vol. 16, No. 4, pp. 237-239.

[12]   Nelson, L.S., 1985, Interpreting Shewhart X-BAR Control Charts, **Journal of Quality Technology**, Vol. 17, No. 2, pp. 114-116

[13]   Pandelidis, I.O., and Kao, J.F., 1990, DETECTOR: A Knowledge-based system for Injection Modeling Diagnostics, **Journal of Intelligent Manufacturing**, Vol. 1, No.1, pp.49-58.

[14]   Pandit, S.M., and Wu, S. M., 1983, Time Series and System Analysis, with Applications, Wiley.

[15]   Rowan, D.A., 1989, On-line Expert Systems in Process Industries, **AI Expert**, August 1989, pp. 30-38.

[16]   Scott, L.L., and Elgomayel, J.I., 1987, Development of a Rule Based System for Statistical Process Control Chart Interpretation, **American Society of Mechanical Engineers, Production Engineering Division (PED)** Vol. 27, pp 73-91.

[17]   Wang, J. H. and Raz, T., 1990, On the Construction of Control Charts using Linguistic Variables, **International Journal of Production Research**, Vol. 28, No. 3, pp. 477-487.

A COMPREHENSIVE APPROACH TO CONTINUOUS PROCESS IMPROVEMENT — CASE STUDY: IDENTIFICATION OF CRITICAL SUBSTRATE PROCESS VARIABLES AFFECTING WIREBONDING

Ingrid J. Vaughan, Brian J. Hogan, and Melvin T. Alexander

Westinghouse, Baltimore, Maryland, U.S.A.

ABSTRACT

Statistical Process Control (SPC), which is an integral part of any Total Quality Management system, has become a general term used when referring to statistical methods relating to optimizing or controlling a process. It is important to realize that traditional SPC utilizing control charts is very useful for monitoring processes that are already stable. However, the real challenge and payoff comes from discovering what the root causes (control factors) of process optimization are, and at what level these factors need to be set to run a controlled process. Only then can proper adjustments be made when a process steps out of statistical control.

To understand and optimize a process, off-line experimentation should be used. Design of experiments (DOE) is one method that has proven very efficient in distinguishing between major and minor contributors of process success and failure. The purpose of design of experiments is to discover the critical variables of a process or product, their effects on variability, and their respective settings.

Off-line experimentation has been used successfully to target and solve hybrid microelectronics assembly problems. Numerous Taguchi Method experiments have been successfully completed to evaluate the wirebond process in the hybrid microelectronics assembly area. The wirebond equipment variables have been optimized, but, much variability remains in the incoming materials being bonded to. This paper specifically discusses the D-optimal experiment designed to identify the critical substrate manufacturing process variables which affect the wirebonding process. In addition, this paper presents an overall discussion of the comprehensive approach used to achieve continuous process improvement.

INTRODUCTION

One of the largest yield detractors in the production of hybrid assemblies is wirebonding. The number of wires on a hybrid assembly greatly exceeds the number of components, thus providing many opportunities for failure. Because of the large number of defect possibilities, the process must be improved to run in the range of parts per million or parts per billion defective. Continuous improvement of the process is essential. To achieve continuous improvement, a well organized approach must be implemented. Figure 1 depicts a four step process to continuous improvement. Each of these steps makes use of various tools and methods, which when used simultaneously, provide a comprehensive approach to continuous improvement. The first step defines the process. The process requirements must reflect the customer's desires and be translated into measurable output characteristics. Tools such as process flow, cause and effect diagrams, and Pareto Analysis are then used to organize the knowledge of the process and to determine the areas of focus. The second step is to determine the process capability. A capability index can be used to compare the current capability to that necessary to run an efficient process. When process capability is below the desired level, the process must be improved.

The third step of continuous improvement is process optimization. A designed experiment is the most economical and most accurate method for performing optimization. A designed experiment will accelerate the learning of the interrelationships of the process variables, determine what variables are critical to the process, and determine at what levels these variables should be set.

Once the process is determined to be capable and controllable, it needs to be monitored. Monitoring the process will not only keep a process in statistical control, but, also show where improvements can be made and when additional experiments are required for further improvement.

The four-step approach has been successfully used to improve the wirebonding process at Westinghouse. The first step was process definition. A multi-disciplined team approach was used to generate the cause and effect diagram shown in Figure 2. A Pareto Analysis of the hybrid assembly area indicated that wirebonding process defects accounted for up to 65% of all defects. The process capability index Cpk was calculated to be approximately 2.0*.

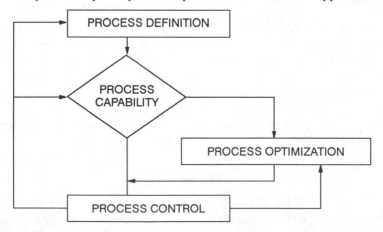

FIG. 1 Four-Step Process to Continuous Process Improvement
* The formula for Cpk is the smaller value of {(USL − MEAN)/3S, (MEAN − LSL)/3S} where USL − Upper Spec Limit, S − Process Standard Deviation, and LSL = Lower Spec Limit.

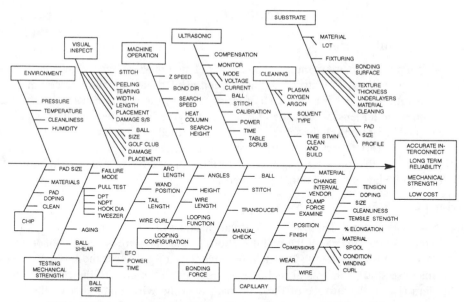

FIG. 2 Thermosonic Ball Bonding – Cause and Effect Diagram

While the process seemed to be capable, the Pareto Analysis indicated additional process improvement was still required. Once these baselines were determined, optimization through design of experiments was performed on the machine specific inputs. These inputs from the cause and effect diagram include machine operation, ultrasonic energy, capillary, bonding force, looping configuration, and ball size. While these experiments achieved significant process improvements, the improvement had been focused on controlling the equipment variables of the wirebonding process. Additional variability reduction was still required. Further Pareto Analysis indicated that the variability of the bonding materials was a major detractor to maintaining a controlled wirebond process. The thick film substrate, gold wire, and chip components were all identified as significant material input factors to the wirebond cause and effect diagram.

In addition, the Pareto Analysis indicated that the majority of wirebond defects occurred at the gold wire/thick film substrate interface (the stitch interface). The thick film substrate was then selected for further investigation. The thick film substrate manufacturing process uses a screen printing operation to apply successive layers of dielectric and gold conductor paste onto a ceramic substrate to form an electrical circuit. The wirebonds are bonded to the gold conductor. Currently, batch acceptance testing for substrate bondability is performed. A sample of each batch undergoes destructive pull tests using go/no go acceptance criteria. This current test does not adequately screen for the requirements truly necessary to achieve parts per billion defect rates. Thus, the goal was established to determine the critical thick film processing parameters that affect the wirebonding process. Once the critical factors are identified and controlled, the substrate variability will be reduced, ultimately improving the wirebond yield.

Following the organized approach, it was also desirable to develop an improved method for monitoring the variability of the substrate material. A preliminary prediction equation was developed which will be used during the batch acceptance testing to compare the actual value to the predicted value. The prediction equation improves the method for detecting inconsistent material over the existing go/no go acceptance method.

CASE STUDY

Thick Film Manufacturing

Thick film substrates are manufactured through batch processing as shown in Figure 3. Layers of conductor and dielectric paste are screen printed onto an alumina ceramic base. The screen printing process is achieved by forcing paste through a stainless steel wire-mesh screen mounted under tension on a metal frame. A polyurethane squeegee applies pressure to the paste moving it through the screen. The screen printed paste dries in a box drying oven at either 100°C or 150°C and is examined for defects. The substrates are subsequently fired in a belt furnace with a peak temperature of 850°C. Successive layers of gold conductors and dielectric insulators form multilevel substrates.

Experiment

Control Factor Selection There are two critical variables of a wirebondable gold pad. First, it is preferred that the surface of the pad is smooth so that the wire is evenly bonded across the pad. Second, it is important to have a high percentage of gold near the surface. The gold paste in this case study contains glass binders which bond the gold to the ceramic base. If these binders rise to the surface of the gold pad, then the wirebondability of the pad decreases.

Figure 4 shows a cause and effect diagram for the thick film process. It was decided to select control factors which had the greatest effect on the surface metal richness. The control factors selected were as follows:

1. Paste Qualification – The gold paste is requalified every three months. When the paste qualification period expires, used paste salvaged from the screen printing process is combined with the unused expired paste and tested for various physical and electrical properties and paste functions. The two levels selected were first qualification paste and requalified paste.

2. Settling Time – The settling time is the time between screen printing the gold paste and placing the substrates in the drying oven. The current minimum settling time criteria is 10 minutes. Since substrates are manufactured through batch processing, the settling time varies for each substrate. The three levels selected were 10 minutes, 30 minutes, and 60 minutes.

3. Drying Time – The minimum drying time for gold paste is 10 minutes. Using only the minimum criteria, some batches may stay in the oven over a break period or longer. The three levels selected were 10 minutes, 30 minutes, and overnight.

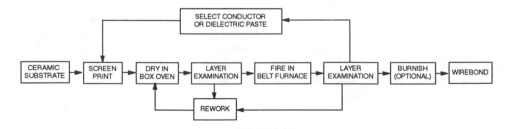

FIG. 3 Thick-Film Manufacturing Process Flow

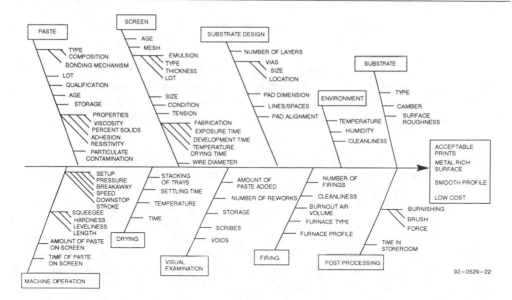

FIG. 4 Thick-Film Manufacturing – Cause and Effect Diagram

4. Furnace Type – The facility has three furnaces which meet the required gold paste firing profile. These furnaces vary in belt widths, belt lengths, and peak temperature tolerances. The three levels selected were Furnace 1, Furnace 2, and Furnace 3.

5. Burnout Air Volume – The air volume mixture in the burnout zone of the furnace is a variable in the firing process. The air volume depends upon such factors as the air flow setting and the number of substrates being fired at one time. An equation calculates the recommended burnout air volume for each furnace [1]. The three levels selected were "less than recommended air volume," "recommended air volume," and "greater than recommended air volume."

6. Refirings – If a defect is found and paste is added to the substrate after one firing, the substrate must be refired. There is currently no limit to the number of times a substrate can be refired. When the gold paste is exposed to many firing profiles, the glass binder tends to rise and reduces the metal richness of the surface. The three levels selected were zero refirings, one refiring, and two refirings.

7. Burnishing – Depending upon the requirements of the substrate, gold pads are hand treated with a burnishing brush. While burnishing smooths the profile of the pad, it also removes gold from the substrate. If too much gold is removed, the surface will contain a higher percentage of the glass binder. The three levels selected were no burnishing, light burnishing, and heavy burnishing.

Design Selection A D-optimal experimental design was used across the seven factors to minimize the fraction of stitch interface failures and to maximize the stitch pull strength. The D-optimal design was chosen because limited funds would not allow conventional experimental design methods to be performed. Budgeted funds limited the study to 40 trials. This limitation placed a major restriction on the type of experimental design used. Taguchi orthogonal array and standard factorial designs were also considered. These designs could not get the desired precision levels nor the degree of orthogonality to satisfy the objectives of the study.

The nearest Taguchi L_n orthogonal array would have been a L_{18} (2^1*3^7), i.e., one two-level factor and seven three-level factors. The actual factors being investigated had one less three-level factor than the factors required by the Taguchi L_{18} array, (i.e., a 2^1*3^6 factorial design). The missing three-level factor made the construction of a pure orthogonal array impossible for the number of trials with the remaining factors. Taguchi method advocates tend to advise against not following his prescribed arrays. Although many of Taguchi's orthogonal arrays were originally developed by statisticians in the 1940s and 1950s, they have been criticized by the statistical community as being flawed. They fail to handle confounding effects of interacting factors properly. Confounding means that estimated effects from factors cannot be isolated, or separated from other factors they interact or mix with. Therefore, it is difficult to tell if the significant effects are from the factors themselves or the factors they are confounded with.

Dr. Taguchi has argued that interactions may purposely be confounded if knowledge of the interactions is known. In practice, knowledge of interactions is rarely known in advance. They may be guessed, but will not be confirmed until experimental results are available. In many experiments, interaction effects are usually added to the experimental design. If confounding occurs, additional experimental runs would be needed to separate the effects. For more information, see [2], [3], and [4].

The smallest standard fractional factorial design would require 243 runs. This was more than six times the 40 runs budgeted for this experiment. The next reasonable experimental design approach was to use an optimal experimental design. The most widely used optimal experimental design method was the D-optimal design. A D-optimal designed experiment uses the degrees of freedom to analyze the desired effects with the least number of runs, with minimum confounding effects, and with minimum correlation between factors. D-optimal designs are used in situations where:

- not all possible combinations of the factor levels are feasible,

- resource limitations restrict the number of experiments that can be performed,

- nonstandard factorial, nonlinear models, or irregularly-shaped experimental regions statistically describe or explain the response characteristics of interest.

To generate the layout of the D-optimal design, the OPTEX procedure of the SAS/QC® software was used. Table 1 lists the factor settings for the 40 runs of the D-optimal design. For more information on D-optimal designs see [5], [6], [7], and [8].

SAS/QC® is the registered trademark of SAS Institute Inc., Cary, NC.

TABLE 1.
Control Factor Settings and Results

Observation	Settling Time (min)	Drying Time (min)	Burnout Air Volume	Furnace Type	Refirings	Burnishing	Paste Qualification	Stitch Pull Strength Run 1	Run 2	Fraction of Stitch Failures Run 1	Run 2
1	60	overnight	greater than	3	1	none	first	7.62	7.71	0.6000	0.5542
2	60	overnight	recommended	1	0	light	requalified	7.94	7.54	0.3500	0.4000
3	60	overnight	less than	2	2	heavy	first	8.02	8.08	0.2815	0.2833
4	60	overnight	less than	1	1	light	requalified	7.77	7.23	0.4167	0.6042
5	60	30	greater than	2	2	light	first	7.76	8.19	0.2292	0.2625
6	60	30	greater than	1	1	none	requalified	5.92	7.02	0.6917	0.6917
7	60	30	greater than	1	0	light	first	3.21	8.56	0.4292	0.3958
8	60	30	recommended	3	1	heavy	requalified	7.90	7.12	0.3333	0.5583
9	60	30	less than	3	2	heavy	first	7.93	8.36	0.3333	0.5208
10	60	10	greater than	2	0	heavy	requalified	3.09	8.30	0.5167	0.4833
11	60	10	greater than	2	2	light	requalified	7.78	7.67	0.3167	0.6000
12	60	10	recommended	2	0	heavy	first	7.20	7.98	0.1000	0.4750
13	60	10	recommended	1	0	none	first	7.55	6.99	0.8750	0.8583
14	60	10	greater than	3	2	none	requalified	6.87	6.85	0.7542	0.8417
15	30	overnight	greater than	3	2	heavy	first	N/A	N/A	N/A	N/A
16	30	overnight	greater than	1	2	heavy	requalified	8.72	8.54	0.3708	0.4208
17	30	overnight	recommended	2	1	none	first	8.26	8.10	0.4583	0.5000
18	30	overnight	less than	3	0	light	requalified	6.77	6.70	0.8583	0.8000
19	30	30	recommended	1	1	heavy	first	7.44	7.70	0.1833	0.5083
20	30	30	recommended	1	2	light	requalified	7.22	7.55	0.4733	0.7208
21	30	30	less than	2	2	none	requalified	7.61	7.13	0.5375	0.6680
22	30	30	less than	2	0	none	first	8.35	8.49	0.3958	0.2833
23	30	10	greater than	3	1	none	requalified	6.09	6.54	0.9328	0.6555
24	30	10	greater than	2	0	heavy	requalified	7.94	7.86	0.5792	0.5750
25	30	10	recommended	2	2	light	requalified	8.59	8.22	0.4958	0.6125
26	30	10	less than	3	2	light	first	7.89	7.66	0.8000	0.5583
27	30	10	less than	1	1	light	first	7.25	8.87	0.3125	0.5292
28	10	overnight	greater than	1	2	none	requalified	8.03	7.03	0.4833	0.8250
29	10	overnight	recommended	3	0	light	first	7.67	7.97	0.5417	0.3000
30	10	overnight	recommended	2	2	none	first	7.88	8.06	0.3958	0.2500
31	10	overnight	less than	2	1	heavy	requalified	8.48	8.47	0.1292	0.0833
32	10	30	greater than	3	0	light	requalified	7.45	6.88	0.6599	0.5208
33	10	30	greater than	2	1	light	first	8.73	8.77	0.1833	0.2250
34	10	30	recommended	3	2	none	requalified	8.70	8.14	0.4542	0.2500
35	10	30	less than	1	0	heavy	requalified	6.32	7.10	0.8333	0.6500
36	10	10	greater than	3	1	light	first	7.46	8.61	0.6120	0.5833
37	10	10	greater than	1	2	heavy	first	7.62	7.04	0.1125	0.7917
38	10	10	recommended	3	1	heavy	requalified	7.86	7.70	0.2917	0.4667
39	10	10	less than	2	1	light	requalified	8.11	7.23	0.3458	0.5667
40	10	10	less than	1	0	none	first	8.33	8.25	0.6250	0.5833

Data Collection A test pattern was designed to minimize the effects of uncontrollable noise factors. A standard pad size of 7 mils by 30 mils was selected with equal number of horizontal and vertical pads. A total of 240 pads were printed on each substrate. In order to get a more representative sample, each run had two substrates. A total of 19,200 wires were bonded and pull tested for the experiment. The pull strength and failure mode for each wire were recorded. Since the purpose of the experiment was to study the effect of thick film processing on the wirebondability of the gold pad, the failures at the gold wire substrate interface were studied. The fraction of stitch failures for each run and the average pull strength of these stitch failures are summarized in Table 1.

Data Analysis The methods used to analyze the data results were Analysis of Variance (ANOVA) and Regression. Regression and ANOVA provided summaries of the data, quantified the relationships between the response variables and factors, and identified the levels of factors. For more information on regression and ANOVA, see [9], and [10].

Results

The statistically significant factors and levels were first paste qualification, heavy burnishing, and Furnace 2 (Tables 2 and 3). The significant negative effect of the requalified paste could possibly be caused by the age of the paste or by contamination from the reused paste (Figures 5 and 6). No burnishing had a significant adverse effect and had more variability than light or heavy burnishing since bonding pads of nonburnished substrates tended to have more irregular profiles (Figures 7, 8, 13, and 14). Furnace 2 had a significant positive effect, but further investigation is required to explain this significance (Figures 9 and 10).

Interaction plots tend to show that light burnishing with first paste qualification had the largest average stitch pull strengths. Heavy burnishing and first paste qualification had the smallest stitch failure fractions (Figures 11 and 12). Burnishing tends to have more significant interactions than the other factors. We believe that burnishing alters the physical substrate surface and, therefore, confounds the response of the other factors being investigated. STATGRAPHICS®, software was used to perform the regression and ANOVA analyses.

Prediction Equation

The primary objective of the experiment was to select factors and levels that gave the lowest fraction of stitch failures and highest pull strengths. Choosing a prediction model was a secondary objective. A first stage regression model was developed. Typical D-optimality designs use a quadratic (second order) regression model of the form:

$$Y = b_o + \sum_{i=1}^{k=7} b_i x_i + \sum_{i \neq j}^{k=7} b_{ij} x_i x_j + \sum_{i=1}^{k=7} b_{ii} x_i^2 + e$$

$$Y = \begin{bmatrix} \text{Pull strength} \\ \text{stitch failure fraction} \end{bmatrix} \text{response vector,}$$

$$X_i \ , \ X_j \ = \ \left[X_{1 \ (Settle)} \ / \ X_{2 \ (Drying \ Time)} \ / \ X_{3 \ (Air \ Volume)} \ / \ X_{4 \ (Furnace)} \ / \ X_{5 \ (Refire)} \ / \ X_{6(Burnishing)} \ / \ X_{7 \ (Paste \ Qualification)} \ / \right] \ ,$$

$b_i \ , \ b_{ij} \ , \ b_{ii} \ = \ regression \ coefficients, \ and \ e = error \ term$

A confirmation experiment would be required to test the validity and predictability of the model. Once validated, the regression model can be used to successfully predict estimated values during substrate batch acceptance testing. Monitoring by this method provides a better indicator of substrate process control compared to the current attributes acceptance testing.

TABLE 2
Analysis of Variance for Stitch Fraction

SOURCE OF VARIATION	SUM OF SQUARES	D.F.	MEAN SQUARE	F–RATIO	SIG. LEVEL
MAIN EFFECTS	1.3542455	7	0.1934636	8.555	0.000
QUAL	0.1992547	1	0.1992547	8.811	0.0045
BURNISH	0.4073459	2	0.2036730	9.006	.0.0004
FURBACE	0.4764291	2	0.2382146	10.534	0.0001
DRY	0.2855355	2	0.1477677	6.584	0.0029
2–FACTOR INTERACTIONS	0.6885978	18	0.0882554	1.692	0.0716
QUAL & BURNISH	0.0891579	2	0.0195790	0.866	0.4267
QUAL & FURNACE	0.399552	2	0.0199276	0.881	0.4204
BURNISH & FURNACE	0.2408643	4	0.0602161	2.663	0.0427
QUALY & DRY	0.0088203	2	0.0044101	0.195	0.8234
BURNISH & DRY	0.2439140	4	0.0609785	2.696	0.0407
FURNACE & DRY	0.1422154	4	0.0355538	1.572	0.1956
RESIDUAL	1.1759589	52	0.0226146		
TOTAL (CORR.)	3.2188022	77			

92-0529-2/BW

TABLE 3
Analysis of Variance for Pull Strength

SOURCE OF VARIATION	SUM OF SQUARES	D.F.	MEAN SQUARE	F–RATIO	SIG. LEVEL
MAIN EFFECTS	7.9153522	7	1.1307646	6.015	0.000
QUAL	2.7077590	1	2.7077590	14.404	0.0004
BURNISH	0.8598343	2	0.4299172	2.287	0.1117
DRY	0.5259356	2	0.2629678	1.399	0.2560
FURNACE	3.5648789	2	1.7824394	9.482	0.0003
2–FACTOR INTERACTIONS	12.050591	18	0.6694773	3.561	0.0002
QUAL & BURNISH	3.286640	2	1.6433201	8.742	0.0005
QUAL & DRY	0.272248	2	0.1361241	0.724	0.4896
BURNISH & DRY	3.697349	4	0.9243373	4.917	0.0019
QUALY & FURNACE	0.491319	2	0.2456595	1.307	0.2794
BURNISH & FURNACE	0.649220	4	0.1623051	0.863	0.4921
DRY & FURNACE	3.200214	4	0.8000534	4.256	0.0047
RESIDUAL	9.7750529	52	0.1879818		
TOTAL (CORR.)	29.740996	77			

92-0529-3/BW

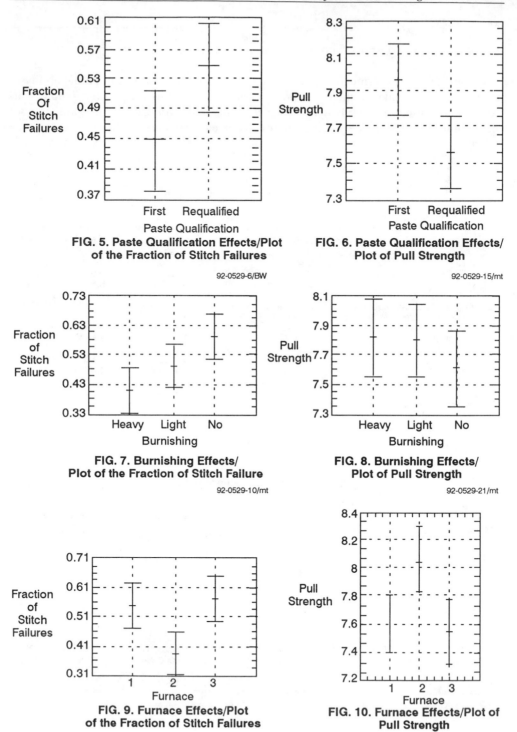

FIG. 5. Paste Qualification Effects/Plot
of the Fraction of Stitch Failures

92-0529-6/BW

FIG. 6. Paste Qualification Effects/
Plot of Pull Strength

92-0529-15/mt

FIG. 7. Burnishing Effects/
Plot of the Fraction of Stitch Failure

92-0529-10/mt

FIG. 8. Burnishing Effects/
Plot of Pull Strength

92-0529-21/mt

FIG. 9. Furnace Effects/Plot
of the Fraction of Stitch Failures

92-0529-7/mt

FIG. 10. Furnace Effects/Plot of
Pull Strength

92-0529-17mt

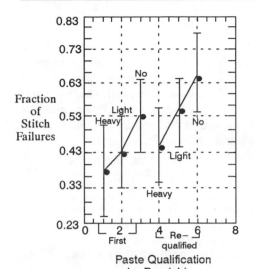

FIG. 11. Interaction of the Fraction
of Stitch Failures for Paste
Qualification by Burnishing

92−0529−11/mt

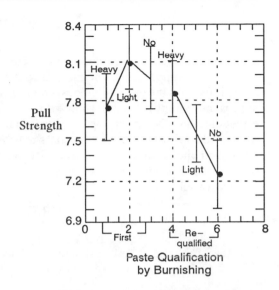

FIG. 12. Pull Strength Interactions
Plot of Paste Qualification and Burnishing

92−0529−18/mt

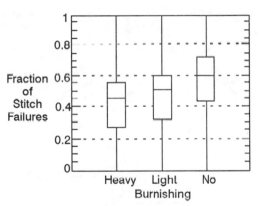

FIG. 13. Burnishing Box Plots of the
Fraction of Stitch Failures

92-0529-9/mt

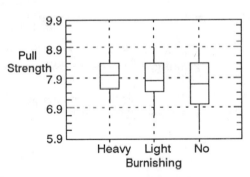

FIG. 14. Pull Strength Box
Plots of Burnishing

92-0529−20//mt

SUMMARY

In order to achieve continuous improvement, a well organized approach must be utilized. The four step approach is one method which achieves continuous improvement. The case study illustrates the many iterations required through the four step approach. The systematic approach is simplified by starting with a well defined process definition using a cause and effect diagram. Design of experiments is then an essential tool to optimize the process and identify the critical process parameters to be monitored. After one set of critical inputs are optimized and controlled, it is necessary to continue improvement by identifying the next set. A prediction equation can then be developed and used as a quantitative measurement of process variability.

REFERENCES

[1]Roffey M, Metke N: "Atmospheres and Their Effects on Air Firing Thick Films," *Hybrid Circuit Technology*, pp. 45–49, November, 1990.

[2]Lochner RH: "Pros and Cons of Taguchi," *Quality Engineering*, 3, pp. 537–549, 1991.

[3]Montgomery DC: "Factorial Experiments and Other Methods for Process Improvement," *Introduction to Statistical Quality Control, 2nd ed.*, New York, John Wiley and Sons, Inc., 1991, pp. 483–547.

[4]Pignatiello JJ, Ramberg JS: "Top Ten Triumphs and Tragedies of Genichi Taguchi," *Quality Engineering*, 4, pp. 211–225, 1992.

[5]Schmidt SR, Launsby RG: *Understanding Industrial Experiments, 2nd ed.*, Longmont, CO, CQG Ltd Printing, 1989.

[6]Nachtsheim CJ: "Tools for Computer-Aided Design of Experiments," *Journal of Quality Technology*, 19, pp. 132–160, 1987.

[7]Snee RD: "Computer-Aided Design of Experiments," *Journal of Quality Technology*, 17, pp. 222–236, 1985.

[8]Mitchell TJ: "An Algorithm for the Construction of D-Optimal Experimental Designs," *Technometrics*, 20, pp. 203–210, 1974.

[9]Berenson ML, Levine DM, Goldstein M: "Selection of an Appropriate Regression Model," *Intermediate Statistical Methods and Applications*, Englewood Cliffs, NJ, Prentice-Hall Inc., 1983, pp. 365–395.

[10]Draper N, Smith H: *Applied Regression Analysis, 2nd ed.*, New York, John Wiley and Sons, Inc., 1981.

THE IMPORTANCE OF ERGONOMICS IN THE CONCURRENT ENGINEERING PROCESS

Robert R. Wehrman, Terry L. Ross,* and Robert J. Marley

University Technical Assistance Program, Industrial and Management, Engineering Department, Montana State University, Bozeman, Montana, U.S.A.

ABSTRACT

Tremendous gains have been realized with the recent implementation of the concurrent engineering process. These include faster lead-times to market, higher quality products, better product designs, lower total life-cycle costs, etc. [1,2]. Unfortunately, the ergonomist seldom has contributed input to the simultaneous engineering process. This suboptimizes the resulting product and manufacturing process in two ways. First, the benefits of proper product and work design are not realized throughout the complete life-cycle and second, this limits the application of ergonomic principles to post-hoc intervention or redesign.

This paper recommends the addition of ergonomics or human factors engineering expertise to the concurrent engineering team. Ideally, this individual should possess the technical and creative design skills obtained through formal engineering training as well as competence in operations management. Coupled with a strong knowledge of occupational ergonomic and man-machine principles, this individual will be a valuable asset in meeting the goals of concurrent engineering. Significant reductions in product life-cycle costs are realized through the application of ergonomic principles. The addition of the ergonomist will amplify the successes that are being achieved through the application of concurrent engineering. This paper discusses the concurrent engineering process as a filtering procedure and describes the advantages of incorporating ergonomic expertise in the design team.

(*) Now with Pacific Northwest Laboratories, Richland, Washington 99352.

INTRODUCTION

Concurrent engineering is a design process in which all aspects of the product from conception to disposal are considered. This is synonymous with simultaneous engineering. Concurrent engineering has established its effectiveness in reducing time to market, manufacturing cost, improving quality, etc. Primarily this is due to the influx of creative thought in the early stages of the design process allowing the design team to filter out inferior solutions before resources are committed to their development. This is critical since much of the total life-cycle costs of a product or service are committed during the design process. T h e selection of the design team is important for establishing an effective concurrent engineering program. Ideally, all disciplines that potentially have a positive impact should be represented on the design team. Realistically, selection is based on the resources of the company and the ability for the team to be managed effectively. This puts management in a dilemma about whom to select. There are some obvious disciplines such as marketing, engineering, management, and production that must be included to form the core of the design team, but there are a myriad of other disciplines that could be included to embellish the team.

The major difference between concurrent versus sequential engineering process is shown in Figure 1 [1,2]. The addition of more disciplines during the design phase infuses more creativity into the total knowledge of the project. This includes creative input from the customer as well as vendors that supply raw materials and components. Thus, it can be seen that the knowledge curve has moved up to better match the impact of the life-cycle cost commitment. Thus, more knowledge is available during the conceptual design phase where most of the life-cycle costs are committed.

This increase in knowledge is used to influence the design of the product before decisions are made that reduce the design freedom. If these are made based on an increased breadth of knowledge, then there will be an increase in the probability of a better decision. Therefore, the goal of concurrent engineering is to increase the knowledge base during the conceptual design phase of the project.

CONCURRENT ENGINEERING AS A FILTER MODEL

Concurrent engineering can be modeled as a filtration system that filters out the solutions that are infeasible and/or suboptimal. Figure 2 shows this model in macro form. Assume that the tank is filled with sand. This sand represents all the possible outcomes of a design. Thus, all of the creative ideas generated by the design team are depicted by the number of grains of sand in the model. An idea's merit is represented by the size of the sand in reverse proportion. Larger grains of sand are ideas or solutions that have not been sufficiently developed. Smaller grains of sand are ideas or solutions that have been developed to a point where the concept is feasible and meets the requirements. The smaller the grain, the better the solution.

It follows that the most preferred solution is the smallest grain of sand in the system. The quantity of sand in the system is a probability function based on the

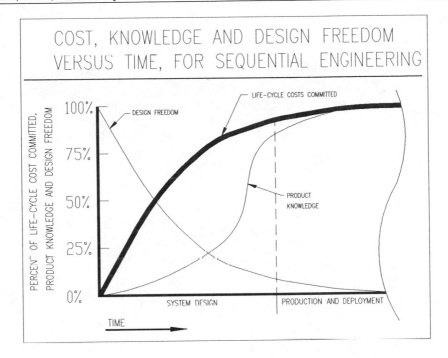

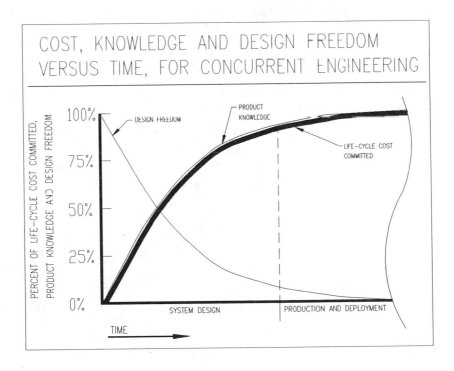

Fig. 1 The theoretical knowledge and design freedom
curves for sequential and concurrent engineering.
Adapted from [1].

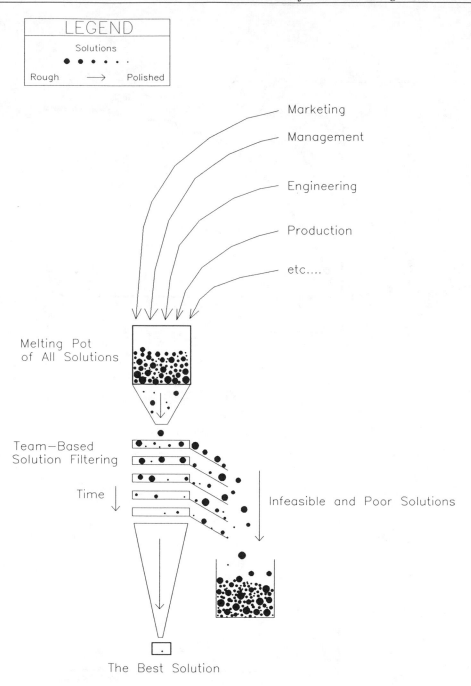

Fig. 2 The basic concurrent engineering filter model.

number of people on the design team as well as their mix of talents. Each member of the team deposits sand into the system as the free flowing ideas are generated in the start of the design process. This can be seen in Figure 2.

Suppose there are an infinite number of members on the team with all the possible disciplines represented. Then the system will be filled with an infinite quantity of sand. This is unrealistic in practice since there are limited resources available as well as a limited amount of ideas generated. The converse is true: if only a few members representing one or two disciplines are on the team, then only a few ideas will be generated. This notion of breadth, discussed earlier, is the inherent flaw of the sequential design process. Management must then face the dilemma about who will participate as a member of the design team to optimize the probability of creating the best solution.

After the design team has generated the ideas (or solutions) that represent the sand in the system, the team proceeds to filter out the ideas that are not practical. This is analogous to filtering the sand with screens of different mesh sizes and configurations. At each phase of the project the team will filter out the solutions deemed inappropriate. In reality, it is virtually impossible to actually find the smallest grain of sand due to all of the resource constraints. One can only develop an environment that will allow the best chance of finding the smallest grain of sand possible. The end result of the design process will result in one solution falling out at the bottom.

Just as each member of the design team contributes a quantity of sand to the system, the team members serve as the fibers of the filter. This has the effect of collectively filtering the sand in a parallel process. Each pass of the filter represents the parallel decision making process at that level of detail, eliminating infeasible solutions. Figure 3 shows a crude representation of this model filter.

Each person on the design team represents a portion of the total knowledge base for the project. The consistency of the resulting filter is a function of the collective knowledge of the team. Any absence of knowledge will result in weak points or holes in the filter. These holes will allow ideas to be accepted by the team that would have been filtered out if that knowledge were available in the team. A consistent and comprehensive team is then the key to the concurrent engineering process. This allows the design team to rapidly reduce the domain of possible solutions to a manageable number before major amounts of resources are committed to solution analyses.

There is a feedback loop in the model that represents the synergistic effect that the team members develop as solutions are filtered out. This feedback loop can be seen in Figure 4. Solutions that are filtered out may spark the team to develop additional solutions that are improvements on the original. This is analogous to crushing the rejected solutions, filtering out the results, and then recycling the feasible concepts back into grains of sand which are then put back into the system. It should be noted that the filter following the crushing process would perform at the same level as the current filter being utilized by the team. This feedback loop will permanently eliminate the ideas that are obviously infeasible, while preserving solutions that still have potential for incorporation.

One could view this crushing of ideas as "getting to the root of the idea" and separating out all but the useful core. For example, Value Engineering attempts to

DEVELOPMENT OF THE CONCEPTUAL FILTER
(Top View)

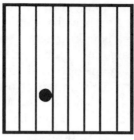

Filtering Limits With
Only Engineering

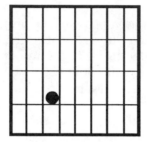

With Engineering and Production

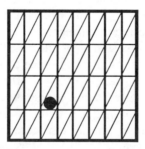

With Engineering, Production,
and Marketing

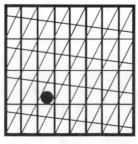

With Engineering, Produciton,
Marketing, and Mangement

LEGEND	
I	Engineering
—	Production
/	Marketing
\	Management

Fig. 3 Generic representation of a filter.

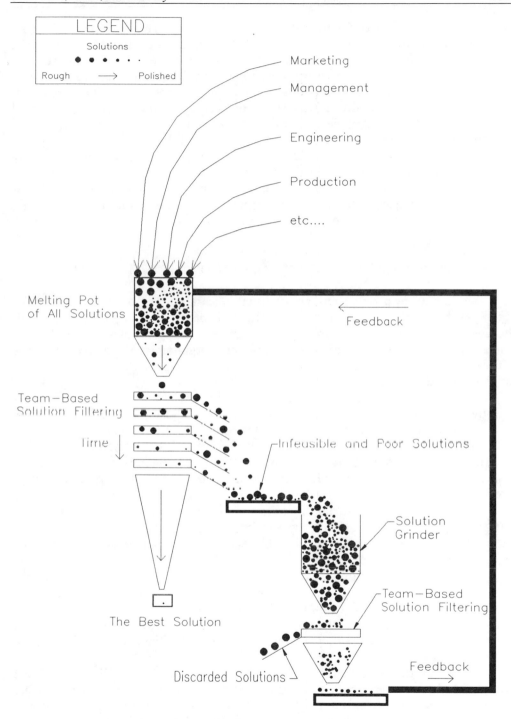

Fig. 4 The concurrent engineering filter model
with feedback.

achieve this by the use of functional analysis [3]. This process illuminates the root functions of a process or physical element. The intent is to only keep the useful functions and eliminate the useless functions as they serve only to add cost.

The final decision between alternatives usually does not fall on the team members. This decision is normally reserved for the next higher level of management. Theoretically, if there are no defects in the previous filters, there will be very little deviation between the utility of these alternatives. This has the effect of reducing the risk of accepting a grossly suboptimal design.

This same process will continue through the implementation and ultimately to the final disposal of the product at the end of its useful life. This portion of the process is not represented in the model. It is easy to see that the team which is responsible for implementation will also generate and filter solutions to unforseen problems throughout the life-cycle of the product.

ERGONOMICS AS AN INTEGRAL PART OF THE DESIGN TEAM

When many people think about ergonomics, they may think about how the dashboard was designed for their car. This type of end-user ergonomics is valid; however, it is only part of the total picture. Ergonomics spans all aspects of human interaction with the physical world. The term ergonomics is derived from the Greek words "nomos" of laws and "ergos" of work. Simply put, ergonomics is the laws of work.

Strictly speaking, an ergonomist is one who specializes in the understanding of how humans can best interact with the physical world. In the context of an industrial setting, an ergonomist specializes in how best to enhance the workers' efficiency in building and distributing a product, as well as enhance the buyers' impression of the product. It is also implied that the ergonomist will decrease the risk of injury and discomfort of both the worker and the user. The ergonomist thus has a positive effect on the profitability of a product from all aspects.

In many cases, industrial ergonomic analysis has been confined to post-hoc "fixes" to problems that already exist. For example, consider a product that is already in production and has assembly stations that put workers at risk for injuries.

From a design standpoint, the above scenario is just the type of problem that has caused so much criticism of the sequential engineering process. Typically, large amounts of resources are committed in the implementation and life-cycle of a project because of the grossly suboptimal design. The concurrent engineering process attempts to address these problems while it is still inexpensive to make the necessary design changes.

The authors of this paper believe that the ergonomic discipline is a critical component in the concurrent engineering process. The design team will most likely have representatives from manufacturing. These representatives will specify the requirements and limitations of the equipment either currently owned or for possible purchase. Any solution would be suboptimal without this representation from manufacturing.

Humans as well as the machines are elements of manufacturing systems. Just as a manufacturing engineer specifies capabilities and limitations of machines, an ergonomist can specify the requirements and limitations of humans. In the context of the filter model of the concurrent engineering process, the absence of the manufacturing representative would be manifested in large holes in the filters relating to equipment specification, for example. Thus too, absence of ergonomic expertise indicates a hole in the filter.

This logic is applicable to the addition of the ergonomist as a member of the design team. With the current state of manufacturing, humans are still a major factor in production, maintenance, and distribution of products and services. The human is still the most intelligent and flexible element in the manufacturing environment.

Historically, there has been a high cost associated with neglecting the ergonomic principles applied to the product and manufacturing process. This requires that a high priority be placed on design and processes to fit humans better if true increases in productivity are to be realized in the future.

Manufacturing engineers strive to design products that are easily assembled by robots. Examples are the use of alternative fasteners, vertical assembly, and reduced material handling. These concepts are consistent with the use of human capability as well.

There is a considerable amount of overlap with respect to what is advantageous for automated as well as human assembly. A preferred solution may be to design a product with the ability for both human and robotic assembly. This will allow management the flexibility to start production with human assembly and progress to robotic assembly as soon as economics show it to be advantageous. This gives management larger flexibility, and faster reaction to market changes.

Other major advantages of designing for the human worker will be the reduction of injuries. There is a definite monetary advantage to reducing worker injuries with a proactive design process that strives to eliminate the hazardous elements of the product design and manufacturing process. This advantage will take the form of reduced workmans' compensation claims, increased worker efficiency, and reduced legal penalties. These serve to increase the profit of the project. More important than the monetary gains are the ethical implications of such actions. All professionals have the responsibility to protect the workers who make a product as well as the end-user. The best way to accomplish this is to consider ergonomic principles in the design stage of the product and process.

With respect to the filter model, an ergonomist will add additional breadth to the design team in three ways. First, the ergonomist can contribute creative solutions to the domain of all possible solutions available to the team. Second, the ergonomist will also serve to fill large holes that exist in the traditional concurrent engineering design team, enhancing the filtration of suboptimal solutions regarding man-machine systems. Finally, the ergonomist will help the team develop the synergistic effect of refining the ideas into more preferred ones. The end result will be a design team that will have a higher probability of finding a solution to the design problem that is closer to global "optimality."

There are pitfalls in selecting ergonomic expertise for the design team. Ergonomics has become a new buzz word in industry in the last few years. That is both positive and negative. It is positive in the sense that people are more aware of the issues regarding efficiency, safety, and quality of life for workers. This benefits companies economically but there are intangible benefits as well, such as industry reputation, labor relations, etc.

The negative aspect is a result of self-proclaimed ergonomists that are out to make money on the wave of interest in the field. Many of these ergonomists make recommendations with a gross disregard for professional ethics. They have no formal training and often do not base their recommendations on facts that were generated through the application of sound scientific principles. The result is pseudo-random solutions that in some cases increase the risk of injury to the customer or worker. The most visible evidence of this is the marketing of "ergonomically designed" items such as chairs, desks, computer equipment, and tools. Many of these items, when scrutinized, do not exhibit any of the accepted principles of ergonomic design.

It is therefore critical that care be given to their legitimate qualifications if ergonomists are to play a positive role in the concurrent engineering design process. To date, there is no standardized professional registration process to regulate the ergonomic discipline. Until such standards are developed, management can use some of the following simple guidelines to help select a qualified ergonomist. In addition, these guidelines can be used to select an appropriate course of action in the professional development of existing staff professionals to serve as qualified ergonomists.

Formal education is the best source for developing skills as an ergonomist. In many cases, specialized ergonomics training is received at the graduate level. Most industrial engineering programs and a growing number of psychology programs offer some amount of coursework at the undergraduate level as well. All formal education in this field must contain applied science methodologies. Ergonomics is the application of science to solve problems relating to man-machine systems. Engineers are uniquely suited for this task as they are already familiar with such principles as engineering mechanics, materials, systems analysis, and stochastic processes. An engineer with a bachelor's degree has had a minimum of 16 semester hours of design credits in addition to 32 hours of engineering science credits (required for any ABET accredited program).

An effective ergonomist should have formal training and/or experience in as many of the following areas as possible (dependent somewhat on nature of particular business):

- Manufacturing and design principles,
- Man-machine principles,
- Biomechanics,
- Work physiology,
- Psychophysics,
- Human perception processes,
- Anthropometry,
- Workstation & tool design,
- Industrial Hygiene,
- Safety engineering.

In addition to these basic topics, an ergonomist should be aware of enforceable standards. OSHA enforces several standards. Knowledge of these standards allow products and processes to be designed to conform to them. This will save the company money in the long run, since costly redesign would be avoided. Use of the National Institute of Occupational Safety and Health (NIOSH) guidelines for manual material handling and cumulative trauma disorders will aid in conformance with future OSHA standards.

CONCLUSION

The philosophy and goals of concurrent engineering provide an important starting point for minimizing the life-cycle costs of a product. The goals of ergonomics are exactly consistent with concurrent engineering. This paper explored in detail the complementary relationship between ergonomics and concurrent engineering.

There exist a finite number of disciplines required to make a satisfactory concurrent engineering team. Examples of such disciplines include engineering, marketing, management, and production. It is proposed that the ergonomic discipline be added to the list of disciplines that comprise the core team. Furthermore, ergonomic expertise may come from one of these, such as engineering.

It was shown how addition of the ergonomic discipline to the concurrent engineering team can result in an increase of creative solutions regarding the man-machine interface as well as earlier rejection of solutions not ergonomically (or scientifically) correct.

It is important to select a knowledgeable and experienced ergonomist to be on the design team. This individual must have a demonstrated ability to solve ergonomic problems correctly and cost effectively. It was noted that many self-proclaimed ergonomists do not to possess the skills needed to contribute effectively to the concurrent engineering process. Thus, management must screen the ergonomist before adding the individual to the team.

Finally, it is recommended that if an organization deems it impractical to hire or purchase legitimate ergonomic services, steps be taken to identify and train appropriate staff members in the principles and application methodologies of ergonomics.

REFERENCES

1. Jack, D., Boeing Company Presentation: "Simultaneous Engineering", Given at Montana State University, 1991.
2. Allen, W. C., <u>Simultaneous Engineering: Integrating Manufacturing and Design, Society of Manufacturing Engineers</u>, Dearborn, Michigan, 1990.
3. Mudge, A. E., <u>Value Engineering: A Systematic Approach</u>, McGraw-Hill, Inc., New York, 1971.

4. Coleman, J. R., "NCR Cashes In on Design for Assembly", Hitchcock Publishing Co., 1988.
5. Goldstein, G., "Integrating Product and Process Design"
6. Stoll, H. W., "Simultaneous Engineering in the Conceptual Design Phase", SME Simultaneous Engineering Conference, 1988.

PHYSIOLOGICAL AND SUBJECTIVE RESPONSES TO ROBOTS IN A NOISY ENVIRONMENT

[1]Jeffrey E. Fernandez, [1]Don E. Malzahn, [1]Brian K. Lambert, and [2]Liang-Sung Wei

[1]*Wichita State University, Wichita, Kansas, U.S.A.*
[2]*Chinese Naval Academy, Tsoying, Taiwan, Republic of China*

ABSTRACT

This study examined the effects of robot arm speed and ambient noise level on the operator's reaction time (RT) and mean heart rate (MHR) response to the moving robot arm. Results indicated that there were statistically significant effects of noise and robot arm speeds on an operator's RT and MHR. The RT varied inversely with robot arm speed whereas MHR varied directly with robot arm speed independent of noise level. A teaching speed of 20 cm/sec is recommended based on both physiological and subjective responses, whereas a teaching speed of 30 cm/sec is recommended based on reaction time.

INTRODUCTION

An increasing number of studies indicate that robots are a hazard in industry. The introduction of robotics has led to work-related accidents and injuries [1-3]. As the use of robots increases, it is important that the safety of operators, maintenance personnel, and technicians be given consideration.

Individuals who program and/or maintain robots, expose themselves to hazards because they work in close proximity to robots. Specifically, the teaching pendant method of robot programming is a job requiring close human-robot interaction. A malfunction or unintentional operation might result in a robot movement that is quicker than the operator's reaction time in releasing the teaching button and/or operating an emergency stop. Clearly, this teaching work poses particular problems concerning safety. Therefore, appropriate guidelines must be provided to make sure that the teaching speed will not endanger the programmer when using the teaching pendant for programming.

The Robotic Industries Association (RIA) [4] has published a standard, ANSI R15.06, which states: "All robots shall have a slow speed while teaching". It also states that the maximum slow speed of any part of the robot shall not exceed 25 cm/sec" (RIA 1984). The teaching speed of 14 cm/sec is endorsed by the Japan Industrial Safety and Health Association [5]. The latter recommendation was derived through a human factors experiment [6]. The RIA 25 cm/sec speed limit is not experimentally based. The choice of appropriate arm speed therefore deserves more research.

Previous studies concerning safe programming speeds were obtained in highly "sterile" laboratory conditions without the presence of noise or other environmental distractions. Distractions might affect the operator's attention and response to the robot speed. Levosinski [7] states that noise, temperature, and smelly air provide powerful distractions in the industrial environment which may result in a decrease in the operator's alertness and mental efficiency. Therefore, the previous studies dealing with robot speed may be misleading since most teaching operations are conducted in a noisy industrial environment.

There are no teaching speed standards that have been accepted and utilized throughout the robotics industry. Therefore, the objectives of this study were:

1. To examine the interrelations between robot speeds and noise levels, and the programmer's reaction time and heart rate response to the moving robot arm.

2. To propose an appropriate teaching speed limit and to determine whether the teaching speed limit of 25 cm/sec proposed by the Robotic Industries Association (RIA) [4] or the 14 cm/sec suggested by the Japan Industrial Safety and Health Association (JISHA) [5] is appropriate.

METHODS

Subjects

Twenty male engineering student volunteers participated in this study. The subjects had an age of 33 $\pm$ 4 years and resting heart rate of 76.8 $\pm$ 10.9 bpm (mean $\pm$ standard deviation). All subjects were given hearing (acuity) tests for the frequency range of 500 to 8000 Hz. Only subjects with hearing thresholds less than 35 dBA for all 7 frequencies measured were selected. Also, all subjects reported having normal visual acuity, corrected or uncorrected. Informed written consent was obtained from the subjects prior to their participation in the study.

In order to reduce the effects of learning and hazards in teaching a robot, all subjects were required to make two 1-hour visits to the laboratory. During these visits all subjects were familiarized with the function of the teaching pendant and then practiced using the teaching pendant at various robot arm speeds.

Procedure

A PUMA Mark II industrial robot, 500 series, with a teach pendant and computer controller was used for this study. During the experimental sessions, each subject was exposed to 30 repeated two-dimensional rectangular movements of the robot arm. Subjects were not told that they were operating outside the working envelope of the robot. The subjects' fingers were required to be placed on the teaching button. Four of the movements, randomly inserted in the control program, were movements toward the subject. The subjects were instructed to release the teaching button and hit an emergency stop button on the top of the teach pendant as soon as they perceived the robot arm moving toward them. Subjects' reaction times and heart rate responses were collected. Six robot arm speeds (10, 15, 20, 25, 30, and 35 cm/sec) and three noise levels (65, 75, and 85 dBA) were studied. The frequency of the industrial noise ranged from 500 to 8000 Hz. A tape recorder was used to present the recorded noise at the three levels. Sound level readings were checked at subject's ear level with a QUEST #215 Sound Level Meter.

Reaction time was measured as the time interval from the start of the robot arm movement toward the subject until the subject pressed the emergency stop button on the top of the teach pendant. This data was automatically collected by a programmable timer in the robot control program. Heart rate was monitored continuously throughout the procedure by telemetry heart rate monitor (UNIQ Heartwatch) and recorded every 5 seconds. Physiological stress of the task was determined by the percent of the observed average heart rate to the predicted maximum heart rate. Maximal heart rate was estimated as 205-(age/2) [8]. Many researchers agree that the interpretation of a stressful task can be made if the heart rate achieved is 85% or more of the predicted maximum [9-11].

At the end of each session, each subject evaluated their perceived level of psychological stress using the checklist of perceived exertion. This checklist was based upon the work of Pearson [12].

Experimental Design

The experimental design was a randomized complete block design with subjects as blocks. The eighteen treatment combinations (6 speeds of robot arm movement x 3 levels of noise) were randomized under each block (subject). The four movements for the data collection per trial were treated as replications in each cell.

RESULTS AND DISCUSSION

The 20 male subjects were randomly assigned to the 18 experimental conditions. The summary of the mean physiological and subjective data for the 20 subjects for the 6 speeds and 3 noise levels is presented in Table 1.

Results of Reaction Time (RT)

In order to test for differences between RT values (milli-sec) versus noise and robot arm speeds, a complete two-way randomized block design for an analysis of variance was specified (subjects as blocks). The analysis of variance for RT is presented in Table 2.

The ANOVA in Table 2 clearly shows that there were significant main effects of subject, noise and speed. The subject effect indicates significant differences in RT among subjects. The noise and speed effects indicate significant differences in RT

TABLE 1.

Mean Physiological and Subjective Data of the Subjects

Responses	Noise level (dBA)	Speeds (cm/sec)					
		10	15	20	25	30	35
Reaction time	65.0	687.0	663.0	645.0	621.0	604.0	620.0
(msec)	75.0	705.0	681.0	662.0	634.0	627.0	651.0
	85.0	724.0	704.0	675.0	652.0	649.0	674.0
Mean HR	65.0	91.2	86.9	85.2	87.2	92.3	96.4
(bpm)	75.0	93.3	89.1	87.4	89.5	95.1	99.3
	85.0	96.2	92.2	90.4	92.7	97.6	103.1
Subjective	65.0	26.6	25.4	24.8	27.5	29.6	33.2
rating	75.0	30.6	30.3	28.9	31.6	34.1	38.6
(score)	85.0	36.6	35.8	33.5	37.3	40.1	42.7

TABLE 2.

Analysis of Variance for Reaction Time

Source	DF	SS	F	PR > F
Subjects	19.0	4.4314	159.04	< 0.01
Noise	2.0	0.3595	122.58	< 0.01
Speeds	5.0	1.0921	148.94	< 0.01
Noise x Speeds	10.0	0.0169	1.15	0.32
Error	1403.0	2.0574		
Total	1439.0	7.9573		

among the three levels of noise and six levels of arm speed, respectively. However, there were no significant interaction effects between noise and speed.

To investigate the differences between the three levels of noise and six levels of speed factors, post-hoc analysis using Duncan's Multiple Range test was performed (alpha = 0.05). Results of this procedure showed significant differences between each of the three levels of noise and each of the six levels of speed. The results above support the hypothesis that RT will vary as a result of different noise levels and robot arm speeds, as shown in Figure 1.

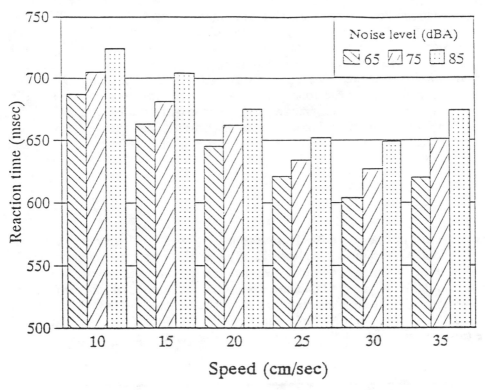

FIG. 1 Reaction Time vs. Robot Arm Speed at 3 Noise Levels

The figure illustrates a curvilinear pattern of RT as a function of robot arm speed. At noise levels of 65 and 75 dBA, RT decreases steadily as speed increases from 10 to 30 cm/sec, then increases after 30 cm/sec. On the other hand, at noise level of 85 dBA, RT decreases steadily from 10 to 25 cm/sec, then appears to be nearly level from 25 to 30 cm/sec and then increases after 30 cm/sec. Robot movements toward the subject produced shorter reaction time at the faster speeds up to 30 cm/sec than at the slower speeds.

The increase in reaction time after 30 cm/sec could be attributed to the increased arousal and reduced acuity at higher arm speed which could delay the human perceptual discriminability. This phenomenon could also be explained by the subjective opinions on items 2 (blurred vision) and 4 (become nervous) of the Checklist of the Perceived Exertion where subjects rated very high scores for the arm speed of 35 cm/sec regardless of noise levels.

The results of this study are consistent with the other experiments which investigated the perceptibility of robot arm movement as a function of speed [13]. To test the learning or practicing effects, a within-trial experimentwise comparison was performed using Duncan's Multiple Range test (alpha = 0.05). There was no significant effect of the within-trial factor for RT.

Finally, a linear relationship was found between the robot arm speed and the distance the robot moved before it was stopped by the emergency stop button (overrun distance) as shown in Figure 2. The overrun distance was calculated by taking 1.6449 standard deviations (for the 5th percentile population) above the mean reaction time shown in Table 1. As can be seen, with noise levels of 65, 75 and 85 dBA, at a robot speed of 35 cm/sec, the mean overrun distance was 24.52, 26.12 and 26.76 cm; and at 10 cm/sec it was 8.22, 8.42 and 8.64 cm, respectively. Although the arm movements toward the subjects were easier to perceive at the faster speeds, the hazard increased with the speed of the robot arm. Based on current study conditions, it is suggested that the overrun distance be a minimum required safety distance of the operator from the end of the robot arm while teaching it under specified conditions. Hence, the corresponding teaching speed could keep the robot from striking a worker if he acted to stop it.

Results of Mean Heart Rate (MHR)

An analysis of variance for MHR is presented in Table 3. There were significant main effects of subject, noise, and speed. The subject effect indicates significant differences in MHR among subjects. The noise and speed effects indicate significant differences in MHR among the three levels of noise and six levels of robot arm speed. There were no significant interaction effects between noise and speed. Post-hoc analysis using the Duncan's Multiple Range test (alpha = 0.05) showed significant differences between each of the three levels of noise and each of the six levels of speed.

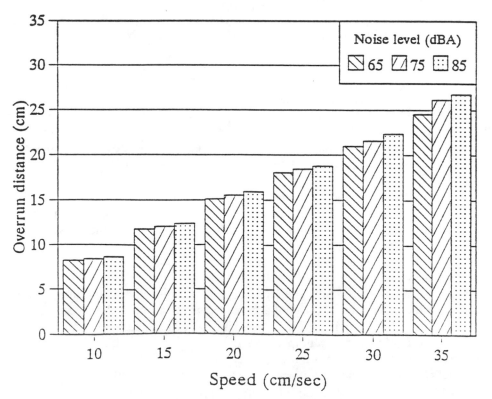

FIG. 2 Overrun Distance vs. Robot Arm Speed at 3 Noise Levels

TABLE 3.
Analysis of variance for Mean Heart Rate

Source	DF	SS	F	PR > F
Subjects	19.0	70546.75	852.06	< 0.01
Noise	2.0	7323.58	840.31	< 0.01
Speeds	5.0	23438.98	1075.76	< 0.01
Noise x Speeds	10.0	73.91	1.70	0.08
Error	1403.0	6113.78		
Total	1439.0	107496.99		

Mean heart rate (MHR) response obtained from this study were below 85% of estimated maximum HR, therefore the task would be considered non-stressful [11]. Plotting MHR against speed, showed a tendency for MHR to decrease as speed increased from 10 to 20 cm/sec, then increase after 20 cm/sec as shown in Figure 3.

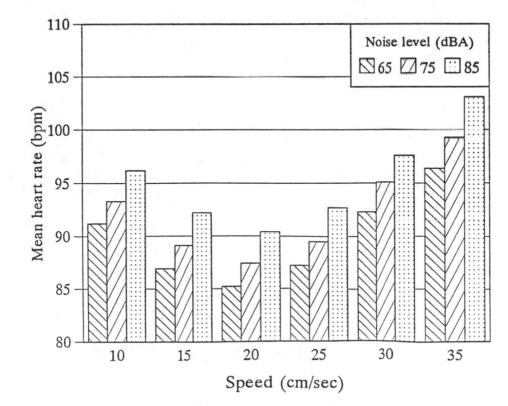

FIG. 3. Mean Heart Rate vs. Robot Arm Speed at 3 Noise Levels

MHR exhibited a roughly U-shape pattern over speeds at all three levels of noise. This might imply that if the robot arm is moved too slowly, annoyance or boredom may result; if the robot arm is moved too fast, nervousness or tension may occur. However, Haider [14] stated that a state of irritation and aversion to activity which provokes annoyance or boredom may put the person involved under increasing internal tension.   Therefore, both extremes of low and high robot arm speeds may lead to higher heart rate when the operator is teaching a robot. It is therefore implied, based on MHR responses, that a 20 cm/sec teaching speed produces the least physiological stress.

Interestingly, the result of this finding (20 cm/sec) is approximately the average of the American Standard (25 cm/sec) and the Japanese Standard (14 cm/sec). It is very likely that neither 25 cm/sec nor 14 cm/sec minimizes the operator's physiological response.

The results also support the hypothesis that MHR will vary as a result of different noise levels and robot arm speeds, as can be seen graphically in Figure 3.

Again, to test the learning or practicing effects, a within-trial experimentwise comparison was performed using the Duncan's Multiple Range test (alpha = 0.05). There was no significant effect of the within-trial factor for MHR.

The results of this study are similar to the findings of the study by Gerber [15] which investigated whether adults displayed alterations of cardiac rate under acoustical stimulus conditions. Likewise, the findings of the present study showed an increase in HR with the presentation of increased noise levels indicating some influence on HR by noise stimuli.

Results of Subjective Opinions

The subjective opinions of fatigue were measured by quantifying the degrees of perceived exertion of 12 fatigue symptoms on the checklist of perceived exertion (Table 4). The ANOVA indicated that there were significant main effects of subject, noise, and speed. Duncan's Multiple Range test was performed (alpha = 0.05) as a post-hoc analysis to investigate the differences between each of the three levels of noise and six levels of speed. Results of this procedure showed significant differences between each of the three levels of noise and six levels of speed.

The total subjective response scores plotted against speed and noise are depicted in Figure 4. It can be seen in Figure 4 that the subjects' responses revealed a slightly decreased level of subjective stress as speed increased from 10 to 20 cm/sec, then an increased level of subjective stress after 20 cm/sec.   Subjects seemed to prefer moderate levels of speed as compared to extremes of low or high robot arm speeds. It is therefore suggested that a 20 cm/sec teaching speed is appropriate based on the minimization of subjective stress.

The results of this subjective evaluation are consistent with the subjects' physiological measure (i.e., MHR) in which the subjects had lower physiological responses at the speed level of 20 cm/sec.

TABLE 4.
The Checklist of the Perceived Exertion

<u>Note</u>: (1) For the questions below, the higher the ranking, the greater the stress. The choices should be rated independently, not additively, for each session.

(2) On a scale from one to five, indicate the degree of stress/fatigue you perceived, during the session you just completed.

		1	2	3	4	5
1.	Heavy eye strain	—	—	—	—	—
2.	Blurred vision	—	—	—	—	—
3.	Feeling of heaviness in the head	—	—	—	—	—
4.	Feeling of nervousness caused by arm speeds	—	—	—	—	—
5.	Feeling of nervousness caused by noise levels	—	—	—	—	—
6.	Difficulty in perceiving risky arm movements toward you	—	—	—	—	—
7.	Difficulty in concentration/ attention caused by noise levels	—	—	—	—	—
8.	Concentration required to perform the task	—	—	—	—	—
9.	Work being dull because of arm speeds	—	—	—	—	—
10.	Dislike work environment because of noise levels	—	—	—	—	—
11.	Dislike work environment because of arm speeds	—	—	—	—	—
12.	Stiffness in the shoulder, arm and hand caused by holding the teach pendant	—	—	—	—	—

Comments:

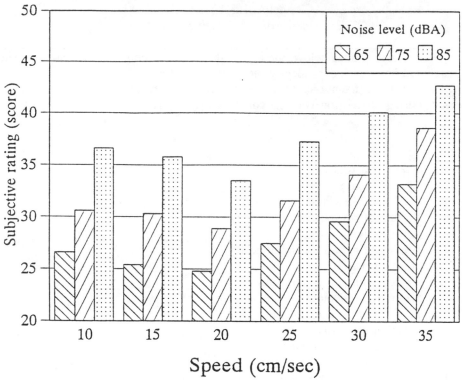

FIG. 4 Subjective Rating vs. Robot Arm Speed at 3 Noise Levels

CONCLUSIONS

The conclusions drawn from the results of this study can be summarized as follows:

1. The environmental noise and robot arm speed produced a significant effect on programmer's reaction time, and mean heart rate response.

2. The reaction time varied inversely with robot arm speed from 10 to 30 cm/sec, then increased after 30 cm/sec. This increase could be attributed to the increased arousal and acuity loss at higher arm speed which could delay the human perceptual discriminability. Therefore a teaching speed of 30 cm/sec is recommended based on minimum reaction time.

3. Mean heart rate response decreased with robot arm speed from 10 to 20 cm/sec, then increased after 20 cm/sec. Therefore a teaching speed of 20 cm/sec is recommended based on the physiological response (i.e., MHR).

4. The subjective responses also showed a slight tendency to decrease with robot arm speed from 10 to 20 cm/sec, then increase after 20 cm/sec. Therefore a teaching speed of 20 cm/sec is recommended based on the subjective responses.

5. The recommended robot arm speed for a particular criteria (RT, MHR, or subjective rating) is the same regardless of the noise level.

6. A linear relationship exists between the robot arm speed and the overrun distance. This indicates that the hazard increases with the robot arm speed for a specified distance away from the robot arm.

REFERENCES

1. NIOSH: Fatal Accident Summary Report: Die cast operator pinned by robot. *Summary Report 84-020.* NIOSH, Division of Safety Research, Morgantown, WV, 1984.

2. Bonney MC: Robot Safety, in Yong YF (ed), Springer-Verlag, Berlin, 1985.

3. Primovic J, Karwowski W: Effects of automation on safety performance. *Proceedings of the 8th International Conference on Production Research.* Cincinnati, Ohio, 1987.

4. Robotic Industrial Association (RIA): Proposed American national standard for industrial robots and robot systems. Dearborn, MI, 1984.

5. Japan Industrial Safety and Health Association: An Interpretation of the technical guidance on safety standards in the use, etc., of industrial robots. Tokyo, Japan, 1985.

6. Sugimoto N, et al.: Collection of papers contributed to conferences held by the Machinery Institute of Japan, No. 8445. 1984.

7. Levosinski GJ: Teach control pendant for robots. *Proceeding of the 1984 International Conference on Occupational Ergonomics*, 1984.

8. Hakki AH, Hare TW, Iskandrian AS, Lowenthal DT, Segel BL: Prediction of maximal heart rates in men and women. *Cardiovascular Review and Reports*, vol 4, ed 7. 1983, pp 997-999.

9. McCarthy DM, Blood DK, Sciacca RR, Cannon PJ: Single dose Myocardial imaging with Thallium-201: Application in patients with nondiagnostic Electro-cardiographic Stress Tests. *American Journal of Cardiology*, vol 43. 1979, pp 899-906.

10. Dash H, Massie BM, Botrinick EH, Brundage BH: The Non-invasive identification of left main and three vessel coronary artery disease by Myocardial Stress Perfusion Scientigraphy and treadmill exercise Electrocardio-graphy. *Circulation*, vol 60. 1979, pp 276-283.

11. Boucher CA, Zir LM, Beller GA, Okada RD, McKusick KA, Strauss HW, Pohost GM: Increased lung uptake of Thallium-201 during exercise Myocardial imaging. *American Journal of Cardiology*, vol 46. 1980, pp 189-196.

12. Pearson RG: Scale analysis of a fatigue checklist. *Journal of Applied Physiology*, vol 41. 1957, pp 186-191.

13. Nagamachi M: Human factors of industrial robots and robot safety management in Japan. *Applied Ergonomics*, vol 17, ed 1. 1986, pp 9-18.

14. Haider M: Ermudung, Beanspruchung und Leistung, Franz Deuticke, Wien, 1962.

15. Gerber SE, Mulac A, Lamb ME: The Cardio-vascular response to acoustic stimuli, *Audiology*, vol 16. 1977.

TRENDS IN THE USE OF PSYCHOPHYSICS IN INDUSTRIAL WORK DESIGN

Robert J. Marley

Industrial and Management Engineering, Montana State University, Bozeman, Montana, U.S.A.

ABSTRACT

The study of psychophysics can be traced to the 1850's when scientists began to consider the quantitative measurement of human sensation as it relates to a physical stimuli. The classical approaches to psychophysical measurement are the method of constant stimuli, the method of limits, ratio scaling, and the method of adjustment. In general, industrial applications of psychophysics make use of the Stevens Power Law, which allows the human worker to make comparative judgements regarding physical stimuli. These stimuli typically involve attributes of a work system. For example, size, shape, texture, color, and mass and inertial characteristics of a load to be handled. Other related examples could include frequency and duration of tasks. This paper discusses previously successful applications of psychophysics in industrial work design. In addition, current trends are explored. These range from improving product design and evaluation, to assessment of upper extremity stress in jobs with risk of cumulative trauma. Special emphasis is placed upon incorporating psychophysical methodologies to help solve modern industrial work design problems.

INTRODUCTION

Psychophysics is the science which is concerned with quantifying a human sensation related to a physical stimuli [1]. Central to early psychophysical investigations was the development of a method of measurement which could describe how sensations differ in both quality and degree. For example, vision is different from hearing and pain different from smelling sensations. But degrees of vision and hearing sensation vary from acute to dull and from loud to soft, respectively. In general, two principles govern sensory reaction to stimuli--the "more than" and the "different from" principles.

By the mid-20th century and due in large part to the work of S.S. Stevens, the method of measurement which had gained greatest acceptance was the psychophysical power law. According to Stevens [1], nature has favored use of the power law in psychophysics because the power function is a common form throughout physics and other sciences. The basic psychophysical formula is stated in Equation 1 as follows:

$$S = kI^n \tag{1}$$

where,   S = Magnitude of human sensation
   k = Constant that depends upon the unit of
       measurement of the stimulus
   I = Intensity of physical stimulus
   n = Exponent which is experimentally determined
       for each sensory continuum.

An alternative formulation for the power function is represented by plotting the power curve in log-log coordinates. This allows using the slope of the straight line to be used as a direct measure of the exponent. This formula is given in Equation 2 as follows:

$$\log(S) = \log(k) + (n)\log(I) \tag{2}$$

where,   S = Magnitude of human sensation
   k = Constant dependent upon unit of measurement of
       the stimulus
   I = Intensity of physical stimulus
   n = Slope of power function when plotted in log-
       log coordinates

Classical Methods of Psychophysics

Gescheider [2] reports that there may be a temptation by some to conclude that sensations can be measured directly as units of sensory magnitude. Rather, sensation must be treated as a concept which is defined in terms of stimulus-response relationships. The classical methods of defining these relationships were outlined by Gescheider [2] and are summarized in the following sections. These methods may be generalized for any stimulus continuum.

Method of Constant Stimuli. The method of constant stimuli refers to a procedure in which a stimuli is presented at between 5 and 9 different levels (equal intervals) of intensity. At the lower end of intensity, the stimulus should be such that it would almost never be detected. At the upper end, the intensity should be detected almost always. As the various stimulus intensities are presented, preferably at random, positive (yes) or negative (no) detections by the subject are recorded. The intensity value which has a proportion of positive detections of 0.5 represents the "absolute threshold."

The method of constant stimuli can also be used to determine the "difference threshold," or more importantly, the "point of subjective equality" (PSE). The PSE is determined by multiple comparisons of a standard stimuli (Sst) and a comparison stimuli (Cst) which changes from trial to trial. Similar to the absolute threshold case, 5 to 9 intensity levels of the Cst are paired (randomly) with the Sst with half of the Cst's being below the Sst and half above. Subjects are asked to state whether the Cst is "greater" or "lesser" than the Sst (even when they cannot determine if a difference exists). When a value of Cst has an equal proportion of "greater than" and "lesser than" subjective ratings, then this value is considered the PSE. Thus, the PSE represents a psychometric function where the value of the comparison stimulus is subjectively equal to the standard stimulus.

Method of Limits. The method of limits is frequently used in determining both absolute and difference thresholds because of it's ease of use, although it is generally considered less precise than constant stimuli techniques. A typical example of the use of the method of limits is in determining the absolute threshold of hearing. Generally, this audiometric technique uses several trials of ascending (A) and descending (D) stimulus (tone signal) generation. The (A) series begins at a intensity level which is certain to be undetected by the subject and is increased until the subject can detect the signal. Conversely, the (D) series begins at an intensity which may be detected without difficulty and is reduced until the subject can no longer detect the signal. The absolute threshold is then considered the average value for signal detection across all trials.

Similarly, in a hearing experiment, the method of limits is used for determination of difference thresholds by allowing the subject to respond "greater" or "lesser" when comparing a standard stimulus intensity with a comparison stimulus in ascending and descending trials. Typically in these experiments, the subject is also allowed to determine an "equal" intensity resulting in an interval of uncertainty in which the Sst and Cst are considered subjectively equal.

Method of Adjustment. The major feature of the method of adjustment is that is allows the observers (subjects) to control the intensity of the stimulus necessary to measure their threshold. This method is generally thought to produce reliable results because of the active participation of the subjects maintains high performance on their part [2].

The absolute threshold determined by the method of adjustment is derived by the experimenter initially setting the stimulus well above, or well below, the estimated threshold and then allowing the subject to increase or decrease the intensity until the sensation "disappears." This process is normally repeated several times and the absolute threshold is considered the mean of these final settings. The difference threshold is determined by adjusting the comparison stimulus until the PSE is reached with the standard stimulus. This is in similar fashion to previously mentioned methods only with the subject afforded direct control over adjustment of the comparison stimulus.

Ratio Scaling. The measurement of physical properties on ratio scales has always been desirable since these scales can contain the characteristics of order, distance, and origin while retaining a strong correlation with a numbered system [2]. The most common forms of ratio scaling involve the methods of ratio production, ratio estimation, magnitude production, and magnitude estimation.

Ratio production, or fractionation, refers to the process of the subject adjusting the comparison stimulus intensity to a prescribed ratio of the standard stimulus such as one-fourth, one-third, or one-half, for example. Ratio estimation is essentially the inverse operation of ratio production. This process requires the subject to estimate the apparent ratio between two stimulus sensations.

Magnitude production refers to the experimenter providing a numerical value of some sensory magnitude and requiring the subject to adjust the stimulus in order to reproduce that value. On the other hand, magnitude estimation refers to the procedure where the subject is asked to provide a direct numerical estimation of a stimulus magnitude. Stevens [1] provides a partial list of exponents which have been experimentally determined. Table 1 summarizes the psychophysical relationship between several stimulus conditions and the sensory magnitude.

PSYCHOPHYSICAL RESEARCH RELATED TO WORK DESIGN

The psychophysical approach has become a widely used methodology in ergonomics, particularly in the determination of acceptable workloads for the manual handling of materials [3,4,5,6]. Genaidy, Asfour, Mital, and Tritar [7] reviewed the psychophysical models which have been developed for manual materials handling (MMH), particularly those which have appeared since the publication of the NIOSH lifting guide [8]. These models were based upon regression equations and fuzzy set modeling.

Stover Snook [6] reported that the design of MMH tasks following psychophysical criteria can reduce back injuries in industry by up to one-third. The advantages and disadvantages of using psychophysical methods for developing criteria for permissible workloads in manual handling tasks were enumerated by Snook [5] and are summarized below in Table 2.

The most common psychophysical approach to the design of MMH tasks is the method of adjustment. Typically, this involves asking workers to perform the task with the objective of achieving the highest level possible while not ". . . straining yourself, or without becoming unusually tired, weakened, overheated, or out of breath" [22]. Thus, a worker performs the task and makes adjustments in the workload over a specified period of time until he/she feels confident that they have achieved the given objective. The adjustable workload parameters are typically load, frequency, duration, or range. The workload that is psychophysically determined is then referred to an "acceptable" workload for the given task conditions.

TABLE 1
Psychophysical Exponents Of Several Stimulus Conditions
Related To Sensory Magnitude

Continuum	Stimulus Condition	Measured Exponent
Loudness	Sound pressure at 3kHz	0.67
Vibration	Amplitude on finger (60 Hz)	0.95
Vibration	Amplitude on finger (250 Hz)	0.60
Brightness	5 degree target in dark	0.33
Brightness	Point source	0.50
Brightness	Brief flash	0.50
Brightness	Point source flashed briefly	1.0
Lightness	Reflectance of gray paper	1.2
Visual length	Projected line	1.0
Visual area	Projected square	0.7
Saturation (red)	Red-gray mixture	1.7
Taste	Sucrose	1.3
Taste	Salt	1.4
Taste	Saccharine	0.80
Smell	Heptane	0.60
Cold	Metal contact on arm	1.0
Warmth	Metal contact on arm	1.6
Warmth	Irradiation of small skin area	1.3
Warmth	Irradiation of large skin area	0.70
Cold discomfort	Whole body irradiation	1.7
Warm discomfort	Whole body irradiation	0.70
Thermal pain	Radiant heat on skin	1.0
Tactual roughness	Rubbing emery cloths	1.5
Tactual hardness	Squeezing rubber	0.80
Finger span	Thickness of blocks	1.3
Pressure	Static force on palm	1.1
Muscle force	Static contraction	1.7
Heaviness	Lifted weights	1.45
Electric shock	Current through fingers	3.5
Angular acceleration	5 second rotation	1.4

Source: Stevens [1].

TABLE 2
Advantages and Disadvantages of Psychophysics
in Manual Material Handling Activities

Advantages:

1. Psychophysics permits realistic simulation of industrial tasks.

2. Psychophysics can be used to study intermittent tasks that are common in industry (steady state physiological responses are not required).

3. Psychophysical results are very reproducible.

4. With the exception of very high frequency tasks, psychophysical criteria tend to agree with physiological criteria for continuous work.

5. Psychophysical results are consistent with the industrial engineering concept of a "fair day's work for a fair day's pay."

Disadvantages:

1. Psychophysical criteria from very high frequency tasks are typically higher than recommended physiological criteria. Thus, permissible loads for such tasks should be based upon the metabolic criteria.

2. Psychophysics is a subjective method relying upon self-report. Therefore, it could be replaced by more objective measures that may be developed in the future.

3. In manual materials handling, psychophysics does not appear to be sensitive to bending and twisting motions of the trunk (though recent research has been successful in establishing guidelines for asymmetrical lifting using psychophysical criteria [14]).

Perceived Exertion

One outcome of psychophysical research that has been widely used in many areas including exercise physiology and ergonomics is the development of scales for perceived exertion [9,10]. Gunnar Borg [11] stated that the rating of perceived exertion (RPE) scale was developed as an attempt to incorporate the known power functions for heavy work and to minimize natural drawbacks of ratio scaling methodologies.

Regarding the former, the exponent for heavy physical work determined in early perceived exertion studies was found to be 1.60 [11]. The main drawback to ratio scaling is that this method often does not provide for good inter-subject comparison. This is because different individuals will typically assign very different values to the same physical stimuli due to differences in experience, sensory acuity, tolerance, etc.. Borg's solution to this problem was to assign verbal "anchors" to the scale. Thus, the resulting 15-grade RPE scale is provided in Table 3.

TABLE 3
Rating Of Perceived Exertion (RPE) Scale

6	
7	Very, very light
8	
9	Very light
10	
11	Fairly light
12	
13	Somewhat hard
14	
15	Hard
16	
17	Very hard
18	
19	Very, very hard
20	

Source: Borg, [9]

As noted earlier, the RPE scale (sometimes known as the Borg scale) has been used extensively in exercise testing, exercise prescription, and in ergonomics research, particularly for physiologically demanding work such as frequent lifting [12,13,14,15,16].

Much of the popularity of the RPE scale can be attributed to its strong relationship with physiological measures such as heart rate, oxygen consumption, lactate concentration in working muscles, and ventilation and respiration rates and therefore its ease of use in clinical settings [10]. For example, heart rate can be approximated by the RPE scale using a factor of 10 (i.e., an RPE of 13 would be associated with a heart rate of 130). However, Borg [11] noted that heart rate is dependent upon many other variables such as age, type of work, and environmental conditions as well.

CURRENT TRENDS IN PSYCHOPHYSICAL MODELING

Industrial Design Applications

McCallam reviewed recent psychophysical methodologies in industrial design applications, particularly consumer product perception [17]. Examples of these include the successful use of magnitude estimation to assess customer ratings of sweetness, tartness, and flavor intensity in an effort to find a product differentiation for a soft-drink manufacturer. Ratings were also used to establish customer preferences for automobile interiors based upon identified physical attributes [17].

McCallam also outlined a nine point structured approach for developing a psychophysical model for complex consumer products [17]: 1) Identify Design Goals and Engineering Constraints (initial delineation of the product attributes and perceptual responses); 2) Determine Range of User Perception (content of user descriptions can be scrutinized); 3) Determine Critical Attribute Ranges (initial estimate of relationship between attribute value and sensory magnitude); 4) Select Product Designs (compromises between the cost of prototype development and psychophysical experimentation must be made here); 5) Measure Consumer Perceptions of Products (typically magnitude estimation used); 6) Measure Physical Attributes of Products (objective criteria for measurement is preferred); 7) Identify Perceptual Factors (reducing all perceptual factors to an adequate set); 8) Develop Psychophysical Model (use results to develop quantitative relationship between physical attribute and perceptual factors); and 9) Develop Design Aids (to be used by future designers and engineers).

Industrial Ergonomics Applications

Recently, the psychophysical approach was proposed as a valid method to examine work involving the upper extremity, analogous to its application in manual handling activities [18]. Initial laboratory investigations have demonstrated that the method of adjustment can be used to find acceptable task frequencies for drilling activities under various wrist postures and force requirements [19,20]. Results from these studies indicate that workers can reliably adjust the frequency of a drilling task based upon the perception of the combined biomechanical and physiological stresses. Psychophysically based models were then developed to aid work designer in establishing acceptable workloads for these activities with the objective of reducing the risk of cumulative trauma disorders in the upper extremities. Such models may be particularly useful when ergonomically preferred task designs cannot be achieved or are not reasonable.

FUTURE TRENDS

It is certainly anticipated that use of the psychophysical approach will continue to grow in years to come, particularly in the arena of industrial ergonomics and work design. An area of application requiring research is in the refinement of RPE scales. Borg's original scales were developed to assess whole-body exertion while reducing inter-subject variation by using the verbal anchors. The standard RPE scale, as well as modified versions (log-scales), have been used to assess perceived exertion in localized body parts with varying degrees of success. However, it is reasonable to question the use of the original anchors for these purposes. It is proposed that research be conducted to update the set of verbal anchors which might more accurately reflect perceptual factors related to localized effort, such as in upper extremity work. As well, the psychophysical exponents for effort in various local muscle groups should be determined (for use in the Power Law equations).

Furthermore, due to the lack of proven objective criteria for the evaluation of cumulative trauma risk to date, it is suggested that the psychophysical approach be further examined as a method to determine population capacities to perform work that contains this risk. Such knowledge of capacity can be used initially as a reference for acceptable levels of stress. This would be analogous to the Job Severity Index (JSI) approach used in manual material handling evaluation [21]. As more objective biomechanical and physiological criteria are developed, they may be incorporated with psychophysical results into comprehensive guidelines for upper extremity work. Further, as these models and guidelines are refined, they will also provide the industrial engineer more information by which to make decisions regarding automation (e.g., when the manual capacity of a worker has been exceeded and redesign for manual performance is not feasible).

CONCLUSIONS

This paper has examined the science of psychophysics as it has been refined and applied to the design of industrial tasks and consumer products. Specific methodologies for psychophysical measurement and the successful use of psychophysics in industrial ergonomics were explored. Current applications for both ergonomics and product design were also outlined in addition to some future needs for psychophysical research. It is believed that psychophysics will play an important role in the design of future end-user and industrial systems including automated and hybrid systems.

REFERENCES

1. Stevens, S.S.: <u>Psychophysics: Introduction to its Perceptual, Neural, and Social Perspectives</u>. New York: John Wiley & Sons, 1975.
2. Gescheider, G.A.: <u>Psychophysics: Method, Theory, and Application</u>. Hillsdale, NY, Lawrence Erlbaum, 1984.
3. Ayoub, M.M.: The problem of manual material handling, in Asfour, S.S. (ed): <u>Trends in Ergonomics/Human Factors IV</u>. North-Holland, Elsevier, 1987.
4. Fernandez, J.E. and Ayoub, M.M.: The psychophysical approach: The valid measure of lifting capacity, in Asfour, S.S. (ed): <u>Trends in Ergonomics/Human Factors IV</u>. North-Holland, Elsevier, 1987.
5. Snook, S.H.: Psychophysical considerations in permissible loads. <u>Ergonomics</u> 28(1):327-330, 1985.
6. Snook, S.H.: Psychophysical acceptability as a constraint in manual working capacity. <u>Ergonomics</u> 28(1):331-335, 1985.
7. Genaidy, A.M., Asfour, S.S., Mital, A., and Tritar, M.: Psychophysical capacity modeling in frequent manual materials handling activities. <u>Human Factors</u> 30(3):319-337, 1988.
8. NIOSH: <u>Work Practices Guide For Manual Lifting</u>, Cincinnati, OH, DHHS(NIOSH), National Institute for Occupational Safety and Health, Technical Report No. 81-122.
9. Borg, G.: Perceived exertion as an indicator of somatic stress. <u>Scandinavian Journal of Rehabilitation Medicine</u> 2(3):92-98, 1970.
10. Mihevic, P.M.: Sensory cues for perceived exertion: A review. <u>Medicine and Science in Sports and Exercise</u> 13(3):150-163, 1981.
11. Borg, G.: Psychophysical basis of perceived exertion. <u>Medicine and Science in Sports and Exercise</u> 14(5):377-381, 1982.

12. Borg, G. and Noble, B.J.: Perceived exertion, In Wilmore, J.H. (ed): <u>Exercise and Sciences Reviews</u>, vol 2. New York: Academic Press, 1974.

13. Chow, R.J. and Wilmore, J.H.: The regulation of exercise intensity by ratings of perceived exertion. <u>Journal of Cardiac Rehabilitation</u> 4(9):382-387, 1984.

14. Garg, A. and Banaag, J.: Maximum acceptable weights, heart rates and RPEs for one hour's repetitive asymmetrical lifting. <u>Ergonomics</u> 31(1):77-96, 1988.

15. Nielsen, R. and Meyer, J.P.: Evaluation of metabolism from heart rate in industrial work. <u>Ergonomics</u> 30(3):563-572, 1987.

16. Purvis, J.W. and Cureton, K.J.: Ratings of perceived exertion at the anaerobic threshold. <u>Ergonomics</u> 24(4):295-300, 1981.

17. McCallam, M.C.: Psychophysics research applications in industrial design, in Karwowski, W. and Yates, J.W. (eds): <u>Advances in Industrial Ergonomics and Safety III</u>. London: Taylor & Francis, 1991, pp. 761-768.

18. Marley, R.J. and Fernandez, J.E.: A psychophysical approach to establish maximum acceptable frequency for hand/wrist work, in Karwowski, W. and Yates, J.W. (eds): <u>Advances in Industrial Ergonomics and Safety III</u>. London: Taylor & Francis, 1991, pp. 75-82.

19. Marley, R.J.: <u>Psychophysical Frequency At Different Wrist Postures Of Females For A Drilling Task</u>. Unpublished doctoral dissertation, The Wichita State University, 1990.

20. Kim, C.H.: <u>Psychophysical Frequency At Different Forces And Wrist Postures Of Females For A Drilling Task</u>. Unpublished doctoral dissertation, The Wichita State University, 1991.

21. Ayoub, M.M., Selan, J.L. and Jiang, B.C.: <u>A Mini-Guide For Lifting</u>. Lubbock, TX: Center for Ergonomics Research, Texas Tech University, 1983.

22. Ciriello, V.M., Snook, S.H., Blick, A.C. and Wilkinson, P.L.: The effects of task duration on psychophysically-determined maximum acceptable weights and forces. <u>Ergonomics</u> 33(2):187-200, 1990.

Part II
Concurrent Engineering Techniques

A CONCURRENT ENGINEERING FRAMEWORK: THREE BASIC COMPONENTS

Janis P. Terpenny and Michael P. Deisenroth

*ISE Department, Virginia Polytechnic Institute and State University,
Blacksburg, Virginia, U.S.A.*

ABSTRACT

Motivated by global competition and rapidly advancing technologies, customers in today's markets have become discriminating. Quality, reliability, service, and product value are no longer optional features to attract sales, but requirements. To remain competitive, manufacturers must minimize total costs while being quick to market with new product introductions. Customers are demanding intelligent, flexible solutions to their needs. Industry, government and university professionals alike, are responding to these new demands by advocating and advancing the practice of concurrent engineering methodologies. These methods promote the consideration, during the product design phase, of the many factors which affect life-cycle cost and product performance. Life-cycle issues to consider concurrently during design may include: manufacturability, assembly, inspection, maintainability, and environmental pollution. The purpose of this paper is to describe and put into perspective the many advancing technologies and methodologies which support a concurrent engineering environment. A framework identifying three components basic to concurrent engineering systems is suggested. These components include: improved design strategies, decision methods, and supporting information and knowledge. Design strategies include general principles for design which enable or improve the function of other life-cycle processes. Decision methods described include: multi-functional teams, classical optimization methods, and artificial intelligence techniques. Supporting information and knowledge includes a discussion on geometric modeling, design by feature, and CIM requirements to achieve an integrated system.

INTRODUCTION

Concurrent engineering is more than simultaneous engineering or life-cycle engineering. It brings to the design phase consideration of the many factors affecting total cost and performance throughout the life-cycle of a product. This involves integrating the many diverse functional areas of an organization into the process of creating a better design when viewed across the entire product life-cycle. In addition to meeting customer functional requirements, considerations during design should include manufacturability (material selection, assembly, process planning), serviceability, reliability, testability, disposability (disassembly, recyclability), environmental pollution, etc.

It has been well documented that at least 70% of the costs which will be incurred over the life-cycle of a product have been committed by the decisions made during the design phase [1], [2], [3], [4]. Moreover, as the product moves beyond the life-cycle phase of design, attempts of making corrections to its definition are met with increased difficulty and expense. Complicating success still further, are traditional industrial business practices which have become counter productive to meeting new market challenges. Functional areas are often encouraged to be more concerned with their own performance measures than the objectives of the business as a whole.

Concurrent engineering methods are countering traditional business philosophy with product designs which consider the complex relationships and trade-off situations of the many cross functional life-cycle factors. Designs created using concurrent engineering methods represent a true case of the whole being greater than the sum of the parts.

To date, the predominant thrust of simultaneous engineering research and its application has focused on factors affecting manufacture, design for manufacturability (DFM), and factors affecting assembly, design for assembly (DFA). While it is logical that manufacturing smarter should offer the greatest and most direct cost savings and reduce time to market, DFM and DFA are, in reality, only subsets of concurrent engineering. Concurrent engineering should, ideally, consider *all* life-cycle factors simultaneously. Tractable solutions to this problem have yet to be found. To gain a better understanding of individual application requirements, other life-cycle factors have identified additional subsets in concurrent engineering. These subsets include design for: automation, test, maintenance, reliability, cost, quality, and environment.

The potential and actual benefits already realized by the practice of concurrent engineering have sparked the interest of many researchers and practitioners in industry, academia, and government. As a result, numerous advancing technologies and methodologies have been developed to facilitate complex multi-objective optimization and communication problems. It is the aim of this paper to describe and put into perspective these evolving technologies and methodologies. To this end, a framework identifying three components basic to concurrent engineering systems is suggested. Basic components include: improved design strategies, decision methods, and supporting information and knowledge. Each of these components is discussed. Future trends and needs in concurrent engineering are also addressed.

A FRAMEWORK: THREE BASIC COMPONENTS

Practicing concurrent engineering is not merely a way of automating, with the use of computers, existing design practices. It involves looking at the design problem in a new way. A way which requires *improved design strategies*. New ways of approaching the design problem with the simultaneous consideration of many life-cycle factors. Additionally, *decision methods* are required to aid the process of selecting from among design alternatives or when maximizing design effectiveness measures. Finally, concurrent engineering requires *supporting information and knowledge* to make these informed decisions. Communication into and out of design is essential to enable concurrency.

As an example, consider computer aided process planning (CAPP). Improved design strategies which consider manufacturing requirements serve to reduce the number of processing operations required. Decision methods assist or guide the designer in identifying and selecting from among feasible design alternatives. Supporting information and knowledge allow for the flow of information between the design and CAPP systems.

A more efficient exchange of information would be one where manufacturing process requirements are present in, or could be readily determined from, the design (drawing) itself. The requirements for information representation should be considered concurrently by both the CAD database design and the CAPP system design to achieve the desired integration. Design by feature and parametric design are strategies currently receiving much attention, with great promise of bridging the design/manufacturing gap. A more complete description of this subject can be found in a later section of this paper, Supporting Information and Knowledge. Refer to Chang [5], Chang, Anderson, and Mitchell [6], Chang and Wysk [7], and Joshi and Chang [8] for more on the integration of design and process planning strategies.

Lu and Subramanyam [9], and Cutkosky and Tenenbaum [10] extend design to simultaneously consider both product and process. This comes closer to a micro analysis of manufacturing systems. The issue to resolve is not just one of integration through information design, but one of designing the process and product concurrently, improved design strategies.

The 'big picture' solution for concurrent engineering should, ideally, consider all interrelated life-cycle functions simultaneously when making design decisions. In reality, the consideration of all factors simultaneously is quite an ambitious task. Methods of enabling and improving large multi-functional design solutions are currently the topic of much research. The three sections which follow describe with more detail each of these basic components.

IMPROVED DESIGN STRATEGIES

Innovative solutions result when knowledge and a true understanding of product and processes are brought together with the intent of finding improved methods. When the product design caters to improving and/or enabling downstream processes in the life-cycle, cost reductions are inevitable. Some general principles which have demonstrated improvements include products designed with [11], [12]:

simplicity - fewer parts and simpler geometries to manufacture, assemble, test, and inventory,

standardized parts and materials - eases design activities, reduces inventory, enables automated handling during fabrication and assembly,

minimum finishing requirements and tolerances - reduces unnecessary processing,

minimum material (both amount and types) - reduced inventory costs and scrap, eases recyclability, eases automation.

easily replaced modules, parts and use of common fasteners - eases service and maintenance during the utilization phase of the life-cycle.

This list is but a small sample of the general principles that can help to guide the design of products which are responsive to other life-cycle needs. For overview and guideline suggestions, refer to Bralla [13], O'Grady and Oh [14], or Boothroyd and Dewhurst [15].

DECISION METHODS

Decision methods are a crucial part of the concurrent engineering process. Solution requirements can vary widely in complexity. These requirements might include anything from evaluating the manufacturability of a design to recommending the 'best' overall product design when consideration has been given to many life-cycle factors. The following methods are commonly used for decision making in concurrent engineering: multi-functional teams, classical optimization techniques, and artificial intelligence.

Multi-functional Team

Teams composed of members across functional areas are formed to bring expertise to the design phase. These groups may be temporary focused task forces for new products or for major redesign projects. Brain-storming innovation into the product design may be a strong point for a team such as this. Alternatively, teams may be permanent groups devoted to a particular product line. This sort of team represents a long term change in the way of doing business ... one which moves away from separate functional measures, and more towards the effectiveness of the product over the life-cycle as a whole. In either case, teams can be quite effective with the right direction and clearly defined purposes. Group dynamics plays a major role in the success of such teams. Team members would most likely include, but not be limited to, representatives from functional areas such as design, manufacturing, quality assurance, marketing, maintenance, cost accounting and information systems.

Classical Optimization

The primary objective of concurrent engineering is to identify the 'best' product design when viewed from the perspective of the product life-cycle. To measure the effectiveness of a design, factors such as functionality, reliability, maintainability, producibility, etc., need to be considered. If all of these design factors were quantifiable, the process of defining optimal designs would be suited to solution using some form of classical optimization procedure.

The concept of finding optimal solutions based on some measure of effectiveness, subject to a system of large and/or complex constraints, is not a new one.   Since the introduction of linear programming by Dantzig [16] in 1949, many practitioners have applied the concept of optimizing over a set of constraints to a wide variety of problems. In 1957, Churchman, Ackoff, and Arnoff [17] introduced the application of evaluation functions to the area of system analysis. Effectiveness, E, a function of controllable, X_i, and uncontrollable, Y_j, variables, could be optimized by determining the appropriate selection and values of the input variables. In this manner, effectiveness is measured in terms of meeting functional requirements, where:

$$E = f(X_i, Y_j)$$

Fabrycky [1] has extended the formulation of system effectiveness functionals to design situations by partitioning system parameters into a design dependent subset, Y_d, and a design independent subset, Y_i. As is most often the case, design is viewed as an iterative process. For any given iteration, design dependent parameters are either determined, predicted, or estimated and optimal values are sought for design variables. For each successive set of values for Y_d, design iteration proceeds through the design space to determine the optimum points for E. The functional is described by

$$E = f(X, Y_d, Y_i); \quad \text{where} \quad g(X, Y_d, Y_i) \leq C \text{ or } g(X, Y_d, Y_i) > C$$

To date, researchers have used classical optimization to help identify and compare design alternatives. Quite often however, the model is either very restrictive in the scope of life-cycle issues considered, complicated by attempts to model relationships which cannot be expressed mathematically, or complicated by multiple measures which are dependent upon one another. Determining how much complexity a model should contain to be effective is an area of study in its own right, see Bell and Taylor [18].

Artificial Intelligence

AI Background Contrary to classical optimization techniques, artificial intelligence (AI) is particularly suited to the solution of problems characterized by incomplete information and variable relationships which cannot be quantified. Additionally, basic to any AI application are its ability to infer new knowledge from existing knowledge and its ability to explain itself.

The most straightforward AI systems, expert systems (ES), apply domain facts to a set of rules searching for a match. When a match is found, the consequent of the rule is enabled. This may be in the form of additional facts which can be added to the knowledge base, in the form of recommendations, or some other enabling action. Search techniques can use either backward or forward chaining, and either breadth first or depth first strategies [19].

AI applications vary considerably in sophistication and complexity. Applications might include techniques which are straight forward rule-based expert systems, systems which use first-order predicate logic and object-oriented programming, constraints, constraint networks, or neural networks.

The level of user interaction required by an AI application can vary widely from one system to another. The user is generally prompted by questions as required by the system. The system may continue to prompt for input throughout the execution of the application or, given enough sophistication, the system may have the capability to continue on its own for certain portions of the reasoning requirements. To date, there are no AI concurrent engineering application systems which are fully automated, i.e. require no user interaction.

AI in Concurrent Engineering In simultaneous engineering, expert systems have been used in design review, as design advisors, to search for standardization opportunities, to search for conflicts in design rules, in diagnostics and in training [5]. Most commonly, applications use manufacturability or design for assembly rules in evaluating the design and/or advising the design engineer. The design may be evaluated upon its completion, or guided during its creation by the AI application. A simple example of a design rule related to hole making might be to check for the length of the hole being no more than six times its diameter .

Many examples of design evaluation/advice systems can be found in the literature. In most instances, applications have been very narrow in the breadth of life-cycle factors considered simultaneously, or are specific to a very small class of problems. The following citations are only a small portion of the many AI systems available or under study: Orady [20], Srihari and Emerson [21], Hernandez *et al.* [22], Lawlor-Wright and Hannam [23], Bao [24], and Mayer and Lu [25]. See

Ghosh and Gupta [26] for a survey of expert systems in manufacturing and process planning.

Advanced AI strategies have been developed which combat some of the inherent shortcomings of expert systems. More sophisticated than rule-based systems, some applications have incorporated the use of mathematically represented constraints and procedures into the AI system. Young, Greef and O'Grady [27] have developed Spark, a constraint network programming language, with frame based inheritance. A constraint network is formed by the linking of constraints with shared variables. In this format, a system can be developed with both quantifiable and unquantifiable measures.

In general, artificial intelligence solution strategies appear to offer a more 'natural' method of approaching concurrent engineering design problems. According to Srihari and Emerson [21], "The human element will continue to be critical in the total design/manufacturing process, but knowledge based systems properly researched and applied will improve the total design cycle."

Much work remains to be done in AI for concurrent engineering applications. Issues to address might include: how to better handle/define the wide variety of life-cycle factors simultaneously, resolving trade-off situations amongst functions (example, manufacturability versus maintainability), and improved knowledge representation schemes for defining the relationships between design and other life-cycle processes.

SUPPORTING INFORMATION AND KNOWLEDGE

The primary purpose of concurrent engineering is to bring information and knowledge to the evaluation and creation of improved product designs. Viewed in this context, finalized product design information serves two functions: 1) to accurately represent product geometric and functional definition, and 2) to support the function of subsequent life-cycle processes. This section will focus on information and knowledge requirements in a concurrent engineering environment. Geometric and application data models are briefly described and contrasted. Requirements for achieving an integrated computer environment to support concurrent engineering are also discussed.

Geometric and Feature Based Modelers

Given current trends in computer aided design (CAD) software development and to facilitate discussion, it will be assumed that the representation of design is accomplished using an intelligent CAD system with solid modeling capabilities.

Geometric Modeling   The two most common methods of representing geometric information in a solid modeling system are Boundary Representation (Brep) and Constructive Solid Geometry (CSG). Briefly, a Brep model represents the solid in terms of the surfaces enclosing a volume. This type of geometric model consists of geometric information (point, line, circle, etc.) and topological objects defined by the hierarchical relationship of elements (vertex, edge, loop, face, etc.). At a higher level of abstraction, CSG models are constructed by the boolean combination of primitive volumes (bricks, cylinders, cones, etc.). Refer to publications by Requicha and Voelcker [28], [29], for a more indepth discussion of these solid modeling methods.

To obtain meaningful information from these models, concurrent engineering applications must rely on feature extraction methods or on the interpretation of neutral geometric files. Difficult to automatically interpret, detailed application data is often stored in the form of textual information.

Design by Feature   Offering an even higher level of abstraction, is the more recent notion of design by feature, sometimes known as parametric design. Inspired by feature extraction applications in manufacturing [30], the method attempts to preserve the designer's intent within the geometric representation. As described by Pratt [31],

> "Finally, the feature recognizer informs the user that it has detected the presence of a cylindrical hole in the part. But this is information which the designer was fully aware of when he created the model; it was lost when the system reduced all input to low-level details of topology and geometry."

Some of the typical features used in design which have been represented explicitly include holes, bosses, cutouts, slots, keyways, and gears. Using group technology (GT) concepts, a given set of features can form the basis for the identification of part families. This approach readily supports the objectives of improved design strategies (standardization, parts minimization, etc.).

The ability to relate surfaces to one another or with other information is accomplished through the use of parameters. In this manner, modifications to one surface would automatically cause adjustments in other related surfaces. This might be particularly helpful, for example, when adjusting the length of a cylinder which has a through hole. Certainly, the length of the hole should change with the length of the cylinder.

The integration of life-cycle considerations into the product design phase can be best achieved when meaningful relationships exist between applications and design data. During the design phase, there must be the ability to analyze other life-cycle factors. Once complete, the design information should be available for input to other life-cycle processes. To allow for this flow of information, the product design data must either be in the correct format for input to other systems or be readily converted to this form. Downstream manufacturing applications requiring design information might include process planning, NC programming, and inspection.

Design by feature would seem to be an ideal method for relating design and application. The definition of this relationship is not clear however, and is currently the subject of much debate.

Features which are meaningful to design may not be the same as those used in manufacturing. In fact, as pointed out by Joneja and Chang [32], there is a possibility that a part created using design features may not be producible within a given manufacturing domain. Two approaches to the use of features in design are to: 1) use design features and convert these to manufacturing features, or 2) use manufacturing features in the creation of the original part design. Either approach has inherent advantages and disadvantages. Issues might include: the amount of manufacturing knowledge required of the designer, flexibility to manufacturing environmental changes, and distortion of design intent. To attain a fully integrated concurrent engineering system using a feature based system, researchers must address many key issues [3]: what features, a taxonomy of these features, how specified, and how related to each other.

Integrated Computer Environment

Recall the primary purpose of any concurrent engineering system: the creation of improved product designs when viewed across the entire product life-cycle. To achieve this objective there must be an integration of life-cycle considerations into the product design phase. Information flow must be input to design for the creation of informed designs and output from the finalized design to enable other life-cycle phases. A truly integrated computer-based concurrent engineering system requires information flow between product databases, knowledge databases, optimization routines, AI applications and CAD systems. Thus, concurrent engineering depends, to a large degree, on the communication of information.

When one considers the many life-cycle factors and the large amount of knowledge required by each, it becomes clear why there is no all encompassing concurrent engineering system in existence. Developers of concurrent engineering systems have used a variety of approaches. A small sample of these include: using a globally accessible database (Subramanyam and Lu [33]), artificial intelligence based approaches (O'Grady *et al.* [34]), and feature extraction methods (Gadh *et al.* [35]). Many other developers have approached the problem by focusing on a specific class of parts (metal formed, plastic, prismatic, etc.).

Weston *et al.* [36] suggest the need for 'soft' integration. The authors describe the usual *ad hoc* interfaces which often exist between applications in a CIM environment. They suggest a concurrent engineering perspective in the construction of CIM architectures. They suggest the use of an integration platform to serve as an intermediary between applications. Conventional approaches require each application to have knowledge of other applications with which they need to be integrated. Contrary to this, soft integration requires only a knowledge of how to use the platform of services and does not require knowledge of other system entities. Application software code generators are proposed. These would lead to applications which interact via standard mechanisms supported by the platform.

Whether supported through some integration platform or through direct communication between applications, a standard for the exchange of data is needed. To this end, product modeling standards are the subject of much research. See PDES [37] and [38].

CONCLUSIONS AND RECOMMENDATIONS

This paper has described and justified what is needed for integrated concurrent engineering. A framework detailing three basic components supporting the practice of concurrent engineering has been identified. Within this framework, techniques enabling concurrent engineering decisions and information flow are described.

Much work has yet to be done before a fully integrated concurrent engineering environment is attained. Confounded by the diverse nature of the many life-cycle factors affecting product costs and performance, it seems unlikely that an obvious, undiscovered solution for concurrent engineering problems exists. Areas which do hold promise for promoting such a system are:

(1) Establishing standards for product data exchange between applications;
(2) The incorporation of other life-cycle concerns during the product design process;
(3) Improving decision making methods to include conflict resolution strategies;

(4) The creation of communication architectures which support the integration of many functional areas in the concurrent engineering process;

(5) A better understanding the design process itself. Creating methods which are supportive of the natural creative processes inherent in the design process;

(6) Identifying new improved design strategies which are likely to reduce costs and improve product functionality.

In conclusion, while implementing a concurrent engineering environment requires considerable effort and a commitment to its success, the long term benefits are impressive when viewed over the product life-cycle. Attention to the advancement of improved design strategies, decision methods, and supporting information and knowledge will prove to be valuable investments.

REFERENCES

1. Fabrycky WI: *Life-Cycle Cost and Economic Analysis*, Prentice Hall, Englewood Cliffs, NJ, 1991.

2. Huthwaite B: Checklist for DFM. *Machine Design*, pp 163-167, January 25, 1990.

3. O'Grady P, Young RE: Issues in concurrent engineering systems, *Journal of Design and Manufacturing* 1:27-34, 1991.

4. National Science Foundation: *Research priorities for proposed NSF strategic manufacturing initiative*, Report of a National Science Foundation Workshop, National Science Foundation, Washington, DC, 1987.

5. Chang TC: *Expert Process Planning Systems*, Addison-Wesley, Reading, Massachusetts, 1990.

6. Chang TC, Anderson DC, Mitchell OR: QTC - An integrated design/manufacutring/inspection system for prismatic parts, in *Proceedings of the ASME Computers in Engineering Conference*, SF, Cal., 1:417-426, 1988.

7. Chang TC, Wysk RA: *An Introduction to Automated Process Planning Systems*, Prentice Hall, Englewood, New Jersey, 1985.

8. Joshi S, Chang TC: Feature extraction and feature based design approaches in the development of design interface for process planning, *Journal of Intelligent Manufacturing*, 1:1-15, 1990.

9. Lu S, Subramanyam S: A Computer-based environment for simultaneous product and process design, in *Proceedings of the ASME Winter Annual Meeting*, vol 31, New York, ASME Production Engineering Division, Nov 1988, pp 35-46.

10. Cutkosky MR, Tenenbaum JM: Methodology and computational framework for concurrent product and process design, *Mechanism & Machine Theory*, 25(3):365-381, 1990.

11. Bancroft CE, et al.: General Product Design, in Bakerjian R (eds): *Tool and Manufacturing Engineers Handbook*, vol 6, ed 4. Dearborn, Society of Manufacturing, 1992, chap 10, pp 1-40.

12. Kalpakjian S: *Manufacturing Engineering and Technology*, New York, Addison-Wesley, 1989, pp 13-29.

13. Bralla JG (ed): *Handbook of Product Design for Manufacturing*, McGraw-Hill, Inc., USA, 1986.

14. O'Grady P, Oh J: A review of approaches to design for assembly, *Concurrent Engineering*, 1:5-11, 1991.

15. Boothroyd G, Dewhurst P: Design for assembly: selecting the right method, *Machine Design*, pp 94-98, Nov 10, 1983.
16. Dantzig GB: Programming of interdependent activities, II, Mathematical model, *Econometrica*, 17(3-4):200-211, July-Oct 1949.
17. Churchman CW, Ackoff RL, Arnoff EL: *Introduction to Operations Research*, John Wiley & Sons, New York, 1957.
18. Bell DG, Taylor DL: Determining optimal model complexity in an iterative design process, in *Proceedings of 2nd International Conference on Design Theory and Methodology*, vol 27, Chicago, ASME Design Engineering Division, Sep 1990, pp 291-297.
19. Rolston DW: *Principles of Artificial Intelligence and Expert Systems Development*, New York, McGraw-Hill Book Company, 1988.
20. Orady EA: Design and development of an expert system for evaluation of the manufacturability of mechanical parts, in *Proceedings of the 15th Annual AMSE Conference*, Brockport, New York, Association of Muslim Scientists and Engineers, Oct 1989, pp 51-66.
21. Srihari K, Emerson CR: A concurrent engineering methodology for PCB assembly, Khalil TM, Bayraktar BA (eds): *Management of Technology III*, Institute of Industrial Engineers, 1992, pp 828-837.
22. Hernandez JA, Luby SC, Hutchins PM, Leung HW, Gustavson RE, De Fazio TL, Whitnew DE, Nevins JL, Edsall AC, Metzinger RW, Tung KK, Peters TJ: An integrated system for concurrent design engineering, in *Proceedings of the Seventh Conference on Artificial Intelligence Applications*, Miami Beach, IEEE Computer Society, Feb 1991, pp 205-211.
23. Lawlor-Wright T, Hannam RG: A feature-based design for manufacture: CADCAM package, *Computer-Aided Engineering Journal*, 6(6):215-220, 1989.
24. Bao HP: An expert system for SMT printed circuit board design for assembly, *Manufacutring Review*, 1(4), 1988.
25. Mayer AK, Lu SC: An Ai-based approach for the integration of multiple sources of knowledge to aid engineering design, *Journal of Mechanisms, Transmissions, and Automation in Design*, 110(3):316-323.
26. Ghosh BK, Gupta T: A Survey of expert systems in manufacturing and process planning, *Computers in Industry*, 11(2):195-204.
27. Young RE, Greef A, O'Grady P: Spark: an artificial intelligence constraint network system for concurrent engineering, in *Proceedings of the First International Conference on Artificial Intelligence in Design*, 1991.
28. Requicha AAG, Voelcker HB: Solid modeling: a historical summary and contemporary assessment, *IEEE Computer Graphics and Applications*, 2(2):9-24, March 1982.
29. Requicha AAG, Voelcker HB: Solid modeling: current status and research directions, *IEEE Computer Graphics and Applicaitons*, pp 25-37, Oct 1983.
30. Lee YC, Fu KS: Machine understanding of CSG: extraction and unification of manufacturing features, *IEEE Computer Graphics and Applications*, 7(1):20-32, January 1987.
31. Pratt MJ: Solid modeling and the interface between design and manufacture, *IEEE Computer Graphics and Applications*, 4(7):52-59, July 1984.
32. Joneja A, Chang TC: Search anatomy in feature-based automated process planning, *Journal of Design and Manufacturing* 1:7-15, 1991.

33. Subramanyam S, Lu SC: Computer-aided simultaneous engineering for components manufactured in small and medium lot sizes, in *Winter Annual Meeting of ASME*, DE vol 21, PDE vol 36, San Francisco, ASME, 1989, pp 175-184.

34. O'Grady P, Young RE, Greef A, Smith L: An advice system for concurrent engineering, *International Journal of Computer Integrated Manufacturing*, 4:63-70, 1991.

35. Gadh R, Hall MA, Gursoz EL, Prinz FB: Knowledge dreven manufacturability analysis from feature based representations, in *Winter Annual Meeting of ASME*, DE vol 21, PDE vol 36, San Francisco, ASME, 1989, pp 21-34.

36. Weston RH, Hodgson A, Coutts IA, Murgatroyd IS: 'Soft' integration and its importance in design to manufacture, *Journal of Design and Manufacturing*, 1:47-56, 1991.

37. PDES, Product data exchange specification: first working draft, PB89-144794, Department of Commerce, Gaithersburg, MD, USA, 1988.

38. PDES/STEP, ISO TC184/SC4/WG1, 1990.

A COMPARATIVE ANALYSIS APPROACH FOR EVALUATING THE EFFECT THAT CONCURRENT ENGINEERING HAS ON PRODUCT LIFE CYCLE COST

Michael D. Reimann and Faizul Huq

Information Systems and Management Sciences, College of Business Administration, University of Texas, Arlington, Texas, U.S.A.

ABSTRACT

American industry is faced today with competition on a worldwide basis. To improve this situation many manufactures are making significant changes in the way they operate their businesses. New techniques such as concurrent engineering, continuous process improvement, quality function deployment, and total quality management are being employed to enhance competitive posture. A successful outcome predicated on these techniques involves philosophical, cultural and technological changes within the company. Well intended changes may lead to the demise of the company rather than its growth. This situation is further exacerbated by the lack of methods for making objective comparisons among possible alternatives. Moreover, once a course of action has been selected, it is difficult to identify and evaluate areas of risk.

Research was undertaken to address these issues. The objective of this research is to develop a methodology for quantifying and comparing alternative product development techniques. The techniques addressed in the work are traditional engineering and concurrent engineering. Having established an accurate comparative model, more sophisticated techniques will be used to define predictive models. Ultimately this approach will be tailored to actual situations in industry.

Well established U.S. Air Force Integrated Computer Aided Manufacturing models provided the basis for defining the entire product life cycle process. This process was transformed into a quantitative model which employed activity based costing as a metric. Implementation of the model required simulation due to the complexity of the product life cycle. Useful insights were gained from preliminary simulation results. Response surface methodology was used to develop a predictive model for a wide variety of product development techniques.

BACKGROUND FOR RESEARCH

Since the end of World War II there has been a dramatic amount of change in the world place for products [1,2]. From a broad perspective changes have occurred in customer preferences, the type of products purchased, complexity and technologies of products, who produces the products, the technologies used to develop, manufacture and maintain the products, etc.

Another facet of change is the rate at which it has occurred. Following World War II, America was the preeminent world producer and largest consumer of products. The American industrial complex was unscratched by destruction, unlike the other principle participants in the war. Moreover, our ability to produce had reached its peak capacity at that time. Many of the major producing countries of the world experienced widespread destruction of their communities and industrial bases. These countries were faced with a substantial if not entire reconstruction effort. The vast capacity and resources of America were used to help rebuild the world community of nations.

It is clear that the dominance of American industries has slipped away. Todays most competitive and advanced producers are from the countries that were the most devastated by World War II. American companies enjoyed a tremendous demand for their products with little offshore competition following the war. This led to complacency, because there was no pressing need for improvement or change. As American production facilities became more obsolete, the Japanese and European countries were building modern manufacturing capabilities. During the last two decades foreign competition has had a significant impact on American industries. Moreover, the pace continues to increase and shows no sign of slowing down.

Productivity in Manufacturing Firms

Many analyses have been done in retrospect to characterize the current situation and trace the evolutionary path that it took to get there [1,2,3]. Most experts concur that American productivity has declined and that the competition from international sources poses a significant threat to American industries. Unfortunately there is no consensus on an identification of the problem that inhibits American productivity. Previous studies of the productivity problem have identified varied and conflicting sources. Among the causes cited are: macroeconomic policies, government policies and practices, shortsightedness of investors, poor management, outdated or inappropriate business practices, lack of innovation, obsolete technology and a lower quality work force [1,2]. All that can be concluded is that the productivity problem in America can not be easily defined and it is probably composed of various interacting factors.

This highly competitive environment is forcing many American firms to reconsider the way they conduct business. To remain competitive requires continual improvements to reduce cost, shorten schedules, satisfy customer needs and minimize risks. A wide variety of new approaches aimed at achieving these goals have emerged. Among the new approaches are, activity based cost management, business performance metrics, concurrent engineering, continuous process improvement, just in time inventories, participative team management, quality function deployment and total quality management. These approaches differ radically from the way American manufactures are accustomed to running their businesses. Moreover the effective implementation of any of these approaches usually involves philosophical, cultural and technological changes within the company.

Research Objective

A review of the literature and other research efforts was unable to uncover any examples of quantitative methods that compare the effects that alternate product development techniques have on product life cycle costs. Even given this fact, many American firms are making radical changes in their businesses. The motivation for change is often driven by a fear of extinction. Subjective and qualitative techniques are used to justify a productivity improvement approach. Figure 1 provides one such example [4]. Relative product life cycle costs are shown for two approaches to product engineering and development. The solid line indicates the costs accumulated through the various product life cycle stages when the traditional approach to product engineering and development is utilized. The dashed line shows the costs incurred at the product life cycle stages when an advanced method for product engineering and development is employed. The advanced approach requires a greater investment in the early stages of specification, engineering and development. However this early investment should result in greater savings in the later stages of the product life cycle. Ultimately, the overall product life cycle cost can be substantially reduced by using advance techniques for specification, engineering and development. The information presented in Figure 1 appears to have a quantitative basis, however, this is not the case. Figure 1 simply illustrates the anticipated cost differences between two alternatives.

Corporate executives must choose between competing alternatives that could potentially provide a competitive advantage for their company or lead to its demise. A quantitative method for evaluating and comparing alternatives would provide objective information for mitigating risk. Unfortunately, this methodology does not exist [3]. Moreover, traditional methods for evaluating and comparing emerging approaches for improving productivity have proven to be inadequate [1,2].

Therefore the objective of this research is to develop a foundation for comparing alternative product development approaches quantitatively.

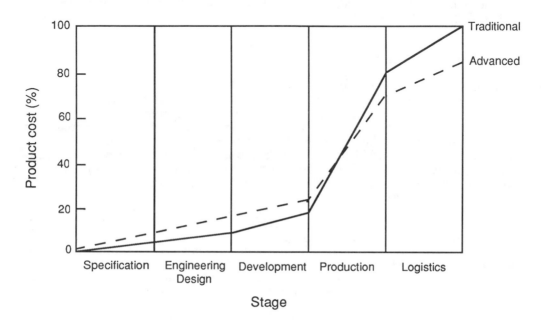

FIG. 1 **Performance Characteristics of Product Life Cycle Costs**

THEORETICAL FOUNDATION FOR RESEARCH

Several techniques were incorporated in this research which previously had not been used together. Quantitative models were derived from functionally oriented process models. Two process models employed in this research were for the traditional engineering and concurrent engineering approaches to product development. Activity based costing served as the common metric for this research. The two engineering approaches and the metric provided the basis for the theoretical foundation in this research.

Traditional Engineering Approach to Product Development

The traditional engineering approach to product development is illustrated in Figure 2 [5,6]. In Figure 2, the product life cycle begins when a product is first conceptualized. Recognition of the need for a product originates from the customer for the product or from the organization that will produce the product. The conceptualized product undergoes a formalization process which produces Customer Requirements for the product. The scope of this research assumes that the Customer Requirements have been generated. The Perform Engineering process utilizes Expertise of the organization to develop a Product Definition. The Product Definition is a technical specification of the product that is intended to satisfy the Customer Requirements for the product. Available Technologies are used to aid the transformation of Customer Requirements into a Product Definition.

In the Produce Product process, materiel is transformed into an Accepted Product. Acceptance of the product is contingent on its conformance to the Product Definition. In the event the Product Definition is inadequate or incomplete, it will not be possible to develop a process for producing the product. When such deficiencies exist in the Product Definition, it will be necessary to change the Product Definition to rectify the deficiencies. Requests for correction to the Product Definition are formally documented as Product Definition Change Requirements. The Perform Engineering process makes the necessary changes to the Product Definition. Then the corrected Product Definition is re-released to the Produce Product process.

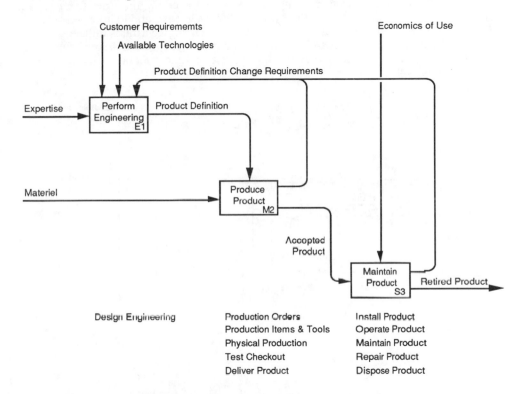

FIG. 2   Traditional Engineering Approach to the Product Life Cycle

Accepted Products are those which satisfy the Customer Requirements. Throughout the operational life of the product, it will undergo various forms of maintenance. Maintenance is scheduled according to policy whereas repair is required when a product malfunctions. Repairability, maintainability and logistical support of the product are affected by its final form, fit and function. Inadequate consideration of these factors during the Perform Engineering process can result in an Accepted Product which is difficult to maintain. Changes to the product will be required to overcome maintenance difficulties. Before the product can be changed, it is necessary to alter the Product Definition to reflect the required changes. Thus Product Definition Change Requirements will be generated to institute the needed product changes. The products are put into service where they are used until they are retired. The decision to retire the product is based on the Economics of Use.

Concurrent Engineering Approach to Product Development

An often cited definition of concurrent engineering comes from Winner, et al. [7]:

> "Concurrent engineering is a systematic approach to the integrated concurrent design of products and their processes, including manufacture and support. This approach is intended to cause the developers, from the outset, to consider all elements of the product life cycle from concept through disposal, including quality, cost, schedule and user requirements."

The objectives of concurrent engineering are to integrate the engineering functions of design, manufacturing and logistics support in order to reduce the time from product conceptualization to production, to reduce product life cycle costs and to improve quality.

Figure 3 shows the product life cycle when concurrent engineering is utilized [5]. This functional model of the product life cycle has many similarities to the traditional engineering approach shown in Figure 2. However, there are two fundamental differences in the engineering approaches that are illustrated by Figures 2 and 3.

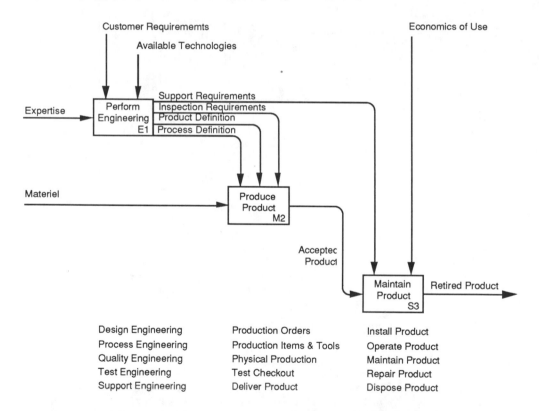

Design Engineering	Production Orders	Install Product
Process Engineering	Production Items & Tools	Operate Product
Quality Engineering	Physical Production	Maintain Product
Test Engineering	Test Checkout	Repair Product
Support Engineering	Deliver Product	Dispose Product

FIG. 3   Concurrent Engineering Approach to the Product Life Cycle

The first difference is because concurrent design must take into consideration the downstream production and maintenance processes. This fact is demonstrated in Figure 3 by the additional lines that emanate from the right side of the Perform Engineering process. In addition to the Product Definition the Perform Engineering process generates Process Definition, Inspection Requirements and Support Requirements. Taken to a theoretical extreme, this new information is sufficient enough to eliminate the deficiencies that were experienced by the downstream processes in the traditional engineering approach.

The second difference between the product life cycles illustrated in Figures 2 and 3 is related to the first difference. In the concurrent engineering approach there is no need for changing the Product Definition. This is because the specifications developed by the Perform Engineering process are sufficient enough so that no inadequacies are experienced by the downstream processes. Thus the Produce Product and Maintain Product processes perform optimally. Under this scenario there is no need to incorporate changes in the Product Definition. Consequently there are no Product Definition Change Requirements feedback loops in the concurrent engineering model shown in Figure 3.

Activity Based Cost Accounting

Advanced technologies have transformed engineering and manufacturing from labor intensive to automation and information intensive processes. This transformation has made existing cost accounting methods inadequate. Automation results in higher proportions of fixed to variable costs. It also alters the proportion of costs that are assigned to direct and indirect cost categories. Traditional cost accounting has relied on direct costs and variable labor as the basis for determining overhead rates. Since the cost of direct labor is no longer a significant portion of a product's total cost it provides little useful information for assigning indirect and fixed costs to the units of production. Direct cost traceability is the key to cost reduction and product profitability decisions. This requires that all costs, even traditional support costs, to be traced to products by directly charging the costs of all activities to the product. This approach to cost assignments is considerably different from traditional cost accounting methods. This orientation of cost accounting requires fundamental changes from historical reporting and variance analysis to a forward view of accounting for each value-added activity in the product life cycle.

During the past decade, a great deal of cooperative research has been undertaken by academia, industry, government and professional accounting organizations to establish improved cost accounting concepts and principles. Activity based costing [4] has emerged as an approach to overcome the deficiencies of traditional cost accounting. Activity based costing identifies activities that consume resources, the cost associated with these activities, and logical links these activities and their costs with specific products. Activity based costing provides an inventory of activities and the associated costs necessary to produce a product. Activity based costing is not a replacement for, but a supplement to, the traditional cost accounting approach.

RESEARCH METHODOLOGY

Various models of the traditional approach to the product life cycle were obtained [5,6]. These models were analyzed for common characteristics and structure. A single process model was developed that represents the overall process of the traditional product life cycle. The process model of the traditional product life cycle served as the basis for the concurrent engineering process model. The concurrent engineering process model was derived by altering characteristics of the traditional model. Basically the modifications reflect the interrelationship of simultaneous activities in the concurrent engineering environment.

Cost attributes for the product life cycle stages were defined based on traditional product cost accounting variables arranged in the context of an activity based cost accounting model [4].

A network model of the traditional product life cycle was built by combining the activity based cost accounting model with the traditional product life cycle process model. A concurrent engineering network model was created in a similar fashion.

It was not expected that product cost data suitable for this research effort could be obtained in the limited time frame of the project. A limited amount of parametric data was obtained. However, many data items were unavailable. This necessitated the creation of hypothetical data.

The network models were implemented as simulation programs in SLAM II [8]. Each experiment was performed by making repeated runs through the simulation programs to obtain product life cycle cost statistics. Based on the statistics the hypothesis of the research was tested. Additional simulations were performed to test sensitivity in the network models. The sensitivity tests altered key parameters of the network models.

PRODUCT LIFE CYCLE MODELS

Three types of models were utilized in this research:
- Process models obtained from industry provided a mechanism for identifying, understanding, communicating and validating the product life cycle stages.
- Transformation models translate the process models into a logical representation.
- Simulation network models translate the logic of the transformation models into a computer program.

Process Models Obtained from Industry Sources

Identification of sources and the selection of product life cycle models were crucial to the successful execution of this research and future expansion of the research. The United States Air Force (USAF) has sponsored numerous research programs during the past fifteen years that were aimed at various aspects of the product life cycle. The results from these programs were the primary source for product life cycle models. The USAF Integrated Computer Aided Manufacturing (ICAM) initiatives provided most of the available information.

All of the process models selected for this project employ a common modeling technique. Specifically the USAF ICAM Definition Language for Activity Modeling ($IDEF_0$) [9] was used. The $IDEF_0$ modeling technique provides a visual way to represent and communicate details about a process. $IDEF_0$ has been widely used by defense contractors since the early 1980s.

No single process model could be found that adequately defined the entire product life cycle. However, basic product life cycle stages were identified and the interrelationships among them were established. Individual models for each product life cycle stage were obtained from existing research results. Thus the entire product life cycle was modeled by combining available models. Distinct components within each stage were delineated and their interrelationships were identified.

The process models selected for use in this research came from major Air Force initiatives [5,6]. Each of these initiatives involved long term cooperative work among numerous and varied organizations. During their development these process models underwent extensive research and development. Final acceptance of the process models required consensus among the contributors.

Even though the original process models were developed for the Air Force, they were targeted for unrestricted technology transfer to industry. Most of the process models used in this research have been used in industry for at least five years. These process models have undergone widespread dispersion and use.

Transformation of Process Models to Logical Cost Models

The process models described in the previous section were originally developed to identify functional manipulations that occur during the product life cycle. The boxes in the models indicate the processes that comprise the product life cycle. The lines that connect the various process boxes serve two purposes. First, the basic relationships between the product life cycle processes are established. Second, the labels associated with the lines identify the type of item that is being transferred between the product life cycle processes.

Inspection of the process models revealed that minimal information concerning costs is conveyed by the models. As they were originally defined, the lines that connect the process boxes are inappropriate for product life cycle costing. However, the basic process boxes that comprise the product life cycle are suitable for costing purposes.

A basic model was defined for the logical manipulation of product cost throughout the product life cycle. The model illustrated in Figure 4 was used to transform the industry process models into logical cost models.

There are two possible inputs for each cost transformation process. Any item that undergoes manipulation by a product life cycle process is either being processed for the first time or is being reprocessed for some reason. When a unit is processed for the first time, cost is added to the unit being processed, the total cost for the process and the total product life cycle cost for all units. In a similar fashion, reprocessing costs are added to the unit being reprocessed, the total reprocess costs for the process and the total product life cycle costs for all units.

When an item is transformed by a product life cycle process, it can have only one of three possible outcomes. A defective design occurs when a process can not be developed that will correctly transform the unit. Defective design will require re-engineering of the product definition so that a suitable process can be developed that will correctly transform the unit.

The second possible outcome of a product life cycle process is when a correct process produces a defective unit. Defective units must be reprocessed to correct any flaws.

The last possible outcome of a product life cycle process is an accepted unit. In other words, a correct process produces a defect free unit.

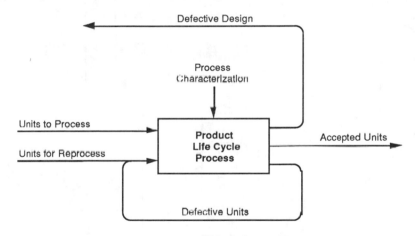

FIG. 4   Process Transformation Model

Process characterization specifies how the process will perform various transformations. Five basic parameters are derived from the process characterization:
- The first time cost of processing a unit.
- The cost of reprocessing a unit.
- The percentage of units requiring rework due to defective design.
- The percentage of units requiring rework due to defective work.
- The percentage of accepted units.

The transformation model shown in Figure 4 was used as the basis to develop logical cost models for the entire product life cycle. The high level model for the product life cycle was transformed into a simplified cost model for the product life cycle that is illustrated in Figure 5. Logical cost models were also developed from the product life cycle process models at lower levels of detail.

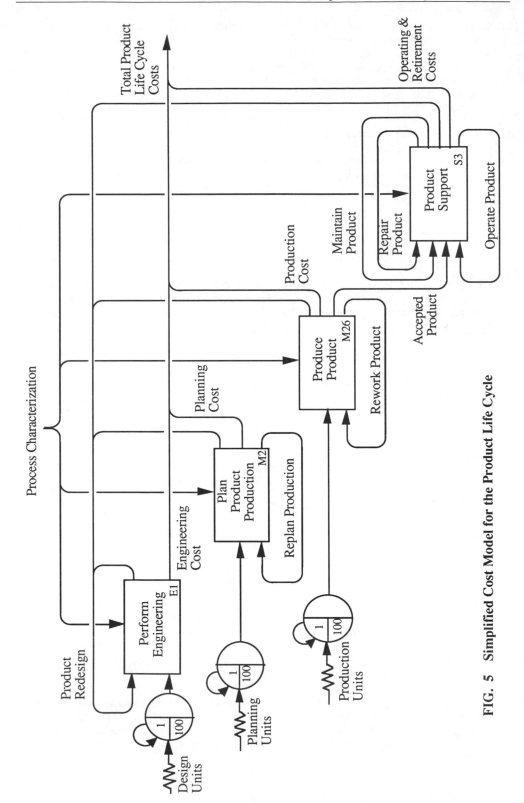

FIG. 5 Simplified Cost Model for the Product Life Cycle

Transformation of Logical Cost Models to Simulation Network Models

The product life cycle process models from which the logical cost models were derived, were not intended for cost considerations of the product life cycle. The logical cost models provided a means to represent, understand and communicate the phenomena under investigation by this research. However, the logical cost models were inappropriate for developing and implementing computer programs. Therefore it was necessary to transform the logical cost models into a graphical representation suitable for simulation on a computer. The Simulation Language for Alternative Modeling (SLAM II) [8] was selected as the graphical technique to represent simulation network models. All of the logical cost models were translated into SLAM network models. Figure 6 illustrates the SLAM network model for a typical subprocess in the product life cycle.

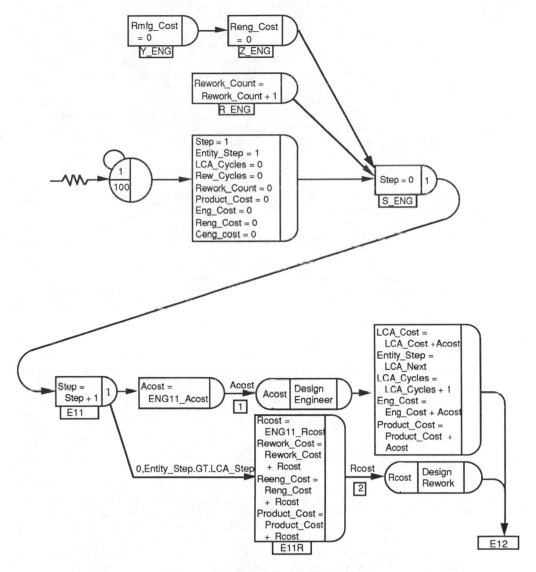

**FIG. 6 SLAM Network Model of the Perform Design
Engineering Subprocess**

RESULTS AND ANALYSIS

This research originally set out to determine if the engineering approach used in product development has an effect on product life cycle costs. Two simulation experiments were conducted to test this hypothesis. A series of additional experiments were conducted to investigate the effects on product life cycle costs that could occur under varying conditions. These experiments produced some startling results which were further analyzed.

Simulation Experiments and Resultant Data

Two simulation experiments were made to establish product life cycle cost data for alternate approaches to product engineering. In these two experiments, process characterization was the independent variable used to quantify the effective level of concurrent engineering. Since the traditional engineering approach to product development does not realize any benefits due to concurrent engineering, a process characterization of 0% was assigned for this situation. At the other extreme, a completely effective implementation of the concurrent engineering approach reaps all the potential benefits. This situation was reflected by a process characterization of 100%.

Various parameters in the simulation models were varied to reflect the effect that the product development discipline would have on the product life cycle. The following parameters were indirectly controlled by the simulation experiments:
- The proportion of total life cycle costs that are allocated to each life cycle stage.
- The relative cost of rework for each life cycle stage.
- The proportion of product life cycle designs that require redesign.
- The proportion of processed units that need rework.

Specific values for these simulation parameters were established based on the extent to which concurrent engineering achieves its theoretical objectives.

A hypothesis test was made to determine if there is a difference between mean life cycle costs of products designed with traditional engineering versus products designed with concurrent engineering. The Behren-Fisher test was used to test this hypothesis. It tests for a difference between the means of two independent normal distributions. The variances of the two means were unknown and not assumed to be equal.

The statistical analysis supports the conclusion that the engineering approach used in product development has a significant effect on the life cycle costs of the product.

Sensitivity Analysis

In actual operational situations neither the traditional engineering or concurrent engineering approaches to product development can be fully achieved. Therefore, three additional experiments were defined and conducted in order to obtain data between the two theoretical extremes. Process characterization levels of 25%, 50% and 75% were used to simulate suboptimal effectiveness of the concurrent engineering approach. For instance, if concurrent engineering can potentially achieve an 80% reduction in rework costs, then at a 75% effectiveness level for concurrent engineering a 60% (e.g. 0.8 x 0.75) true reduction in rework costs will be realized.

The results from all five experiments are illustrated by the notch plots shown in Figure 7. Each plot provides insight into the descriptive statistics and distribution of the data from each experiment. The leftmost and rightmost notch plots depict the experimental data used to test the null hypothesis. The leftmost plot reflects the total product life cycle cost resulting from the traditional engineering approach to product development. A median cost of 135,000 is indicated by the narrowest horizontal line at the notch. The rightmost plot shows a median product life cycle cost of 115,000 for the concurrent engineering approach to product development. Inspection of these two plots leads to the rejection of the null hypothesis. This is readily apparent because the notches of the two plots do not overlap one another.

The three notch plots in the middle of Figure 6 depict the product life cycle costs for concurrent engineering effectiveness levels of 25%, 50% and 75%. The results for these three levels were unexpected and alarming. At concurrent engineering effectiveness levels of 50% or less, the product life cycle costs are significantly higher than they were for the traditional engineering approach to product development. Even at a 75% effectiveness level for concurrent engineering the product life cycle costs were only slightly less than they would have been based on the traditional engineering approach.

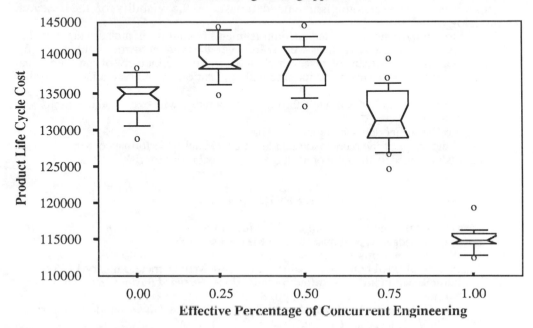

FIG. 7 **Sensitivity Analysis of Product Life Cycle Costs Based on Different Levels of Concurrent Engineering**

Summary of Research Accomplishments

This research was originally undertaken to establish a methodology for comparing alternative techniques that can be applied to problem solving. The methodology has general applicability to a broad class of problems due to its flexibility. The methodology may be modified to fit a particular area of concern.

In order to test the feasibility of the methodology a specific problem was identified. The problem under consideration was to evaluate the efficiencies of two alternative techniques used for product design. The scope of the problem addressed the product life cycle stages from product definition, through production and product use and support.

Analytical models previously had been developed which quantitatively assessed particular stages of the product life cycle [5,6,9]. Typically these analytical models measure the efficiency of a single design technique in an operational situation. However no quantitative models were found which address the entire product life cycle. Also no quantitative approaches which compared alternative design techniques were identified.

Thus this research is unique in three respects:

- The research defines and tests an approach for comparative analysis of alternative problem solving techniques.
- The scope addressed by the particular project in this research is broader than any known analytical model.
- The comparative analysis approach also provides a means to conduct sensitivity analysis of the alternative problem solving techniques.

Recommendations for Further Research

Several interesting conclusions were drawn from this research effort. The most important finding came from the sensitivity analysis that was conducted. It revealed that inefficient implementations of concurrent engineering will result in an unexpected and undesirable increase in product life cycle costs.

Several important accomplishments of a more general nature were realized as a result of the research. These accomplishments demonstrate the viability of the research methodology.

Even though the conclusions and accomplishments from this research are theoretically oriented, they have practical value as well. Moreover, there are numerous implications for further research. The stream of research should combine refinement of the theoretical methodology with an expansion of its practical application. Candidate areas for further research include:

- Refinement of the network models through collaboration with experts and consensus of their opinion.
- Validation of the models with empirical data.
- Expand the scope of the models to include time and quality performance factors.
- Embellish the models to incorporate subjective probabilities for risk analysis.

REFERENCES

1. Dertouzos ML, Lester RK, Solow RM: *Made in America: Regaining the Productive Edge*, Cambridge, Massachusetts, The MIT Press, 1989.
2. The Packard Commission: *A Quest for Excellence, Final Report to the President*, President's Blue Ribbon Commission on Defense Management, January 1986.
3. Kaplan RS: Must CIM be justified by faith alone, *Harvard Business Review*, Volume 64, Number 2, March-April 1986, pp. 87-95.
4. Berliner C, Brimson JA: *Cost Management for Today's Advanced Manufacturing - The CAM-I Conceptual Design*, Boston, Massachusetts, Harvard Business School Press, 1988.
5. *Geometric Modeling Applications Interface Program (GMAP), Functional Requirements and Architecture of a Product Modeler, Technology Transfer Document*, TTD 560240003U, Wright-Patterson Air Force Base, Ohio, Materials Laboratory, Air Force Wright Aeronautical Laboratories, Air Force Systems Command, August 1989.
6. Norwood DL, Moraski RL, Snodgrass BN, Kelly SJ, Ward L: *ICAM Conceptual Design for Computer-Integrated Manufacturing - Conceptual Framework Document*, Report Number MMR110512000, Wright-Patterson Air Force Base, Ohio, Materials Laboratory, Air Force Wright Aeronautical Laboratories, Air Force Systems Command, February 1984.
7. Winner RI, Pennell JP, Bertrand HE, Slusarczuk MMG: *The Role of Concurrent Engineering in Weapons System Acquisition*, IDA Report R-388, Alexandria, Virginia (available through the National Technical Information Service, Springfield, Virginia), Institute for Defense Analysis, December 1988.
8. Pritsker AAB: *Introduction to Simulation and SLAM II*, 3rd Edition, New York, John Wiley & Sons, 1986.
9. *Integrated Computer-Aided Manufacturing (ICAM) Architecture Part II, Volume IV - Composite Function Modeling Manual (IDEF$_0$)*, Report Number AFWAL-TR-81-4023 Volume IV, Wright-Patterson Air Force Base, Ohio, Materials Laboratory, Air Force Wright Aeronautical Laboratories, Air Force Systems Command, 1981.

DESIGN CONSIDERATIONS FOR CNC MACHINING IN CONCURRENT ENGINEERING

Vijayakumar Pandiarajan

Industrial Engineering, West Virginia University, Morgantown, West Virginia, U.S.A.

ABSTRACT

Designing a product in the perspective of CNC machining is a crucial function in the concurrent engineering domain. In this paper, an architecture is presented to enhance the interaction between various functional groups involved in the product development. The goals and the needs of the functional groups are presented. The goal programming approach is suggested to produce globally satisfiable design as the design evolves in the concurrent engineering environment. The interaction effect in making a CAD based design more producible in the CAM environment is discussed through illustrative case studies.

INTRODUCTION

In the modern day production scenario, it is commonly seen that Computer Numerically Controlled (CNC) machines play a significant role. The predominant input supplied to such machines to accomplish the set goals is largely the Numerical Control (NC) tape. The preparation of such numerical control tapes does not always guarantee problem-free machining on the machine tool. Cutter interference problems are addressed while machining sculptured surfaces and remedy is presented through algorithmic approach [1]. A method through simulation to detect and eliminate cutter tool assembly collisions with work surface is presented [2]. In the case of compound surface numerical control machining, the occurrence of gouging problem is addressed and approaches to remedy the occurrence is reported [3]. The gouging problem in numerical control machining of sculptured surfaces is reported and attempt has been made to alleviate the problem by generating an offset surface which avoids the intersection loop [4]. A methodology to determine the optimal sequence of operations on a CNC machining center for the production of automotive engine blocks is presented [5]. In numerical control machining, an attempt has been made to reduce the cutter travel time by exploiting the orientation of the workpiece with respect to the machine axis[6]. Hence there is a need to foresee the potential problems during machining, while designing the product and making the numerical control tapes. Manufacturing excellence can be attained only by designing products and processes to address potential problems before they occur [7]. There are some efforts to promote the concept of concurrently designing a product with parallel interaction with manufacturing group. A system has been developed called 'First-cut', which can support the design refinement, as it evolves from scratch [8]. The development of a system, using the structure of process plan together with a catalogue of manufacturing alternatives to generate suggestions to improve the design is reported [9]. A methodology is proposed for simultaneous product and process design of components manufactured in small and medium lot sizes, for the lowest possible cost [10]. In this paper, the goal programming approach is suggested to bring in concurrency while designing a product.

1.1 Concurrent Engineering

Concurrent Engineering is defined as "a systematic approach to the integrated, concurrent design of products and their related processes, including manufacture and support". This approach is intended to cause the developers, from the outset, to consider all elements of the product life cycle from conception through disposal, including quality, cost, schedule, and user requirements [11].

1.2 Concurrent engineering in CAD and CAM

The general definition of concurrent engineering given in the previous section portrays the early interaction of all the functional groups involved in the product development. In the perspective of computer aided design and computer aided machining, if we focus only on the subsystem level which is involved in the design and physically building the part, the following architecture can be thought of to bring in the issues pertinent to the CNC machining at the very early stage [12]. The functional groups involved in the part design and manufacture are the geometrical design group, various engineering analysis groups, system integration groups, production scheduling groups, production tooling and fixturing groups, process planning groups, shop floor production groups and inspection groups. These groups have the need to maintain open window policy, to receive the comments from other functional groups in the network, to be able to make the part design more producible. The proposed architecture is shown in FIG.1.

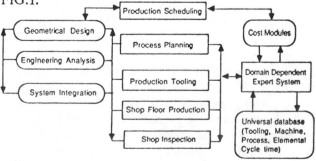

FIG.1 Concurrent Engineering Architecture

2.0 Functional group goals

The goals guide the task groups in designing the guidelines. In the scenario presented, there are several groups which are expected to work in parallel. The irony is the fact that, though every functional group is committed to produce the physical part, the individual interests differ to meet the sub goals which are predominant in the domain in which they operate.

2.1 Design group goals

The design group is normally committed to producing a design which can deliver maximum performance under the specified operating conditions. They often recommend stringent technological specifications. These imply asking the manufacturing group to produce a part with close tolerances on the part dimensions, while having a total disregard to the available production capacity in terms of delivering the tight specifications economically.

2.2 Process Planning goals

Process planning groups are often blamed for increased production time and shop rejection in the manufacturing domain. This is because they are at the receiving ends traditionally accepting the given design and engineering drawing, and figuring out the possible operational procedures. For new materials, and new geometrical shapes, identifying and recommending the process steps is a time consuming process. They often tend to prefer the selection from the known process plans which have proved their ability in delivering the required dimensional attributes.

2.3 Scheduling goals

Scheduling is often an ignored and less recognized functional group except during emergency situations when there are heavy rejections in one or more production centers. They will be posed with a problem of scheduling the parts at short notice to meet the delivery schedules. In reality, this group is found to have fixed constraints in terms of machine tools which can do the desired operation. In a situation where there are more machining centers with medium capacity limits, this functional group prefers to have similar design or at least features on design which can be interchangeably produced on these machines. The primary domain goal in this case is to load the parts on the available machines to satisfy the global goal of meeting the desired production quantity which is a function of time.

2.4 Tooling goals

In the manufacturing subsystem level, the existence of this group is very critical due to several obvious facts. In a multi product, multi project production situation, one is often faced with totally dissimilar products having differential engineering specifications. The supply of cutting tools and work holding fixtures to satisfy varied designs is a very complex problem. This group often tends to have a sub goal of providing the same fixtures for similar work pieces. The similarity is expressed often in terms of occupying space, part location facilities, clamping arrangements and tool approach directions.

2.5 Shop floor production goals

The shop floor gets engineering drawings from the design group which dictate the operational sequence in numerical control tape. This group has some preset goals such as, reducing the set up time by machining larger batch quantities, loading similar components, and performing as many operations as possible before the part is disturbed from the fixture. These goals may often be constrained by decisions taken by production planning and control, scheduling and design groups.

2.6 Inspection groups

The internal functioning structure of an inspection group often depends upon the kinds of components manufactured and the order quantities. In a mass production company, the inspection groups are entrusted with inspecting one particular kind of component. Whereas in a flexible production set up the same inspection group is assigned to inspect parts of a different nature. This poses some practical problems such as loading the inspection control tapes and establishing inspection set up in a coordinate measuring machine. In reality, the inspectors accord priority to components with larger batch size, in order to increase their groups net efficiency and reduce throughput time. But, again this may conflict with the goals of other groups.

3.0 GOAL PROGRAMMING APPROACH

In the previous section, the functional goals of various task groups are briefly highlighted. In the process of making a component, satisfying individual interests and still meeting the common goal of producing parts to perform to the functional specification should be the major objective. In the proposed architecture, the interactions of functional groups are facilitated to make the design right the first time. In concurrent engineering, the focus is to satisfy the individual interests while making the final product design. The problem of product development in the new environment can be mathematically formulated through goal programming.

Let

P1 = design the parameters to deliver required performance

P2 = produce the part at reduced cost

P3 = make the design more producible, by specifying the material
and technological specifications which can fit the
available in-house resources.

P4 = make the design such that maximum number of features can be machined in
one set-up. This implies that, for instance provide features in clusters which can
be easily approached by machining center spindle either in one orientation or
multiple orientations of several faces of the component in reference to the
spindle axis.

P5 = provide similarity between features.

P6 = provide layout of dimensions keeping common absolute reference.

Where P1, P2, P3, P4, P5 and P6 are the priorities or the goals of individual groups.

X1 = the performance achievement through design parameters

X2 = the estimated product cost for the given design

X3 = the percentage utilization of the in-house resources in producing the physical
part for the design

X4 = the # of set-ups in which the given design can be machined in machining
centers

X5 = the degree of feature resemblance

X6 = the percentage satisfaction of dimensions layout

$DP1^{+}$ be the over achievement of priority 1

$DP1^{-}$ be the under achievement of priority 1

$DP2^{+}$ be the over achievement of priority 2

$DP2^{-}$ be the under achievement of priority 2

$DP3^{+}$ be the over achievement of priority 3

$DP3^{-}$ be the under achievement of priority 3

$DP4^{+}$ be the over achievement of priority 4

$DP4^{-}$ be the under achievement of priority 4

$DP5^{+}$ be the over achievement of priority 5

$DP5^{-}$ be the under achievement of priority 5

$DP6^{+}$ be the over achievement of priority 6

$DP6^{-}$ be the under achievement of priority 6

The objective function can be expressed for this scenario as follows.

Minimize

$$P1(DP1^{-}) + P2(DP2^{+}) + P3(DP3^{-}) + P4(DP4^{+}) + P5(DP5^{-}) + P6(DP6^{-}) \quad (1)$$

Subject to the constraints

$X1 - DP1^+ + DP1^- = $ Rating for full performance R_p
$X2 - DP2^+ + DP2^- = $ Preset cost factor C_f
$X3 - DP3^+ + DP3^- = $ Resources Utilization level R_u
$X4 - DP4^+ + DP4^- = $ Set-ups upper bound S_u
$X5 - DP5^+ + DP5^- = $ Preset similarity index S_i
$X6 - DP6^+ + DP6^- = $ Preset dimensions index D_i
$P1 >= P2 >= P3 >= P4 >= P5 >= P6$
$X1, X2, X3, X4, X5, X6 >= 0$
$DP1^+, DP2^+, DP3^+, DP4^+, DP5^+, DP6^+ >= 0$
$DP1^-, DP2^-, DP3^-, DP4^-, DP5^-, DP6^- >= 0$

In the mathematical form, the problem is reduced to simply solving for the variables which make the objective function optimum. This mathematical expression can be extended to accommodate the interests of all the groups involved in the product development.

4.0 CASE STUDIES

The design can be made more producible by satisfying all the groups involved in the product development in the concurrent engineering environment. As illustrated in the goal programming approach, the focus now is to minimize the variables indicated in the objective function. The following case studies illustrate how a product becomes more producible by the enhanced interaction.

Case study 1:

An example part is presented as shown in FIG. 2. This is a gear box casing, also called banana casing, which goes into the gear box of super-sonic aircraft turbo engine. This gear box casing has several kinds of holes on the mating face, such as type 1 and type 2 on a PCD (Pitch Circle Diameter), four type 3 holes for location, and several type 4 holes. The body of the casing has oil holes at the top and an inclined face F1 with equi-distant holes on its face. Also the type 3 holes are through holes with counter bores on either side. In the case of type 1, 2, 4, the holes are blind and do require counter sinks of various sizes in relation to the hole diameters.

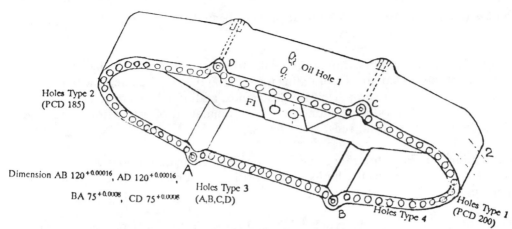

FIG. 2 Gear Box Casing

This component is normally made of a magnesium alloy, which comes for machining in the form of casting. If we were to take this component for machining on a machining center with four degrees of freedom, then this would have to be done in three different set-ups. In set-up 1, side 1 can be machined except Face 1, its corresponding holes, and oil holes. In set-up 2, using tilting fixture, the inclined face (Face 1) has to be machined. Similarly, in set-up 3, the oil holes can be machined using another inclined fixture. The last set-up may be performed in conventional machine instead of machining center. If we pay close attention to the drawing, it will be evident that, we can not use the same subprogram for type 1 and type 2 holes as they are located on different PCDs. Also because they are of different diameters, it calls for usage of different size drills. As type 4 holes are of different diameters, they need to be produced using different diameter drills. In addition to these requirements, measurement of counter sinks of these holes take more time in inspection, as they are of different sizes and having differential depths. The dimensional requirements for the holes A, B, C, and D pose another problem. As they are too closely placed in dimensional tolerance, they need to be maintained in jig boring operation either prior to CNC machining or after CNC machining.

In the new concurrent environment, if all the groups are brought together, the individual group goals can be highlighted and design for this drawing can be modified as follows, subject to the functional requirements.

Modification 1 : Holes Type 1 and Type 2 can be made to the same size, including the counter sinks on a common PCD.

Effect realized : This enables usage of same drill sizes, reduces the number of tool dispatches, and reduces inspection time. Also, this permits the use of the same subprogram in locating the tools on these PCDs once the tool is positioned at the center of the PCD through incremental numerical control program.

Modification 2 : Oil holes which are at present located at the top of the casing, can be brought to one of the sides.

Effect realized : This modification, reduces the total number of set-ups by one, as the holes can now be drilled, just by orienting the table to the required angle with respect to the reference axis of the part. Also this avoids, additional fixture to position the part. As this is now produced in one set-up, accumulation of errors in the multiple set-ups, is controlled within reasonable limits.

Modification 3 : Close tolerance specifications for linear dimensions AB, AD, BA and CD can be relaxed to the limits of the positional tolerances offered by the CNC machining center.

Effect realized : This change, avoids additional set-up in the jig boring machine and the requirement of corresponding fixture. Also, this avoids, possible dimensional deviations that could result in other hole dimensions, if they are produced in different set-ups.

Modification 4 : The Face 1 orientation and the associated feature details can be placed in line with side 1. This implies, making Face 1 parallel to the mating face.

Effect realized : The new change, permits the machining of the Face 1 along with other features on the side 1. This reduces the total number of set-ups by 1. Also, this facilitates the inspection in checking the face details, as they are now in a plane parallel to the mating face.

In the above case, the goals of various functional groups which influenced the design change, confirm with the objective function presented in equation 1. In the same lines, we can bring in several other goals which can play a significant role in altering the design, in making it more producible.

Case study 2:

In this case study, another example part is presented, which is often encountered in electronic assembly of communication equipments in aircraft. This is called outer cover and is made out of special grade aluminum material. The three dimensional drawing is given in FIG. 3. In this part, it is seen that, there are four uniform holes at the top and there are sixteen equi-spaced holes on the front end side. The flange sections are indicated as part of the side walls. As a requirement, inner base must be a machined face and the walls need to be machined to the dimensions.

As such, the producibility of this part is on the average side due to many reasons. In the concurrent environment, if the same part is produced and individual group goals are met, the same part can be remodeled taking into account the following modifications.

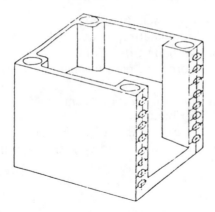

FIG. 3. Electronic Assembly Unit

Modification 1 : The overall height needs to be reduced, as it prevents the cutting tool with spindle head from reaching the bottom floor of the unit. This is essential, in order for the machining to take place without fouling with the top surface of the part.

Effect realized : This modification obviates the need to use special extensions for the tool holder. Also it produces good surface finish at the floor, and permits machining to take place at increased speed and feed rates.

Modification 2 : This is regarding the flange sections which are part of the walls. As we see, on three sides, the wall thickness is very small compared to the rest of the wall thickness. During machining, this kind of design, will pose numerous problems in maintaining uniform wall thickness. In addition to this, one could expect tapered wall surface, which is undesirable in the assembly stage. The modification suggested to avoid this problem is increasing the wall thickness to an acceptable assembly requirement level. As there are sub-assemblies to be placed inside the cover unit, there is an upper limit for the wall thickness. Also the overall weight and heat dissipation characteristics depend on the wall thickness, and need to be respected.

Effect realized : The modifications brought to this unit, in terms of increased wall thickness, enables machining at a higher speed and feed rates. This facilitates easy machining and enhances the ability to maintain uniform wall thickness, avoiding possible assembly problem.

Modification 3 : This is regarding the tool vibration, possible cutter breakages and the eventual poor surface finish on the inner walls. It is evident from the drawing that, the corner radii at two places are very sharp. If this is insisted, this requires additional set-up on a slotting machine to maintain the specified radii. Moreover, in the rough machining, if we approach the final radii in CNC machining, this will result in heavy tool vibration and tool chatter, as we need to use as small tool diameter as possible, with more tool overhang from the tool holder. Hence, it is suggested to increase the corner radii in order to plan machining with larger size cutters.

Effect realized : Good surface finish can be maintained at the corners, allowing uniform cutting forces during machining. Another, important factor is, reduction in tool breakage and the use of less number of cutting tools. The number of set-ups get reduced by one, which results in improved cycle time and the associated cost.

5.0 Conclusion

In this paper, the real concerns of various functional groups in product development are discussed and the need to respect the individual goals in the fulfillment of the global goal is emphasized. The problem is mathematically formulated using a goal programming technique and the objective function is constructed. The architecture that promotes mutual interaction in the product development is proposed to facilitate the development of globally agreeable design. Design refinement is illustrated through case studies, in the perspective of CNC manufacture. It is suggested that, all the functional groups need to be interconnected by networks to facilitate exchange of drawing information and relevant technical advice. A black board concept needs to be brought into the system, enabling posting of information to the appropriate agency. Tools like domain dependent expert system can be embedded in the system architecture, to work in an advisory mode, when there is a lack of coordination in the organizational structure.

REFERENCES

1. Choi, B.K., Jun, C.S : Ball-end cutter interference avoidance in NC machining of sculptured surfaces, Computer Aided Design, Vol 21, No 6, July/Aug 1989, pp 371-378.
2. Anderson, R.O : Detecting and eliminating collisions in NC machining, Computer Aided Design, Vol 10, no 4, July 1978, PP 231-237.
3. Choi, B.K., Lee, C.S., Hwang, J.S., Jun, C.S : Compound surface modelling and machining, Computer Aided Design, Vol 20, 1988, pp 127-135.
4. Chen, Y.J., Ravani, B : Offset surface generation and contouring in computer Aided design, American Society of Mechanical Engineers, Journal of Mechanics, Transmissions, and Automation in Design, Vol 109, 1987, pp 133-142.
5. Bisschop, J.J., Stegeman, H., Striekwold, M.E.A : The sequencing of point operations on a CNC machining center using a microcomputer, International Journal of Production Research, Vol 26, No 8, 1988, pp 1375-1383.

6. Yang, D.C.H., Chou, J.J : Part set-up angle for optimal machining time in CAD/CAM, Robotics and Computer Integrated Manufacturing, Vol 6, No 2, 1989, pp 125-131.
7. Heldenreich, P : Designing for Manufacturability, Quality Progress, May 1988, pp 41-44.
8. Cutkosky, M.R., Brown, D.R., Tenenbaum, J.M : Extending Concurrent Product and Process Design Toward Earlier Design Stages, Concurrent Product and Process Design, The Winter annual meeting of the ASME, Dec 10-15, 1989, pp 65-72.
9. Hayes, C.C., Desa, S., Wright, P.K : Using Process Planning Knowledge to make Design suggestions concurrently, The Winter annual meeting of the ASME, Dec 10-15, 1989, pp 87-92.
10. Subramanium, S., Lu, S.C.Y : Computer-Aided Simultaneous Engineering for Components Manufactured in Small and Medium Lot sizes, The Winter annual meeting of the ASME, Dec 10-15, 1989, pp 175-183.
11. Winner, R.I., et al : The role of concurrent engineering in weapon system acquisition, IDA report R-338, Institute For Defence Analysis, Alexandria, VA, Dec 1988.
12. Pandiarajan, V : Computerized Numerically Controlled Machining of Product Designs with Complex Geometries in Concurrent Engineering, Internal report, Industrial Engineering Department, West Virginia University, 1992.

AN APPROACH TO OVERALL DESIGN CONSISTENCY IN CONCURRENT ENGINEERING

James S. Noble

*Industrial Engineering Program, University of Washington,
Seattle, Washington, U.S.A.*

ABSTRACT

The design process typically used by manufacturing firms reflects a hierarchical organizational structure. It has been well documented that this approach to design is overly complex and inefficient in today's rapidly changing, consumer-based market. A proposed solution to this problem is to integrate the design activity within product development. This approach, termed concurrent engineering, presents new opportunities for efficient design activity.

Effective, timely communication is an important element in concurrent engineering. This paper will address one aspect of this communication, design decision consistency. Design consistency is achieved when design decisions from each different design function/area are integrated and communicated within an organization in a consistent manner.

This paper presents a conceptual framework for the design consistency problem that is based on an overall perspective of the manufacturing firm. The design consistency component of the framework is implemented using a rule-based production system for a specific company's product design project. An example of the design consistency component applied to a specific design scenario is presented, and issues that must be addressed in an implementation of a complete design consistency framework are discussed.

1.0 INTRODUCTION

Traditionally, manufacturing firms have used a hierarchial organizational structure in their design process. This implies that design has been viewed as a combination of processes which are treated sequentially along company defined function boundaries. The design process is then conducted by iterating between these independent functions. As a result, the development and implementation of a design is extremely time consuming because the corresponding channels of communication are not directly integrated. This approach to design is overly complex and inefficient, especially in light of the current dynamics of the competitive market environment.

In recent years the size of manufacturing organizations and the amount of information associated with them has grown, causing the organizational complexity to increase as well. Historically, complexity has been dealt with by increasing the level of specialization within the organization. This has caused the design function to become a department of the organization, which is then typically broken down into further specializations. As a result, design can end up competing with other departments within the company rather than working with them to achieve a marketable product. This division of labor has created a more sequential design process, with more recognizable iteration [1, 2].

Two main assumptions underlie the concept of organizational specialization. The first assumption is that smaller parts of a larger problem when solved independently result in a solution to the larger problem. This is called problem decomposition. The second assumption is that the problem of coordination and communication due to fractionalization can be dealt with by using traditional management organization and techniques. The validity of both of these assumptions has been questioned. Lardner addresses the error in these assumptions in the following statements: 1) "optimization of each of the parts inevitably results in unacceptable sub-optimization of the whole," and 2) "the cost to manage the problem begins to increase at some exponential rate, while the effectiveness of the management effort plunges" [1, p. 37]. Schmidt echoes the words of Lardner when he says "Performance has not been optimum because we have carried specialization too far"[2, p. 6]. The need he suggests is for "shared responsibility" in design.

There are several possible approaches to the problems resulting from organizational specialization. All require a greater degree of integration of the whole organization which implies that the design process should take place in an integrated fashion as well. One approach suggests that this type of integration be implemented through the formation of design teams that consist of individuals from each of the functional areas of an organization. A variety of other integration approaches include the development of a common database, specific design tools that present integrated information to the designer, and expert systems for post design analysis that suggest areas to improve the design with respect to X (where X can be assembly, cost, manufacturing, inspection, testing, etc.)

Yet another approach to integrating the design process is to provide integrating information on the different design processes to the associated departmental functions so that communication is improved. This can be achieved by providing the ability to see the effect of individual design function/area decisions on the "different" functions/areas and on the organization as a whole. The result of integrating the different function/area decisions is that the final design is consistent. This ability to see the effects of different design decisions and how they interrelate from a broad consistency perspective is crucial if there is to be integration of the design process so that overall organization objectives are attained.

This paper will first review the general topic of concurrent engineering providing the background for the problem of overall design consistency. Next, a conceptual framework for overall design consistency will be described. Finally, an implementation of the overall design consistency component of the framework will be presented for a specific design problem.

2.0 REVIEW OF CONCURRENT ENGINEERING AND DESIGN CONSISTENCY

The concept of concurrent engineering was initially promoted in the early 1980's. However, it wasn't until 1986 that the term "concurrent engineering" was first coined by the Institute for Defense Analyses to describe a "systematic method of concurrently designing both the product and its downstream production and support processes" [3]. The concept has taken a variety of names such as simultaneous engineering and integrated product design, yet all have as their intent to provide an integrated design ability.

All though the concept has been around for more than a decade, it has been said that behind the jargon little of substance has been developed [4]. However, recently a variety of techniques have been developed that make the concept usable and practical. The basic approaches behind the techniques can be broken down into three major areas: team organization, structured procedures, and consulting systems. Representative work in each area is described as follows:

    1. Team organization - approaches have focused on the development of design teams that include representatives from all aspects of an organization [5, 6],

    2. Structured procedure - approaches provide either a list of questions to be considered

[3] or specific design analysis procedures [7], and

3. Consulting system - approaches have ranged from rule-based manufacturing advisors [8, 9] to various constraint based approaches [10, 11, 12]. This area has been the focus of much interest recently and appears to have high practical value.

(Note: A more extensive review concurrent engineering approaches can be found in O'Grady and Young [13].)

The design consistency approach discussed in this paper can be classified as a consulting system type. The following briefly reviews the concurrent engineering literature that address three issues associated with design consistency: constraints, information, and integration scope.

The constraint based concurrent engineering approach has received a considerable amount of attention and is most analogous to the concept of design consistency. The origin of the constraint based work is constraint programming, which has been researched since the 1970's with the purpose of developing procedures for which the goal is the satisfaction of a set of constraints. However, Bowen et al. [11] comment on the inappropriateness of pure constraint satisfaction to the concurrent engineering problem due to the complexity of the issues and problems associated with concurrent engineering rendering it an infeasible and undesirable approach. Constraints were also addressed by Ishii et al. [10] through the development of a design compatibility analysis that explores the appropriateness of the different components of a design and the design alternatives. It then assigns a design rating based upon the level of compatibility. A progression of constraint programming languages have been developed to represent constraint networks so that resulting advice systems can consider life-cycle engineering issues. Bowen et al. [11] describe the specifics of such a constraint language applied to the analysis of printed wiring board holes. O'Grady et al. [12] present an implementation of a constraint network that notes which constraint is violated and which parameters caused the violation.

Another issue associated with design consistency is the availability of concurrent information from a broad range of decision functions and tools. Preiss et al. [14] noted that inefficient information flow is the primary problem associated with integrating design and manufacturing. Salzberg and Watkins [15] addressed the information problem by discussing the issues related to the development of a heterogenous design database that would make design information accessible to those who need it. Mullineux [16] proposed that a blackboard structure, a central data area accessible to all functions, would provide for the integration and storage of design data. He further states that decision intelligence can be added to the blackboard by including the ability to make decisions through the analysis of constraints. The addition of decision intelligence is an area of research that addresses the shortcomings of the constraint satisfaction methodologies mentioned previously.

The majority of the work on concurrent engineering has concentrated on integrated product and process design issues. However, the scope of integration should be expanded if design decisions are to be totally consistent with the overall organizations objectives. Morely and Pugh [17, 18] have both proposed that it is necessary to consider all factors of the organization within the product design task. In what Pugh calls the "business design activity model" the viewpoints of research, purchasing, marketing, development, manufacture, finance and sales are all considered as part of the product design process. Their premise can be summed up by their statement that, "What is needed above all else in the context of design is the use of systematic methods which provide a structure so that disagreements converge productivity onto solutions all can understand and can accept" [17, pp. 219-220].

The framework for overall design consistency presented in this paper draws from the constraint network approaches, information issues and an overall organization viewpoint to provide a comprehensive approach to a key aspect of the concurrent engineering problem.

3.0 APPROACH TO OVERALL DESIGN CONSISTENCY

There are two general scenarios in which a tool for design consistency can be utilized in the design process. The first situation is one in which an individual who has decision responsibility for a specific function needs to consider the impact of their decisions on others associated with the design effort. In this case the individual would need to be able to access the most current input of each of the other design functions/areas. If the design effort was still in the preliminary stages and at least one pass at the design had been taken by each area, it would be necessary to obtain "most likely" design information from a database. Once the data is obtained the individual could experiment and explore trade-offs associated with the decision that they are responsible for. The second situation for which the design consistency concept would be applied is within a concurrent design team meeting. In this case, the need is for a basis from which to discuss the inherent trade-offs between different design decisions. A design consistency tool enables team members to see the point where the design inputs have become inconsistent and which input is responsible for the inconsistency. The ability to explore design trade-offs using the consistency approach removes some of the subjectiveness and potential miscommunication that can occur within a design effort.

Interviews with personnel with different responsibilities on design projects provided the basis for the development of the overall design consistency approach. They also confirmed the application of design consistency in the two general scenarios mentioned. It was also noted that in organizations where a concurrent design approach had not been formally adopted, that the use of the design consistency approach would serve to introduce the concurrent design concept and would provide benefit in the transition from a traditional sequential design effort.

The design consistency concept depends on the participation of a representative from each area/function that has input to the design process. Representative areas/functions and the perspective they bring to the design process are presented in table 1.

Table 1: Design Function Responsibilities

Functional Area	Design Impact/Function
Management	Provide perspective on goals
Finance/Accounting	Provide understanding on capital investment budgets, product costs
Marketing	Provide information on customer needs and preferences in terms of cost, quality, quantity and timing.
Production Engineering	Provide information on manufacturing capability and costs.
Product Engineering	Provide information of materials, function, and form.

Observation of the design process reveals that no single design function has total control or input into a single design factor. Rather, each design factor is influenced by several areas. Table 2 shows how each functional area contributes to each design factor.

The underlying concept behind the design consistency framework is to provide an environment where the design information factors from different departments within an organization can interact and see how they affect other departments. The result of the interaction is to show which design information is consistent with other departments and which is inconsistent or conflicting with what another department has specified.

Table 2: Design Factor Input

Design Factor	Functional Area
Fixed Investment (amount and cost)	Finance, production engineering, product engineering
Material (amount and cost)	Product engineering, production engineering
Labor (amount and cost)	Management, production engineering, product engineering
Output (amount, cost, & ability)	Marketing, production engineering
Profit, Investment	Management, Finance

The general framework of the design consistency concept contains seven major components. Figure 1 illustrates the interaction between the different components. In the first component the specific design problem is defined. At this point the problem is broken down in a logical manner so it can be addressed from the various departmental views in the organization. In the second component, design groups are developed from the overall organizational structure. A decomposition algorithm such as the one developed by Kusiak [19] could be used to form logical design functions that are problem specific. The third component is made up of the actual departmental/design functions that will interact with each other throughout the design process. The fourth component is a data board where the results of each decision made by a design function are posted and information is exchanged in a consistent manner. Underlying the data board is the fifth component which is a general organizational data base used to provide early data before specific decisions have been made. The sixth component is the actual design consistency tool that checks for consistency of the design inputs and suggests where inconsistencies might be resolved. The final component is a consistency resolution tool that not only resolves inconsistencies, but also provides a means of conducting a decision trade-off analysis of the variety discussed by Noble and Tanchoco [20].

4.0 IMPLEMENTED DESIGN CONSISTENCY RULE-BASED SYSTEM

The implemented design consistency rule-base was developed using data obtained from a company that designs and manufactures electrical metering devices [20]. The implemented design consistency component is for the design of an EMI (electromagnetic interference)/RFI (radio frequency interference) shield for an electrical metering device. This

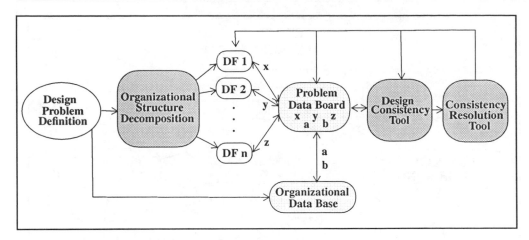

Figure 1 Overall Design Consistency Conceptual Framework

product became necessary after recent advances in electrical metering devices resulted in the use of solid state electronics as the measuring mechanisms rather than mechanical components. These meters are typically used in environments that are characterized by large amounts of electromagnetic and radio waves which affect the solid state electronics by altering the specified performance characteristics. Therefore, to ensure accurate electrical metering, a shield for the metering device became necessary.

The major function of the shield is to reduce the EMI/RFI in the metering circuitry. The effectiveness of such a barrier relies primarily on the conductivity of the material used. Two ways in which sufficient conductivity can be obtained in the shield are 1) use of an engineering plastic / conductive material composite, or 2) use of an engineering plastic coated with a conductive material. The design process for an EMI/RFI shield consists primarily of selecting the performance requirements and shape, structural materials and associated forming process, and the shielding material with its associated application process.

The design consistency system was implemented using the KES-PS software, version 2.4. KES-PS is an expert system building tool that utilizes a production rule inference engine. KES-PS allows for the inclusion of certainty factors for attributes provided to the system. However, due to the nature of the rules and attribute values needed, the ability to utilize certainty factors within the implementation was somewhat limited.

When using the design consistency system, information related to five (5) different areas is requested. These five areas are described below and their interaction with the design parameters is illustrated in figure 2.

Marketing (customer)
 desired performance - EMI protection, deflection temperature, quality
 desired quantity and timing - annual demand, early due date, first batch size,
  expected product life
 desired cost - unit production cost, market price

Finance/Management
 minimum rate of return, maximum payback period, capital interest rate, project
 capital budget

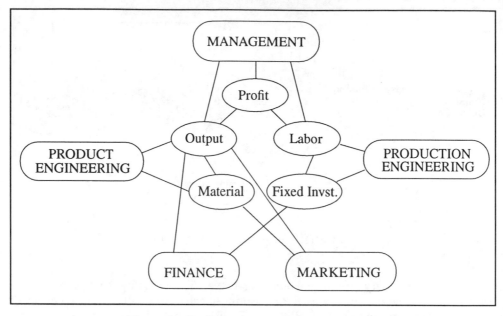

Figure 2 - Basic Design Parameter Interaction

Product & Production Engineering
    shield weight, shield area, plastic material type, shielding type, shielding amount
Production Engineering
    forming process, shielding application process

In this implementation the following inputs can be given a certainty factor: EMI protection, deflection temperature, early due date, maximum payback period, plastic material type, forming process, shielding type, and shielding application process.

From the above data the production system infers the following:
    - actual EMI protection, actual deflection temperature, actual product quality
    - capital investment cost and shield unit cost (material cost, shielding cost, shielding
       application cost, forming capital cost, shielding application cost)
    - project rate of return, project payback
    - process consistency
    - consistency of each department/area
    - overall project consistency
    - recommendations to correct inconsistencies.

The overall departmental consistency rules implemented are given in figure 3. The rules are only illustrative of the types of rules that would be included in a design consistency rule base. Figures 4 and 5 show one example of the type of data input and the results of applying the design consistency rules for EMI/RFI shield design. It should be noted that the suggestions given in figure 5 provide a point where there is a design trade-off, since the inconsistency can be addressed by altering the data input of several different design functions/areas.
    The implemented rule-based production system approach has its advantages and disad-

```
MARKETING CONSISTENCY:              PRODUCT ENGR. CONSISTENCY:
   if                                  if
      customer temp. <= engr. temp.       process check is consistent
      and customer EMI <= engr. EMI       and coat check is consistent
      and customer quality <= engr. quality   then
      and customer cost <= final cost        product engineering is consistent.
   then
      marketing is consistent.
                                       PRODUCTION ENGR. CONSISTENCY:
                                          if
FINANCE CONSISTENCY:                         customer due date >= prod. due date
   if                                        and customer quantity <= prod. quantity
      capital cost <= capital budget      then
      and max. payback >= project payback    production engineering is consistent.
      and min. ROR <= project ROR
   then
      finance is consistent.
```

Figure 3 - Overall Design Consistency Rules

vantages. An advantage is that it is easy to express design requirements rules. A disadvantage is that it can become a tedious process to capture all of the various design interactions. There are also logical limitations to the rule based approach that make the constraint programming languages more robust and desirable.

The implemented design consistency tool fits into the overall framework as represented in figure 1. In particular, the implementation issue of data integrity is dealt with through the data board that maintains an up-to-date representation of design decisions.

There are other issues that must be addressed before a complete organization consistency environment is implemented. The first is the development of the general relationships for overall organization. Considerable effort is required to accurately capture and represent the organizational interactions. Although there is a degree of generality common to each organization, the relationships must still be tailored to the specific organization. The second issue relates to the development of specific product and process data modules. Once again there would be some commonality between different products and processes. Depending upon the level of detail desired in the consistency analysis, the effort involved in developing the modules would vary.

6.0 SUMMARY

It is clear that the product and process design processes must be improved and integrated for products to compete successfully in a dynamic market environment. A major barrier to this effort is the availability and quality of information. The approach termed concurrent engineering has explored various organizational and informational approaches to improving the design process. This paper has presented an approach that addresses the overall design consistency from an entire organization perspective. A framework was proposed that addresses the development of design functions, ensures data availability and integrity, analyzes the design consistency and explores the resolution of design inconsistencies. The framework enhances communication in two different concurrent engineering scenarios. First, it allows for communication and integration of decisions between differing views of the product and manufacturing system design processes within a broad-based design team meeting. Second, it provides a methodology whereby an individual decision maker can experiment with their specific design factors within the context of the overall

```
****************************************
Welcome to the EMI/RFI Design Consistency
System
****************************************
```

The following is information needed from the marketing group at your company.

Enter <RETURN> to continue:

Enter the customers requirements for EMI protection in DB, include certainty (i.e. 10 <0.8>)
```
    =? 80 <0.8>
```

Enter the temperature requirements for the shield, include certainty factor (i.e. 100 <0.8>)
```
    =? 400 <0.7>
```

What quality level does the customer require? Enter a decimal value (i.e. 0.90 => 90%) [constraint: cquality <= 1.0]
```
    =? 0.95
```

What is the estimated annual demand for the shield?
```
    =? 25000
```

In how many months does the first lot need to be ready? Include certainty factor (i.e. 1 <0.8>)
```
    =? 1 <0.75>
```

What is the size of the first lot?
```
    =? 15000
```

What should be considered an upper limit for the final cost of the shield?
```
    =? 2.00
```

What is the estimated selling price of the shield? (in dollars)
```
    =? 4.00
```

What is the estimated life that this shield will be in demand? Enter value in years:
```
    =? 5
```

The following is information needed from the finance group at your company?

What is the maximum allowed payback on this project in years? Include certainty factor (i.e. 1 <0.8>)
```
    =? 3 <0.85>
```

What is the minimum rate of return on this project? Enter a decimal value (i.e. 0.10 => 10%)
```
    =? 0.2
```

What is the cost of financing this project? Enter a decimal value (i.e. 0.10 => 10%)
```
    =? 0.12
```

What is the initial investment budget allocated for this type of project? Enter a dollar amount ($)
```
    =? 100000
```

The following is the design information required from the engineering and production groups.

What kind of shielding is the shield utilizing?
```
                1. coat
                2. fill
    =? 1
```

The following is the design information required for the coat option.

What is the weight of the shield in pounds? (enter a number)
```
    =? 0.5
```

What is the area of the shield in square inches? (enter a number)
```
    =? 35.0
```

Which engineering plastic is the shield make of:
```
                1. nylon
                2. HH ABS
                3. polycarbon
                4. polysulfone
    =? 3
```

Which forming process is to be used for form the shield?
```
                1. injection mold
                2. compression mold
                3. thermoform
    =? 1 <0.95>
```

What type of conductive material is to be used to coat the shield?
```
                1. silver
                2. copper
                3. zinc
                4. nickel
    =? 2
```

How many mils of coating does the shield require? (enter a number)
```
    =? 3
```

What type of coating process will be used to apply the coating?
```
                1. spray
                2. electroplate
                3. vacuum plate
    =? 1
```

The following areas of the design are inconsistent:

```
        FINANCE <0.85>
        PRODUCTION <0.75>
        MARKETING <0.70>
```

Figure 4 - Example Design Consistency Session

The following are marketing consistency problems:

> Product quality too low <0.70>
> Temperature requirement violated <0.70>

The following are finance consistency problems:

> The project is over budget <0.85>

The following are production consistency problems:

> Can not product enough shields to meet due date <0.75>

The following are my recommendations to correct the inconsistencies, for:

TEMPERATURE REQUIREMENT VIOLATED:

- Marketing should recheck environmental requirements, or

- Engineering should change to a higher deflection point material,
  i.e. T > 220 => HH ABS or polycarbon, T>300 => nylon66 or polysulfone).

PRODUCT QUALITY TOO LOW:

- Marketing should examine customer quality requirements, or

- Engineering should try different forming and/or coating processes.

PROJECT IS OVER BUDGET:

- Finance can either increase the budget, or

- Engineering can reduce the initial cost by considering either,
  1) a different forming process, or
  2) a lighter weight shield.

CAN NOT PRODUCE ENOUGH SHIELDS TO MEET DUE DATE:

- Marketing can check to see if customer can change due date and/or quantity
  requirements, or

- Production should try different forming and/or coating processes.

The justification for this conclusion is:

> overall consistency = inconsistent <0.85>

Figure 5 - Example Design Consistency Session (continued)

design effort and can post their decisions for others to consider.

There are still implementation issues that need to be addressed in the overall design consistency framework. However, the framework as presented provides a starting place for developing an approach to a problem that is in need of continuing attention.

ACKNOWLEDGEMENT

The Landis and Gyr, Inc. of Lafayette, IN is acknowledged for providing the product design information and insight to design organization needs.

REFERENCES

1. Lardner, J.R., "Integration and Information in an Automated Factory," *Proceedings of the 1984 SME Autofact 6 Conference*, Anaheim, CA, October 1-4, pp. 34-44, 1984.

2. Schmitt, R.W., "Design-Centered Innovation," *Design Theory '88*, S.L. Newsome, W.R. Spillers, and S. Finger, Eds, Springer-Verlag, New York, NY, pp. 2-7, 1989.

3. Carter, D.E. and B.S. Baker, *Concurrent Engineering: The Product Development Environment for the 1990s - Volume 1*, Mentor Graphics, Bellevue, WA, 1991.

4. Lee-Mortimer, A, "The Reality Behind the Jargon," *Integrated Manufacturing Systems*, vol. 1, no. 1, pp. 11-14, January 1990.

5. Dean, Jr., J.W. and G.I. Susman, "Organizing for Manufacturable Design," *Harvard Business Review*, pp. 28-36, January-February 1989.

6. Evans, S., "Implementation Framework for Integrated Design Teams," *Journal of Engineering Design*, vol 1, no. 4, pp. 355-363, 1990.

7. Boothroyd, G., C. Poli, and L.E. Murch, *Automatic Assembly*, Marcel Dekker, New York, 1982.

8. Lu, S. C-Y and S. Subramanyam, "A Computer-Based Environment for Simultaneous Product and Process Design," *Advances in Manufacturing System Engineering - 1988*, M. Anjanappa, D.K. Anand, Eds., ASME, New York, pp. 35-46, 1988.

9. Lim, B.S, "An IKBS for Integrating Component Design to Tool Engineering," *Expert Systems*, vol. 4, no. 4, pp. 252-291, 1987.

10. Ishii, K., R. Adler, and P. Barken, "Application of Design Compatibility Analysis to Simultaneous Engineering," *Artificial Intelligence for Engineering Design, Analysis and Manufacturing*, vol. 2, no. 1, pp. 53-65, 1988.

11. Bowen, J., P. O'Grady, and L. Smith, "A Constraint Programming Language for Life-Cycle Engineering," *Artificial Intelligence in Engineering*, vol. 5, no. 4, pp. 206-220, 1990.

12. O'Grady, P., R.E. Young, A. Greef and L. Smith, "An Advise System for Concurrent Engineering," *International Journal of Computer Integrated Manufacturing*, vol. 4, no. 2, pp. 63-70, 1991.

13. O'Grady, P. and R.E. Young, "Issues in Concurrent Engineering Systems," *Journal of Design and Manufacturing*, vol. 1, no. 1, pp. 27-34, September 1991.

14. Preiss, K., R.N. Nagel, and K. Krenz, "Design and Manufacturing in an Information-limited Environment," *Journal of Design and Manufacturing*, vol. 1, no. 1, pp. 17-25, September 1991.

15. Salzberg, S. and M. Watkins, "Managing Information for Concurrent Engineering: Challenge and Barriers," *Research in Engineering Design*, vol. 2, pp. 35-52, 1990.

16. Mullineux, G., "A Blackboard Structure for Handling Engineering Design Data," *Engineering with Computers*, vol. 7, pp. 185-195, 1991.

17. Morely, I.A. and S. Pugh, "The Organization of Design: An Interdisciplinary Approach to the Study of People, Process and Contexts," *Proceedings of the 1987 International Conference on Engineering Design*, W.E. Eder, Ed., ASME, New York, pp. 210-222, 1987.

18. Pugh, S., *Total Design*, Addison Wesley, New York, NY, 1991.

19. Kusiak, A. and K. Park, "Concurrent Design: Decomposition of Design Activities," *Proceedings of the 2nd International Conference on Computer Integrated Manufacturing*, IEEE, Los Alamitos, CA, pp. 557-563, 1990.

20. Noble, J.S. and J.M.A Tanchoco, "Concurrent Design and Economic Justification in Developing a Product," *International Journal of Production Research*, vol. 28, no. 7, pp. 1225-1238, 1990.

MANAGEMENT OF COUPLING IN DESIGN

Sridhar S. Condoor, Srinivasa R. Shankar, and Christian P. Burger

Institute for Innovation and Design in Engineering, Texas A & M University, College Station, Texas, U.S.A.

ABSTRACT

The philosophy of concurrent engineering requires the designer to consider various life-cycle issues early in the design process. This leads to highly coupled design problems, in which a single design parameter strongly influences two or more requirements. It is argued that a reduction in the degree of coupling leads to shorter product development time and improved quality.

This paper addresses some key issues in the management of coupling in design. In the paper, coupling is defined and different types of coupling are characterized. The process of identification of coupling and its reduction are illustrated through examples. It is demonstrated that coupling which is associated with a particular configuration can be eliminated by modifying the configuration. In some problems, the requirements may be coupled irrespective of the configuration. In such cases, a compromise in the final solution is necessary, unless the problem is reformulated.

INTRODUCTION

Any design problem has several requirements which must be met. The term requirements encompasses both functions which the product must perform and constraints which it must satisfy. The attributes of the design, such as length or thickness, which affect the requirements are referred to as parameters. A parameter can also be a key variable which is independent of the physical form, such as conductivity. A requirement can be influenced by several parameters to a varying degree.

Coupling is defined as the interdependence between two or more requirements. This interdependence is the result of a parameter (or a set of parameters), which influences these requirements differently. Thus, a certain change in the parameter may have a positive (desirable) effect on some requirement, but a negative (undesirable) effect on other requirements. Such a parameter is called a *coupling parameter*.

In a concurrent engineering approach, manufacturing and other life-cycle issues are considered very early in the design process. Therefore, the complexity of the interrelationships is much greater. Individual design teams working on different aspects of a problem may not be free to vary some parameters, without imposing new constraints which significantly affect the work of the other teams. Resolving these issues requires a highly iterative design process and lengthy negotiations. For effective management of the design process it is essential to identify the parameters which couple the problem. Technical conflicts can therefore be anticipated and effectively managed. This will result in improved quality of the design solutions and shorter lead times for the product development.

BACKGROUND WORK

The idea of reducing coupling in design has been addressed by Suh [1] who divides design solutions into three groups : coupled, decoupled, and uncoupled. The design is represented in terms of a set of functional requirements (FR's) and a set of design parameters (DP's), which affect the FR's. In an uncoupled design, it is possible to vary a particular FR by varying just one DP, without affecting other FR's. In decoupled designs, the order in which the parameters are varied determines whether or not all FR's are satisfied. Finally, in a coupled design, varying any DP always affects more than one FR. Rinderle [2] has developed measures of coupling based on the earlier work by Suh.

Eppinger [3] discusses the importance of coupling in concurrent engineering. The emphasis is on the partitioning of a complex design problem into a series of tasks which may be performed in sequence or in parallel. He views increasing coupling as a means to improve the quality of designs. The partitioning method used by Eppinger is based on the work of Steward [4]. Essentially, it is a method for examining the precedence relationships between different tasks and analyzing the flow of information between these tasks.

Shankar et al [5] have provided a detailed discussion of coupling in design with composite materials. Their paper identifies fiber orientation as a key coupling variable which affects strength and manufacturability in the design of a horizontal stabilizer of an aircraft. The importance of coupling in design for manufacturability has also been addressed by Shankar [6].

The abundance of examples in a wide range of engineering applications which exhibit some form of coupling has been a major motivation for this research. This research establishes a framework for coupling. It is recognized that constraints imposed on a design problem can also cause coupling. It is further argued that increased coupling in concurrent engineering is a result of the consideration of various life cycle issues simultaneously, as opposed to Eppinger's view wherein increased coupling is seen as a design strategy. The various strategies for managing coupling presented in this paper, all stem from the desire to reduce or eliminate coupling.

STRUCTURE OF COUPLING

The phenomenon of coupling is common in design. To illustrate this point, two simple examples are presented. Two more examples are presented later which discuss coupling and demonstrate its impact on the design process.

Example: Design of a car door

In the design of a car door, the requirements from a customer's perspective are:
* provide access to the car;
* easy to open and close; and
* avoid damage due to collision with adjacent vehicles during opening.

The parameters which link these requirements in the common hinged car door are the size of the door (S) and the angle of opening (α). Table 1 gives the desired magnitudes for these parameters. There is a conflict between the desired magnitudes shown in the column corresponding to the parameter S, where two of the requirements demand a small door while greater access is provided by a larger door. A similar conflict is also observed with respect to the desired values of α. The parameters which couple the requirements are dependant on the configuration, since a different configuration, such as a sliding door, does not exhibit coupling between these two requirements.

TABLE 1.
Desired Magnitudes for the Design Parameters for a Car Door.

Requirements	Parameters	
	Size (S)	Angle (α)
Accessibility	Large	Large
Ease of operation	Small	Small
Damage	Small	Small

Example: Design of an automobile headlight

The second example addresses the design of an automobile headlight. The headlight must:
* provide increased visibility for night driving; and
* avoid blinding of on-coming traffic.

Increased visibility can be obtained by increasing the brightness of the headlight. However, it is attained at the expense of inconvenience to on-coming traffic. Therefore, the parameter which couples these requirements is the brightness of the lights. At this level of analysis, the configuration of the headlight is not a concern. All headlights will exhibit coupling between the above requirements.

From a study of these and many other examples, it was observed that most design problems can be classified into two types, namely configurational coupling and conceptual coupling. This classification is patterned after the concept-configuration space model of the design process [7]. In *configurational coupling*, parameters in the configuration cause the coupling. Therefore, configurational coupling is solution dependant. A different solution to the problem may not exhibit the same coupling. The hinged car door is an example of configurational coupling. *Conceptual coupling* is a form of coupling in which the coupling parameters relate to the concepts and not to specific configurations. All configurations which stem from the same basic concept will exhibit coupling. The degree to which this coupling affects the requirements will depend on the quality of the design. In the headlight example, the coupling parameter is brightness which is not an attribute of the physical configuration, and therefore the coupling is conceptual in nature.

MANAGEMENT OF COUPLING

A powerful strategy to handle complex design problems is to divide them into simpler sub-problems. This division helps designers to focus their attention on specific aspects of the problem at any given time. Also design teams can work independently on the sub-problems thereby reducing the total design time. However, such a partitioning will not be very effective in a highly coupled system, due to the inability of the different teams to make independent design decisions. The identification of coupling limits the information flow between design teams to key variables and facilitates design decomposition. It also leads to a better understanding of the problem and a greater awareness of the trade-off decisions which will dictate the quality of the final solution and the design time.

Identification of Coupling

Coupling may be identified by looking for parameters which influence some requirements positively and others negatively. This identification procedure should be performed several times during the design process. At the initial stages the purpose is to discover conceptual coupling in the problem. During the later stages if it is observed that an attempt to satisfy one requirement by varying the configuration adversely affects another requirement, it is an indication of configurational coupling. The examples below illustrate the process of identifying coupling.

Example: Injection Molding

Injection molding is a complex process which requires design decisions to be made at various stages. These decisions involve the design configuration (e.g. determining wall thickness at different locations in the part), material selection (e.g. selection of a material with desired viscosity and shrinkage), and mold design (e.g. choosing a suitable gate location). These decisions, which are often made by different teams (part design team and mold design team), determine whether or not the design satisfies the requirements. The two requirements used in this example to illustrate coupling are low injection pressure and high part quality. During each machine cycle, several parts are molded simultaneously in different mold cavities. The total clamping force required for a mold depends on the number of cavities and the pressure required to fill a cavity. The maximum clamping force is determined by the machine size. By designing a part to have a low injection pressure, a greater number of cavities can be built into a single mold, thereby increasing the production rate. Therefore, maintaining low injection pressure can be viewed as a design requirement. Part quality in this example is viewed mainly as a consequence of the amount of warpage in the part. Figure 1 and Table 2 show coupling between the requirements of low injection pressure and high part quality.

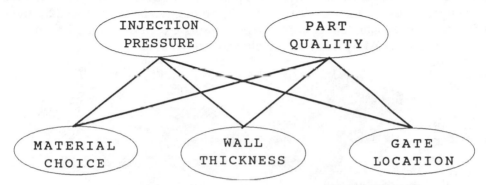

FIG. 1. Design requirements for an injection molded part (top row) and the parameters that affect these requirements (bottom row).

TABLE 2
Coupling in injection molding

Requirements	Parameters		
	Material choice	Wall thickness	Gate Location
Low injection pressure	Low viscosity	Large	To reduce flow length
High part quality	Low shrinkage	To allow uniform shrinkage	To improve fill pattern

Coupling due to material choice occurs only if there is no suitable material available which has a low viscosity and exhibits low shrinkage. Coupling due to wall thickness occurs because of the conflicting requirements of maintaining a large wall thickness in order to reduce injection pressure, and reducing wall thickness at certain locations to improve the fill pattern. Coupling due to the gate location is illustrated in Figure 2. In this figure the fill pattern and part warpage of an injection molded tray are shown (for two different gate locations). In Figure 2a, the center-gated mold has the shorter flow length but a poorer fill pattern. The fill pattern is good if the material orientation is reasonably constant throughout the part. In the center-gated mold the flow first reaches the two side walls which are closer to the gate and is then forced to change direction. Therefore the material orientation varies considerably at various locations. In the end-gated mold, the flow is more unidirectional and a better material orientation is achieved. However the end-gated mold has a much larger flow length, thereby increasing the required injection pressure. Due to the poor fill pattern in the center-gated mold, this design exhibits considerable warpage. Figure 2b shows that the center-gated design buckles due to the stresses set up during cooling. This design is therefore unacceptable because the final geometry is unpredictable. The end-gated design exhibits some warpage but does not buckle. Warpage in this case can be compensated by minor modifications to the mold dimensions. Deflections have been exaggerated for display and have been obtained by using the simulation package Moldflow.

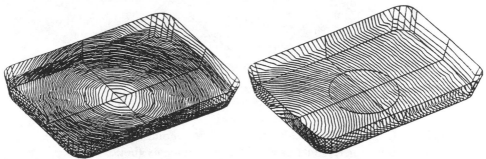

FIG. 2a. Effect of gate location on fill pattern
(Left: center-gated; Right: end-gated).

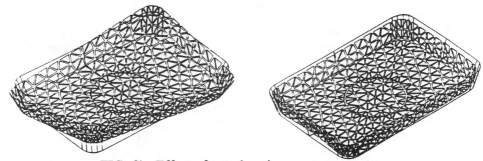

FIG. 2b. Effect of gate location on part warpage
(Left: center-gated; Right: end-gated).

Example: Cable for undersea robotics

A typical undersea robot or Remotely Operated Vehicle (ROV) includes a manipulator for accomplishing the desired tasks, and a vehicle for transporting the manipulator. The ROV is unmanned because it is expensive and unsafe to maintain a human at large depths. Due to the limitations of machine intelligence, most undersea robots are not fully autonomous and need to constantly communicate with a human operator on the ocean surface.

Usually the robots are connected to the surface by a tether which carries power to the vehicle and allows communication between the vehicle and the surface. The tether is designed to be neutrally buoyant, but is a burden on the limited energy resources, because of increased drag. It also limits the effective operational range of the robot.

Such a tether needs to satisfy four requirements:
* have a high communication rate;
* carry the required amount of power;
* have a low drag force; and
* allow a large operational range.

Two parameters that define the tether are its length and diameter. Table 3 shows the desired magnitudes of the length and diameter of the tether for each of the requirements. It is clear that the designers of the tether will be under strong pressure from the designers of the power and communication subsystems to increase the diameter. However, to reduce the drag force the tether designers prefer small diameters. As tether length is increased, the drag force increases which in turn increases the motor power requirement which calls for a larger diameter power cable causing more drag.

TABLE 3
Desired Magnitudes for the Cable.

	Parameters	
Requirements	Length (L)	Diameter (d)
Communication	Short	Large
Power	Short	Large
Drag force	Short	Small
Operational range	Long	Small

Strategies to Handle Coupling

After the key coupling parameters in a design problem have been identified, the designer can use any of several strategies to reduce the undesirable effects of coupling. Through a reduction in configurational coupling, the designer can achieve superior designs which satisfy most of the requirements to a high degree. Reduction of conceptual coupling facilitates the effective partitioning of a complex design problem into sub-problems which can be concurrently tackled by different design teams. In either case, the designer should first attempt to eliminate or reduce the coupling in the problem. If this fails, a compromise should be made in the degree to which some requirements are satisfied.

The strategies for handling coupling are:

1. Identifying new parameters This strategy is indicated when the coupling is caused by having a greater number of functional requirements than design parameters [1]. Therefore, making the number of parameters the same as the number of requirements can either decouple the problem or reduce the influence of coupling. In the head light example, the spread of the light beam may be another parameter along with brightness which can affect both visibility and blinding of on-coming traffic. Now, the designer has two parameters which s/he can vary to find a better solution to the design problem.

2. Re-examining the assumptions When conceptual coupling is encountered, it is important to re-examine the implicit assumptions which have been made in defining the problem. Implicit assumptions can artificially constrain the problem and also cause coupling. Some potentially good concepts may have earlier been rejected because they violated some artificial constraints. Upon re-examination of the assumptions, these concepts may emerge as attractive alternatives. Also new concepts may be generated. This reformulation will result in a shift in the focus on the design effort. In the undersea robot example, one of the assumptions is that the cable should be long to achieve the desired operational range. However, an umbilical cord which can bear heavy loads can be used to carry a launcher to launch the robot. The size of the cord is not a constraint on power transmission and communication, because the robot does not drag the cord. Therefore, the effective length required to maintain the operational range requirement is greatly reduced. As a consequence of adopting this strategy, the design team has the additional tasks associated with design of the launcher system.

3. Transforming the coupling This strategy stems from the realization that most design teams have certain areas of expertise and other areas in which they are less experienced. If the coupling involves an area of lesser expertise, a deliberate effort can be made to shift the coupling to an area in which the team has greater experience or resources. In the undersea robot example, the vehicle design team may be much more experienced than the tether design team. In such a case, the importance of power as a requirement for the tether design can be artificially reduced, thereby reducing the coupling and making the task of the tether designers easier. The task of the vehicle designers is now more difficult because they have to design the vehicle for a lower power consumption. This transformation is deliberate and the coupling in the area of weakness is transformed into coupling in an area of strength.

4. Adopting new technologies An innovation in a different domain can sometimes be beneficial in solving the problem at hand. However, there is a need to be aware of both the desirable and undesirable side-effects of the new technology. For example in the undersea robotic cable, the use of fiber-optics makes the diameter required for communication extremely small. If the power cable can be eliminated by the use of on-board power systems, the overall cable diameter is negligible and the issue of drag with respect to its effect on motor power is no longer critical.

5. Using coupling parameters as information links In this strategy, the coupling parameters act as the main means of communication between various design teams. The teams negotiate and arrive at a range of acceptable values for these parameters. This is done in advance of any detailed design so that the possibility of conflict and subsequent iteration is reduced. For example, in the design of injection molded parts, gate location is one of the information links between the part designer and the mold designer.

CONCLUSIONS

The main contribution of this paper is the development of a framework for managing coupling in design. Several examples from widely different domains have been used in developing this framework, thereby establishing the universal role of coupling as an important issue in design management. It has been demonstrated that configurational coupling may often be reduced through a modification in the design configuration, whereas a reduction in conceptual coupling typically requires a change in the underlying concept itself. Although a restatement of the design problem can lead to a radically different solution, constraints acting on the designs often prevent a radical change to the configuration. In such situations, a compromise in one or more of the requirements is a must. Early recognition of coupling and the source of coupling can significantly improve both quality of the design and the time taken to complete the design.

REFERENCES

1. Suh NP: *The Principles of Design.* NY, Oxford University Press, 1990.
2. Rinderle JR, Suh NP: Measures of functional coupling in design. *Journal of Engineering for Industry.* 104:383-388, 1982.
3. Eppinger SD: Model-based approach to managing concurrent engineering. Hubka V (eds): *Proceedings of International Conference on Engineering Design.* Zurich, Heurista, 1991, p 171-177.
4. Steward DV: The design structure system: A method for managing the design of complex systems. *IEEE Transactions on Engineering Management* EM-28:71-74, 1981.
5. Shankar SR, Sharkawy A, Burger CP, Jansson DG: Designing with composites: A study of design process. Hui D, Kozik TJ (eds): *Composite Material Technology.* ASME publication PD-Vol. 32: 1-6, 1990.
6. Shankar SR: *Generalized methodology for manufacturability evaluation.* Ph.D. dissertation, Texas A&M University, College Station, 1992.
7. Jansson DG: Conceptual engineering design. Oakley M (eds): *Design Management.* Basil Blackwell, Oxford, 1990.

CONCURRENT ENGINEERING COOPERATION USING IDEF$_0$ AND PARTITION LATTICES

Albert W. Small and Elizabeth A. Downey

Downey and Small Associates, Inc., Kensington, Maryland, U.S.A.

ABSTRACT

Concurrent engineering can provide benefits in efficiency and effectiveness. The benefits come from properly identifying, understanding, and communicating the details of the functions to be integrated throughout the engineering enterprise. Individual operating groups are thereby enabled to cooperate in structuring the overall concurrent process to best meet the enterprise objectives.

In complex enterprises, process understanding and communications can be enhanced by the use of the graphic process analysis language, IDEF$_0$. To address areas of overlapping interest, IDEF$_0$ can be supplemented with partition lattice techniques.

I. INTRODUCTION

A. Concurrent Engineering Environment

Companies determined to succeed in the competitive global market place are striving for improved quality, lower costs, speed to market and customer satisfaction. *Concurrent engineering* is widely discussed as a strategy for achieving these goals. This paper suggests tools, and outlines the way they can be used, to develop a concurrent engineering process.

Technology has advanced more rapidly than our engineering management capability to utilize it. Concurrent engineering is an effort to address this situation by integrating the activities of various groups to run simultaneously, thereby increasing competitiveness through more efficient utilization of resources. The concept of concurrent engineering may be narrowly applied to designing the product and its manufacturing process simultaneously, or may be broadened to include any or all of the other activities of an enterprise which relate to its product. When concurrent engineering reaches a certain breadth of scope it is more often termed *enterprise integration*.

B. Concurrent Engineering Approach

An essential requirement for concurrent activities and integration is the provision of the right information to the right people at the right time for them to make the best possible decisions. For example, when activities are performed sequentially information is not communicated from further along the process at the time it could have reduced expensive errors made at the front end. All activities can benefit from knowledge of opportunities and constraints from other activities that relate to them.

Information and communication gaps and confusions are both the greatest problem and offer the greatest area of opportunity for improvement in a concurrent engineering process. Computing technology can support but by itself cannot provide a solution. Methods to manage information and structure activities which produce and use information to best meet objectives are necessary. This paper outlines a method to build an activity and information framework in which people of diverse background and expertise can participate and communicate to develop the concurrent engineering process.

C. Contribution of $IDEF_0$ and Partition Lattices

A major obstacle to achieving effective concurrent engineering in a complex organization is the misunderstanding of processes among the different operating groups. In order to work effectively together in an integrated way, these groups need to develop a shared understanding of the overall process in which they operate. They need to understand the activities within the process, the interdependence among the activities, and how and why the information constraining these activities is produced and used.

An effective methodology for supporting diverse groups in understanding and communicating process concepts is known as $IDEF_0$, a graphic hierarchical modeling methodology used for process analysis. The method can display processes with compelling graphic clarity that can be widely understood, but its full power requires the integration of multiple process models with overlapping scopes and different decompositions. In general practice multiple models are often avoided, probably due to the perceived difficulty of integrating them.

The application of $IDEF_0$ to the development of a concurrent engineering process is described in Section II. An intuitive approach to the multi-model integration problem is presented with Venn diagrams using Boolean logic, in Section III. The Venn diagram approach is consistent with the partition lattice treatment in Section IV, where a technique used for the functional decomposition of digital switching circuits is extended and shown applicable to the resolution of enterprise process integration problems. By using partitions to characterize the enterprise functional decompositions in $IDEF_0$ models, the properties of lattices of partitions become available for addressing the integration of multiple $IDEF_0$ models.

For background reading, Ross [1], Marca and McGowan [2] are recommended for structured analysis (predecessor of $IDEF_0$); Rutherford [3], for lattices; and, Brown [4], for Boolean logic. The U.S. Air Force manual [5] should be consulted for $IDEF_0$ syntax.

II. DEVELOPING A CONCURRENT ENGINEERING PROCESS

The application of $IDEF_0$ and partition lattices to the development of a concurrent engineering process in an operating enterprise is presented in this section. It is assumed that high level management is giving essential support to the project and has established objectives. All areas of the enterprise conducting activities within the scope of the process should participate in its development.

A. Process Situation

A concept of *sequential process* is portrayed in Fig. 1. In this process, activity A is completed before activity B begins. The work done by activity A is handed-off to activity B. There is no feedback from B to A, nor is there advanced work started by B before A is completed.

A concept of *concurrent process* is shown in Fig. 2. In this situation the activities have been modified to permit activity B' to start before activity A' is completed. Two other features are also included. Two-way interaction provides appropriate information at key stages of both A' and B' thereby facilitating simultaneous work. Some shared activity in which both groups P and Q participate provides dual expertise where needed for higher quality work with less rework.

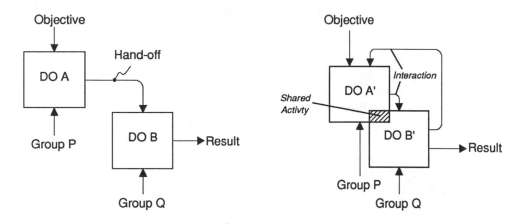

Fig. 1 Sequential Process   Fig. 2 Concurrent Process

Design of a concurrent process requires that participants have a more detailed understanding of each activity area than is required for a sequential process. The activities must be decomposed to a level of detail where individual interaction interfaces can be connected. In the shared areas, functions must be decomposed in a manner that all participating groups can understand. This requires special tools and effort for analysis and communication. Thus, management commitment and cooperation are called for.

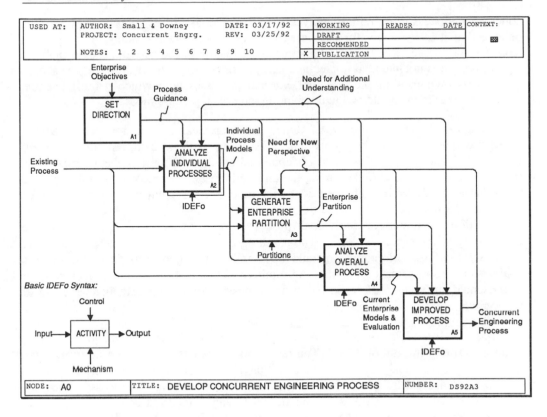

Fig. 3 IDEF$_0$ Diagram of Proposed Approach

B. Proposed Approach

The approach proposed for developing a concurrent engineering process is presented graphically in Fig. 3, as an IDEF$_0$ diagram. The approach has five basic activities: **A1**, Set Direction; **A2**, Analyze Individual Processes; **A3**, Generate Enterprise Partition; **A4**, Analyze Overall Process; **A5**, Develop Improved Process. IDEF$_0$ is a suggested mechanism for **A2**, **A4**, and **A5**, while partitions are recommended for **A3**.

A1, Set Direction. Understanding the enterprise objectives is essential for setting proper project direction. The project direction will guide the selection of analysis scope and viewpoint (**A2**, **A4**), and ultimately the focus of improvements to the process (**A5**).

A2, Analyze Individual Processes. A unique feature of the proposed approach is the initial analysis of individual process areas, separately. When an organization has made the decision to increase integration it is usual to begin with a joint effort to build the concurrent process using representatives from the various process areas. However, we believe that it is important for each area to have a clear understanding of the process from its own perspective before attempting participation in the overall process analysis. This approach is not intended to preclude analysis of context activities. Each area should develop its own view of the principal external processes with which it must interact. These context views should be included in the individual process models.

A3, Generate Enterprise Partition.   Once the individual models are developed the enterprise partition can be generated.   Partition lattices using straight forward mathematics can be applied to areas of functional overlap. Discussion between all areas (two at a time) is needed to determine the underlying enterprise task set and the specific tasks that are included in each activity of each model. Agreement between every pair of areas must be reached.

Additional analysis of individual processes may be required to complete the partitioning to the necessary detail. The resulting partition will be an enterprise building block set used for communication and cooperation during overall enterprise process analysis and improvement (A4, A5). An intuitive example of the building block concept is presented in Section III, followed by a discussion of the underlying mathematical theory in Section IV.

A4, Analyze Overall Process.   The individual process models are next analyzed together by an enterprise integration team. The team should include people who performed the individual process analyses (A2). Information from the individual models is used to develop overall models. In a complex project, analysis of the overall process will require development of several models having different viewpoints and scopes. All models should be explicitly correlated with the enterprise building blocks, either by top-down decomposition or by bottom-up aggregation. This is required for maintaining correlation among models.

It is expected that the analysis of the overall process will reveal issues and opportunities that require new viewpoints to examine. Whenever a new perspective is needed that was not present in any of the individual process models, the specific need should be fed back to revise the enterprise partition (A3).

A5, Develop Improved Process.   After evaluation of the overall current process, the enterprise integration team formed in A4 is prepared to develop process improvements. This activity will involve development of new and modified process models, and may also require new perspective. The new concurrent engineering viewpoints should be fed back and the enterprise partition refined as appropriate (A3).

When the improved process has been developed, each activity node should be clearly assigned to one responsible party. The generation source and destination of all information is also shown. When the process model is finalized the roles and responsibilities of each group should be clearly understood and agreed to by all.

III. ENTERPRISE BUILDING BLOCKS

Individual process models developed in the proposed approach (Fig. 3, A2) will have different viewpoints, functional scopes and decompositions. This is necessary and should not be discouraged. However, the differences can cause confusion when discussing a functional area of interest to two or more groups. For this reason, the groups must join to generate an enterprise partition of building blocks (Fig. 3, A3).

The *enterprise building blocks* are those process details required to provide a common basis for communications among individual groups in the enterprise. Details should address both the cross-functional interaction and the shared function areas anticipated for the new concurrent engineering environment (Fig. 2).

A Venn diagram method for identifying the enterprise building blocks is presented in this section by means of an example. The example will assume that two areas share a common process interest. A mathematical definition is provided in Section IV.

A. Identifying Building Blocks

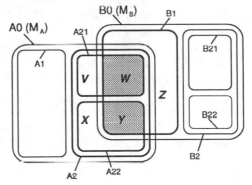

Consider the $IDEF_0$ models, M_A and M_B, with functional decompositions represented by Venn diagrams in Fig. 4. For simplicity some liberty is taken with the $IDEF_0$ language notation and decomposition conventions. Also, the gaps between subsets in the diagram exist only to facilitate visual distinction of the subsets.

The two process activities A0 and B0 include common tasks (within A2 and B1). However, the different viewpoints used for decomposition in models M_A and M_B have resulted in the two models having no

Fig. 4 Building Block Example--
Two Models

common activity. Further decomposition of A21, A22, and B1 is needed to specifically identify the tasks common to the scopes of both models. That is, A21 must be decomposed into *V* and *W*, A22 into *X* and *Y*, and B1 into *W*, *Y*, and *Z*. The tasks common to both areas will then be totally contained in the union of *W* and *Y*.

The most detailed activities from both areas are assembled in a single *building block set*, {A1, *V*, *W*, *X*, *Y*, *Z*, B21, B22}. The detailed activities in the building block set provide the correct level of detail for integrating the two models, M_A and M_B.

To the extent that models include context activities (e.g., A-1.2), those activities should be included in identifying the building block set. When a group includes context activities in its model, those context activities are usually the external activities having important interfaces with the group's individual process.

In complex projects where more than two areas must be integrated to improve concurrency, a single enterprise building block set is needed which will support all areas of the project. The universal set can be created by first generating the building block set, for any pair of models, M_A and M_B. Additional models are then incorporated one at a time, refining the building blocks and increasing scope as required.

B. Using Building Blocks

IDEF$_0$ activity hierarchy trees (also called *activity node trees*) will now be used to demonstrate some properties of the enterprise building block set. Fig. 5 shows the activity hierarchy trees for M$_A$ and M$_B$. The figure also contains the additional decomposition of A21, A22, and B1 into *V, W, X, Y, Z*. The relationship between the two models and the new smaller building blocks, *V, W, X, Y, Z*, is shown. The combined building block set for M$_A$ and M$_B$, namely {A1, *V, W, X, Y, Z*, B21, B22}, contains the lowest level activities of the combined tree.

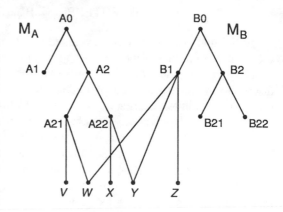

Fig. 5 Building Block Example--
Combined Hierarchy Tree

Alternative models are needed for analyzing specific process issues in an enterprise level, and for defining new improved processes. Alternative models can now be developed using the building block set, {A1, *V, W, X, Y, Z*, B21, B22}. The models can have different scope and viewpoint and will be capable of integration, so long as they draw upon the same building blocks.

In Fig. 6 there are two alternative models (left and center) which use only building blocks from the set {A1, *V, W, X, Y, Z*, B21, B22}. The third model, however, requires additional decomposition of *Z* into *Z1* and *Z2*. For this third model to be communicated among groups, the enterprise building block set must be refined by agreement to include *Z1* and *Z2*.

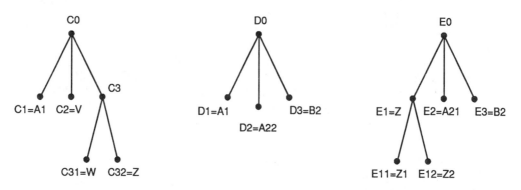

Fig. 6 Building Block Example--Alternative Models

Observe that building blocks address the overlap between model scopes. They do not show interfaces between models. Interfaces between areas are shown as standard arrow constraints on IDEF$_0$ context diagrams or higher level models. A mathematical treatment of the building block concept is presented next in terms of sets, partitions and lattices.

IV. TASK PARTITIONS

The approach to concurrent engineering proposed in this paper includes a specific step for generating the *enterprise partition* to support communication among divergent functional areas. A notional method for identifying the *building blocks* of the partition was presented in the preceding section. In this section, a mathematical characterization of the enterprise building blocks is presented using the properties of lattices of partitions.

A. Trees and Tasks

The *activity-hierarchy tree*, H_A, for the IDEF$_0$ model M_A is a Hasse diagram representing a particular decomposition of the overall process activity A0 and of selected context activities (e.g., A-1). An activity node AX positioned beneath another activity node AY and connected by one or more edges signifies that AX is a *sub-activity* of AY; that is, activity AX is performed as part of AY. Fig. 7 shows an example of the activity-hierarchy tree for a simplified abstract model.

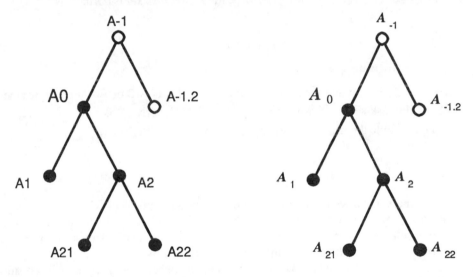

Fig. 7 IDEF$_0$ activity-hierarchy tree, H_A Fig. 8 Task-set tree, $(A, \prec)$

The notion of *task* is used here to represent the smallest distinguishable activities (functions) relevant to a given enterprise. The *universal task set* for a given enterprise is defined as the finite set

$$T = \left\{ t \mid t \text{ is an enterprise task with no relevant distinguishable sub task} \right\}. \qquad (1)$$

This subjective definition places tasks at a level which is more detailed than normally used by management, but not so detailed as to have infinite cardinality. The purpose is to provide a granularity fine enough to analyze any functional decomposition of interest.

The *task-set tree* for $IDEF_0$ model M_A, denoted $(\mathcal{A}, \prec)$, is defined as a Hasse diagram with node set $\mathcal{A}$ and edge relation $\prec$ which are isomorphic to H_A. The *node set* $\mathcal{A}$ is defined as the collection of all *task sets* on M_A, where *task set* A_X is defined as

$$A_X = \{t | t \in T, \text{ t is a sub task of activity } AX \text{ in } M_A\}. \tag{2}$$

The edge relation $A_X \prec A_Z$ (read A_X *is a component of* A_Z) signifies that the set-tree node A_X is beneath and directly linked to node A_Z by an edge. The *edge relation* is defined for $A_X, A_Y, A_Z \in \mathcal{A}$:

$$
\begin{aligned}
&A_X \prec A_Z \text{ if and only if} A_X \subset A_Z \\
&\text{and there is no } A_Y \text{ such that } A_X \subset A_Y \subset A_Z
\end{aligned} \tag{3}
$$

Given the task-set tree $(\mathcal{A}, \prec)$ for model M_A, the *model task set* is defined as

$$T_A = \bigcup_{\mathcal{A}} A_X. \tag{4}$$

Observe that $T_A = A_0$ when the highest level node in H_A is A-0 or A0. Fig. 8 provides an example of the task-set tree corresponding to the $IDEF_0$ activity-hierarchy tree, H_A, in Fig. 7. In the example the highest level activity is A-1; hence, $T_A = A_{-1}$.

B. Partitions and Task Distinction

An $IDEF_0$ analyst considers the objective of the modeling project when setting scope and making decomposition decisions so that the model will include the needed tasks and will illuminate selected distinctions among those tasks at each level. This is essential for the model to communicate the desired information about a process.

In the following discussions let the $IDEF_0$ models be decomposed so that no two sub-functions at the same level include the same task, and that a task is included in a particular function if and only if the task is included in an immediate sub-function. Then the ability of the models to distinguish among tasks can be specified by partitions.

The collection $p = \{p_1, \ldots, p_n\}$ is a *partition* on the set S if and only if

$$\bigcup_i p_i = S \text{ and } p_i \cap p_j = \Phi(i \neq j). \tag{5}$$

The system $(P(S), \leq)$ containing the collection $P(S)$ of all partitions on S together with the partition-inclusion relation $\leq$ is a *lattice of partitions*.

Given M_A with activity hierarchy tree H_A, and task-set tree $(\mathcal{A}, \prec)$, the *activity collection*, α_X, the *atom collection*, α, and the *extended partition*, π_A. are defined:

$$\alpha_X = \left\{ A_{Xj} \middle| A_{Xj} \prec A_X \right\}; \tag{6}$$

$$\alpha = \left\{ A_X \middle| A_X \in \mathcal{A}, \text{ no } A_Y \prec A_X \text{ in } (\mathcal{A}, \prec) \right\}; \tag{7}$$

$$\pi_A = \alpha \cup \left\{ T \cap \overline{T_A} \right\}. \tag{8}$$

The three collections are partitions, but on different sets: α_X is a partition on A_X; α is a partition on T_A; and, π_A is a partition on T. Each partition specifies the tasks that can and cannot be distinguished. Two tasks can be distinguished if and only if they belong to two distinct blocks of the partition.

The activity collection α_X specifies the tasks that can be distinguished by the components of A_X. For the example in Fig. 8, let $T_A = \{a, b, c, d, e, f, g\}$, with $A_1 = \{a, b\}$, $A_{21} = \{c, d\}$, $A_{22} = \{e\}$, $A_{-1.2} = \{f, g\}$. The activity collection $\alpha_2 = \{\{c, d\}, \{e\}\}$ specifies that the components of A_2 can distinguish tasks c and e, d and e, but not c and d.

The atom collection α specifies the tasks in T_A that can be distinguished by the model M_A. Observe that two models having the same atom collection but different decompositions can distinguish tasks differently at particular levels and branches, but exactly the same for the overall models.

The extended partition π_A specifies all tasks in T that can be distinguished by M_A; this extension of α allows comparison of partitions for all models of the enterprise .

C. The Enterprise Partition

Let $\{M_A, ..., M_N\}$ be a collection of IDEF$_0$ models developed independently for a given enterprise. The *enterprise partition*, π, for the collection is the *greatest lower bound* of the extended partitions, $\pi_A, ..., \pi_N$, of models $M_A, ..., M_N$:

$$\pi = \prod_{X=A}^{N} \pi_X \tag{9}$$

A model M having atom collection π can distinguish among all tasks that can be distinguished by any model used to generate the enterprise partition. The enterprise building blocks needed for integrating $M_A, ..., M_N$ are the blocks of π. A similar result has been demonstrated for decomposing of digital logic functions [6], and visual methods for calculating the *greatest lower bound* of two partitions have been developed [7], [8].

V. CONCLUSION

Before a concurrent process can be developed to integrate the activities and information of an enterprise, a shared understanding must be created. In this paper we have suggested a method to build this common understanding using the graphic process analysis modeling technique, $IDEF_0$, supported by the mathematics of partition lattices. Using $IDEF_0$ and partitions, people with diverse backgrounds and expertise can communicate and reach detailed understanding of their own and others' processes.

A concept that this paper emphasizes, and that we believe has not been widely recognized, is the need to analyze the individual process activities, separately, prior to the integration effort. The use of partition lattices is the second concept that we believe is uniquely applied here to concurrent engineering. Partition lattice theory is used to identify the activity breakdown required to show the correct functional detail in areas of shared interest. The task blocks that are generated are used to develop process models for analyzing overall current operations and synthesizing a concurrent integrated process.

ACKNOWLEDGMENT

We thank John A. Small for the helpful discussions regarding applicability of the enterprise partition to a political science project at the University of Maryland. We also appreciate the support provided by E. Claire O'Hanlon during development of this paper.

REFERENCES

1. Ross, Douglas T: Structured analysis (SA): a language for communicating ideas. *IEEE Transactions on Software Engineering*, Vol SE-3, No 1, pp 16-34, 1977.
2. Marca, David A and McGowan, Clement L: *SADT™ Structured Analysis and Design Technique*, (™SADT is a trademark of SofTech, Inc). New York, McGraw-Hill, 1988.
3. Rutherford, DE: *Introduction to Lattice Theory*. New York, Hafner, 1965.
4. Brown, Frank M: *Boolean Reasoning--The Logic of Boolean Equations*. Boston, Kluwer, 1990.
5. Integrated computer aided manufacturing (ICAM) function modeling manual ($IDEF_0$), report UM110231100. U.S. Air Force Materials Laboratory, Wright-Patterson AFB, Ohio, 1981.
6. Small, Albert W: Partitions and functional decomposition, in Lin, S (ed): *Proceedings Fourth Hawaii International Conference on System Sciences*. Honolulu, University of Hawaii, 1971, pp 698-700.
7. Ore, O: Theory of equivalence relations. *Duke Math Journal*, Vol 9, pp 573-627, 1942.
8. Small, Albert W: Partitions and edge-weighted pair-graphs. *IEEE Transactions on Computers*, Vol C-17, No 11, p 1089, 1968.

Part III
Computer Integrated Manufacturing

WAYS OF UTILIZING REFERENCE MODELS FOR DATA ENGINEERING IN CIM

Christian Kruse, Alexander Hars, Ralf Heib, and August-Wilhelm Scheer

University of Saarbrücken, Saarbrücken, Germany

ABSTRACT

Conceptual models are of paramount importance in the process of developing and maintaining integrated information systems in the heterogeneous CIM environment. Reference models alleviate and accelerate the creation of conceptual models significantly. This paper, resulting from the ESPRIT-project 5499 CODE (Computer supported enterprise-wide data engineering), discusses the feasibility of the reference model approach for a production control application. General properties of reference models are defined and their potential benefits discussed. It is shown how data reference models - described in an EERM notation - may be integrated into a broader information system engineering context. Reference models may be used for requirement elicitation and -modelling as well as for the customization process and re-engineering applications. The utility of the reference model approach is demonstrated by a real-life example for the field of production control.

CIM AND DATA ENGINEERING

Computer-integrated manufacturing (CIM) spans all activities of the product life cycle such as product design, planning and manufacturing [1]. It provides a means to integrate heterogeneous data originating from various sources through computers and computational techniques. The distributed heterogeneous computer environment often found in hierarchically structured CIM systems with data being stored and processed locally, necessitates data engineering approaches that facilitate and support the integration and sharing of data among different component systems.

The role of data engineering in CIM is to uncover semantic properties of the underlying systems and make them explicit in the data definition. Data engineering comprises methods and procedures for the design and development of conceptual data models, their administration and the implementation by data base management systems. Data engineering allows to share conceptual data models by people at different enterprise levels. Hence it enforces a common perception of the data and operations of a company, establishes a way of communication and leads to conceptual standardization and integration.

Various approaches for integrating data-related aspects of CIM have been described in literature [2]. Achieving conceptual integration by an enterprise-wide data model [3] has proven to be both feasible and affordable.

NOTION OF REFERENCE MODELS

The notion of reference models can be found in many scientific domains. However, the interpretation of what a reference model really is and how it can be used differs significantly. On the one hand, reference models are seen as 'de-facto standards'. According to this interpretation reference models should be theoretically proven and have an imperative character. It is possible to measure the deviation of other models from the de-facto standard. Contrary to this rather rigid view of the term reference model, a more flexible approach is proposed in this paper. Particularly, the further utilization of reference models shall be demonstrated. Accordingly, the description of a car in a catalogue - for instance - can be regarded as a reference model for the car a customer wants to buy. By making certain alterations to motor power, interior design, exterior design, colour etc. the customer can model his desired car. The catalogue description is used by many customers and can be regarded as a starting point for their decision-making process leading to the description of their customer-specific car.

The characteristics of a reference model - which universally apply to the car reference model as well as to data reference models - are
-        general applicability,
-        variability, and
-        feasibility.

Reference models are general, type of trade or type of industry-oriented models, which conform to the requirements of many potential customers. They do not focus on a particular enterprise but on the structures which are typical for a set of enterprises that may be classified and clustered according to common features. With respect to the variability, reference models are adapted to the specific situation of the user, a process which is termed customizing. This customization further emphasizes the distinction between reference models and standards since the latter may not be altered. Reference models, on the other hand describe a viable solution which is open to changes. The customizing process has to be followed by a verification and validation process to ensure the feasibility of the customized model. In the car example a four wheel drive would have to be rejected if motor power were too low etc. The reference model itself has to describe a feasible solution, which is consistent within itself, complete, and can serve as an implementation description without modifications or enhancements (which in the car case implies that the catalogue car can directly be ordered).

The concept of reference model as outlined above can beneficially adopted for developing software and information systems in CIM. In the ESPRIT project 688 CIM-OSA (i.e. Computer Integrated Manufacturing - Open Systems Architecture), reference models are embedded in the so-called CIM-OSA architectural framework [4] on a level denoted as partial. The framework is represented by a three dimensional cube containing 36 building blocks. It is distinguished between four generation views (function, information, resource, organisation), three instantiation levels (generic, partial, particular) and three derivation levels (requirements definition, design specification, implementation description).

The ESPRIT- project CODE explores and validates the reference model approach for the partial level of the information view on both the requirements definition and design specification level. Procedures to develop data reference models and to incorporate them into the complete software development life cycle are explored. The approach of incorporating conceptual models in the software design process is particularly emphasized by the shift towards the early phases of the software development life cycle i.e. the analysis and design.

REFERENCE MODELS FOR DATA ENGINEERING

Syntax of the data reference model

Reference models have to use a widely accepted, proven syntax allowing an easy identification and interpretation of their structure and content. To ensure a general applicability of reference models this syntax must support a simple transformation into different notations and has to be open for modifications.

There are many representation forms for conceptual data models [5]. The Entity Relationship Model (ERM) proposed by CHEN [6] serves as a basis for representing conceptual data reference models because of its close links to the network and the relational data models [7]. The original model is extended by several constructs resulting in a so-called Extended Entity Relationship Model [8]. The modelling constructs applied for data reference models are a subset of the CIM-OSA Entity-Relationship-Attribute approach. This approach comprises popular extensions to Chen's model such as the generalization/specialization relationship and relationships between more than two entity-types (n-ary relationship). Another extension allows the aggregation of relationships into new entity types. This construct - denoted as aggregated relationship [9] - provides a means to model relationships between relationships. Since many complex facts are essentially relationships between relationships this construct vastly enhances the clarity of the model. Cardinalities of relationships are represented by the notation of Schlageter/Stucky [10]. Attributes are shown in circles with primary keys being underlined.

Structure of the data reference models

Data reference models could either be enterprise-wide data models [11] for specific type of industries or domain-specific data models (e.g. marketing data reference model, production data reference model). In the latter case, an enterprise-wide data model could be constructed by merging domain-specific reference models for different areas of an enterprise. For each reference model the functional scope and possible relationships between area-specific reference models have to be defined. Depending on the purpose and scope of reference models three levels of modelling details can be distinguished.

The majority of existing enterprise-wide models are macro-level models. They contain only few, so-called kernel entity types. Not all relationships between kernel entity types but only the most important ones are described. These relationships are often information-bearing with descriptive attributes assigned to them. On the medium and micro level these relationships often are detailed into a number of other relationships which also involve additional entity types. Hence, a specification of cardinalities is of limited value on the macro level. A macro-level data model describes the master data of an enterprise. The medium-level data model is a detailed representation of the data structures of an enterprise on the object type level. It incorporates cardinalities and primary keys for all entity types and relationship types. The relationships are more specific and their cardinalities are precisely specified. A micro level data model is a project-specific detailation of a medium level model comprising implementation-oriented aspects. Attributes are added to all entity and relationship types and functional dependencies between data elements are considered.

Data reference models of different levels of detail necessitate not only different functional requirements for their support but also additional needs for the description of the reference models themselves. A typology for the selection of reference models has to be established and there certainly has to be a textual description to allow a person to understand whether a selected model could be appropriate. Furthermore constructs for the solution of typical problems (e.g. recursive relationships, header-position

relationships, planned-actual comparisons) should be provided. A comprehensive reference model approach therefore requires the following components which are an
- extended Entity Relationship Model constructs,
- domain- and enterprise-wide data models,
- data structure frames and a
- classification typology of models.

Based upon these components, a library for reference models can be established and administered.

Impact of data reference models on the data engineering process

Data reference models are primarily applied in the early phases of the software development life cycle, especially for requirements modelling and elicitation. They lead to modifications and extensions in the process of developing conceptual data models and implementation descriptions for CIM information systems. This is illustrated in figure 1.

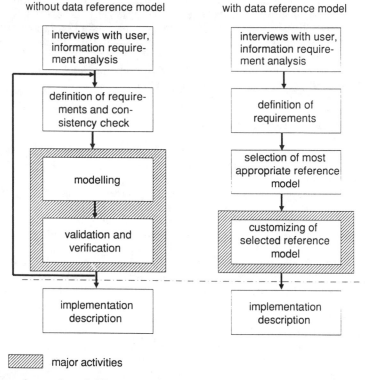

Fig. 1: Comparison of modelling procedures

The first two steps of conducting user-interviews and performing an information requirement analysis are incorporated into both approaches. However, there is a substantial difference to what extent these processes are structured. Utilizing pre-determined reference models facilitates the interactive user requirements analysis considerably. Both the interviewer and the user are guided and controlled by the reference model. After the requirements have been defined the conceptual models have to be developed. Without reference models a modelling process takes place transforming from scratch the user requirements into a model according to the syntax of the applied description language. This model has then to be validated and verified. If a model does not meet the specified requirements it is rejected and the modelling process has to be repeated. When applying data reference models the modelling process varies significantly. First, the most appropriate data reference model is chosen from the library with the help of the classification typology. This data model is then customized to the needs specified in the user requirements phase. The emphasis of the modelling process has shifted from an error-prone unstructured modelling to customizing.

In figure 2 it is illustrated how data reference models are utilized in the process of developing and integrating different models of implementation descriptions.

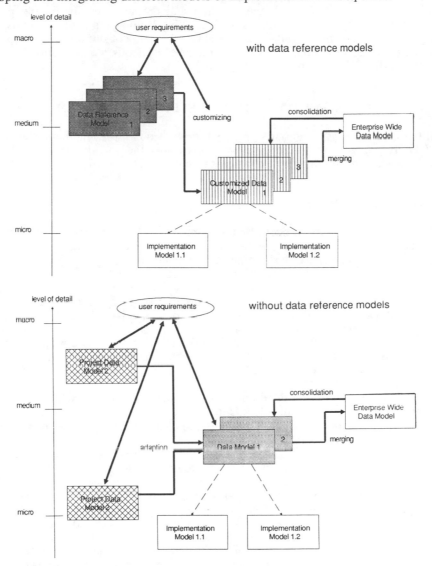

Fig. 2: Utilization of reference models

Applying area-specific data reference models of medium level detailiation the user requirements are represented in a standardized description language and homogeneous with respect to the level of detailiation. The requirements models expressed as data reference models are the starting point for a customizing process that comprises a modification and further detailiation of the reference models. The customized data models represent aspects of enterprise-specific application areas. In a stepwise procedure they are merged, consolidated and integrated in an enterprise-wide data model. The merging and consolidation process includes the removal of redundancies and an abstraction process. Each consolidated and customized data model may serve as a documentation basis for different, function-oriented implementation description models. These models focus on implementational aspects and reflect function-specific requirements. The described modelling process is contrasted by a modelling approach without reference models. Without reference models, the requirements modelling and elicitation process is very much unstructured and depends heavily on the project-specific approach chosen. No standardized modelling language is applied and the level of modelling detail may vary considerably. A time-consuming and expensive adaptation process is required to transform the heterogeneous requirements definitions into a standardized, homogeneous description.

APPLICATION OF DATA REFERENCE MODELS

In the following the customizing process is described in more detail and exemplified by a CODE project test partner application. Beside, the incorporation of the data reference models in a re-engineering process is delineated.

Reference data models for customizing

To start the customizing process first an appropriate data reference model has to be selected from a library of existing reference models. This requires that a
-        structured enterprise description and, based upon that
-        a selection of the most appropriate data reference model
has to be conducted prior to the customizing. The enterprise-specific situation is classified and described according to a pre-defined typology. Classification criteria include the type of production (e.g. make-to-order vs. make-to-stock), the type of planning, the product structure, the degree of product standardization, the organization of the sales activities etc. Based upon this description a set of suitable reference models is derived. The user analyzes the selected reference models, compares them with the enterprise situation and finally chooses the most suitable data reference model. The customizing process itself contains
-        an iterating adaption of the domain-specific data reference model,
-        an evaluation of customized data model and
-        the merging, consolidation and integration of the customized data model into an
         enterprise-wide data model.
The data reference model is iteratively adapted to the enterprise-specific situation by thoroughly analyzing each component of the reference model. The adaption process contains formal and syntactical changes (i.e. removal of inconsistencies), the addition of structures which have not yet been part of the reference model but are relevant for the enterprise, and the deletion of structures from the reference model which are not relevant for the enterprise. The semantic evaluation of the modified data model is an important task during the customizing process.

After the adaption process the customized reference model has again to be checked for consistency. This comprises syntactical correctness as well as a semantic cross-check by a third person that was not involved in the customizing process. This consistency check can lead to new modifications of the customized reference model until the model is finally accepted.

Customized area-specific data models may be merged to enterprise-wide data models which are very important means to enforce enterprise-wide integration according to the CIM paradigm. The enterprise-wide data model has to be validated and tested again. The major task is the computer-supported recognition of synonymous constructs and subsequently the elimination of redundancies. Checks are oriented towards completeness and uniform level of detail. These checks might as well include third persons to increase the objectivity of the model.

A sample customizing application

The following example is taken from the CODE testbed Pilkington PLC, a British glass manufacturing company which operates world-wide. Pilkington is currently prototyping a production planning and scheduling system for one of its subsidiaries, the Triplex Safety Glass Company Ltd., which produces window glasses for vehicles. The CODE reference model approach is explored and validated by developing a data model for the production control area of Triplex's Eccleston factory, which is situated in St. Helens. The data model is generated by using the customizing technique based on the reference model introduced by Scheer [12]. The Scheer reference model is an enterprise-wide data model which contains about 300 entity- and relationship types. The model has evolved from experiences gained in academic research and from implementation projects in industry. The example presented focuses on the primary data of the production control area.

The part of the Scheer reference model which was used as starting point of the customizing process is depicted by the left part of figure 3. After the first analysis of the data reference model the entity types PART, PERSONNEL, TOOL and EQUIPMENT and the relationship type HIERARCHY were transferred to the customized data model. The distinction between EQUIPMENT GROUP and EQUIPMENT was not necessary for Triplex. Therefore, the EQUIPMENT GROUP was dropped and replaced by EQUIPMENT, i.e. EQUIPMENT participated in all relationships which belonged to EQUIPMENT GROUP. The relationship types PERSONNEL ASSIGNMENT and TOOL USE as relationships between PERSONNEL and EQUIPMENT and TOOL and EQUIPMENT were also transferred to the customized model. The relationship TOOL ASSIGNMENT was not necessary and dropped. A relationship type between the entity types TOOL and PART was not part of the reference model and had to be added to describe correctly the Triplex production control area. The entity type PERSONNEL had to be further detailed. Therefore, it was specialized into the entity types OPERATOR, FOREMAN and CHARGE HAND. Up to this point, no substantial modifications of the data reference model were necessary. Most of the entity- and relationship types could have been taken over from the reference model or had to be modified only slightly.

The fast generation of a first draft version of the data model helped to focus on the most important or most critical parts of the data model. Modelling the relationship type OPERATION and the entity types belonging to it was more difficult. Although the Scheer reference model already presented a feasible solution, the Triplex engineers wanted a solution which fits better to their particular situation. After discussing different alternative solutions, the most suitable was selected and integrated into the data model. The data model which was developed by customizing the reference model is depicted by the right part of figure 3.

The experiences gained by the application at Triplex showed that the usage of a reference model for data engineering in CIM is a feasible approach which is leading to tangible results in a short time. The customized data model presented in figure 3 is only a small part of the comprehensive data model developed at Triplex. The entire data model of the Triplex production control area contains 26 entity types and 39 relationship types.

The data model developed for Triplex reflects knowledge gathered during the customizing process which is not only important for the purpose of designing a database. The model led to a better understanding of the Triplex production site. Some of the findings were important to improve the scheduling functionality and pinpoint weaknesses of the current situation.

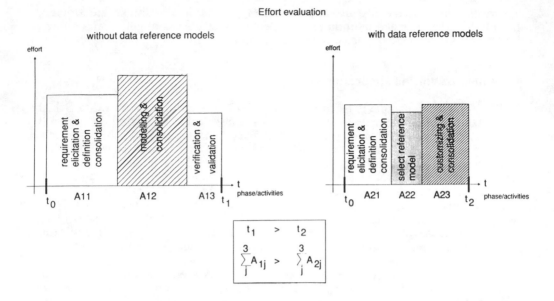

Fig. 3: Customizing process for primary data of production control area

Reengineering using data reference models

The reengineering approach addresses the task of documenting, restructuring and enhancing existing but not adequately documented and structured software. In the data modelling context, the task of reengineering is to analyze the logical data structures of existing application systems and to derive a sound conceptual schema for it which represents the data structures of the analyzed business application. The process of reengineering may also incorporate the restructuring of programs and computer code. This paper, however, only deals with the reengineering of data. Yet, there is a close connection between the reengineering of computer code and data.

Any data engineering approach aiming for the development of integrated information systems has to take the existing structures into account. Reengineering is a prerequisite for the migration of currently isolated and redundant structures to integrated systems. The reengineering process is supported by a classification system. In the following, the steps of reengineering are presented in figure 4. It is demonstrated how data reference models may be used to support the reengineering of data structures.

The reengineering process comprises activities such as
- the analysis of existing data structures,
- the classification of data elements,
- the documentation of the links between existing data structures and the data reference model,
- a remodelling of the existing data structures and
- a restructuring of the existing applications.

The informational basis for the analysis could be data dictionaries (if available) or data catalogues. The analysis phase is an interactive process which essentially needs to include the participation of human knowledge. The analysis should focus on the recognition of data fields, member types, sets, pointers and indexes (primary keys, foreign keys). Result of this first step is the documentation of existing data structures.

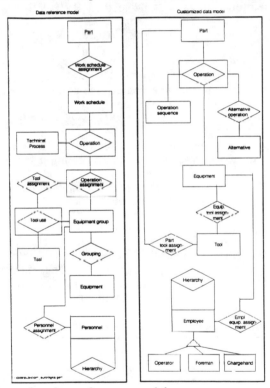

Fig. 4: Reengineering using data reference models

The documentation of the existing data structures is completed by a classification of its information objects. This classification supports the recognition and removal of synonyms and homonyms within the data base. The classification plays an important role because it implements the interface between the documented data structures of the existing application systems and the data model which elements are classified by the classification system.

The documented and classified data structures are compared to a reference data model which is most appropriate to describe the application area. The selected reference data model may already be customized to the enterprise-specific situation. In the following, the elements of the existing data structures can be linked to the elements of the selected reference data model by using the functionality of the classification tools. Synonyms of the existing data structures can be found by the retrieval component of the classification tool. Elements which are classified in the same way are synonym candidates and have to be analyzed by the user. In the case of existing synonyms, the links between these elements have to be documented. Based on the defined links between the existing data structures and the reference model, a data model can be developed which considers the structures of the existing systems. Redundancies can be removed and new aspects may be inserted into the data model.

Based on the new findings documented by the data model new information systems can be developed or existing information systems can be restructured. Thereby, the administration of the links between the existing structures and the data model is a prerequisite for the definition of migration strategies from existing, isolated and redundant information systems to new integrated information systems.

The process of reengineering in combination with the usage of data reference models has the potential of creating a platform for the enterprise which is stable and independent of technological changes. The CODE project validates the reengineering process using data reference models at its testbed RWE-DEA, a large German oil-processing company.

Benefits of data reference models

The effort to create such an integrated enterprise-specific data model is very large. Data reference models provide substantial support for an efficient and correct design of enterprise-wide data models. Experience gained in large modelling projects shows that data reference models serve as a catalyst in the requirements definition phase. They provide a methodologically sound and well-structured framework for focussing on the substantial aspects of the CIM system under study. Being a feasible solution that has been validated and verified, reference models reduce the effort for validation and verification significantly. Furthermore, they prevent the development of too a detailed and implementation dependent solution, a problem often encountered in many software design projects.

Applying reference models avoids disadvantages of the development from scratch as there are the lack in expertise within the enterprise with several risks associated and the tendency to model the current situation of the enterprise with all its positive and negative aspects. Tangible benefits of applying data reference models in CIM are
- a reduction of costs and efforts for the design on an integrated CIM system,
- the utilization of proven and verified input for the requirements modelling and design stages,
- an improved structure of the information system and
- a documented migration path from currently isolated towards integrated systems and
- the incorporation of data engineering knowledge.

In figure 5 these benefits are illustrated.

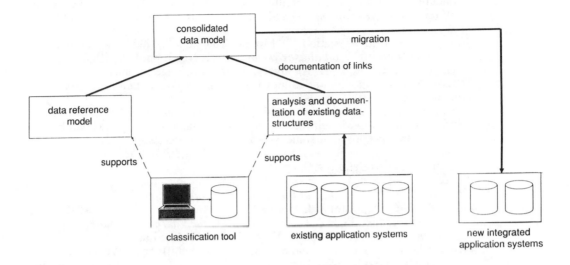

Fig. 5: Benefits of data reference models

CONCLUSION

The concept of applying reference models for data engineering in CIM proves to be both feasible and beneficial. The main benefits of applying data reference models are an improved quality of software and data bases, reduced efforts for the maintenance of data bases, a faster implementation of information systems and the availability of off-the-shelf reference models with proven, documented and tested data structures.

The data reference model approach will be further investigated in the course of the CODE project. A computer-supported tool for the classification process is currently being developed. For both the customizing and the reengineering process a computer-based system will be developed in cooperation with a renowned software vendor.

REFERENCES

1. Scheer AW: *CIM Der computergesteuerte Industriebetrieb.* 4.Aufl. Berlin, Springer 1990, p 11.
2. Dilts DM, Wu W: Using Knowledge-Based Technology to Integrate CIM Databases, in *IEEE Transactions On Knowledge And Data Engineering*, Vol.3, No.22, pp. 237-245, 1991; Krishnamurthy V et al.: IMDAS - An integrated manufacturing data administration system, in *Data & Knowledge Engineering*, Vol.3, pp. 109-131, 1988.
3. Scheer AW: *Enterprise-wide data modelling - Information Systems in Industry.* Berlin, Springer 1989, p 537. Also Scheer AW, Hars A.: Enterprise-wide data modelling - the basis for integration, in *Proceedings of the 7th CIM-Europe Conference*, 1991.
4. Jorysz HR, Vernadat FB: CIM-OSA Part 1: total enterprise modelling and function view, in *International Journal of Computer Integrated Manufacturing*, Vol.3/4, pp.144-156, 1990.

5. For a comparison see: Hars A, Heib R, Kruse C, Michely J, Scheer AW: Concepts of Current Data Modelling Methodologies. - A Survey -, in Scheer, AW (ed.): *Veröffentlichungen des Instituts für Wirtschaftsinformatik*, Vol.84. 1991.

6. Chen PP: The Entity-Relationship Model - Toward a unified view of data, in *ACM Transactions on database systems*, Vol.1, No. 1, pp.9-36, 1976.

7. Mylopoulos J, Levesque H: An Overview of Knowledge Representation, in Brodie ML, Mylopoulos J, Schmidt JW (eds.): *On Conceptual Modelling*. Berlin, Springer, 1984, pp 3-17.

8. No consensus has yet been reached for the standardization of the Extended Entity Relationship Model; for a discussion: Spencer R, Teory J, Hevia E: ER Standards Proposal, in Kangassolo H (ed.): *Proceedings of the 9th International conference on ENTITY-RELATIONSHIP APPROACH*. Lausanne, Switzerland, 1990, pp.405-412.

9. Scheer AW, Hars A: Enterprise-Wide Data Modelling - the basis for integration, in *Proceedings of the 7th CIM-Europe Conference*, 1991; see also: Rochfeld A, Morejon J, Negros P: Inter-Relationship Links in E-R-Model, in Kangassalo, H (ed): *Proceedings of the 9th International conference on ENTITY-RELATIONSHIP APPROACH*. Lausanne, Switzerland, 1990, pp.143; and Scheuermann P, Schiffner G, Weber H: Abstraction Capabilities and invariant properties modelling within the Entity-Relationship-Approach, in Chen P (ed): *Entity-Relationship Approach to System Analysis and Design. Proceedings of International Conference on ENTITY-RELATIONSHIP APPROACH to System Analysis and Design*, Amsterdam, Netherlands, 1980, pp.121

10. Schlageter G, Stucky W: *Datenbanksysteme: Konzepte und Modelle*. 2. Aufl. Stuttgart, Teubner 1983, p. 50.

11. Vernadat F: A conceptual schema for a CIM database. *Proceedings of the 6th Autofact Conference*, October 1-4 1984.

12. Scheer AW: *Enterprise-wide data modelling - Information Systems in Industry*. Berlin, Springer 1989, p 165.

AN ARCHITECTURE FOR COMPUTER GENERATED INSPECTION PROCESSING PLANNING

Michael D. Reimann and Joseph Sarkis

*Information Systems and Management Sciences, College of Business Administration,
University of Texas, Arlington, Texas, U.S.A.*

ABSTRACT

The intent of computer integrated manufacturing (CIM) is to link together the primary design and manufacturing functions of the enterprise. For the most part industry and research has been concerned with the integration of computer-aided design (CAD), computer-aided manufacturing (CAM), and computer-aided process planning (CAPP). Little effort has been directed toward the linkage of quality control and automated inspection in the CIM environment [1,2,3,4,5].

Equipment that can fill this void are numerically controlled dimensional measuring equipment (DME), of which, coordinate measurement machines (CMM), Flexible Inspection Systems (FIS), optical comparators, robotic measuring devices, and laser based measuring devices, are examples. To inspection, the automated DME bring the same advantages that numerical control (NC) machines bring to manufacturing. A properly integrated system can enhance flexibility, increase throughput, reduce setup time, minimize operator error, improve accuracy, improve product quality and lower costs. To take full advantage of DME, CIM requires that flexible inspection centers no longer act as islands of automation, but that they be integrated with the other elements of automation, including CAD, CAPP and CAM. Integration of inspection tools is one of the issues that will be addressed in this paper.

We also present a framework for generating automated inspection process plans based on CAM-I's advanced numerical control (ANC) processor design. The elements that compose this framework will be discussed. The framework along with its elements are referred to as the Expert Programming System - One (EPS-1) [1,2].

AUTOMATED INSPECTION AND DME

The philosophies of Total Quality Management (TQM) and Total Quality Control (TQC) are gaining greater acceptance from domestic manufacturers. An underlying concept of TQM is for quality at the source. This concept points to the need for monitoring the product throughout all processing stages. In an integrated framework the inspection process can be run simultaneously with the actual machining processes of the product, where direct results and measurements of specifications can be determined while processing. Thus, a framework similar to the automated development of numerical control programs for processing equipment (i.e. ANC) can be used for the development of programs for automated inspection equipment.

In many instances DME are the appropriate devices for data collection and data processing. Automated CMM's are capable of collecting dimensional data at rates approaching one measurement point per second. This makes CMM an ideal choice for gathering Statistical Process Control (SPC) data. Moreover these inspection devices make it possible to obtain a near real-time SPC system. Reverse engineering, or design for manufacturability, are also greatly influenced by DME, where design and redesign times can be dramatically reduced [6].

Similar to generative process planning for CAPP, the framework developed here would be classified as a generative process planning approach for automated inspection.

OPERATIONAL ENVIRONMENT OF THE EPS-1

An operational environment for linking CAD systems with DME has been defined. The Expert Programming System-One (EPS-1) was developed through the cooperative efforts of the CAM-I Advanced Numerical Control (ANC) program. Figure 1 illustrates the environment of the EPS-1. The elements of this environment are described in the remainder of this section.

Geometric Modeler

A geometric modeler is used to delineate, identify and define the physical characteristics of a part. Various geometric entities can be described by their boundaries. For instance, a straight line can be defined by its end points. In the same way, solids can be defined by the surfaces that enclose them. A relatively simple geometric modeler [7] was used in the EPS-1 . This modeler employs boundary representation as its standard means for defining solids. The geometric modeler provides only swept generation solids. Objects are defined by first developing a profile on a general plane. This is specified in the standard form of a normal vector and the distance the plane is from the origin. A profile may be a simple circle or any combination of arcs and/or straight lines which comprise a closed figure. In order to define a portion of an arc or line, a user selects circles and lines in sequence around the profile. This is repeated until the closed figure is completely specified. There are five methods for creating straight lines:
- Line through two specified points
- Line through a specified point in a specified direction
- Line through a point parallel to an existing line
- Line through a given point and tangent to a specified circle
- Line tangent to two specified circles

There are two methods for defining circles:
- Specified center point and radius
- Circle with a specified radius tangent to two specified lines

The profile is then swept outward a specified distance normal to the profile plane to define a solid. Additional profiles may then be defined on any of the planar surfaces of the object and swept outward to define protrusions or swept inward to define holes or slots.

Another definition method provided by the geometric modeler is referred to as reflect and join. This is accomplished by selecting a face on a planar object, then, the object is reflected at that face. A "glue" operation is performed to make a symmetrical object.

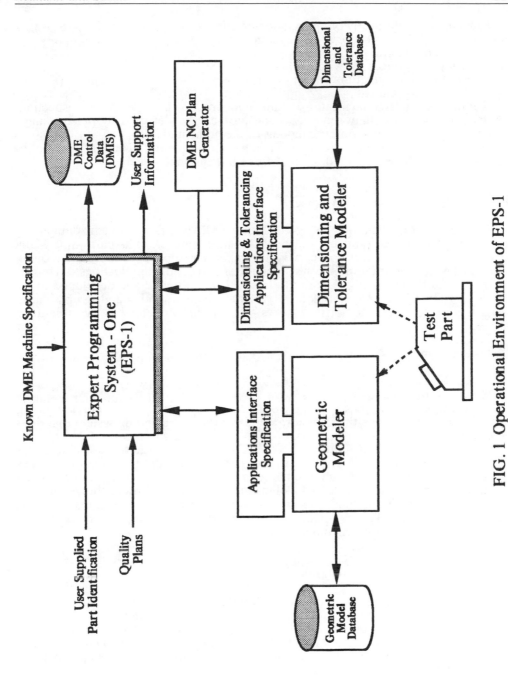

FIG. 1 Operational Environment of EPS-1

Entry points into the geometric modeler allow external applications to use its functional capabilities. These are categorized into four groups:

- Primitive operators
- High-level operators for whole objects
- Interrogation/verification
- Additional procedures

The geometric modeler facilitates certain functions that are specific to inspection. In particular, collision between probe and part can be detected by performing a Boolean intersection on geometric models of the probe and part.

Even though the geometric modeler is limited to planes, cylinders, quadratics and cubics, it could be extended to other three dimensional objects.

Dimensioning and Tolerance Modeler

Inspection of physical objects by DME and CMM's requires specific data about the dimensions and tolerances for the geometric features of a part [8]. Many existing geometric modelers do not have the capability to produce the dimensioning and tolerancing information that is needed by inspection applications. Thus, a Dimension and Tolerancing (D&T) Modeler [9] was designed and developed to fill this void. The D&T Modeler defines various tolerance nodes and assigns these nodes to one or more geometric features. The D&T Modeler has five tolerance classes:

- Location
- Orientation
- Size
- Form
- Surface Finish

It is necessary for the D&T nodes to remain linked with their corresponding geometric features. Since the geometric modeler is a free standing system that does not utilize measurement information, it is the responsibility of the D&T Modeler to insure that the nodes and features are aligned with one another. Coordination is accomplished in the D&T model of the part by reference to feature names previously assigned by the geometric modeler. After geometric features have been added to a part model, the D&T Modeler must be used to augment the geometries with specific dimensioning and tolerancing data. Likewise, geometric features that are removed from the geometric model must also be removed from the dimensioning and Tolerancing model.

Applications Interface Specification

Due to unique capabilities that are offered by various geometric modelers, an inspection application may require the use of several modelers. To avoid the necessity for developing unique interfaces for each combination of application and modeler, a standard interface similar to IGES has been developed. The Geometric Modeling Program of CAM-I has defined the Application Interface Specification (AIS) to satisfy this need.

The AIS is a dynamic interface between modelers and applications software. Modelers and applications that have been designed and implemented to conform to the AIS can interface directly with one another. Non-compliant modelers and applications require modifications to make their input and output compatible with the AIS. For example, if an application requires a modeler function such as the evaluation of a point on a surface, the applications software can issue a subroutine call requesting a solution from the modeler. This is straight forward provided the modeler has the ability to perform the function requested. If the function is not available in the modeler, then the application interface will be expected to perform the task.

The AIS defines a core set of capabilities that should be available from all modelers. If a particular modeler has the capability to perform a specified function, then the interface acts only as a communications media. On the other hand, when the function is not supported by a modeler, the application interface would be required to perform that function. The core set of capabilities consists of about fifty entries or subroutines covering various categories which include:

- Creation/Deletion
- Naming
- Modification
- List Handling
- Low-level Interrogation
- High-level Interrogation
- System Utilities

Dimensioning and Tolerance Applications Interface Specification

The AIS allows the user to retrieve geometric data about a part and also enables applications to utilize the functional capabilities of the geometric modeler. The D&T model expands the geometric model of the part to include data that are necessary for dimensioning and tolerancing. Unfortunately, the AIS does not accommodate dimensioning and tolerancing data, nor does it enable external applications to access the functional capabilities of the D&T Modeler. Therefore it was necessary to extend the AIS to create, read, modify and delete nodes in the D&T model. The Dimensioning and Tolerancing Applications Interface Specification (DTAIS) [10] was created to fill the void.

The DTAIS is similar in concept to the AIS in that it is used to construct the D&T model. The DTAIS constrains the types of nodes that may be associated with the D&T model. The valid nodes include:

- Feature nodes
- Tolerance nodes
- Datum Reference Frames (DRF) nodes

Dimensional Measuring Interface Specification

There are three classes of interface for linking CAD systems with DME [11].

One-to-one interfaces require users to develop their own communication linkages that work with specific devices in their system. The one-to-one interface typically involves only one type of CAD system and one type of DME.

One-to-many interfaces allow one CAD system to interface with numerous DME and CMM's. A characteristic of this type of interface is that the CAD system usually produces commands in a generic or neutral language and then a post processor, unique to each DME, converts the neutral language into DME specific commands.

Many-to-many interfaces allow the user to select the most appropriate CAD systems and DME which satisfy their particular needs. Users are able to mix and match systems due to the standardization of the interfaces between the devices.

Because there is a wide variety of CAD systems and DME available to industry, it is desirable that a standard interface exist for linking these devices. Ideally, this standard interface would accommodate as many existing products as possible, thus allowing for many-to-many matches of CAD systems and DME. Standardization of a many-to-many interface occurred quite recently with the adoption of the

Dimensional Measuring Interface Specification (DMIS) [12] by ANSI as an American National Standard in February 1990.

DMIS consists of a neutral command file, a neutral data file and protocols developed for transferring information. CAD systems communicate with the DME using these commands, data structures and protocols. While primarily designed for communication between automated equipment, DMIS is both man-readable and man-writable, allowing inspection programs to be written and inspection results analyzed without the need for computer aids.

The DMIS vocabulary is similar in syntax to the APT NC programming language. The vocabulary is a structured programming language which is designed to be processed sequentially by a DME. There are two basic types of DMIS statements; process oriented commands and geometry-oriented definitions. Process commands consist of motion commands, machine parameter commands and other commands which are unique to the inspection process itself. Definitions are used to describe geometry, coordinate and tolerancing systems, and other types of data which may be included in the CAD database.

FUNCTIONAL STRUCTURE OF THE EPS-1

The basic architectural model of the EPS-1 processor is illustrated in Figure 2. This diagram shows the relationships among nine functional modules that form the EPS-1 architecture. The EPS-1 requires a part identification number and an overall process plan as inputs. The identification number is used to retrieve the geometric model and the dimensioning and tolerance model of a part.

The DME NC plan generator uses the overall plan to produce DMIS control data and support information for a specific DME.

Obtain Operation Plan

The user provides required data to the EPS-1 for initiating its process. This module determines the scope of the inspection process to be performed on the indicated part. Input needed by the EPS-1 include:
- A valid part identification number as defined in the geometric part data base.
- The type of inspection (e.g., courtesy, in-process, or final).
- An inspection process strategy.
- If known, a predetermined DME class or specific DME machine.

All of these inputs are required except for DME Class/Machine. Values provided by the user are validated to ensure that corresponding data files exist in the appropriate external databases. Once this information is validated, a header data record is created in a Work Element Table and assigned the values provided by the user.

Task Decomposition

The task decomposition module determines the evaluated dimension and tolerance (EDT) features that are to be measured by the inspection process. EPS-1 automatically determines this based on information provided from the previous module. The part identification and the inspection type uniquely identify the geometric and D&T models of the part. An entire D&T model is obtained through DTAIS subroutine calls to the D&T Modeler.

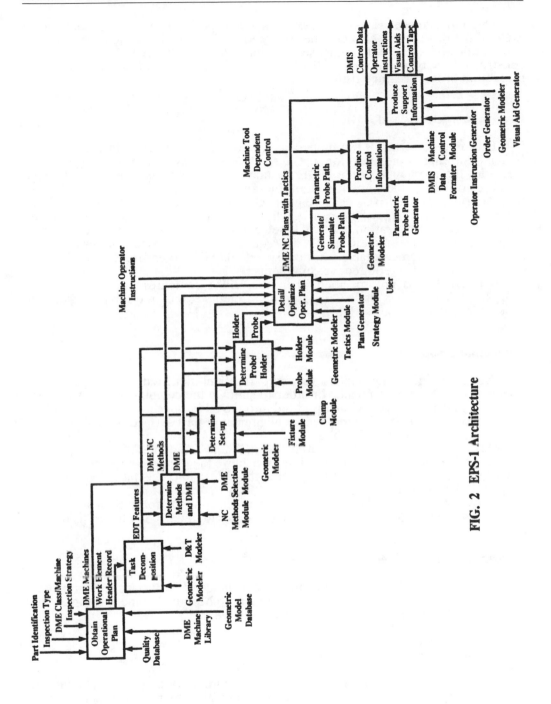

FIG. 2 EPS-1 Architecture

The D&T model is used by the task decomposition module to construct a skeletal list of the work elements necessary to perform the inspection process. For each node in the D&T database, an individual work element line item is constructed. Subsequent steps in the EPS-1 use this information to determine other inspection process characteristics.

Determine Methods and DME

This module performs two functions. The first function determines appropriate NC measurement methods for the inspection process. The second function identifies the most appropriate DME for taking the actual measurements. For the purposes of EPS-1, the combination of tolerance class and sub-feature class, is used to establish individual NC measurement methods. Once the methods have been determined, a DME machine can be chosen. If a DME machine was specified during process initialization, the DME selection function will be bypassed. In this case, the NC measurement methods required for the inspection process are compared with those available from the specified DME, to ensure that the DME is capable of taking the required measurements.

If a DME was not specified, then the EPS-1 compares the capabilities of available DME against the NC measurement methods required for the inspection process in order to choose a DME.

Determine Setup

Part setup generally applies to the part orientation and fixture and clamp selection in a NC environment. In order to simplify the complexity of this process, fixtures and clamps were not incorporated into the initial specification of the EPS-1.

Part orientation is accomplished by alignment of part surface normals with the DME measurement axes. The goal is to:
- Maximize the number of measurements in any given orientation.
- Minimize the number of orientations to achieve the required inspections.
- Eliminate possible collisions between the DME and the part due to improper orientation.

The above strategy requires the geometric modeler to perform Boolean algebra intersections on the part and probe models in order to detect potential collisions.

Determine Probe/Holder

Combinations of measurement probes and holders for the inspection process are determined by this module. Selection of a probe can become a complex task when the following factors are taken into considered:
- DME technologies
- Probe relative orientation
- Multiple probe setups
- Probe shape

Therefore, candidate probes must be identified through a process of elimination. The capabilities of the available probes are compared to required measurement characteristics of the inspection process. A probe is determined for each sub-feature on the part. Therefore, several probes may be needed to measure the entire part. The strategy at this point may be to select probes based on minimizing the number of probes for the inspection process. Another probe selection strategy might be to maximize the number of measurements obtained from the individual probes.

Detail/Optimize Operation Plan

Inspection process information which was developed in previous modules, is used by this module to complete the operation plan. EPS-1 finalizes the sequence of work elements and determines the corresponding movement tactics for the inspection process. Three activities are performed by this function:
- NC plan generation
- Determination of NC tactics
- Review and modification of the NC plan

Sequencing of the individual work element items is determined in this module. The NC strategy previously specified by the user forms the basis for the ordering process. Based on this strategy, an appropriate set of logic is used to establish the specific sequencing of the work elements. The types of logic used is dependent on the sequencing strategy. Mixtures of heuristic, optimizing and decision table logic may be employed. Relative positioning of the probe is also determined for every sub-feature measurement to be made.

All the processing up to this point has been automatically carried out by the EPS-1. The Work Element File, NC plan, and NC tactics are now complete and the user can intervene to make any adjustments. The user may review the graphical representation of the inspection process sequence. This will indicate potential collision and coverage problems. The user makes necessary corrections to eliminate any problems. Simple changes can be made by the user without any further processing by the EPS-1.

Generate/Simulate Probe Path

There are two major functions performed by this module. These are to generate locational and parametric probe path data and to visually simulate that data. The primary input to this module is the NC plan with tactics and the primary output is an internal representation of the probe path data.

The NC plan with tactics contains an ordered list of the work elements required to inspect the identified EDT features. The work elements are processed sequentially and contain the sub-features class, the tolerance class, and the sub-feature name for each entry.

The measurement process involves the generation of initial and departing probe positions and parametric probe path motion. Positioning of the probe for measurement is determined by generic routines which are geometry specific. EDT sub-feature class/tolerance class combination dictate the appropriate routine. The routines use specific D&T information to establish actual probe position.

Parametric probe path data is mathematically determined for each measurement. The probe path data is comprised of two curves that are parametrically synchronized. One of the curves represents the probe end, while the other represents the probe axis. Once the parametric probe path is created, it is used to graphically illustrate the movement of the probe in relation to the part. The user can review the movements of the probe over the part to determine coverage and detect collisions.

Produce Control Information

Numerically controlled DME inspection instructions are generated by this module. These instructions conform to DMIS. The NC plan with tactics provides the necessary tolerancing information. The coordinates of the actual points for every movement and measurement are retrieved from the internal representation of the probe path. All of this information is combined to produce DMIS instructions.

This module also allows DME independent control information to be included in the DMIS file.

Produce Support Information

Information that supports the actual NC inspection operation is developed in this module. It is defined in such a way that it can be tailored to support the specific environment of individual sites. Information for the DME operator is generated as well as for other users indirectly involved in the NC inspection operations. These individuals may include quality assurance engineers, shop supervisors, foreman, or tool crib operators.

DISCUSSION

To fully leverage the potential benefits of CIM it will be necessary to incorporate quality considerations of the product and process along with the more traditional computer-aided tools used in CIM. In an integrated framework the inspection process can be run simultaneously with actual manufacturing processes. Thus the results from measurements can be used to correct the manufacturing process in a real time fashion. A framework similar to advanced numerical control serves as the logical mechanism for automating the inspection process.

This paper has briefly described the EPS-1 framework. The framework addresses both the operational environment and architecture of the EPS-1. The EPS-1 links automated inspection process planning to the rest of the CIM environment. Moreover, it identifies the required elements for automating the inspection process. Considerable development work has been undertaken to define and refine the EPS-1 architecture. Proof-of-concept has been demonstrated for the EPS-1 and the results of this work will be presented in a more detailed and technical publication. A prototype system [3] that utilizes the EPS-1 as its foundation has also been implemented. The results produced thus far point to numerous areas for further research which include:

- Inspection processing planning requirements
- Inspection information requirements
- Heuristic selection techniques for DME, probes and holders
- Object oriented process structure
- Blackboard mechanisms for controlling the EPS-1 process
- Optimization of inspection paths

REFERENCES

1. Reimann MD, Fowler JW, The Expert Programming System - One (EPS-1), in the proceedings of the *Fourth International Conference of the State-of-the-Art on Solids Modeling*, sponsored by CAD/CIM Alert and CAM-I, Boston, Massachusetts, May 1987.
2. *The Expert Programming System - One (EPS-1), Task One Design Review*, Computer Aided Manufacturing - International, Inc., Arlington, Texas, R-87-ANC-01, 1987.
3. Brown CW, IPPEX: An automated planning system for dimensional inspection, in the proceedings of *The 22nd CIRP International Seminar on Manufacturing Systems*, Enschede, The Netherlands, June 11, 1990.
4. Merat FL, Radack GM, Automatic inspection planning within a feature-based CAD system, *Robotics & Computer-Integrated Manufacturing*, Vol. 9, No. 1, Pargamom Press, pp. 61-69, 1992.

5. ElMaraghy HA, Gu PH, Expert system for inspection planning, *Annals of the CIRP*, Vol. 36, No. 1, pp. 85-89, 1987.
6. Owen JV, CMMs for process control, *Manufacturing Engineering*, Vol. 107, No. 2, pp. 39-41.
7. *CAM-I Test Bed Modeler*, Computer Aided Manufacturing - International, Inc., Arlington, Texas, PS-85-GM-01, 1985.
8. *ANSI Y14.5M, Dimensioning and Tolerancing*, The American Society of Mechanical Engineers, ANSI Y14.5M - 1982.
9. *CAM-I D&T Modeler Version 1.0 - Dimensioning and Tolerancing Feasibility Demonstration Final Report*, Computer Aided Manufacturing - International, Inc., Arlington, Texas, PS-86--ANC/GM01, Nov. 1986.
10. Johnson RH, *Dimensioning and Tolerancing Final Report*, Computer Aided Manufacturing - International, Inc., Arlington, Texas, R-84-GM02.2, May, 1985.
11. Zink JH, Closing the CIM loop with CMM, *Automation*, Vol. 36, No. 1, pp. 48-50, 1989.
12. *ANSI/CAMI 101-1990 Dimensional Measuring Interface Specification*, Computer Aided Manufacturing International, Arlington, Texas, 1990.

SHOP FLOOR DATA COLLECTION — THE CORNER STONE OF CIM

[1]Thomas D. Halpin, M. Munir Ahmad, and [2]Donald Walshe

[1]*Computer Integrated Manufacturing and Engineering (CIME) Research Centre,*
University of Limerick, Limerick, Ireland
[2]*Symnet International, Shannon, Co. Clare, Ireland*

ABSTRACT

Today's business problems are complex, the manufacturing industry is faced with the problems of decreasing product life cycle, increased demand for customisation and increasing complexity, together with increased international competition to reduce total cost, lead times and to improve due date performance. Firms are facing questions, the answers to which could affect their survival, good solutions require information to be readily available from many sources within an organisation.

Considering a large multi-national, the information contained in each department, and associated software applications, is often isolated and not readily available to people across the organisation. Decisions taken to the benefit of one department may have a detrimental effect on others. Thus it is essential, that the decision makers, from the operators through to the executives, have available to them the information they require to make the correct decisions. This is the realm of Shop Floor Data Collection (SFDC) systems. They should collect information as close to the operational source as is possible and present it to the relevant personnel, in a timely manner, suitable to their function and consequently aid the decision making process through the organisation. This will ensure that integrated decisions are made, and that the decisions made do not conflict with the organisation's objectives. This paper addresses the role of SFDC systems in a multi-national organisation, the functions to be provided and the implementation strategy.

INTRODUCTION

A CIM project is being undertaken, by the Computer Integrated Manufacturing and Engineering Research Center (CIMERC) at the University of Limerick for a multi-national industrial client. The general manufacturing and engineering facility of the organisation has been chosen as a pilot area for the implementation. The general engineering facility is driven by individual orders, which are scheduled into the system dependent on customer priorities, material availability, machine availability and manpower. The general engineering facility is in effect a job shop with lead times of upto one year, currently the information processing functions are performed manually.

ROLE OF IT STANDARDS

It is clear since the advent of the PC that we are in the midst of a technological revolution, and that there are dangers of implementing technology for technology's sake and not for clearly defined business reasons.   This approach could have catastrophic effects on a multi-national organisation when the time comes to integrate the various systems toward attaining the objective of the global factory.   There is therefore a need for standards defining the hardware, operational software, application software, communications and system management approaches for use within an organisation. The intelligent use of such standards permits a multi-national organisation to apply, tried and tested, low cost technology throughout the organisation, to provide a common frame of reference toward which the various business functions may move or branch from.   It is important, however that once the standards are set, the organisation keep track of international developments, and ensure that the standards are reviewed periodically.

Standard Pitfalls

However there are a number of important points to be realised in relation to standards, by default in today's rapidly changing environment, a computing standard will nearly always trail the leading edge of technology by at least one year if not more. It is extremely unlikely that a particular standard will provide an ideal solution to the business function it addresses; more often than not the standard will be perceived within the organisation to lack essential functions. There is bound to be considerable resistance to what people perceive as being forced to use set computing tools to solve their unique problems. This resistance must be overcome, in view of the alternative of trying to integrate the various stand alone systems which would propagate through the absence of standards, this is a complex task within a single company let alone a multi-national.

Standards in the Company

The company have over the last three years set in place a wide range of corporate standards in relation to computer technology.   Standards now exist defining the selection of hardware platforms, mainframes, mini's and PC's, operating systems, application software such as word processors, LAN and communication software. Now that these standards are in place, implementations of applications within individual companies and departments may take place within the corporate CIM plan of integration.

WHERE TO START ?

Strategy by default must originate at the top, many papers state that it is essential to have the support of the board for the planning and implementation of CIM. However it is universally recognised that CIM should be installed from the bottom up, by the engineering staff of the shop floor. Defining the link between strategy and the shop floor, is one of the most important steps toward a successful CIM implementation, this however is not without difficulties as evident from the questions Nicholson [1] poses

"But who is responsible for the result ? Is it up to the shop floor to implement the strategy or is it up to the strategic statement to go into such detail, that it is obvious to implement it ?"

The answer lies in a marriage of the ideas. In a multi-national organisation the managers responsible for the various departments must be free to act to solve their business problems. However any action should only be taken with the guidelines for CIM as defined by a CIM Implementation team. Once the project specifications have been agreed between the two parties, the individual managers supervise the implementation using either internal or external resources. The progress should continually be reviewed by the CIM team to ensure that the long term objective of corporate integration is achieved.

Where to Direct the Technological Focus ?

The trend in the manufacturing industry of throwing the latest lathe or CNC machine at the problems of the shop floor, is in our opinion, seriously misplaced. This is evident upon walking in to any batch manufacturing environment and asking the following two questions

> What is the lead time ?
> What is the actual manufacturing time ?

There will always be an alarming difference between the two, this difference can be summarised in one word "confusion", the initial aim of CIM should therefore be to eliminate this confusion and the incurred costs. For a number of years the technology of automatic identification, more commonly known as barcoding, has been over shadowed by swarf generating technology. The latest CNC and CAD technology has been installed and has been seen in the words of Hutchin [2]

"as the path toward profitability and security through the nineties"

When it is in effect further reducing the actual manufacturing process time, the smallest portion of the lead time, in our opinion the focus should be directed toward reducing the time that components are sitting around waiting for manufacturing operations. If this time can be reduced by even as much as one half day per week, the knock on effect throughout an organisation would be considerable. At the Computers In Manufacturing '91 exhibition held in Birmingham, England, it was interesting to see that in excess of 30% of the stands present were in the field of barcoding technology and shop floor data collection (SFDC) systems. It is the opinion of the authors, that the area of SFDC has greatest potential initially for savings in the manufacturing lead time.

Linking SFDC with the Manufacturing Plan

SFDC systems, however do little in isolation, they must be conceived in response to an overall manufacturing plan. The manufacturing plan may be viewed as being comprised of three elements

- external constraints, originating from other parts of the organisation i.e. orders.
- internal constraints, i.e. setup and processing times and the type and sequence of production.
- specified objectives, which define the goal of the organisation.

Requirements of SFDC in the Company

The company have identified a need for an SFDC system, the driving factor for this requirement is control, this is a fundamental requirement of any business. If a company is to make a profit, it must know where it stands, to know this it must have reliable information and a manner of collecting information that is accurate and cost effective. The following requirements were deemed to be essential toward the justification of an SFDC system.

1. Collect data as close to source as is possible, in the most efficient manner, and in doing so reduce the paper flow to and from the shop floor, and the time and effort spent in maintaining it.
2. Present the information to the relevant personnel in a manner suitable to their function as they require it.
3. Maintain and improve existing information flow to other departments i.e. accounts.
4. Provide a means of identifying and tracking a job through the manufacturing process, enabling a complete history of the job, from stock to manufacturing through to QC and finally to the customer, to be compiled.
5. Provide employee monitoring, to ensure employee accountability for work.
6. Provide an interface to a scheduling package, to ensure that the most up to date information is used for scheduling.

7. Provide the companies staff with accurate and detailed historical information to base future decisions upon.
8. Have the facility to be expandable, as new processes and operations come on line.
9. Adhere to the corporate IT standards

Role of the Database in SFDC Systems.

In order to establish a proper link between the SFDC system and the manufacturing function, a plan is needed, for the plan to succeed three pieces of information, are required

- the current status of the manufacturing organisation, how it is structured, how it functions and the existing technology.
- the future objectives and desired status of the organisation.
- the "roadmap" or plan, necessary to make the journey.

Once the SFDC system is installed and operating, all production and operating knowledge will be maintained as a single entity. Production information will no longer be solely distributed in the memories of the organisation's personnel. In the words of Sobczak [3]

" The computer becomes the collector of production trivia. In time this trivia becomes the basis for important savings ... When we collect data it is of no specific use. The utility is added by assembling like items and orienting (sorting) them so that they can be properly used. Having assembled and oriented the data it must be monitored."

It is evident from the preceding quote that the use of a SFDC system requires the compilation of an extensive database of manufacturing information. The database is the core of any SFDC system. The database should include the following information categories

- historical and performance related information
- WIP or status related information
- instruction related information

It is important that every happening on the shop floor is logged to the database as it occurs. This database will in time become the most important tool at the organisations disposal, consequently it should be readily available to all personnel in a manner suitable to their function and thus act as a catalyst to decision making process throughout the organisation. As an example, consider the installation of an SFDC system in the general engineering facility of the company, the planning people would now have access to accurate historical information to base quotes and schedules upon, shop floor supervisors will have up to the minute shop floor status information available at their fingertips and QC will have access to the manufacturing record of the components at the inspection stage. It is in this functionality that the main benefits of IT standards are realised. If no standards were present for the selection of systems, departments within an organisation would independently implement information systems, which would result in the birth of a number of local databases. These local databases would not be on-line to one another, creating un-necessary technical and cost barriers to the work of data and systems integration. The prime operational disadvantage to this being, that there is limited inter-departmental access to knowledge, resulting in inconsistencies between data stores, thus decisions can be taken in one department to the detriment of other departments within the same site or within the entire organisation. The resulting inefficiencies ensures the continual but un-necessary existence of departments and their interfaces.

By adapting a database standard, systems are put in place with a common frame of reference, at or near the point at which the operations occur. Thus providing the management and organisational staff with access to on-line information from the databases at the dispersed locations. So decisions can be taken for the corporate good, rather than solely to the benefit of the department making the decision.

IMPLEMENTATION

Having identified the need for standards and the company driven requirement for SFDC systems to act as information base, upon which applications such as scheduling and costing systems may operate, the question remains as to where the implementation is started ?

Pilot Area

In determining where to commence, it must first be understood what is required from the implementation. At a corporate level, the company requires a generic SFDC system, offering flexibility where they as an organisation require it, at the operational level. The final set up is required to act as an internal corporate SFDC reference site, designed and implemented using the corporate IT standards, which managers considering similar implementations may visit. The pilot area chosen, the general engineering facility simply requires control, that requirement is considerably more profound than it initially seems. It is important in this instance to note that the pressure for the implementation in this instance originates from the pilot area, thus reducing the risk of failure of the implementation, as it forms a solid link between the shop floor and the strategy.

The general engineering facility is ideally suited for use as a pilot area, it is a highly definable section of the organisation, while still having the links to all the organisational services such as accounts, stores and purchasing etc. The advantage to this is that the system can be worked out in total for the pilot area and any problems encountered can be solved for the pilot area, resulting in efficient organisational damage limitation. Once the problems have been solved for the pilot area, the implementation can be expanded to other areas of the organisation on demand. A cautionary note in that if the pilot area is not truly indicative of the organisation all the problems may not be encountered.

Implementation Issues

Implementation implies change, to change one must first understand what one is changing from, what one wishes to change to and how it is intended to supervise the change. Among the implementation issues to be addressed are :

> Current modus operandi of the area.
> Effect we hope the change to have on the area.
> Market review of vendors and products.
> Physical planning and installation.
> Development of bespoke software interfaces and applications.
> Software and installation.
> Communication with corporate systems.
> Employee training and industrial relations.
> Initial setup and parallel running.

Current Practice

To understand the current operating practices of the pilot area, a functional specification was developed by the CIMERC. This defined the functions and responsibilities of each member of the staff and the information flow within the area and the links with other corporate areas.

Desired Practice

From the analysis of the functional specification and through extensive consultation, with the staff of the pilot area, a system specification was developed by the CIMERC. The system specification outlines what the eventual goals of the CIM project are. It is the added bonus of ensuring that everyone is aware of the scope of the change being planned, thereby reducing the risk of technological shock.

It is important to note that both the functional specification and the system specification, were open to discussion and criticism at all stages by the personnel of the pilot area. This discussion and eventual agreement is essential to the success of the program, as if the end users are not happy with the proposed changes at the conceptual stage, serious difficulties would be encountered when the ideas are actually being physically implemented.

Vendor Selection.

As we are considering a pilot area, extensive time and effort has been spent sourcing and evaluating solutions. There are three sets of criteria for the selection of a system:

1. Ability to meet the requirements of the pilot area.
2. Ability to meet the corporate requirements.
3. Adherence to corporate IT standards.

It has been suggested, by Dunne[4]

"The buying of IT requires the street sense of a cattle-trader and the skills of a brain surgeon."

We would endorse that statement, as the company objectives and the vendors objectives are totally different. The following points are intended as an aid to the navigation of the vendor invested waters of the outside world:

1. Define exactly what is required, via documents like the functional and system specifications.
2. The best place to start the search, is within ones' own organisation. Enquire as to whether another site has got a similar system in place. If so evaluate the system in relation to the technology that is available today and re-establish contact with the supplier.
3. See as many vendors as is possible, it is useful to visit trade shows and exhibitions, as a large number of vendors are accessible under the one roof.
4. Define your requirements to the vendors and do not have the vendors defining your requirements otherwise you may inadvertently find what you perceive to be an ideal solution.
5. Use the corporate IT standards, to reduce the list of vendors to a manageable size.
6. Enter into detailed discussions with the short listed vendors, determine the underlying principles of their systems, how they operate, their hardware and software skills etc.
7. Allow the vendors to visit the pilot area and talk the personnel, do everything possible to ensure the vendor gets an extensive understanding of your requirements.

8. Ask for reference sites, visit the reference sites to see the proposed system working and to get an independent assessment of the system, from your peers at the reference site.

9. Only once you are satisfied, that each of the vendors has the capability to offer a solution, in relation to the selection criteria, invite them to offer a proposal in relation to the pilot area.

10. From the proposals received, select the system which a consensus of people within the organisation perceive to be the most suitable proposal.

The ten point vendor selection plan, outlined doesn't guarantee that the best system is selected, it however can generally be assumed that the success of the selection is generally proportional to the time and effort spent in the selection process.

Physical Planning and Installation.

Once the vendor has been selected, the task of planning the installation commences. In relation to the hardware the major issue to be addressed is the type and the location of terminals. The type of terminal is governed by the proposed environment, the type of transactions, the rate of transactions and whether interfaces to electronic devices such as PLC's or weighing scales are required. The terminals should be located as near to the operational activity as possible and to ensure flexibility at the operational level one terminal should be allocated per machining station. The physical planning also covers tasks like: cabling, power supplies, plug-in points etc.

Development of Bespoke Software.

Device drivers should be developed, if non standard transactions are required, such as an interface to a CNC controller, for the automatic downloading of CNC programs.

Software Development and Installation.

The transaction and reporting software, should if the correct system was selected, require just minor configuration changes before installation. If any changes are required, they should be agreed by both the vendor and the company and included in the final contract.

Communication with Corporate Systems.

The pilot area has interactions with many other areas on site and within the entire organisation. These links with these systems once developed for the pilot area can effortlessly be expanded to other areas on demand.

Employee Training and Industrial Relations.

Any form of change causes concern, people worry as to how the change will effect their jobs. It is important that the change is handled correctly, this requires early awareness and involvement of the shop floor in the change. The training of the workforce is essential, this includes the entire staff of the pilot area, from the operators of the shop floor terminals to the manager of the area. The training should focus on the new work practices that the new system will enable, the new practices should have been agreed long before the installation with the employee representatives to ensure that industrial relations do not cause un-necessary problems.

Initial Setup and Parallel Running

Once the system has been installed and is ready to go live, run the system in parallel with the old manual methods for a trial period. This is to ensure that the old techniques are not abandoned until one is sure that the new system is fully functional.

CONCLUSIONS

1. The importance of IT standards, in a multinational organisation striving toward the objective of the global factory can not be over emphasised.
2. CIM implementations, should initially be focused toward reducing the manufacturing lead time, rather than the manufacturing process time.
3. SFDC has a key role to play, in reducing manufacturing lead time, but the initiative must be conceived in response to the organisations overall manufacturing plan.
4. The most important component of an SFDC system is the database, which will in time, if it is properly structured and maintained become the most important tool at the organisations disposal.
5. Implementation should commence with a pilot area, that is indicative of the organisations production philosophy.
6. Once the implementation is complete in the pilot area, it will provide a solid, and reliable information base upon which required computer applications, i.e. scheduling, can be built.
7. The final setup will act as an organisational SFDC reference site, designed and implemented using the organisation's IT standards.

ACKNOWLEDGEMENTS

The authors would like to thank the company and EOLAS - the Irish Science and Technology Agency for funding the project, under the Higher Education Industry Co-operative (HEIC) Research Programme. Special thanks to the management, project team and to all the other people involved in the project from the company. We are also grateful to all those people who are working in the pilot factory. Without their support and cooperation, the implementation of the project would be impossible.

REFERENCES

1. Nicholson TAJ: Strategy and Shopfloor: A One-way Initiative ?. *International Journal of Operations & Production Management*. Vol 11 No. 1, pp. 41-50, 1991.
2. Hutchin T: Training for Flip-Flop Manufacture - The Road to Continuous Improvement Manufacture: *Proceedings of the Eight Conference of the Irish Manufacturing Committee*. University of Ulster, Ireland. Harris, Sept.1991.
3. Sobczak Dr. TV: Productive Use of Information: *Applying Barcodes for Industrial Productivity, Ed. 1*. Society of Manufacturing Engineers, Michigan, Sobczak, 1985, pp. 123-125.
4. Dunne SD: Making IT Work. *Computerscope*. November 91, p 17.

KIMS: KNOWLEDGE-BASED INTEGRATED MANUFACTURING SYSTEM

[1]Moonke Min (*), [2]Thomas J. Schriber, and [2]Kathryn E. Stecke

[1]*National Institute of Standards and Technology, Gaithersburg, Maryland, U.S.A.*
[2]*University of Michigan, Ann Arbor, Michigan, U.S.A.*

ABSTRACT

This paper proposes an information systems architecture for integrated planning and control of manufacturing systems. An implementation of the architecture for production planning and control of a typical flexible manufacturing system (FMS) is presented here and called KIMS (Knowledge-based Integrated Manufacturing System). KIMS builds on the previous work; the knowledge-based hybrid modeling approach to planning problems in FMSs, which was proposed by authors of this paper.

In KIMS, knowledge-representation methods are used to describe the FMS. Especially, frames are used to represent FMS structural entities and rules are used to represent the behavior of the entities. The formal and explicit rule-representation of hierarchical communication behavior among various levels of managers (planners and controllers) provides modularity and flexibility, which is vital to maintain a global systems viewpoint in the analysis, design, and operation of a manufacturing system.

The simulation component of KIMS also provides FMS managers with dynamic observations regarding the effect of using various planning decisions and control strategies on FMS performance. KIMS is written with GoldWorks, a sophisticated knowledge-based system building tool on Macintosh computers.

INTRODUCTION

The body is a unit, though it is made up of many parts;
and though all its parts are many, they form one body.
1 Corinthians 12:12

A system is a set of related objects (subsystems) that have common goals. A simple model of a general system [1] in Figure 1 (a) shows two subsystems: the operation system and the management system. The operation system is a set of operators which transform input to output; the management system plans and controls the operators with the goal of maximizing system effectiveness. Operationalized goals are called objectives. System effectiveness is defined as the degree to which these objectives are achieved.

(*) Moonkee Min is a guest researcher at the National Institute of Standards and Technology.

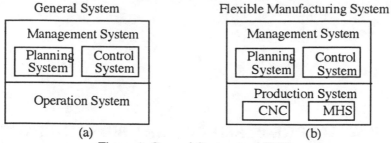

Figure 1 General System and FMS

An FMS consists of several computer-numerically-controlled (CNC) machines interconnected by automated material handling systems (MHSs), as shown in Figure 1 (b). In CNC machines, a digital computer controls the position and paths of cutting tools. A machine may have three, four, or five axes of motion, which allows complex precision operations. Some typical machine types are mills, drills, and vertical turret lathes (VTLs). These machines are versatile: they are capable of varied operations. An operation may require a set of tools. The tools needed to produce part types are loaded in the tool magazine of each machine before production begins.

A premise of systems thinking is that a system, a set of interrelated parts, should be viewed as a whole. From the systems perspective, intelligence of a system is not just summation of intra-intelligence of each subsystem but it also involves inter-intelligence among subsystems. Recently in manufacturing area, the flexibility has been considered as a strategic weapon for a company to survive in an increasingly competitive market. However, most current planning and control software still tends to focus on specific problems without incorporating a global systems viewpoint [2]. Thus, the potential of flexible machines which are capable of a concurrent production of a variety of part types, is not fully achieved. The purpose of this paper is to present an information system architecture which emphasizes the systems perspective and allows a flexible design of manufacturing planning and control systems.

REVIEW OF LITERATURE

This section reviews some of concepts in cybernetics theory, simulation, and artificial intelligence which are employed in the design of KIMS.

Cybernetics Theory

According to cybernetics theory there are three basic elements of a control system: detector, comparator, and effector. A control object is the variable of the system's behavior chosen for monitoring and control. The control system keeps the systems output within certain predetermined limits via a self-regulating mechanism consisting of a detector which constantly monitors the changes in output over time, a comparator which compares the sensed output against a set of standards, and an effector which decides whether deviations (i.e., the output of the comparator) warrant corrective action to be taken against the operating system's input function. In cybernetics theory, there are three types of feedback systems: automatic goal attainment, automatic goal changer, and reflective goal changer [1].

Simulation Modeling

Computer simulation is a method for predicting the future state of a system by constructing a realistic model of the system. A discrete-event simulation model is one in which the system state changes only at a set of discrete points in time called event times. Fishman [3] presents the following figure which demonstrates the concepts of event, process, and activity. An event signifies a change in the state of an entity. A process is a sequence of events ordered in time. An activity is a collection of operations that transform the state of an entity. These three concepts give rise to the following three most prevalent approaches to discrete-event simulation modeling.

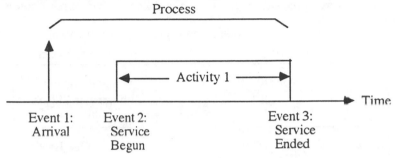

Figure 2 Event, Process, and Activity.

Event Scheduling Approach   The event scheduling approach emphasizes a detailed description of the steps that occur when an individual event takes place. Each type of event has a distinct set of steps associated with it. Whenever an event is scheduled, a record identifying the event and the time at which it is to occur is filed in a special list. When the instruction to select the next event is encountered, the simulation searches this list to find and perform the event with the earliest scheduled time. Simulated time is then advanced to this scheduled time, thus skipping the "dead" time. This procedure is the essence of the next-event approach to simulation.

Activity Scanning Approach   The activity scanning approach emphasizes a review of all activities in a simulation to determine which can be begun or terminated each time an event occurs. An activity is usually defined by two events; one marking the beginning of the activity, the other marking the end. Each activity contains test conditions and actions. The activity will execute when the test conditions are satisfied. The actions specify what is to be done to accomplish this activity when the test conditions are true. Whenever time is advanced to the next event, all activities are scanned to determine which can be begun or ended. This process is repeated until the simulation terminates.

Process Interaction Approach   The process interaction approach combines the event scheduling feature of the event scheduling approach with the concise modeling power of the activity scanning approach. A process is a collection of events that describe the total history of a job's progress through the shop. Since jobs arrive at different times, it is clear that system behavior is described by a collection of processes, some of which may overlap. The process interaction approach provides a process for each entity in a system; this approach emphasizes the progress of an entity through a system, from its arrival event to its departure event. In modeling a system, interaction between the processes is stressed. Schriber [4] models an FMS with GPSS/H (General Purpose Simulation System/H) [5]. "Transactions" (entities) simulate parts moving through the system. Each transaction moves through a model segment corresponding to the type of part it represents.

Artificial Intelligence (AI) Modeling

   In an AI approach, a problem is formulated as a search problem and problem-solving is viewed as a search process. This involves the design of a problem space model which consists of:
- States (collections of features that define a situation)
- Operators (the actions which transform one state into another)
- An evaluation function (which rates each state in the problem space)

In search, problem solving means finding a sequence of operators that map the initial state to one of the goal states. One aspect of intelligence is applying problem-specific knowledge to the search process. Thus, search and knowledge representation (KR) are two core concepts in AI.

   Figure 3 (a) shows a general system and Figure 3 (b) shows a knowledge system.

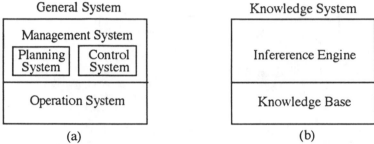

Figure. 3 General System and Knowledge System

   The knowledge system (KS) consists of two subsystems: the knowledge base (KB) and the inference engine. The KB corresponds to the operation system of the general system, and the inference engine corresponds to the management system. The KB represents states, operators, and an evaluation function. An inference engine is a mechanism that plans and controls the sequence of operator applications to solve a problem (to achieve one of the goal states from the initial state).

Constraint Satisfaction Problem (CSP) Search CSPs are a type of search problem. CSPs have three components: variables, values, and constraints. The goal of a CSP is to assign values to variables subject to a set of constraints. Since the inception of AI as a distinct field, CSPs have been studied intensively by many researchers. The importance of CSPs is due to the wide range of practical problems they can model. Applications of the standard form of the problem or a close relative have included such diverse areas as theorem proving, belief maintenance, graph problems, and machine vision.

Knowledge Representation Methods Rule-based KR centers on the use of "IF condition THEN conclusion/action" statements [6]. When the current problem situation matches the condition part of a rule, the conclusion/action part of a rule is performed. The conclusion may add a new fact and the action may direct program control (e.g., cause a particular set of rules to be tested and applied). This matching of rule IF portions to the facts can produce inference chains. The use of rules simplifies the job of explaining what the program did or how it reached a particular conclusion.

   A frame provides a structured representation of an object or a class of objects [7]. Frames are a network of nodes organized in a hierarchy, where the topmost nodes (frames) represent general concepts and the lower nodes represent more specific situations of those concepts. Frames low in the hierarchy automatically inherit properties of higher level frames. Each frame is defined by a collection of attributes and the values of those attributes, where the attributes are called slots.

KIMS

This section describes the proposed KIMS. KIMS is the management system which plans and controls the FMS resources. The management system is a set of managers which is organized in a hierarchy. Figure 4 shows a generic manager in the hierarchy. The manager has two subsystems: knowledge base and inference executive. Knowledge base contains a planner, a controller, and a knowledge-based simulation model.

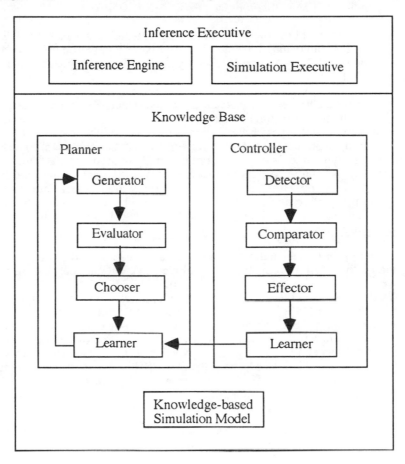

Figure 4   Generic Manager

Planner

A planner in KIMS is decomposed into four components: generator, evaluator, chooser, and learner.

Generator   We define a generator as a mechanism which finds candidate solutions. In KIMS, this mechanism is comprised of a generation rule and a CSP search. Whenever a planning problem needs to be solved, a generation rule is applied to invoke a CSP search.

The invoked CSP search follows the structure suggested by Nadel [8]. It starts a tree search and checks constraints at each node. Only the constraints local to each node are applied to check whether the generated solution is feasible or not. In KIMS, rules represent procedures regarding which constraints are applied and how they are applied during the CSP search. Stecke, Min, and Kim [9] [10] suggest that KR methods can be employed to utilize integer programming (IP) models and a simulation model to solve the part mix problem. For example, rules specify which IP model to use to generate candidate solutions, depending on FMS situations.

Evaluation   We define an evaluator as a mechanism which assesses candidate solution alternatives. In KIMS, an evaluation rule and a rule-based simulation model constitute an evaluator. When a candidate solution is found by a generator, an evaluation rule is applied to execute the simulation model.

Choice   We define a chooser as a mechanism which selects a solution alternative from among candidate alternatives. In KIMS, a set of choice rules constitutes a chooser. Rules can easily represent various types of choice strategies which might be difficult to formulate in analytic modeling approaches, since they rely on an analytic function(s) to be a chooser. Although some analytic models (e.g, goal programming) can incorporate multi-objective choice strategies, they have to adhere to assumptions inherent in analytic modeling approaches. As Lee and Hurst [11] point out, rules are appropriate methods for formulating multi-objective decision-making problems which have qualitative as well as quantitative objectives. Choice strategies regarding trade-offs among multiple objectives can be easily incorporated in rules.

Learning   After evaluating a candidate solution by a simulation model, we can analyze the performance of the heuristic constraints employed in the CSP search. Based on this analysis, we can refine the constraints to improve the candidate-solution generation process. We define a learner as a mechanism which refines heuristic constraints based on evaluation results. In KIMS, a set of learning rules constitutes a learner. Through the refinements, unnecessary CSP searching space is pruned more closely. Learning rules can be specified not only to refine but also to dynamically add or remove heuristic constraints. These rules can be obtained from the experts' experiences as well as from theory; e.g, the program by Seliger et al. [12] uses knowledge of FMS design experts as learning rules to refine the initial design. In KIMS, learning rules are provided by humans. A more sophisticated KIMS would have a machine learning mechanism which can replace the humans' role by automatically generating such learning rules.

Controller

Based on cybernetics theory, a controller in KIMS is decomposed into four components: detector, comparator, effector, and learner. A detector is a scanning system. The function of a comparator is to compare the current performance against the predetermined plan and schedule. An effector evaluates alternative courses of corrective action and chooses one. By having a learner as a component, the controller can be a self-organizing system, which is a kind of a reflective goal changer. While the learner in a planner is based on analytic inference, the learner in a controller utilizes the empirical results of control. Active values or when-modified-slot concepts in AI can play a role as a detector and a comparator. The effector uses the knowledge-based simulation model in evaluating alternative control actions. As the behavior of the controller is represented with rules, the inference executive executes these control procedures and sends command messages to its subordinates.

Knowledge-based Simulation Model

FMS Structure Represented as Frames   Figure 5 shows frames that represent an FMS structure. A frame corresponds to an object constituting an FMS structure.

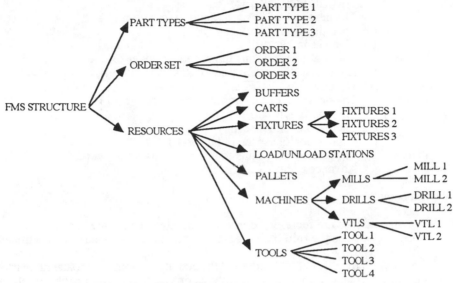

Figure 5 FMS Structure Represented with Frames

Rules to Represent FMS Behavior   Figure 6 shows rules representing the dynamic behavior (activities) of an FMS.

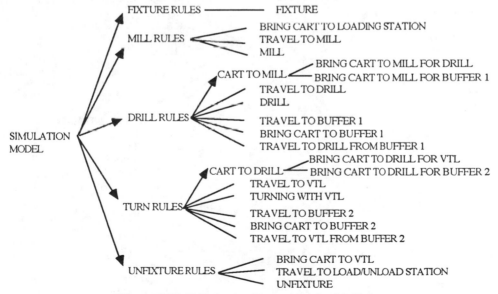

Figure 6 FMS Behavior Represented with Rules

Each rule represents an activity required to transform the raw parts into finished parts. Rules (activities) are grouped into five classes: FIXTURE, MILL, DRILL, TURN, and UNFIXTURE. Each of the five rule classes has members denoting a series of subtasks needed to finish the task. Because the rules are partitioned, only the relevant rules are scanned in the simulation process. Examples of the contents of rules are below:

BRING CART TO MILL FOR DRILL
```
(IF      (   The next task of this transaction is cart-to-mill )
         (   There is an idle cart )
         (   There is an idle drill )
THEN     (   The future task of this transaction is travel-to-drill )
         (   Advance future event time of this transaction )),
```
BRING CART TO MILL FOR BUFFER-1
```
(IF      (   The next task of this transaction is cart-to-mill )
         (   There is an idle cart )
         (   There is not an idle drill )
         (   There is an idle space in buffer-1 )
THEN     (   The future task of this transaction is travel-to-buffer1 )
         (   Advance future event time of this transaction )).
```

Inference Executive

When all activities are represented in rules, an inference engine can play a role as a simulation system, because the rule-based reasoning process corresponds to the simulation process. This is basically the same as the activity scanning approach to discrete-event simulation modeling. Bruno et al. [13] adopted the activity scanning approach in their rule-based system for FMS scheduling. However, they concede that the approach is inefficient due to the need to scan all rules during the simulation process.

Min [14] proposes a simulation executive that allows one to build a rule-based simulation model with the process interaction approach (see Figure 7) and named KGPSS (Knowledge-based General Purpose System).

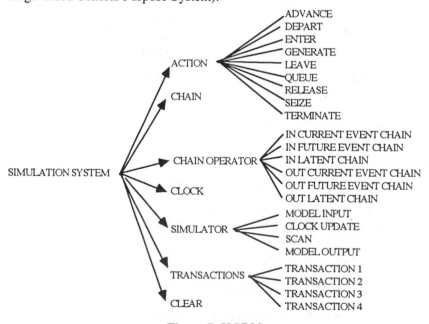

Figure 7 KGPSS

KGPSS, having an event scheduling feature, helps an inference engine scan the action rules efficiently in performing rule-based simulation. KGPSS is based on frame data structures which can store various types of information and functions effectively.

As shown in Figure 7, KGPSS consists of a set of frames. The frame TRANSACTIONS has child frames such as TRANSACTION 1, TRANSACTION 2, TRANSACTION 3, and TRANSACTION 4. The slots (attributes) of the frame TRANSACTIONS are inherited by the child frames. The inherited slots include "name" and "next task." A transaction models each entity in a system; here it represents a part.

The frame CHAIN has slots such as "future events chain," "current events chain," and "latent chain." The latent chain contains a list of transactions to be used in simulation. When a new part is introduced to an FMS, a transaction is taken from the latent chain and is placed in the future events chain. When a transaction (part) finishes production and leaves an FMS, the transaction goes to the latent chain again. The frame CLOCK has a slot that keeps a record of simulation time.

Frames such as ACTION, CHAIN OPERATOR, and SIMULATOR store functions (procedures) in the slots. The functions stored in ACTION and its child frames such as ENTER and ADVANCE are similar to simulation-modeling blocks in GPSS/H. These functions can be embedded in the rule representation of activities: we can write the rule BRING CART TO MILL FOR DRILL mentioned in the previous section as follows.

```
BRING CART TO MILL FOR DRILL
(IF        (    The next task of this transaction is cart-to-mill )
           (    ENTER cart )
           (    ENTER drill )
THEN       (    The future task of this transaction is travel-to-drill )
           (    ADVANCE )).
```

ENTER returns "truth" when there is an idle resource to be used. Otherwise, the transaction goes to the queue of the resource and updates queue statistics. ADVANCE not only advances the future event time but also moves the transaction out of the current events chain and places it in the future events chain.

Simulation starts when a message is sent to the frame SIMULATOR which has a procedural slot named *start*. In KIMS, the action part of Evaluation Rule-1 sends a message to SIMULATOR when a feasible solution is found by a CSP search. The procedure of the *start* is as follows.

```
(function start
        (model input)
        (clock update)
        (while (terminated transactions < number of transactions for simulation)
           (scan)
           (clock update))
        (model output)).
```

In the above procedure, MODEL INPUT generates the initial transaction. (e.g, If part input sequence is (x_1, x_3, x_4), TRANSACTION 1 is taken from the latent chain to model the first part (x_1), and is placed in the future events chain.) The slots of frame TRANSACTION 1 include following:

```
TRANSACTION 1
    Name          :        ORDER 1
    Next Task     :        FIXTURING
```

The frame data structure has an advantage in that the value of a slot can be a pointer. The value of slot "Name" is a pointer to a frame ORDER 1, and it can access all information about ORDER 1 without storing it in the transaction. For example, the part type of this transaction is obtained by stating:

(The part-type of (The name of TRANSACTION 1)).

Simulation cycles of CLOCK UPDATE and SCAN continue until the number of terminated transactions satisfies the objective in this regard. CLOCK UPDATE updates the simulation clock, collects data (e.g., resource-utilization statistics), removes imminent transactions from the future events chain, and places them in the current events chain. SCAN infers the future task of each transaction in the current events chain using an inference engine. The procedure of SCAN can be stated as:

(function *scan*
(what is the future task of this transaction using rules for the next-task?)).

An inference process, backward reasoning, is performed to conclude the future task of each part. When the next task of a transaction is CART TO MILL, rules for the next task are BRING CART TO MILL FOR DRILL and BRING CART TO MILL FOR BUFFER-1. These two rules are child rules of the parent rule CART TO MILL. Only these two rules, rather than all rules, will be selected to infer the future task of the transaction. When there is an idle cart and an idle drill, the rule BRING CART TO MILL FOR DRILL is applied and the future task becomes TRAVEL TO DRILL. When there is an idle cart and an idle space in buffer-1 but not an idle drill, the future task becomes TRAVEL TO BUFFER-1 after applying BRING CART TO MILL FOR BUFFER-1. If the condition of a rule is met as such, the transaction moves along to the future events chain. Otherwise, the transaction remains at the current events chain, waiting for the next CLOCK UPDATE and SCAN. When the number of terminated transactions becomes the same as the number of transactions for simulation, the simulation stops and MODEL OUTPUT prints out statistics such as mean tardiness and system utilization.

KGPSS employs the process interaction approach to discrete-event simulation modeling. In the simulation model of an FMS, a transaction represents a part, only transactions in the current events chain are scanned, and only a subset of rules is used to infer the future task. Thus, the current KGPSS helps an inference engine perform rule-based simulation efficiently, and automatically collects data such as resource utilization by allowing operators such as ENTER, ADVANCE, QUEUE to be embedded in rules.

CONCLUSIONS

In this paper, we proposed KIMS which emphasizes the flexibility in information systems. Several other advantages over previous approaches are also discussed.

- Accuracy in evaluation: KIMS can predict the effect of a solution alternative on FMS performance more accurately than analytic modeling approaches because it uses realistic simulation modeling of the system. As an evaluator, analytic modeling approaches employ an analytic equation(s), which makes it difficult to incorporate some performance attributes (e.g, meeting due dates) and detailed behavior (e.g., transient-state behavior).
- Consistency in representation: In KIMS, rules represent both human problem-solving behavior (heuristic rules) and FMS behavior (a simulation model).
- Shareability of data structure and simulation model: In KIMS, heuristic rules and a simulation model share a data structure represented as frames. In addition, it is not necessary to create a new simulation model to solve other

planning problems. The same rule-based simulation model developed to solve a planning problem can be used for other problems.

- Contingency approach to management can be easily considered (e.g., incorporating of strategic planning rules into operational planning and control [15].

- Efficiency in rule-based simulation: We have designed a simulation executive, KGPSS, to allow one to build a rule-based simulation model with the process interaction perspective. This approach is more efficient than the one based on the activity scanning perspective.

- Separation of domain knowledge from an executor: An inference engine and the KGPSS constitutes the inference executive, which is a domain independent and a general purpose executor of plans, schedules, and control procedures.

The current KGPSS automatically collects data such as resource utilization. To model more complex manufacturing activities in rules, and to collect the associated simulation statistics, however, more operators like those found in GPSS/H would have to be developed. Otherwise, users themselves should represent data collection procedures as well as manufacturing activities. The current KIMS uses a deterministic simulation to evaluate alternative plans: it does not take into account the random arrival of new orders and random processing times in decision making. Later, the operators such as GENERATE and ADVANCE in KGPSS need to incorporate random behavior.

REFERENCES

1. Schoderbek, Peter P., Schoderbek, Charles G., Kefalas, Asterios G., Management Systems: Conceptual Considerations, Richard D. Irwin Inc., p. 12, 1990.

2. Biemans, Frank P. M., "Manufacturing Planning and Control: A Reference Model," Manufacturing Research and Technology, Vol. 10, 1990.

3. Fishman, George S., Concepts and Methods in Discrete Event Digital Simulation, John Wiley & Sons, Inc., 1973.

4. Schriber, T. J., "The Use of GPSS/H in Modeling a Typical Flexible Manufacturing System," Annals of Operations Research, Vol. 3, pp. 171-188, 1985.

5. Schriber, T. J., An Introduction to Simulation Using GPSS/H, John Wiley & Sons, Inc., 1990.

6. Hayes-Roth, Frederick, "Rule-based Systems," Communications of the ACM, Vol. 28, No. 9, pp. 921-932, September 1985.

7. Fikes, Richard and Kehler, Thomas, "The Role of Frame-based Representation in Reasoning," Communications of the ACM, Vol. 28, No. 9, pp. 904-920, September 1985.

8. Nadel, Bernard A., "Tree Search and Arc Consistency in Constraint Satisfaction Algorithms," Search in Artificial Intelligence, Kanal, L. and Kumar, V. (Editors), Springer-Verlag, 1988.

9. Stecke, Kathryn E., Kim, Ilyong, and Min, Moonkee, "A Knowledge-based Approach to Part Type Selection Considering Due Dates in Flexible Manufacturing Systems," Proceedings of the 12th IMACS World Congress on Scientific Computation, Paris, France, July 17-22, 1988; reprinted in the IMACS Transactions on Scientific Computing, Vol. 2, Huber, R. et al. (editors), J. C. Baltzer, Basel, Switzerland, pp. 321-325, 1989.

10. Stecke, Kathryn E., Min, Moonkee, and Kim, Ilyong, "A Hybrid Model-Based Approach for the Production Planning of FMSs," Proceedings of the Third International Conference: Expert Systems and the Leading Edge in Production and Operations Management, pp. 281-291, May 1989.

11. Lee, Jae Kyu and Hurst, E. Gerald Jr., "Multiple-Criteria Decision Making Including Qualitative Factors: The Post-Model Analysis Approach," Decision Sciences, Vol. 19, pp. 334-352, 1988.

12. Seliger, G., Vieherger, B., Wieneke-Toutouai, B., and Kommana, S. R., "Knowledge-Based Simulation of Flexible Manufacturing Systems," Proceedings of the Second European Simulation Multiconference, Vienna, Austria, pp. 65-68, 1987.

13. Bruno, Giorgio, Elia, Antonio, and Laface, Pierto, "A Rule-based System to Schedule FMS Production," Computer, Vol. 19, No. 7, pp. 32-40, July 1986.

14. Min, Moonkee, "A Knowledge-based Hybrid Modeling Approach to Planning Problems in Flexible Manufacturing Systems," Ph. D. Dissertation, Business Administration, University of Michigan, Ann Arbor, MI, 1990.

15. Kochen, Manfred and Min, Moonkee, "Intelligence for Strategic Planning," Seventh International Workshop in Expert Systems and Their Applications, Avignon, France, May 1987.

APPLICATION OF A EUROPEAN INTEGRATING INFRASTRUCTURE TO THE NIST MSI DEMONSTRATION SYSTEM

Paul J. Gilders

Department of Manufacturing Engineering, Loughborough University of Technology, Loughborough, U.K.

ABSTRACT

Since June 1991, work has been undertaken at NIST (National Institute of Standards and Technology) at Gaithersburg, to bring together two CIM projects. The first of these projects is the Manufacturing Systems Integration (MSI) project at NIST. Here the concern is focussed on developing standard functions, interfaces and informational requirements of an automated, distributed manufacturing control system. The second project is one that has been conducted in the Manufacturing Engineering Dept. at Loughborough University of Technology (LUT). This project has been based on producing an integrating infrastructure called CIM-BIOSYS. This software provides a range of basic services to the integrator and separates general system knowledge from the individual software processes within a system.

The work reported in this paper has involved establishing mechanisms by which the MSI demonstration system can utilise the services provided by the CIM-BIOSYS software. The software engineering effort involved in the system implementation was minimal and the resulting flexibility of the system allows for automated configuration and an extremely flexible environment for experimentation and alteration. The ideas encapsulated in this proof of concept system would provide the small to medium-sized batch manufacturer with a means of implementing highly dynamic control systems and an ability to incrementally extend the overall scope of integration.

INTRODUCTION

This paper reports upon a collaboration effort between two research groups in the area of computer integrated manufacturing. The collaboration was based around the merging of two demonstration systems used by the two groups to experiment and verify new concepts in the integrated manufacturing environment. The first of the research groups involved was the Loughborough University of Technology (LUT) Systems Integration (SI) Group, who have been involved in research in the integration arena for a number of years. The focus of the group's research has been to develop a systematic approach for the migration towards computer integrated manufacturing (CIM) through the identification of the methods and processes necessary within an integrated system and to separate out this functionality to provide a definite system structure. The approach to the separation of concerns is believed by the group to yield great advantages to the system implementor, enabling flexibility to external change and reconfiguration without considerable redevelopment of application software.

The other research group involved was the National Institute of Standards and Technology (NIST) Manufacturing Systems Integration (MSI) Group. This group is part of the large research initiative into new concepts in automated manufacturing known as the Advanced Manufacturing Research Facility (AMRF) project which has been ongoing since 1981. The MSI group itself is concerned with research into the systems and issues in the control and management of automated and semi-automated manufacturing environment, from the shop order entry level to the controllers of automated machinery.

The initial collaboration was initiated from a desire from both groups to expand and test their demonstration systems and to gain more experience in the automated manufacturing environment though this activity. The initial collaboration was to be undertaken over a six month period whereby the author (a member of the Loughborough Systems Integration Group) was seconded over to NIST in Gaithersburg to work with the MSI group and merge the two projects.

This report firstly details the aims of the two projects and how they were merged to produce a new demonstration manufacturing control system. The report then proceeds to analyse some of the details of the implementation and some of the technical issues involved in the amalgamation that the author sees as being useful and advantageous to the manufacturing community.

THE PROJECTS

The MSI demonstration system

The MSI group has two main aims in the research and development of integrated manufacturing concepts. The first is to develop a testbed for the experimentation of different concepts in an integrated environment, both by implementing proprietary elements and specifically developed functional units into the system. The second aim was to try and develop some suitable candidates for standards for integrated manufacturing control systems. The activity was aimed at the specification of functionality, interfacing and informational requirements of such component elements as configuration management, process planning, production scheduling, manufacturing data preparation and shop floor control.

In 1990, the MSI group developed its first operating demonstration system, which introduced a number of concepts in manufacturing system organisation and functionality. The main concepts demonstrated in the initial system included a hierarchical control architecture and the development of compatible interfaces between the system components. The project is not involved in the specification of the internal mechanisms of the functional units but rather their external interaction in the manufacturing system and their interface requirements to enable appropriate system performance. A brief schematic arrangement of the demonstration system is shown in Figure 1. The diagram emphasises

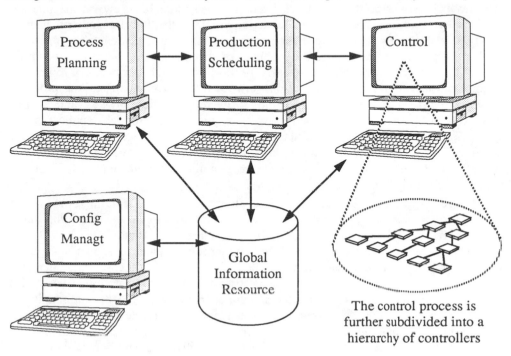

The control process is further subdivided into a hierarchy of controllers

FIGURE 1: Diagram to Show the distributed functionality and interfacing requirements of the MSI Demonstration System

the highly distributed nature of the system, especially with the further subdivision of the control functionality into the hierarchy of partially generic controllers.

The MSI demonstration system, from a pure communication perspective, can be viewed as a heterarchy of processes. With this in mind, the MSI group developed a common interaction mechanism between the participant processes. The interaction mechanism originally introduced by the MSI group utilised a common memory environment together with a set of monitoring and interaction access routines known as SAP (Service Access Point) routines. These concepts will be described in a later section.

More details of the concepts and components involved in the previous and current version of the MSI project can be found in [1-5].

The CIM-BIOSYS Integration Platform

The main thrust of research of the LUT SI group has been similar to that of the NIST MSI group: to decompose functionality involved in the integration of manufacturing processes [6]. However, the effort has been focussed more intensely on the practical problems of enabling processes to interact rather than the development of standard mechanisms and protocols for information exchange and communication (more the stance of MSI). This research first tackled the problems of direct messaging between applications and developed a consistent interaction basis for this purpose. The problems of lower level communication and system configuration between different computer-driven processes was separated from the interface provided for software applications. The advantage of the separation of interaction functionality was in the ability for all applications to communicate through a consistent set of application services via a common service provider. This reduced the number of different interfacing mechanisms between applications and enabled simple system reconfiguration since all applications utilised identical interfacing primitives.

The result of these ideas led to the production of an actual integrating infrastructure by the group which has been the subject of ongoing development since 1985 and is currently known as CIM-BIOSYS (CIM-Building Integrated Open SYStems)[7-9]. The functional division of the CIM-BIOSYS integrating infrastructure is shown in Figure 2. The platform is conceptually viewed by applications as a single network-wide interaction tool for all system processes. In reality it consists of a number of integration servers (CIM-BIOSYS servers) distributed across the integrated network linked by communication drivers. In this diagram the software applications are shown to reside above the integration platform, plugging into the services provided by the service manager. Below the platform, communication between different physical hardware platforms and across networks is provided by communication drivers, which utilise the CIM-BIOSYS services. The services provided by the platform have been extended over the life of the platform and now include the following categories:

Basic Interaction services   These services provide the software applications with the ability to establish and disconnect links with each other, request the status of a particular connection and send data to other applications.

Communication Services   The services provided to the communication drivers are similar to those provided by the application services, however, they are provided at a more basic level. The communication services are also used to communicate to drivers providing an interface to non conformant, proprietary processes.

Configuration Services   The configuration services offer the ability for applications to alter the arrangement of applications over the physical devices in the participating CIM-BIOSYS system.

Information Services   The information services provided by CIM-BIOSYS offer mechanisms to access information held on databases and in files within the system. These services include the retrieval of information, update of information, creation of information units and the deletion of information units. The information services also include the transfer and management of files held within the system. CIM-BIOSYS manages the location and access mechanisms used for a particular data resource.

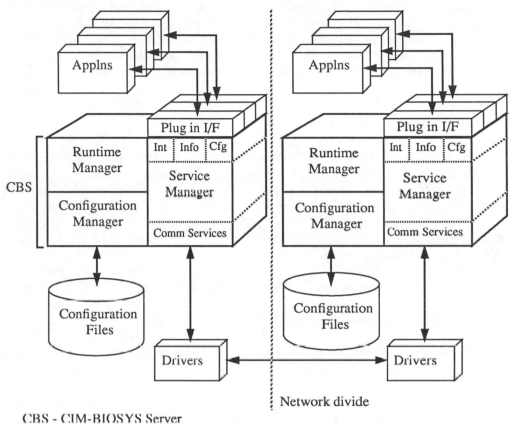

CBS - CIM-BIOSYS Server

FIGURE 2: Diagram to Show the Division of Functionality using CIM-BIOSYS

In addition to the management of the consistent interaction mechanisms, the platform also provides two other functions. The first of these involves a runtime manager which handles such tasks as buffering I/O, event handling, process monitoring and manual intervention. Secondly, a configuration manager deals with the locations of applications with the system and passing messages along the appropriate route.

The CIM-BIOSYS integrating infrastructure is also accompanied by a set of tools and application libraries to aid in the transformation of systems to the requirements of the CIM-BIOSYS platform. The majority of these tools were developed for a particular environment (using the UNIX operating system running on SUN workstation hardware) and are therefore not included as part of the platform. The tools include SUN to SUN ethernet drivers, template applications, and testing applications (most of which are written in C). The library routines also aid implementors in situations where event handling routines are required and CIM-BIOSYS message packeting is required (these libraries are also currently written in C).

An earlier basic version of the CIM-BIOSYS integration platform had previously been used in industry by a medium sized electronics manufacturer in the UK. This implementation, although useful, (and still being utilised) was relatively restricted, only involving the interaction between three applications in one cell. The reconfiguration capability and extensibility was still considered to be of great benefit in this relatively simple application, but more extensive testing was thought to be necessary.

Details of the Specific application

The application of the CIM-BIOSYS platform to the MSI demonstration system required a clear understanding of the interaction requirements of the demonstration system. It was clear at the outset that the overall system was of an extremely distributed nature, especially with the further distribution of the control hierarchy which used direct interaction as a mechanism for conveying task execution requests, administrative status reports and requests for planning. This meant that the platform would have to cope with high volumes of messaging which, although could not be classed as real-time, still remained time dependant for the appropriate performance. The system would also have to be capable of running over a generated configuration of hardware devices and hence the physical distribution of the system was not fixed. This meant that adjustments to the physical hardware configuration would be required.

The implementation of the CIM-BIOSYS platform was executed in two stages. The first stage was to implement the configuration manager within the system and the second was to alter the interaction mechanisms for the other participant applications.

Application of the Configuration Manager   The original MSI demonstration system included a configuration manager which was used to monitor and control the configuration of the control hierarchy of the system. In addition it was also used to allocate a suitable initial physical and logical layout of the hierarchical control system. Using information about the hardware available, a physical factory definition model and a baseline logical configuration, the process derived an appropriate physical and logical distribution of controllers and caused their associated execution.

In order for the controllers to run over the CIM-BIOSYS platform it was necessary for the configuration information to be transferred to the network of CIM-BIOSYS servers. CIM-BIOSYS does not provide a centralised configuration manager for the distributed system, since the central configuration management problem was considered to be outside the initial scope of the integrating infrastructure. However, each CIM-BIOSYS server within the participating system is required to have a local view of the application configuration to enable correct message routing. The configuration manager was therefore included as a participating application in the CIM-BIOSYS environment and used to set the local CIM-BIOSYS views of the configuration using the CIM-BIOSYS configuration services. The configuration manager was also required to execute the controller processes, which was accomplished using the "connection establish" service.

Alteration of the communication mechanisms   All other participating applications within the original MSI system had utilised a common interaction mechanism known as common memory. The common memory environment provided applications with appropriate input and output mailboxes for connections between processes. The management of these mailboxes was undertaken by the common memory manager which would provide suitable locations to address and retrieve messages. In addition to this common memory environment, applications used procedures called SAP (Service Access Point) routines to advertise and request application services. These routines were used to provide a common interaction mechanism for applications that resembled the OSI view of communications. The routines included the activities of buffering incoming messages, monitoring the current connections between applications and making the appropriate requests to the common memory manager. The situation is shown in Figure 3(i). It was clear that the code construction used by the original system was already very modularised and mapped very closely to the requirements of the CIM-BIOSYS interaction mechanisms and library routines. The application specific code was not affected by the change since it was provided with the same interface as before. Much of the functionality of the SAP routines also remained, since the monitoring of connections and the event handling procedures could be used in the same manner. The main alteration was in the routines called by the SAP routines to interface to the CIM-BIOSYS platform. These original routines required substitution by the basic interaction services of CIM-BIOSYS. The schematic representation of the change is shown in Figure 3(ii).

ANALYSIS OF THE COLLABORATION

The application of the CIM-BIOSYS platform as a basis for integrating the MSI demonstration system did not go without its problems. However, most of the difficulties were technical issues concerned with inexperience in the combining of event management systems and the more general problem of debugging the code of asynchronous processes. However the two projects were also very compatible in that the division of functionality in the original MSI applications was similar to the division of code required by CIM-BIOSYS. This ensured that a relatively simple substitution of the interaction routines could be achieved and kept the changes to software minimal and localised. A good understanding of the modularisation and relationship of the application code was found to be essential so that substitution of the original interaction routines would not cause restrictions in more specific circumstances. In addition, well structured and modularised code provided an improved incremental conversion of the interaction mechanisms which is an important ability for the purpose of debugging.

The implementation of CIM-BIOSYS did not provide great advantages over the previously existing common memory system. The main advantage in the implementation was that the platform provided a more extensive distributed environment. The common memory system meant that all process required access to the common memory process using a tcp or similar link. CIM-BIOSYS provided a more localised link to applications using distributed servers, only causing remote linking when required by the particular configuration. This negates the need for all messages to be transported via a single centralised process on the network, keeping local messages local and directing messages

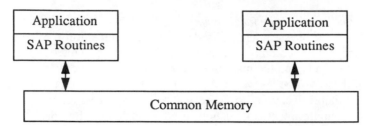

FIGURE 3(i): Modularity of Interaction Provided by Common Memory

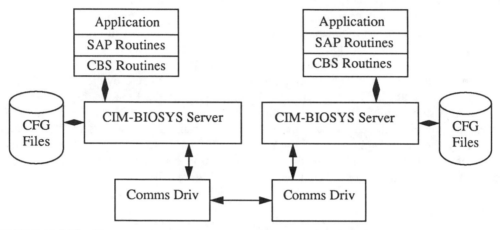

FIGURE 3(ii): Change in Modularity of Interaction using CIM-BIIOSYS

via other services according to the required connection and configuration. This system enables messages to be forwarded a number of times across the system and hence no centralised access is required. The only requirement of CIM-BIOSYS applications is that they are able to send appropriately constructed packets to a UNIX file descriptor and receive packets from the CIM-BIOSYS server. Hence, applications are freed from the networking capability which can be altered without affecting the view of the application. Therefore reconfigurability of a system is further extended without the requirement for any regeneration of application software. Another slight difference between the two interaction methods is that CIM-BIOSYS uses only one port for input and output for each application. This means that message monitoring is relatively simple, but may cause problems when large numbers of messages use a single port.

The basic services provided by the two interaction mechanisms were very similar, in that the basic primitives of connection request, disconnect request, and send data were used in both cases. However, some discrepancies in the service confirmation primitives were evident, causing a mixed layer of services. This was not a problem but made some of the code untidy.

The use of an integration platform is very useful concept for the implementation of distributed manufacturing systems. However, the level of support provided is a very difficult issue to address. If too much support is given this tends to impose restrictions on the method by which integration is tackled. Conversely, too little support requires significant design procedures and code generation on the part of the implementor. CIM-BIOSYS provides a very basic set of services to the user, but even this implementation causes the implementor to view integration from a particular perspective (i.e. using sockets as a communication mechanism, using a particular message buffering procedure, using service response primitives, using a particular message packet primitive). The importance of CIM-BIOSYS is not in the particular implementation, but in its insistence on modularisation and functional decomposition. The application of CIM-BIOSYS to the MSI project worked well since the MSI requirements were met by the services provided by CIM-BIOSYS. However, this may not always be the case and hence alteration of the services and management functions provided should be available and easy to alter. The usefulness of CIM-BIOSYS in its current form is that it comes as a particular implementation of an integrating infrastructure which can be used by industry together with a set of tools such as communication drivers which provide particular parts of the essential functionality for implementors. The CIM-BIOSYS platform can be considered as a baseline implementation of the set of functions that should be included within an integrating infrastructure and although this particular flavour of implementation may not suit all circumstances, the functional components and their relationships remain applicable.

Some of the services provided by CIM-BIOSYS were not utilised by the MSI demonstration system, since they involved areas not currently being addressed by the group. The CIM-BIOSYS integration platform includes some basic information retrieval services that enable the blind retrieval and alteration of information from distributed information sources and files. This is considered by the LUT SI group to be of great benefit to the flexibility of CIM systems, since applications query and retrieve information from the CIM-BIOSYS platform which is then able to map the requests onto particular physical data resources. Research is also currently being undertaken to map a conceptual information model onto a distributed system of data resources running under the CIM-BIOSYS platform.

The MSI group processes were mainly resident on SUN workstations using the UNIX operating system. Although this system made this particular implementation easier since SUN to SUN drivers had already been developed, the system did not demonstrate the ability to reconfigure the processes to run over a variety of hardware and using a range of operating systems. This capability is more appropriate to the needs of industry who often begin the climb towards total integration though the linking of software which runs on a variety of "fixed" hardware bases.

Application of the projects to the needs of industry

Whilst the CIM-BOSYS integration platform does not attempt to hide many of the complexities involved in the implementation of integrated manufacturing, it does provide a platform on which to base the incremental introduction of CIM. It also encourages implementors to approach the problem of integration with flexibility as an essential part of system design. Even if the CIM-BIOSYS services become unsuitable for future needs, then the functional decomposition of the code ensures that alterations can be introduced with the minimum of conflict with other functions. Although the LUT SI group realises the importance of work in standards for integrated manufacturing, the philosophy behind the CIM-BIOSYS platform has been to enable the incorporation of standards rather than to prescribe them. In this way the platform is less restricted and implies a more long term usage. The platform is therefore a concept useful for manufacturing industry today and whilst remaining flexible for the future.

The MSI group is, however, more involved in the development of standards for manufacturing industry, providing a greater level of guidance for particular implementations. Many of the activities still require further testing and development before manufacturers will be prepared to accept the more radical change to highly automated distributed manufacturing control.

One aspect that both projects clearly stressed was the need to clearly define and separate out functionality in integrated manufacturing systems. This makes future development and alteration more readily incorporated by the implementor and provides great savings in the time and effort involved in application.

SUMMARY OF ACHIEVEMENTS

The application of CIM-BIOSYS to the MSI project has enable the evaluation of an integrating infrastructure, which has proved to be very flexible to the requirements of a distributed computer controlled manufacturing system of the MSI project. The application has shown that even in the most complicated manufacturing systems, the integrating infrastructure was generic and flexible enough to achieve the required functionality. The application did, however highlight the need for the platform itself to be capable of alteration and reconfiguration to enable differences in system performance.

ACKNOWLEDGEMENTS

The author would like to thank the members of the MSI group who helped during the collaboration. The members included Mark Luce, Sarah Wallace, Kate Senehi, Ed Barkmeyer, Evan Wallace and Steve Ray.

REFERENCES

1. Senehi MK, Wallace S, Barkmeyer E, Ray S, Wallace EK: Control entity interface document. *NISTIR 4682* September 1991.

2. Senehi MK, Wallace S, Luce ME: An architecture for manufacturing systems integration. *Proceedings of the Manufacturing International Conference,* Dallas, TX, 1992

3. Senehi MK, Barkmeyer E, Luce M, Ray S, Wallace EK, Wallace S: Initial architecture document. *NISTIR 4682* September 1991.

4. Ray SR: Using the ALPS process planning model. *Proceedings of the SME Manufacturing International Conference*, Dallas, TX, 1992.

5. Libes D: A clock for manufacturing systems integration. *NISTIR 4666*, Sept 91.

6. Weston RH, Gascoigne JD, Sumpter CM, Hodgson A: A generic framework for systems integration. *CIME*, Vol 7, No 3, pp48-58, Dec. 89.

7. Weston RH, Gascoigne JD, Sumpter CM, Hodgson A, Rui A, Coutts I: Configuration methods and tools for manufacturing systems integration. *Special Issue IJCIM*, Vol 2, No 2, pp 77-85, 1989.

8. Weston RH, Hodgson A, Coutts I, Murgatroyd IS: Soft integration and its importance in design to manufacture. First Issue of *Int. J. Des. & Manuf.*

9. Weston RH, Hodgson A, Coutts I, Murgatroyd IS, Gascoigne JD: Highly extendible CIM systems based on an integration platform. *Proc. Int. CIMCON Conf. '90*, NIST, Gaithersburg, MD.

PROBE METROLOGY AND PROBING WITH COORDINATE MEASURING MACHINES

[1]Howard H. Harary, [2]Jacques-Olivier Dufraigne, Patrice Chollet, and Andre Clement

[1]*National Institute of Standards and Technology, Gaithersburg, Maryland, U.S.A.*
[2]*Laboratoire de Mécatronique, Institut Supérieur des Matériaux et de la Construction Mécanique, Saint Ouen, France*

ABSTRACT

A harmonic frequency approach to the modeling of form in manufactured holes is discussed. The utility of sampling theory as a rational basis for choosing a sampling density in discrete measurement cases is suggested. The possible usefulness of Fourier analysis is demonstrated in the context of a harmonic model for form errors and an example presented. A series of holes produced by ten common machine shop processes will be measured at very high sampling densities with roundness measuring instrumentation. This data will be analyzed for frequency content of form error and sampled for the calculation of functional parameters of the hole. The possibility of under-sampling in the measurement process, while achieving acceptable accuracy in the calculated hole parameters, will be investigated using various probe stylus radii, both experimentally and theoretically.

BACKGROUND

Researchers from the Institut Supérieur des Matériaux et de la Construction Mécanique, a graduate engineering school in France, and the Precision Engineering Division at NIST have been working together for several years on topics of mutual interest in metrology. Our successful collaboration results from complementary strengths in theoretical and experimental research. Specifically, we have developed a new metrology system at NIST for the characterization of analog probes used on coordinate measuring machines [1], and applied software error correction mathematics developed in the Laboratoire de Mécatronique improving the accuracy of measurement of these probes by as much as a factor of 30 [2].

We are now beginning a new project designed to explore the application of signal analysis and sampling theory to probing strategies for measuring actual part features at a limited number of touch locations. Simply put, how many touches are necessary to characterize a feature? The answer depends on (1) the type of feature; (2) the characteristic form errors of that feature; (3) the functionally important characteristics of that feature and their acceptable tolerance bands; (4) the accuracy of the measurement.

INTRODUCTION

When a coordinate measuring machine (CMM) is used to measure a geometric feature of a part, the metrologist or quality inspector must decide on a strategy for inspection. The vast majority of CMMs use touch-trigger probes for sensors. Each time the probe touches the part the CMM records the coordinates of that point by recording the current value of its scales. A picture of the size and shape of the part is then built up from a series of these discrete measurements. In the discussion that follows, we limit ourselves to sampling issues related to the measurement of nominally circular holes.

PRINCIPLES

Sampling

When inspecting a hole we wish to determine its radius, center coordinates, and deviation from ideal form. The key parameters [3] in the inspection strategy of a hole are the number and distribution of inspection points. We limit our discussion to the case in which the sampled points are equally spaced around the circumference of the hole. A decision must be made as to the number of touches (data points) that will sufficiently characterize the form, radius and position of the hole. The value of time generally dictates that the inspection be accomplished with as few inspection points as possible. When more than the minimum number of inspection points (three) necessary for mathematically computing the parameters of the equation of a circle are available, most CMM software routines use a least-squares method [4,5]. The circle so determined is called a substitute element; it is a model that "stands in" for the originally measured hole.

The residuals are the deviation between the substitute element and the actual data points. These residuals can take the form of random noise or systematic deviations. When the number of data points is in excess of the minimum for the definition of a geometric feature, the least squares, and other fitting criteria tend to minimize the effect of random noise on the fit parameters. The effect of random noise on the fit parameters is inversely proportional to the square root of the number of data points.

A Harmonic Model For Form Errors

Systematic residuals are evidence that the circular mathematical form originally postulated to represent the feature is inadequate. Then, errors in the fit parameters of the hole may occur when we force the data to fit a circle. In some cases the magnitude of the form error can be comparable to the tolerances demanded of a measurement. In those cases, fitting the data to a more complicated model may be beneficial. If the systematic residuals are periodic an appropriate model might be a harmonic model in which the form is represented as a sum of sinusoids [6]. Each additional periodic term added to the fundamental term, r_0, is termed a harmonic. The equation for the form of the hole is then:

$$r = r_0 + r(\theta) \tag{1}$$

$$r = r_0 + \sum_n a_n \cos(f_n\theta + \phi_n)$$

$$r = r_0 + a_1\cos(f_1\theta + \phi_1) + a_2\cos(f_2\theta + \phi_2) + \dots \tag{2}$$

where:

r_0 is the least-squares radius of the circle-like feature,

f_n is an integer equal to the frequency of the nth harmonic,

a_n is the amplitude of the nth harmonic,

and ϕ_n is the phase of the nth harmonic

It is interesting to note that the first harmonic term ($n=1$) contains information regarding a mis-centering of the hole in the measurement coordinate system, (r,θ).

There is good evidence that for real parts the higher order form errors, $r(\theta)$, are well represented by a small number of relatively low frequency harmonics ($n < 10$). Note that the form of the new equation which represents the feature is the sum of a substitute element, a circle, and a group of higher order terms $r(\theta)$ for $n > 1$, which we call a "substitute form error".

Shannon's Sampling Theorem

The common formulation of Shannon's sampling theorem [7] states that a sinusoidal signal can be unambiguously recovered if sampled at a rate f_N such that:

$$f_N \geq 2 f_h \tag{3}$$

where

f_N is the Nyquist rate

and f_h is the highest frequency present in the signal.

In fact, for a general periodic signal composed of sines and/or cosines with arbitrary phase, it can be shown [8] that the minimum sampling rate (see Appendix 1 for a graphical explanation) should satisfy the condition:

$$f_N > 2 f_h \tag{4}$$

We intend to use a Nyquist rate, f_N, defined by this somewhat stricter sampling criteria.

The Fourier Transform

The mathematical tools used in electrical signal analysis can be applied to our problem by recognizing the analogy between an electrical signal and a representation of the form of a circle-like feature as a function of θ (equation 1). The Fourier transform [9,10] can be applied to determine the amplitude and phase of each sinusoid necessary to adequately represent the form of the hole:

$$r = r_0 + \sum_n a_n \cos(f_n \theta + \phi_n)$$

$$FC = FT\{r\} \tag{5}$$

$$FC = FT\{r_0 + a_1 \cos(f_1 \theta + \phi_1) + a_2 \cos(f_2 \theta + \phi_2) + \ldots\}$$

where

FT is the Fourier transform of the function r,

and   FC are the Fourier components.

The Fourier component consists of a real and imaginary parts for each frequency (harmonic) present in the form. The amplitude, a_n and phase, ϕ_n, can be calculated from the Fourier components as follows:

$$\text{for } n=0: a_0 = |FC_0| = (Re_0^2 + Im_0^2)^{(1/2)}$$

$$\text{for } n \geq 1: a_n = 2|FC_n| = 2(Re_n^2 + Im_n^2)^{(1/2)} \tag{6}$$

and $\phi_n = TAN^{-1}(Im_n/Re_n)$

where FC_n is the nth harmonic Fourier component, and Re_n and Im_n are the real and imaginary parts, respectively of that Fourier component.

Sampling at less than the required Nyquist rate, f_N, provides insufficient information to reconstruct the form error. This under-sampling can lead to indeterminate, or incorrect results. Frequency components in the form present at greater than half the actual sampling frequency appear at lower frequencies and are confused with genuine low frequency components; this phenomenon is called aliasing (see Appendix 2). In electrical signal analysis a low-pass filter is used to limit the high frequency components (bandwidth) of a signal. In metrology applications the averaging effect of the stylus radius could serve an analogous function [11,12,13]. Increasing the stylus radius increases the area of contact with the part surface; the averaging acts as a low-pass spatial filter.

A Simulated Fourier Analysis of Measurement Data

An example of the Fourier analysis of a form is presented for a form with second and third order harmonic form errors. The equation for this form is:

$$r = r_0 + a_2\cos(f_2\theta+\phi_2) + a_3\cos(f_3\theta+\phi_3) \qquad (7)$$

where $\qquad r_0 = 1$ and $a_2 = 0.001 \quad \phi_2 = \pi/5$
$$a_3 = 0.001 \quad \phi_3 = \pi/2$$

A "population" data set of 512 points was generated using equation (7). The data set is plotted in Figures 1a and 1b. The magnitude of the Fast Fourier Transform (FFT)- an efficient, widely available implementation of the Fourier transform - as applied to the population data set is shown in Figure 2. The second and third harmonics are visible. Figures 3a and 3b show the results of sampling the population data set at 8 equal intervals. The magnitude of the FFT of the sampled data set is shown in Figure 4; note that it is identical to the FFT of the population data set. The complete Fourier coefficients are shown in Tables 1a and 1b. The coefficients are identical, demonstrating that we have recovered the exact form and amplitudes, phases, and frequencies of the population data set from the sampled data set.

ILLUSTRATION OF THE PRINCIPLES WITH THE MEASUREMENT OF REAL PARTS

Ten different processes were used to generate 25.4 mm holes in machined aluminum 2024 plates (152 mm square by 19 mm thick). The processes were chosen to provide a wide range of hole quality (see Appendix 3). It is our working hypothesis that the processes will have characteristic form errors that are a signature of that process. The feeds and speeds used were close to maximum production conditions as recommended in the Metcut (Cincinati, OH) machinability database. A separate blank was used for each hole in order to minimize inter-hole interactions; each process was repeated four times in four different blanks. The fixture used to hold the blanks provided kinematic location, good support and even, reproducible clamping pressure.

Proposed Measurements

We intend to measure the holes in our manufactured parts using a roundness measuring machine at a sampling rate of 512 samples per revolution. This data set will serve as our sampling population. The sampling theorem predicts that we should be sensitive to frequency components < 256 cycles per revolution; we consider this to be over-sampling given the relatively low frequency components expected in our parts. The data will then be analyzed with the FFT. The magnitudes and phases of the frequencies determined, we will then sample our data set at rates starting above the Nyquist rate set by the highest frequency component present in the Fourier transform of the population

Figure 1a

Figure 1b

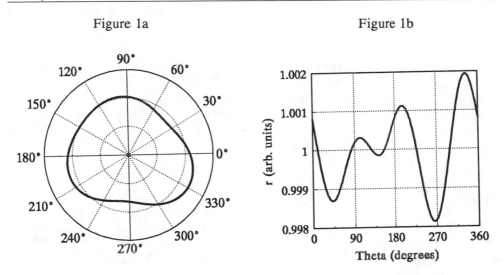

Figure 1. Polar (1a) and rectangular coordinate plots (1b) of the example form error given in the text (see equation 7) for the population data set. In Figure 1a the form error is amplified by 100; the interval between the dotted circles is 0.005 units.

Figure 2

Population Data Set

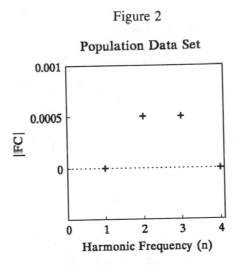

Figure 2. The magnitude of the Fourier components of the population data set.

Figure 3a   Figure 3b

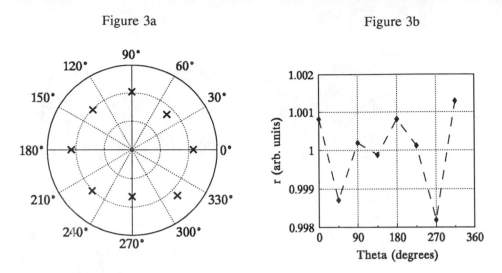

Figure 3. Polar (3a) and rectangular (3b) plots of the example form error given in the text (see equation 7) for the sampled data set. The data was sampled at 8 equal intervals. In Figure 3a the sampled form error was amplified by 100; the interval between the dotted circles is 0.005 units.

Figure 4

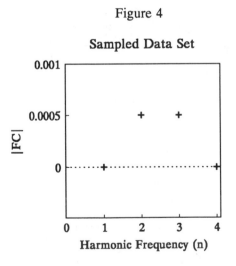

Figure 4. The magnitude of the Fourier components of the sampled data set.

Table 1a

Fourier Components Of The Population Data Set[*]

n	RE (Real Part)	IM (Imaginary Part)	|FC| (Magnitude)	ϕ/π (Phase)
0	1	0	1	0
1	0	0	0	-
2	$4.0451 \cdot 10^{-4}$	$2.9389i \cdot 10^{-4}$	$5 \cdot 10^{-4}$	5
3	0	$5i \cdot 10^{-4}$	$5 \cdot 10^{-4}$	2
4	0	0	0	-

Table 1b

Fourier Components Of The Sampled Data Set[*]

n	RE (Real Part)	IM (Imaginary Part)	|FC| (Magnitude)	ϕ/π (Phase)
0	1	0	1	0
1	0	0	0	-
2	$4.0451 \cdot 10^{-4}$	$2.9389i \cdot 10^{-4}$	$5 \cdot 10^{-4}$	5
3	0	$5i \cdot 10^{-4}$	$5 \cdot 10^{-4}$	2
4	0	0	0	-

[*]Fourier components shown rounded to four decimal places

data set. We will then compare the form reconstructed from the FFT of the sampled data to the form given by the original sampling population. The presence of random noise in the measured form might require a sampling rate somewhat greater than the predicted Nyquist rate; only real data can tell us this. We intend to repeat our measurements using stylus diameters ranging from 4 mm to 25 mm.

DISCUSSION

A knowledge of the signature form errors produced by different manufacturing processes should allow the determination of a more realistic function to represent that feature. In the case of a hole, the usual substitute element of a circle is supplemented by higher order harmonic terms called substitute form errors. The Nyquist rate provides a guide for the proper sampling of the feature, including its substitute form error. The form error function could also prove useful in finding a data fit called the maximum inscribed circle [4] - the maximum radius circle that will fit inside of a nominally circular feature. The maximum inscribed circle is used to determine a useful functional characteristic of a hole - the maximum radius of the cylindrical pin that will fit in it. The substitute form error could help this determination by predicting the peak values of the substitute error independent of the phase of the sampling data. Also, the substitute form error may lead to a more efficient mathematical method or computer algorithm to find the maximum inscribed circle. Finally, note that the averaging provided by the radius of the probe stylus is biased towards detecting the peaks of the form error, a property also beneficial to finding a more accurate maximum inscribed circle with limited inspection data.

The goal of our studies is to develop analysis methods which can be used on real part data. Eventually, we hope to present a data base which relates process parameters (tool type, feeds, speeds, materials, etc.) to recommended inspection strategies.

CONCLUSION

The frequency analysis of the form error of a circular feature should lead to a rational basis for choosing the sampling frequency required in an inspection of that feature. In addition, a more complete model of the hole which includes a substitute form error could prove useful in improving the accuracy of algorithms used to find the maximum inscribed circle. Increasing the radius of the probe stylus should limit the high frequency components of the form error and permit lower sampling rates.

Acknowledgements: We thank the Navy ManTech program for support and are grateful to W.T. Estler, G. Stenbakken, T. Hopp, T. Doiron, and N. Oldham for helpful discussions.

The identification of commercial materials and instruments is given only for the sake of clarity. In no instance does such identification imply endorsement by NIST, nor does it imply that the particular material or equipment is necessarily the best available for the described purpose.

REFERENCES

1. Harary H, Crcismcas P, Chollet P, Clement A: A new method for the characterization of analog probes for coordinate measuring machines. in *Proc. 5th Intl. Precision Engineering Seminar.* Monterrey, CA. 1989, pp 155-157.

2. Harary H, Clermont D, Lavigne C, Chollet P, Clement A: The characterization of analog probes for use on coordinate measuring machines. in *Proc. American Society for Precision Engineering Annual Conference.* 1991 p 32

3. Caskey G, Hari Y, Hocken R, Palanivelu D, Raja J, Wilson R, Chen K, Yang J: in *Proc. NSF Design and Measurement Systems Conference*, 1991 p 779-786.

4. Feng SC, Hopp TC: A review of current geometric tolerancing theories and inspection data analysis algorithms. *National Institute of Standards and Technology Internal Report*, NISTIR 4509, February 1991.

5. Forbes AB: Least-square best-fit geometric elements, *National Physical Laboratory Report*, DITC 140/89, Revised February 1991.

6. Damir M: Approximate harmonic models for roundness profiles. *Wear* 57:217-225, 1979.

7. Stanley WD, Dougherty GR, Dougherty R: *Digital Signal Processing.* 2nd Ed., Prentice-Hall, Reston, 1984.

8. Hamming RW: *Digital Filters*, Englewood Cliffs, Prentice-Hall, 1977, pp 141-143.

9. Tretter S: *Introduction to Discrete-Time Signal Processing.* Wiley, New York, 1976.

10. Weaver HJ: *Applications of Discrete and Continuous Fourier Analysis.* Wiley, New York, 1983.

11. Radhakrishnan V: Effect of stylus radius on the roughness values measured with tracing stylus instruments. *Wear* 16:325-335, 1970.

12. Church EL, Takacs PZ: Instrumental effects in surface finish measurement. *Proc. SPIE* 1009:46-55, 1988.

13. Church EL, Takacs PZ: Effects of the non-vanishing tip size in mechanical profile measurements: *Proc SPIE* 1332: 504-514, 1990.

Appendix 1

Two first harmonic form errors are illustrated in Figures A1a and A1b. The form errors are sampled at twice their frequencies. The cosine form error (Figure A1a) is properly sampled; the function could be properly reconstructed from the sampling data. However, the sine form error (Figure A1b) is not properly sampled; it contains no amplitude information.

Figure A1a   Figure A1b

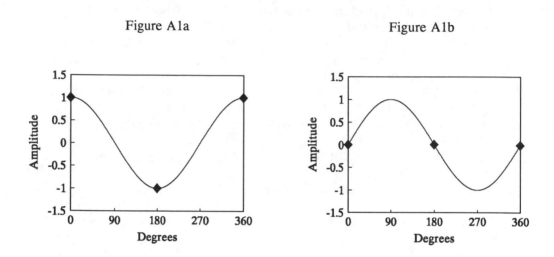

Figure A1. Illustration of sampling of a cosine (Figure A1a) and sine (Figure A1b) form error. The sampling points are indicated by the symbol ◆ .

Appendix 2

An example of aliasing is shown in Figure A2. The form shown has a fourth harmonic form error and is sampled with phases of sampling differing by $\pi/4$ (45°). The data from the two samplings would result in computed radii differing by twice the amplitude of the fourth harmonic form error.

Figure A2

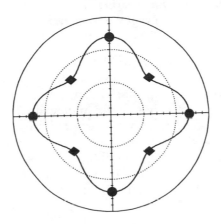

Figure A2. Demonstration of Aliasing. The fourth harmonic form error is sampled twice, indicated by the symbols ● and ◆.

Appendix 3

Ten processes used to generate the 25.4 mm holes using a Hurco (Indianapolis, IN) BMC-40 machining center are listed below.

1. Drill (short*)
2. Drill (long**)
3. Drill (short) Undersize Pilot Hole [12.7 mm, (1/2")], re-drill to size
4. Drill (short, [25 mm (63/64")] and Ream - Spiral
5. Drill (short, [25 mm (63/64")] and Ream - Straight
6. Drill (short, [25 mm (63/64") and Bore
7. Drill (short, [24.6 mm (31/32")], Bore [(25.27 mm (0.995")] and Ream - Spiral
8. Drill (short, [25 mm (63/64")] and Mill - 2 Flute
9. Drill (short, [25 mm (63/64") and Mill - 4 Flute
10. Drill (short, [12.7 mm (1/2")] and Counter bore (1/2" pilot)
 * screw machine length twist drill
** standard length twist drill

DESIGN OF OPEN AUTONOMOUS MANUFACTURING CELLS

Kim Siggaard, Arne Bilberg, and Leo Alting

*Institute of Manufacturing Engineering, Technical University of Denmark,
Lyngby, Denmark*

ABSTRACT

Cellular manufacturing has been proven as a good concept in realizing flexible manufacturing. This paper deals with the problem of developing computer based cell controllers for fully or partially automated cells.

It describes how to design flexible, autonomous and open cell control programs in an efficient way using a set of structured modelling techniques and discrete computer simulation for modelling of the applied logistic concept.

Through functional modelling (IDEF0) a generic model has been developed identifying the common functions in a cell controller, that has to be open for integration with production planning and control systems. It is shown how a relational client/server database architecture gives an open and flexible way to configure and integrate the cell controller with surrounding systems.

Finally, the paper will discuss how to benefit from using a relational database as the integration platform in manufacturing cells.

INTRODUCTION

Cellular manufacturing has been accepted as a rational way to organize the production. The Japanese have presented excellent manufacturing competitiveness by applying a cellular and group oriented organization of the production, e.g. implementation of JIT production and quality circles.

In connection with automation of production (e.g. FMS), the cellular concept has been adapted as the only feasible path. Flexible manufacturing cells (FMC) developed as autonomous production units are used as the basic building block of advanced manufacturing systems. Much research and development has the last decade been put into the development of control programs for FMCs because these were the basic building blocks needed to realize the "FMS dream" with one-of-a-kind production and still achieving a mass-production productivity.

Today, we can not say that FMS has been a second industrial revolution. Most of the successful FMS implementations are more like a new kind of automated mass-production with a limited additional flexibility. One can ask why FMS has not fulfilled its promises?

Looking at the market for enabling tools for implementation of cell controllers, it is clear that a large supply has come on the market (e.g. The FIX, FactoryLink, Alert, RealFlex, CIMCell, PlantWorks, CellWorks, EasyMAP and many others). All very excellent enabling tools for developing cell controllers to nearly any manufacturing cell /9/. This paper discusses two possible reasons why integration of the shop floor in the job shop is going quite slow despite the availability of enabling tools.

First, there is a lack of a well-established design and implementation methodology. When it comes to shop floor control systems, the traditional software design methods are not efficiently coping with dynamic multi-agent environments at the shop floor. The industry has regarded the work, done in developing structured design and analysis methods for the manufacturing environment, as a rather academic occupation.

Secondly, there has been a tendency for the available enabling tools to be rather closed systems not directly compatible with e.g. production planning systems and other management systems. This can very easy result in automation islands and not the intended use of cells as building blocks for more overall integrated manufacturing systems.

SYSTEM DEVELOPMENT METHODOLOGY

Developing computer programs it is always important to have a good set of structured system modelling techniques. Especially, when control systems for flexible manufacturing cells are to be developed, because these type of systems are combined of developments of the logistic, the management and the control system. Therefore a methodology has to be established, telling how to organize the system development process as well as defining what modelling tools to use at the different phases in the project.

In a shop floor control system development project five important phases can be identified: *Objective setting, specification, design, development* and *implementation / running-in.* Figure 1 illustrates these phases and the activities we propose to perform to go through these five phases.

"Objective sitting" is the task of converting company strategic goals to operation goals and milestones for the manufacturing system in question. It is important that this conversion is done with respect to the economical, technological and social conditions in the organization. When modernizing an existing manufacturing facility this phase will have to include analysis of the existing system. The result of this phase is an overall work-plan, goals and milestones for the project.

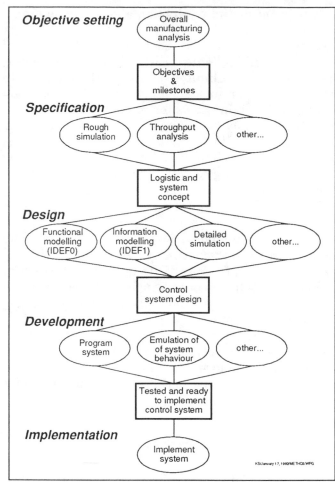

Figure 1 *Shop floor control system development framework.*

The "Specification" phase is done on the basis of the overall workplan making detailed specification of the developments/changes needed in the manufacturing system to accomplish the workplan. The specification phase shall produce the conceptual frame for the logistic concept to be applied when controlling the cell, as well as the control system concept covering how the system must be integrated with other systems within the company, the functionality of the control system (e.g. user interfaces) among others. If an external company is going to develop and implement the system the specification documents need to be so detailed that it can be used to contract a subcontractor. In making the specifications, some analysis will be needed, we recommend the use of throughput analysis /7/ for investigating the logistics to determine where important problems are located. This should tell us what to focus on in the new manufacturing concept. Computer simulation is an efficient tool which can be applied for making rough simulations of alternative logistic concepts and thereby give an objective basis for deciding which logistic concept to select. Of course other analysis can be performed if needed.

The "Design" phase is converting the specifications to actual designs of computer systems, databases, production layouts, special or customized production equipment. The outcome from this phase is documented design specifications ready for development and implementation. A set of design techniques has to be applied. In Figure 1 it is proposed to use IDEF0 and IDEF1 for respectively functional modelling and information modelling, and detailed computer simulation for evaluating layouts (e.g. material flow or collision test of robot systems) and the logistic performance. It is also important to test as many alternative designs as possible at this early stage. This could be by prototyping computer applications to validate functionality and performance.

The "Development" phase is converting the designs to physical systems. This can be machines/equipment for the production system as well as the needed computer systems to realize the control system. A set of enabling tools has to be put in use, embracing database systems, cell control software as FactoryLink, CellWorks, PlantWorks or SIMCON (a system developed at the Technical University of Denmark). What tools to choose is naturally depending on the actual control task and preference of the developer. An important issue here is to perform as much off-site test and validation of the developed systems as possible. In developing computer based control systems, emulation is an important tool, the application of emulation is covered in /1//2/.

The "Implementation and running-in" phase covers all the developed components in the factory. Of course changes always have to be made, not all bugs can be found in the previous test and validation. Therefore, it is important to prepare a final system documentation on the same form as the detailed design specifications, and the final documentation should also document that the implemented system meets the objectives for the system.

IDEF0	Functional modelling
IDEF1	Information modelling
IDEF2	Simulation modelling
IDEF1x	Data modelling
IDEF3	Process description capture
IDEF4	Object-oriented design
IDEF5	Ontology description capture
IDEF6	Design rationale capture
IDEF8	User interface modelling
IDEF9	Scenario-driven info sys design spec
IDEF10	Implementation architecture modelling
IDEF11	Information artifact modelling
IDEF12	Organization modelling
IDEF13	Three schema mapping design
IDEF14	Network design

Figure 2 *Next generation IDEF methodes. /6/*

The IDEF[*] techniques are among the most mentioned and used system design methodologies for manufacturing applications and are also used in this paper. IDEF proposed 3 modelling techniques needed to model and analyze the behavior of manufacturing systems. IDEF0 - for modelling the static functional relations between subfunction within the manufacturing system /3/. IDEF1 - for modelling the static information structure, relations between information entities needed to operate the manufacturing system. IDEF1 results in an information structure that can be implemented in a relational database. IDEF2 - for modelling the dynamic behavior of the manufacturing system, general speaking this is computer simulation of material and information flow within the manufacturing system.

---

[*]   IDEF (<u>I</u>CAM <u>D</u>efinition method) was developed in the ICAM program - a ten years U.S. Air Force program for <u>I</u>ntegrated <u>C</u>omputer <u>A</u>ided <u>M</u>anufacturing started in 1975.

The problem is that the IDEF techniques do not cover all the needs for modelling tools. This has also been realized by the IDEF user group which has defined a next generation of IDEF methodes that have to be developed the next years. In Figure 2 a summary of these next generation IDEF methodes are presented /6/. One can argue that the huge number of techniques will make structured modelling look even more academic to industry.

DESIGN OF AN OPEN CELL CONTROLLER

During the last 5 years the Institute of Manufacturing Engineering at TUD has worked with development of a new kind of cell controller based on the idea of using a simulator as decision maker in the control program /4/. This work has resulted in the discussion of how we develop a cell controller open for integration with other systems. The result has been a generic functional architecture that can be applied in all types of cellular manufacturing systems.

For describing the generic functions and architecture of control systems, IDEF0 modelling is applied. In Figure 3 the A-0 diagram is showing the context of a cell controller as it serves the purpose of this modelling. There is not included any material flow in the analysis but only information flow with the external environment and the flow internally between modules within the cell controller.

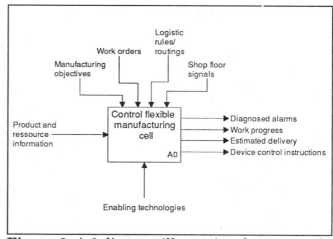

Figure 3 *A-0 diagram illustrating the context of a cell controller,*

Figure 4 shows an A0 sheet with the main functions in a cell controller and their mutual communication. It is a decomposition of the A-0 diagram in Figure 3. The diagram identifies 5 generic functions:

o   Introduce manufacturing data
o   Schedule production sequence
o   Dispatch work orders
o   Manage job execution
o   Monitor cell condition

The functionality of these five functions will be discussed briefly.

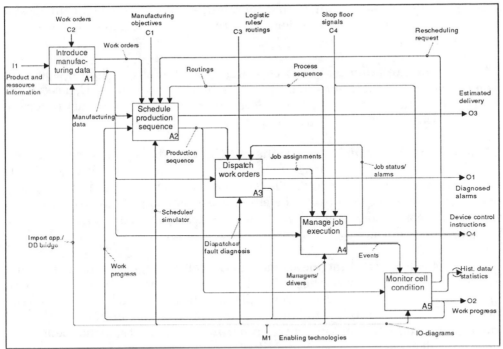

Figure 4 *Main function in a cell controller.*

"Introduce manufacturing data":   is the function of feeding the cell controller with the necessary manufacturing data to carry out the production. The function is order driven meaning that "work orders" received from the overall scheduling system or entered manually are controlling the activities. When a work order is introduced there has to be performed a validation of whether or not the needed manufacturing data (e.g. part programs) are available. If manufacturing data are missing the work order has to be rejected or the missing data must be requested.

"Schedule production sequence":   is determining in what sequence the work orders should be handled in order to optimize the fulfillments of manufacturing objectives. The type and freedom in the scheduling activity vary from cell to cell. The trend is that more and more decisions are distributed down in the organization, which in a cellular manufacturing means that scheduling authority for the cell is given to the employees managing the cell. The most important output is the "production sequence", but it is additionally proposed that the scheduling function could produce and feedback an "estimated delivery" time to the higher level of planning function.

"Dispatch work orders":   is controlling the actual manufacturing of work orders. This activity is primarily controlled by the scheduled production sequence and the status feedback from the shop floor. The function is breaking down work orders into jobs needed for producing the order (e.g

transportation from machine A to machine B, an operation on machine B and so on). The "dispatch work order" activity has to coordinate all these jobs within the cell.

"Manage job execution":   is covering execution of the different jobs on the individual workstations. E.g. to process part A on machine C, a sequence of activities has to be carried out: Setup machine, download part-program, start operation and take down the part.

"Monitor cell condition":   is the continuous registration of activities within the cell and monitoring of how work orders progress. One function is to report work progress to higher level of planning and control in the factory. Another function is to support the employees operating the cell, they need a system for continuously monitoring how the cell is operating to decide when actions are needed.

The described functional model of the cell management task is generic in the way that each of these five functions to some extent have to be performed in any cell, even a manual manufacturing cell with-out use of computer based control programs. If we concentrate on the problem of developing computer based control programs for manufacturing cells, four essential demands have been identified:

First: the cell must be flexible for configuration of system layouts and control logic within the cell. In this area a simulation program is used as decision maker, this research is discussed in /4/.

Second: there has to be an open architecture for integrating shop floor devices, e.g. machine tools, bar-code reader etc.. OSI standards and MAP network are solutions which are offering such an open integration architecture towards the production equipment.

Third: when cells are going to be used as building blocks in advanced manufacturing systems they have to be open for integration with overall production planning and control systems.

Fourth: cell controller systems should be prepared for allowing continuously improvements and modification, e.g. regarding data collection, manufacturing statistics and graphs.

It is the last two demands for open cell controllers that are the topic of the remaining part of this paper. The idea to be discussed is the use of an integration platform based on a relational database.

ESTABLISHMENT OF AN INTEGRATION PLATFORM

The term integration platform means a computer hardware and software configuration, which enables integration of the cell controller with external systems, as well as it enables the user to develop his own application for controlling or monitoring the cell activities. The key issue in this problem is storing and retrieval of information in a flexible manner.

Using a relational SQL database as the integration platform offers a number of advantages:

o   The relational model gives a very flexible way of storing information so they can be used for multiple purposes.

o   Through use of LAN, multi user application of the database is possible.

o   SQL has been established as a market standard in DBMSs and for making queries in relational databases.

o   Using well-established databases (as Oracle), the system can grow by using a distributed architecture or transfer the database server to a higher computer platform.

o   One can utilize the continuously development and improvement made by the DBMS supplier in the future releases.

It has been argued that relational databases are too slow to be used in cell controllers on the shop floor. It is right that data access in relational databases can not compete with various types of record files, but performance of computers and query managers in the DBMSs are increasing rapidly. Therefore, performance of the database transactions will be a less and less critical issue.

For investigating the transaction time for a relational database in a cell controller a performance test has been made, testing the performance of a Client/Server configuration using 386-PC's as server and client. The database server was a DOS SQL-Base from Gupta Technology, Inc. and it was interfaced through a C Application Programmers Interface. The result of the test was that on average a database transaction took 0.1-0.2 sec. (seen from the client application).

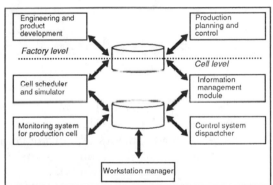

Figure 5 *Database as integration platform in a manufacturing cell.*

This is an important information to remember when designing how the database should be used in the cell controller.

Figure 5 shows how the database is used as integration platform within the cell and as the interface to surrounding systems. The integration upwards is done by sharing a database between the cell controller and the overall PPC system. Through this interface the cell controller can receive work orders, part information, process plans etc. In many factories direct interface to the PPC database is not possible or only a part of information can be imported through this interface, why manual entering or file import of data is an important feature. In the cell controller a local database is used for integrating the different modules of the cell controller (the functions described in previous chapter) as well as being the platform on which the user can develop his own applications for special purposes, e.g. monitoring, data collection, quality control etc..

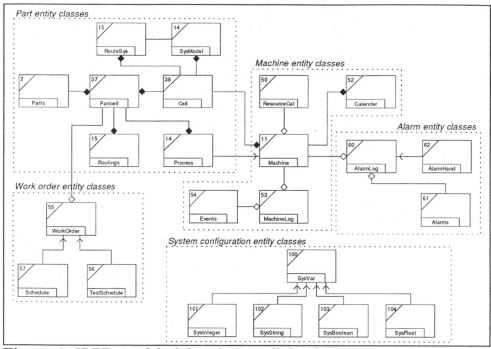

Figure 6 *IDEF1 model of the relation cell database.*

Figure 6 shows the initial information model (IDEF1) for our implementation of the SIMCON/2 cell controller at TUD. The developed information model consists of 5 groups of entity classes: *"Part entity classes"* storing information about parts defined in the cell and process plans for manufacturing the parts. *"Machine entity classes"* storing information about all the machines within the cell. *"Work order entity classes"* for managing work orders and scheduling. *"Alarm entity classes"* for alarm and fault management. *"System configuration entity classes"* used for setting up the cell because all the configuration parameters for the different modules are stored in the database.

Without going into detail with the database design, a few facilities can be pointed out: The structure of the "part entity classes" is designed in such a way, that the same database can be used for multiple cells (a central database server shared between several cells). This is made possible by the entity class "PartCell" connecting a part entity with a specific cell entity telling where the part can be produced. Sharing a database server is an important feature because it would be difficult to justify a separate database server within each cell. The "Work order entity class" is designed to handle a distributed scheduling in the cell. The database can store the actual operating schedule and a set of candidate schedules, which is under preparation.

APPLICATION OF THE OPEN INTEGRATION PLATFORM

To demonstrate how one could exploit the open integration platform of the SQL based database server, the problem of making a customized monitoring facility in the cell controller is considered. This is an important area because production management needs the possibilities for making logistic analysis through statistics and trend analysis. The problem is that the wanted/needed analyses differs from company to company or even between departments within the same company and often requirements for new types of analysis arise in the course of time.

Recording manufacturing history in the SQL database (as by the database architecture shown in Figure 6) enables one to use standard software packages for making analyses and manufacturing reports, e.g. spreadsheet systems as Excel[**] enable one to make analyses and reports working on-line on a SQL database. Such facilities give a new level of freedom to customize the logistic monitoring of the manufacturing process.

In the area of quality control, the integrated use of relational databases on the shop floor offers excellent opportunities for tracing products through the manufacturing processes. Quality data can be combined in new ways, e.g. product specific production documentation certifying all process steps, which is a demand to many manufactures in the 90s.

Only a few commercial factory automation products have been built on an integrated use of a relational database. One product, EasyMAP[***] has been developed with the philosophy of using standardized data exchange formats /8/. For communication with shop floor devices the MAP/MMS standard is used, and for storing internal information within the cell controller an Oracle database is used. For making manufacturing reports and monitoring function a spreadsheet (MicroSoft Excel) is used. This extensive use of standard software products and standards has reduced the new developments within the control software to modules for definition of displays, control logic and the MMS kernel enabling a distributed control architecture.

CONCLUSIONS

It can be concluded that the use of structured system design methods is crucial to successful development of shop floor control systems and there is presented a methodology for handling the complex task of developing a shop floor control system.

This paper provides an overall generic control architecture that has

---

[**]   Excel as product and trademark from Microsoft inc..

[***]   EasyMAP is a Danish factory automation software packages from PROCOS A/S running on PCs under the OS/2 operating system.

been applied for development of a simulator-based cell controller open for integration with various PPC systems. The architecture specifies 5 generic functions to be handled in any cell: *Introduce manufacturing data, schedule production sequence, dispatch work orders, manage job execution* and *monitor cell condition.*

It is described how a standard relational database can be used as integration platform for the functions within the cell as well as for integrating the cell with other functions in the manufacturing enterprise. It is shown how a database gives flexibility for continuously improving and customizing the cell controller functionality.

REFERENCES

/1/   Kim Siggaard and Leo Alting: Methodology for test and validation of shop floor control systems. *Conference proceedings for FAIM '91. March 13-15, 1991.* Limerick, Ireland. Page 952-960.

/2/   Kim Siggaard: Design and implementation of open and flexible shop floor control systems. *Ph.D. thesis from the Institute of Manufacturing Engineering, The Technical University of Denmark.*
       (TO BE PUBLISHED IN SEPTEMBER 1992)

/3/   SoftTech Inc.: Integrated Computer Aided Manufacturing (ICAM), Architecture part II, Volume IV - Functional modelling manual (IDEF0). 1981. 161 pages.

/4/   Arne Bilberg: Simulation as a tool for developing FMS control programs. *Ph.D. thesis from the Institute of Manufacturing Engineering, the Technical University of Denmark.* 1989. 258 pages.

/5/   J.E. Cooling: Software design for Real-Time systems. Chapman and Hall. 506 pages. First published 1991.

/6/   Richard J. Mayer and Michael K. Painter: Roadmap for enterprise integration. *Conference proceedings for AUTOFACT '91. November 10-14, 1991, Chicago, Illinois.* 26 pages.

/7/   H. P. Wiendahl: The throughput diagram - An universal model for the illustration, control and supervision of logistic. *Annals of the CIRP Vol. 37/1/1988.* 1988. Page 465-468.

/8/   PROCOS A/S: EasyMAP - Extensive application system for manufacturing and process control. *General description.*

/9/   Peter Bubbel Jensen: Computer integrated manufacturing systems, communication and control. *Ph.D. thesis from the Institute of Manufacturing Engineering, the Technical University of Denmark.* 1991. 157 pages.

AN INTEGRATED INTELLIGENT SHOP CONTROL SYSTEM

Michael P. Deisenroth and Yaoen Zhang

Industrial and Systems Engineering, Virginia Polytechnic Institute and State University, Blacksburg, Virginia, U.S.A.

ABSTRACT

A new method for the designing of a shop control system in a computer integrated manufacturing environment is presented in this paper. This method uses two basic techniques, expert systems and object-oriented programming. Using this method, the control system model will have three concentric layers. The innermost layer is an expert system of which the main functions are scheduling and routing. The two outer layers perform shop floor monitoring, manufacturing planning, and communication functions. Two main characteristics of this method are modularity and flexibility. As an example, a shop control system model designed with this method is described and analyzed.

INTRODUCTION

Presently there is a trend in manufacturing systems design from stand-alone, automatic operations to Computer Integrated Manufacturing (CIM) systems. The success of the integration depends largely on the performance of shop control software. The objective of CIM is to develop a cohesive database that integrates manufacturing, design, and business functions. Information is sent on demand and as needed to the largest number of intra- and interdisciplinary groups as possible. The ideal CIM controlled system will design and manufacture without disruption from raw material to the finished products. The typical CIM functions include: business planning and support, engineering design, manufacturing planning, manufacturing control, shop floor monitoring, and process automation [1].

In implementing CIM, an organizational structure and a communication mechanism are required which tie the different processes together. To put it another way, a CIM system involves linking together numerically-controlled machines, robots, material handling equipment, and other automatic equipment. To integrate these machines under one control system is certainly not an easy job. A few companies have done this using a hierarchical control philosophy. The organizational structure involves both partitioning the control problem into smaller, more manageable and less complex control modules and arranging them in a hierarchical fashion, so that the control system becomes easier to implement and modify.

RELATED WORKS

McLean and his colleagues [2] have designed a test-bed control hierarchy on the Automated Manufacturing Research Facility (AMRF) at the National Bureau of Standards and they are striving to develop standards in this area. The test-bed control hierarchy is composed of five major levels: facility, shop, cell, work station, and equipment. Each level has one or more control modules that are further broken down into sub-levels or modules.

A similar approach can be seen in the document, "Reference Model of Production Systems," jointly prepared by Digital Equipment Corporation (DEC) and Philips Bedrijren BV [3]. In this document, the NBS hierarchy is expanded to seven levels. In this model, one more level, production facility, is added to the top of the factory level. The DEC/Philips model divided the equipment level on the NBS model into two levels: Automation Module and Device. The functions performed at the other levels are quite similar to those performed at similar levels in the NBS hierarchy.

In designing the control system for a specific level in the system control hierarchy, considerable efforts have been expanded in order to develop flexible and efficient models. For example, the DEC/Philips [3] cell control model consists of three major functions: goal or task decomposition, sensory data processing, and world data representation, storage, and retrieval.

Another approach for the cell control system can be found in the research on AMRF at National Bureau of Standards [4]. In this research, state tables were utilized for the control of a manufacturing cell at the AMRF. The left side of the state table is the input table which consists of a list of internal states, all commands and feed back inputs that are possible at the given level. The right side of the state table is the action table which specifies the appropriate action to be taken by the controller at a particular given state.

Limited research has been done on using expert systems in shop control software. In the scheduling domain, an example on using expert systems in a job-shop environment is ISIS [5]. This system which is written in SRL (a knowledge representation language) uses a heuristic search approach in a constraint environment to achieve its objective.

Object-oriented problem solving and object-oriented programming (OOP) represent a way of thinking and a methodology for computer programming that are quite different from the usual approaches supported by structured programming languages [6]. The main objective of OOP is to improve software reusability and extensibility. The OOP defines the unit of modularity so that programmers produce sub-components, instead of complete solutions. The sub-components are controlled by standards and can be interchanged across different products. These reusable sub-components are named class in an OOP environment. An object is a variable declared to be of a specific class. Such an object encapsulates state by duplicating all the fields of data that are defined in the class definition.

CIM FACILITY / CONTROL STRUCTURE

The control system described in this paper is used to control the flexible manufacturing system presently hosted in Whittemore Hall at Virginia Polytechnic Institute and State University (VPI&SU). It consists of two processing workcells and a material handling and storage system.

The machining workcell has two CNC milling machines, a SCARA robot, a local storage buffer, and an AT&T 6386 PC which serves as its workcell controller. The main function of this cell is to perform all machining of individual work pieces. The assembly and kitting workcell has a second SCARA robot, a local storage buffer, assembly fixtures and parts feeders, and an AT&T 6386 PC to serve as the workcell controller. It performs two system functions. First it places all the parts that are necessary to make an individual product on an empty pallet. This is referred to as "kitting". Secondly, this workcell is used for assembling the parts after they have been machined into a finished product. The AT&T personal computer serves to coordinate and control all the work within the workcell. The last workcell is a material storage and handling sub-system which consists of an automatic storage and retrieval system (AS/RS), an automatic conveyor system, a machine vision system and an AT&T 6386 PC workcell controller. A shop control computer is connected to the three workcell computers through a local area network (LAN) to coordinate and control system level activities.

Several functions are required to manage a manufacturing system. Teicholz [1] groups these functions into six categories: business planning, engineering design, manufacturing planning, manufacturing control, shop floor monitoring, and process automation. For a particular case, it is not necessary or feasible to include all functions. Rather, different manufacturing systems require different sets of functions depending on the system configuration and the product structures. The control system considered in this research focuses on the areas of business planning and support, manufacturing planning, and production control in a make-to-order fashion.

The hierarchical approach was used to design the system control structure with the individual workcell controllers linked to the shop controller. The shop control functions and the workcell control functions were implemented in a DOS environment on four AT&T 6386 computers, each with 4 Megabytes of RAM memory and a 60 Megabyte disk drive. The communication between the shop controller and the workcell controllers is carried via an AT&T STARLAN network with a 10 megabit transmission speed. A fifth AT&T 6386 PC running Unix is used to serve as the network host and the file server. For the communication between the cell controllers and the devices, several kinds of communication devices are used. To communicate, the machining cell controller uses a built-in serial port for the robot, a four-port serial card for the two CNC milling machines, and a parallel I/O card for discrete input/output to both the robot and the two CNC milling machines. In the assembly cell, the built-in serial port and a parallel I/O card are used for communication with the robot. To communicate, the material handling controller uses the built-in serial port for the programmable controller, a second serial port for the machine vision system, and a 64-pin I/O parallel card for the programmable controller and the machine vision system.

THREE-LAYER INTEGRATED METHODOLOGY

A three-layer integrated methodology, was developed to use in controlling the CIM facilities. This methodology is generic and can be applied to any level of the control structure. As illustrated in Figure 1, the methodology has three concentric layers. The innermost layer is called the "Decision Layer" because of the functions performed. An expert system is used in this layer. The objective of the expert system is to use all available data from the automatic production facility (environment knowledge) and human expertise and experience to create an intelligent control mechanism. This intelligent control mechanism manages the production system and generates real-time responses to problems that may occur in the production system.

The middle ring is the "Action Layer" which is composed of several "action objects." The algorithmic knowledge and part of the state database reside in this layer. This layer is responsible for generating commands for the production system and for interacting with the controllers of the next level to carry out the commands. This layer also gathers information necessary for the expert system to make decisions from the controllers or machines of the next lower and higher levels.

The "Service Layer" is the outermost ring of the control system. This layer has some generic objects which help the action objects in the middle layer to do some calculations and operations. The term "generic" means that each service object is used by several action objects. The service objects usually don't interact directly with the decision layer.

SHOP CONTROL SYSTEM

To design the shop control system for the VPI&SU CIM facility, the three-layer integrated approach was used. The functions identified earlier, the communication functions and other related functions were analyzed. Those functions which were associated with the decision-making process were incorporated into the expert system, and others were implemented as independent objects. Then the conceptual form of the shop control structure was developed and is given in Figure 2.

At the top layer of shop control, the central layer consists of a main driver and an expert system. The function of the main driver is to initialize the control system. The expert system performs the following functions: shop scheduling and routing, bill of material processors, machinability data system, shop order flow system, machine performance monitoring, in-process inventory management, and material storage monitoring. Using object-oriented problem solving methodology to solve a problem requires that the problem space be mapped onto a set of objects and modules in the solutions space. The problem space for the shop system includes the manufacturing planning and control functions and the functions which handle the interfaces between the shop controller and workcell controllers. The system objects are the following:

1. Inventory management;
2. Order management;
3. Production report;
4. Parts program management;
5. AS/RS management;
6. Interface to the machining cell (MCS);

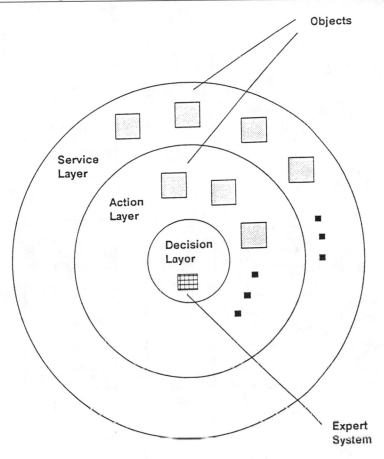

Figure 1. Graphical illustration of the three-layer integrated method.

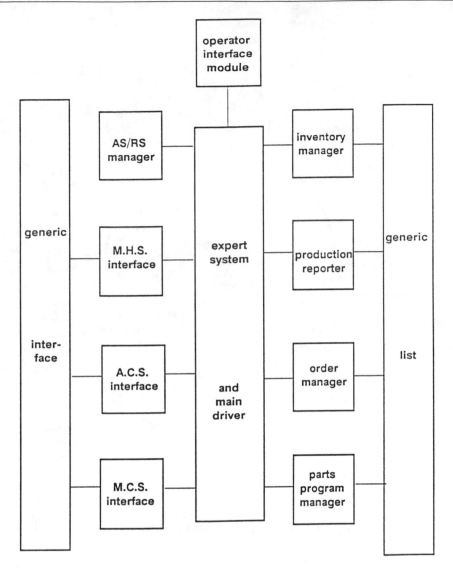

Figure 2. The conceptual design of the shop control system.

7. Interface to the assembly cell (ACS);
8. Interface to the material handling system (MHS); and
9. Operator interface.

All of these software objects constitute the middle layer of the control methodology. Only two major entities, generic list and generic interface functions, were put into the outer layer. Two classes are declared for these two entities, and the objects of these two classes composed the outer layer.

Decision-Making Function

The main function of the shop control system is to schedule and route parts through the production system according to the available information about the state of the system. This function was implemented through an expert system. The expert system was meant to put human expertise into a series of production rules and then to utilize these rules and knowledge of the state of the system to make decisions. This kind of expert system is often called a rule-based system. The expert system has three main parts: working memory, production rules, and rule interpreter.

Working memory is the part of the expert system which represents a particular problem situation (system state). This part provides information about machine availability, local storage, AS/RS status, and the order queue and is represented by integers and integer arrays.

The production rules are the system representation of the human expertise. The rule base is represented in a simple IF-THEN form, and it is written in C language. All rules in the knowledge base can be put into three main groups: operator interface rules, system maintenance rules, and system operation rules. Operator interface rules relate to the events which require an operator interface. Whenever an action requires an operator interface, one of these rules will be fired, and the control system is put into a waiting state. System maintenance rules are used to maintain the control system. For example, the system has to save status periodically to prevent accidental power off. A product report should be generated at the end of the day or by request. If the order buffer is empty, the control system may go to the order queue to get an order. The rules in this group and the group above have one thing in common. If one of these rules is fired, the system does not end the control cycle and it continues to search for matching rules in the third group. System operation rules are associated with machine operations like milling or assembling on shop floor. These rules are the part of the knowledge base which do the scheduling and routing.

In this system, the rule interpreter has three operations to perform. First, it interprets the feedback messages from cells and the human operator, and updates the working memory. Second, it selects rules to apply by the matching technique. If several rules have their conditions satisfied, these rules are all selected. Third, if more than one rule is selected, the rule interpreter resolves conflicts by selecting one rule for activation. Two strategies are used to resolve conflicts. Rule ordering is the strategy in which the rule that occurs earliest in the rule base has highest priority. The other is context limiting in which the rules are put into groups and a procedure for activating and deactivating groups is provided.

Action Objects

The middle layer of the shop controller is the action layer which contains objects which serve the decision layer directly. The objects in the middle layer provide interface between the expert system and the shop floor. And they also do bookkeeping and computations. If algorithmic knowledge is added to this system, it can be translated into objects which reside in this layer. These action objects can use objects in the outermost layer to do some specific operations. A class has been declared for each action object and will be described separately in the following section.

The inventory manager class contains all the information about raw materials before entering the production process. This includes component type, quantity, minimum ordering level, and maximum storage level. The in-process inventory management function is handled by the expert system in the decision layer.

The customer order is the basic driving force of the production system. "No order" means no production. The class order manager is developed to process customer orders. The main responsibility of it is to maintain a prioritized list of all active customer orders for input to the production control functions.

The class production reporter contains all the information about the products produced. The main task of this class is generating a production report which shows the finished orders and the total numbers for each product.

Class parts program manager has the responsibility of managing all part programs for the machines in the system. It may help to load new programs, remove the old ones, and to provide information about programs in storage. In this system only one object is used to manage all parts programs. An alternative approach is to declare an object of this class for each cell.

The class AS/RS manages the automatic storage and retrieval system. There are 124 storage positions in the rack. Several different types of pallets can be stored on the rack. So the pallet name is used as an identifier for storing and retrieving. Two rules are used in the management. First, whenever a pallet enters the storage system, the AS/RS manager files the first empty position which is near the entry point of the conveyor. Second, whenever there is need to withdraw a pallet, the AS/RS manager finds one near the entry point.

Class MCS, ACS, and MHS are interface modules which carry the message back and forth between the shop controller and the cell controllers. MCS stands for machining cell system, which handles the Assembly Cell interface, and MHS stands for material handling system, which handles the interface of the material handling system. Since these functions have something in common, that is, basic I/O functions, a generic interface class was designed. The three cell interface classes are declared to be the subclasses of the generic interface class.

The main function of class operator interface is to translate the messages between the shop controller and the operator. There are three types of interaction between the operator and the shop controller. One is supervising activity such as order input, another is maintaining activity, and the last is loading activity.

Service Objects

Service objects are those objects in the outermost layer of the control system model. In this shop controller, only two kinds of objects are grouped into this category. Two classes are declared for each kind. One of them is class generic list. The reason to declare the generic list as a class is that it is used by many other objects such as the order manager, the production reporter, the inventory manager and the part program manager.

The class generic interface is used by each of the three cell interface classes in the middle layer. Basically what the three cell interface classes do is the message I/O function, but each class has its own private data. The declaration of the generic interface class simplifies the development of the three cell interface classes. With the generic interface class the program developer can concentrate on message processing rather than on both the message processing and I/O function.

It is important to notice that class parts program management and class generic interface are designed specifically for the AT&T STARLAN network. These two classes may not work with other networks. But all the other classes are portable.

CONCLUDING REMARKS

The CIM approach has emerged as a leading alternative for the manufacturing companies to compete and survive in the world market; the success of the approach depends largely on the design of the shop control software. In this paper we presented a systematic approach for designing CIM system control software through an example. Some common techniques like group technology, hierarchical philosophy are incorporated in this approach. Using the systematic approach, it starts with the analysis of the physical facility layout and the communication equipment necessary for the information integration. Then the required manufacturing functions, communication functions, and other related functions are determined. Next, using a hierarchical philosophy, the system control structure is designed. Finally after analyzing the required manufacturing functions for a specific level in the system control hierarchy, the control system for the underlying level can be designed using the three-layer integrated method of any other approaches.

A strategy for the integration of a variety of manufacturing processes and functions is also devised. This strategy is the three-layer integrated methodology which can be used to design the control system for any level in the system control hierarchy. This method has several advantages:

1. Software intelligence. By its very nature, a CIM system can provide a vast amount of data regarding the past and present states of the system. The problem of a system controller is how to use the data in controlling the system. The three-layer integrated approach attempts to include human expertise in the system via the central layer. Adaptation and modification of this layer can be used to improve system performance.

2. Modularity. The production functions of the middle and third layers of the control system are modeled as a set of objects. Each object is independent and the expert systems are used to integrate these objects into a working unit. Hence, it is easy to modify the control system structure designed with this approach. For

example, when a new cell is added, the programmer will need to design a new interface object for the new cell using the generic I/O class, and change some rules in the expert system to accommodate the changes. Changes to other parts of the code would be minimal, if any.

3. Reusability. Program reusability is enhanced by object-oriented programming because of the use of the encapsulation concept. With OOP, the programmer needs to understand the behavior of a class as specified by the methods; the programmer does not have to worry about its implementation. Thus, if a class exists which can perform the desired functions, the programmer can incorporate it into his/her software in a new situation.

4. Maintainability. Program maintainability is enhanced by OOP. Changes in data and procedures can be localized to the classes or objects which implement the data and procedures. If the class can keep the existing interface format, there will be no fall-out effects on other data and procedures in the software system.

The drawback of the three-layer integrated approach is the size of the software system. Using OOP makes the software code larger than using a more traditional method. This problem can be remedied by using efficient languages such as C, or by keeping the production unit the controller manages small. Additionally, as the speed and size of the control computers increases, the effects of this disadvantage are reduced.

REFERENCES

1. Teicholz, E, "Computer Integrated Manufacturing," *Datamation*. Vol. 30, No. 3, March, 1984, pp. 169-174.

2. McLean, C., M. Mitchell and E. Barkmeyer, "A Computer Architecture for Small-Batch Manufacturing," *IEEE Spectrum*. Vol. 20, No. 5, May 1983, pp. 59-64.

3. DEC/Philips, *Reference Model of Production Systems*. Digital Equipment Corporation and Netherland Philips, Bedrijven BV, 1987.

4. Albus, J.S., C.R. McLean, A.J. Barbera and M.L. Fitzgerald, "An Architecture for Real-Time Sensory-Interactive Control of Robots in a Manufacturing Facility." *Proceedings of the 4th IFAC Symposium on Information Control Problems in Manufacturing Technology*. Gaithersburg, MD, Oct. 26-28, 1982.

5. Fox, M.S., B. Allen and G. Strohm, "A Computer Aided Decision System." *Management Science*. Vol. 15, No. 10, June 1969, pp. 550-561.

6. Wiener, R.S. and L.J. Pinson, *An Introduction to Objective-Oriented Programming and C++*. Addison Wesley, Reading, Massachusetts, 1986.

7. Zhang, Y., "An Integrated Intelligent Shop Control System," Master Thesis, Department of Industrial Engineering and Operations Research, Virginia Polytechnic Institute and State University, 1989.

8. Arai, Y., S. Hat, T. Imakubo and K. Kikuchi, "Production Control System of Microcomputers Hierarchical Structure for FMS." *Proceedings of the 1st International Conference on Flexible Manufacturing Systems*, Brighton, UK, Oct. 20-22, 1982, pp. 259-268.

9. Ben-Arieh, D., "Knowledge Based Control System for Automated Production and Assembly," Ph.D. Thesis, Department of Industrial Engineering, Purdue University, 1985.

10. Charniak, E. and D. McDermont, *Introduction to Artificial Intelligence*. Addison Wesley, Reading, Massachusetts, 1985.

11. Cox, B.J., *Objective-Oriented Programming*. Addison Wesley, Reading, Massachusetts, 1986.

12. Gershwin, S.B., R.R. Hildebrant, R. Suri and S.K. Mitter, "A Control Perspective on Recent Trends in Manufacturing Systems." *IEEE Control System Magazine*. Vol. 6, No. 2, April, 1986, pp. 194-199.

13. Ham, I, K. Hitomi and T. Yoshida, *Group Technology*. Kluwen-Nijhoff Pub., Boston, 1985.

14. Jones, A.T. and C.R. McLean, "A Cell Control System for the AMRF." *Computers in Engineering*. Vol. 2, The Society of Mechanical Engineers, 1984.

15. Mesarovic, M.D., D. Macko and Y. Takahara, *The Theory of Hierarchical Multilevel Systems*. Academic Press, New York, 1970.

16. Nau, D.S., "Expert Computer Systems." *Computer*. Vol. 16, No. 2, Feb. 1983, pp. 63-85.

17. Rolston, D.W., *Principles of Artificial Intelligence and Expert Systems*. McGraw-Hill, New York, 1988.

18. Stroustrup, B., *The C++ Programming Language*. Addison Wesley, Reading, Massachusetts, 1986.

A METHODOLOGY FOR INTEGRATING SENSOR FEEDBACK IN MACHINE TOOL CONTROLLERS

Frederick M. Proctor, John L. Michaloski, and Thomas R. Kramer

National Institute of Standards and Technology, Gaithersburg, Maryland, U.S.A.

ABSTRACT

A reference model architecture for real-time hierarchical control systems has been proposed by researchers at the National Institute of Standards and Technology, and has been implemented on a variety of computing platforms for manufacturing and vehicle control applications. A fundamental aspect of this architecture is the notion of nested control loops, which incorporate sensory feedback in a hierarchy whose cycle times decrease in frequency as planning becomes more abstract. The nested control loops provide a hierarchy in which to model command and control. This architecture was formalized during work with the National Aeronautics and Space Administration on the Flight Telerobot Servicer for the space station, and is known as the NASA/NBS Standard Reference Model, or NASREM. Although NASREM was intended to serve as a guideline for space application robot control, it has applicability to a wide range of real-time control applications. This paper adapts the NASREM reference model architecture to a machine tool control model. A computational architecture will be presented that describes expected behavior at each layer. A functional analysis will outline a baseline task tree vocabulary. The task tree vocabulary is given by a set of command verbs for each layer and is a critical component of task description within a hierarchical control system.

INTRODUCTION

The accuracy of a machine tool can be greatly increased through the use of high-resolution position sensors such as glass slides, or calibration of components such as lead screws or gears. Unfortunately, much of the inaccuracy in finished parts is due to dynamically varying quantities, such as tool wear, chatter, or thermal expansion, which cannot be predicted in advance. Methods have been developed which rely on sensors to measure these quantities in real time as the part is being machined, and modify the position of the machine tool to compensate for these disturbances. Incorporating these methods into machine tool control brings benefits in failure prediction, surface finish improvement, and accurate machining, all of which improve the quality and reduce the cost of manufacturing. The limit to the effectiveness of these methods lies not with the engineering principles, but in the practical effort required to interface to proprietary machine tool controllers with closed architectures.

The answer is an open architecture that integrates sensor feedback into the control structure. At NIST, an architecture for integration of sensor and control has been applied to numerous projects. This paper will review the architecture as outlined in the NASA/NBS Standard Reference Model for Telerobotic Control System Architecture (NASREM) [1] as it applies to machining. NASREM describes a system architecture to handle the specific responsibilities of the Space Station Flight Telerobot (FTS), a multi-armed manipulator intended to assist service and construction of the Space Station Freedom. NASREM proposes a hierarchical control methodology to decompose FTS functionality from high-level Space Station directives into low-level physical actions. NASREM is not restricted to space robotics. It is a general real-time control model and architecture, adaptable to a broad range of applications.

The goal of this paper is to present canonical or base-class vocabularies of machine tool commands stratified along levels of operation. The stratification of operation subscribes to the NASREM notion of hierarchical, feedback control systems. The canonical vocabularies drew from machining experience from the Automated Manufacturing Research Facility (AMRF) at the National Institute of Standards and Technology (NIST). The AMRF serves as a testbed for developing techniques and standards for automated manufacturing. The paper is organized as follows. The first section will describe the basic NASREM architecture and present the enabling architectural design concepts. The second section will study the concept of NASREM-style commands and content of command vocabularies. The third section will present canonical commands for a machining hierarchy. Finally, an example of the sensor-integrated machining architecture will be presented.

BACKGROUND

NASREM defines an application control system as a hierarchical collection of sensory-interactive, controller nodes. These controller nodes reflect the fundamental aspect of the replicated architectural structure. A *controller node* is composed of sensory processing (SP), world modeling (WM), behavior generation (BG) and appropriate human interface (HI) components. Behavior generation determines control activity and sends either actuator signals or commands to subordinate controller nodes to effect that activity. Sensory processing obtains and processes feedback from system sensors and subordinate controller nodes. World modeling interprets sensory processing data to maintain an internal model of the world. World modeling is the part of the system which mediates sensory processing and behavior generation activities. The key enabling architectural design concepts are:

- *hierarchical organization*—levels of control are derived from a hierarchical decomposition of control functionality, and hierarchical composition of sensory-feedback knowledge.
- *well-structured*—all controller nodes have the same structure and data formats
- *cyclic execution*—node execution provides a predictable model using feedback of command and status.
- *inherently concurrent*—nodes are concurrent. NASREM controller nodes support concurrent threads of execution.
- *sensory-interactive*—closed loop control is possible. SP external world sensing in conjunction with WM internal predictions provide robust feedback.
- *one master rule*—all subordinate control processes obey a single superior control process at each instant in time. A superior may control multiple subordinates. This gives a hierarchical control tree decomposition.

Of great importance to sensor-based machining is the concept of sensor and control integration. NASREM integration of control and sensor feedback comes out of the tradition of servo mechanisms and state-space control. Each of the controller nodes is a sensory-interactive servoed feedback loop. Each controller node accepts task commands that define goals (i.e., set-points, or attractor sets). Each controller node regularly samples sensors, computes the state of the world, and generates actions designed to reduce the difference between the current state of the world and the goal state. The overall architecture is a multilevel, hierarchical, nested control system. An implementation would typically employ periodic "servoing" or data sampling as opposed to "event-driven" interrupt processing in order to cut down on processing overhead and at the same time ensure deterministic and verifiable behavior (in terms of execution time and response time) particularly at the lowest levels of the architecture. At higher levels of an implementation—and lower performance bandwidths—it is often convenient to transition to an asynchronous technique.

The fundamental NASREM design principle is the use of hierarchical task decomposition with stepwise refinement to reduce a larger problem into more elementary steps, as well as bottom-up aggregation of controller nodes based on equipment composition. Hierarchical aggregation of control nodes leads to the definition of a *control level* as the collection of controller nodes operating at the same spatial and temporal level of execution. NASREM decomposes control actions into levels both temporally and spatially. Temporally, each successively higher level depends on completion of a lower level task, much like increasing digits of magnitude on an odometer. Spatially, levels are similar to different magnifications of a microscope—lower levels have a higher degree of magnification, but observe less of the total picture.

As an example of this architectural structure, consider the hierarchy shown in Figure 1. This diagram depicts an abridged view of a machining cell consisting of a workstation containing a machine tool and a robot consisting of one arm with a simple gripper, and a camera with pan, tilt, zoom, focus and iris control. Applying NASREM stratification leads to a hierarchy of six levels. The path from the cell level to the robot arm in this hierarchy, for example, implements the following functionality:

- Level 6: CELL decomposes the factory orders into a sequence of workstation action commands.
- Level 5: WORKSTATION decomposes actions to be performed on batches of parts into tasks performed on individual objects.
- Level 4: TASK decomposes actions applied to object task commands into sequences of elemental moves defined in terms of motions and goal-points.
- Level 3: ELEMENTARY MOVE (EMOVE) decomposes elemental move goal-points into paths.
- Level 2: PRIMITIVE (PRIM) decomposes paths into smooth, dynamic trajectories.
- Level 1: SERVO transforms trajectories into coordinated joint motion space, using either position, velocity or force parametrization.

NASREM COMMAND VOCABULARY

Controller processes within the NASREM hierarchy communicate through message passing. These messages are defined by a language that will be referred to as the neutral messaging language, or simply NML[*]. NML is not a programming language; it is a type of non-procedural language. Further, NML is not a general purpose grammar. Instead, a project specific language is developed that defines the set of messages into and out of each controller node. Of course, projects that share many characteristics could in fact use an identical NML. Thus, dialects of NML can be developed for all or portions of a specific control applications. As an application language example, NASA has developed vocabularies for space robotics [2]. As a portion of a control application example, a PRIM controller node may communicate to a SERVO controller node with the NML defined by the RS-274 machine tool programming language.

The method to develop the grammars is to analyze the functionality of each controller node and develop verbs which describe appropriate commands. The set of NML verbs defines controller node input commands or task names. An NML message is then a verb together with a list of controller node attribute values defining the task goal, object, and parameters. Command names should use the active connotation of a verb since a verb has a rather descriptive and characterizing quality of the intended action. Hence, command names

---

Compare with the Neutral Manufacturing Language specified by the Next Generation Controller program. The Neutral Messaging Language specifier was chosen to evoke the connotation of the Neutral Manufacturing Language, generalized as a messaging language to more accurately reflect its suitability for control in non-manufacturing applications, such as vehicle control.

will be termed as the *command* or *task verb*. The collection of all task verbs in a controller node will define a *task vocabulary*. Other vocabularies for status and world model communication are also necessary, but will not be covered in this paper.

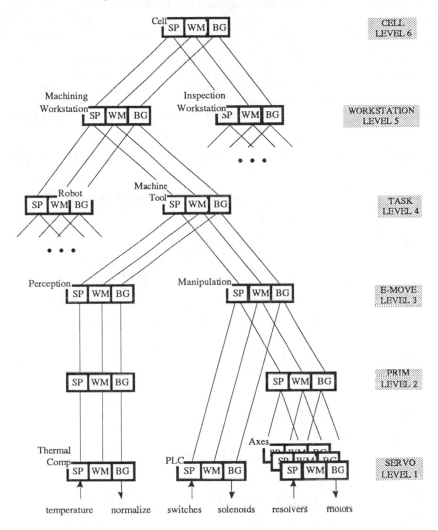

Figure 1. Machining Cell Control Hierarchy

Each controller node should have a baseline set of verbs describing the atomic actions of that control node. Verbs should be unambiguous and descriptive. Thus, verbs such as "do," which don't provide the necessary level of description, should be avoided. Defining a controller node task vocabulary with a set of verbs is a fuzzy linguistic process [3]. One must match the needs of the task to the capabilities of the controller node. What makes the linguistic process fuzzy is that there exist numerous equally valid grammars that can define the task vocabulary. For example, one can specialize the task vocabulary to match specific project goals or one can use a smaller, general vocabulary that assumes extra steps. For example, one could have the task verb CHANGE-OUT, if this were a specific need of the project, or one could use more general verbs, such as REMOVE and INSTALL. In the second case, a CHANGE-OUT can be accomplished with a composition of REMOVE and INSTALL tasks.

To define controller node vocabularies, one must stratify functionality into the NAS-REM hierarchy. Some functionality transcends levels within the hierarchy. This leads to the concept of an internode function or task. For example, all levels of the hierarchy which exhibit motion could have the verb MOVE. Redundancy of the verb MOVE across levels is necessary to capture the different time and spatial capabilities. To illustrate the stratification of motion control within the functional model, a partial verb vocabulary for a single-chain of controller nodes is presented in Figure 2.

In this diagram, TASK communicates task commands to EMOVE with the verbs AP-PROACH, DEPART, MOVE, and GRASP. APPROACH and DEPART are obviously moves, but imply some special parametrization that is required as one nears contact. Further, the PRIM to SERVO interface vocabulary contains only a single command verb. Different motion requirements in this interface do not translate nicely into different verbs, so that a single verb with different sets of parameters is required. Each level issues new commands based on feedback from the lower level. Such new commands and parametrization can be issued upon a simple acknowledgment of task completion or a complicated adaptive control modification based on sensed feedback.

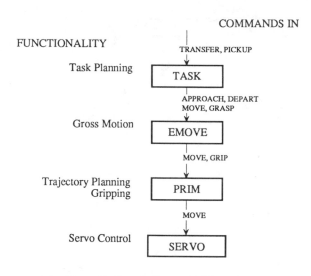

Figure 2. Sample Command Hierarchy

CANONICAL COMMANDS FOR 3-AXIS MACHINING

For the general case, a CELL would be capable of making parts itself and would contain both machining and inspections workstations. A turning, vertical machining, and inspection workstation, for example, might comprise a CELL. A single part might be routed through all three workstations. Each WORKSTATION (Level 5) would contain one or more pieces of equipment, each of which would have a controller at the TASK level (Level 4). The machining workstation of Figure 1 is typical of this view. The activity of a workstation would be to issue commands to subordinate equipment to get it to transform the workpiece according to the capabilities of the workstation.

For a machining workstation (vertical or horizontal, and any number of axes) the command to the workstation would result in the partial or complete machining of a part, starting from a piece of stock or a partially machined workpiece, and ending with a partially machined workpiece or a completely machined part. The vocabulary of CELL and WORK-STATION commands greatly depends on allocation and scheduling strategies that will not be addressed in this paper. Instead, command vocabularies will be presented for TASK and subordinate controllers that are invariant of time and scheduling considerations.

A given controller might decompose a command from a superior by simply taking a pre-built process plan off the shelf and constructing commands from the process plan, or it might plan and build commands in real time. It is expected that a command would be very similar to the step of a process plan from which it was built. The command would have a variety of information relating to timing and sequencing not found in the process plan, and some process plan information (such as expressions evaluated by the controller to get explicit values) would not be contained in the command.

A command might be a macro that decomposes into a series of commands for the same controller. In machining, this is already done with existing controllers in implementations of RS-274-D, for example, (commands g80 through g89 are macros which decompose into g0, g1, g2, g3 and other elementary commands). The discussion of canonical vocabularies will concentrate on commands for making parts and not with commands for setting up or taking down a hierarchy of controllers, dealing with error conditions, etc. Some method for doing these tasks would, of course, be required but is out of the scope of this paper. These issues might or might not be handled similarly to the handling of manufacturing tasks.

The remainder of this section will outline a list of proposed canonical commands for a 3-axis machining center. Most of these commands should be suitable for machining centers with more than three axes. Commands at Level 4 (TASK), Level 3 (EMOVE), and Level 2 (PRIMITIVE) of the NASREM hierarchy are given. For each command, the name of the command and a description of the effects of the command are given. Parameters are given in italics for Level 4 and Level 3 commands. Parameters for Level 2 commands are generally required for each command and are included in a follow-up discussion on control and data parametrization.

TASK Level

Each workstation would have its own component equipment, and each would require its own set of task level commands. In a machining workstation, there might be controllers for a robot which would load parts, unload parts, move tools in and out of the machining center, and adjust passive fixtures. There may also be separately controllable equipment, such as a powered adjustable fixture. For machining, the task level is responsible for interpreting the geometry of a part and then generating a process plan to machine that part.

The REMOVE_VOLUMES (*plan_id, design_id, material_removal_volumes_id, setup_id, workpiece_id, fixture_id*) command causes a set of material removal volumes to be removed from a workpiece. The command parameters include a process plan identifier (*plan_id*), a design identifier (*design_id*), an identifier for the set of material removal volumes referenced in the program (*material_removal_volumes_id*), an identifier for setup instructions (*setup_id*), an identifier for the workpiece that will be machined (*workpiece_id*), and an identifier for the fixture to be used (*fixture_id*). This command is used to do the work done in a single fixturing of a workpiece. The workpiece may start as a piece of stock or as a partially machined workpiece. The workpiece may end as a completely machined part or as a partially machined workpiece.

EMOVE Level

The elemental move level is the transition point between the abstract notion of geometrical part description and the physical notion of machining. EMOVE is responsible for feature interpretation and generation of tool or motion paths to achieve removal of some volume of material. Each machine tool requires its own geometric and physical model of volume removal and tool path generation.

The BORE (*tool_type_id, material_removal_volume, spindle_speed, feed_rate*) command results in a hole being bored. The cutter must be a boring tool.

The CENTER_DRILL (*tool_type_id, material_removal_volume, spindle_speed, feed_rate*) command results in a small starter hole being made with a center drill cutter by a single-stroke plunge into the material.

The COUNTERBORE (*tool_type_id, material_removal_volume, spindle_speed, feed_rate*) command results in an existing hole being enlarged.

The END_PROGRAM (*no parameters*) command indicates the end of a program has been reached. It may cause activities such as spindle retract, return to home position, cleanup of the world model, resetting machine parameters, etc.

The FACE_MILL (*tool_type_id, material_removal_volume, spindle_speed, feed_rate, pass_depth, stepover*) command results in the material removal volume being machined away by a face mill cutter.

The FINISH_MILL (*tool_type_id, material_removal_volume, spindle_speed, feed_rate, stepover*) command results in the removal with a finish end mill (with cutter nose geometry suitable for the material removal volume) of any material in the material removal volume, so that the resulting surfaces meet some desired quality specification. Only a small thickness of material should be removed in this operation. The *stepover* parameter is required for milling with the flat portion of the nose of the cutter, where an area larger than the area of the nose of the tool is being finished.

The FLY_CUT (*tool_type_id, material_removal_volume, spindle_speed, feed_rate, pass_depth, stepover*) command results in the material removal volume being machined away by a fly cutter.

The INITIALIZE_PROGRAM (*program_name, design_id, material_removal_volumes_id, setup_id, workpiece_id, material, fixture_id, program_x_zero, program_y_zero, program_z_zero*) command initializes the controller to be ready to accept additional Level 3 commands, all of which, up to an END_PROGRAM command, are logically parts of a single program for machining a single workpiece using a single fixture. The command identifies the name of the program (*program_name*), a design identifier (*design_id*), an identifier for the set of material removal volumes referenced in the program (*material_removal_volumes_id*), an identifier for setup instructions (*setup_id*), an identifier for the workpiece that will be machined (*workpiece_id*), the name of the type of material being machined (*material*), an identifier for the fixture to be used (*fixture_id*), and the location of the program zero in machine coordinates (*program_x_zero, program_y_zero,* and *program_z_zero*). This command causes no motion in the machining center. This command is not used to bring the machining center task controller to a ready state from a cold start; that must be done before an INITIALIZE_PROGRAM command is given.

The MACHINE_CHAMFER (*tool_type_id, material_removal_volume, spindle_speed, feed_rate*) command results in an edge being chamfered. The cutter must be a chamfer tool (tool profile is a cone, possibly truncated).

The MACHINE_COUNTERSINK (*tool_type_id, material_removal_volume, spindle_speed, feed_rate*) command results in a hole being countersunk with a countersink cutter.

The MACHINE_ROUND (*tool_type_id, material_removal_volume, spindle_speed, feed_rate*) command results in an edge being rounded. The cutter must be a rounder (side of tool profile is an arc of a circle).

The PERIPHERAL_MILL (*tool_type_id, material_removal_volume, spindle_speed, feed_rate, pass_depth, stepover*) command is for milling an exterior or interior contour by milling at the periphery only. Unlike ROUGH_MILL, it may not contain any plunging or slotting. The cutter must be an end mill (with nose geometry suitable for the material removal volume).

The REAM (*tool_type_id, material_removal_volume, spindle_speed, feed_rate*) command causes a small amount of material to be removed from the inside of an existing hole. The cutter must be a ream. The material cut away must be a very small thickness around the surface of the material removal volume.

The ROUGH_MILL (*tool_type_id, material_removal_volume, spindle_speed, feed_rate, pass_depth, stepover*) command results in the milling with an end mill of the designated material removal volume. It is expected that requirements on the surfaces created by this operation will be such that no consideration needs to be given to surface quality in determining machining methods. The operation may include plunging, slotting, and both conventional cutting and climb cutting. The cutter which is used may be a rough end mill or a finish end mill (with nose geometry suitable for the material removal volume).

The SET0_CENTER (*tool_type_id, near_x, near_y, x_offset, y_offset, near_diameter*) command results in a probe cycle being run in which a hole with its axis parallel to the z-axis is probed. Program x_zero and y_zero are set at the center of the hole or by offsetting from the center. The tool must be a probe.

The SET0_CORNER (*tool_type_id, near_x, near_y, x_offset, y_offset, corner_type*) command results in a probe cycle being run in which a corner is probed. The corner must be formed by two planes parallel to the z-axis. Program x_zero and y_zero are set at the corner or by offsetting from the corner. The tool must be a probe.

The SET0_Z (*tool_type_id, x_location, y_location, z_offset*) command results in a probe cycle being run in which a surface parallel to the xy-plane is probed. Program z_zero is set at the surface or by offsetting from the surface. The tool must be a probe.

The SLOT_MILL (*tool_type_id, material_removal_volume, spindle_speed, feed_rate, pass_depth*) command results in a slot being milled. The shape of the material removal volume will be such that it may be produced by having the tool follow a path (simple or complex) in which the tool will generally be cutting across its full width to form a slot.

The TAP (*tool_type_id, material_removal_volume, spindle_speed, feed_rate*) command results in the inside of an existing hole being threaded. The cutter must be a tap.

The TWIST_DRILL (*tool_type_id, material_removal_volume, spindle_speed, feed_rate, pass_depth*) command results in a hole being drilled. The pass_depth parameter is used if the user's TWIST_DRILL strategy is to perform peck drilling. It may be desirable to split this command into three commands: PLUNGE_DRILL, PECK_DRILL, and SMALL_HOLE_DRILL.

PRIMITIVE Level

The PRIMITIVE level is responsible for generating a time sequence of closely-spaced goal states from a static description of a desired motion. As such, it generates primitive trajectories or motion profiles. The output commands of the EMOVE level are time-independent descriptions of motions, for example static position or position and force paths, or directional fields. In the first case, the position and force commands are in the form of parametrized paths to be followed. In the case of a directional field description, the command takes the form of position-dependent fields which indicate the desired direction of motion or force application. In addition to these types of commands, EMOVE can also simply specify a set of termination conditions, or goal states, along with an algorithm specification which determines the strategy to be used to achieve them. This type of command is useful for sensory-interactive algorithms.

The ARC_FEED (*first_axis_coordinate, second_axis_coordinate, rotation, axis_end_point*) command describes a move in a helical arc from the current location at the existing feed rate. The axis of the helix is parallel to the x, y, or z axis, according to which one is perpendicular to the selected plane. The helical arc may degenerate to a circular arc if there is no motion parallel to the axis of the helix. If the selected plane is the xy-plane, *first_axis_coordinate* is the axis x-coordinate, *second_axis_coordinate* is the axis y coordinate, and *axis_end_point* is the z-coordinate of the end of the arc. If the selected plane is the yz-plane, *first_axis_coordinate* is the axis y-coordinate, *second_axis_coordinate* is the axis z-coordinate, and *axis_end_point* is the x-coordinate of the end of the arc. If the selected plane is the xz-plane, *first_axis_coordinate* is the axis x-coordinate, *second_axis_coordinate* is the axis z coordinate, and *axis_end_point* is the y-coordinate of the end of the arc. The *rotation* parameter represents the number of degrees or radians in the arc. *Rotation* is positive if the arc is traversed counterclockwise as viewed from the positive end of the coordinate axis perpendicular to the currently selected plane. The radius of the helix is determined by the distance from the current location to the axis of the helix.

The DWELL (*duration*) command indicates a pause in motion for the amount of time specified by *duration*. The ability to dwell is useful in finish cutting.

The PARAMETRIC_2D_CURVE_FEED (*first_function, second_function, start_parameter_value, end_parameter_value*) command describes a move along a parametric curve in the selected plane. We will call the parameter *u*. If the selected plane is the xy-plane, *first_function* gives x in terms of *u* and second_function gives y in terms of *u*. Analogous assignments are made if the selected plane is the xz-plane or the yz-plane. Allowable functions should include at least cubic polynomials and elementary trigonometric functions. The current position of the spindle must be at coordinates corresponding to the *start_parameter_value*. The final position of the spindle is at coordinates corresponding to the *end_parameter_value*.

PARAMETRIC_3D_CURVE_FEED (*x_function, y_function, z_function, start_parameter_value, end_parameter_value*) describes a move along a parametric curve in three dimensions, where x, y, and z are each functions of the parameter *u*. Allowable functions should include at least cubic polynomials and elementary trigonometric functions. The current position of the spindle must be at the coordinates corresponding to the *start_parameter_value*. The final position of the spindle is at the coordinates corresponding to the *end_parameter_value*.

The SPINDLE_RETRACT (*no parameters*) command describes a retract at traverse rate to fully retracted position.

The STRAIGHT_FEED (*x, y, z, probe*) command describes a move in a straight line at existing feed rate (or using the existing z-force) from the current point to the point given by the *x*, *y* and *z* parameters. If z-force is enabled, the values of *x* and *y* must be the same as those of the current point. The *probe* parameter is optional. If the *probe* parameter is present, the feed motion will stop when the probe is tripped or when the endpoint is reached, whichever happens first.

The STRAIGHT_TRAVERSE (*x, y, z*) command describes a move in a straight line at traverse rate from the current point to the point given by the *x*, *y* and *z* parameters.

Other PRIM Physical Activities   The FLOOD_OFF, FLOOD_ON, MIST_ON, and MIST_OFF commands enable or disable flood or mist coolants. The START_SPINDLE_CLOCKWISE and START_SPINDLE_COUNTERCLOCKWISE commands turn the spindle in the appropriate direction at the currently set speed rate. The STOP_SPINDLE_TURNING command stops spindle rotation. The LOCK_SPINDLE_Z command locks the spindle against vertical motion. The UNLOCK_SPINDLE_Z command unlocks the spindle to permit vertical motion. The SET_SPINDLE_FORCE (*force*) sets the force with which the spindle is pushed in the z-direction, potentially useful in tapping operations. This command also specifies whether or not to use the spindle force. The CHANGE_TOOL (*tool id*) command indicates a tool change. The TURN_PROBE_ON and TURN_PROBE_OFF commands turns the machine probe on or off. The ORIENT_SPINDLE (*orientation, direction*) command turns the spindle to the given orientation and direction at the current spindle speed, then stops the spindle.

PRIM Data and Control Parametrization   In a normal machining operation, the execution is open-loop: a series of commands are issued by one level and interpreted at a lower level. This execution mode does not accommodate servo update parametrization. For example, it may be desirable to specify the feed rate with each commanded motion path every servo update. Traditional parametrization prefers to minimize communication, and only send this parameter once with a set command. However, to effect the servo-update mode of command feedback, all relevant parameters should be included within every command, with default parametrization established with the SET command. This leaves the following parameters as either command parameters or isolated set data commands:

FEED_RATE (*feed*) sets the feed rate that will be used when the spindle is told to move at the currently set feed rate. TRAVERSE_RATE (*rate*) sets the traverse rate that will be used when the spindle traverses. SPINDLE_SPEED (*speed*) sets the spindle speed that will be used when the spindle is turning. CUTTER_RADIUS_COMPENSATION (*radius*) sets the radius value to be used in cutter radius compensation, and enables or disables cutter radius compensation when executing spindle translation commands. The PROGRAM_ORIGIN (*origin*) parameter specifies an absolute or relative origin. TOOL_LENGTH_OFFSETS (*offset transformation*) parameter specifies normal, modified or no tool length offsets. SPEED_FEED_SYNCHRONY imposes or cancels the requirement that feed and speed rates be synchronized exactly.

EXAMPLE

This section describes a prototype architecture for sensory-interactive machining. Figure 3 illustrates the architecture consisting of 4 levels of controller nodes. A job in this architecture performs the following steps. First, a command to manufacture a single part or batch of parts is sent to the TASK level, which interprets the part CAD boundary representation to produce a Constructive Solid Geometry (CSG) model of the part. With the CSG model, part features are extracted and a process plan is generated to cut each feature. Then, part features from the process plan are passed to the EMOVE level to determine the proper machining paths. Feature path geometries are sent to the PRIMITIVE level, which generates path segments. The SERVO level interprets these path segments to produce actuator signals. As commands cascade down the BG leg of the hierarchy, feedback percolates up the SP hierarchy. This section will review the use of sensor feedback at applicable levels of operation [4]. As status feedback filters up the hierarchy, a superior may use lower-level feedback to alter its control strategy. For example, EMOVE may use lighter cuts if the SERVO senses tool vibration.

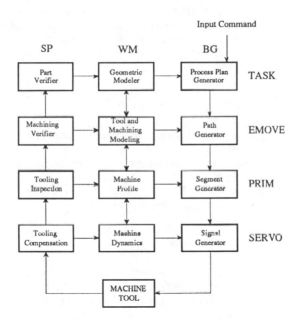

Figure 3. Machining Functional Hierarchy

The TASK level will analyze geometric models of workpieces, fixturing and tools. A typical command is to take the model of an "as-is" blank workpiece and generate a process plan describing feature volumes to be removed that will generate the "to-be" model of the workpiece. Another responsibility will be to test whether the swept volume of a tool following a path intersects the volume occupied by fixturing. For sensor feedback, the TASK level monitors the progress of the job to check deadlines, and expected worksystem failures. For example, if a machined part has exceeded some overall tolerance limit, then the part is discarded and a new blank part and alternative process plan must be undertaken.

The EMOVE level will interface the feature-based machining to the physical machine tool. EMOVE will do Tool Path BG to generate the tool path required to carry out a step from a process plan. It will also identify canned cycles (such as peck drilling) if that is appropriate to the plan. The Tool Path Generator will undertake geometric analysis of the cutting situation in order to generate efficient paths. For example, it will find entry points for peripheral milling, and it will avoid cutting air where material has been removed by earlier operations. The EMOVE BG will use world modeling to determine tools, spindle speeds, feed rates, stepovers, pass depths.

For sensor feedback, EMOVE can do machining verification. If an inspection probe is available, feature tolerance verification can be performed on the part. Out of tolerance machining could result in an additional machining step in the process plan to bring the part within tolerance. EMOVE could also perform tool wear detection and replacement by measuring tool sharpness with either a torque, temperature or acoustic sensor. A camera sensor could do visual inspection to detect excess chip accumulation and invoke a clean-up procedure.

The PRIM Segment Generator will translate EMOVE paths into straight line, arc or helix path segments. PRIM validates that the tool is proper for its machining use. Then, Machine Profile conditions the path to check for suitable machine limits, spindle speed, feed rate, cutting depth, and horizontal stepover. For sensor feedback, PRIM can inspect cutter radius to compensate for different sizes of the cutter. Further, torque feedback on the tool shaft can be used to adjust the feed rate.

The SERVO level will apply a control law to transform path segments into the necessary actuator signals. For feedback, SERVO is responsible for minimizing machining error through such techniques as chatter detection and thermal compensation [5]. SERVO feedback can detect chatter by sensing tool vibration with an accelerometer and then reduce the feed rate, or change the spindle speed. Thermal compensation can fix positional errors due to machine thermal expansion [6].

SUMMARY

A reference model for hierarchical control has been applied to the control of a machine tool workcell, emphasizing the requirement for sensor feedback in areas other than traditional position control. As part of the development of this reference model, a candidate set of messages for three-axis machining has been generated. This messaging language for machine tools grew from consideration of a scenario involving the four lower levels of a machine tool hierarchy, described in the final section.

REFERENCES

1. Albus, J.S., McCain, H.G. and Lumia, R., "NASA/NBS Standard Reference Model Telerobot Control System Architecture (NASREM)", Technical Note 1235, National Institute of Standards and Technology, Gaithersburg, MD, June 1987.
2. "Flight Telerobotic Servicer, Task Analysis Methodology," Ocean Systems Engineering, McLean, VA., NASA contract NAS5-30897, April 1991.
3. Meystel, A. "Representation of Descriptive Knowledge for Nested Hierarchical Controllers," Proceedings of the IEEE 27th Conference on Decision and Control, Austin, Texas, December 1988, pp. 1805-1811.
4. Donmez, M. A., "Progress Report of the Quality in Automation Project for FY90," National Institute of Standards and Technology Internal Report 4536, March 1991.
5. Gavin, R.J., and Yee, K.W., "Implementing Error Compensation on Machine Tools," Southern Manufacturing Technology Conference Proceedings (Sponsored by NMTBA, The Association for Manufacturing Technology), Charlotte, NC, May 1989.
6. Donmez, M. A., Blomquist, D., Hocken, R., Liu, C., and Barash, M., "A General Methodology for Machine Tool Accuracy Enhancement by Error Compensation," Precision Engineering, Publication No. 0141-6359/86/040187-10, 1986.

INTEGRATION TESTING OF THE AMRF

William G. Rippey

*National Institute of Standards and Technology,
Gaithersburg, Maryland, U.S.A.*

ABSTRACT

This report presents recommendations that aim to improve the effectiveness and efficiency of developmental testing of large manufacturing systems. This type of testing can require up to half of the total development effort of large systems. [14]   The recommendations apply to distributed systems in general and are not limited to manufacturing systems.

The recommendations are based on testing principles that were learned and applied during the development of the Automated Manufacturing Research Facility (AMRF). Use of these integration principles and guidelines for testing will help manage complexity involved in the development of manufacturing systems like the AMRF. Developers can produce a higher quality system, in less time, if they prepare for the complexity through modular subsystem design and good practices of development and testing. In terms of human resources, preparation can reduce time and effort needed to get the system working, reduce stress on the developers, and increase satisfaction with the results.

The AMRF is a major U.S. national laboratory for research in automated manufacturing. It consists of automated workstations and the control rooms and computer equipment necessary to operate them. The facility was developed by NIST as a testbed where scientists and engineers from industry, academia, and the federal government work together on projects of mutual interest. Their research concentrates on the interfaces and measurement techniques needed for successful computer integrated manufacturing.

The material will be of interest primarily to people involved in detailed design and implementation of computer integrated manufacturing systems. Others who may benefit are people responsible for performing, directing, and scheduling system testing.

INTRODUCTION

NIST is investigating the issues of integrating many discrete components into complex manufacturing systems. This report deals with the initial development, from 1982 to 1987, of NIST's Automated Manufacturing Research Facility (AMRF).

There are four major sections in this report:
- Brief Description of the AMRF
- Issues of Integration Testing - discussion of what is meant by the term integration testing, and a description of some AMRF testing concepts.
- Recommendations for Integration Testing - these recommendations form the conclusion of the report.
- An Example of AMRF Integration Testing - this illustrates some of the design recommendations.

Brief Description of the AMRF

The AMRF is a research testbed that models automated production of small batches of machined parts. In 1987 manufacturing tasks were performed by six groups of equipment components, called workstations. Three workstations machined parts by using robots to tend numerically controlled (NC) machine tools. A two-robot workstation deburred parts produced by the machining workstations. The last process in part manufacture was a robotically tended inspection workstation that used a coordinate measuring machine (CMM) and surface roughness sensor system to inspect parts. A materials handling workstation transported part blanks, using a wire guided cart, from a central storage to each of the AMRF workstations. The cart also transported finished parts and NC tooling. Data preparation and NC programming is done by a variety of methods. [6]

To reduce complexity the AMRF hierarchical control architecture modularizes the system by using levels of control. This modularity aids system testing. Command-status interfaces tie AMRF control levels together. The six workstations were coordinated by a cell controller which was at the highest level of the AMRF. Groups of equipment components --robots, machine tools, and fixtures-- were coordinated by workstation controllers.

AMRF manufacturing data is distributed throughout the factory both physically and logically. The information that must be exchanged between processes is managed by the Integrated Manufacturing Data Administration System (IMDAS) [6] .

ISSUES OF INTEGRATION TESTING

What is Integration Testing?

Integration testing is the step-wise testing of a system that comprises discrete, interconnected subsystems. Interactions between the subsystems are rigorously tested under controlled conditions that reduce the complexity of system debugging. Its goal is to confirm that a system can function according to design requirements. In progressing toward this goal, the testing must detect system errors, and assign corrections or changes to specific subsystems.

The quality and efficiency of developmental testing has a great effect on:
- system reliability
- how soon a new system becomes operational and useful, i.e. how much time is required for testing.
- how soon an existing system can be revised for more advanced system capabilities or for addition of new subsystems
- the amount of human effort required for system development
- the sense of satisfaction the project staff gain from system development.

A new system that is tested in progressive stages increases in complexity at each stage. Effective system testing constrains the increases in complexity of system operation. Failure to manage testing complexity results in the system becoming too complex to test or operate.

Complexity impedes peoples' ability to understand system operation, and has a direct effect on system reliability. These four factors determine the complexity of any system: number of elements, attributes of the elements, interactions between elements, inherent degree of organization. [10]

The first factor means that when there are more elements in a system, there is more complexity. 'Attributes' are properties or possible states of an element. 'Interactions' are relationships between elements in which they exchange outputs and process them. 'Organization' describes the extent to which predetermined rules, which guide interactions or describe attributes, exist. The overall complexity of a system is determined by the relationship among the four factors.

Testing and the System Development Life-Cycle

The development of a system can be divided into four phases: design, implementation, testing and maintenance, shown in Figure 1. Examples of techniques to ensure and improve the quality of the system at each phase are shown below the boxes. The recommendations in this report apply to the design and testing phases.

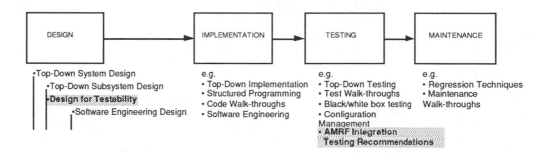

Figure 1. Steps in the life-cycle of software/hardware systems. The bullets denote techniques that can be used to improve quality of the system. The shaded areas are addressed by this report.

Adapted from reference [13].

AMRF Integration Testing

In the AMRF each subsystem is a distributed hardware, software, or hardware-and-software system; e.g. robots, machine tools, communications systems, database systems, intelligent sensors, controllers. The goal of AMRF integration testing is to achieve progressively advanced levels of operation of the automated factory.

Control System Configurations for Integration Testing   Configuration is the internal logical structure and operating mode of a controller or control system. During integration testing, AMRF control systems were tested in different configurations for the following reasons:
   • to permit step-wise development of controllers. Controllers were tested incrementally, to limit complexity, by integrating untested sub-processes one-by-one into a rigorously tested configuration
   • to simplify integration testing. Complexity of the integrated system can be decreased by reducing complexity in individual controllers.
   • to support concurrent development of subsystems. It was often desirable to test interactions between two separate subsystems before either subsystem was complete. Selected modules were run to exercise interface functions: other subsystem functions were not performed for the test. In the AMRF, the most frequent tests run in this fashion were communications tests and IMDAS access tests.
   • to aid troubleshooting. If a controller failed during an integration test, its internal configuration was simplified or diagnostics were added, and the test was rerun.
   • to conduct a system test when a subsystem failed. The objective was to operate and test an integrated system that contained a failed controller. If a control system component failed, its functionality was emulated by human interaction or by use of a simplified software module.
   The number of possible configurations of a controller contributes directly to testing complexity by increasing the number of attributes of the elements in the system. Multiple

configurations provide the advantage of flexibility, if each possible configuration is thoroughly tested and operators are skilled in manipulating the controller.

Table 1 shows the controller configurations used for testing and for operation. A control system operating unattended, fully integrated, is in all five modes on the left. Testing progresses from step-wise modes to integrated modes.

TABLE 1
Control System Configurations for Integration Testing

Integrated Operation Modes	Step-wise Development Modes
Integrated	Stand-alone
AMRF IMDAS	Local database
Lower level connected	No lower level subsystems connected
Minimal or no diagnostics	Diagnostics or trace on
No operator interaction	Operator interaction

Other configuration factors
Environment , change in operator performance.

The first three step-wise modes in Table 1 require the emulation of processes that interact with a controller: specifically, the higher level controller, IMDAS, and lower level controllers. Thus the emulation modules become temporary parts of the system. Figure 2 shows the interfaces involved.

AMRF testing policy requires the stand-alone capability for all AMRF subsystems, in which the higher level controller is emulated. Operator utilities produce command information inputs, and display the status information outputs.

AMRF controllers must be able to switch from accessing AMRF IMDAS information to using their own local database. There are three uses for this configuration change:
• Use of local data is a usual step in individual system development. The first tests are with simplified data and data access protocols.
• By using previously stored data, manufacturing operations can be performed even when communications and/or the IMDAS are not available.
• Complexity of integration testing can be decreased. The subsystem operation is simplified and more reliable in local mode. System-wide, there is reduction of Network traffic and IMDAS computer loading.

In the no lower level subsystems configuration command-status interactions and sometimes the physical tasks of lower level subsystems (e.g. robots, machine tools, fixtures, automatic guided vehicles, automated storage and retrieval systems) are emulated or simulated. Integration testing benefits from increased subsystem reliability and faster subsystem response when physical tasks are not performed. This mode is also used in stepwise subsystem development. The use of emulated lower level subsystems for testing is currently being emphasized in the testbed of the Manufacturing Systems Integration project. [11]

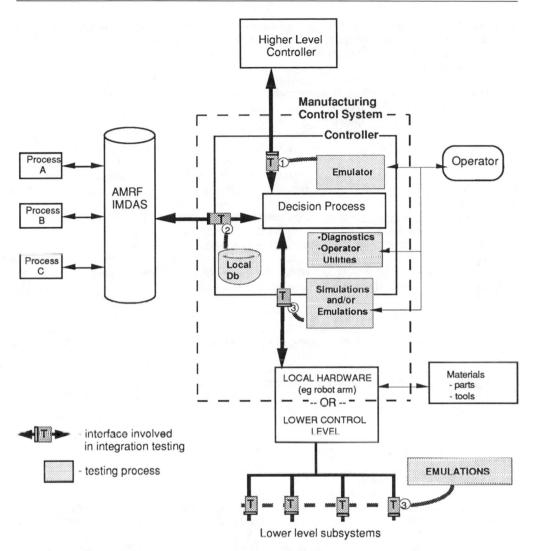

Figure 2. Manufacturing control model showing configurations for integration testing. Testing modes that involve the labeled interfaces are: 1) Stand-alone 2) Local database 3) No lower level subsystems.

Diagnostic testing is the use of extra software routines to capture information during process operation. The use of diagnostics changes the configuration and can affect the operation of a system. Adding or removing diagnostic software statements in source code, and turning procedures on and off, can affect code execution paths, timing, and the behavior of the subsystem at interfaces to other processes. Some of these changes in system performance may not be detected in stand-alone testing.

A change in one process that can produce an apparent change in the behavior of another is an environmental configuration change. A common change in the AMRF was to change to newer versions of operating systems or compilers. Other examples of environmental configuration changes for a controller "C" are: a changed version of controller "H" that issues commands to "C", a changed version of controller "L" that receives commands from "C", and a change to IMDAS.

Phases of AMRF Integration Testing   The testing of a large system is done in phases to manage complexity and to permit concurrent development of the subsystems. AMRF testing is done in four phases: interface testing; standalone subsystem testing; step-wise integration testing; and regression testing.

Configuration control is important in proceeding from one phase to the next. Complexity is reduced when tested conditions and data are carried forward to the next phase. The ability to isolate changes in the system as testing progresses is a key to achieving repeatable system tests.

Interface testing exercises the subsystem functions that interact with other subsystems. Typical AMRF functions are communications protocols, and exchange of command and status information. Selected modules of subsystems are run without involving the processing of control programs and control inputs. Interface testing simplifies the system testing by simplifying subsystem operation. It also permits system testing to be started before all subsystems are complete.

Standalone subsystem testing is performed in two phases. First, a controller is run while all interfacing subsystems are emulated. The objective is to thoroughly test internal functions of the controller and the interfaces between control modules and external functions. Decision making and processing of control information by the controller is tested. Next, in the hierarchical systems of the AMRF, lower-level subsystems and their interfaces to the controller are added. The goal is to aggregate complete control systems; i.e. a controller plus all of its lower-level hardware or control processes. The higher level interface to the controller is exercised via operator utilities.

The goal of step-wise integration testing is to connect the control systems that were tested in standalone mode by replacing operator utilities at high level interfaces with the actual controller. If the global data system was not previously tested, it is included at this stage.

System regression testing is required when subsystems are changed, usually to improve them via operating system upgrades, partial redesign, or when they are replaced with a new implementation with the same functionality.

Designing Manufacturing Subsystems for System Testability

Modular subsystem structure is essential for effective system integration testing. There are three steps for incorporating testability needs in subsystem design.

1) Modularize the subsystem to accommodate the four phases of system testing. Specific requirements are stated in recommendations 1-3. Verify that the design is correct by planning a scenario of the four phases of system testing. Some checks on the design are: Can interfacing functions be run independently of control processes? Can information exchanged between functions be examined and/or manipulated? Have interfaces to operator utilities been designed?

2) Incorporate the features described in recommendations 4 and 5.

3) Generate sets of information to be used in system testing. See also reference [4]. Step 3 is a cooperative effort with staff from other subsystem projects. This is not strictly a design task, but it involves the same personnel and should be done immediately after design is complete, and before implementation is begun. This step contributes to the overall understanding of system operation and of the communications and control protocols between subsystems. The early establishment of test data contributes to effective configuration control throughout the four testing phases.

RECOMMENDATIONS FOR INTEGRATION TESTING

The theme of these recommendations is to isolate and constrain system complexity. Two measures of success in efficient testing are: tests that can be run with repeatable results, and the ability of developers to predict test results.

Design Principles and Features

Some measures to control testing complexity affect the design of system components. The key is to anticipate system test methods. Subsystems operating in an integrated environment must have certain configuration change capabilities to support stepwise development and testing.

1) Incorporate into subsystem design the ability to select the testing configurations listed in Table 1. Don't wait until after modules are coded and then patch in changes to accommodate testing.

2) Modularize controller functions that are linked to integration interfaces. Data exchanged between modules should pass through buffers that are accessible to operator utilities. This structure, plus the ability to manipulate buffer data, allows configuration changes to be made more efficiently, and tests of different configurations will be more repeatable. See the example.

3) Put configuration information used for subsystem initialization in data files, not in source code. It must not be necessary to compile a different version of source code to reconfigure a system. The operator procedure for startup should be to select and set all configuration parameters in the file, then load the subsystem software. These initialization files must be covered by configuration management.

4) These operator capabilities for controllers are useful testing features:
- Pause the controller--examine status information--resume, with minimal operator action needed.
- Pause--manually or automatically manipulate outputs--then resume.
- Switch between local and remote database systems dynamically (i.e. without reloading, restarting, or recompiling software).
- Manually initiate module tasks, including external communications
- Assess controller status and activity, especially with a continuous display that does not require keystrokes to invoke.
- Initialize and activate subsystems using automated procedures (rather than step-by-step manual actions).
- Change controller configuration via operator interface while the controller is running. This is the most useful and versatile technique for changing configuration.

5) Controller developers must provide tested software utilities that support the configuration changes listed in Table 1. These utilities emulate or simulate the interactions of other controllers and subsystems.

Integration Testing Procedures

- Before integration testing begins -
6) Test subsystems in stand-alone mode, to verify their start-up and performance.

7) Stand-alone testing must exercise subsystem integration interfaces and protocols. That is, emulated input information should be expressed in the format specified by the integration interface--stand-alone testing using internal representations of interface information is not adequate. See the testing example at the end of the report. The integration buffers are B1 and B2, the internal buffer is B3.

8) Confirm that the performance of emulations of other subsystems used in standalone testing conform to interface specifications. Note: the deliberate generation of

bad interface information is a useful technique for testing subsystem error detection and correction.

9) Specify the design of interfaces with documents that are based on a well-written narrative description of subsystem interactions and information formats. Program listings of procedures and/or data definitions can be used to supplement the narrative, but should not be the sole mechanism for conveying interface information to people.

10) If a tested subsystem is modified (hardware or software) retest it in stand-alone mode before it is included in further integration tests. Also retest when initialization data changes, even if source code did not change.

11) Consider choosing a project-neutral test coordinator to coordinate test activities, centralize configuration control, and provide impartial judgment during troubleshooting.

12) Provide for and enforce central configuration control for subsystem versions of software, hardware, and initialization data. Developers often cannot "see" effects of changes to their subsystems on system operation.

13) Design integration tests before the test is begun. Elements of a good Test Design are:

- Purpose of the test: what will we find out by trying the test; by successfully completing the test?
- Definition of the system to be tested, configuration of the subsystems, and description of the environment. Use drawings or sketches in addition to narrative.
- A functional scenario, including description of initial conditions
- A technical scenario: how functions are done and what interfaces are involved.
- Criteria to use in determining a test's success. These form a prediction of the results of a successful test. Criteria include technical results as well as operational procedures (e.g. the criterion "The subsystem performs well with a non-expert operator.") Quantify criteria when possible, e.g. the robot will process and perform four different commands, the workstation controller will perform three different types of IMDAS transactions.

14) Use a Test Plan to describe tests that are a subset of a given Test Design. In this way, ambitious, longer-range Test Designs can be generated that do not cover details of individual tests. Smaller scope, step-wise tests can be run without generating a new Test Design each time. A Test Plan can reference specific sections of the Design, and note exceptions if applicable.

15) Use a standard form for Test Design and Test Results documents. The formats and titles should be easily recognizable. Make forms easy to read and to complete.

16) Perform upgrades of subsystem computer operating systems or compiler versions after reliable system operation has been achieved. Do not make changes when some progress has been made, but more testing is needed. Notify all subsystem staff before changes are made.

17) Experts on interface protocols and system design should publish examples of transactions or data encodings for use by other project staff. It is especially important to illustrate transactions that affect the most subsystems: in the AMRF these were Network and IMDAS transactions.

18) Generate tools to produce and manage diagnostic trace information. Use "save" files to preserve data in case it is needed later. Develop, document, and use procedures for cleaning up and reclaiming file space after tests.

- Procedures for Integration Testing -

19) All subsystem modules that will run in integration testing must first be successfully run in the stand-alone test.

20) Activation procedures for subsystems and the integrated system must be in written form or directed by interactive software utilities--not remembered by operators. Include activation procedures in configuration control.

21) The following documentation must be on-hand during tests:
• software listings
• hardware wiring diagrams and manuals
• interface specifications (not just well documented software listings)
• a test plan or test design
• standard initialization and startup procedures
• an integration testing policy (recommendations 1 through 30).

22) Dedicate shared resources such as multi-user computers to integration testing, at least in the early stages, so that baseline performance can be assessed.

- Regression testing (retesting of a previously tested system)

23) Retest a subsystem if any of its modules is changed.

24) Retest the <u>system</u> when the configuration of any <u>subsystem</u> has changed, including the apparently harmless removal of diagnostics.

25) Provide utilities for each subsystem for examining interface information for inputs and outputs. Note: continuous display of important system status parameters, without need for keystrokes, is a very desirable capability.

26) Do not train operators of subsystems or the system during integration tests.

27) Do not change operators during critical testing phases. Operator performance may change, and a new operator may not be aware of the latest changes in procedures and environmental conditions.

28) Develop, test and use procedures for orderly system shutdowns. This includes orderly process exits and file space cleanup. Failure to do so can produce conditions that can corrupt subsequent tests.

29) The test coordinator should not operate a subsystem during an integration test.

- After Testing -

30) Issue reports of test results to the participants. This is their feedback; to know if the test was successful, to learn how other subsystems performed, what the overall problems were, what progress was made, and what additional testing may be necessary.

Human Factors

31) To help communication among test personnel, develop scenarios of system operation, system test procedures and desired results of system operation. These scenarios are elements of good Test Designs.

32) Plan milestones and goals in steps. Early and intermediate successes help build confidence and morale. There are some differences between integration testing and testing of small projects regarding personal factors. Integration testing is a team activity in which individuals' satisfaction is linked strongly to cooperative success. Also, results of the cooperative testing are visible to members of the team. In smaller project development intermediate results may only be known to that project's staff.

33) Visual displays
• Make operator interfaces instructive, not terse. Eliminate distracting information on visual displays.
• Don't leave displays blank during process execution. Use positive feedback to indicate a process's state, such as "Process Begun", "Requesting Data", "Loading Part". Also see [12], p. 98, 213, 223, 308, 311, 312, 325.

34) When applicable, identify and document a common high-level methodology to be used in system and subsystem design by all personnel . This provides a common ground that simplifies interpersonal communications and produces more consistent designs of subsystems and interfaces. The example in the AMRF is the concept of hierarchical control and characteristics of all command-status interfaces.

AN EXAMPLE OF AMRF INTEGRATION TESTING

This example shows the flexibility that should be designed into controllers to allow the configuration changes required for step-wise development and for integration testing. The configuration change is between integrated and stand-alone modes. The most important design principle is to modularize controller functions. The controller configuration is then changed by controlling module execution and manipulating data that is exchanged between modules, rather that by compiling different versions of software.

Figure 3 shows the internal modules and information buffers of a controller. In integrated operation a Network mailbox conveys commands from a higher level controller. The Common Memory Protocol (the AMRF communications paradigm) module copies the information into an internal buffer. The conversion module processes the standard AMRF format information into an internal representation of commands and parameters that is used directly by the controller's decision process.

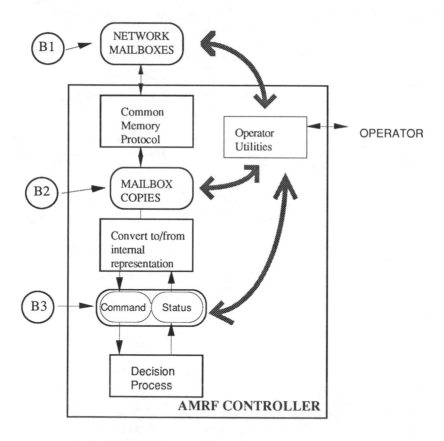

Figure 3. Processing modules and information buffers of a controller.

In stand-alone configuration an operator generates the command used by the controller. The operator utilities can produce information that is inserted into any of three information buffers, either the Network mailbox B1, or the internal copy buffer B2, or the internal representation buffer B3. The utilities also display and provide ability to change all information in the buffers, including commands and the status information produced by the controller.

There are four testing activities that benefit from the modular controller structure and the use of buffers for information exchange:

1) Changing controller configuration. In integrated configuration a higher level controller produces command data in buffer B1. In stand-alone mode the operator utilities would produce the command information--the configuration of internal modules is identical to that of integrated operation. If the absence of the Network dictated that buffer B1 was not physically present, then the operator utilities would generate data for buffer B2.

2) Stepwise controller development can be performed by first testing the decision process via B3, then adding the conversion module by using B2, then adding mailbox communications functions by using B1.

3) An effective diagnostic technique is to isolate the actions of modules by manipulating input data in buffers, running the module, and observing results. Modules that exchange data can also be run in "single-step" mode so testers can trace controller internal performance.

4) Error recovery can be performed via operator intervention in some cases by pausing the controller, clearing the local error and replacing the error status report in buffer B3 with a status report of DONE. The controller can be restarted, it will report its DONE status to the higher control level, and be ready for the next command.

REFERENCES

1. Barbera, A.J., Fitzgerald, M.L., Albus, J.S., "Concepts for a real-time sensory interactive control system architecture", Proceeding for the Fourteenth Southeastern Symposium on System Theory, April 1982, pp. 121-126.

2. Barbera, A.J., Fitzgerald, M.L., informal notes, "Demo suggestions", June 1984.

3. Basili, Victor R., Perricone, Barry T., Software Errors and Complexity: an Empirical Investigation, Communications of the ACM, Volume 27, Number 1, January 1984.

4. Branstad, Martha A., Cherniavsky, John C., Adrion, Richards W., "NBS Special Publication 500-56", U.S. Department of Commerce, National Bureau of Standards, February 1980, reprinted in Structured Testing, Thomas J. McCabe, ed., IEEE Computer Society, 1983: see page 84.

5. Furlani, C.M. et al., "The Automated Manufacturing Research Facility of the National Bureau of Standards", Proceedings of Summer Computer Simulation Conference, Vancouver, B.C., July 1983.

6. Libes, Don, Barkmeyer, E., "Integrated Manufacturing Data Administration System - An Overview", Vol. 1, No. 1, International Journal of Computer Integrated Manufacturing, January, 1988.

7. Nanzetta, Philip, Weaver, Asenath, Wellington, Joan, Wood, Linda, Publications of the Center for Manufacturing Engineering Covering the Period January 1978 - December 1988, NISTIR 89-4180.

8. Nanzetta, Philip, Hutchins, Cheryl, Wood, Linda, Publications of the Manufacturing Engineering Laboratory Covering the Period January 1989 - September 1991, NISTIR 4713.

9. Rybczynski, S., et al., <u>AMRF Network Communications</u>, NBSIR 88-3816, June 30, 1988.

10. Schoderbek, Peter P., Schoderbek, Charles G., Kefalas, Asterios G., <u>Management Systems, Conceptual Considerations</u>, Plano Texas, 1985, 383 pp.

11. Senehi, M.K, Barkmeyer, Ed, Luce, Mark, Ray, Steven, Wallace, Evan K., Wallace, Sarah, <u>Manufacturing Systems Integration Initial Architecture Document</u>, NISTIR 4682, September 1991.

12. Smith, Sidney L, Mosier, Jane N., <u>Guidelines for Designing User Interface Software</u>, Bedford, MA., The Mitre Corporation, 1986, 478 pp. NTIS document number AD A177 198.

13. Walsh, Thomas J., "A software reliability study using a complexity measure", p. 111, reprinted in <u>Structured Testing</u>, Thomas J. McCabe, ed., IEEE Computer Society Press, 1983.

14. Wasserman, Anthony I., "Information system design methodology", Journal of the American Society for Information Science, January 1980, reprinted in <u>Tutorial on Software Design Techniques</u>, IEEE Computer Society, 1983, edited by Peter Freeman and Anthony I. Wasserman.

CURRENT AND FUTURE TELECOMMUNICATION NETWORKS

Seamus O'Shea

Department of Computer Science and Information Systems, University of Limerick, Limerick, Ireland

ABSTRACT

This paper describes the current scene in public telecommunication networks and sets out some of the shortcomings of having multiple service-specific overlay networks. A service independent, integrated, universal network, commonly called the Broadband Integrated Services Digital Network,(BISDN), which can carry all services is described in this paper. The technology on which the BISDN will be built is described, especially the Asynchronous Transfer Mode (ATM) of switching and multiplexing, together with the Synchronous Digital Hierarchy (SDH) based architecture for the transport network.

INTRODUCTION

This paper is concerned with the emerging trends in public telecommunications networks. In particular, it describes the two core technologies which will serve as the infrastructure for a future universal service independent network. They are: 1) the Asynchronous Transfer Mode (ATM) of switching and multiplexing, and 2) the Synchronous Digital Hierarchy (SDH) which is the intended architecture for future transport networks. Firstly it describes the current position of many separate overlay networks corresponding to separate services,(e.g telephony, telex, X.25 etc), and outlines the shortcomings of this type of development for future requirements. The demands of the information age in which we live, are giving rise to many new services, (facsimile, , high speed data transport, videotelephony, multimedia), and it is not feasible to build separate networks for each new service. It is very desirable to have one universal network, which is service independent, and therefore future proof, which can transport any service, irrespective of its required bit rate.

This paper considers the essential characteristics of a generic service, and sets out its requirements on a transport network. It then describes a transfer mode and a transport architecture which can serve as the building blocks of a future universal service-independent network. The paper describes the technology relating to the transfer mode and to the transport architecture in some detail. This universal network is currently being planned in many countries throughout the world. It is referred to as the Broadband Integrated Services Digital Network (BISDN), and will be a close successor to the

Narrowband Integrated Services Digital Network (NISDN) currently being introduced in many countries.

AN OUTLINE OF THE CURRENT SCENE

There are many varieties of Telecommunication Networks in existence today. Each is almost entirely service specific. For example:

(1) Telex service is carried over the Telex Network, which is intended to carry text messages at very low speeds, typically 300 bps.

(2) Telephone Service is carried over the Telephone Network and allows interactive two way speech communication.

(3) Computer data is primarily carried over dedicated X25 networks. However, computer data can also be carried over the telephone network (even though this use of the telephone network was never intended) or dedicated circuit switched data networksspecially built for data transport. In the private domain, special Local Area Networks (LANs) carry computer data over limited distances.

(4) TV signals are transported over separate broadcast networks, or over cable networks.

The current situation is characterised by:

(1) No sharing of resources, i.e idle resources (e.g bandwidth) in one network cannot be used in another. Peak hours in the Telephone network occur during normal office hours, while peak hours for the CATV networks are in the evening. However no sharing is possible.

(2) Each network type is dimensioned according to the characteristics of the service carried, this normally means that the worst case expected must be catered for.

(3) Inflexibility, in the sense that new developments in technology may not always be easily incorporated into existing networks and put to advantage. For example, the current generation of telephone switches are designed to carry voice traffic at 64kbps, however progress in speech coding algorithms means that speech traffic can now be carried at half that bit rate, i.e 32kbps. This means that switches designed for 64kbps operation may be very inefficiently used when better coding techniques are implemented.

(4) The existence of multiple networks means that it is expensive to design, install, maintain and plan such networks, where many resources may be duplicated. The network management task is complicated by the existence of multiple networks (e.g interworking). It is clear from the foregoing that a future evolution strategy based

on multiple overlay networks is not the way to proceed. The infrastructure on which BISDN will be built is optical fibre, which has already been proven to have almost unlimited bandwidth capacity. The only limitations in this regard concern the opto-electronic devices to insert, switch, and recover the message that is transmitted on fibre optic cables. The key requirements on the BISDN are that it must be:

(1) Flexible, i.e its mode of operation must be independent of any particular service or bit rate. This is important not alone within the BISDN itself but also in terms of interfacing to existing service specific networks.

(2) It must have almost unlimited capacity/bandwidth, not alone in the backbone network but also in the local customer access network. Fibre Optic technology is capable of meeting this demand.

(3) Resource sharing must be possible. Statistical sharing of resources leads to greater efficiency in a network. A new switching and multiplexing scheme, called Asynchronous Transfer Mode (ATM), which has been adopted by CCITT as being the basis for BISDN, will be the technology by which such resource sharing will be possible. ATM is described in detail in a later section of this paper.

THE CHARACTERISTICS OF A SERVICE.

Every service is characterised by a natural information rate, i.e the rate at which a source (e.g voice, video) generates information independently of any considerations relating to transport over a network. The natural information rate depends on the coding and compression algorithms used to express the information from the source. In general, the natural information rate of a source is a stochasitic quantity, which fluctuates over time. Of particular importance from the transport point of view is the peak information rate, together with the average information rate of a source.

Service	Average	Burstiness
Voice	32 Kbps	2
Interactive Data	1–100 Kbps	10
Bulk Data	1–10 Mbps	1–10
Standard Video	20–30 Mbps	2–3
HDTV	100–150 Mbps	1–2
Video Telephony	2 Mbps	5

Table 1. Service Characteristics

A further parameter, the 'burstiness' of a source is defined as the ratio of the peak to average information rate. Some typical values for the peak and average information rates for a range of services are shown in Table 1. Those values are taken from [3](DePrycker).

It is clear that there is great variation in burstiness from service to service. Furthermore, no service has a burstiness value of 1, i.e peak rate equal to the average.

If the transfer rate (i.e the rate at which information can be transmitted, switched and multiplexed) in the network is greater than the peak information rate of a source, then there is wasted capacity. On the other hand, if the transfer rate is less than the peak bit rate, then there is degradation of quality, since bits will be discarded at times.

The bursty nature of many services can often be put to advantage, for example in the IEEE 802.3 Medium Access Control layer, and of course in packet switching. It is shown in [3] that significant advantage is gained by statistically multiplexing many bursty services over a given link using ATM.

There are two further service characteristics which are important. The sensitivity of a service to network errors, and to network delay often determine if a given network can be used to carry a given service. For example, Telephony is not sensitive to network errors, but it is very sensitive to network delay, and delay jitter. On the other hand, data transport is very sensitive to transmission errors, but is not very sensitive to network delay. The ability of a network to transport information without error is referred to as 'semantic transparency', while the ability of the network to transport information with acceptably low end-to-end delay and delay jitter is referred to as 'time transparency'.

In summary, the important characteristics of a service for the purposes of this paper are:

(1)   The Average bit rate
(2)   The peak bit rate
(3)   The burstiness of the service
(4)   Sensitivity to transmission errors
(5)   Sensitivity to transmission delay

THE ROLE OF TECHNOLOGY.

The core technologies for BISDN are the Asynchronous Transfer Mode (ATM) for switching and multiplexing, and the Synchronous Digital Hierarchy (SDH) for the transport network. ATM is a particular type of fast packet switching which has the capability to allocate any bit rate on demand. Its switching capability is also independent of any particular bit rate which may be associated with any service. The SDH (Sonet in the US) is a high speed transport architecture ratified by CCITT. A key feature of SDH is that it can create higher speed interfaces by byte interleaving streams from lower speed channels. This simplifies add/drop multiplexing compared to the current plesiochronous systems. Within SDH systems, there is provision for ample operation and maintenance data associated with transport paths. SDH is proposed to become a worldwide standard. Both ATM and SDH will be described in more detail in the next sections.

The Asynchronous Transfer Mode

Earlier X.25 packet switching networks have, in general, been built from low quality transmission links, where complicated protocols are needed on a link-by-link basis to recover from error conditions. Such networks are not able to meet the stringent end-to-end delay requirements of some services (e.g telephony). However with the emergence of NISDN, higher quality optic fibre based communication links are now being used, with much lower error rates. This in turn makes it unnecessary to repeat many functions (e.g error control, flow control) on a link-by-link basis within the network, instead such functions can be removed to the edges of the network, i.e to the end stations. This means that within the network the switches have less processing to do, and therefore better end-to-end performance can be achieved. If the overall delay, plus the delay jitter, can be reduced sufficiently, then delay sensitive services like telephony and video can be transported over packet switched networks. Figure 1 shows the position in current X.25 networks.

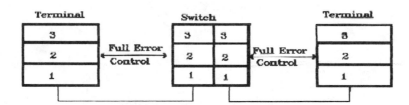

Fig. 1 X25 Operation

The data link layer level, includes functions such as flow control, frame delimitation, bit transparency, checksum generation, and error recovery. Figure 2 shows a communications architecture where only a subset of the above functions are performed on a link-by-link basis. This mode of operation is referred to as Frame Relay.

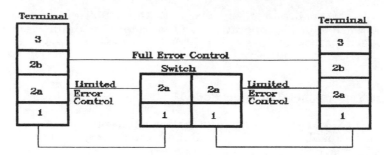

Fig.2   Frame Relay Operation

In the figure, layer 2 is divided into two sublayers, layer 2a contains only core functions like frame delimitation and data transparency. It operates on a link-by-link basis. Layer 2b operates on an end-to-end basis and contains functions like error correction and flow control. Figure 3 shows the operation of an ATM based network. In this scheme the core functions of layer 2a have also been removed to the edges of the network, while layer 1 is subdivided into a physical medium dependent layer (layer 1a) and an ATM layer (layer 1b).

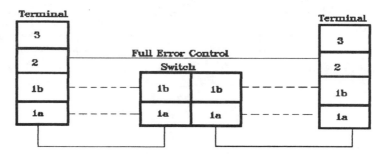

Fig. 3   ATM Operation

In ATM networks, error control is not performed within the network, only at the end stations. This means that in ATM networks only very limited functionality exists in the switching nodes, leading to minimal complexity allowing very high speed switching. (~600Mbps).

From the foregoing it is clear that an ATM based network can deliver very low end-to-end delays, and delay jitter. It also exploits the fact that some traffic sources may not require to send information at a constant rate for the duration of an entire call, i.e the burstiness of certain services can be exploited to achieve statistical multiplexing. ATM does not allocate capacity statically as in synchronous switching, it allocates only on demand, thereby achieving flexibility. This leads to enhanced efficiency with acceptably low error rates. Because ATM is a packet switching technology, all information to be transported and switched in the network is organised into fixed length packets or cells. Each cell consists of a 5 byte header together with 48 bytes of user data. Each cell can be addressed independently to a destination via a label in the header which associates a cell with a particular virtual channel. Allowing each cell to be allocated a different destination address is equivalent to being able to allocate any bit rate on demand, the required number of cells are allocated according to the current needs, of course subject to the overall capacity of the link in question. It is from this property that ATM gets its name, i.e cells are asynchronously allocated to services on demand. There are no periodically appearing frames as in the case of synchronous switching. However, ATM can emulate fixed rate channels since cells can be allocated at a fixed or variable rate. When services requiring Constant Bit Rate (CBR) channels are adapted to ATM cells, it is necessary to buffer the arriving information, place it into the ATM cells, transmit it to the appropriate destination, and then convert it back to the original CBR stream.

The switching fabrics to support ATM will be based on CMOS, ECL, and Gallium Arsenide components in the near term. Switching speeds in excess of 100Mbps can be achieved at present using CMOS technology, and it is anticipated that speeds of up to 300Mbps will be possible in a few years. User network interfaces and switching elements will be based on CMOS technology at least in the near term. Silicon Bipolar technology based devices can operate at higher speeds (~600Mbps at present) and it is expected that this technology will also be used in the user interface for the distribution of TV channels. Within the backbone network, where much higher speeds (~3Gbps) are required, it is anticipated that Gallium Arsenide (GaAs) technology will be used. (i.e high speed switching).

The performance of ATM as a switching and multiplexing mechanism has been studied both from the viewpoints of semantic and time transparency. Semantic transparency relates to the ability of the network to transport information to its destination without error. Time transparency relates to the capacity of the network to meet the requirements of very low end-to-end delays, and very low jitter levels on the delay. Results relating to the additional traffic levels in the network arising from the use of ATM end-to-end protocols are shown in the following diagram:

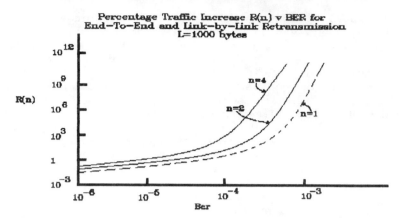

Fig. 4 Traffic increase v Ber

In this diagram the percentage increase in traffic over the network is plotted as a function of the Bit Error Rate (BER) of the Network. Results are presented where the end-to-end path consists of 4, 2, and 1 links, and where the packet length is 1000 bytes. It emerges that for BER values less than 10^{-6}, the increase in traffic is very small, and it remains small even up to a BER of 10^{-4}. Similar results are shown in the following diagram where the packet length in this case corresponds to 48 bytes, the payload of an ATM cell.

It follows that, in the case of the shorter packet, the increase in traffic remains small up to a BER of 10^{-5}, after which it increases exponentially. In the case of the longer packets, the traffic increase is insignificant up to a BER of 10^{-4}, after which an exponential growth sets in. The shorter packet is more suited to the ATM type network.

The above results demonstrate the feasibility of the ATM concept. It is shown that there is no significant overhead in terms of increased traffic due to retransmissions where error recovery is done on an end-to-end basis over the normal case of link-by-link recovery if the BER is less than 10^{-4}. A typical BER over optical fibre is 10^{-8}.

A detailed study relating to time transparency in an ATM network is reported in [3] where a typical end-to-end route is selected,(8 ATM switches, end-to-end distance of 1000km, with a traffic loading of 80% on the links) and the overall end-to-end delay is calculated for packets of lengths 16, 32, and 64 bytes respectively. The overall delay increases with packet size, however in all cases studied, the total end-to-end delay was less than 25msec. This is the value specified by CCITT in the context of telephony, above which echo cancellors must be fitted. This is partly the reason why CCITT approved a fixed cell size of 53 bytes.

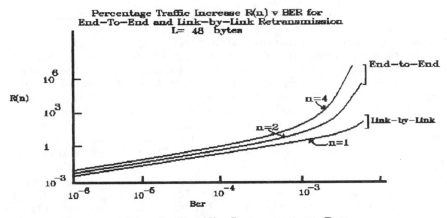

Fig. 5 Traffic Increase v Ber

In conclusion, this section has demonstrated that the underlying principles of ATM in terms of shifting many network functions to the ends of the network has no adverse effects on the semantic or time transparency requirements of services. It has also shown that considerable flexibility is inherent in the ATM mode of operation, whereby capacity is allocated on demand, and the fact that services do not always require their peak bit rate is exploited using statistical multiplexing. This results in far greater transport efficiency. In addition of course, it follows that cells can be allocated at a fixed rate, to emulate synchronous switching. Therefore ATM can be considered as a very general switching and multiplexing technique which includes within it a synchronous switching mode.

THE SYNCHRONOUS DIGITAL HIERARCHY.

SDH is a synchronous transport network. It uses a frame structure to perform basic synchronization. Its operation and interfaces are standardised by CCITT. The plesiochronous digital hierarchies as defined by CCITT in 1972 led to different standards in Europe, North America, and in Japan. The SDH is intended to achieve one worldwide transmission standard, where byte interleaving is used for economical add/drop multiplexing, and where ample provision for network management overhead is made. The standardised optical carriers, or Synchronous Transport Modules (STMs) defined in SDH are shown in the following figure:

SDH level	SDH signal	Bit rate
1	STM-1	155.520
4	STM-4	622.080
16	STM-16	2,488.320

It shows that 4 x STM-1 is equivalent to 1 x STM-4, while 4 x STM-4 is equivalent to 1 x STM-16. Higher modules can be created using the same interleaving structure. Only the three rates shown are being considered for near term applications. The bit rate of STM-1 is the fastest rate at which CMOS electronics can be applied over the next few years. The bit rate of STM-4 is base don ECL or bipolar technologies. The bit rate of STM-16 is based on the use of Gallium Arsenide (GaAs) devices. CCITT envisages the combined use of the ATM switching and multiplexing scheme with the synchronous nature of the SDH type of networks, whereby the payload of SDH frames is organised via ATM multiplexing. The ATM cells are carried to their destinations as payload within STMs where the payload is then recovered and the virtual channels separated from each other. In fact two options are retained by CCITT for the user network interface, one being a pure ATM interface without the synchronous structure of SDH, and the other being as described above i.e ATM carried as payload in SDH frames.

SUMMARY AND CONCLUSIONS.

This paper has described the current position of many public overlay networks, most of which are almost totally service specific. This type of evolution has a number of associated problems, which are outlined in the body of the paper. Reasons are given why a universal integrated service independent network is required for future developments. The paper discusses at some length the technology advances which makes this universal network feasible. The key technological building blocks are a flexible form of fast fixed size packet switching known as asynchronous transfer mode, together with a CCITT ratified transport network architecture known as the synchronous digital hierarchy. The paper demonstrates that the combined use of the asynchronous transfer mode and the synchronous digital hierarchy leads to a service independent network, where its mode of operation is not dependent on any specific bit rate. Valuable benefits in terms of resource sharing and flexibility are inherent in this type of network.

References

[1] CCITT Recommendations G.707 Synchronous Digital Hierarchy bit rates.

[2] CCITT Recommendation G.709 Synchronous multiplexing Structure.

[3] DePrycker, M. Asynchronous Transfer Mode, Solution for Broadband ISDN. Ellis Horwood Publishers. 1991.

LOCAL AREA NETWORKING FOR MANUFACTURING ENGINEERS

M. Munir Ahmad and Declan M. Melody

Department of Mechanical and Production Engineering, Limerick University, Limerick, Ireland

ABSTRACT

The current debate between the supporters of the de jure OSI-based networking standards and the users of the de facto proprietary standards is intense and complex. This paper attempts to clear up some of the confusion surrounding the networking issue by discussing the technologies available for the networking of the factory.

The paper investigates the options in hardware and software for networking. It outlines the popular cabling, topology and media access methods, and the most common network operating systems, including the newest implementation of DECnet. The paper discusses the feasibility of the OSI solution in the face of the newest attempts by major manufacturers such as DEC and HP to incorporate Open Systems into their networking strategies and examines solutions for a predetermined industrial application based upon both OSI-dependent and OSI-independent equipment.

INTRODUCTION

CIM is an approach to manufacturing which uses computer technology to enhance productivity through the collating and processing of information from all phases of manufacturing. This view of CIM provides a broad definition which includes not only all of the engineering functions but also the business functions of the firm. The ideal CIM system applies to all of the operational and information processing functions in a factory from order receipt through design and production and finally to product shipment.
The Computer and Automated Systems Association (CASA) of the Society of Manufacturing Engineers (SME) defines CIM as having three foundations
i) The establishment of a plantwide information network
ii) The establishment of an overall data and information flow architecture
iii) The simplification of the manufacturing function. [1]
The purpose of this paper is to concentrate on the design of a plantwide Local Area Network for a given facility, discussing the options available to the typical medium sized company which wishes to initiate or further it's CIM strategy through the implementation of an 'Open' networking environment,

Local Area Networking in Boart Europe

This work has been carried out in association with the Management Information Systems and Technical Departments of Boart Hardmetals Europe. Boart Europe Ltd. started operations in Shannon, Ireland in 1961. It has grown steadily to become one of the leading rock drilling and accessories manufacturers in the world. Boart Europe employs 230 people and comprises three divisions, all with modern production facilities, these divisions are ;

i) The Mining and Construction Division (MCD) which manufactures a full range of drill steels and bits for percussion and rotary drilling, down the hole bits, adapters and couplings to fit all types of rockdrills.

ii) The Tungsten Carbide Division (TCD) which provides tungsten carbide inserts for drill bits and produces a wide range of cutting and forming tools and wear resistant parts for a broad spectrum of industries.

iii) The Diamond Division (DD) which meets the demand of sister companies involved in diamond bit manufacture.

MCD is the largest of the three facilities, and is located in it's own building, TCD and DD share an adjacent but separate facility. In installing a network (distributed between two buildings) to serve the data communications needs of the factory, it is sought to lay the foundation for a system that will be self managing, flexible, capable of being installed on a modular basis, with the associated advantages from the point of view of capital expenditure, and which will be as "Open" as possible.

Existing Systems :- Figure 1.0 illustrates the different systems presently in place in the factory. These are :

i) A customised 'Applicon' DEC VAX 11/750, which supports the CAD editor and database.

ii) A 'Cromenco' minicomputer with a 'Jarogate' machine as backup which runs the Unix derivative 'Cromix' operating system. The purpose of this Cromix system is to provide instruction to the CNC machine controllers connected to the Cromenco through RS422 cabling. NC code for this purpose is generated on a dedicated Texas Instruments PC.

iii) A HP3000 series 52, running HP's proprietary MPE/V operating system. This machine supports the business and production management applications for the factory.

iv) 18 IBM-compatible PC XT/AT 286 HP Vectra PCs located in the MCD, another 10 386 PCs located in the TCD and DD divisions. Uses, personal data and word processing.

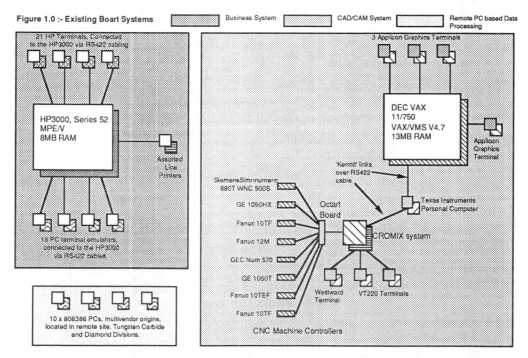

Figure 1.0 :- Existing Boart Systems

Requirements :- IDEF0 analysis of the existing systems revealed networking requirements at four levels, these were to be addressed by a solution divided into four corresponding modules, the requirements and respective modules are described on the following page;

i) Transfer of data and electronic documents between the local site and other members of the Boart Group worldwide and customers or vendors. Access to the national Public Data Network (PDN) 'Eirpac' was to be provided using a network accessible, X.25 gateway.The solution at this level was to be provided by an enterprise-wide module

ii) Transfer of data among the minicomputers and between the minis and the DOS based machines present in the factory. This requires a plant-wide module.

iii) Transfer of data and resource, file sharing among the Boart PCs, also office automation in the form of automatic PC backups, internal mail and fax facilities etc.This requires an area module.

iv) Providing the basis for control of existing and planned shopfloor equipment.This requires a shopfloor module.

At this point a decision was taken to concentrate on the enterprise to area modules of the solution, where benefits would be most immediately apparent. The existing shopfloor control system was judged to be adequate for the company's needs.

Figure 2.0 :- Solution Modules

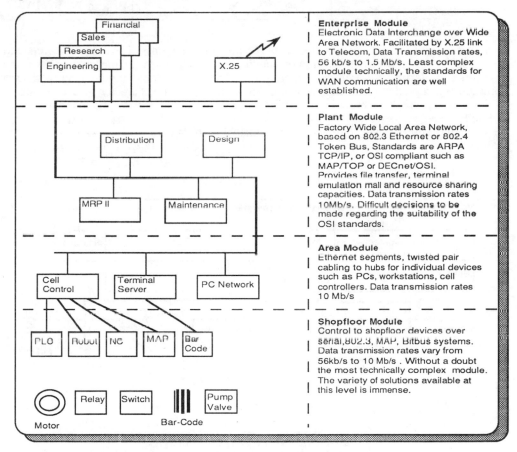

NETWORKING THE FACTORY

A computer network can be considered to be an interconnected system of autonomous computers, interconnected because they are capable of sharing information, autonomous because no one machine is capable of being started, stopped or controlled by another [3]. The connection is enabled both logically through the commonly accepted standards relating to the meaning of the signals carried via the network and physically through the mechanism used to build the network and link the different machines together.

Network Standards : Figure 3.0 illustrates the network standards. These are agreed reference models which regulate combinations of networking options and machine-machine communication protocols and can initially be divided into two separate camps ;
• Standards conforming to the International Standards Organisation (ISO) Open Systems Interconnection (OSI) model, referred to in this paper as OSI-Dependent Standards
• Independent standards, usually developed by one manufacturer or organisation for a specific purpose, set of equipment or application. These will be referred to as OSI-Independent standards.

Figure 3.0 :- Network Standards

OSI Dependent Standards				OSI-Independent Standards		
MAP	Mini Map	TOP	DECnet Phase V	X.25	TCP/IP	DECnet Phase IV	Novell Netware

Network Elements : Figure 4.0 illustrates the network elements. These are the 'nuts and bolts', so to speak from which the communication mechanism for the network is to be assembled. In terms of the physical medium used to connect the computers, networks are commonly accepted as being defined according to three separate characteristics, these are ;
• The mechanism by which the machines gain access to the network (Media Access Method)
• The medium used to physically connect the machines (Network Media)
• The manner in which the machines are connected to the network (Network Topology)

Figure 4.0 :- The Network Elements

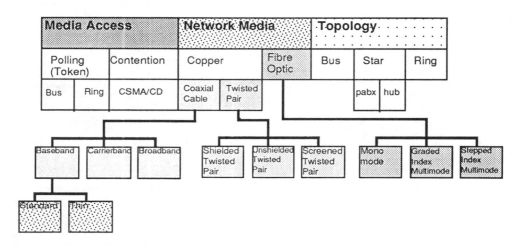

OSI AND NON-OSI SYSTEMS

Existing standards for device communication are of two types ;
• Standards developed before the OSI-model was conceived to provide a particular type of operating system with network capabilities. Examples are Novell Netware and MS LANmanager based on NetBIOS for DOS, or Advanced Research Projects Agency (ARPA) Transmission Control Protocol / Internet Protocol (TCP/IP) for UNIX systems. These standards gradually found their fields of use extended to include different operating systems. Netware for example will now support UNIX, OS/2, MAC and DOS. TCP/IP is supported by virtually every operating system

• Standards created for a specific purpose, such as the interconnection of a particular vendor's devices. Typical examples include protocols such as DECnet for machines designed and manufactured by the Digital Equipment Corporation and the Systems Network Architecture (SNA) for IBM product.

Figure 5.0 :- 7 Layer Open System Interconnection Protocol Stack

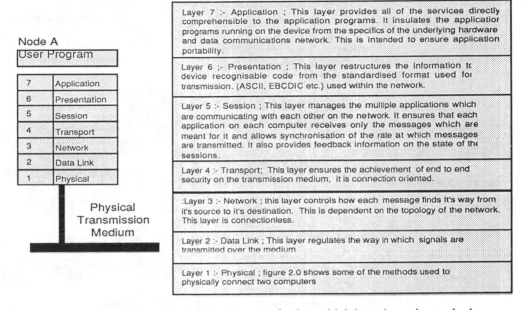

The need to standardise communication systems further which has given rise to the by now familiar OSI model (figure 5.0) came about for two reasons :

• A widening rift was developing between what was possible in network communications and what amounted to the current practice.Consider the most 'open' of OSI-Independent systems in terms of multi-vendor communications, which has proven to be the TCP/IP standard. The wide area of application of this standard has added considerably to the inertia associated with changing it, any upgrade in the functionality necessarily implies a decline in the amount of systems to which communication can be supported. This is also true of the vendor proprietary systems, furthermore, the responsibilities associated with ensuring that new versions of these systems were backward compatible, compounded by the difficulties involved in extending the functionality of such systems to include other vendor's equipment had made upgrading such systems a matter of extreme complexity.

• The lack of truly 'Open' multi-vendor communications perpetuated the restrictions associated with information exchange between different manufacturer's systems. Enabling the user to choose that system which was most suited to requirements without having to worry about information exchange with other manufacturer's devices was obviously a priority among machine consumers, if not among the manufacturers themselves.

Attempts at standardisation could quite easily have been stymied had individual vendors or other interested parties tried to force their particular solution on users as an international standard. The user community sidestepped this obstacle by creating the International Organisation for Standardisation (known not by the acronym, IOS, but by the Greek prefix ISO, which means 'same') from the national standards organisations of 89 different countries (e.g. ANSI - USA, BSI - U.K., DIN- Germany)

Since 1978 the ISO has been attempting to achieve consensus on machine-machine communication based on it's seven-layer Open Systems Interconnection (OSI) reference model protocol stack which borrows from many different systems, especially at the lower levels, but which seeks to provide for more advanced applications at the higher levels. The establishment of this completely new standard was intended to provide a reference model, to which the myriad different systems would conform in order to enable communication between them. A common communication platform for differing operating systems was thus to be provided, independent of vendor and machine.

As OSI protocols were devised by consensus rather than directly, through the modification of existing systems, they were to exist in theory long before they existed in fact, they are in effect however the officially recognised or de-jure standards. At present the base of systems installed make use of what, since the international acceptance of the OSI model have come to be known as the de-facto standards.

The seven layer OSI reference model illustrated in figure 5.0 may or may not be the optimum manner in which to structure network communications, it does however make an excellent way in which to compare different networking systems. Tables 1.0 and 2.0 illustrate the composition of some of the OSI dependent and OSI-Independent protocols stacks mentioned in figure 3.0, they are described in terms of the seven layer model.

Table 1.0 :- OSI Dependent Protocol Stacks

OSI LAYERS		MAP/TOP	EPA	DECnet Phase V	X.25 Messaging Protocol
7	Application	File Transfer Access and Management (FTAM) - (ISO 8571/1.4) Job Transfer Access and Management (JTAM) - (ISO 8831/2) Virtual Terminal - (VT) - (ISO 9040/1 Association Control Service Elements (ACSE) - (ISO 8649/50) Manufacturing Message Standard (MMS) - (ISO/DIS 9506/1/2) MHS X.400	Selected MAP Application Services are Provided Here	DECnet Application, User Application or OSI Application. Examples : ISO 8571 - FTAM (File Transfer Access and Management) ISO 8649 - ACSE (Association Control Service Element) ISO 9506 - MMS (Manufacturing Message Specification)	X.400 Messaging Specification X.500 Directory Services
6	Presentation	Abstract Syntax Notation One (ASN/1) - (ISO 8824/5) Presentation Service/Protocol Spec (PS/PS) - (ISO 8822/3)		OSI Presentation or DNA Session Control, examples : ISO 8822/23 - PS/PS ISO 8824/25 - ASN/1	
5	Session	Session Service/Protocol Specification (SS/PS) - (8326/7)]		OSI Session or DNA Session Control ISO 8326/27 - SS/PS (Session Service / Protocol Specification)	
4	Transport	Transport Service Protocol (TP4) - (ISO 8072/3)		NSP - From DECnet Phase IV ISO TP0,TP2,TP4 - (Transport Protocols) ARPA - TCP/IP	
3	Network	Internetwork Protocol (IP) - (ISO 8473/8348)		ISO 8473 - IP (Internet Protocol) ISO 8880 - OSI Network Service Connection and Connectionless	Packet Level
2	Data	Medium Access Control 802.3 (TOP) MSC 802.4 (MAP) Logical Link Control 8802 (LLC)	Logical Link Control 8802 (LLC) Medium Access Control 802.3	OSI Data Link ISO 8802-2 - LLC (Logical Link Control) ISO 9314-2 - FDDI MAC	Frame Level
1	Physical	802.3 Broadband/Carrierband/ Baseband CSMA/CD - (TOP) 802.4 Carrierband Token Bus - (MAP) 802.4 Broadband Token Bus - (MAP/TOP) 802.5 Baseband/Token Ring - (TOP) There are also stds. in development for Fibre Obtic Cabling of MAP/TOP	Cell Level Physical Connection for MiniMAP	ISO 8802-3 (IEEE 802.3) ISO 9314-1 - FDDI	Physical Level

FACTORY SOLUTIONS

The modular approach to networking the factory described in the introduction is to provide for the fulfilment of both the immediate enterprise to area requirements and for the expansion of any installed system into the shopfloor area or future refinement of any of the other systems. What is sought is not just temporary solution but a solid foundation flexible enough for easy expansion and stable enough for long term use. The following approaches were investigated, OSI-Dependent based on MAP/TOP and DECnet Phase V, and OSI-Independent based on TCP/IP incorporating Novell Netware.

Table 2.0 :- OSI Independent Protocol Stacks.

OSI LAYERS		TCP/IP	DECnet Phase IV	Novell Netware V 2.2
7	Application	File Transfer Protocol - (FTP) Simple Mail Transfer Protocol - (SMTP) TELNET Internet Name Server - (INS) Trivial File Transfer - (TFT)	Network Management : MOP,NICE, Loop Back Mirror Protocol, Event Log Protocol etc. Network Application : Data Access Protocol Command Terminal Protocol Terminal Communication Protocol	Workstation Shell :- Adaptable to 286, 386 and 486 DOS machines, also MAC, OS/2 and UNIX
6	Presentation			
5	Session		Session Control Protocol - (SCP)	Net Bios
4	Transport	Transmission Control Protocol - (TCP) User Datagram Protocol - (UDP)	Network Service Protocol - (NSP)	SPX :- Connection Oriented, Virtual Terminal Method
3	Network	Internet Protocol - (IP) Internet Control Message Protocol - (ICMP) ARP	Routing Protocol	IPX :- Connectionless, Datagram Method.
2	Data	TCP/IP can be used with many different types of physical and data link conditions	X.25, DDCMP Ethernet	
1	Physical		Line Controller Twisted Pair, Coax, Fibre Optic	

OSI-Dependent Solutions

MAP/TOP Solution : The MAP and TOP standards are basically application specific subsets of the ISO protocols designed to meet the needs of factory and office automation respectively. They are complementary standards that work together in separate but mutually supportive sectors of the same multi-vendor computer network [4].The function of MAP is specifically geared towards the control of and communication with shopfloor devices. TOP on the other hand was designed with the intention of enabling communication between office machines. MAP is a broadband or carrierband token bus communications protocol. at the physical layer, the other layers are as illustrated in table 1.0. TOP differs from MAP only at the physical layer, where it is based on baseband Ethernet technology.The advantages of MAP technology are fundamentally concerned with the sophisticated application layer service, the Manufacturing Message Specification (MMS). MMS is a draft standard, the purpose of which is to enable communication with all types of shopfloor equipment, including PLCs, NC controllers, and robots. Machines communicating through MMS are capable of stopping, starting and otherwise controlling client mechanisms such as those mentioned above. A diagram illustrating the application of MAP/TOP technology to networking in Boart Europe is illustrated in figure 6.0. Several points should be noted about this solution ;
• There are no products available to date which would enable the existing Hewlett -Packard or DEC equipment to communicate through MAP, nor, due to the relatively advanced age of these systems are there likely to be in the future.

• The HP machine would have to be replaced by, or interfaced to, the network through a HP9000 series 300 or 800 machine running HP-UX, (HP's version of AT&T's System V Unix). This system would be connected to the network through the HP OSI Express MAP 3.0 product. HP's version of MAP version 3.0 (HP MAP 3.0) would then be used to communicate with any other MAP compatible systems on the network [5]

• The DEC machine would also have to be replaced by, or interfaced to, the network through a newer machine, for example a microVAX 3600. Alternatively and as shown, the present system could be replaced entirely by an arrangement of DEC workstations running the newest OSI-compatible DECnet, Phase V. The physical and data-link layers of the OSI model are implemented in the hardware used to connect such a device to the broadband cable. VAX DEC MAP 3.0 software implements the network to application layers. VAX DEC MAP 3.0 is a multivendor broadband plant floor device interconnection for the manufacturing environment [6]. This machine could make use of DECs implementation of MMS, known as DEComni (DEC OSI Manufacturing Network Interconnect) which includes utility programs for creating and managing stored definitions of application data structures. This would provide the basis for replacing or upgrading the present shopfloor control system.

• Wide Area Network communications would be enabled through X.400 messaging on top of the TOP protocols for PC networking or through the HP9000/X.25 link, running on the HP MAP 3.0 stack and accessible over the LAN

• PC Local Area Networking would not be possible with the existing equipment.TOP stacks and hardware are not available for the Vectra. IBM supply the Concord MAPware series of equipment for connection of MCA and XT/AT compatible equipment to MAP networks but this would be of limited use.

Figure 6.0 :- Local Area Networking in Boart Europe according to MAP/TOP

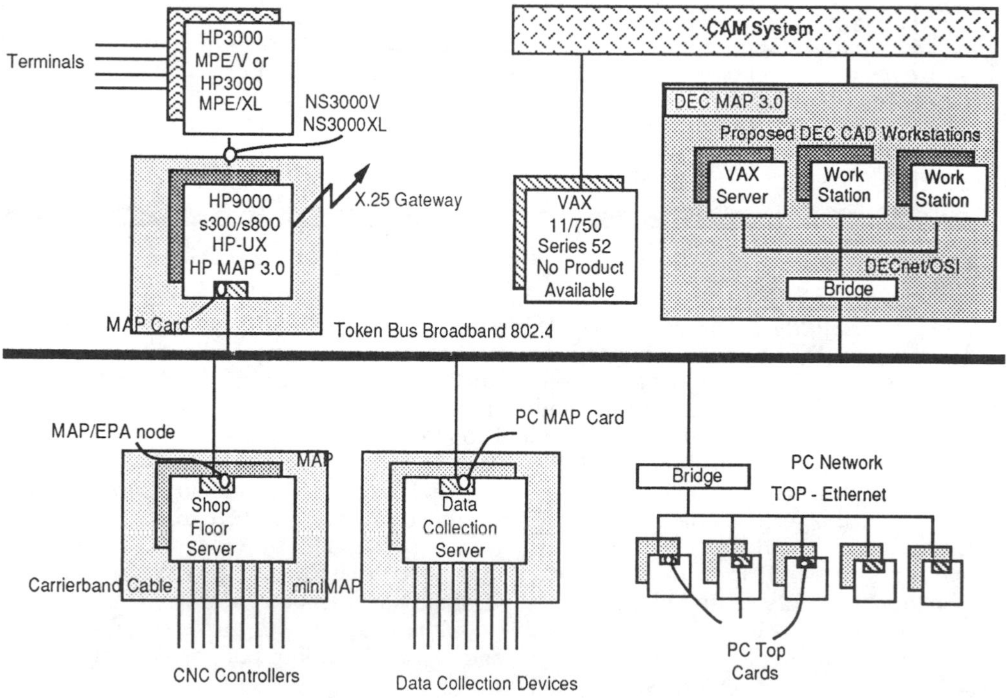

Assuming that the products necessary for PC communication under TOP were to become available and that the network was to be expanded to include shopfloor control and shopfloor data collection equipment running under the reduced MAP stack of miniMAP (See table 1.0), the network could be expected to resemble the illustration in figure 6.0 (Note that proposed equipment is surrounded by shaded boxes)

This system is not feasible for the following reasons
• The cost involved in upgrading or replacing the existing systems simply to get the network to function is exorbitant. Furthermore the technology associated with MAP Broadband products is already well beyond the budget of most medium sized companies.
• PC networking using TOP is unlikely using the present systems. Firstly because of the fact that little or no products or support are available, secondly because of the fact that running the entire seven layer OSI stack on a machine with a maximum of 640K available RAM is likely to be a slow if not impossible process.
In conclusion, MAP/TOP systems, while becoming more available, as evidenced by the support which companies like HP and DEC are providing for the systems on their newer models, are expensive and in the case of equipment designed and built before the widespread acceptance of OSI principles simply impossible. Implementing MAP in anything other than a start-up situation therefore remains a highly dubious proposition.

DECnet/OSI Solution : DECnet/OSI, also referred to as DECnet phase V is part of the newest implementation of the Digital Network Architecture (DNA). DECnet Phase V supports the integration of OSI protocols while enhancing DECnet and providing compatibility with both preceding versions of DECnet and TCP/IP systems [7]. Table 1.0 illustrates this version of DECnet and shows how it incorporates the proprietary OSI and ARPA protocols, this can be compared with the last OSI-Independent release illustrated in table 2.0. DECnet/OSI is based on standard or thin baseband coaxial cable, although fibre-optic cabling can also be used as the network backbone. As stated DECnet/OSI is compatible with both MAP and TCP/IP and theoretically seems to offer the best solution in terms of preserving the 'openness' of the network. The fact that DECnet runs on Ethernet makes it possible to avail of some of the more enticing features of MAP technology (MMS) without incurring the expense of purchasing broadband communication equipment. The DECnet/OSI solution is illustrated in figure 7.0. Some points to note are as follows ;
• Minicomputer communications can be effected through TCP/IP. DECnet Phase V will not, unfortunately, run on the existing VAX, consequently, as with the MAP solution a new system is necessary, once this system is in place however, DECnet/OSI will interface with any TCP/IP stack. The existing shopfloor control system could theoretically be replaced by an arrangement of workstations, all running DEC MAP 3.0 and communicating over the DECnet through the File Transfer Access and Management protocol (FTAM) included in DEC MAP 3.0 and DEComni, DEC's aforementioned implementation of MMS.
• The HP3000 system, be it the existing MPE/V or a replacement MPE/XL could make use of HP's ARPA services or independent TCP/IP stacks such as those manufactured by the Wollongong group (WIN TCP/IP) for HP systems. PCs could also avail of the TCP/IP relationship through the use of various ARPA services for DOS applications. (Clarkson PC TCP/IP, WIN TCP/IP for DOS)
• PC networking could be accomplished within the DECnet architecture through the use of a novel approach by DEC and the Microsoft corporation known as 'Teamlinks'. This is a DEC compliant application of the microsoft LANmanager, known as the Personal Computer Systems Architecture (PCSA) version 3.0. VAX or microVAX servers, (including the VAX 11/750 present in the factory) can run 'VMS services for PCs' server software and the PCs themselves run DECnet PCSA client software.Physically, the PCs are to be connected to a 'Chipcom' management Hub using Unshielded Twisted Pair cabling (UTP) and from there to the backbone. The advantages of such a system will be outlined in the next section.
• X.25 Wide Area Network communications are provided through a PC gateway, resident on the network, communicating with the other elements through MS-Mail (Microsoft's implementation of the very successful 'Courier' mail system) and connected to the local telephone exchange through a Packet Assembler Disassembler and X.25 Modem. Alternatively X.400 communications are available through the use of DEC's All-in-1 application, which resides on the network server.
This solution is certainly more viable than the previous pure MAP solution, connectivity is easily achieved to the ethernet-based network and all of the solution modules can be catered for by connection to the backbone. DEC's move to OSI compliancy has been much

Figure 7.0:- Local Area Networking in Boart Europe according to DECnet/OSI

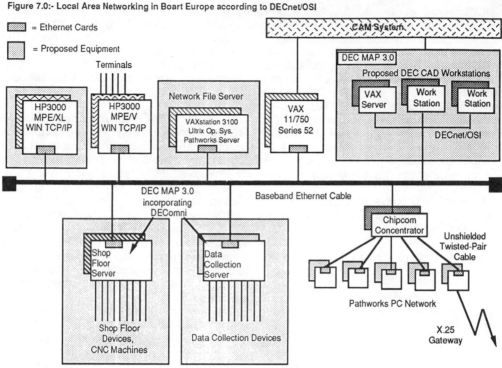

postponed but now that it has finally arrived it seems an attractive solution. The joint ventures between DEC and microsoft and DEC and Chipcom mark a step away from the strictly mainframe/minicomputer area of the market which has been DEC's traditional area of expertise and into the most up-to date end of the PC networking field. This type of collaboration can be expected to be more and more prevalent as the older companies seek to° adapt their products to the new distributed computing environments. The solution is feasible and for established DEC customers it offers a relatively painless approach to OSI compatibility.

OSI-Independent Solutions

Ethernet Solution : A networking solution which can be developed independently and yet still retain a degree of 'openness' and the possibility of migration to OSI compatibility is an attractive option to most companies. Such a system is likely to be easier to maintain once installation has taken place, and certainly cheaper than a solution provided by companies such as DEC or HP. It has the added advantage of being able to select equipment on it's own merits, and not being restricted to products endorsed by the largest vendor. What follows and is illustrated in figure 8.0, is a description of a solution devised around the application of existing standards to the network, with an eye to future refinement and expansion. The following points are worth noting.
• Minicomputer Communications. Minicomputer communications are to be provided by the ARPA TCP/IP protocols, versions of which are available for virtually every operating system and machine. Typically Wollongong TCP/IP could be used to provide communication between the MPE/V, VMS and DOS operating systems. Use of a third party vendor such as Wollongong in this case, is actually cheaper than buying direct from DEC or HP. TCP/IP systems are likely to be sufficient in terms of Open systems in the short term , however a migration path to OSI compatibility will have to be provided in the long term. At present there are several options for existing TCP/IP users or those planning to use TCP/IP when it comes to providing communications with OSI equipment or using OSI applications. Some are protocol based, some are service based. At present RFC 1006,

ISO transport services on top of the Transmission control protocol allows Internet users to run OSI applications (typically MMS and FTAM) over a TCP/IP stack. However RFC does not provide communication with the full OSI stack. A gateway is required for this purpose, eventually however, due to the massive installed base of TCP/IP systems, OSI communication layers are expected to replace the communications of the TCP/IP protocol, established TCP/IP users should then be able to upgrade their systems to full OSI compatibility.

• PC Networking. This is to be achieved using the system introduced in the previous section of Ethernet over Unshielded Twisted Pair (UTP) cable. Such systems, based around the IEEE 10BASET standard are rapidly superseding the older coaxial cable based 10BASE5 and 10BASE2 Ethernets [8]. The reasons for this are three fold, firstly, UTP is a cheaper cable for networking requirements, typically the building to be networked is 'flood wired' with an abundance of RJ45 nodes, to which users can be connected as the needs arise, this endows a greater degree of flexibility to the network. Secondly, managing the network through the use of a central hub is simplified through the use of the Simple Network Management Protocol (SNMP). Such a device constantly polls the attached nodes for evidence of malfunction and shuts them down should a discrepancy occur. This non-disruptive method of network management has many advantages over the older coaxial system where cable breaks could cause havoc by bringing down the entire network and being exceedingly difficult to locate. Thirdly, the standard in development for data transmission rates of up to 100Mbps over UTP ensures that the network will be capable of supporting traffic from virtually any type of future node.

Fig 8.0 :- Ethernet Solution

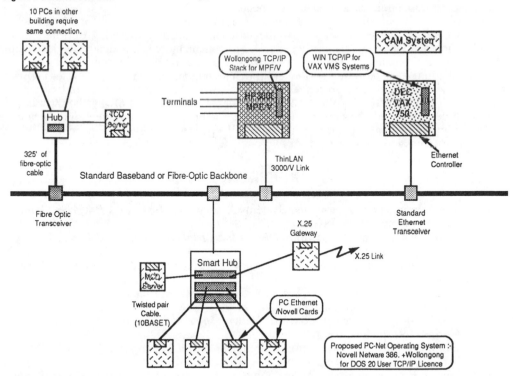

• Wide area networking can be accomplished through the use as in the DECnet solution of a PC based gateway, resident on the network. This gateway communicates through the server with it's sister nodes on the network using an internal mail package (MS-mail for example), then communicates with the global X.25 network through a Packet Assembler/Disassembler Device (PAD) and X.25 modem resident on an X.25 card through the local telephone exchange.

• The present shopfloor control system is not to be adjusted, however should the need arise, further nodes can be connected to the hub or directly to the network backbone. These nodes could communicate using TCP/IP and if necessary the RFC 1006 facility outlined above.

This solution contains many of the newest approaches to plant networking, Figure 8.0 illustrates these, this diagram does not stress the incorporation of proposed systems but these could be included as in the discussion of minicomputer communications. The diagram does illustrate the distribution of the network between the buildings, an application to which intelligent hubs are particularly suited [9]. Several points are worth considering about this solution, these are ;

• The cost of this solution is the least of all of the systems considered.

• Modular installation of such a system is easily managed. This involves the initial installation of the structured cabling and hub necessary for PC communications, coupled with a facility wide backbone in MCD and fibre optic connection to the remote network in TCD.This provides the foundation for future development of the network to include communication with the minicomputers and the Shopfloor.

CONCLUSIONS

The final solution has been selected as being the most suitable for the following reasons:

• TCP/IP is a proven networking solution with a developing migration path to OSI compatibility.

• The structured cabling approach enhances the flexibility of the system, augments the management capabilities and is relatively cheap.

• The solution is well suited to modular development, laying an initial foundation which fulfils many of the requirements of the enterprise to area modules outlined in the introduction and which can be easily expanded to contain more sophisticated devices within the enterprise/plant/area or to make contact with machines on the shopfloor.

In conclusion it must be stated that even though an ethernet based TCP/IP system remains the most viable solution at present, and in particular when the relatively old systems in question in this particular application are considered, this may not always be the case. As has been demonstrated, the newest models manufactured by the large vendors are capable of OSI communication, the choice between OSI-Dependent and OSI-Independent networks will not be so easy to make in the future.

REFERENCES

[1]   The Computer and Automated Systems Association of SME : *A Program Guide for CIM Implementation*, ISBN 0-87623-206-7

[2]   Digital Equipment corporation : *Digital Industrial Networks Guidebook* , 1990, pp 1-8, 1-9

[3]   Tanenbaum Andrew S : *Computer Networks* , Prentice Hall International 1988, pp 2-3

[4]   B.Tangney, D. O'Mahoney, '*Local Area Networks and their Applications*', Prentice-Hall International, 1988, pp 94-99.

[5]   Hewlett-Packard Corporation, '*HP Networking and Communications Specification Guide 1991*' pp 94-98

[6]   Digital Equipment Corporation, '*Digital Industrial Networks Guidebook*', 1990, pp 4-7 - 4-8

[7]   Digital Equipment Corporation, ' *Open Systems Handbook, A Guide to Building Open Systems* ', 1991, pp 10-5, 10-6

[8]   M. Johnston, 'Troubleshooting UTP for LANs ', *Data Communications International*, September 1991, pp 73 - 79

[9]   J. Herman, ' Smart LAN Hubs Take Control ', *Data Communications International*, June 1991, pp 66 - 79

Part IV
CAD/CAM Databases and Applications

FLEXIBLE REFERENCE MODEL AS BASIS FOR SIMULTANEOUS ENGINEERING IN AN INTEGRATED ENVIRONMENT

Carlos F. Bremer and Henrique H. Rozenfeld

LAMAFE-EESC-USP, São Carlos, São Paulo, Brazil

ABSTRACT

The application of CIM concepts in industries results in a new approach interrelating the design and the process areas. In this new situation the Simultaneous Engineering philosophy plays an important role in enterprise integration. The São Carlos CIM Center - USP develops much research related to CIM, mainly in the areas of enterprise modelling and process planning. For CIM planning and implementation, a flexible reference model has been developed, which allows the modelling of an enterprise according to its real necessities. For the link between the design and process areas, the model is used as basis for function and data integration, as well as for other models defined by the enterprise. A Computer Aided Tool for the modelling, based on the flexible reference model is in the design phase. This model is applied in the Pilot Integrated Factory of the São Carlos CIM Center and also in an initial phase at a local industry.

INTRODUCTION

The world context of market and business demands an increasingly higher adaptability on product design and manufacturing within industry. New products launching time, as well as the delivery time for one job production, is nowadays a crucial competitive factor.

This way, functions that were developed in a divided and sequencial way, are now done in an integrated and simultaneous manner (<u>Fig. 1</u>). The results are related by Putnam, showing that the launching time of a company is reduced from 8 weeks to 6 weeks [1].

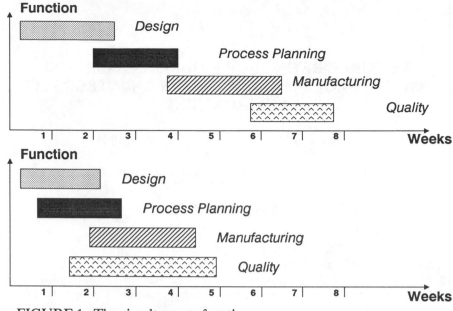

FIGURE 1 - The simultaneous functions

Simultaneous Engineering (SE) is a result of integration between the design and process planning areas [2], where other ones such as quality assurance, marketing, purchase and so on must be taked into account. Integration of different areas is not the actual SE aim, but rather that determinated functions of each of these areas should be integrated [3, 4].

For sucessfull SE introduction and operation, not only technical measures must be applied, but also other ones such as personal qualification, reorganization and so on [5]. The utilization of a reference model for planning and operative SE is of great value in order that all of the other variations be considered. However the integration of the enterprise as a whole should be taken into consideration in this context.

This paper describes the development of a flexible reference model at São Carlos CIM Center - USP and its application in an integrated environment, the Pilot Integrated Factory. The model servs as a theoretical framework for integration, and it is presently still applied in a manual way. Its migration to computer aided tool is already in the design phase. Implementation in an enterprise is also being performed at a machinery industry with one job production, specifically in the product development and specification and process planning areas.

The São Carlos CIM Center - USP

Created in 1987, the São Carlos CIM Center - USP develops 28 projects with a group of almost 70 researchers basically in the areas of CAD/CAE, Finite Scheduling, Shop Floor Control and Enterprise Modelling.

The project that links all the others, with the target of demonstrating an integration solution is the Pilot Integrated Factory. Based on the reference model, a complete integrated solution, from product development up to its manufacture has been developed. The system that guarantees the integration is the CIM-Manager, that uses the Metadata Base concept [6].

THE DESIGN FOR MANUFACTURING

One of the requirements that industries need in order to make themselves competitive is the product launching time. Much work has been developed in design for manufacturing, targeting to reduce the development and manufacturing time. Figure 2 shows the situation of design and process planning.

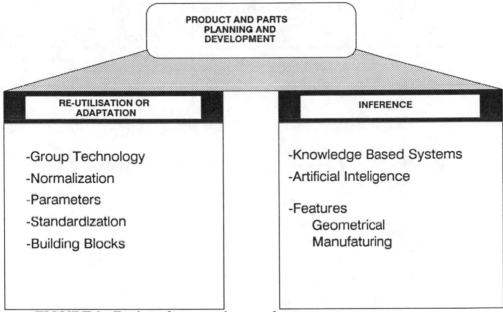

FIGURE 2 - Design of new products and parts

A new product can be designed using an existing one, through the re-utilisation of some parts or their adaption. For this, techniques such as Group Technolgy can be applied. Another aspect is the search to utilise as many possible parts that are already manufactured or bought. In these cases the parametrisation and normalisation is an important effort.

If re-utilisation is not possible or planned, design for manufacturing can be reached through the automatic inference, based on features. These features are used in the process planning for the automatic generation of a process plan. There is much research in this area, and one of the trends is to apply automatic inference on restricted knowledge sectors, for example in gears manufacturing.

In reality, the aspects described above have been introduced with a view to designing for easier manufacture. However it is necessary to achieve an integration of different functions, beginning in the product specification, targeting the product planning. Product planning means that by the product specification and design, inspection procedures, assembling dispositives, and so on are already being decided. This results not only in a lower launching time, but also in a higher product and manufacturing quality and reduced general costs, increasing the competitivity of the enterprise.

DESIGN AND PROCESS PLANNING INTEGRATION ASPECTS

The integration between product and process planning can be considered as one of the first integration islands that has been realized, as shown in Figure 3.

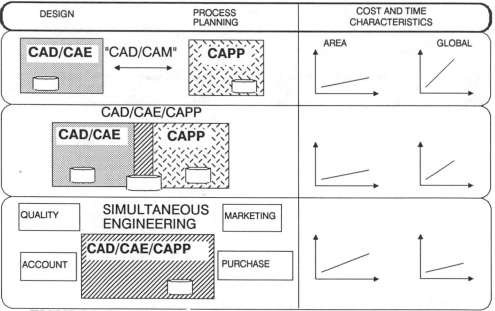

FIGURE 3 - Integration evolution between product and process planning

With the development of NC machines, the first CAM or NC programming systems were applied. In a further stage, with the increasing use of CAD/CAE systems, the term CAD/CAM represents for the integration between these systems. This integration was produced by automatic NC-Programm generation from the geometry produced on the CAD system. The date of both systems although were stored at their own databases, caracterizing in reality an interfacing between the systems. The time and costs of these areas were lower and the total time and product cost were not greatly reduced.

A more effective integration was reached with the development of common databases between the design and process planning and with the utilisation of product models, in which, along with the geometrical data, also technological, managing and other kind of data are associated. The time and costs of each area is reduced, but the most important reducing factor is applied for global time and costs.

Nowadays, the integration is more relevant, through model products that have more associated data and use a CIM view. In the context of design and process planning Simultaneous Engineering adopts integration with other areas, like Quality, Purchase, Marketing and so on. This greater relationship between different areas results in a management and operational problem, since more functions are involved. Figure 4 exemplifies the integration of functions.

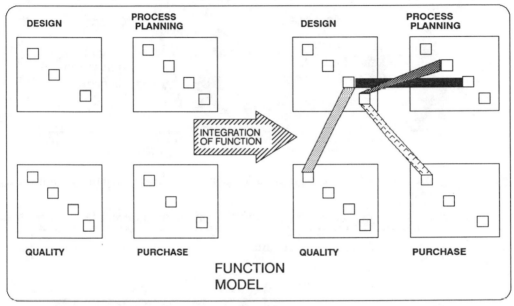

FIGURE 4 - Function integration

With the integration of areas, some functions of each sector can be eliminated, modified or added. As an example, the function of material determination is divided into three functions: resistance verification, material specification and weight determination. In an integrated environment, the function of material determination can be modelled in four functions: resistance verification, consultation of the purchase database, material specification as in the internal inspection procedure and weight determination. The purchase consultation function is new, and based on the functions already performed by purchasing and quality control, the function of material specifications is then modified, targetting to guarantee the material quality and to provide process planning with an inspection norm.

A function modification results in a new input or output of information . This function can, although be realized by the same group, which will need, for instance, a new qualification, more motivation and a computer aided tool to manipulate greater quantity of information. Another alternative is that the group receives more people or that they be transferred. In this case a reorganisation is necessary and consequently a

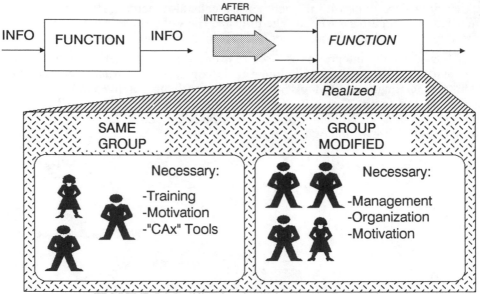

FIGURE 5-Modification in an integrated scenario

new management form, as well as a motivation plan for an interdisciplinary teamwork. Figure 5 represents this scenario.

In both cases, a series of decisions must be defined on different levels. The lack of a global view of the modification may occur with the change of integration function and can be predjudicial for the implementation of Simultaneous Engineering, as seen in Figure 6.

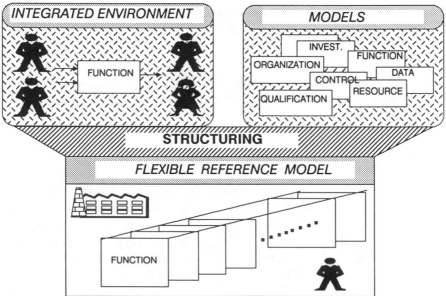

FIGURE 6 - Lack of a global view in an integrated environment

The implementation of new procedures, without a structure can result in a certain amount of chaos in the integration. To implement and operate the Simultaneous Engineering in a structured way a reference model is proposed, which is flexible, targetting to satisfy the reality and requirements of the enterprise.

THE REFERENCE MODEL FOR INTEGRATION

There are already many proposals for enterprise integration. Part of the proposals describe as necessary a reference model, that represents and abstracts the reality of the enterprise and serves as a basis for the integration, since this is done in phases. However the existing reference models are fixed, determining to which models the enterprise must adopt. The CIM-OSA of Project ESPRIT uses four models for derivation: function, information, organization and resources [7]. The Integration Infrastructure Model adopts the following models: function, data, application systems, data storage, networks, hardware, staff and qualification and organisation units [8]. It can be assumed then that the model to be adopted should be flexible, defined by the reality and requirements of the enterprise [9].

Based on the aspects described, a reference model for integration is proposed. This model has as kernel the function, information and integration models, being complemented according to the integration concept adopted by the enterprise. An example of other models can be seen in figure 5.

Most of these models are, based on the existing functions and their information, it is possible to characterize an existing and an idealized enterprise situation, which achieves the enterprise targets. Since the models form a complete reference model, the consequences of a function modification can be verified.

As an example, if a function executed manually is planned to be aided by a software, the existence of a qualification model, oblies the enterprise to verify which are the necessary training activities.

An important utility of the reference model is the possibility to navigate into the processes, allowing the identification of their relationsships with the other models adopted. With the utilisation of this model by the people involved in an integration management and operation, they all share from an updated and common situation and information, it being an important tool for the decision-making process. These people can also generate reports to help in the execution of common functions, meetings or even auditories like ISO-9000.

The reference model described in this paper is a theoretical framework for integration. Its optimized utilization depends on developing it as a software, like a computer aided tool. The specification has already been done and it is now in the design phase.

With the model in the enterprise the following benefits in Simultaneous Engineering can be obtained :

- a base for the planning of new procedures or technologies;

- process control as a whole;

- elimination of redundancy and incosistency;

- clearer definition of the business process;

Next the function, information and integration model will be detailed, that constitute the kernel of the reference model.

Function Model

This model is the nucleous of the enterprise model. It contains all the existing function structured in a hierarquical form. Each function is interrelated with the others through information, that represents as documents as well as files or displays. Information that interconnect the functions are objects of the information model.

As well as this control and resources elements are determined. Each function block is identified also by detailing levels through the simbols + (yes) and - (no). All the other models adopted are specified in the function block, as seen in Figure 7.

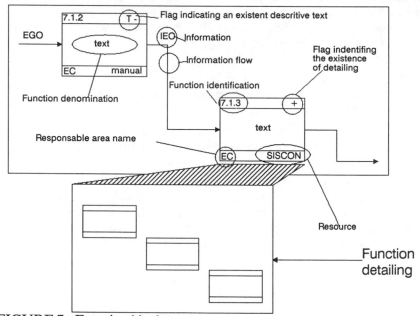

FIGURE 7 - Function block

Information Model

In this model a data dictionary with the register of all exixting documents, files and displays (views) and their attributes fields exists . The specification of the data model is in this model.

The logical model used is the entity/relationship, where the connection with the function model is an information dictionary.

Integration Model

This model has the specification of the network, data base implementation and of the communication norms, interconnecting the both other models. This model characterizes the physical implementation of the former models.

REFERENCE MODEL APPLICATION

The first application of the reference model has been performed at the Pilot Integrated Factory of the São Carlos CIM Center - USP. The reference model has, as well as its of its kernel, the following models:

Qualification Model

This model describes what knowledge is required for the realization of the specific function. This model serves in a further stage for the acquisition of knowledge with the view to the development of a knowledge based system.

Hardware Model

Here hardware elements are specified which the function utilize. If it is manually executed, nothing is specified. This model servs for investiment planning, network projects and so on.

Software Model

This model orientates which systems execute the functions. It allows other people to know and access other systems dependent on their priority in the enterprise.

The utilization of this model makes possible the harmonic development of a first scenario integration. In this first scenario it a simple part is manufactured, that goes through many phases of the enterprise, like sales, product planning, process planning, NC programming, finite scheduling and a flexible cell control system. The CIM Manager system, based on the Meta Data Base concept, guarantees the real integration. The quality control, material administration and shop floor control functions are executed without a computer system.

With the model as basis a second scenario is planned, with the inclusion of new hardware, software and machines. For the functions realized without a computer system, software is being developed or bought.

Another application of the model is in the initial stages in a machinery enterprise, which already has a complete E/R model. The macro bussines functions has been modeled and also the functions of product development and specification and process planning, with the target of implementing Simultaneous Engineering. The reference model has the resource and organization models, besides its kernel.

CONCLUSIONS

The utilization of the flexible reference model in the São Carlos CIM Center - USP has demonstrated that it is an important tool for a common understanding of the business process and as a normalized language for the different research areas involved in the integration at the Pilot Integrated Factory. It allows groups, made up by multidisciplinary people, to use it for the definition of new integration scenarios, consequently of new functions and their reflections in the other models adopted .

With the knowledge already acquired by its implementation in the integration of design and process planning areas, the flexible reference model has enabled:

-navigation through the functions, verifing the consistency and updating of the information flow;

-that the CAPP system works only with updated designs;

-greater adaptability in new products design, as well as lower general costs;

-verifation of the documents and displays during the function navigation;

-creation of many reports, such as: documents flow, functions realized by one area, functions that use an specific software and so on;

-creation an open integration island, already structured for CIM.

To obtain all the possible advantages that the flexible reference model allows, it will be necessary to migrate it to a computer aided tool. After its application in the Simultaneous Engineering, it must be extend for the other enterprise areas.

REFERENCES

1. Putnam AO: A redesign for engineering. Harvard Business Review :139-144, May-June, 1985.

2. Garret RW: Eight steps to Simultaneous Engineering. Manufacturing Engineering :41-47, November, 1990.

3. Spur G: Integrierte Producktentwicklung. ZWF-CIM, 3:CA14-CA14, 1991.

4. Pedersen MA, Alting L: Life-Cycle design of Products - An Important Part of the CIM Philosophy. Proceedings of AUTOFACT, :2.1-2.15, 1991.

5. Painter MK, Mayer RJ, Cullinane TP: The Many Faces of Concurrent Engineering. Proceddings of AUTOFACT, :17.1- 17.25, 1991

6. Rozenfeld HH, Takahashi S: Management System for Computer Integrated Manufacturing with Different Automation Levels. Proceedings of FAIM, 1991.

7. Stotko, EC: CIM-OSA. CIM-Management, 11:9-15, 1989.

8. Mertins K, Suessenguth W: Integrated informartion modelling for CIM. Computer-Integrated Manufacturing Systems, 4(3):123-132, August, 1991.

9. Blinn T, Mayer RJ, Cullinane T: A Service Approach to Information Integration in Concurrent Engineering Environments. Proceedings of AUTOFACT, :7.27-7.38, 1991.

PROMOD: A NEXT GENERATION PRODUCT DESIGNER SYSTEM

Bartholomew O. Nnaji

Automation and Robotics Laboratory, Department of Industrial Engineering and Operations Research, University of Massachusetts, Amherst, Massachusetts, U.S.A.

ABSTRACT

In today's design method, there is a lack of communication between the designer and the manufacturing engineering. This lack of communication is more enhanced by the inability of today's designer (CAD) systems to capture and propagate the designer's intentions for a product. The goal of this paper is to present a method by which a designer's intents for a product can be captured and propagated to all the planning modules. This approach allows for a tight coupling of design and manufacturing of a product as well as all its life-cycle considerations. We will present *ProMod* which is an embodiment of this type of designer system developed at the University of Massachusetts at Amherst.

INTRODUCTION

As companies strive to remain competitive in the global market, there is an increasing realization that reduction in time from product conceptualization to marketing is crucial. There is also a realization that this reduction can only come if the time from design through manufacturing is reduced. This means that the product design and process planning for manufacture must be tightly coupled. A second area of concern is the cost of a product. For the finished product to be competitive, it must carry competitive price tag and have high quality.

It has become clear that the factories of the future must be automated in order for companies to compete and generate the required profit margin and at the same time maintain high quality. The success of the Japanese industries, in large measure due to automation, has underscored the need for this technological direction. Automation means that individual processes must be computerized and integrated and that such processes must be capable of performing their transactions with other processes without dedicated need for human intervention. It also means that some machine information processors must become "intelligent" in dealing with their tasks and the environment where they perform these tasks. This role has traditionally stayed with the human decision maker.

The major bottleneck in the integration of design and manufacturing is that designer systems generate information which must be interpreted by humans in order to develop a manufacturing process plan. For example, the computer-aided design (CAD) system can only carry geometry information which do not tell the planner what *features* exist in the product, the function of the product *features* and how to make the part. Although a designer normally comes up with these sorts of information, the current CAD system does not have any capability to propagate these most crucial data.

The implication of this problem is that since a CAD system cannot capture, represent and deliver design information to the planner, the one way to plan is to use another human who knows the needed manufacturing processes to re-generate and represent the design data in a manner that can be used for manufacturing planning. This has been the traditional approach.

One clear solution to the problem is to develop a new type of CAD system which will be capable of capturing both the usual design data (gragphic drawing, and dimensions) as well as the designer's intentions (how the parts fit together in terms of spatial relationships; tolerances; bill of materials; surface finish; etc.).

This paper will discribe *ProMoD*, a new generation designer (CAD) system developed at the Automation and Robotics Laboratory at the University of Massachusetts at Amherst which is capable of capturing all the design data as well as the designer's intentions for the product. It is also capable of evaluating the product for life-cycle issues (to assure quality and manufacturability); and propagating the data to planning modules. This system is able to deal with all geometric reasoning issues. Since this type of CAD system cannot be developed for all purposes, we have chosen three areas for which we have developed a complete designer system: sheet metal, mechanical (and robotic) assembly and machining. This paper will discuss the principles of such a designer system.

BACKGROUND

Recent efforts in industry and research laboratories across the United States, Europe and Japan have attempted to integrate the design work with other life cycle issues, including the evaluation of the product for producibility, inspectibility, serviceability, etc. This means that the design of the product and the manufacturing system are executed concurrently.

Nevins et al stressed that engineering schools teach a fairly straightforward version of how something is designed and illustrated this approach in Figure 1 [1]. It can be seen that product and process designs are sequential in this approach. Needs are determined, product specifications are done, trial designs are produced, prototypes are produced for bench testing, and then final designs are produced. After all these then the required manufacturing process plan is developed.

The traditional relationship between the designer and the manufacturing personnel has always been one of initiator and implementer. This relationship merely required the designer to dictate what should be done without much care about if it could be achieved. Thus, designs are created and it is the responsibility of the manufacturing engineer to determine whether or not the product is manufacturable. If the product violates manufacturing rules, then either the manufacturing environment is redesigned to accommodate the new product or a redesign of the product by manufacturing engineers, in consultation with design engineers, will be essential. Violations can occur in several ways. Available tools and equipment might be incapable of fabricating the product to the desired tolerance, or the general cost of making the product might be beyond the budget of the company.

It is clear then that design of a relatively complex high-technology product will require a lot of analysis, investigation of basic physical processes, experimental verifications, complex trade-offs, and difficult decisions [1].

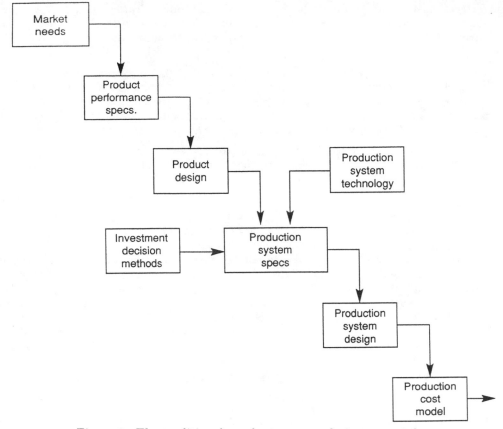

Figure 1: The traditional product-process design approach

The lack of communication between the designer and the manufacturing engineer can be bridged by:

a) training the design engineer in manufacturing principles,

b) making the manufacturing engineer a design engineer, or

c) having manufacturing engineers work with design engineers during the design of a product.

The first two options are difficult to achieve since neither the design engineer nor the manufacturing engineer would want to take on the full responsibility of the other. It seems plausible that the manufacturing engineer and the design engineer could work together to produce a manufacturable product.

The state of the evolution of product and process design interaction is as shown in Figure 2 which is adapted from Nevins et al (1989).

This figure shows the depth of interaction between the product design process and the process design. Since no designer has the total knowledge necessary to evolve the concurrency in design and product, and process, a team approach is the advocated norm today. It means that people whose roles normally come later in the life of the product evolution are involved early in the design process.

The method of effecting this cooperation can vary from company to company.

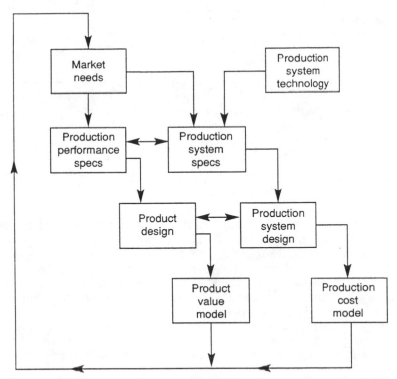

Figure 2: The strategy for concurrent product and process design

But in general, the manufacturing engineer would provide some manufacturing rules which become design constraints for the designer. A designer can submit a "draft" design to a manufacturability evaluator, which can be an expert system or just an automated evaluator. Design for manufacturability requirements will vary from one manufacturing application to another. That is, it will be different for metal machining than for assembly processes.

Designing for manufacturability requires the fusion of the designer's intentions for a part with the manufacturing engineer's manufacturing requirements for the product. The designer's intentions are a set of functions which the product will provide or require. The functions are related to the features of the product [2]. For instance,

   a) What is the function of a given feature on the part?

   b) What operation is implied by this feature?

   c) What tool can be used to make this feature?

In actuality, the design engineer is usually totally responsible for answering question (a). He communicates the responsibility of a feature entity on a part for life cycle consideration. It could be that the feature is on the part only to satisfy a temporary function, for instance ease of assembly, or that the feature is there to serve a life-long function of the part or the product. In some cases, a surface may only be on a part to provide support or to provide solid requirement for the part. In the case where a surface provides support, it may be required to satisfy certain spatial relationships with other part surfaces or with features of other parts.

Question (b) is partially answered by the design and partially by the manufac-

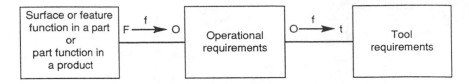

Where: F is the feature, surface or part funtional responsibility
O are the operational requirements
f is an algebraic mapping function
t are the tool requirements

Figure 3: The mapping relationship of object and feature form and function

turing engineer. Although this question derives partly from the functional intentions of a part or a part feature with regard to its intermediate and ultimate use in some part or product, the operational requirements in large measure derive from geometric considerations and material properties. For instance, while knowing that a hole is to be made may be a designer's prerogative, knowing that a drilling operation will yield the geometric configuration which we call a hole is an operational consideration which the manufacturing engineer stipulates. While the design engineer stipulates the tolerance requirements for the part, the manufacturing engineer uses operational requirements planning to assure that the desired tolerance is achieved. The manufacturing engineer goes the next step to stipulate which machines and/or tools can be used for the drilling operation and establishes not only the operational sequence but the operational tool requirements as well (Figure 3).

In answering question (c), the manufacturing engineer takes the designer's intentions, which can include the temporary functional objectives of the feature as well as the operational requirements for making the part, to generate a workable process plan which will obey the capacity constraints of the manufacturing environment of the organization.

Designing for manufacturability allows the designer to work with a set of manufacturing constraints and requirements in the design on the product.

It is clear from the above discussion, that design for manufacturability is achievable by establishing some design rules in the form of knowledge engineering which can be used to guide the design to obey manufacturing requirements.

Computationally, predicate logic, semantic networks, production rules or frame theory can be used to represent the manufacturing knowledge. These methods can represent data in a uniquely specified structure which can be augmented with deductive reasoning [3]. The more common method for representing manufacturing knowledge has been frame theory. In this application, frames can be used to capture the various manufacturing rules as well as tool requirements for various processes. Certain parameter constraints in the process planning activities such as process selection, machine selection, process sequencing, tool selection, and jig and fixture selection can be captured in the manufacturing knowledge base. It will be useful to use manufacturing application areas to illustrate the design for manufacturability concept in an application area. Then we will present the product modeler to show how future CAD and manufacturing relationships will work.

The basic rules in designing for manufacture are [4]:

  a) Use standard components whenever possible,

5

b) Take advantage of the work material's geometric shape in order to design the components,

c) Use previous designs whenever possible,

d) Minimize required machining whenever possible,

e) During geometric shape design, consider the ease of material handling, fixturing, machining and assembly,

f) Avoid over-specified tolerances and surface finish specifications, and

g) Consider the kinematic principles during the initial steps of a design.

Nevins et al (1989) outlined a guide to strategic design in the concurrent engineering approach:

1. Determining the character of the product to understand what kind of product it is and how to develop appropriate design and production methods. The character of the product is articulated with the resulting consequences. For example, complex items, with no model mix which are used by untrained personell, must have 100% reliability when used once and thrown away. Examples of such products are missiles hand grenades and fire extinguishers. The consequence of such a product character is that the product should be made of high quality parts, glued or welded together. It should not be designed for repair after production.

   On the other hand, if the product is a complex item, with model mix, the user has options. Also, if this user is untrained and the product is intended to last for years, then the consequence is to make the product with high quality parts, screwed together and to provide replacement parts and accomodate field repair service.

2. Subjecting the product to a product function analysis to assure that the design is made rationally.

3. Carrying out a design for producibility and usability evaluation to determine if the product's producibility and usability can be improved without impairing its desired functions.

4. Designing a fabrication and assembly process for the product that takes into account its character. This task can be accomplished by determining the appropriate assembly sequence, identifying subassemblies, integrating a quality control strategy with assembly, and designing each part so that its functional tolerances and tooling tolerances are compatible with the assembly method and sequence, and its fabrication cost is compatible with the product's cost goals.

5. Factory design should fully involve the production workers, assure that inventory (including in-process inventory) is minimal, and is integrated with the procedures and capacities of the vendor.

THE PRODUCT MODELER

The Product Modeler is capable of creating a product in engineering and directly using the basic elements of the model throughout the life cycle of the component. Figure 4 contains the system architecture of the product modeler. The life cycle can range from engineering to manufacturing to quality to support. The product modeler is thus expected to possess an intelligent database which documents all the product information. This database supports many architectures. This figure contains the conceptual environment of the Product Modeler.

The concept of concurrent engineering whereby some aspects of the process plan are evolved at the time of design is a goal of the product modeler. Some of the key development areas which will play significant roles in the evolution of the product modeler are the generalized feature characterization allowing for design with features and feature based reasoning for process plan generation; and development of expert systems.

The ProMod [2] systems illustrate well the product modeling type of CAD system. The ProMod system embodies the attributes of the future CAD system as enunciated above. As can be seen, the CAD systems of today produce drawings on paper or geometric models on the computer as a result of a design. Since they are only capable of producing the geometrical characteristics of the part, they are merely able to assist in the final stages of the design process which is the design documentation (e.g. drawings). The *product modeler* can capture the designer's intentions throughout the whole stage of the design process. The *product modeler* is concerned about the process which generate the designer's intentions to assure preservation of the intention information. For instance, the physical and geometric properties of a part are normally conceptualized at the early stages of design – the conceptual design and detail stage rather than the drawing stage. In conventional design/manufacturing systems, the production planner must supplement the raw geometric data with additional manufacturing information such as dimension tolerances, surface finish, straightness of surface, spatial relationships between surfaces, etc. In addition, the functions of features and surfaces are incorporated by the planner.

Design for Automated Assembly

In the past, designers have produced designs for assembly in which little or no attention was paid to how the product was made. As mentioned in the previous section, what is needed is a design system which accommodates the manufacturing requirements. Therefore, in the assembly application, the goal will be to design products which are not only assemblable but also cost efficient as products. In addition, they should be easy to disassemble.

It is essential for assembly to be considered at three levels, including *product variety, product structure*, and the *product*. This consideration allows for product rationalization. Productivity is influenced by product variety. If a product is not designed to accommodate the variations in product lines, then costs will increase due to non-flexibility. Also, since frequent production starts and stops increase cost and production time, fewer components should be desirable. Therefore, similarity of parts as well as minimization of the number of components needed for a product should be an objective. The product structure consists of the components which aggregate to yield the product. Figure 4 illustrates the method of capturing product structure.

In designing for ease of automated assembly, two general objectives can be seen. The first one is to advise the designer of the difficulties in automated handling of the parts during assembly [5]. The second is the estimation of handling cost for parts being designed.

In the design, four major solution stages have been identified [6]:

a) Feeding of parts,

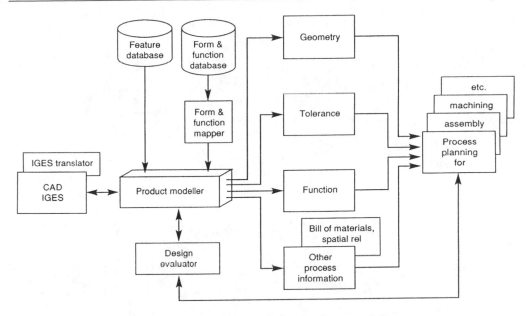

Figure 4: The schema of the product modeller

b) Orientation of parts,

c) Presentation of parts, and

d) Handling system cost.

The feeding of parts segment checks whether or not the part could be fed automatically. The result of this check could be that (i) there is no suitable feeding method for the part in its current design; (ii) the part could be handled by incurring some extra tooling time [6], or; (iii) that the part could be handled without extra penalty in time or cost. In the first two cases, some redesign changes would have to be made to stay within the required design constraints. Some design rules in the form of using the *production rules* could be used to deal with feeding problems. These rules will deal with such issues as tangling, flexibility, weight, size, and quality [6].

Orientation shows the general attitude of the part or feature with reference to some other parts or features in the assembly task. In automatic assembly to date, this problem has been largely solved using the α and β symmetry approaches of Boothroyd and Dewhurst [7]. The symmetries are illustrated in Figure 5. In "intelligent" CAD systems such as ProMod [2], the symmetry of a part can be automatically inferred from the spatial relationships which have been specified by the designer. In any case, the orientation problem is normally solved to determine the complexity of orientation of a part as depicted in the tooling time and efficiency or orientation.

The presentation problem is normally solved to determine the best approaches to presenting parts for the orientation area, and presenting the oriented parts to the assembly area.

Finally, handling system cost can be classified into the basic feeder cost and the cost of tooling [6]. The handling device could be a vibratory feeder [7], in which case cost will be for the basic feeder (including drive unit and bowl) and the cost of tooling the bowl.

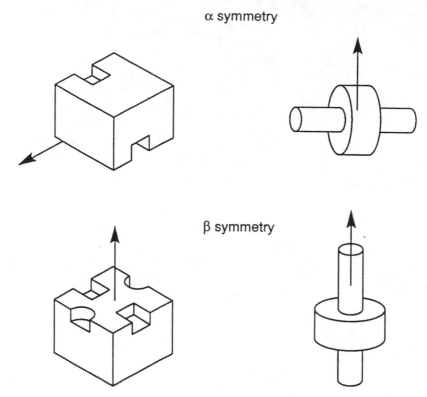

Figure 5: Symmetry of Objects

Product Assembly Modeler

The objective of product assembly modeling is to illustrate the product modeling concept in the assembly domain of application. One major stumbling block in the development of the automatic process planning for assembly is the loss of parts relationships among the assembly objects. Although standard systems such as PDES may eventually provide the framework for transferring such information from the designer to the process planner, not much work has gone into this area. Whitney et al [8] have shown how to automatically generate and evaluate alternative assembly sequence coupled with economic analysis tools for designing assembly systems. It is a bold effort in addressing the shortcomings of today's design process. In addition to the assembly sequence problem, there is need to automatically generate key process information directly from the designer. This product assembly model (PAM) serves as an interface between designer and process planner.

The product assembly model works as follows:

1. A designer uses a modeler (e.g. CATIA, GEOMOD, etc.) to represent a product.

2. The designer initiates the product assembly model which provides a list of all the components which make up the product.

3. The product assembly model provides a menu system through which spatial

relationships can be established for each pair of components. The system also automatically labels the faces and features of each component [9].

4. The system will then arbitrarily choose a part from the list and exhaustively pair the other parts against this part. Each pair is presented on the CRT screen along with a list of spatial relationships, such as *against, fits, attached, coplanar, aligned, etc*, which the designer can use to establish the relationship which must exist between the two objects, Figure 6.

5. The designer uses the menu system, (employing a mouse or the keyboard), to establish spatial relationships between each pair of components. The output of a work session is shown in Figure 6.

The product assembly model is also capable of building the bill of materials tree. A bill of materials can be built by using the physical relationship **part_of** to connect parts and establish inheritance attributes (where they exist) in **ancestor/descendant** manner. When all the parts in the list are linked in the bill of materials, an upper or a lower triangular matrix will thus be established for the parts and their relationships. A bill of materials graph can be obtained from this and the number of part i which directly go into the making of part j will be used to complete the product. Similarly, the product assembly modeler automatically establishes feature matrices needed for mating. The feature matrices work the same way as the above parts relationships.

One of the most significant benefits of building this product assembly model is seen in automatic assembly planning. A planner must establish these relationships in form of expressions. For instance if part A is against part B. The expression

$$against(PartA, Feature(face), PartB, Feature(face))$$

can be derived; where *against* is a developed spatial relationship protocol [10]. With a product assembly modeller, we can now automatically generate the same expressions based on the entries into the triangular matrix. It is now easy to develop and solve the huge algebraic problems inherent in solving the spatial relationships of an assembly manipulator with its work environment. The latter is represented in a world model.

There is another major advantage. The establishment of relationships among fixed objects in an assembly world will be easier. This task in world modeling has been quite cumbersome and therefore discouraging to even the most dedicated automatic programming workers. Once the entities in a facility plan have been identified, the object spatial relationship approach can be used to relate them in the world.

The product assembly model is an interactive module through which a designer or process planner can enter the functions between assembly parts and specification requirements. This information contains the designer's intent in carrying out assembly process planning. The designer creates the part on a CAD system with *Boundary Representation* or *CSG* as its mode of representing the CAD data. Through the product assembly model, the geometric information combines with mating functions, and design specifications. The kernel of the product assembly modeler is based on the spatial relationships among the parts being assembled. While these spatial relationships are being assigned, attributes such as mating functions, assembly tolerances, required torque for screwing, the contact force exerted on each part which signals the robot to stop (using the information carried via force sensors) etc. are requested.

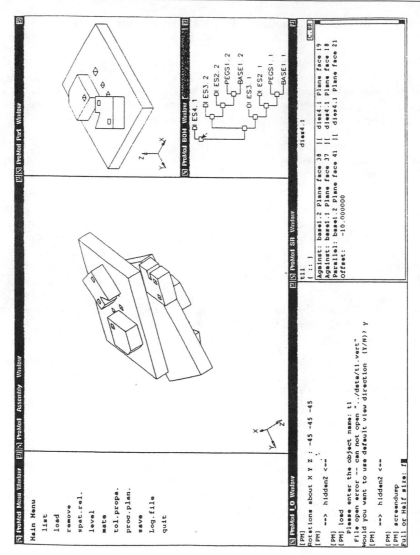

Figure 6: Product Assembly Modeler

CONCLUSIONS

In this paper, we have presented a more contemporary perspective of design which embodies the advocacy of concurrency in product and process design. Since today's designer systems are not capable of carrying a designer's intent for a product, or producing the needed information for process planning of the product, we have presented a product modelling system which is representative of the desired futuristic designer system. It is envisaged that designer systems of this kind will be capable of severely reducing the time and cost of production. Along these lines, we have discussed the progressive issues of standards and considered a product's entire life-cycle during the design process.

REFERENCES

[1] D.E. Whitney Nevins, J.L. and et. al. *Concurrent Design of Products and Processes: A Strategy for the Next Generation in Manufacturing.* "McGraw-Hill Publishing Company, New York", 1989.

[2] B.O. Nnaji. CAD-driven machine programming. In *"Proceedings of JAPAN-USA Symposium on Automation and Robotics"*, "Kyoto, JAPAN", "July" 1990. "ASME".

[3] A. Takeshige Ando, K. and H. Yoshikawa. An approach to computer integrated production management. *International Journal of Product Research*, 26(3):333–350, 1988.

[4] G. Boothroyd. *Fundamentals of Metal Machining and Machine Tools.* "McGraw-Hill, New York", 1975.

[5] K.G. Swift. *Knowledge-Based Design for Manufacture.* "Prentice Hall, Engle wood Cliffs, NJ", 1987.

[6] A. Kusiak. *Intelligent Manufacturing Systems.* "Prentice-Hall, Englewood Cliffs, NJ", 1990.

[7] G. Boothroyd and P. Dewhurst. Design for assembly: A designer's handbook. Technical report, Dept. of Mech. Eng., Univ. of MA- Amherst, 1983.

[8] D.E. Whitney, T.L. DeFazio, R.E. Gustavson, S.C. Graves, T. Abell, C. Cooprider, and S. Pappu. Tools for strategic product design. In *"Reprints of the NSF Engineering Design Research Conference"*, " Amherst, MA ", "June " 1989. "U. of MA - Amherst ".

[9] B.O. Nnaji and H. Liu. Product assembly modeler. In *"Proceedings of 4th Int'l Conference on CAD/CAM"*, "Delhi, INDIA", "December" 1989. "IIT".

[10] B.O. Nnaji and H. Liu. Feature reasoning for automatic robotic assembly. *International Journal of Production Research*, 28(3):517–540, 1990.

NEW-MODELLER: CONCEPTS FOR AN INTEGRATED PRODUCT DEVELOPMENT ENVIRONMENT

Edison Da Rosa

Mechanical Engineering Department, Universidade Federal,
São Carlos, São Paulo, Brazil

ABSTRACT

This work presents some concepts for a new approach of an integrated product development environment, based on a multimodelling and on a design by features strategies. Initially, it is presented the general data structure, for use in the integrated system, which describes the product under consideration as a set of model files. These model files are related with the conceptual product design, the geometric product definition, the product engineering analysis models and so on, with the associated manufacturing analysis models. One of the main objectives of this research is to permit a straight forward feature recognition. The methodology adopted also would help the product analysis procedures, the product assembly planning and others analysis kinds. The product multimodelling concept is just the proposed strategy to permit the use of the power of the actual available hardware, in an efficient way.

INTRODUCTION

With the ever increasing use of computers in the industry, terms like CAE, CAD, CAM, CAPP, CIM, CIE, and others are usual. They imply computers helping the product development, including its design, analysis, and manufacturing. Usually the product development cycle starts with a detected market need. In a simple way, the product development cycle can be sketched as shown in figure 1.

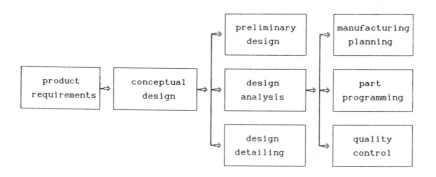

Figure 1. Product development cycle.

The detected need must be carefully identified and defined, permitting that the product specification can be clearly written and understood. These specifications are used to start the product development, beginning with the conceptual design. In the conceptual design phase some alternative configurations are proposed and tested to define the basic product guidelines, [24]. This is a very personal step and the designer ingenuity is put into proof. Some alternative options are proposed, in this step, with a valuable help provided by a CAD system, to get some product sketches. These sketches can be used to obtain a geometric model or a product analysis model. Obviously, these models are very crude, with little information about the product definition, but very useful to get some performance data about one or other product configuration. This performance data are very convenient to restrict the possible product configurations to a small number. With the definitive product configuration, a basic geometric model can be created, usually as a graphic file on some CAD/CAM system. The geometric model will grow as new product data are introduced into model, along the product design detailing phase, [14,16,25].

After some time, say two or three months after the product development starts, the design team need some information about the probable product field behaviour, material behaviour, or any other kind of data. Then, the natural step, for products with responsibility, is an analysis by finite or boundary element method, for some critical parts of the product. To do that analysis, a convenient part model is needed, compatible with the analysis software. This type of model is formed by a geometric idealization of the part, a load modelling and a set of constraint conditions, [6,10].

THE MULTIMODEL CONCEPT AND CONCURRENT ENGINEERING

In a traditional CAE/CAD/CAM system concept we have, for a product, some models associated with it, as the geometric, the graphic, the FEA, the NC, etc. When the user detect the need of a FEA, or a NC program, he starts to derive the model, from the geometric one. This way of doing the job, typical for an actual commercial software, imply usually in effort duplication to generate new models.

On the other hand, with the high speed of hardware development, in the PC and in the workstation segments, with a price/performance index in the range of US$ 400 / MIPS to US$ 250 / MIPS, the work methodology must be changed. What we need is a better approach to use efficiently the processing power of the hardware, and not only to overload the CPU with a lot of job, to produce a very detailed ray traced image, or a two million degrees of freedom dynamic non linear finite element analysis. The change involves a new approach to software design, with smart systems helping the designer in his mental work for product development, [11,12,18,24]. These smart systems can be build around some product class, with a knowledgement data base aggregated, which helps the designer in his decision making design process, [1,2,4,5,7].

In the usual engineering work the hardware is poorly used, mainly if we are working in the design of a product through an interactive modelling software. The reason is that the user spends 90% to 99% of his time thinking how good is the result, to decide which is the next step. Then, the CPU is awaiting by seconds or minutes. For a 20 or 50 MIPS machine this isn't reasonable. What is the answer for a better usage of the hardware? It is the multitasking of the design activities, since only in a few situations we have a big job running in background, that makes full use of the CPU power of the workstation. The multitasking in the design activity is obtained with the use of a multimodelling concept for the product development, as required by an concurrent engineering environment. This concept is derived from the assumption that if we have an adequate processing power, the simultaneous processing of product models is feasible. The product models can be the functional or conceptual model, the geometric model, the analysis model, the visualization models, the drawing models, the process planning model, the part programming model, etc. In the proposed software architecture some of the above models are updated as soon as the product is changed by the designer. In this sense, as the product is detailed, not only its functional and geometric files are updated, but also the others product model files selected by the user. If the designer wants, also the analysis model can be solved by the hardware, starting a new process, in low priority level. In this case, as the designer makes some design change, he gets the results from this change some seconds later. This is very useful for a true interactive design, where the predicted performance of the product is available to the designer as soon as he changes the design and requests an analysis.

Therefore, the product multimodelling concept is just what is needed to use the full power of the hardware, in an efficient way. This concept cannot be confused with a cluster of static product models, since it is the update of the models dynamically. The figure 2 bellow illustrates the idea of the multimodelling concept, where the product models are divided in functional or conceptual models, in geometric models, in analysis models, documentation models and manufacturing models. The figure also shows the corporative data bases used by the product models.

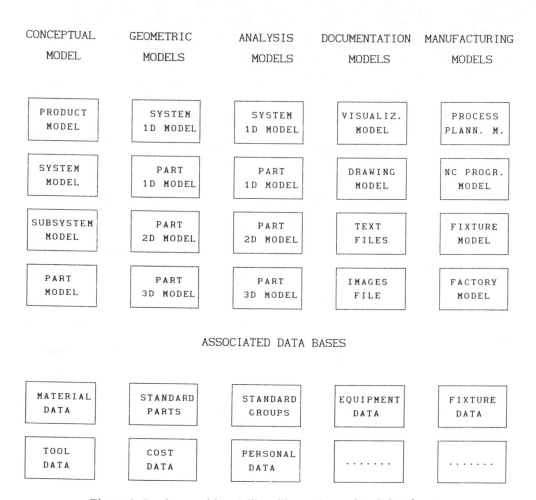

Figure 2. Product multimodelling files and associated data bases.

Conceptual Model

The conceptual model, or product functional model, are a very crude product definition, but are the core of the product models. If the functional model are properly processed, the geometric models are correctly generated and then any other kind of model. The product model starts by the product hierarchy, from product as a whole, to parts, involving the product systems and sub systems. Figure 3 is a sketch of the hierarchy in the definition and of the data to be stored in this type of file. For each hierarchical level, there is a set of parameters, derived from the level immediately above, stored in the P vector. The attributes are stored in the A vector. Finally the I vector stores the calculated indices, like performance, cost indices, and so on. The model calls predefined parts, stored in a library in the associated data base. In some cases the user can model the part using standard groups, constructive blocks and geometric elements.

Geometric Model

This model defines the geometric product data, with explicit element definition, with their faces, edges and points. For each product part we have one geometric model.

Visualization Model

The visualization model is a derived model, which use the geometric model to create relevant data for a wire frame product visualization, or with hidden lines/surfaces removal, or a shaded model.

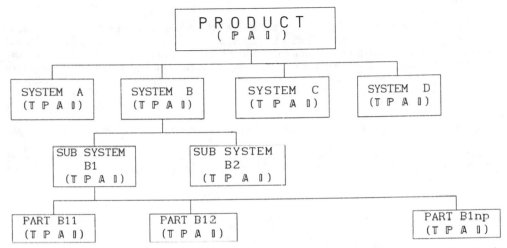

Figure 3.   The structure of the product conceptual model. P, A and I are, respectively, the product parameters, attributes and indices vectors. T is the transform matrix needed to locate a part.

1D Analysis Model

This model is useful for an initial CAE analysis, as a kinematics, dynamics, stress, fluid flow, heat transfer, electromagnetic field, etc. This type of model stores data in a global sense, as the gravity center location, the inertia matrix, the part stiffness matrix, etc. The model may be for a part or for the whole system, assembled by many parts.

2D Analysis Model

This model is a more detailed part description for analysis, generated as the designer starts to put details in the part that cannot be represented in a 1D model, like a filet, a hole, etc. For a structural analysis, the 1D model uses beam elements, while in the 2D model, plate, shell or plane strain elements are used.

3D Analysis Model

This model describes the full geometric part information, using solid elements to make the model. It is expensive to analyze, if every kind of details are modeled.

Process Planning Model

The process planning model extracts only the relevant data from the product functional model and from the product geometric model. Those data are directly used for process planning.

NC Programming Model

This model gets its data from the product functional model, from the geometric model and from the process planning model, to generate the NC part program.

Result Files

When some kind of analysis is realized, its results need to be stored for further analysis and display.

These files, but not all of them, are updated as the design is altered or refined and, if the designer wishes, the analysis file can be run as soon as it is updated. Then, in the usual interactive product development, when the designer evaluates and appreciates his work, the CPU is underloaded and can be used to run the analysis file. In the proposed software configuration, the user commands needs to have a higher priority, over the priority for other software actions, running concurrently, to assure a good interactive environment. In a networked configuration, some analysis effort can run on a server, with bigger RAM memory and disk area for storage.

On the other hand, as the design is more and more detailed, some aspects can't be put in the 1D analysis model, but can be put in the 2D model and obviously in the 3D model too. The kind of details that are added to the product model determines which analysis model need to be updated, using some A.I. rules. The evolution of the analysis model as the design goes step after step is showed in figure 4.

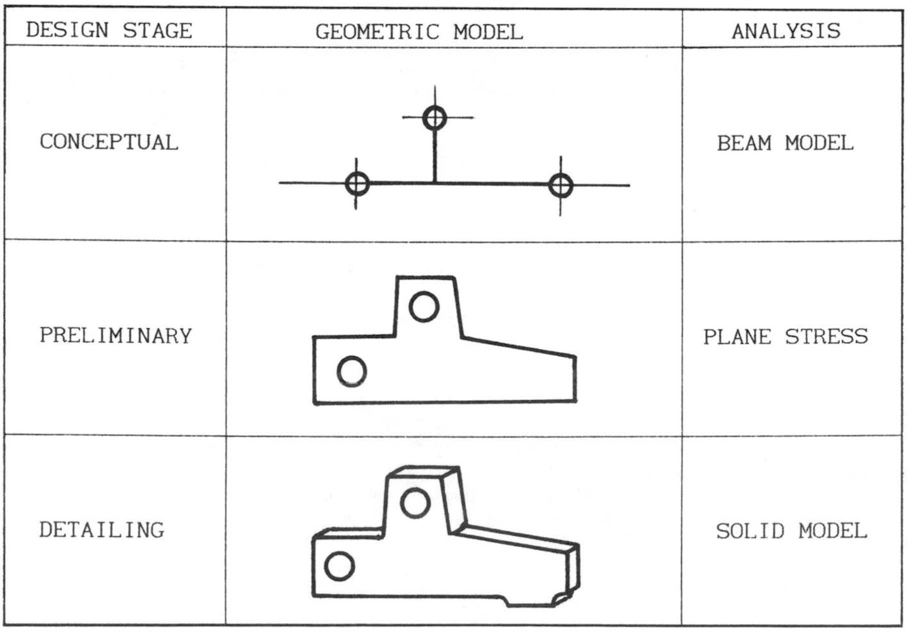

DESIGN STAGE	GEOMETRIC MODEL	ANALYSIS
CONCEPTUAL		BEAM MODEL
PRELIMINARY		PLANE STRESS
DETAILING		SOLID MODEL

Figure 4. The analysis model evolution, in three steps, from a 1D to a 3D model.

THE PRODUCT DEFINITION

In the proposed concept, the product is hierarchically defined by systems, subsystems, and parts. Each has its specific data, which may be like that illustrated in figure 5. The product or part code, the product or part GT code, the parameters, the attributes, indices and subcomponents codes are read from correspondent tables. They define what is the code and the allowable range of inputs. These tables can increase in size as a new type of class of attributes, indices, etc. is needed. Then this new type can be created and incorporated in the table. The associated code is valid only for the application software that can correctly interpret that code. A list of probable parameters, attributes and indices associated with a typical mechanical product may be written as follows:

Dimensional:
Length, width, thickness, radius, distance, diameter, angle.

Tolerance:
Dimensional, form, properties.

Manufacturing:
Manufacturing time, cost, necessary equipments, process sequence, preferred finishing pattern, material and treatment, surface treatment.

Assembly:
Adjacent parts, plug in parts, fixtures, assembly cost, assembly time.

Inspection:
Fixtures, gages, cost, inspection time, approval criteria.

Performance:
Weight, maintenance, load bearing capability, stiffness, stress factors, total cost, expected life, life curve, reliability.

Visualization:
Active viewport, active window, projection data, surface color, surface type, border visibility, shading type.

Drawing:
Line type, line thickness, line color, visibility, text.

Simulation:
Inertia matrix, stiffness matrix, stress matrix, damping matrix.

H	PRODUCT (PART) NAME	
E	PRODUCT (PART) CODE	
A	PRODUCT (PART) GT CODE	
D	DATE	
E	RESPONSIBLE	
R	STATUS	
	TEXT FILE POINTER	
	IMAGE FILE POINTER	
	DATA POINTER	
TRANSFORM MATRIX	Ti j	VALUE
PARAMETERS VECTOR	P_CODE	VALUE
ATTRIBUTES VECTOR	A_CODE	DATA
INDICES VECTOR	I_CODE	VALUE
SUBCOMPONENTS CALL	S_CODE	PARAMETERS

Figure 5. The data structure for a product definition file.

THE PART DEFINITION

The product definition ends with the part definition, but then it is necessary to go deeper, to define completely the part. This is done with the modelling hierarchy, which determines the part geometry. The geometric definition of the part is done by calling some standard constructive groups from a library. The group concept is a set of geometric data with some functional characteristics, for example a cylindrical shaft end, with a keyway, or with a spline, or an internal hole and so on. Together with the geometric data, the group have a set of other data, concerning its dimensions, surface finishes, dimensional and form tolerances, load capability, etc. This set of non geometric data are stored in the form of group attributes and group indices, in the group definition file, with a format like that of figure 5. The called subcomponents by a group is what are named as blocks. A block is a small set of geometric data, useful in the design detailing, as for add a hole in the part, or a chamfer or a filet. Figure 6 shows some groups and some blocks.

A very useful concept is that of geometric activity of a block. If the block is set as active, its geometric details are inserted in the part geometric model, but if the block is set as inactive, only its non geometric data are added to the part definition file, without any geometric change. This is illustrated in figure 7 where a small chamfer is added to the part and this needn't be modelled, but must be present in the product data for the part manufacturing analysis, as also for its drawing generation.

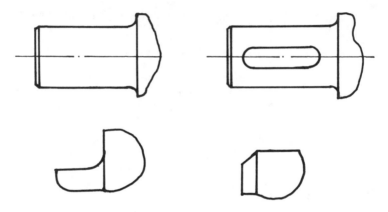

Figure 6. Examples of groups and blocks.

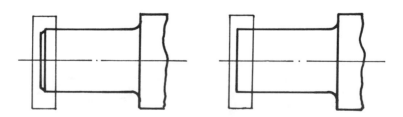

Figure 7. A inactive block.

The block has its geometry defined by a set of geometric elements, the smallest geometric data manipulated by the system. The geometric element is a line segment or an arc segment, for a 1D model. For a 2D model, the geometric element is a four sides polygon. For a 3D model, it is a hexahedron, as figure 8 shows.

Figure 8.   The geometric elements considered. They can have linear (planar) edges (faces) or not.

The geometric elements have their faces treated as planar, but in a more general application, the faces can be considered as a NURBS surface, with the data specifying the mesh point coordinates and the knot vector, for each parametric direction, [19].

The groups and blocks are formed not only by geometric elements, but also by others kinds of elements. Another type of element is the reference element. This element doesn't define the part geometry, but define some kind of associativity or relationship between groups in a part or between parts in the product. The associativity and the relationship are useful in the definition of an assembled set of parts, specifying which part is attached to another and how it is attached. The reference elements can be used also to define kinematic constraints between assembled parts. The last type of elements is the non geometric elements, needed in the drawing files. Some non geometric elements are center lines, witness lines, text, hatches, symbols, etc.

The product definition file has the whole product hierarchic structure, starting with the product and ending with the groups or blocks calls. Each system, subsystem, part, group or block has, in addition to its geometric data, the data associated with it, stored in the parameters vector, in the attributes vector and in the indices vector. When the product definition file is processed, it generates the geometric model of the product parts, and from this, the other product files.

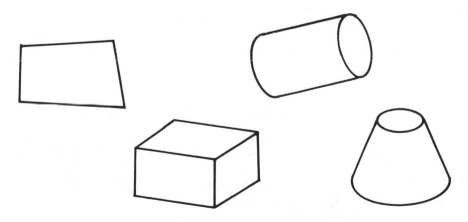

Figure 9. Basic constructive blocks.

On the other hand, when a block, built by geometric and reference elements, is called by a part, its elements are transferred to the part geometric definition. The block parameters are adjusted accordingly, [15,17,18], as a function of the calling parameters. The block attributes are inherited from the part attributes and processed by some inheritance rules, [9]. The blocks are divided in basic constructive blocks, or geometric primitive blocks, and in detail insertion blocks, parametrically defined blocks, with a specific topology of elements, to represent some detail, [15]. A detail block change the already defined blocks, opposite to the constructive blocks, which only add to the geometric data, without any change on the existing blocks. Figures 9 and 10 illustrate the behaviour of the two block types discussed.

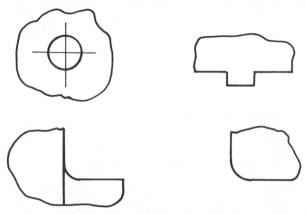

Figure 10. Detail insertion blocks.

Some detail blocks may eventually not change the part geometry, changing only parameters and attributes for the region where the block is inserted. This kind of detail is, for example, a better surface finish, by a milling or a grinding operation. Another possibility is for a special surface treatment, like a shot peening or an induction hardening. Figure 11 shows an example.

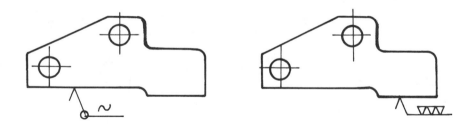

Figure 11. Insertion of a detail block, to modify the surface finish.

The concept of groups is twofold. Initially, it helps the designer in the part modelling, since a great percentage of details are stored in a group library. The user also can define his special library, customized for some application. The group concept is also useful as it contain aggregated data, which tells what is the group and its properties. That kind of information is essential for any attempt to build a smart CAE/CAD/CAM system around a data base, [2,5,8,12,13]. The group concept is analogous of the feature concept, [3,8,20,21,22,23], since both aggregate geometric data in a functional approach and non geometric data that characterizes the group.

In a part design, the user starts with a part sketch, usually as a line with some added information. This corresponds to the conceptual design of the part. As the designer goes to the preliminary design stage, he adds new geometric data and the part model becomes more sophisticated, as an axisymmetric shaft model for example. In the design detailing, the model incorporates some features like a keyway, a threaded hole, a lubrication hole, etc. The details that are added to the geometry are from predefined blocks. Some two dimensional blocks are showed in figure 12 and for a three dimensional geometry in figure 13.

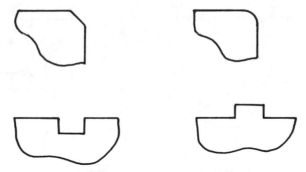

Figure 12. Details for a 2D modelling.

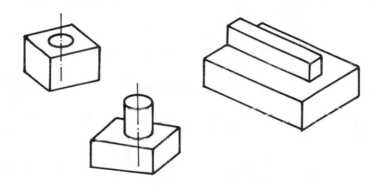

Figure 13. Details for a 3D modelling.

CONCLUSIONS

As already established, the geometric modelling is based on a few very simple constructive geometric elements, as shown in figure 8. One reason for that choice is to build up a simple modeller software, without great sophistication. This has as result a relatively small package and fast algorithms. Another reason to use simple geometric shapes is the easy by which an analysis mesh can be generated, for a FEA software for example. Then, with that consideration in mind, the concept of a multi modelling for a product, with simultaneous on line processing of the relevant models, can be effectively implemented. When the designer begins his work on a product, the product concept is transferred to the CAD system as a sketch, interactively refined, as more and more details are added to the design. The details that are added to the design can be classified in some basic geometric forms, such that illustrated in figure 12, for a 2D geometry and in figure 13, for a 3D geometry.

The easiness in detailing the design is concerned with the types of available details, in the system data base. The details are build up by one or more geometric elements, parametrically defined in the call for that detail. Figure 14 illustrates some two dimensional solutions for detail insertion in the geometric data base.

The concepts discussed in this work are being implemented in a software prototype, using a Windows environment for user interface and programs manager, a PHIGS graphic library to execute all the graphic data manipulation and a cellular boundary representation as the solid modelling technique, using NURBS surfaces as the geometric algorithm for each cell face.

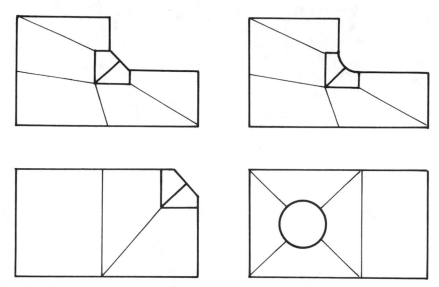

Figure 14. Some solutions for a 2D detail insertion problem.

REFERENCES

1. Ahmad, M. M., M. O'Callaghan, J. O. McEnery - The role of CAD/CAM Database in Computer Integrated Manufacturing. Proceedings FIAM 91, CRC Press, pp 743-757.

2. Bhat, S. K., D. D. Beagan - Feature Based Data Management. Mechanical Engineering, March 1989, pp 68-72.

3. Butterfield, W. R., M. K. Green, D. C. Scott, W. J. Stoker - Part Features for Process Planning. CAM-I 1988.

4. Cattell, R. G. G. - What are Next-Generation Database Systems? Communications of the ACM, October 1991, pp 31-33.

5. CIME Staff, - Automating Data Management. Mechanical Engineering, March 1989, pp 73-76.

6. Cook, W. A., W. R. Oakes, Mapping Methods for Generating Three-Dimensional Meshes, Computers in Mechanical Engineering, August 1982, pp 67-72.

7. Develly, M. - CAD/CAM Integration: The Product Data Management System. Proceedings FIAM 91, CRC Press, pp 713-728.

8. Faux, I. D. - Reconciliation of Design and Manufacturing Requirements for Product Description Data using Functional Primitive Part Features. CAM-I 1986.

9. Graves, G. R., S. Banerjee - Object-Oriented Programming in CAD/CAM Integration. Proceedings FIAM 91, CRC Press, pp 639-648.

10. Hamilton III, C. H. - CAD/CAM Techniques for Generating Finite-Element Meshes, in Computer-Aided Design, Engineering, and Drafting, Auerbach 1984.

11. Henderson, M. R.; S. Musti - Automated Group Technology Part Coding From a Three-Dimensional CAD Database. Journal of Engineering for Industry, August 1988, pp 278-287.

12. Jakiela, M. J., P. Y. Papalambros - Design and Implementation of a Prototype 'Intelligent' CAD System. Journal of Mechanisms, Transmissions, and Automation in Design, June 1989, pp 252-258.

13. Klein, A. - A Solid Groove. Mechanical Engineering, March 1988, pp 37-39.

14. Machine Design, CAD/CAM Industry Report: Changing the Way Engineers Work. Machine Design, May 23, 1991, pp 48-55.

15. Mantyla, M. - An Introduction to Solid Modeling, Computer Science Press, 1988.

16. Mantyla, M. - A Modeling System for Top-Down Design of Assembled Products. IBM J. Res. Develop. September 1990, pp 636-659.

17. Mantyla, M. - Solid Modeling: A State the Art and the Next Steps. MICAD 87, Paris.

18. Menon, U., T. Yang - Integration Framework Using Parametric Design Solid Modeling. Proceedings FIAM 91, CRC Press, pp 323-332.

19. Rogers, D. F., J. A. Adams - Mathematical Elements for Computer Graphics, 2nd Ed. McGraw Hill 1990.

20. Shah, J. J., P. Sreevalsan, M. Rogers, R. Billo, A. Mathew - Current Status of Feature Technology. CAM - I, 1988.

21. Shah, J. J., M. T. Rogers - Functional Requirements and Conceptual Design of the Feature-Based Modelling System. Computer Aided Engineering Journal, February 1988, pp 9-15.

22. Shah, J. J., A. Mathew - Experimental Investigation of the STEP Form-Feature Information Model. Computer-Aided Design, May 1991, pp 282-296.

23. Shah, J. J. - Assessment of Features Technology. Computer-Aided Design, June 1991, pp 331-343.

24. Yoshikawa, H., E. A. Warman, Eds, Design Theory for CAD. Proceedings of the IFIP WG 5.2 Working Conference, Tokyo, October 1985. North-Holland 1987.

25. Zhang, Z., S. L. Rice - Conceptual Design: Perceiving the Pattern. Mechanical Engineering. July, 1989, pp 58-60.

OBJECT-ORIENTED PARADIGM FOR CAD FRAMEWORKS

Hyeon H. Jo, Jian Dong, and Hamid R. Parsaei

Department of Industrial Engineering, University of Louisville,
Louisville, Kentucky, U.S.A.

ABSTRACT

In the area of computer-integrated manufacturing (CIM), *concurrent engineering* (CE) has recently been recognized as the manufacturing philosophy for the 90's, to develop high quality products and bring them to the highly competitive global marketplace at a lower price and in significantly less time. However, practicing the CE philosophy requires a large number of design tools during the design phase, to examine the entire aspects of a product's life-cycle in a coherent way. The entire design process has consequently become very complicated. A large number of highly specialized design tools have to be developed to meet the technological needs of complex design tasks, and integrated into a computer-aided design (CAD) framework to provide better service to the users. This paper identifies the needs and requirements of such *CAD frameworks* for concurrent engineering environment and discusses the *object-oriented paradigm* which has several salient features applicable to the development of sound CAD frameworks.

INTRODUCTION

The objective of developing a product may be to introduce the product with maximum *quality* at the lowest *cost* within the shortest *time*. As a more systematic approach addressing the objective, increasing awareness is being directed to the products design stage. It has been recently acknowledged that the best opportunity to rebuild the manufacturing leadership could be found in design. *Design must be viewed accordingly, as the central pivot of all manufacturing activities, and the designers' role should be redefined and understood.*

As technological tasks in manufacturing environment have become progressively complicated, today's designers are faced not only with increasing complexity of product designs but also with a constantly increasing number and sophistication of specialized design tools. Consequently, the entire design process has become very complicated. A large number of highly specialized design tools has to be developed to meet the technological needs of complex design tasks, such as design analysis for manufacturability, assemblability, cost estimation and engineering analysis. Moreover, these design tools should be integrated into a CAD framework to provide a variety of functions and services to the design community.

This study proposes that *the concepts of CAD frameworks should be of special interest to those who try to practice the philosophy of concurrent engineering* (CE), because a large number of design tools are required during the design stage to examine the influence of the design upon all aspects of a product's entire life-cycle in a coherent way.   The literature survey on CE indicates the need and lack of study in the field of CAD frameworks, particularly for the mechanical design community. Nearly all aspects of a product's life cycle have been examined in "*isolated*" fashions based on local criteria, frequently yielding locally satisfied designs.   Therefore, higher-order framework services will be needed to fully support concurrent engineering practices.   This paper focuses on three issues: a structural model for the computer-based approach to practicing the CE philosophy, the basic concepts and the needs for a CAD framework to fully support the CE practices, and the object-oriented paradigm which has several salient feures useful for the development of such CAD frameworks.

CAD FRAMEWORKS FOR CONCURRENT ENGINEERING

Background

Concurrent engineering (CE) has recently been recognized as a more integrated approach to develop high quality products and bring them to a highly competitive global marketplace at lower price and in lesser time.   This is stemmed from the recognition that design decisions made early in the product development cycle have significant impacts on manufacturability, quality, cost, time-to-market, and thus the ultimate marketplace success of the product.   It implies that all the information pertaining to a product's life cycle should be used to augment the information for design decisions in order to achieve the *globally optimized product design* for manufacture.   This is the philosophy of CE which entails the concurrent consideration of product designs and all their related processes within an organization to address all of the above issues.

The two basic approaches to implementing the CE practices are team-based and computer-based approaches.   CE practices have been mostly done by team-based approaches.   Team members are selected for their ability to contribute to the design of a product and processes by early identification of potential problems and timely initiation of corrective actions to avoid a series of costly reworks [1].   Though the team-based approach can be readily implemented and is being adopted in industry, some shortcomings seem to arise [2].   As more sophisticated computer tools emerge constantly, the team-based approach is being enhanced by the computer-based approach in which the CE philosophy is woven into the internal logic operations, enabling *design justification or optimization with respect to all the aspects of a product's entire life-cycle*.   The figure 1 shows a product development cycle employing the CE philosophy [3].

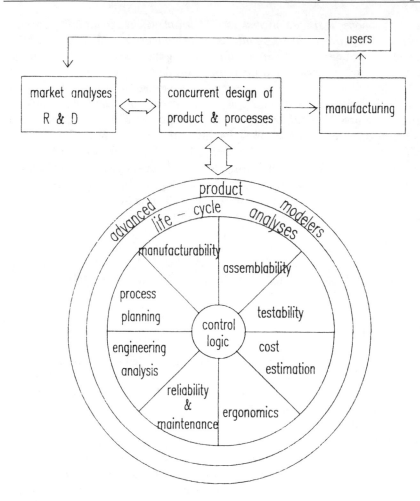

Fig. 1   Product Development Cycle Employing
'Concurrent Engineering Wheel'

The outer layer of the 'concurrent engineering wheel' is the advanced product modelers and the core is the control logic which involves the steering of various CAD tools to provide a variety of services, finding a globally satisfied design. It must be emphasized that it is the core of the wheel for which we badly need the *scientific theories of the design process*. Between the layers is the functional layer that comprises various life-cycle analysis tools.

As for the implementation of the CE philosophy, it can be inferred that computer-based approaches will be predominant. A large number of CAD tools will be required to examine the entire aspects of a product's life cycle from conception to disposal.   Consequently, what we are expecting to see is an integrated design environment in which *all the CAD tools interact and cooperate to find a globally optimized or compromised design.*   This design environment may consist of sets of various CAD tools and an underlying CAD framework.

CAD Frameworks

A CAD framework is *"a software infrastructure that supports the application environment, providing a common operating environment for CAD tools"* [4].   It is a logical extension to the general computing environment that provides the functions and services demanded by the design community.   The framework is intended to be a well-organized software infrastructure which properly guides designers through the entire design process so that they can reach a design with maximum quality which can be produced at the lowest cost, and in the shortest time.   A CAD framework is actually layers of code between the operating system and application environment, which provides a variety of functions and services to designers.   A notable improvement in design productivity can be accomplished, because designers using the framework can concentrate on the main design rather than on other tasks such as interface.

In general, today's CAD tool users seem to be satisfied with the performance of their design tools.   However, a great deal of time that could have been spent on the design is often spent in integrating these diversified tools into a single, working CAD environment [4]. The domain of the integrated CAD environment has emerged to address this problem, particularly from the area of electronic circuit design. The design process of the area has been more automated than those of any other products. The designers in the domain have been able to spend less time directly manipulating design data and more time managing CAD tools that manipulate the design data.   This results in the need for an improvement in the integrated CAD environment which consists of sets of numerous application specific tools and an underlying framework.

The concepts of CAD framework are not new.   Traditionally, the term CAD framework has been used to indicate a set of common software facilities that provides services to make it easier to build, use, and manage the CAD environment.   In 1986, a notable effort was made in the domain of electronic circuit design to consistently and uniformly integrate a broad spectrum of CAD tools into a framework based on an object-oriented database system [5].   Several attempts that address similar issues have been noticed thereafter [6, 7, 8, 9, 10].   However, most attempts to develop the CAD frameworks have been made without concurrent engineering practices in mind. Consequently, those frameworks have revealed some inappropriate characteristics for CE environment [11, 12, 13].

In 1988, the CAD Framework Initiative (CFI) was formed by the major companies in the electronic industry to develop a set of standards for such frameworks, because several products under the name of framework may be different to one another.   It was hoped that the set would allow CAD tool developers to focus on enhancing tool functions rather than laboring on the interfaces. Smith and Cavalli [13] briefly described the evolving stages of the framework concept.

As for the mechanical design community, the concept and the value of a CAD framework have not yet been fairly understood and appreciated. However, as more and more sophisticated CAD tools required for the life cycle analyses become available, CAD frameworks should be recognized in the community as a requisite for improving the design productivity. It should be also noticed that the advances in CAD frameworks may influence the development of individual CAD tools. That is, *the synergistic relationship between the developments of a CAD framework and individual tools* should be understood.

It has been believed that the development of sound CAD frameworks that fully support concurrent engineering practices will not be trivial and will require some fundamental redesign of the framework architectures [13]. A CAD framework should contain several facilities which coordinates the entire design process that complies with CE philosophy. Bretschneider et al. [14] describe three components that may form a CAD framework: a consistent database management system, a uniform graphical user interface, and a design management facility.

Since today's technological milieu changes rapidly, the frameworks should be constantly updated. To ease the burden from the CAD system integrators or designers, the frameworks are intended to be of *open* architectures that make the frameworks more domain independent, easier to use and update, easier to integrate the tools, and more capable of supporting a large population of tools and corresponding data models [11]. A methodology for implementing '*soft*' integrated systems that facilitate the adaptation to changing environments is described by Weston et al. [15]. van der Wolf et al. [16] also discuss the principles of the architecture in terms of designer orientedness, openness, configurability, simplicity, and efficiency.

OBJECT-ORIENTED PARADIGM

A CAD framework is itself a complex software system whose capabilities can impact the success of a product design. Object-oriented paradigm is one technology that offers, substantial help in simplifying the design and implementation of CAD frameworks. In this section, two issues related to the paradigm, language and database, will be discussed, followed by the discussion of an approach to control CAD tools within a CAD framework.

Object-oriented Languages

Generally, object-oriented languages allow a complex software system to be built out of much simpler entities called objects, each representing a physical entity, a concept, an ideal, or some aspect of interest to various applications. Major prominent features of object-oriented languages are built upon *data abstraction* which is the key to reliable and flexible software design. Data abstraction and hiding capability of the languages enable the designer, who uses various design data structures and CAD tools, to have access *only to their abstraction or interface parts, and not to the complete implementations,* thereby providing significantly increased modeling power. This allows programmers to simplify the design and implement sound CAD frameworks. This advantage indicates the necessity for a clear distinction between the storage structure associated with *data* and the logical structure of the *information* to be used. Other critical features that object-oriented languages support include *inheritance* and *polymorphism.* The former implies that an object type automatically has all the properties/operations of its parent type, and the latter refers to the situation in which objects belonging to different classes can respond to the same call, usually in different ways. As a consequence, inheritance capability may allow a programmer to assign default behavior to objects, and polymorphism can dispatch a call to an appropriate class according to the type of parameters used. It is believed that languages providing all these feaures are effective tools for the construction of complex and sound CAD frameworks.

Database Management

Traditional CAD systems based on file systems to strore the design data reveal numerous shortcomings. As a remedy to ease the problems, database management systems have been introduced, which provide a central storage facility, a uniform data interface, and a means for controling data redundancy. However it has been argued that the traditional record-oriented data models - relational, network, and hierarchical- are inadequate in dealing with complex applications such as found in engineering, office information system, and multimedia databases. Obviously, conventional data modeling schemes require considerable effort to represent real world objects as fixed programming constructs, resulting in a *semantic gap between the way data are strored and the way the information is used.*

Particularly, the objects needed in CAD systems are too complex to be efficiently handled with the traditional data models. Design data have many different, very complex structures connected by many relations which change dynamically. As a consequence, a new class of database management systems called object-oriented database management sysem (OODBMS) is being developed. It allows each entity of the real world to be very closely represented by exactly one data object which may contain complex attributes structure. For instance, any subassembly, component, fastener between components that exist in the real world of manufacturing can be modeled as an explicit object. The interrelationshiops between objects also can be aligned very closely to the real world interrelationships by using *is-a, is-instance-of,* and *is-part-of* relationships that can provided by OODBMSs. Object-oriented databases provide persistent objects that we can organize to hierarchical classes. Notice that the design has an inherent hierarchical nature. They also provide more flexible data access and retrieval mechanisms. The database objects as well as the retrieval/update procedures are defined by classes whose implementation determines how data is accessed. Moreover, object-oriented database encourages modeling of the behavior of objects, making it more frequently used in applications with strong dynamic components like those in CAD frameworks.

CAD Tools Control

At the heart of a CAD framework is the task of invoking and controlling numerous CAD tools, which remains as an error-prone and a largely unsolved problem [17].   The CAD tool communication and control mechanism is perhaps the most important technological part of the framework.   It provides the communication backbone for a large amount of design data, and controls various design events that occur throughout the entire design processes.   Invoking a large number of tools manually is difficult or at best tedious and time-consuming.   The issue of CAD tool control is critical for solving many problems that have plagued CAD frameworks. A few attempts have been made to resolve the problem by encapsulating the tool execution and management in a CAD framework.

A design management system that controls the design process using a network of fifty engineering workstations has been discussed [12].   Scripts were recommended to describe and serve as the master description for each tool.   The management system accommodated a set of software tools and allowed multiple designers to work on a design simultaneously.   Six main control and communication facilities which constitute a suitable design management system were discussed ; version control, control of database hierarchy, process control, validation, notification, and user interface.   However, it is known that scripts lack portability and support for parallel tool execution [18].   Scripts usually make explicit reference to CAD tools, requiring the existing scripts to be rewritten, as new tools are introduced into the framework.

By taking an object-oriented view of a large number of CAD tools in a framework, the efficiency of the CAD tool control mechanism may be significantly improved, because the data abstraction capability enables an object to separate an *interface part* from an *actual implementation portion*.   Daniel and Director [11, 19] described an approach to CAD tool control within a framework in which each tool was represented as an object-oriented entity.   By taking an object-oriented view of CAD tools, they tried to derive the modeling mechanism that provided a coherent view of each tool.   Each original tool was replaced by the CAD tool knowledge object that was intended to fulfil several requirements ; represent the capabilities of the tool to the designers, manipulate the low level details associated with the tool, and furnish a consistent control scheme to make the framework more flexible, easier to control, use, and maintain.   A framework, Cadweld, has been implemented to test the proposed approach in the domain of VLSI design.   The volunteer-based blackboard architecture was used as the central mechanism to support communication between the tools, designers, and other modules in the framework.   The control mechanisms based on object-oriented paradigm will serve as the basis of a new way to develop CAD frameworks that accommodate a large number of CAD tools to support CE.

CONCLUSIONS

A survey of concurrent engineering has shown that practicing the concurrent engineering philosophy is *a huge project spanning all aspects of a product's entire life cycle*.   Furthermore, concurrent engineering is essential to attain the *computer optimized manufacturing* (COM) system that may be the next generation beyond the present computer integrated manufacturing (CIM) system, and will be a challenging task for manufacturing society to implement its philosophy.

A significant improvement in the design productivity may be accomplished by using a CAD framework that provides a variety of functions and services. The design community, especially *mechanical design community*, should be aware of the synergistic relationship that exists between the developments of CAD framework and individual tools.

Because of its many desirable feaures, the *object-oriented paradigm* will be a promising candidate for the develoment of sound CAD frameworks that support the CE practices.

REFERENCES

1 Pennel, James P., and Winner, Robert I., 1989, "Concurrent Engineering: Practices and Prospects", IEEE GLOBECOM '89, Part 1 (of 3), Dallas, TX, Nov. 27 - 30, pp. 647 - 655.

2 O'Grady, P., and Young, R.E., 1991, "Issues in Concurrent Engineering Systems", J. of Design and Manufacturing, Vol.1, pp.27-34.

3 Jo, H.H., Parsaei, H.R., and Wong, J.P., 1991, "Concurrent Engineering: the Manufacturing Philosophy for the 90's", Proc. of the 13th Ann. Conf. on Com. & Ind. Eng., Orlando, FL, pp.35-39.

4 Graham, A., 1991, "The CAD Framework Initiative", IEEE Design & Test of Computers, September, pp.12-15.

5 Harrison, D.S., Moore, P., Spickelmier, R.L., and Newton, A.R., 1986, "Data Management and Graphics Editing in the Berkeley Design Environment", Tech. Pap. of the IEEE Int'l Conf. on CAD, Santa Clara, CA, Nov.11-13, pp.24-27.

6 Martin, C.C., and Hutchinson, K.K., 1989, "Computer Aided Concurrent Design for Printed Wiring Boards", Proc. of the 2nd Int'l Conf. on Indus. & Eng. Appl. of AI & Exp. Sys., Tullahoma, Tennessee, June 6-9, pp.493-499.

7 Birmingham, W.P., Kapoor, A., Siewiorek, D.P., and Vidovic, N., 1989, "The Design of an Integrated Environment for the Automated Synthesis of Small Computer Systems", Pro. of the 22nd Ann. Hawaii Int'l Conf. on Sys. Sci., Kailu-Kona, Hawaii, Jan.3-6, pp.49-58.

8 Pendleton, J.M., and Burns, C., 1989, "An Integrated CAD Environment for System Design", Proc. of the 22nd Ann. Hawaii Int'l Conf. on Sys. Sci., Kailu-Kona, Hawaii, Jan.3-6, pp.39-48.

9 Srinivasan, R.V., Agrawal, R., and Kinzel, G.L., 1990, "Design Shell: A Framework for Interactive Parametric Design", in Advances in Design Automation-1990, Vol.1, pp.289-295.

10 Bingley, P., and Wolf,P. van der, 1990, "A Design Platform for the NELSIS CAD Framework", Pro. the 27th ACM/IEEE DAC, Orlando, FL, June 24-28, pp.146-149.

11 Daniel, J., and Director, S.W., 1989, "An Object Oriented Approach to CAD Tool Control Within a Design Framework", Pro. of the 26th ACM/IEEE DAC, Las Vegas, June 25-29, pp.197-202.

12 Cooke, D., Swan, G., Sirott, J, Kane, R., Stevens, P., Yang, J., and Chen, D., 1989, "Design Management in a Workstation Environment", Pro. of the 22nd Ann. Hawaii Int'l Conf. on Sys. Sci., Kailu-Kona, Hawaii, Jan.3-6, pp.111-117.

13 Smith, Roy, and Cavalli, A., 1990, "Building a Fourth-Generation Framework", High Performance, June, pp.63-67.

14 Bretschneider, F., Lagger, H., and Schulz, B., 1989, "Infrastructure for Complex Systems - CAD Frameworks", in EROCAST '89, Pichler, F., and Moreno-Diaz (eds), pp.125-133.

15 Weston, R.H., Hodgson, A., Coutts, I.A., and Murgatroyd, I.S., 1991, " 'Soft' Integration and Its Importance in Design to Manufacture", J. of Design and Manufacturing, Vol.1, pp.47-56.

16 van der Wolf, P., Bingley, P., and Dewilde, P., 1990, "On the Architecture of a CAD Framework: The NELSIS Approach", Proc. of the EDAC, Glasgow, Scotland, Mar.12-15, pp.29-33.

17 Fiduk, K.W., Kleinfeldt, S., Kosarchyn, M., and Perez, E.B., 1990, "Design Methodology Management - A CAD Framework Initiative Perspective", Proc. of the ACM/IEEE DAC, Orlando, FL, June 24-28, pp.278-283.

18 Allen, W., Rosenthal, D., and Fiduk, K., 1990, "Distributed Methodology Management for Design-in-the-Large", IEEE Int'l Conf. on CAD, Nov.11 -15, Santa Clara, CA, pp.346-353.

19 Daniel, J., and Director, S.W., 1991, "An Object Oriented Approach to CAD Tool Control", IEEE Trans. on CAD, Vol.10, No.6, pp.698-713.

CAD "AN INTEGRAL PART OF MANUFACTURING"

M. Munir Ahmad, Edward M. O'Mahony, and [2]James O. McEnery

[1]Department of Mechanical and Production Engineering,
University of Limerick, Limerick, Ireland
[2]Boart Europe, Shannon Industrial Estate, Clare, Ireland

ABSTRACT

The persistent pursuit for high quality products in the manufacturing environment today has been a major contributor to the rapid move towards CAD/CAM systems.

This paper deals with one such system and the further development it has to under go to keep pace with technology. Topics covered include the initial development of the CAD/CAM database, the development of an expert parametric design package and its knowledge-base. The system design will take into account the manufacturability of the product and impose limitations on the designer by utilizing an interdisciplinary approach to the engineering design. The parametric design system will be the front end to the original CAD/CAM system and will be integrated into the design database, thereby eliminating the possibilities of design proliferation.

The eventual aim is to develop an easy to use CAD/CAE system capable of encapsulating the designers experience, and performing retrieval and optimisation routines to control the interactive process of design modification through rules and practices.

1.0 INTRODUCTION

The research and development work discussed herein is being conducted in Boart Europe in conjunction with the CIME Research Centre of the University of Limerick. The company started operation in Shannon Industrial Estate in 1961 with just 20 people, but today employs approximately 250. They are now one of the major manufacturers of high quality rock drilling consumables, (Bits, Rods, Shanks, Couplings and Hammers), for the mining, quarrying and civil engineering industries.

More than 90% of Boart Europe's products are manufactured for the export market due to its geographical position. This has necessitated the need to keep pace with technology and improvements in machining methods. So in 1986 the company decided to introduce a Computer Integrated Manufacturing Strategy. The first step decided on was the introduction of a CAD/CAM system. After a feasibility study was conducted which highlighted the benefits that a CAD/CAM system could yield the company, one of the few completely integrated CAD/CAM system on the market at the time was selected and implemented [1,2].

This system so far has afforded many benefits, improved quality, increased customer satisfaction, a gain in work completion, design and manufacture cycle reduction, better reliability and better utilization of capital resources. But these benefits are difficult to determine financially as they do not directly influence the product value.

2.0 PHASE 1: THE INITIAL DEVELOPMENT OF THE CAD/CAM SYSTEM

2.1 Overview

Prior to the introduction of the CAD/CAM system there was no real communication between the design office and the manufacturing side of the organisation. As a result, this led to a duplication of work, as the design office issued a complete drawing to manufacturing, where the CNC programmer had to digitize it into the CAM system. The drawings were full engineering working drawings and contained more information than the CNC programmer required, as he was only interested in the product profile to produce the cutter paths. This led to an obvious loss in man hours, data redundancy, to a loss of security. The integrity of data was also effected due to unauthorised editing.

2.2 CAD/CAM Configuration

The configuration of the CAD/CAM system implemented in Boart Europe is shown in Fig(1). When the CAD system was successfully selected and implemented, the next question that had to be addressed was how to structure the database and the existing information, in a method the designers could access quickly and easily [3,4,5,6]. As the implementation of the CAD/CAM system was to be the first foundation stone for the implementation of a totally computer integrated manufacturing system, a considerable amount of research was conducted into the structure of the CAD/CAM database and its eventual integration into the CIM environment [7,8].

2.3 CAD database structure

The initial requirements of the CAD database were, to store all the necessary information and limitations the designer needs for the successful completion of his/her work. Therefore it was decided to divide the main CAD database into 3 databases Fig(2).

2.3.1 The working database

As the name implies this is the working database. Each designer is assigned one and this is where he/she stores the designs he/she is working on and all the information of relevance to them. When the designer is satisfied with the completed design it is stored in the pending release directory "PENDREL.DIR" to await the chief designers inspection and authorization for production.

2.3.2 The local database

There are two local databases i.e. STDPARTS for the design department and NCREL for the production department.

The STDPARTS.DIR is used for storing standard elements that are used in the construction of the product drawings. These elements are drawn, dimensioned and stored in libraries. Therefore if the designer needs a standard element he/she can instance/retrieve it into the drawing in his/her working directory.

The NCREL.DIR is used to store the product profile, which is drawn in the design office and down-loaded to the CAM office, where the CNC programmer produces the CNC code for down-loading to the machines via a direct numerical control (DNC) link.

2.3.3 *The global database*

This is the permanent database and is used to store all the drawings which have been authorized by the chief designer. It contains two directories,

(1) PRODUCTS.DIR: This is where all the product drawings are stored. The directory is divided into sub-directories for the different product ranges and further divided in relation to product codes, as seen in Fig(3).

(2) TOOLING.DIR: All the drawing for specialised tooling is stored in this directory. The tooling could be made up in house or contracted out.

Logical names have been set up to allow the designers to retrieve information from the lower levels of the VMS directories in a fast and convenient fashion as shown in Fig(4).

3.0 PHASE 2: ANALYSIS OF THE PRESENT SITUATION AND TECHNOLOGY

3.1 Overview

The first step has been successfully completed. The CAD/CAM system has now been in operation in the factory for the past 5 years, and has to some extent, simplified the tasks facing the designers. The major benefits on the design side are; it speeded up and standardized the process of entering new drawings and simplified the task of modifying drawings already in the system.

On the CAM side of the organisation, as the current practice now is for the designers produce the product profile and down-load it to the CAM terminal where the CNC programmer requires it, has eliminated the need for duplication of work, increased drawing security, ensured that the integrity of the data is maintained and that unauthorised editing is eliminated.

Even with the benefits outline above the CAD system is still being used like an electronic drawing board. The products continue to be drafted in a similar manner. This can be accounted for by a number of reasons.

(1) Like most companies that have been in operation for a long time that decide to invest in a CAD/CAM system the majority of their product drawings are not in the system due to the vast quantities that have been produced throughout the years. Boart have less than 10% of their total product drawing in the system and as a result have a very limited set of standard parts in the libraries. The time saving of having these elements in the libraries could be in the region of 500% to 800% in the drafting of a typical drawing.

(2) The designers have little time to start entering the old drawing into the system.

(3) Another problem is the amount of time the designer spends doing clerical work, such as, photocopying drawings and answering queries.

The aim of this work is to reduce if not eliminate the first two problems outlined above, thereby speeding up the process of producing new drawings leading to a reduction in lead time for preproduction.

3.2 Preliminary Investigations

To identify the best route forward on the CAD side of the organisation it was necessary to identify the activities being performed in the design office i.e. information flow, the usage of the designers time and the present CAD facilities. Also an investigation was performed into different expert systems and their shells [9].

3.2.1 Information flow through the design office

As expected the main requests for new products came from the sales department. When a customer makes an order for a new product, the sales department raise a notification form and deliver it to the design office, where it is placed in a queue in order of priority. The designer design the product, produces the drawing, places the product code into the material manager system and enters the bill of materials. The information proceeds then as shown in Fig(5).

Another important area from which requests originate is product development. This is handled in a similar fashion as the sales request and is detailed in Fig(6).

The design department also receives requests from various other areas of the factory, these include production, quality assurance, administration etc. A complete list of these and the information they require is provided in Fig(7).

3.2.2 Utilization of the designer's time

One of the most important factors to analyze is how the designer's time is spent. To do this a time study was conducted with the full co-operation of the designers. The results can be seen in Fig(8). As expected it highlighted the fact that the designers were only using the CAD system 51% of their working day.

3.2.3 Present CAD facilities

The present CAD system is similar to most present day CAD systems with the same drafting capabilities. When a designer needs the drawing of a product he/she creates it on the screen in the usual fashion by entering a series of commands i.e. Add line, Add arc etc. If there are parts that are used in several drawings the designer will create these parts and store them in libraries. The problem with this method of construction is the parts are constructed to a set dimension, therefore, if a different size parts is required the designer has to construct it from scratch, working through the same series of commands. This method is time consuming and tedious for the designer and as a result can lead to errors in the completed work.

3.2.4 Part duplication and proliferation

Like most companies involved with manufacturing Boart produce a variety of products. Over a period of time these products undergo alteration in their design i.e. proliferations (slight variations of the existing product design). With the result that within each of Boart's product ranges, there are components that only vary in size. For most companies as the part number is not adequate for design retrieval methods, the designer must rely on memory and experience to locate a drawing. When memory fails, or experience is lacking, redundant designs are created.

Many studies have been performed which show that in an average company 40% of new designs already exist, 40% are variations on old designs and only 20% are actually completely new designs. In Boart's situation, as the technology and designs of the drill bits have not changed much over the past 30 years, less than 20% would actually be new designs. Therefore, a fast retrieval system is necessary to reduce duplication and proliferation.

3.3 The Expert System and its problems

At the outset of this work it as decided to use an expert system [10,11]. In short, an expert system is a set of computer programmes that provide expert level solutions to difficult problems. They are normally,

(1) Heuristic in nature (use rule of thumb in their reasoning).
(2) Transparent (give an explanation for their solutions).
(3) Flexible (allows knowledge to be integrated into their knowledge base).
They normally contains 3 sections:
(1) The user interface, which allows the user to communicate with the expert system.
(2) The inference engine, which drives and controls the system.
(3) The knowledge base, which is the most important section and contains the rules, facts and questions from which the expert system formulates its result.

To develop a complete expert system would be prohibitive by cost and time. So an investigation into different the shells available was conducted. An expert system shell is an expert system who's knowledge base is empty [12,13]. These shells are normally subsets of existing expert systems that have been designed to handle particular problems, where the solution can be derived ie. it assumes the solution can be located somewhere within the knowledge base. This approach is not suitable for engineering design as it is a creative activity and not a deductive one. Therefore, there has to be a formulative approach taken to find a solution. So to find a shell that would satisfy all the constraints outlined in Section (3.4), and be capable of dealing with the problem in hand would be an impossibility [14,15,16].

3.4 Limiting factors to further development
(1) The present CAD system operates on a MicroVAX II that supports a VMS platform. Therefore all software must run under VMS.
(2) The software must be compatible with the current CAD system eliminating the need to reenter information already in the system.
(3) Information is required to be entered in graphical, menu and textual format as this is what the designers are familiar with.
(4) The user interface must be similar to the existing CAD package, thereby reducing the learning curve encountered by the designers.
(5) The software must be capable of searching the existing database in a fast efficient manner.
(6) Most importantly it must simplify the designers job.
(7) Should allow further development by the designer ie for new product groups.

4.0 PHASE 3: THE FURTHER DEVELOPMENT OF THE CAD SYSTEM

4.1 Proposed solution
All of the preceding factors plus the fact that Boart manufacture family-of-elements led to the investigation of parametrics for the solution.

4.2 Parametric design
One of the most common ways to automate a process is through standardizing similar or redundant segments in a process. This is called family-of-parts or parametric programming and is usually done with a form of macro language. Parametric design is a relatively new approach when it comes to CAD. It operates on the principle that instead of drawing each part in a family-of-parts individually, one part is drawn and variables are used instead of constant numerical values. By using this approach, the geometric data can

be separated from the image generating process, therefore with a single programme similar geometric shapes can be produced by simply changing the values of the parameters [17].

4.3 Parametric programming for shank adaptors

Initially it has been decided to develop the software for only one of Boarts product ranges, to establish the benefits and the viability of the development work. The language used for this work is IAGL (Interactive Applicon Graphic Language) which was developed by the CAD software vendors. With the use of this language, it is possible to customise the system by writing small application macros. These macros are developed within the system and can be executed at any time during a session.

When the programmes are executed the designer is prompted for the main dimensions of the shank adaptor. The programme then reads in relevant data from available files on tooling etc., for the sizes and informs the designer, if for instance, specialised tooling may be required. When all this information is gathered the system calculates the value of each parameter and then proceeds to drawing the component.

The main advantage of using the macro language, is that it is easy to learn due to its English like format. Also, as it is supplied with the CAD system, it allows for the interface to be developed in the same manner as the original CAD package, thereby eliminating a steep learning curve for the designer. The designer can also develop his own parametrics, as time goes on thus eliminating the need for a specially trained person to maintain or extend the system.

4.3.1 Advantages of parametric programming
1) Time saving of up to 1000% in the production of a product drawings.
2) Standardization in the way the information is presented on the drawings.
3) Reduction in repetitive work.
4) Utilises the CAD package while the designer is doing other work ie. the system will produce the drawing once the initial information is entered.
5) Elimination of errors in the drawings once the program is correct.

4.3.2 Disadvantages
1) Requires time to develop the programmes.
2) If the program is aborted before it is finished the CAD setup could be changed

4.4 End Product

The end product will be a package capable of creating drawings through the use of parametrics for the complete shank adaptor range and linking these drawings to those already in the system. It also provides the designer with a mix and match facility for the construction of product drawings. The package will also take into consideration various factors, for example, standard tooling etc. causing the designer to be more standard in his/her design work.

4.5 Plans for the future

The initial parametric programmes have been developed in IAGL and have been tested and verified to ensure they run correctly. All modifications have been conducted and the software documentation produced. The future plans for this software is to develop it in VAX 'C' thereby allowing the software to be compiled and executed externally to the

CAD system. This will allow the designer to run the software from the text screen without entering into the CAD system until he/she wants to view the completed drawing.

5.0 CONCLUSIONS

1) This system will provide a front end to the CAD/CAM system already in the company. It will complement the designer's job, as all the information required for a design will be automatically retrieved after the initial questions are answered.
2) The designs will not have to be created from scratch, the amount of standard elements in the libraries will increase and will increase the speed for the production of drawings.
3) The system will reduce the time spent by the designer doing repetitive drafting jobs which will allow him concentrate on the design of the products, thus leading to greater job enrichment.
4) This system will lead to the reduction of the lead time for the drawings going through the design office and aid the factory in its aim of reducing the total product lead time to six weeks.
5) It is vital to maintain a good information management system to eliminate part duplication and proliferation.

6.0 ACKNOWLEDGEMENT

The authors would like to thank Boart Europe and EOLAS - the Irish Science and Technology Agency for providing the funds under the Higher Education Industry Co-operation Scheme to carry out the research work presented in the paper. The authors would also like to thank the company's management, John and Tom in the design office and Gerry McInerney the chief designer for their support and co-operation.

7.0 REFERENCES

1. Ahmad M.M., O'Callaghan M. and McEnery J., Selection, Implementation and Management of a CAD/CAM System , *Proceedings of the 2nd World Basque Congress Part B (Advanced Tech. in Des. and Manuf.)* Bilbao, Spain, 14-18 Dec.,1988, 295-312.
2. Ahmad M.M., O'Callaghan M. and McEnery J., A Case Study on the Selection of a CAD/CAM System, *Proceedings of the Irish Manuf. Committee Conf.,* University of Limerick, Ireland, Sept, 1987, SP194.
3. Managaki M., Kawagoe K., and Naniwada M., A Model and Its implementation in a Practical CAD/CAM Database, *Computer Ind.* Vol. 5, No. 4, Dec, 1984, 319-327.
4. Eberlein W. and Wedekind H., A Methodology for Embedding Design Database in Integrated Engineering System, *File Structure and Databases for CAD,* edited by Encarnacao, J. and Krause, F.L., North Holland, 1982, 3-37.
5. Ulfsby S., Mean S. and Oian J., *A DBMS or CAD/CAM System,* 335-346.
6. Gardenas A.F., *Database Management Systems,* Allyn & Bacon, New York, 1979.
7. Ahmad M.M., O'Callaghan M. and McEnery J., The Role of CAD/CAM Databases in CIM, *Proc of the Factory Automation and Information Management Conference*, FAIM 13-15 March, 1991, University of Limerick, Ireland.

8. Ahmad M.M., O'Callaghan M. and McEnery J., Engineering Database Technology and it's impact on Computer Integrated Manufacturing, *Proc. of the Design Productivity International Conference*, Honolulu, 3-9 February,1991.

9. Ahmad M.M., O'Mahony E. and McEnery J., The Development of a CAD Synthesis System, *Proc. of the Irish Manuf. Committee Conf.,* University of Ulster, Northern Ireland, Sept, 1991.

10. Winfield M.J., Expert Systems: An Introduction for the Layman, *Computer Bulletin*, Dec., 1982, 6-7/18.

11. Alty J.L., Use of Expert Systems, *Computer Aided Engineering Journal*, Vol. 2, Feb., 1985, 2-9.

12. Born G. and John R., Expert system shells for development & delivery, *Knowledge Based Systems Proc. of the Int. Conf.*, London, July, 1986, 273-283.

13. Simmons M.K., Artificial Intelligence for Engineering Design, *Computer Aided Eng.* Vol. 1 No.3, Apr., 1984, 75-83.

14. Maher M.L., Longinos P., Development of an Expert System for Engineering Design, *Int. Jour. of Applied Engineering Education* Vol. 3 No. 3, 1987, 279-286.

15. Mausbach W., Carwile B. and Dorris R., Design for Manufacturing Use of an Expert System, *Autofact '90 Conf. Proc.*, Detroit, Michigan, November 12-15, 1990, Section 21, 25-40.

16. Krouse J., Ten Trends for the 1990's, *CAE, Computer Aided Engineering*, Vol.7 No.10 Oct. 1988, Varing pages.

17. Unny M., Yang T., Integration Framework Using Parametric Design Solid Modeling, *Factory automation and Information Management Conf. Proc.*, University of Limerick, Ireland, March 13-15 1991, 323-332.

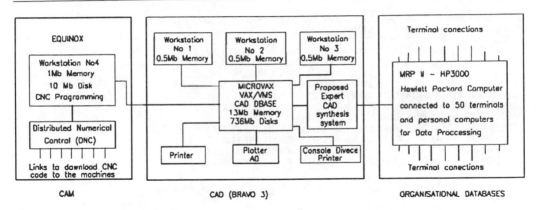

FIGURE 1: The Configuration of The CAD/CAM System

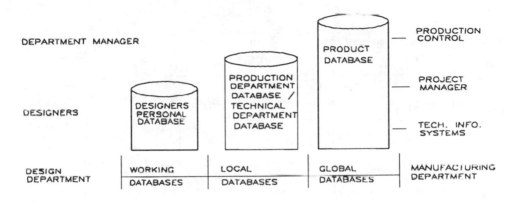

FIGURE 2: The Database Structure

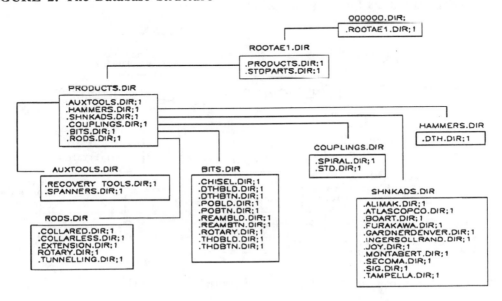

FIGURE 3: Directory Structure For The Products

```
SHK$DIN     FDA1:[ROOTAE1.STDPARTS.SHANKS.THREAD_DIN]THREAD_DIN
SHD$ENDS2   FDA1:[ROOTAE1.STDPARTS.SHANKS.BACKEND2]BACKEND2
SHK$ENDS3   FDA1:[ROOTAE1.STDPARTS.SHANKS.BACKEND3]BACKEND3
SHK$ENDS4   FDA1:[ROOTAE1.STDPARTS.SHANKS.BACKEND4]BACKEND4
SHK$ENDS8   FDA1:[ROOTAE1.STDPARTS.SHANKS.BACKEND8]BACKEND8
SHK$GD      FDA1:[ROOTAE1.STDPARTS.SHANKS.THREAD_GD]THREAD_GD
SHK$HM      FDA1:[ROOTAE1.STDPARTS.SHANKS.THREAD_HM]THREAD_HM
SHK$ROPE    FDA1:[ROOTAE1.STDPARTS.SHANKS.THREAD_ROPE]THREAD_ROPE
SHK$SEALS   FDA1:[ROOTAE1.STDPARTS.SHANKS.SEALS]SEALS
SHK$SLOTS   FDA1:[ROOTAE1.STDPARTS.SHANKS.SLOTS]SLOTS
SHK$SPLS2   FDA1:[ROOTAE1.STDPARTS.SHANKS.SPLINES2]SPLINES2
SHK$SPLS3   FDA1:[ROOTAE1.STDPARTS.SHANKS.SPLINES3]SPLINES3
SHK$SPLS4   FDA1:[ROOTAE1.STDPARTS.SHANKS.SPLINES4]SPLINES4
SHK$SPLS5   FDA1:[ROOTAE1.STDPARTS.SHANKS.SPLINES5]SPLINES
```

FIGURE 4: Logical names used for Shank Adaptor libraries

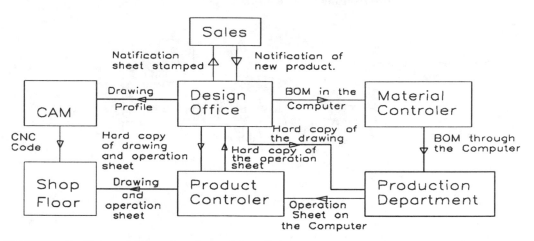

FIGURE 5: Requests for drawings from Sales.

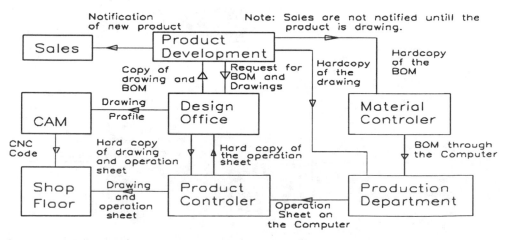

FIGURE 6: Request for drawings from Product Deveopment

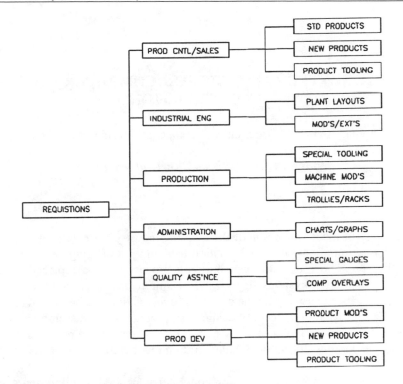

FIGURE 7: Requisitions sent to the design office

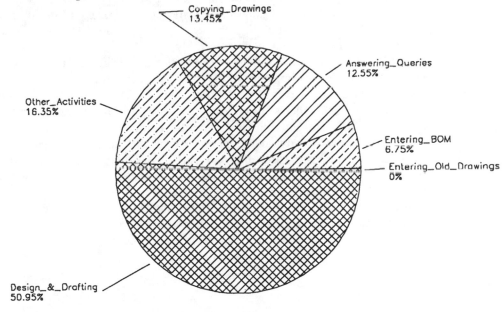

Note : Total for all the designers
This study was conducted over a peroid of 383 working hours.

FIGURE 8: Time usage in the design office

OBJECT ORIENTED CONCEPT FOR COMPUTER INTEGRATED CONSTRUCTION

Thomas A. Bock, Friedrich S. Gebhart, and Kunibert H. Lennerts

University of Karlsruhe, Karlsruhe, Germany

ABSTRACT

The process of industrialization of the building industry has entered a new dimension by the introduction of advanced information technologies offering new possibilities for the building process. The traditional building process will be replaced by computer integrated construction systems. By means of simultaneous planning, production and assembly, the capacity as well as the quality of construction can be improved.

To realize this parallel concept of design, pre-production and assembly, the computer integration of all related processes is necessary. Re-design, re-programming and re-simulation are possible during the whole building process, to handle the changes as they occur.

To show the possibilities of realizing computer integrated building, the assembly of masonry was chosen as a case study. Within this study, an integrated concept of the information flow was developed, to connect all parts of the process from computer aided architectural design of the building through automated pre-fabrication of the building components to the robotized on-site assembly.

To realize a continuous information flow, highly flexible interfaces are necessary. They must be integrable, extendable, reuseable and compatible. To meet all requirements, an object-oriented approach was chosen. Through object-oriented interfaces, an object-oriented database provides all processes with the necessary data.

For this purpose, different activities are currently carried out in our laboratory. Especially, the interfaces between the components of the integrated concept are identified and realized, to enable the different participants immediate access to all necessary information. Also, components for advanced implicit robot programming tools and an integratable advanced robot control are realized.

With these prototyped realizations of an advanced integrated concept, we begin to meet the needs of the construction industry for flexible and rationalized one-of-a-kind production which includes complex management and logistic tasks.

STATE OF THE ART

The progress of productivity in the construction industry is relatively slow compared to other fields of the manufacturing industry. Reasons are mainly single piece production, complex tasks, and constantly changing environments with every new building task; e.g., the very individual designs of apartment houses with angled and curved walls or bay windows.

The shortage of skilled workmen in Germany due to difficult working conditions, time and cost pressures, and increasing demands for quality are leading to a need for more rationalization on the building-site. The labour shortage is also captured in the "pyramid of age", for example, 50% of skilled workmen in Germany on construction-sites are over 50 years old. The labour shortage can be compensated for by flexible automation of the building process.

Another factor is that the maximum lifting weight allowed for construction workman will be more and more limited: In the Netherlands, for example, the limit is already set at 18 kg. In order to obtain higher productivity, it is necessary to automate the construction process through computer integrated planning, construction and execution of building projects. Robotic systems can be considered a key technology for construction automation.

Conventional Masonry

The development of building systems in Europe in the last twenty years showed that masonry gained more and more market share in comparison to other building techniques, (e.g. large prefabricated panel systems). In 1988, in Germany approximately 85% of all newly constructed housing was erected using masonry; the masonry usage rate for other building types was 50%. The total production of bricks amounted to 18 million m^3. Due to immigration and increased demands from the former German Democratic Republic, the building of 500,000 flats/year is expected for the next 10 years.

The reason for the wide market acceptance of masonry is because of positive experiences in solving the problem of individualized building tasks. This can only be accomplished by using small building elements, which are highly flexible and fulfill all requirements for free planning. In addition, a large array of additives can be used to create blocks with different biological and physical features for a wide field of applications.

The conventional masonry is still characterized by high work intensity, difficult working conditions, and low productivity. To define the tasks, which must be fulfilled by an automated system, one first has to identify the conventional method of masonry construction. The traditional building materials are bricks and mortar. The bricks must meet the requirements of load bearing capacity, heat and noise insulation, and protection against other environmental influences. The mortar is basically used as a levelling layer because of the inaccuracy of the bricks. The traditional mason tools are bubble level, plumb, batter cord and a hammer, which are used to get a vertically and horizontally correct wall at the right place with the right length. The building method is to level out each brick in a mortar bed with these tools and, if necessary, to cut the brick with the hammer into the right shape. These materials, tools, and methods have remained unchanged for centuries.

From Mechanized to Advanced Masonry

The conventional masonry process requires highly skilled workmen in order to achieve sufficient and consistent walling quality. This labour intensive construction process results in relatively high costs. To achieve higher rationalization and humanization, different approaches are currently being tested.

One of the approaches, which is currently used by the sandlime brick and cellular concrete industry in Europe, is the enlargement of the building block sizes. The larger dimensions of these blocks, however, coincide with their increased physical weight of up to 300 kg. Due to larger dimensions and heavier weights these building blocks are not ergonomically desirable and, therefore, various mechanical aids such as hydraulic balancers or minicranes with counterweights are used.

A second approach is the pre-fabrication of customized block in the factory. This requires positioning plans for the on-site assembly which are generated with the help of CAD/CAM-systems. The assembly information was also used to program the in-factory cutting equipment to customize the blocks. The pre-cutting and the assembly plans enhance the working speed on-site, significantly, and also help to reduce the complexity of the masonry task.

Another tendency, which is steadily gaining momentum, is the use of more exact, plane-parallel blocks, which are layed in thin-bed mortar/glue or, as shuttering blocks, are assembled dry and filled with concrete after the assembly of the whole wall. These methods result in better physical properties because of the reduction of walling operations because of no mortar joint, enabling higher working speed and easier use.

The differing accuracies of below grade and of walling work are usually compensated for by a first layer of smaller blocks in a more or less thick mortar bed. By this method, the difficult alignment, which is required for adjusting the position of each block, can be reduced. Even though an increased working speed can be achieved while employing less qualified workers on the construction site, the present and future share of cost due to labour will be constantly increasing.

Further rationalization can only be achieved by reduction of labour and construction time. Since the above mentioned mechanised methods for masonry have reached their systems limits, they can not contribute to further effectiveness enhancement. Therefore, a certain innovative leap is required by a systems approach that synchronizes existing construction material with new information technology (IT) enhanced and semi-autonomous robotic technologies which are designed for construction site application.

The system approach deals mainly with the construction oriented modification of existing technologies and with closing the gaps between them through intelligent interfaces and IT-tools, in order to provide the necessary flexibility for one of a kind building production, robust design, and user friendly programming.

The advantage of on-site mechanization and future on-site automation compared to a stationary pre-fabrication is higher flexibility relative to materials, damaged elements and (important for the building industry) short term market changes.

Conventional Information Flow

The implementation of CAD programs into architectural offices has been a very disjointed process. Many different systems are currently in use in the bigger offices, while the smaller ones are still working by hand to produce the design plans.

In Germany, right now, in almost all building projects the following information flow prevails. The hand/CAD made design plans are used by the construction companies to order the necessary materials from the building material manufacturers and to cut, trim and assemble the blocks on-site.

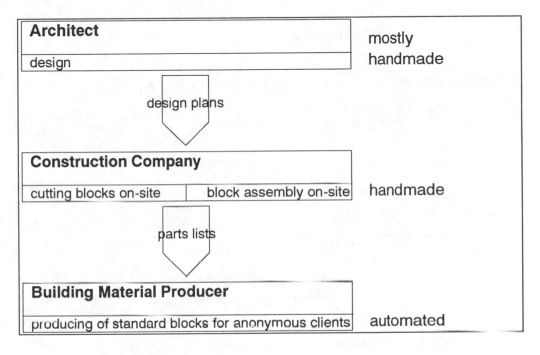

Fig. 1: Conventional information flow

The whole construction process is mostly done by hand, because of the heterogenous structure (hand, different CAD´s) of the design plans and, thus, it is almost impossible to process all the information automatically. In this above mentioned context CAD means computer aided drafting rather than computer aided design!

Advanced Conventional Information Flow

Today, computer aided design gains more and more importance. The use of CAD in architectural offices is increasing rapidly and with this the availability of electronically processed data. An additional trend can be observed in that building material producers are working together with service companies to provide different factories with the necessary production data.

The third process taking place today is the automation of the cutting and customizing process in pre-fabrication by NC-cutting equipment. To rationalize the construction process, the pre-fabrication companies changed the information flow and get the information they need directly from the architects instead of the construction companies. As mentioned above, the information is used to customize the blocks and to generate assembly plans for the construction companies.

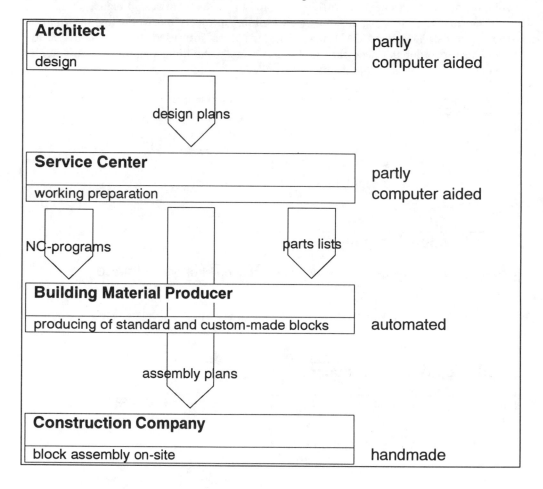

Fig. 2: Advanced conventional information flow

The automated pre-fabrication procedure has some advantages for the whole process:
- The rationalization and humanization of the cutting and customizing process through in-door automation.
- The focusing of the information flow due to the reduced number of building material producers.

COMPUTER INTEGRATED CONSTRUCTION (CIC)

To integrate the construction process, every task must be based on electronic data processing. The design plans should be generated by CAD at the architect´s office. The preparation of work for pre-fabrication and assembly on-site should be done by computers at the service centers or building material producers. Customized blocks should be produced automatically in the building material factories using NC-equipment. Construction robots should assemble the customized materials on-site. But, how does the information get to the different stages of this production chain?

CIC Information Flow

The CAD-based architectural information must include a database of the service center or producer. Here the necessary data for pre-fabrication and on-site activities will be generated: Data for the production planning of the standard blocks, NC-programs for the cutting equipment, and robot programs for the on-site assembly.

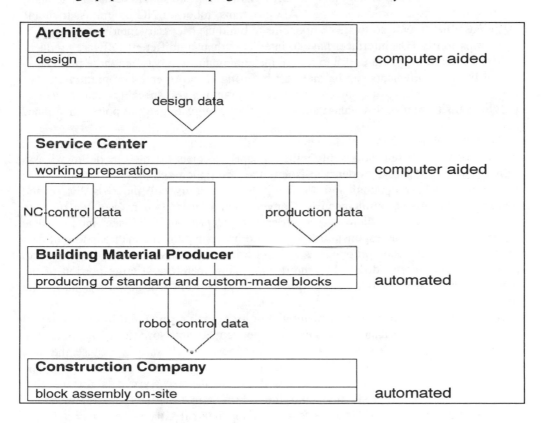

Fig. 3: CIC information flow

To provide all the participating processes with the necessary data, different interfaces are necessary, which are identified and described in the next section.

CIC Interfaces

Having acquired the needed CIC information, the interfaces should fulfill some important requirements:
- Flexibility: Construction is a dynamic process. Because we are dealing with many different persons and an always changing environment, a highly flexible system is necessary.
- Integration: Data integration is needed because the CIC process consists of different soft- and hardware products. Therefore, an overall integratable system is required.
- Extendibility: CAD/CAM, robotics, etc. must be considered rapidly evolving. Therefore, it is important to build interfaces, which ensure that an adapting to a changing environment (e.g. new CAD programs) can be done easily.
- Reuseability: The interfaces have to fulfill many tasks and should be useable to many different software products, e.g., CAD programs, robot, or NC programs. It should be possible to reuse software components to build the necessary interfaces.
- Compatibility: The interfaces have to interact with many different software products. Abstract data types are useful to realise the communication between these products.

All these requirements can be met best by using object-oriented programming. The basic characteristics of object-oriented programming, especially encapsulation, polymorphism, hierarchies, inheritance and genericity guarantee the points mentioned above.

Object-oriented database systems are distinguished by the integration of conventional database functionality with concepts of object-oriented modelling. It is a primary aim of database systems to isolate the description and administration of data from application programs, and to facilitate a solid use of the data by several application programs, simultaneously. Conventional database systems isolate data from application programs which makes programs independent from the internal data representation. But the responsibility for a solid interpretation and manipulation of data rests with the application programs. An object-oriented database, on the other hand, can process highly structured data like the design plans; therefore, a large portion of the work that normally the interface programs had to perform can now be done by the database.

The correct interpretation and manipulation of complex structured information will be guaranteed by the database system, independent of any specific application. The object-oriented database system approach is especially well suited, because the data encapsulation by operations supports the transfer of data into the database system. The object-oriented interface programs that will extract the necessary data for the NC-equipment, the block production, and the robots, can be easily programmed and changed, because they are independent from the internal data representation in the object-oriented database. The often encountered problem of changing data formats can be easily handled by the object-oriented database. Thus, an object-oriented-database helps to avoid many problems occurring with relational databases.

Instead of programming every interface between the partners mentioned above, a programmed object-oriented-interface will be like a magnifying glass focusing different CAD software to an object-oriented-database (see fig. 4). In summary, the object-oriented-database can be considered as the main part of the integration process.

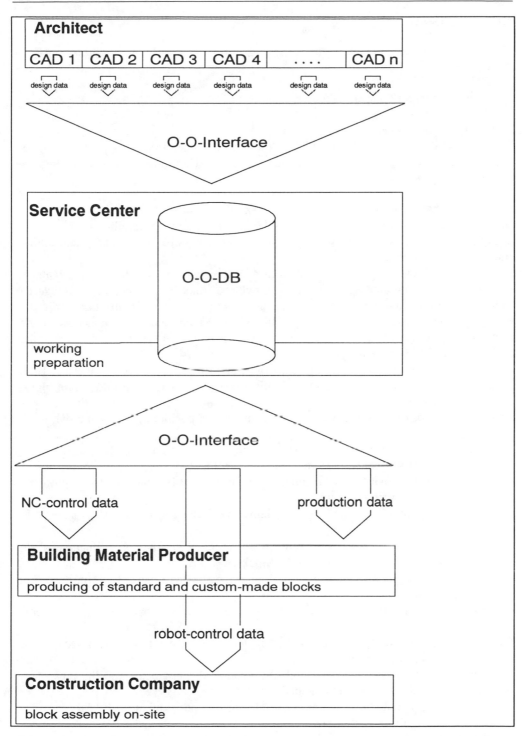

Fig. 4: CIC interfaces

CIC Components

From the integration aspect, the following information technology components can be identified:
- CAD
- Building component generation
- NC program generation
- robot program generation

The first three points: CAD, building component generation/work preparation, and NC program generation are state of the art, today. Therefore, in the following, the emphasis will be on the last point:

Robot program generation　In one-of-a-kind production like the construction business, the programming is a critical key factor of the rentability of automated equipment. Therefore, an easy to use and as highly as possible automated programming interface must be provided.

Because of the single piece production and constantly changing requirements, normally used programming of robot work cycles cannot be applied. To minimize the individual robot programming efforts, an automatic generation of control data from the existing geometrical data is necessary. This requires completely new concepts of off-line programming, which must integrate as much pre-planning as possible and as much on-site flexibility as necessary.

The whole implicit programming task can be subdivided into several subtasks. Three major positioning tasks can be identified after the wall partitioning into single blocks and the block cutting in the factories: (1) The position of the blocks on the transport pallets, (2) the position of the palets on the site, (3) the positioning of each block in the wall.

Different restrictions and requirements of the overall system must be considered by the solution of each task. Therefore, we chose an integral approach where an object-oriented-database provides the common platform to solve the different problems identified above by different application programs.

For the first application of block palletizing, the following requirements have to be considered:
- Every block must be reachable taking into consideration the robot kinematics, the grippers, and the assembly order of the blocks.
- Due to the low price of the building components, a load capacity optimization of each palet is necessary.
- The position accuracy of the blocks on the pallets must be considered with respect to gripping as well as transport security.

The second application (positioning of the pallets on-site) has to take into account the following requirements and restrictions:
- Dynamic obstacle avoidance through the generated robot paths and the previously assembled walls must be implemented.
- The robot path should also be minimized by proper positioning of the pallets on the site accounting for the assembly order.

For the third application, emphasis is placed on the following key factors to meet the requirements of the construction site:

- Sensor integration is a key approach for the uncertain environment on construction sites.
- In order to allow hybrid control strategies, robot-control modularity must be provided.
- Fault tolerance of the system is necessary, especially in case of inaccurately positioned or damaged blocks.

The above mentioned advantages of an object oriented approach can also be used in the applications programming. The properties of object orientation like data encapsulation, polymorphism, inheritance and genericity are also useful to solve the optimization, positioning and sensor integration problems occuring in the implicit robot programming.

CURRENT AND FUTURE RESEARCH ACTIVITIES

Currently, different pilot implementations are being carried out in our laboratory. We examine different CAD systems, which are available on the market, with respect to their functionality and their interfaces. Also, existing wall partitioning systems for the off-line building element generation are examined with respect to their integratability into an integrated information concept.

In the laboratory different robot platforms and simulation systems are available to identify the requirements on the construction site and to test implicit robot programming approaches on- and offline.

To gain intial experience with object-oriented programming, we chose the high level programming task of the layout of the construction site. As the programming tool, we chose the object-oriented programming language Eiffel. Eiffel has been taken instead of C++, for example, because the language keeps the involved people to analyse and design the problems in an object-oriented way. Also, Eiffel fulfills all the requirements mentioned above.

The emphasis of future research activities will be on the further development of the applications and the integration of them based on an object-oriented-database to a prototype CIC chain stretsching from architectural planning to the robotized assembly of building blocks on-site.

REFERENCES

1. Bock, T.A.: *Concept of building system for robotization - A systematic approach for development and design*, Fourth International Symposium on Robotic & Artificial Intelligence in Building Construction, Haifa, Israel, 1987.
2. Bock, T.A.: *Robot Oriented Design*, Shokokusha, Tokio, 1988 (in japanese language).
3. Meyer, B.: *Object-oriented Software Construction*, Prentice Hall International (UK) Ltd., 1990.

CMES: COST MINIMIZATION AND ESTIMATION SYSTEM FOR SHEET METAL PARTS

Sagar V. Kamarthi, Paul H. Cohen, and Edward C. De Meter

Industrial and Management Systems, Penn State Univesity, University Park, Pennsylvania, U.S.A.

ABSTRACT

This paper describes a system which estimates the manufacturing cost of sheet metal parts in conjunction with their process plans. In order to make the optimal use of product information, both the process planning and the cost estimation tasks are integrated with an attribute based group technology system. Here, a process plan is represented by an acyclic multistage digraph. In this graph, the initial node represents the product raw material, the goal node represents the final product, and an intermediate node represent a <feature-machine> pair. A connected path from the initial node to the goal node is considered a feasible process plan that transforms the raw materials into a product. Each arc in the graph is assigned a cost based on the actual machining and material handling cost required to move the product from its head node to tail node. For a given product, all possible alternative process plans are evaluated and the process plan with the minimum cost is chosen for manufacturing the product. This approach provides both cost optimal process plans and product cost at the same time.

INTRODUCTION

This paper describes a computer based cost estimation system CMES (Cost Minimization and Estimation System). CMES serves two functions. First, it is used to manage product, process and cost information. Second, it is used to generate optimal process plans and estimate respective product costs.

CMES has been developed for a relatively small focused sheet metal product company. The cost estimates provided by CMES are used for submitting bids, revising quotations, evaluating alternatives, controlling manufacturing expenses, and establishing selling price.

This company has a job shop manufacturing environment. The shop floor has burning, punching, bending, drilling, grinding, and tumbling machines. This company produces a variety of products in its job shop facility, often, in small quantities to meet the orders from different customers. In this situation, unlike in mass production environment, the product cost estimates cannot be based on standard product costs and instead should be directly related to its process plan. In order to obtain an optimal process plan, process planning and cost estimation are performed concurrently in CMES. At the same time, these two tasks are integrated to with a group technology system to make an effective use of product information.

In the early stages of computer technology in the 60s, computers were simply employed for faster computation of manual procedures for cost estimation. In the late 70s and in 80s, a new generation of cost estimation systems integrated the advances in computer technology such as personal computers and database techniques with traditional cost estimation methods [1]. Some of the well known systems of this class are COSTIMATOR, MICAPP System, CACE System, AM Estimator, and E-Z Quote System [2, 3, 4, 5]. The latest trend in cost estimation systems is their integration with design and manufacturing databases such as Computer Aided Design, group technology, methods-time measurement, and accounting. Examples of such cost estimating systems are Automatic Cost Estimating System [6], Integrated Manufacturing Cost Estimating System (IMCES) [7], Totally Integrated Manufacturing Cost Estimating System (TIMCES) [8], and Cost Estimating System for Concurrent Engineering [1].

CMES is a similar system which not only provides accurate production cost estimation but also permits product cost minimization during process planning. Though focused, the development of this system gives insight into the feasibility, utility of such a system for similar industries where different varieties of products are produced in small quantities. The following sections present the details of its implementation, function, and a sample running session.

DESIGN AND IMPLEMENTATION OF CMES

CMES contains three major subsystems: a group technology subsystem, a process planning subsystem, and a cost information subsystem. The division of CMES into these subsystems serves to achieve modularity in implementation. All three subsystems are implemented using the Clipper database language. Clipper works with dBASE-compatible databases and allows the development of stand-alone systems. The architecture of CMES is shown in Figure 1.

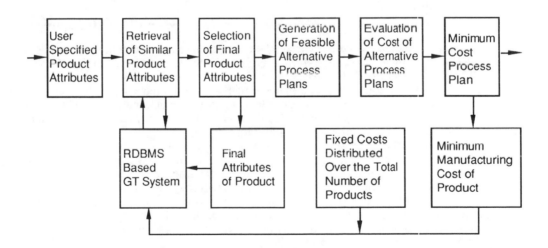

FIG. 1 Structure and information flow in CMES.

The system starts with user supplied product attributes. These attributes are used to retrieve product data of similar description from the group technology system. This retrieved data is used for selecting the final product attributes. For a given set of product specifications, all feasible alternative process plans are generated and evaluated to determine the minimum processing cost. The integration of process planning and cost estimation with a group technology system is one of the features that distinguishes this system from the other cost estimation systems reported in literature. The group technology system in the system contributes to the standardization of product designs and provides a source of information useful for process planning and cost estimation. The following sections present the details of the subsystems.

Group Technology Subsystem

Classification of production engineering information is one of the first steps of manufacturing integration. Over the past ten years, group technology has been applied to the machining industry in order to establish generic part families for part design and manufacturing cell design [9]. In practice, classification and coding is the most widely used method for implementing a group technology system.

A classification and coding scheme involves allocating to each part a code number by means of a classification system which covers the important design features and attributes of the part. Drawings numbers and other design details of parts with similar codes are stored together. Whenever a new part is to be designed, data on similar parts are retrieved and used in the new part design. However, in the present work an attribute based relational database scheme is used instead of a classification and coding scheme.

The choice of the attribute based relational database scheme is motivated by limitations and shortcomings of the traditional classification and coding scheme. Maintenance of a uniform and coherent coding system over a wide range of parts is difficult. The subjective decisions which must be made while coding a product increase the risk of similar parts having different codes. In the process of coding, important attribute information may be lost since each digit or alphabet in a code represents a range of values of an attribute. Often, code similarity may not correspond to the true similarity in part attributes and process plans because code numbers do not carry sufficient part information [10].

The advent of relational data base management systems, where items of data can be linked according to special relationships, has provided an alternative method for building group technology systems. These relational database management systems allow the storage of not only the data and information but also the knowledge of the production system [11, 12]. Therefore, in this application the group technology subsystem is developed using a relational database management system called Clipper instead of a classification and coding system. The excellent query retrieval capabilities of Clipper improves the ability of the group technology subsystem to store and retrieve product data, and eliminate the duplication and proliferation of designs.

The group technology subsystem has two main databases: a part database and a feature database, and several support databases. The part database contains one record for each part. Each record stores a set of part attributes: part name, customer name, customer part number, drawing number, total number of parts required, revision number, revision date, CAD file name, NC file status, raw material form, material, external shape, peripheral shape, overall dimensions, tolerances, special tool and fixture requirement information, part complexity code, programing complexity code, inspection complexity code, part weight, and other part-specific information.

The feature database contains one record for each feature of a part. In this work four types of features are considered: periphery, hole, slot, and bend. Here it is assumed that a sheet metal part has only one periphery, though the part could have multiple counts of holes, slots and bends. For the convenience of developing process plans, any two features of the same type but with different geometric or technological attributes are treated as separate features. A record in the feature database stores feature dimensions, position and size tolerances, and the count of features. The records of holes and slots also store information as to whether they are threaded, countersunk, defined from a bend, defined from an edge, and/or internal. A many-to-one relationship exists between the feature database and the part database records. The part database and the feature database are linked by the part number field common to both databases.

The support databases are used for the selection of the customer number, raw material form, material, paint and coatings, external shape, and peripheral shape for each part. Each item in every support database is assigned a standard code. These individual codes link records in support databases with the records in the part database. The group technology subsystem allows the user to add, delete, and edit part records and feature records using a full screen format. The user can also query the system for specific parts. Queries can be based on any part or feature attribute stored in the part and feature databases. The system also allows the user to browse and modify all support database records.

Process Planning Subsystem

Process planning subsystem has one rule base, two main databases: machine capability database and machine status database, and many auxiliary databases. This subsystem matches part and feature attributes from group technology subsystem with the machine capability information to make selection of machines. The subsystem uses the forward chaining rule-based strategy to do matching. Once the machine selection is complete, a graph theoretic cost optimization model is used to find an optimal process plan.

The machine capability database contains one record for each machine available on the shop floor of the sheet metal company. Each record stores machine attributes: machine name, features that can be produced, minimum and maximum feature dimensions that can be handled, various tolerances that can be achieved, machine tonnage that can applied, and the information as to whether the machine can produce features that are threaded, defined from a bend, defined from an edge, and/or countersunk. In addition, each machine has a separate auxiliary database which contains a list of codes of raw material forms, materials, and peripheral shapes that can be handled by the machine. Finally, the machine status database contains one record for each machine and stores the machine status information.

The rule base of the process planning subsystem has about 35 rules which matches the part and feature attributes from the part database and feature database with the machine capability information found in the machine capability database. These rules are applied at two levels to qualify the machines. At the first level, rules check if a machine status is 'on', and if a machine is capable of processing a part's raw material form, material, and peripheral shape. If so, rules at the second level check whether a machine can produce a feature considering its type, dimensions, tolerances and tonnage. In this project, features are created by punching, burning, shearing, drilling, or bending operations. At the end of this machine qualification process, each feature has a set of machines that can be used to produce it. This information is passed to the next step for determining an optimal cost process plan.

The next step is to create an acyclic multistage digraph representing a part processing and find a shortest path from the start node to the end node. Figure 2 gives an example of such a graph for a sheet metal part with periphery, holes, slots, and bends. In this graph, the start and end nodes represent the initial raw material and the final product respectively. A node in between the start and end nodes represents a feasible <feature-machine> pair. These <feature-machine> pairs are arranged in levels according to a predetermined precedence of features. A level corresponds to a feature and each node in

Start   Periphery   Holes   Slots   Bends   End

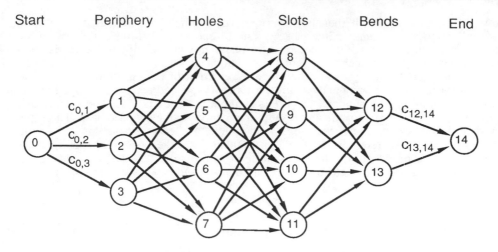

FIG. 2 Part process representation graph.

the level corresponds to a machine that can produce the feature. For a given feature, all feasible machines have already been identified in the previous step. By completely connecting the nodes in consecutive levels with arcs, an acyclic multistage digraph is created. Each arc in the graph is assigned a cost. The cost of an arc represents the cost of moving a part from its head node to tail node plus the cost of processing the part at the tail node. The cost of an arc is determined as follows:

$$C_{ij} = d_{ij} + SC_j [S_j + (n * m * P_j)] \qquad (1)$$

C_{ij} = cost of generating a feature on machine j when part is currently located at machine i
d_{ij} = travel and material handling cost between machine i and machine j
SC_j = standard machine rate for machine j
S_j = standard setup time for machine j
n = number of parts being produced
m = number of features of a particular type
P_j = processing time for a feature on machine j

A separate program is written in the C programming language to construct and solve this feature digraph for the least cost path. This C program is interfaced with the Clipper program. In the above equation the quantities d_{ij}, SC_j, S_j, P_j, comes from the cost information subsystem. The graph representing the part processing has the special property of being completely connected between consecutive stages only. Dijkstra's well known label setting algorithm [13] is used to find the least cost path. If the feature precedence is not imposed in building the graph based on the practical consideration, it would result in a completely connected digraph and the search problem would belong to a class of computationally complex NP-complete problems. From the practical consideration, the fixed sequence of generation of features is: periphery, pre-bend holes, pre-bend slots, bends, post-bend holes, post-bend slots. The detailed description of the process planing subsystem is given in [14]. Unlike the existing systems which use assumed weights and heuristics search strategies to obtain a feasible plan, this system directly uses cost data and optimization methods to generate an optimal plan.

Cost Information Subsystem

Cost information subsystem is primarily responsible for providing cost information necessary in the process planning subsystem. This subsystem generates cost information by two different methods. In the first method, the subsystem retrieves cost information from databases. In the second, it estimates cost information from rules and equations.

This subsystem contains machine cost databases, machine operation rate databases, a travel matrix table, and miscellaneous tables. Each machine cost database captures the information related to machine rate, major consumable and utility rates, labor rate, part loading and unloading cost, and machine set up cost. A machine operation rate database is a look-up table which gives variable machine operation rates such as burn rates, punch rates, bend rates, for different ranges of part or feature attributes. For example, burn rate is specified as a function of sheet metal thickness. Similarly, punch rate is specified as a function of punch travel codes (punch travel codes measure the complexity of pattern of slot and/or hole), and bend rate as a function of the number of bends. The travel matrix table is an M X M matrix where M is the total number of machines in the shop floor. Each element t_{ij} in this matrix gives the cost of moving a part from machine i to machine j (it includes material handling cost also). Finally, the miscellaneous tables contain cost items such as special tool or fixture cost, programming cost, and first piece inspection cost. The programming costs and the first piece inspection costs are specified as functions of programming codes and first piece inspection codes (these codes can be obtained from the part database).

By using cost estimation equations and information from the machine cost database and the part database, the values of variables SC_j, S_j, in Equation 1 are estimated. Similarly, the value of P_j is estimated using the information from the machine operation rate database, the part and feature databases. The value of d_{ij} for a part is determined as a function of t_{ij} from the travel matrix table and the weight of the part.

After solving the part processing graph for the least cost path, CMES generates the minimum cost process plan and the corresponding apparent minimum manufacturing cost. To get the actual minimum manufacturing cost, certain fixed cost items such as special tool or fixture requirement cost, programming cost, first inspection cost (distributed over the total number of parts) are added to the apparent minimum manufacturing cost. In addition to the minimum cost process plan and the minimum manufacturing cost, CMES also provides a report which contains a step by step trace of rules fired, necessary explanations for elimination or selection machines, and intermediate cost figures etc. This report is very useful for a design engineer in identifying the cost influencing factors and in conducting the Process Value Analysis [15].

These three subsystem are integrated such that the information in CMES interacts as shown in Figure 1. Next section illustrates the operation of CMES by running it on an example sheet metal part.

EXAMPLE RUN ON CMES

This section presents a sample session with CMES. The illustration of the session is restricted to only a few important screens. Figure 3 gives the drawing of the part used for the illustration of the session. This part is a L-Bracket with 8 hole, 2 slots, 1 bend (90° angle), and 1 periphery. The detailed attributes of the part are shown in the part header stored in the group technology subsystem.

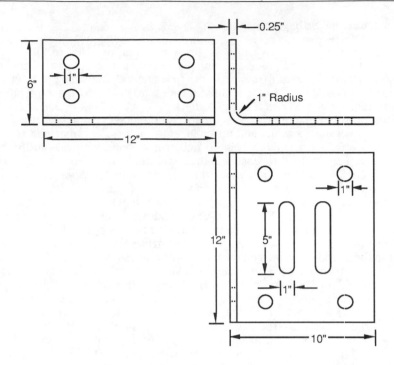

FIG. 3 Sample part used to illustrate a session with CMES.

1. System starts with the following menu:

```
        Add Part
        Edit Data
        Query System
        Delete Parts
        Print Parts
        Database Utility
        Process Planning
        Cost Estimation
        Exit System
```

2. The user can either enter the part number and the item number or retrieve a desired part through a query process.

```
        Enter the part Number P1ØØØ        Item Number ØØØ
```

3. If the part is found in the part database, the part information (or part header) is displayed in the format shown below. User has a provision to modify the data displayed. If the part is not found, user should enter the data in the same format. In

the first screen part attributes are displayed. In the second, feature attributes are displayed. (Note that these attributes belong to the sample part shown in Figure 3.)

Part Number **P1ØØØ** Item **ØØØ** Name **L-Bracket**

Customer **1ØØ** Cust. Part No. **PØØØ** Drawing No. **P1ØØØ**
Rev. No. **1** Rev. Date **Ø2/27/92** CAD File Name **P1ØØØ** NC File Exist? **N**

Raw Material Form **P** Material **51** Paint/Coating **ØØ1** External Shape **7**
Surface Finish **1ØØ** Out-of-Shop Process **51** Dim. Tolerance **Ø.Ø3ØØ**

Per. Shape **1** Length **16.ØØØ** Width **12.ØØØ** Thickness **Ø.25Ø**
Per. Inches **48.ØØØ** Estimated/Actual (E/A) **E** Machined Bevels **N**

F-factor **1.ØØ** S-factor **1.ØØ** Shear Length **12.ØØ** Parts Per Sheet **1**
Total Number of Parts **2ØØ** Part Weight **5Ø** Customer Supplied Blank **y**

Special Tools/Fixtures Required **N** Punch Travel Code (A/B/C) **A**
Programming Code (A/B/C) **N** First Piece Inspection Code (A/B/C) **A**

Grinding or Tumbling (G/T) **G** Shear Hits Required Per Piece **1**

Features: Holes **8** Slots **2** Bends **1**

View Features (Y/N)? **Y**

Bends - 1
 Dimensions Tolerances

Radius	Angle	Dim.	Pos.	Number
1.ØØØØ	90.ØØØØ	2.ØØØØ	Ø.Ø3ØØ	1

Holes - 8
 Dimensions Tolerances

Diameter	Depth	Dim.	Pos.	THR	CTS	DFB	Num.	Int.
1.ØØØØ	Ø.25ØØ	Ø.Ø3ØØ	Ø.Ø3ØØ	N	N	N	8	Y

Slots - 2
 Dimensions Tolerances

Length	Width	Depth	Dim.	Pos.	THR	CTS	DFB	Num.	Int.
5.ØØØØ	1.ØØØØ	Ø.25ØØ	Ø.Ø3ØØ	Ø.Ø3ØØ	N	N	N	2	Y

Do You Want To Plan This Part (Y/N)? **Y**

4. The system displays the feasible machines for each feature. This information is actually used by the system to construct the acyclic multistage graph. First column gives feasible machines, the second feature types (P=Periphery, H=Hole, S=Slot, B=Bend), and the third feature precedence numbers in the graph.

Machine: 1	Type: P	Feature Number: 1	Record: 534
Machine: 3	Type: P	Feature Number: 1	Record: 534
Machine: 1	Type: H	Feature Number: 2	Record: 959
Machine: 2	Type: H	Feature Number: 2	Record: 959
Machine: 3	Type: H	Feature Number: 2	Record: 959
Machine: 5	Type: H	Feature Number: 2	Record: 959
Machine: 6	Type: H	Feature Number: 2	Record: 959
Machine: 7	Type: H	Feature Number: 2	Record: 959
Machine: 8	Type: H	Feature Number: 2	Record: 959
Machine: 9	Type: H	Feature Number: 2	Record: 959
Machine: 1	Type: S	Feature Number: 3	Record: 960
Machine: 3	Type: S	Feature Number: 3	Record: 960
Machine: 5	Type: S	Feature Number: 3	Record: 960
Machine: 6	Type: S	Feature Number: 3	Record: 960
Machine: 7	Type: S	Feature Number: 3	Record: 960
Machine: 8	Type: S	Feature Number: 3	Record: 960
Machine: 9	Type: S	Feature Number: 3	Record: 960
Machine: 10	Type: B	Feature Number: 4	Record: 961
Machine: 11	Type: B	Feature Number: 4	Record: 961
Machine: 12	Type: B	Feature Number: 4	Record: 961

Any key to continue . . .

5. The system will ask for the number of parts in the batch because the batch size will affect the fixed costs that are to be distributed over the entire batch.

Part number ∅∅∅   Item ∅∅∅ : **L-Bracket**

Enter Number of Parts **5∅**

6. The system displays the shortest path as shown in the window below. (Note that the sequence is displayed from bottom to top, i.e. the operation "Generate Periphery" is actually the first operation in the processing sequence.) The contents on the right hand side of an arrow give a feature description and those on the left hand side give the recommendation of the machine to produce the feature.

```
Shortest Path:
.........................
                                                      ==>   End of Processing
                                Bend with radius 1.00  ==>   Pacific 300 ton
                      Slot with dimension (1.000 by 5.000)  ==>   Metal Muncher
                          Hole with diameter 1.0000  ==>   Strippit (punch)
                              Generate Periphery  ==>   Strippit (burn)


Total cost for 5Ø parts  =  223.23


Cost per part  =  4.46


Any key to continue  . . .
```

7.   This completes a cycle of process planning and cost estimation for a specified part.

CONCLUSIONS

The paper described a new cost estimation model which is useful for both cost estimation and cost minimization. The utility and feasibility of the system was demonstrated by developing CMES for a small sheet metal industry. Though the paper identified the part and feature attributes, cost elements, and various databases need for implementing the model for a specific project, this work should provide guidelines for developing similar systems for other projects.

REFERENCES

1.   Wong JP, Parsaei HR, Liles DH: Design of a Cost Estimation System for Concurrent Engineering, in Ahmad MM, Sullivan WG (eds.), *Factory Automation and Inf. Manag.*, Boca Raton: CRC Press, pp 600-607, 1991.
2.   Malstrom EM: Cost Estimating Handbook, New York: Marcel Dekker, 1984.
3.   Whealon I: Sound Basis for Computer Aided Cost Estimating, *Machine and Tools Blue Book*, pp 54-58, 1985.
4.   Casy J: Digitizing Speeds Cost Estimating, *American Machinist and Automated Manuf*, 131/1, pp 71-73, 1987.
5.   Goldberg J: Rating Cost Estimating Software, *Manf. Eng.*, 98/2, pp 37-41, 1987.
6.   Lee S, Ebeling KA: Automating Cost Estimating Systems, *Comp. in Ind. Eng.*, 13, pp 356-360, 1987.
7.   Kamrani AK, Wong JP, Parsaei HR, Tayyari F: Design of an Integrated Manufacturing Cost Estimating System (IMCES), *Proc. of 1989 Inst. of Ind. Eng. Integrated Sys. Conf.*, Atlanta, pp 599-604, 1989.
8.   Wong JP, Imam IN, Kamrani AK, Parsaei HR, Tayyari F: A Totally Integrated Manufacturing Cost Estimation System (IMCES), in Parsaei HR, Mital A (eds.), *Economic Analysis of Advanced Prod. and Manuf. Sys.*, London: Taylor and Francis, 1990.
9.   Oliver CR., Quesnel LH: Group Technology: An Approach to be Used by an Existing Sheet Metal Industry for CIM Implementation, *SME Tech. Paper*, MS87-762, 1987.

10. Filho EVG: Computer-Aided Group Technology Part Family Formation Based on Patter Recognition Techniques, Ph.D. Dissertation, Department of Industrial and Management Systems Engineering, The Pennsylvania Sate University, University Park, 1988.
11. Kumara S, Ham I: Database Considerations in Manufacturing Systems Integration, *Robotics and Computer-Integrated Manuf.*, 4/3/4, pp 571-583, 1988.
12. Schafer H, Brandon JA: The New Face of Group Technology in the Light of Recent Developments in Information Technology, in Ahmad MM, Sullivan WG (eds.), *Factory Automation and Inf. Manag.*, Boca Raton: CRC Press, pp 292-301, 1991.
13. Ahuja RK, Magnanti TJ, Orlin JB: Network Flows, Sloan Working Paper No. 2059-88, 1988.
14. Smith JS, Cohen PH, Davis JW, Irani SA: Process Plan Generation for Sheet Metal Parts Using an Integrated Feature-Based Expert System Approach, to appear in *Int. J. of Prod. Res.*, 30, 1992.
15. Kammlade JG: The Cost Management System: A Critical Link Between Design and Manufacturing, *SME Tech. Paper*, MS89-746, 1989.

AUTOMATED DECISION MAKING FOR PROCESS PLANNING

Antony R. Mileham and Elfatih A. Rustom

School of Mechanical Engineering, University of Bath,
Bath, U.K.

ABSTRACT

BEPPS-GSCAPPP is a generative System of Computer Aided Process Planning for Prismatic components under development at the University of Bath. It has been designed for planning prismatic components on conventional machine tools in a batch manufacturing environment. This paper describes the program structure and details two main modules. Firstly the interactive input stage is described, which is designed to elicit component feature and other general manufacturing information. Secondly it focuses on the automatic stage concentrating on the automatic selection of raw material from a restricted list of standard material types stored in the material data base and follows by describing how feature ordering of the 7 common features has been automated by grouping them into flat and cylindrical groups and then prioritising them in a novel way to achieve a best order for machining.

1. INTRODUCTION

Process planning is an important activity that links the design and manufacture activities in the component cycle. It involves the selection and sequencing of the operations and processes required to convert a component design economically and effectively into a finished.

Traditionally process plans are generated manually and are documented on route sheets that specify both the processes and the machines to be used. This function is usually carried out by an expert planner who is highly skilled in the decision making aspects and has great experience of shop floor operations. The major disadvantage of manual process planning is inconsistency [1]. It is not unusual for different planners to specify different routes for the same part, each expressing their own preference. Further there is no way of being sure that any route is optimal and thus the level of planning proficiency will affect the efficiency of manufacturing.

The computer offers potential for reducing routine clerical work, and at the same time, it is capable of calculating complicated formula and analysing logic rules in a much faster time. Process planning systems which are assisted by computer power are called computer aided process planning (CAPP) systems.

Literature reveals that four typical approaches have been put forward for computer aided process planning (CAPP). (1) The *Constructive* approach in which the information of the materials, machines, tools, operations ...etc. are held in separate menus in the computer data base. Typically, the planner has to specify the sequence of operations, machines, tools and materials to be used to produce a component. Using a menu structure, the planner selects the relevant page from which to choose the appropriate material, machine, tool and operations from the screen. Once the machine type has been selected, the system will often automatically choose appropriate cutting conditions and then calculate the machining time, and finally output the process planning sheet. (2) The *Variant* approach creates process plans for parts which are related to a specific composite part in a computer data base. The composite part is then retrieved and modified to suit the new part and hence a process plan is created. (3) The *Generative* approach generates a new process plan for a given part from first principles. The system uses information which is available in a manufacturing data base that contains the part description data and technological information. Using expert process decision logic the computer program manipulates the data in order to automatically generate a process plan. (4) The *Expert System* approach is a new form of generative process planning that uses an expert system program structure to make the planning decisions. Expert planning systems are currently being researched [2], [3].

2. GENERAL STRUCTURE OF GSCAPPP

BEPPS-GSCAPPP is a knowledge-based Generative System of Computer Aided Process Planning for Prismatic components. It has been designed with a modular structure. The general structure of BEPPS-GSCAPPP is illustrated in figure 1 [4]. Basically it contains four options; (1) User's help, (2) Process Planning, (3) Decision logic modification and (4) Data base file modification. The user's help option provides general guidance on how to use the system at the initial stage. In option (3) and (4), the user can have access to both decision logic and data base files to enable updating and/or modification whenever it is required. The main option is process planning which is divided into three further stages; *Interactive* stage, *Automatic* stage and *Output* stage.

2.1 Interactive Stage:

In this stage, the planner provides the system with the input data required to generate the process plan. Input data is subdivided into five sections. i. General information data input, ii. Component classification and coding, iii. Component type, iv. Machine availability, and v. Feature data input.

2.1.1 General Information Data Input: In this section the planner is asked to input general information about the component and the production plan. The component information includes; component's name, number, material and shape envelope (length, width and depth). The planning information includes the batch type (discrete or continuous), the planner's name and date.

2.1.2 Component Classification and Coding: The system at present only considers only standard raw material forms with plate, flat, and square bar forms in selective sizes only being included. It is necessary for the planner to be familiar with the system devised for coding both the planes and edges that form the shape envelope in which the component lies. As the system is designed for prismatic components, it is important to code the surface planes in a certain way, as this enables the planner to input the features in a distinct order for each plane.

2.1.2.1 Plane Coding: Generally, a plane is named with reference to the axis to which it is normal i.e. (X-plane, y-plane, or z-plane). The six surface planes of the component are divided into two types; *Datum Planes*, and *Opposite Planes*. A datum plane is a plane in which one corner is set at (x=0, y=0, and z=0). An opposite plane is a plane which is parallel to the datum plane at an x, y, or z position appropriate to a specific component. Figure 2 shows the six plane surfaces with their codes.

2.1.2.2 Edge Coding: The edge code is used to recognize the position of each feature and for determining machining direction. Edges in GSCAPPP are coded according to their plane positions. For example, the edges of the x-axis, are named as x-edges and coded as EX0 for the original x-axis, then moving in an anti-clockwise direction for the next edge EX1, ...etc. The same procedure is applied for the original y-axis and z-axis. Figure 3 illustrates the edge codes for the component envelope.

2.1.3 Component Type: Furthermore, the entire component is classified according to it's shape, particularly, the flat features required. This classification relies on; (1) The cross-sectional profile of the component, and (2) The machining direction for the flat features. Using this classification, prismatic components in BEPPS-GSCAPPP are considered as belonging to one of the three types; (1) *Totally*

Constant Cross-Section component (TCX-SEC), (2) *Partially Constant Cross-Section* component (PCX-SEC), and (3) *Non-Constant Cross-Section* (NCX-SEC).

A component is of totally constant cross section (TCX-SEC) if each of the surfaces that required machining have a constant profile in any plane direction. The partially constant cross section component (PCX-SEC) is a component which, has at least one surface, of those requiring machining with a constant profile in any one plane direction. A component is of non-constant cross section (NCX-SEC) if none of the surfaces requiring machining have a constant profile in any one plane. A component is inputted to the system initially as either a constant or non-constant component. A component of partially constant-cross section is considered as a non-constant component at first. After the system has interactively obtained information about the types of plane surface present each plane surface is then classified as either of constant or non-constant cross-section. Any component containing both is subsequently classified as PCX-SEC by the system.

2.1.4 Machine Availability: The machine tool data in BEPPS-GSCAPPP has been limited to a vertical milling machine, a horizontal milling machine, a pillar drill, a radial drill, a vertical boring machine, a surface grinder, and an internal grinding machine. Actual machine tools have been selected and these in turn impose size constraints on the components that can be machined by the system. The system displays the machine tools (names and codes) so that the planner or production control system are able to delete machines that are currently occupied with other jobs.

2.1.5 Features in BEPPS-GSCAPPP: The system considers a range of machined features that are commonly produced on conventionally machined prismatic parts. The simplified research versions uses 7 features namely; flat surface, pocket, slot, plain hole, stepped hole, countersink and thread.

2.1.5.1 Feature Classification: These seven features are divided into two groups; (1) *Flat* group, and (2) *Cylindrical* group according to the tool geometry and motion required to machine them. The flat group includes faces, pockets, and slots, where as the cylindrical group includes plain holes, stepped holes, countersinks and threads. Each group is then subdivided into Basic and Secondary features as shown in figure 4. The basic feature represents a primary form of the feature and the secondary feature represent deviations from this primary form. This classification has been designed to give a much simpler feature ordering decision logic and to group features requiring the use of the same machine tool type.

2.1.5.2 Feature Data Input: The feature data for each surface that requires machining is inputted to the system interactively via system prompts. Initially the planner is asked to choose one of the three main options after studying the component to be planned. These are; (a) Only flat features are required, (b) Only cylindrical features are required, and (c) Both flat and cylindrical features are required. Once the choice has been made the system displays the range of features within the group for the planner to choose the appropriate feature set. In the case of choice (c), information on the flat features is requested prior to that for cylindrical features. For each feature, the planner is asked for a variety of parameters including; feature code, location, dimensions, tolerances, surface requirements, ...etc. This data is stored in a component data base file that can be retrieved and processed by several modules. At this stage, if the component is constant and the features required for any plane are flat only, the planner is required to input them in a particular order according to the *"Top-To-Bottom"* technique (TOP-TO-BOT).

The TOP-TO-BOT technique is designed to input a feature's information with reference to it's position on the plane. This means that the feature on the top level (greatest z value for x-plane , ...etc.) has priority over the bottom ones. An example of this technique is shown in figure 5. This technique is applied only for flat feature types. If more than one feature exists on the same level, then the feature input sequence is left to the planner's judgement, or alternatively, the *"Scoring"* technique (SCORE) can be applied. The scoring technique is an automatic feature ordering technique designed for all features required on a non-constant cross-section plane surface. This technique is further discussed in the automatic stage.

2.2 Automatic Stage:

Once the input of data has been completed, the system stores the information in a file named by the component number so that it can be either retrieve for modification or be used to generate a process plan automatically. The automatic stage is divided into 8 modules; 1. Raw material selection from stock, 2. Feature recognition and ordering, 3. Operation determination and sequencing, 4. Machine tool selection, 5. Cutting tool selection, 6. Cutting conditions selection, 7. Total time calculation, and 8. Workpiece holding device consideration. This paper concentrates on modules 1 and 2.

2.2.1 Automatic Selection of Raw Material from Stock: The system considers three of the most common materials used for prismatic parts in batch manufacturing factories. Those are; Mild steel, Carbon steel, and Aluminium. Standard shapes and sizes for these materials have been used rather than castings or forgings. GSCAPPP is based on a small batch working shop that only machines components from stock and only keep a small range of standard shaped bars, ...etc. This information is contained within the raw material data base of the system. When during the input stage the shape envelope dimensions are requested, the system checks to establish what if any material allowances are required in order to achieve the tolerances and finishes specified. This is carried out as part of the process planning module and uses expert logic to make the required decision. If any allowances are required, then a fixed allowance is added to the specified dimension.

By using this metal addition technique the system changes the shape envelope of the finished component into the minimum shape envelope of the required raw material and details which surfaces if any are sufficiently accurate to negate machining.

The data base of standard forms assumes at present that each bar, ..etc. is substantially longer than any component that the system can accommodate, hence only X and Y cross-section dimensions are important. A *"best"* fit comparison is carried out in order to match a component's shape envelope with the material X Y dimensions. Basically the component's Length/Width, Length/Depth and Width/Depth are compared against all X Y material dimensions of the stock held. If there is an exact match, the matching algorithm is stopped and the material stock designated as the *"Ideal Form"* i.e. no excess machining is required.

If no ideal form is available, then the data base is searched to find the nearest fits. These are displayed on the screen together with the most appropriate stock size. The most appropriate size is arrived at automatically by taking into account; (a) The minimum volume of excess metal to be removed, (b) The minimum contact area for the machining, and (c) The method of removing the excess metal.

By using these three factors that the choice of the most appropriate stock size is based on a combination of minimum volume and economics. Figure 6 shows the structure of the material selection module.

2.2.2 Automatic Feature Recognition and Ordering Module: As discussed earlier, the SCORE technique is designed to reorder flat and cylindrical features on any non-constant cross-section plane surface as well as cylindrical features on any constant cross-section plane surfaces. In the scoring technique each feature is given a score based on it's *"priority"* for machining. The flat feature group for example typically has priority over the cylindrical group. This will result generally in plans showing *'Mill'* and then *'Drill'* i.e. a hole will not be machined unless the face on which it is located has been completed.

The system retrieves the feature input data from the component data base file and checks the plane surface type. If the plane surface is of non-constant cross-section, firstly the system reorders the flat features and secondly the cylindrical features according to their basic score. If the plane surface is of constant cross-section type, the system reorders only the cylindrical features. Figure 7 shows the structure of the scoring technique.

Once the feature data has been organized into the correct order for processing, the remaining modules are used to automatically determine the required operations and their sequence, the machine tool set, the appropriate cutting tools, cutting conditions and the machining and non-machining times.

CONCLUSION

A generative CAPP system, BEPPS-GSCAPPP has been developed for the automatic planning of prismatic components. It is able to generate a fully documented process planning sheet for the shop floor using automatic planning modules which have been designed for easy modification.

A raw material selection module has been developed that will choose the most appropriate stock size available using decision logic that takes into account the economics of the operation. It is considered that advantages could also be gained by using this module as a stand alone package in a design for manufacture environment.

Two techniques used for feature ordering have been developed and successfully applied to components that can be described using 7 basic feature types. The feature ordering technique (TOP-TO-BOT) is based on the relative height of the flat features present, where as the scoring technique relies on allocating a priority to individual features depending on their basic scores.

REFERENCES

[1]   Chang T., and Wysk R, "An Introduction To Automated Process Planning Systems", Prentice-Hall, Inc., 1985.

[2]   Groover, M. P., "Automation, Production Systems, and Computer Integrated Manufacturing", Printice-Hall International, 1987.

[3]   Muthsam H., And Mayer, C., "An Expert System for Process Planning of Prismatic Workpiece", Proceeding , 1st Conference, Artificial Intelligence and Expert Systems in Manufacturing, March 1990, pp. 211-220.

[4]   Rustom E. A., Mileham A. R., "The Development of a Generative Computer Aided Process Planning System for Prismatic Components", Advanced Manufacturing Technology, Proceedings of the Fifth National Conference on Production Research, Huddersfield Polytechnic, U.K. September 1989, pp. 259-263.

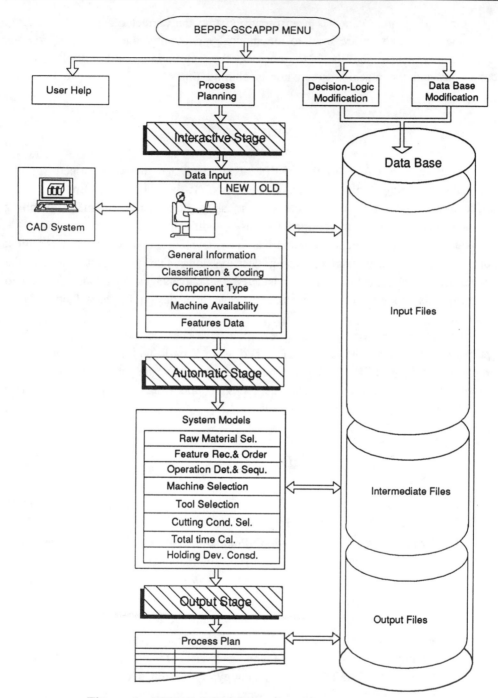

Figure 1: BEPPS-GSCAPPP General Structure.

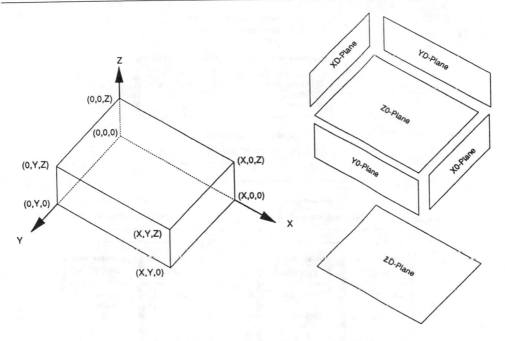

Figure 2: Datum and Opposite Planes of a Component.

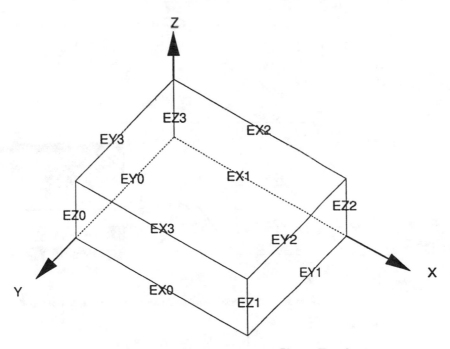

Figure 3: Edge Codes of a Component Shape Envelope.

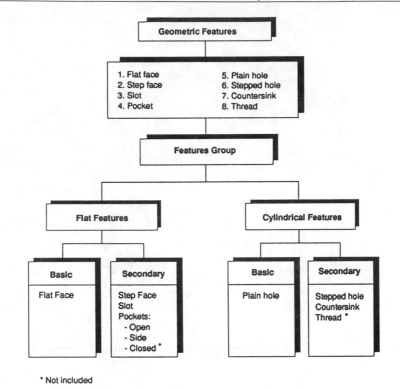

Figure 4: BEPPS-GSCAPPP Features Classification.

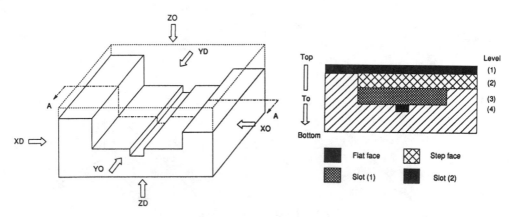

Figure 5: Top-To-Bottom Technique Applied for ZO Plane Surface.

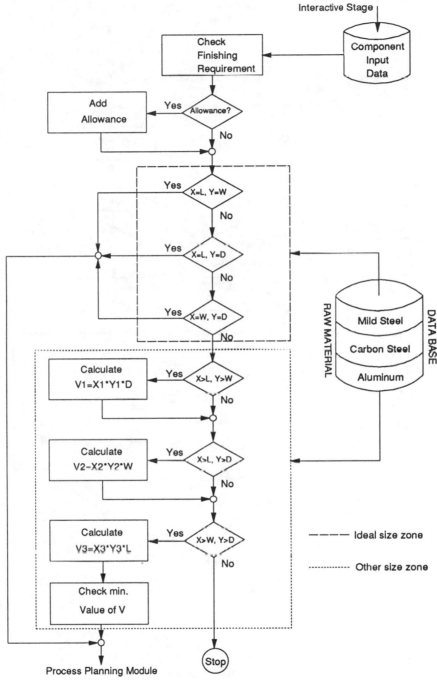

Figure 6: Structure of Raw Material Selection Module.

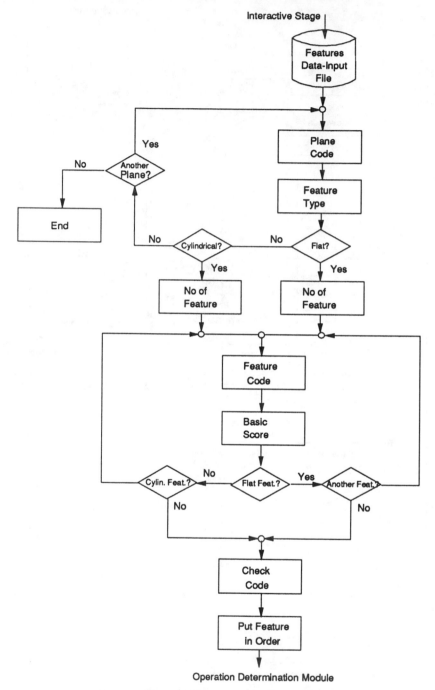

Figure 7: Scoring Technique Structure

CONCURRENT PROCESS PLANNING AND SCHEDULING USING NON-LINEAR PROCESS PLANS

[1]Bernd C. Schmidt and [2]Jochen Kreutzfeldt

[1]*CIM-Fabrik Hannover, Hannover, Germany*
[2]*Institute for Production Engineering and Machine Tools, Hannover, Germany*

ABSTRACT

The conventional approaches to Production Planning and Control (PPC) are being reconsidered with respect to the shift in production goals from maximum output at minimum cost to high flexibility, quality, reduced throughput-time and high schedule observance. On the level of workshop control, the implementation of a generated schedule in the shop is impeded by unavoidable disruptions. Since process planning is usually performed a long time before manufacturing, the load situation and the availability of machines in the shop is not taken into account. Investigations have shown that in many small and medium-sized metal-working companies, 20 to 30% of the total load has to be redirected to other resources and that only a small amount of workshop orders actually complies with the original short-term schedule. This situation can be improved significantly by using flexible process plans that comprise manufacturing alternatives. By using these flexible or non-linear process plans it is possible to reconsider the final manufacturing route only shortly before or during the workshop scheduling process, when the actual situation in the workshop is known. On the other hand it facilitates rescheduling of orders to other manufacturing resources in case of unforeseen disruptions.

This paper describes developments achieved within the ESPRIT Project 2457 FLEXPLAN. FLEXPLAN develops an integrated process planning and scheduling system based on 'non-linear' process plans (NLPPs). These process plans are represented as Petri nets that resemble a net of feasible operation sequences. The system comprises knowledge-based automatic process planning functions for a limited spectrum of workpieces, a graphical editor for non-linear process plans and an hierarchical workshop scheduling system that utilises the NLPPs to increase flexibility in the workshop. The workshop scheduling system comprises a load-oriented module for medium range scheduling, a finite scheduling module with graphical user interface and schedule evaluation functions. The optimisation algorithm used in the finite scheduling is based on a rule-based execution of petri-nets.

1. INTRODUCTION

1.1 Starting Point

Some factors can be identified that currently impede cost efficient small batch production in classical job shop environments. The scheduled flow of orders is impeded by the unavoidable disruptions, such as machine breakdowns or rush-orders so that a significant percentage of the total load has to be redirected to other resources [BECKEN 91]. This requires rushed and improvised replanning of urgent orders to other machines. Due to the rapid technological progress many workshops now have machines with widely differing characteristics. These differing machines require more and more machine-specific process plans and NC-programs. This has restricted the former flexibility of workshop scheduling in levelling the load between different machines and reacting on disruptions. The frequent replanning also inhibits the controlling of time, cost, and quality in the shop which is becoming more and more important. The traditional functional division between process planning and workshop scheduling prevents often a closer cooperation of both functions, that is required to align the decision criteria in process planning with the actual load and cost situation in the workshop. The decisions taken in process planning do not consider the actual load situation and machine availability and consequently often do not meet the requirements of workshop scheduling. Since a significant reduction of disruptions in the shop cannot be achieved with cost efficient effort in the near future, development has to focus on an improvement of the other above-mentioned deficits.

1.2 Required Actions

The flexibility in workshop production has to be regained. New approaches of simultaneous engineering require a closer cooperation between workshop scheduling and process planning. The growing effort for process planning and NC-programming has to be decreased by improved computer support of process planning. This requires a thorough integration of process planning with CAD and CAM. This can only be achieved by a new integrated approach towards process planning and workshop scheduling.

1.3 General Approaches

One approach is to combine process planning and workshop scheduling into one single optimisation task. This known as 'dynamic process planning'. The current research efforts focus on application for flexible manufacturing systems. Due to the complexity of the planning task this approach seems unfeasible for job shop production.

A less radical solution does not consider all shop orders simultaneously, but still generates order dependent process plans that account for the actual workshop situation [KHOSHN 90]. This also requires a process plan generation during workshop scheduling. Considering the usual throughput times in job shop production there is still a considerable time span between process plan generation and the completion of the last operation in the shop. The conditions based on which process plan was made may have changed. The still necessary replanning in case of disruption is not supported by that approach.

Another possible approach is a two-level process planning, where the operation are only roughly determined in the first step. The detailed planning and the selection of resources is then done at a very late point shortly before manufacturing.

2. THE FLEXPLAN APPROACH

The FLEXPLAN project pursues another approach. Process planning and workshop scheduling remain distinct functions with respect to time [TÖNSHO 87], [TÖNSHO 89a],[TÖNSHO 91]. Only one process plan is generated for each workpiece. This process plan is independent from an specific order. The required flexibility for scheduling is achieved by including manufacturing alternatives or possible changes in manufacturing sequence already in the process plan. The usually linear process plan sequence is enhanced to a net structure that contains the feasible operation steps. Several alternative routings or feasible sequences of operations are combined in one Petri-net structure, so called non-linear process plans (NLPPs).

FLEXPLAN also tries to achieve a better integration of the process planning and scheduling task by using a common data base that centralizes all data that is relevant for planning and scheduling.

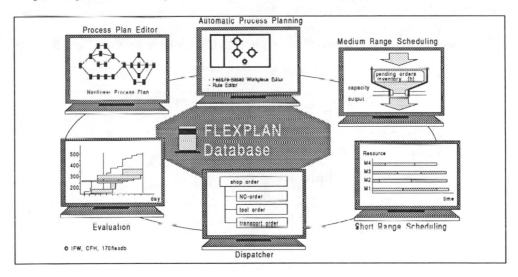

Figure 1: Main components of the FLEXPLAN system

2.1 System Architecture

The system comprises two main modules: The FLEXPLAN Process Planner (FPP) and the FLEXPLAN Workshop Scheduler, that can be subdivided into modules for Medium and Short Range Scheduling, Evaluation and Monitor/Dispatch (Figure 1).Non-linear process planning requires to describe more possible operation steps and thus increases the process planning effort. FLEXPLAN therefore supports nonlinear process planning with a graphical process plan editor and functions for automatic generation of NLPPs for a limited spectrum of workpieces. FLEXPLAN also comprises a function that converts linear process plan information as included in PPC systems into NLPP format to allow a step-wise introduction of the NLPP concept.

An hierarchical workshop scheduling system uses the NLPPs for a medium range, load-oriented scheduling of orders, that recognizes capacity bottle-necks at an early stage and considers manufacturing alternatives. The finite scheduling of shop orders follows a rule-based algorithm and also utilizes the flexibility of NLPPs. The graphical user interface is similar to modern computer planning board systems (Leitstands).

Additional manufacturing alternatives can be generated 'just-in-time' using the graphical editor when required by workshop scheduling. Information about workshop load and bottle-neck resources is supplied to process planning, so that the process planner can focus on generating the most important alternatives with respect to the load situation in the shop. The common data base enables comprehensive production evaluation functions.

2.2 Information Model

All essential data is stored in a common relation data base (Fig. 2). The underlying entity-relationship model links the following data:

○ An order model, that comprises quantity and due date information, the current schedule and auxiliary orders.

○ A technical workshop model that represents the available resources and the shift model.

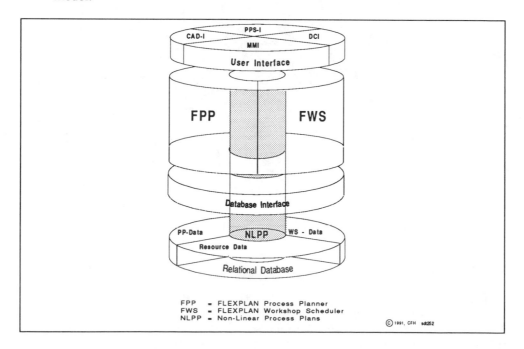

Figure 2: An architectural view of the FLEXPLAN system

O A workpiece model that describes the part that are referenced by the orders and also includes technical features required by the automatic process plan generation.

O A process plan model that contains the non-linear process plan representation and information required for process plan generation

3. THE FLEXPLAN PROCESS PLANNER

The most important feature of the FLEXPLAN process planning system is its ability to handle so called non-linear process plans. In these plans a set of alternative manufacturing sequences is represented in an AND/OR graph. These graphs allow to represent technological manufacturing alternatives as well as alternative manufacturing sequences. Alternative manufacturing options are modelled by OR-knots. Manufacturing operations that can be exchanged in their sequence are modelled by AND-knots.

The process plans remain partially structured. Rules and constraints that apply to manufacturing operation modelled in one path can be assigned to each decision node (AND/OR split). The manufacturing operations are only fixed when required by manufacturing technology.[TÖNSHO 89b, BECKEN 91]. The representation of alternative manufacturing operations in process plans is also investigated in several other research projects [CATRON 89, CHRYSS 091, KALS 91, KHOSNE 90, KRAUSE 89].

The FLEXPLAN process planning system offers support for interactive process plan generation as well as for automatic process planning function for a limited workpiece spectrum. The generated process plan format can directly be used by the FLEXPLAN workshop scheduler.

A graphical NLPP-editor has been developed to support the interactive planning (Figure 3). With this editor operations and knots can be placed and connected on the screen. After a process plan net has been created, the operation information (setup and process times, tools etc.) can be entered and stored. Apart from the interactive graphical editor the FLEXPLAN project developed a planning module for the automatic generation of non-linear process plans for a limited spectrum of parts [DETAND 90]. The result is a prototype of an automatic system for generating NLPPs for hydraulic blocks. The system essentially supports the planning of milling, drilling and boring processes. The planning module requires that the workpieces have been described beforehand using technical features. The generation of feature based workpiece descriptions would be much easier if the geometrical and technological workpiece information could be transferred from a CAD system, most likely a feature-based CAD system, in the future.

4. THE FLEXPLAN WORKSHOP SCHEDULER

Workshop schedules are usually based on linear sequences of operations as fixed in a conventional process plan. They do not provide the necessary flexibility to react on the irregularities of job shop production in terms of alternative routings of a job.

Recent attempts to implement highly sophisticated finite scheduling systems did not get into industrial applications [MADEMA 89], [CALIER 85], [FOX 87].

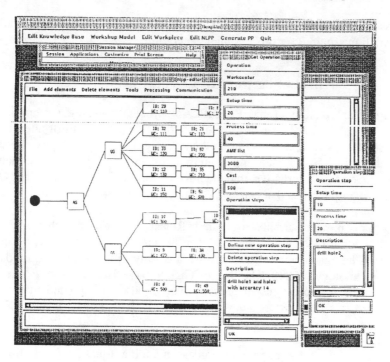

Figure 3:   The NLPP Editor [screen dump from the system prototype]

On the level of workshop control the new concept of the 'Computer Leitstand' has gained great attention and is already widely accepted and implemented in industry. A 'Computer Leitstand' visualizes the situation in the workshop and for the experienced workshop controller provides insight into the flow of orders and schedule performance in the shop. On the other hand it requires a high effort for attendance and data maintenance and an highly skilled operator.

The FLEXPLAN Workshop Scheduler (FWS) is designed to support and facilitate this process. The FWS is capable of accessing information on alternative routings in a non-linear process plan and utilizing it for rescheduling in order to harmonize the loading in the workshop and to improve schedule observance. It provides for scheduling with respect to different goal systems with differently weighted objectives and criteria. The FWS additionally performs an evaluation of the generated schedules with respect to the different goal systems and visualizes the results using characteristic values and diagrams. The functionality of the FWS also comprises the dispatching of shop orders and auxiliary orders to the shop and the monitoring and analysis of feedback data and especially of disruption messages from the shop. To perform these functions the FWS consists of five main functional blocks. The purpose of the fundamental functions and the logical flow of information between these functions is represented in Figure 4.

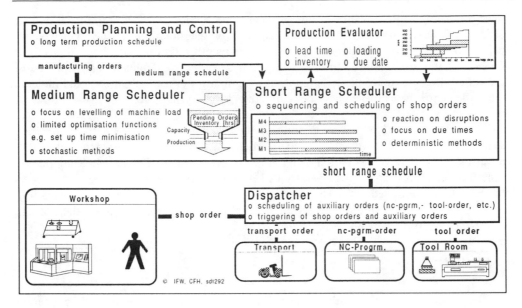

Figure 4:   Main functions of the FLEXPLAN Workshop Scheduler

4.1 Medium Range Scheduling

The **Medium Range Scheduling** (MRS) module schedules manufacturing activities (Shop Orders) in order to comply with the due date of the corresponding manufacturing order. From the manufacturing and shop orders of the PPC system a new resource allocation with limited accuracy and completeness is generated. The MRS carries out a capacity check to determine possible bottle-necks and thus accounts for the actual load situation in the workshop. The MRS usually follows a first-in-first-out strategy but it will also rearrange orders within a specified time horizon with respect to setup optimisation. The applied method for medium range scheduling was developed at the Institute for Production Plant Engineering (IFA) of the University of Hannover [WIENDA 92]. It is based on the use of realistic operation lead times for shop orders and the consideration of the actual load range of workcenters. Realistic mean operation lead times are calculated and periodically updated for each workcenter from a monitoring system and stored for each workcenter in the Flexplan database. The 'load range' of a workcenter is determined by the earliest date at which capacity on that resource is available for new orders. The available capacities for each day are cumulated and compared to the cumulated capacity requirements, which are given by the existing load from already scheduled shop orders and the backlog from unfinished and delayed orders. The load range is now determined by the first date where the cumulated load is lower than the cumulated workcenter capacity.

A backward scheduling for the shop orders, starting from the delivery date, is performed with the realistic operation lead times. Comparing the scheduled dates of shop orders with the dates of free capacity of the assigned workcenters results in a classification of manufacturing orders. If a scheduled date of a shop order is earlier than the date of free capacity of the workcenter, a bottle-neck and therefore a critical order is detected. The MRS now checks whether the situation can be improved by using an alternative manufacturing route in the corresponding NLPP. If this is not successful the amount of additionally required capacity is calculated and displayed to the scheduler. If extra capacity is required from many workcenters, this would indicate that the capacity profile assumed by the PPC system for the generation of manufacturing orders is not realistic.

The task of the MRS is to harmonize the loading coming down from the PPC system with respect to the actual load situation. This is necessary since the SRS requires a levelled load situation to be able to perform its task. The MRS provides a schedule and a resource assignment with limited accuracy and horizon as an input to the Short Range Scheduling. The applied time horizon for medium range scheduling depends on the order structure and should usually comprise once or twice the average lead time of a typical manufacturing order.

4.2 Short Range Scheduling

The **Short Range Scheduling** (SRS) module performs the function of scheduling manufacturing activities (Shop Orders) on the shop floor by specifying the detailed timing of operations in order to comply with due dates, priorities, availabilities of resources etc. Short range scheduling takes place on the basis of the medium range schedule and generates a complete and detailed resource assignment. Important auxiliary resources such as special tools and fixtures are also considered. The planning horizon is limited. The function provides a list of operations that is sequential with respect to time as an input to the Dispatcher.

The SRS applies conventional scheduling strategies using Petri-Net algorithms. The employed method of allocation of operations to resources is either backward or forward scheduling or a combination of both. The method of selection of an order or operation to be scheduled next can be either order based (sequential scheduling of activities of one order) or resource based (parallel scheduling of operations on the next available resource) (Figure 5). The choice of methods is performed by means of heuristics that are based on priority rules. The heuristics are applied mostly according to attributes of the orders, such as lead time, due date or used resource, and defined according to the actual goals of manufacturing.

Bottle-neck situations that have already been indicated by medium range scheduling are handled in two different ways within SRS. One method to alleviate bottle-neck situations is to schedule operations require a bottle-neck resource in a combined forward/backward strategy. That means, that all operations that have to pass through the bottle-neck resource are allocated first. Subsequently, all earlier operations are scheduled backwards and all later operations are scheduled forwards. This effects a high utilisation and short lead times at the bottle-neck resource. A problem is that sufficient time must be available for the remaining operations of the orders that had to pass through the bottle-neck resource.

A completely different approach is offered by the use of non-linear process plans (NLPPs). The bottle-neck resources are identified during the load calculation using the 'optimal routing' of the process plan. A rule is then dynamically introduced into the rule system whereby the bottle-neck resource is 'relieved' by choosing an appropriate alternative routing in the NLPP during scheduling.

The SRS employs a reactive scheduling philosophy that reacts on disruptions and leaves as much of the old schedule intact as possible to avoid unnecessary scheduling effort and disturbances in the workshop to keep an existing schedule as stable as possible. A disruption may be a rush-order or the unavailability of a resource. The emphasis is put on finding immediate solutions to disruptions. The scheduler will first look at local conditions and determine whether an obvious solution exists, such as the allocation to an identical machine. If this fails, a new schedule that considers all manufacturing alternatives from the NLPPs will be generated using the above described methods. This rescheduling starts of course from the actual state of orders. It will be determined whether an alternative routing of the NLPP can be used. It will also be determined how delays affect the start times of subsequent operations in the NLPP. If the alternative route cannot be accomplished in the same time than the previously scheduled route, the succeeding operations must also be released for rescheduling.

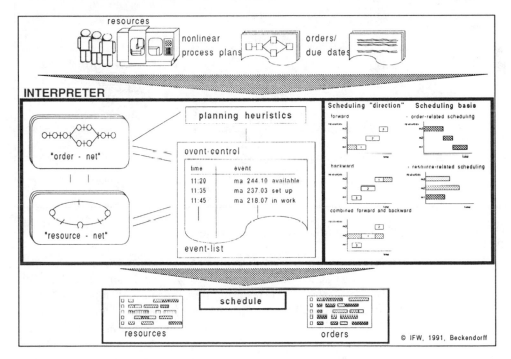

Figure 5: Workshop scheduling by means of Petri net interpretation [according to Beckendorff]

4.3 Schedule Evaluation

The **Evaluation** module calculates the performance of a proposed schedule according to the active goals. This enables the Scheduler to reason about the definition of goals and strategies and about the results of the scheduling activities. This is achieved on the basis of the so called 'funnel model' developed at the Institute for Production Plant Engineering.

The first option will be a visual representation of the order flow by means of a 'throughput'-diagram. It represents a complete description of the load situation and output of a resource for a given period of time. The data for the generation of a throughput'-diagram are obtained from a list of events, that records input and output of orders for a resource. It also allows for the calculation of important key values. This diagram can be generated for single resources or for groups of resources. It thus provides the necessary transparency about the flow of orders to the human scheduler (Figure 6).

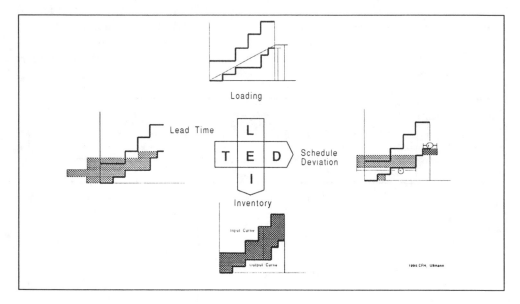

Figure 6:   Use of Throughput-diagrams for evaluation

The second option is the relative evaluation of different candidate schedules, calculated by the short range scheduler or demanded by the human scheduler. This allows to choose the best schedule with respect to different goals and strategies.

An overall evaluation facility will be implemented that recognizes an overload that is caused mainly by the PPC system that generates the manufacturing orders and due dates. This allows for determining the responsibility for the actual load situation in the workshop. To be able to evaluate the effect of using NLPPs, cost evaluations will be performed to be able to differentiate between results with respect to cost and with respect to logistics.

4.4 Monitor and Dispatch

The **Monitor** module performs the feedback functions and therefore allows actualization of the whole FLEXPLAN Workshop Scheduler according to shop floor events. These events include the registration of the normal order progress and additionally the unpredictable disruptions such as tooling failures, machining failures, unavailabilities of resources etc. It collects data from production data registration systems and compares them to the schedule. An analysing function performs the examination of 'scheduling-sensitive' feedback-data from the Monitor module in order to generate a proposal for necessary scheduling activities to the short range scheduling.

The **Dispatch** module performs the distribution of an evaluated and established schedule by transmitting it either by generating the required printouts of order papers or via a production data distribution system to the operators on the shop floor. It also controls auxiliary activities such as for example NC-programming, tool pre-setting or transport.

5. INTEGRATION OF FLEXPLAN IN EXISTING ENVIRONMENTS

The FLEXPLAN system is designed to be integrated in existing software environments. The installation of the FLEXPLAN system requires a PPC system that performs the explosion of bills of materials and determines the due dates of parts or subassemblies. The information on released orders from PPC is accessed by an interface program and transfers them into the FLEXPLAN data base. The PPC orders are then processed by the Medium Range Scheduler. The required changes of order release dates depends on the quality of the supervising PPC system. Investigations of PPC data at the pilot end user have shown that plausibility checks and far reaching changes in release dates may be required. The development of an interface to CAD was not included within the scope of the project. The feature-based workpiece description that is required for automatic process plan generation has to be performed manually. The Dispatch and Monitor module releases orders for production in the shop and communicates the NC-programming, tool management and data capturing systems.

6. HARD AND SOFTWARE PLATFORM

An important project objective was to develop a system for open, standardized and easily portable hard- and software platforms. To achieve that, different software environments had to be linked. The rule-based planning modules are coded in LISP, all other software modules are coded in ANSI-C. The common data base was implemented with ORACLE. The graphical user interface is coded in OSF/Motif. The system uses DEC VAXstation 3100 running under the Ultrix 4.1 operating system.

A system prototype has been displayed on the ESPRIT conference in Brussels in November 1991. The project will be continued together with professional software vendors in order to develop a commercial software tool within the next three years.

For additional information please contact:
Dipl.-Ing. B. Schmidt,   Tel. +49-511-27976-42
CIM-Fabrik Hannover,   Fax.+49-511-17976-88
Hollerithallee 6
3000 Hannover 21
Germany

7. REFERENCES

[BECKEN 91]   Beckendorff, U.: Reaktive Belegungsplanung für die Werkstattfertigung. Dr.-Ing. Dissertation, Universität Hannover, Germany 1991

[CALIER 85] Calier, J.; Peters, J.: MOPS, A Machining Centre Operation Planning System. Annals of CIRP, Vol. 34/1/1985, S. 409-412

[CATRON 89]   Catron, B.A.; Ray, S.R.: ALPS - A Language for Process Specification (Draft). National Institute of Standards and Technology, Gaithersburg, USA 1989

[CHRYSS 91] Chryssolouris; G., Pierce, J.; Dicke, K.: MADEMA: An Approach for allocating manufacturing resources to production tasks. Journal of Manufacturing Systems, Vol. 10, 1991

[FOX 87]   Fox, S.F.: Constraint directed search: A Case Study of Job-Shop Scheduling, Ph.D. Thesis, Carnegie-Mellon University, Pittsburgh (PA), published in: Research Notes in Artificial Intelligence, Pitman, London, Morgan Kaufmann Publishers Inc, Los Altos, California 1987

[KALS 90]   Kals, H.J.J.; van Houten, F.J.A.M.; Tiemersma, J.J.: CIM in small Batch Part Manufacturing. Proceedings of the 22nd CIRP International Seminar on Manufacturing Systems, major theme Computer Aided Process Planning, 11-12 June 1990, University of Twente, Enschede, Netherlands

[KHOSN 90] Khoshnevis, B.; Chen, Q.: Integration of Process Planning and Scheduling Functions. Journal of Intelligent Manufacturing, Vol. 1, pp. 165-176, 1990

[KRAUSE 89]   Krause, F.L.; Altmann, C.: Arbeitsplanung alternativer Prozesse für flexible Fertigungssysteme. ZwF 84 (1989) pp. 228-231, Carl Hanser Verlag, München

[MADEMA 89]   N.N.: MADEMA: Product Description and Applications, hrsg. durch Manufacturing Software Inc., Cambridge (USA) 1989.

[TÖNSHO 87]   Tönshoff, H.K.; Beckendorff, U.; Schaele, M.: Some Approaches to represent the Interdependence of Process Planning and Process Control. Proceedings of the 19th CIRP International Seminar on Manufacturing Systems, S. 257-271, Pennsylvania State University, USA 1987

[TÖNSHO 89a]   Tönshoff, H.K.; Beckendorff, U.; Anders, N.: FLEXPLAN - A Concept for Intelligent Process Planning and Scheduling. Proceedings of the CIRP International Workshop on Computer Aided Process Planning (CAPP), S. 87-106, Hannover, 21./22. September 1989, herausgegeben vom Institut für Fertigungstechnik und Spanende Werkzeugmaschinen (IFW) der Universität Hannover, 1989

[TÖNSHO 91]   Tönshoff, H.K.; Becker, M.; Hellberg, K.; Kreutzfeldt, J.: A Planning Strategy based on advanced Modelling Techniques for CAPP in Mechanical Engineering. Proceedings of the Seventh CIM-Europe Annual Conference, S. 197-210, Turin Mai 1991, Springer Verlag.

[WIENDA 90]   Wiendahl, H.-P.: Fundaments and Experiences with Load Oriented Manufacturing Control, Proceedings of 33rd APICS Conference, New Orleans, Oct. 1990

[WIENDA 92]   Wiendahl, H.-P.: Load Oriented Manufacturing Control, Springer, London, 1992

A METHOD FOR SUBASSEMBLY STABILITY CHECKING IN ASSEMBLY PLANNING

[1]Antonio Armillotta, Paolo Denti, and [2]Quirico Semeraro

[1]*CNR — Istituto di Tecnologie Industriali e Automazione, Milano, Italy*
[2]*Politecnico di Milano, Milano, Italy*

ABSTRACT

Assembly planning is a key problem towards the effective CIM implementation. Focusing on the problem, the generation of the assembly sequences is a necessary step and should be faced considering an automated assembly context. Automated assembly of mechanical products requires the handling of parts and subassemblies for placing and fastening operations. A subassembly can be handled by an automated device (robot, pick-and-place, conveyor) only if it is stable under the action of disconnecting loads. Therefore, a robust planning strategy is likely to use the subassembly stability as a criterion for selecting candidate subassemblies and finding precedence constraints among operations.

The present paper proposes a method for subassembly stability evaluation in CAD based assembly planning. The method consists of verifying that no part is free to translate along any direction in space, because of the presence of other parts. For every part an ordered sequence of geometric tests is executed, each test requiring only information about the geometrical entities involved in the contacts (normals to planes, axes of cylindrical and conical surfaces, etc.). The type of relations among components helps in driving and simplifying the stability evaluation, decreasing the number of required tests. The information required by the method are just the solid models of the parts, the model of the whole assembly and the contact relations among parts (possibly extracted by a CAD system).

The proposed method is applied to an industrial case and the obtained results are discussed in terms of effectiveness and time efficiency.

INTRODUCTION

Mechanical assembly processes are carried out in current automated systems (robotic cells and lines) as a sequence of operations. Each of them consists in picking a part or a subassembly from either a feeding or an intermediate storage location and placing it on another part or subassembly clamped in a fixture. So each operation involves two items: a fixtured one and a handled one; in order the operation to be feasible, the handled subassembly needs to be stable, that is, not to be able to disconnect itself under the effect of gravity and the other forces acting on it during the manipulation.

Most of the assembly planning systems found in literature do not care for stability requirements. In fact, the aim of these systems is the generation of geometrically feasible assembly plans. If a plan is geometrically feasible, i.e. no physical interference between components occurs during the operations, we can say it is certainly executable by a human-operated system, because dexterity and handling flexibility allow it to deal with unstable subassemblies as well; however, it might be impossible for that plan to be carried out by an automated system. Therefore, an assembly planner has to be able to check the stability of any subgroup of the product to be assembled: such a capability provides criteria to select the most convenient subgroups and to find out feasible assembly sequences for each of them.

The present work proposes a method to check subassembly stability within an assembly planner able to generate assembly sequences for mechanical products. The application of the method requires a limited amount of information, which can be easily extracted by the boundary representation of the parts and by the assembly structure model provided that the relations of contact between parts are known.

The paper is organized in such a way: we start briefly reviewing the previous work in the field; then we show the assumptions on which the method is based. We later formally describe the problem and the method we propose for the solution of the stability checking problem. After giving some implementation notes, we draw some conclusions.

RELATED RESEARCH

As said before, only a few methods among those proposed for solving the assembly planning problem take the stability criteria into account for selecting subassemblies and searching hierarchies among assembly operations. Moreover, it is possible to find in literature different definitions of stability; according to them, a subgroup is considered as stable if:

- no part is free to translate relative to other components, no matter of forces acting on the subgroup [1, 2, 3];
- no part can translate under the effect of gravity, taking friction with other components into account [4];
- no part is free to move (translate or rotate) relative to others under the effect of loads acting during assembly operations [5].

Clearly the first and the second definition help detecting stability conditions during the handling of the subgroup, whilst the third one helps in testing if the subgroup can be easily assembled, no matter of later manipulations. In all cases shown above, however, the stability checking of a subassembly consists in verifying if every component is enough constrained with respect to the others.

Stability can also be defined in a relative way, by computing heuristic evaluation parameters applying to the whole subassembly [6]: in this case only comparisons between the stability of two alternative subgroups can be made, while checking subgroup stability has no sense.

Focusing on an absolute definitions of stability, the various approaches differ in the way they deal with type and geometry of contacts among parts.

In some cases [2, 3] stability of each part is checked processing only information about kinds of contacts, and the test consists in searching contacts, or patterns of them, viewed as stable according to a set of rules. This kind of test leaves the geometry of contacts out of consideration.

In other cases, on the contrary [1, 4, 5], stability is considered as independent on kinds of contacts, and affected only by their geometry. In the latter case the whole of the contacts corresponds to a set of constraints that bounds the feasible movements of the components. The test consists in searching, for each part, a direction of motion satisfying all geometric constraints; if at least one component can move along some direction, the subassembly is unstable. Constraint conditions take a simple mathematical formulation by assuming particular hypotheses on geometry of contacts: the most common ones allow only planar [1, 5] or bidimensional [4] contacts.

MAIN ASSUMPTIONS

A method to test subassembly stability has to meet some needs in order to be applicable within an assembly planner:

1 the definition of stability has to be consistent with physical context of the problem and planning requirements;

2 the hypotheses about kinds and geometry of contacts between parts have to cover a wide range of products.

With regards to item 1, we assume to consider a subassembly as stable if no part can translate relative to others, no matter what forces act on it. In other words, a subassembly is stable if it can be handled as a single component. According to this definition a subassembly is stable even if some of its components can rotate on one or more axes. To justify this simplification, note that, in most cases, the existence of components with rotational degrees of freedom neither affects the manipulation of the subgroup, nor prevents them from mating with other subgroups. More precisely, we can observe the following:

- usually a mechanical part cannot rotate on more than one axis;
- if a part has one rotational degree of freedom relative to the remainder of the subgroup, the latter can be handled by grasping surfaces of other components;
- when a subassembly S, containing a rotating part A, is manipulated through a surface belonging to another part B, A can easily mate with components of different subgroups if it is symmetrical about its rotating axis (as an example we can grasp an electric motor by its housing and couple it with an operating device); otherwise mating is yet possible by adopting particular handling tools.

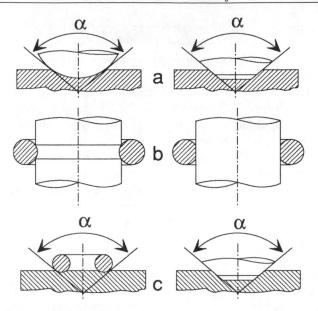

FIG. 1   Equivalences among different kinds of relations

With regards to item 2, the kinds of relation we consider (capable of describing a very wide set of products) are the ones listed in table 1.   This set of relation types can be simplified if we consider that the difference among contacts, for stability purposes, is the way each of them affects the motion of the components involved. From the stability point of view, note that:

- relation 6 (splined joint) is equivalent to a cylindrical contact, because, as said before, we do not take into account relative rotations between components;

- relation 7 (screw-hole coaxiality) can be turned into a cylindrical contact;

- relation 10 (spherical contact) can be approximated to a conical contact (figure 1a);

- relation 11 (torical contact) can be approximated, under different conditions, to either a cylindrical press fitting (figure 1b) or a conical contact (figure 1c);

- relations 12, 13, 14, 15 and 16 (rectilinear contacts) are not interesting for stability evaluation, as they do not usually help to constrain the motion freedom between the mated parts (e.g. in the case of the contact between the pitch surfaces of gear wheels).

THE PROBLEM

Let P be a set of n objects forming a subassembly and let we also suppose that the geometric description of the contacts existing between the parts is available; let us also suppose that the technological information describing the kind of contact

TABLE 1
Feasible relations

ID	Relation name	Geometrical entities involved
1	planar contact	two planar surfaces
2	planar bonding	two planar surfaces
3	cylindrical contact	two cylindrical surfaces
4	threaded contact	two cylindrical surfaces
5	cylindrical press fitting	two cylindrical surfaces
6	splined joint	two cylindrical surfaces
7	screw-hole coaxiality	two cylindrical surfaces
8	cylindrical bonding	two cylindrical surfaces
9	conical contact	two conical surfaces
10	spherical contact	two spherical surfaces
11	torical contact	two torical surfaces
12	plane-cylinder rectilinear contact	a planar surface and an external cylindrical surface
13	plane-cone rectilinear contact	a planar surface and an external conical surface
14	cylinder-cylinder rectilinear contact	two external or internal cylindrical surfaces
15	cylinder-cone rectilinear contact	an external cylindrical surface and an external conical surface
16	cone-cone rectilinear contact	two external conical surfaces
17	cone-plane point contact	a planar surface and the vertex of a conical surface (the axis of the cone is perpendicular to the plane)

(bonding, screwing, ···, simple contact) is joined to each relation. The kinds of relations taken into account in the previous section can be further subdivided into two classes:

- blocking relations (bonding, press fitting), which will be called *attachment* in the sequel;
- simple geometrical relations (contacts between surfaces), which will be called *simple contacts*;

Table 2 describes the information required for each kind of either simple contact or attachment.

In the following, we will describe a method, for checking the subassembly stability, which needs just the information listed in table 2.

The whole of necessary geometrical data needed can be easily extracted from the boundary representation of the subassembly parts; besides, let us suppose that, during the assembly modelling by CAD, all technological attributes of the contacts are joined to the features involved.

TABLE 2
Geometrical info required

Kind of relation	Data required	Kind of contact
1	normals to planes	simple contact
2		attachment
3	axes of cylinders	simple contact
4		attachment
5		attachment
8		attachment
9	axes of cones and taper angle	simple contact
17	normal to plane	simple contact

The contact relations existing between the parts are not usually explicitly available in the representation of a CAD model; such information should be extracted analyzing the assembly model.

FORMAL DESCRIPTION OF THE PROBLEM

Let us now describe the problem in a more rigorous way.

Let $P = \{p_1, p_2, \cdots, p_n\}$ be the set of the subassembly parts.

Let $G = \{g_1, g_2, \cdots, g_m\}$ be the set of the simple contacts existing between the parts $p_1, p_2, \cdots, p_n$; every element $g \in G$ is a 4-tuple (p_i, p_j, s_k, s_l) where s_k and s_l are the surfaces, respectively belonging to the parts p_i and p_j, in contact. Note that, for every couple of elements in P, more than one element might be in G.

Let $T = \{t_1, t_2, \cdots, t_o\}$ be the set of the attachments that act on the parts in P. Every element $t \in T$ is a couple of elements in P and represents the parts joined by the attachment. The elements in T are not necessarily disjoint; that is because a part $p \in P$ might belong to more than one attachment.

A subassembly model is represented by a 3-tuple:

$$Sub = \langle P, G, T \rangle.$$

The subassembly in figure 2 is represented by the following structure:

$$
\begin{aligned}
Sub \ = \ & \langle \{p_1, p_2, p_3, p_4, p_5, p_6\}, \\
& \{(p_1, p_2, s_2, s_3), (p_1, p_6, s_1, s_{18}), (p_2, p_3, s_5, s_6), \\
& \ (p_2, p_6, s_4, s_{18}), (p_3, p_4, s_7, s_{10}), (p_3, p_4, s_8, s_{11}), \\
& \ (p_3, p_6, s_9, s_{19}), (p_4, p_5, s_{12}, s_{14}), (p_4, p_5, s_{13}, s_{15}), \\
& \ (p_5, p_6, s_{16}, s_{19}), (p_5, p_6, s_{17}, s_{20})\}, \\
& \{(p_1, p_6), (p_3, p_4), (p_4, p_5)\}\rangle
\end{aligned}
$$

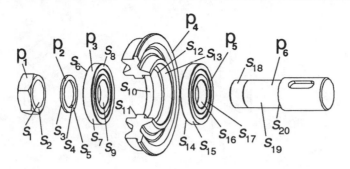

FIG. 2   A subassembly example

The parts joined by attachments can be considered, for stability purposes, a single part. We can manipulate the structure Sub removing the set T of the attachements, adding new parts to the set P and modifying the set G of the contacts in the way explained in the next section.

ATTACHMENTS SET REMOVAL

Let $Sub = \langle P, G, T \rangle$ be a subassembly; let us define $P_t = \bigcup_{t \in T} t$. Note that P_t is a subset of P and it represents the set of the parts belonging to at least one attachment.

Definition 1 *Let us define a binary relation* $\xleftrightarrow{\text{blk}} \subseteq P_t \times P_t$, *called* blocking relation, *as follows:*
let p_i, p_j be parts in P_t, then:

- *if $i = j$ then $p_i \xleftrightarrow{\text{blk}} p_j$*
- *if $(p_i, p_j) \in T$ then $p_i \xleftrightarrow{\text{blk}} p_j$*
- *if $\exists p_{k_1}, p_{k_2}, \cdots, p_{k_l} \in P_t \mid$*
  $((p_i, p_{k_1}) \in T) \wedge ((p_{k_l}, p_j) \in T) \wedge ((p_{k_h}, p_{k_{h+1}}) \in T, 1 \leq h < l)$
  then $p_i \xleftrightarrow{\text{blk}} p_j$

$\xleftrightarrow{\text{blk}}$ relates parts linked by an attachment chain of arbitrary length. It is easy to verify that $\xleftrightarrow{\text{blk}}$ is an equivalence relation.

Let $Q = \{Q_1, Q_2, \cdots, Q_p\}$ be the set of the equivalence classes inducted by $\xleftrightarrow{\text{blk}}$; each equivalence class Q_i represents the set of the parts joined by one or more attachments and it can be considered like a single part; let $q_1, q_2, \cdots, q_p$ be so defined:

$$q_i = \bigcup_{q \in Q_i} q \quad 1 \leq i \leq p$$

Let $f : P \rightarrow P \cup \{q_1, q_2, \cdots, q_p\}$ be so defined:

$$f(p) = \begin{cases} q_i & \text{if } \exists Q_i \mid p \in Q_i \\ p & \text{otherwise} \end{cases}$$

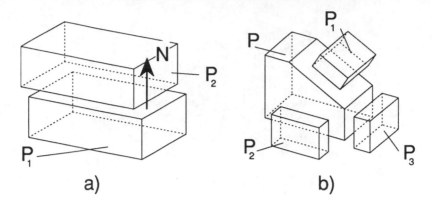

FIG. 3   Examples of planar contacts

Let us build a new subassembly $Sub_g = \langle P', G', \emptyset \rangle$ where P' e G' are defined in the following way:

$$P' = \bigcup_{i=1}^{n} \{f(p_i)\}$$

$$G' = \bigcup_{(p_i, p_j, s_k, s_l) \in G} (f(p_i), f(p_j), s_k, s_l), \quad f(p_i) \neq f(p_j)$$

The subassembly Sub_g is derived from Sub considering all parts linked together by one or more attachments like a single part; the stability checking of subassembly Sub_g, equivalent to Sub, can be carried out taking into consideration just geometrical information.

GEOMETRIC STABILITY FOR PLANAR CONTACTS

Let P_1 and P_2 be two objects; let a planar contact be the only one acting between them like in the example in figure 3a. We want to compute the set of the directions along which P_1 is free to translate.

Let π be the plane lying on the faces of P_1 and P_2 which define the planar contact. Let N be the normal to π pointing outwards from P_1. The set D_{free} of the directions along which P_1 is free to translate is so defined:

$$D_{\text{free}} = \{w \in \mathbf{R}^3 \mid N \cdot w \leq 0\}$$

Note that if w is a direction along the contact surface then $N \cdot w = 0$.

If an object P has b planar contacts (figure 3b) then the set of the feasible directions for P is the solution of the system of linear inequalities

$$\begin{cases} w \cdot N_1 & \leq & 0 \\ w \cdot N_2 & \leq & 0 \\ \quad \vdots & & \\ w \cdot N_b & \leq & 0 \end{cases}$$

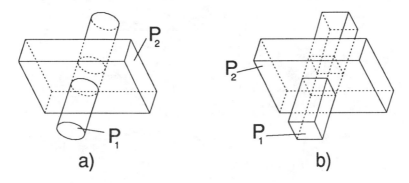

FIG. 4   Cylindrical contact and equivalent planar contacts

The trivial solution $(0,0,0)$ is always admissible; if it is the only admissible one then P is fixed; if every part of the subassembly is fixed then the subassembly is stable.

Each inequality defines a halfspace generated by a plane passing through the origin of the cartesian system. The solution of the system represents a region in $\mathbb{R}^3$ which is a polyhedral cone whose apex is in $(0,0,0)$. If the apex is not the only admissible solution, then at least one of the following objective functions

$$\min x, \max x, \min y, \max y, \min z, \max z$$

is not limited; otherwise, every function reaches its optimum in $(0,0,0)$ and its value is 0. The problem can be turned into an optimization problem and it can be solved using well known linear programming techniques [7]

TURNING EACH CONTACT INTO A PLANAR ONE

The method shown above is applicable to subassemblies involving planar contacts only; however, we also take into account curve contacts. It is possibile to substitute each non-planar contact previously described with some planar contacts turning the subassembly into an equivalent one, from the stability point of view; in the sequel we will analize every situation involving non-planar contacts.

Cylinder-cylinder contact   Figure 4a shows an example of a cylinder-cylinder contact. This kind of contact prevents one of the cylinders from translating along directions which are not parallel to their own axes; figure 4b shows a set of planar contacts which obstructs the same set of directions. Note that we do not consider the rotations of the cylinders around their own axes.

Cone-plane contact   Figure 5a shows an example of the cone-plane contact; the presence of cone P_2 (usually a screw or a dowel pin) prevents P_1 from moving along directions having a positive component along the normal N to the contact plane. The same set of directions is obstructed by the planar contact shown in figure 5b. This is equivalent to the former contact, because P_2 can only translate along N due to the other parts constraining it

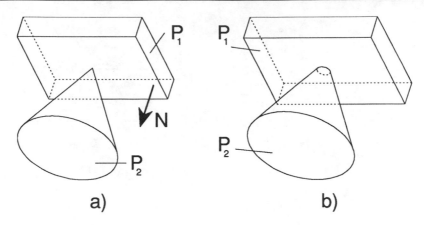

FIG. 5   Cone-plane contact and equivalent planar contact

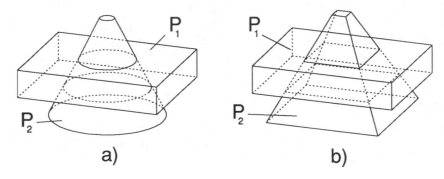

FIG. 6   Conical contact and equivalent pyramidal contact

Cone-cone contact   For this kind of contact note that it does not exist a finite set of planar contacts equivalent to the conic contact. We can just approximate this contact to a pyramid-pyramid contact (figure 6); the accuracy of the approximation depends on the number of side-faces of the pyramids: when this number tends toward infinity, the pyramid tends to become a cone and the error tends toward zero. More precisely, the error made by approximating the conical surface of contact to a pyramid, consists in considering feasible the translations within the region included between the side-faces of the cone and the pyramid which, in fact, are not. It is easy to realize that the size of such a region does not affect the stability checking response, even when dealing with a square pyramid, which is a very loose approximation to a cone.

Note that all the unstable subassemblies, involving conical contacts, are still considered unstable after the substitution with pyramidal ones.

IMPLEMENTATION NOTES

The proposed method was implemented in a computer program in order to evaluate its efficacy. The program, coded in C++ language on Sun SPARC2 workstation,

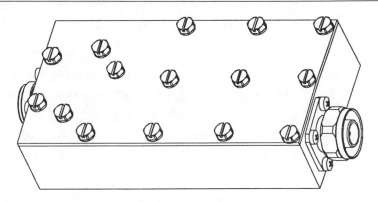

FIG. 7   Microwave filter

is able to check the stability of each subgroup of a given product.

The program was tested with some assemblies covering the allowed set of relation types. Correct checks were reported for all problems with short computation times. As an example, figure 7 shows a microwave filter, utilized in telecommunication devices. The assembly is made by 92 parts connected by 196 relations of 5 different types (planar, cylindrical and threaded contacts, bond cylindrical joints and hole-screw coaxiality). Using interactively the program, more than 30 stable subassemblies were found (excluding subassemblies differing for components belonging to the same pattern). The maximum size subgroup, namely the whole assembly, has been checked in a CPU time of nearly 0.6 seconds.

CONCLUSIONS

In this paper a method for checking subassembly stability has been presented. It is based on turning the set of contact relations into a new one, involving only planar contacts; a geometrical check is then performed on the modified set. Our approach is applicable to a wide set of contact types even though computation can be carried out in a very short time. So the proposed algorithm can be easily integrated in a computer-aided planner, in order to restrict the set of feasible sequences.

ACKNOWLEDGEMENTS

This work has been funded by CNR - Progetto Finalizzato Robotica, contract no. 90.00399.67. We gratefully thank Mr. B. Malnig and Mr. G. Birindelli of Alcatel Telettra company for the technical support provided.

REFERENCES

[1] A. C. Sanderson and L. S. Homem de Mello. Automatic generation of mechanical assembly sequences. In *Geometric modeling for product engineering*, pages 461–482, Elsevier, 1990.

[2] A. Delchambre. A pragmatic approach to computer-aided assembly planning. In *IEEE international conference on Robotics and Automation*, pages 1600–1605, 1990.

[3] A. Delchambre and A. Wafflard. An authomatic, systematic and user-friendly computer-aided planner for robotized assembly. In *IEEE international conference on Robotics and Automation*, pages 592–598, 1991.

[4] R .H. Wilson and J. F. Rit. Mantaining geometric dependencies in an assembly planner. In *IEEE international conference on Robotics and Automation*, pages 890–895, 1990.

[5] N. Boneschanscher, H. van der Drift, S. J. Buckley, and R. H. Taylor. Subassembly stability. In *Proceedings AAAI conference*, pages 780–785, 1988.

[6] S. Lee and Y. G. Shin. Assembly planning based on subassembly estraction. In *IEEE international conference on Robotics and Automation*, pages 1606–1611, 1990.

[7] M. S. Bazaraa and J. J. Jarvis. *Linear Programming and Network flows*. Wiley, 1977.

LASER BASED SYSTEM FOR REVERSE ENGINEERING

Yasser Hosni and Labiche Ferreira

University of Central Florida, Orlando, Florida, U.S.A.

ABSTRACT

The reverse engineering process involves the use of 3D data to manufacture a part. A crucial element in the reverse engineering process is obtaining 3D data of the part or it's CAD model. The paper describes a laser-based system for reverse engineering. The laser scanning system performs the crucial role of capturing the geometric details of the object. This data can then be processed by routines to accurately define the object. This definition can then be used to manufacture a replica of the part or scaled object. Also addressed in the paper are issues for integrating the laser scanner with other stages in the product development and manufacturing cycle.

This paper is an outcome of ongoing research aimed at manufacturing heat resistant tiles for the space shuttle to fit cavities created due to damaged or lost tiles after each space mission.

INTRODUCTION

An essential step in the reverse engineering process is the creation of a CAD model for the object to be manufactured. The creation of a CAD model from a part or prototype is referred to as *Reverse Engineering*[1]. The CAD model serves as an input for subsequent analysis such as planning, tooling and NC tool path generation. An important element of the reverse engineering process is the need of a geometric database. Creation of such databases is one of the most complex elements of a computer based design and analysis system. For objects with geometric regularity, it is practical to generate them analytically, using one of the several geometric modelling methods - such as solid modelling. However there are many areas in which there is a need to create a database by extracting an object definition from a complex real life object; i.e. one that already exists and which may not have regular geometric properties. Examples of these are found in the field of medicine (prosthesis, plastic surgery), in industrial applications such as measurements of castings and production of dies (having a model of a product). Another important component of the system besides the geometric database is a link to a computer controlled metal removal machine such as: milling, turning, etc.; which permits the replication of the scanned object.

The creation of a CAD model may be necessary under the following conditions:

1.   *modification of existing design:* modifications on a part can be best evaluated

using a CAD model.

2.   *replicate the part/rapid prototyping:* in situations where the part has to be reproduced.

3.   *manufacture a die or mold for the part:* the CAD model of the part created can be used to generate a model of it's "negative" or the mold.

4.   *inspection:* the part manufactured can be scanned using a scanner or measurements can be taken using a CMM (coordinate measuring machine) at specific points. The model of the newly manufactured part can be compared with it's original CAD model (model subtraction) to determine inaccuracies in the manufactured part.

The use of a contact measuring technique (such as a CMM) could be used to obtain X-Y-Z information and then create a CAD model. In case of complex shapes such information would be difficult to obtain due to the inability of the probe to capture the minutest detail of the objects surface. In these cases the use of non-contact measuring methods such as photomechanics: namely laser beam scanning or the Moire method could be more efficient.

The outcome of scanning (mapping) yields X-Y-Z data of the object scanned (mapped). This 3D data of the object may then be used for: generating 2-D drawings, rapid prototyping, producing a modified version of the existing object, producing a negative of the object, i.e. to produce a mold or die, inspection and measurement verification and maintenance purposes.

We describe here an integrated system for object manufacturing using a non-contact measuring technique, such as the laser-based scanner. The system is experimental in nature and currently being explored in the laboratory of the Department of Industrial Engineering and Management Systems at the University of Central Florida (UCF). The paper starts with a brief description of surface mapping technologies, laser beam scanning in particular, followed by object representation and the experimental setup at UCF.

SURFACE MAPPING TECHNOLOGIES

A scanning arrangement can be used to scan the part (if no or old 3D data exists), and then transform the scanned data into a solid model. The model could then be used to manufacture the mold or the die for the part or replicate the part on a milling machine. Many ways exist for capturing the geometric information of a surface. An obvious way of obtaining such data is to use a CMM. The object could be placed on the table and by the use of a probe the coordinates of the surface can be obtained. However, the limitations of a CMM are:

1.   The size of the object that can be measured is restricted by the size of the CMM table.

2.   The accuracy of a CMM limits its use where dimensions have to be measured within close tolerances.

3.   The measurement of complex profiles may be impossible. This is due to the fact that the probe may not be in a position to reach sharp corners, grooves and peaks.

Modern measuring systems are optical based. These techniques include:

1.   capturing a picture of the object. The image of the objects surface is then

analyzed through the use of image processing algorithms to extract pertinent data.

2.   projecting fringes on the objects surface (Moire Method). A picture of the distorted fringes on the objects surface is captured and analysis of the fringe pattern results in topological information of the surface.

3.   scanning a laser beam across the objects surface. The output is an X-Y-Z representation of the object scanned.

This paper addresses the use of a laser scanner in obtaining 3D data of an object and the different issues related in using this data in the manufacturing cycle of the product. In the following section the concept of laser beam scanning and the components of a laser scanner are described in detail.

Laser Beam Scanning

The industrial environment adds constraints and limitations, such as need for non-contact, real-time measurement, hostile environment, cost, and compactness. Active methods where a beam of light, such as a laser beam is superimposed to the naturally lightened scene simplify a lot of the signal processing to be done in order to recover distance information (Z coordinate). Besides this advantage, the use of a laser beam for surface mapping provides a number of unique advantages. The brightness of the source ensures a good signal-to-noise ratio in most applications. The most significant advantage is the ability to capture data in realtime, thus enabling applications such as machine control, inspection and others[2].

The basic elements of a system using triangulation for surface mapping (scanning) consists: of a light source that is usually a laser, a scanning mechanism to project the light spot onto the object surface and a position sensor with a collecting lens looking off-axis for the light spot. Distance measurement is done by trigonometric algebra applied to the projection direction (scanner angular position) and the detection direction made by the light spot position on the sensor with the principal point of the collecting lens.

Synchronized scanning[9] eliminates the shadow effects and permits continuous profile scanning. The basic idea is to synchronize the projection and detection in a way that the detected light spot on the position sensor keeps it spatial position stable when the projected beam is scanning a flat surface (scanning being parallel to the surface). The implementation of a synchronized scanning mechanism can be done electrically or geometrically.

Figure 1 shows an arrangement to electrically synchronize two galvanometer driven motors G1, G2. In this case the source signal is a ramp, then the output signal from a position sensor seen on the oscilloscope provides a direct profile reading for which the time axis is proportional to displacement along the X axis. Amplitude of the deflection is proportional to departure of the object surface from the reference plane. In order to get a surface profile measurement a third galvanometer driven motor (indicated by G3 on figure 1) is used to deflect perpendicularly to the page both the projected and received beams.

Figure 2 shows an arrangement that is geometrically synchronized. Both mirrors F1 and F2 are fixed when the device is in operation. Angular adjustment of those mirrors sets the position of the reference plane in space. Synchronization is

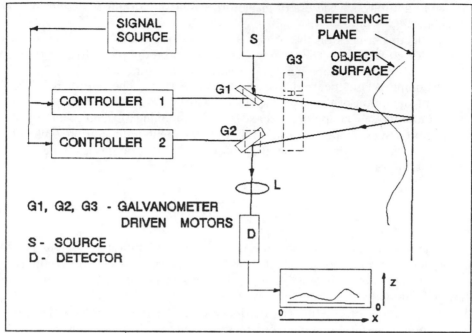

Figure 1.   An example of an electrically synchronized scanner[9].

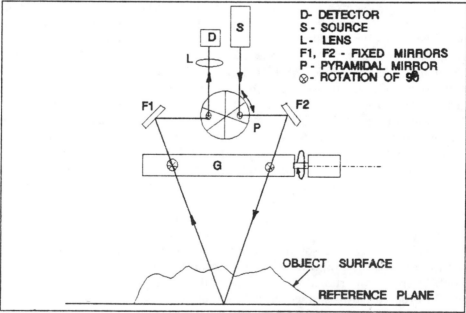

Figure 2.   An example of a geometrically synchronized scanner[9].

realized using a pyramidal rotating mirror(P). With mirror G, the scanner becomes a two-axis scanner. By geometrical analysis, synchronization can be achieved by using two opposite facets of the scanner. The arrangement is simpler to implement than electrical synchronization and is also more precise and stable. Another feature is the ability of rotating mirror scanners at very high speed (approx 10,000 lines/s) compared to galvanometers (approx 100 lines/s).

As seen from figure 2, the major components of the scanning arrangement are : polygonal scanner (mirror), motor for rotating this mirror, light source and a position sensor. The following sections discuss the various aspects of making the scanned data useful at all stages of the manufacturing cycle of the product.

OBJECT REPRESENTATION AND MANUFACTURING

As mentioned earlier, a CAD model is a crucial element of the reverse engineering process. From the 3D data obtained after scanning an object, a model of the object has to be created. The need for a CAD model for manufacturing a part is crucial in all stages of the product development and manufacturing cycle. A CAD model is needed for designing, inspecting, prototyping and modifying the product. This model could then be transferred to a more complex CAM package to develop process plans and manufacturing routines. Advanced CAD/CAM packages such as CATIA, CADAM, and PRO ENGINEER can create a solid model of the object and generate tool path programs for numerically controlled programs. The use of these packages depends on the availability of product drawings prior to the start of model creation. In situations where product data on hand may not be valid or replication of existing product is desired, the role of contact or non-contact (surface mapping) instruments for measurement becomes evident. A complete system would include the integration of data capturing equipment, object representation, tool path generation and a numerically controlled machine.

Most of the surface mapping techniques including the laser beam scanning technique generate a huge amount of 3D data of the object mapped. This data could then be used to create a model of the object which could be used in all the stages of the development and manufacturing cycle of the product. Since this model will be used by design engineers, quality department and process planners, the accuracy of the model created from the scanned data is important. Besides this, the model created should be compatible with CAD/CAM packages. In some cases the development of a CAD interface would be necessary to transform the data into formats that are compatible with specific CAD/CAM packages.

A considerable amount of work has been done in the area of surface definition from 3D data and the use of this data to manufacture the object. In all methods the ultimate goal is to create a model of the object. Most of these methods use the basic curve and surface fitting techniques developed over the years. Description of these methods can be found in the references listed.

An important application of a laser scanner is it's use in obtaining topological data of complex profiles. This data can then be used to manufacture a replica of the part. In some cases the object representation can be modified through a suitable CAD interface before manufacturing. An important issue at this stage is the accuracy of the object definition from the 3D data obtained. Most laser scanners are marketed with an interface or software package to represent the object before manufacturing it. However, the scanned data could be imported into any CAD package, such as AutoCAD or CADKEY before it is manufactured. This CAD file or the 3D data file can then be imported into a CAM package via the DXF format. The 3D data file contains a huge number of points of the object scanned. The amount of points are in excess to that needed for definition of a surface. The points will have to go through a "smoothing algorithm" to reduce the number of points before creating surfaces or splines and generating the necessary machining instructions(tool path).

A crucial issue in the use of a laser scanner is the compatibility of the output data with CAD/CAM packages. In some cases the output data will have to be modified to a format acceptable to that of the post-processing package/software.

EXPERIMENTAL SETUP AND PARTIAL RESULTS.

This paper is an outcome of ongoing research aimed at manufacturing heat resistant tiles for the space shuttle to fit cavities created due to damaged or lost tiles after each space mission. The research group at the Department of Industrial Engineering and Management Systems at UCF surveyed the market for 3D scanning systems. In order to familiarize ourselves with the technology and the issues involved a relatively less expensive but capable laser probe was selected. This section contains the description of the hardware acquired and its capabilities. The laser triangulation system chosen for initial study and development is the precimeter system distributed through Candid Logic Inc. The precimeter system contains a controller and a gauge probe (scanner). Figure 3 shows the experimental set up of the scanning arrangement using the laser probe and the motion system. Described below briefly are the features of the laser probe, motion system and the results of initial experiments at UCF.

Laser probe (Point Scanner)

The controller contains the power supply and necessary electronics to operate the scanner. The scanner contains a diode laser and a linear detector with optics. The distance information is available from the controller by way of an RS-232 serial port, parallel port or visually from the LED read out. The probe has a measuring range and standoff distance of 4.7 inches. This probe has the lowest accuracy of all the precimeter line at just under .01 inches.

Motion System

The motion system is made up of individual slides and tables each moved with a stepper motor. This approach allows modifying the set up of the different axis to vary the way in which the probe or object are to be moved. It consists of two linear slides mounted on a rotational table. This gives linear motion in the vertical and horizontal planes and rotational motion around a vertical axis. The movement is accomplished through the use of stepper motors. The motors have 400 steps per revolution, each with it's own power supply. With the associated gearing one step

Figure 3.   Experimental setup of a scanning arrangement using a laser probe and a motion system.

equates to one 4000th of an inch linear travel and one 200th of a degree rotational movement.

The power supply receives movement instructions from the PC23 controller via an adapter box. The PC23 controller sits inside of the personal computer. It supports many complex movement commands and need only receive instructions from the computer. The PC23 sends logic signals to the adapter box to control motor movement. A ribbon cable connects the PC23 and adapter box. Custom cables run from the adapter box to the motor power supplies and again from the power supplies to the motors.

Interfacing

An IBM compatible Intel 486 machine is used to coordinate the interaction between the precimeter and the PC23 controller. This enables 3-D data to be collected. Other functions the computer serves is to display, store, and manipulate data. The system uses a custom program developed in the UCF laboratory. The motion system with probe attached are mounted on an optical breadboard along with the object being scanned. The use of an optical breadboard minimizes the effects of vibrations and provides a simple method of mounting hardware for experimentation.

```
SUB scan.3d.first
INPUT "Enter filename for output (.xyz will be appended) ", filename$
INPUT "How many planes? ( 0.2 inches vertical seperation ) ", plns%
filename$ = filename$ + ".xyz"
aspectratio = (4 / 3) * (350 / 640)
OPEN filename$ FOR OUTPUT AS #5
PRINT #5, plns%      'plns% data blocks
VIEW (0, 0)-(639, 250)
CLS 1
LINE (0, 0)-(639, 250), 2, B
LINE (320, 20)-(320, 180), 2
LINE (240, 100)-(400, 100), 2
FOR planes% = 0 TO plns% - 1
PRINT #5, 360    '360 data points
FOR theta% = 0 TO 359
distance% = get.distance
IF distance% = 999 THEN GOTO jhg
x% = SIN(3.141593 / 180 * theta%) * distance%
y% = COS(3.141593 / 180 * theta%) * distance%
xprint% = x% * .4 + 320
yprint% = y% * .4 * aspectratio + 100
PSET (xprint%, yprint%), planes% + 1
jhg:
clockwise ("cw 01.0")
delay .5
xydata%(0, theta%) = x%
xydata%(1, theta%) = y%
PRINT theta%, distance%, x% / 100 + 5, y% / 100 + 5
NEXT theta%
up ("up 0.2")
FOR x% = 0 TO 359
PRINT #5, USING "#.##"; xydata%(0, x%) / 100 + 5;
PRINT #5, " ";
PRINT #5, USING "#.##"; xydata%(1, x%) / 100 + 5;
PRINT #5, " ";
PRINT #5, USING "#.##"; planes% * .2
NEXT x%
NEXT planes%
CLOSE #5

END SUB
```

Figure 4.   Program listing used to coordinate data collection and the motion system.

Partial Results

Many objects have been used to test the capability of the probe. Code was written in Quick Basic incorporating the standard motion statements to control the movement of the laser probe past the object. Figure 4 shows one of the programs used to coordinate the laser probe and the motion system. The 3D data obtained after scanning is then imported into AutoCAD for representation or could be imported to CAD/CAM package such as SURFCAM™. The digitized data can then be used to generate splines and surfaces. After the definition of the surfaces; the required NC code can be generated for it's manufacture. Figure 5 shows the object representation after importing it into a CAD/CAM package. Some problems have yet to be a resolved; these are:

1.   Coordination between different scans. Despite the fact that the 3D laser scanning technique is capable of capturing complex profiles; different scans of the same object will be needed to fully capture the objects geometric detail.

This problem could be resolved by identifying a reference point on the object and using it as a reference or datum point for other scans.

2. Formatting data for CAD/CAM packages. Most CAD/CAM packages available can import digitized files (files containing the X-Y-Z data) and display the object. However, the format has to be as required by the package. The data file in some cases will have to be modified to indicate the start of a new scan line.

3. Process and sequence of machining the object. The object scanned could be machined by a number of different processes (milling, turning, EDM) and a different sequence. Determining the correct process and proper sequence need to be resolved. Research may lead to routines which may identify the proper process and machining sequence.

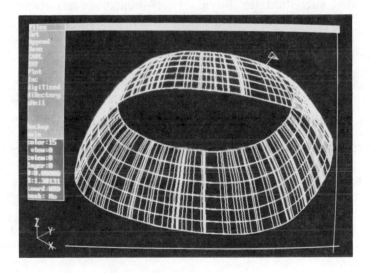

Figure 5.   Object representation after importing data file into a CAD/CAM package.

CONCLUSION

The use of a laser scanner for obtaining 3D data is promising. It is clear that with the use of a laser scanner, the lead time for producing the final product can be substantially reduced. The development of an integrated system such as the one described above is feasible. The experiments conducted at UCF validate the potential of a laser based scanner for reverse engineering. The experiments helped identify possible problems and identify new research avenues. The crucial task of integrating the various elements of the system has not yet been perfected and still remains a challenge. As the development of automated systems and many forms of required software continues this challenge will someday be met.

REFERENCES

1. Bidanda B, Motavalli S: "Reverse Engineering: A survey of prospective non-contact technologies and applications in manufacturing systems." *University of Pittsburgh, Technical Report Series No. 90-20.*

2. Boulanger P, Evans K, Rioux M, and Ruhlmann L: "Interface between a 3-D Laser Scanner and a CAD/CAM System." *Proceedings of the 5th CAD/CAM and Robotics Conference.* Toronto, Canada, June 1986,: 731.1-731.7.

3. Ferreira L: "Surface Mapping and Automatic Tool Path Generation." *Thesis,* University of Central Florida, Spring 1991.

4. Hosni Y: Photomechanics Based System, Proposal and Final Report for submitted for Florida High-Technology and Industrial Council (FHTIC), Tallahassee, Florida, 1991.

5. Hosni Y: Productivity Techniques, Report to NASA, Kennedy Space Center, Florida, 1991.

6. Hosni Y, Jueng-Shing H, and Ferreira L: "Tool Path Generation from Surface Mapping of an Object." *Proceedings of PROCIEM '90,* Tampa, Florida, November 11-13 1990: 23-27.

7. Hosni Y, Pax T, Ferreira L: "Integration of a Laser Scanner in the Product Development and Manufacturing Cycle", *Proceedings of The Conference on Systems Integration,* Orlando, Florida, October 6-8, 1991.

8. Nasser D: "Non-Contact, Three Dimensional Object Digitizing Systems." *Thesis,* University of Central Florida, Fall 1989.

9. Rioux M, Blais F, Beraldin J, Boulanger P: "Range Imaging Sensors Development at NRC Laboratories." *Proceedings of the Workshop on Interpretation of 3D Scenes,* Austin, Texas, Nov 27-29 1989: 154-160.

RAPID PROTOTYPING IN A COMPUTER SYSTEM POWER AND PACKAGING DESIGN PROCESS

Marto J. Hoary

Digital Equipment Corporation, Boxborough, Massachusetts, U.S.A.

Abstract

Introducing change in large global business organisations has obvious difficulties. Yet in order to remain competitive, particularly in todays Information Technology business it is essential to be able to anticipate and respond quickly changing market trends. Smaller companies tend to be much more dynamic than large ones. But success breeds success and the small companies grow into larger more diverse organisations and begin to suffer from the impaired dynamism of large complex organisations. (Ref. [5]). This paper describes aspects of the impact of introducing change in a global engineering organisation. It relates to the impact of deviating from a rather standardised design process in pursuit of a more aggressive development cycle in the face of accelerating competitive pressures.

The main focus is the transition from CAD to CAE technology and the development of interfaces to world wide manufacturing to meet the competitive challenge. Other elements of this transition were building a highly motivated team with a proven background and simplifying the ground rules. The development effort outlined here served to extend the benefits of Euclid throughout the entire product development cycle and into volume production.

It useful to understand some background to traditional use of CAD in order to put the benefits of migrating to a CAE system with a relational type data structure in true perspective. By way of introduction it deals with the initial impact of CAD. Progressing to the migration form CAD to CAE and extending the application of CAE to support the broader aspects of the design process such as prototyping forms the main thrust of the study.

1 The CAD Revolution

The evolution of computer applications for design and manufacturing engineering functions has progressed from assistance with numerical computation in the fifties and sixties, through the introduction of computer graphics to Computer Aided Drafting applications in the mid seventies. The late seventies saw the emergence of turnkey CAD systems offering applications in 3D wireframe and surface modeling, machining, basic finite element modeling,. The proliferation of those systems was relatively slow initially because of the poor cost performance ratio. Only the larger R&D intensive corporations such as Aircraft designers invested. The emergence of low cost 32 bit hardware and improvements in the reliability and functionality of CAD software saw an expansion in the use of CAD.

1.1 The Legacy of Computer Aided Drafting.

Turnkey systems offered a selection of diverse applications and functionality to assist the design process but, for the most part CAD was applied to improve the productivity of the drawing office. This is still true today even though systems have made vast improvements all-round. I see the reasons for this as follows;

- The productivity of using CAD for creating drawings in 2D became apparent quickly and the migration to the process from the drawing board was simple.

- Working in 3D wireframe models had some disadvantages when it came to drawing productivity. So unless there was a specific need for a 3d model there was no advantage to creating one. Some of the disadvantages were;

  - No hidden line removal made visualisation difficult due to part disorientation in distinguishing foreground from rear and wireframe fog on complex parts.

  - Hidden lines had to be erased or fonted manually in each view on a drawing often with a slow screen refresh for each. This tended to be a frustrating exercise because of multiple lines in a single location.

  - Creating a wireframe involved much more effort that creating a 2D drawing.

  - The size of the part in the database was much greater. Mass storage was expensive and large models were susceptible to becoming unstable or corrupt.

- Surface models had nothing to offer the drafting function. They were used for computation of mass property data and to support other CAD tools and applications such as Machining and Finite Element Modeling.

Consequently, even though the use of CAD began to proliferate as cost/performance improved and functionality increased with the introduction of Solids modeling it was largely as a drafting system that it continued to be applied. This is not surprising because after all drawings are the key means of communication between engineering and the rest of the world for product design. Furthermore, standards on the use of CAD tended to focus only on the drafting function thus further ingraining the restricted application of CAD/CAM. In the last three years or so developments in client serving applications, relational or object oriented database type CAE systems, clustering technology and high speed communications have lead to the emergence of highly integrated CAE/CAD/CAM systems which make automating the entire product development cycle feasible. Making the transition to true Computer Aided Engineering whereby drafting is a back end application after the design work is complete has significant risk and is especially daunting where the standards and procedures and expectations of using CAD are already well established. Productivity must be further increased sufficiently to justify the new investment. Associated with the transition also is a change in the work routines of the regular users. This paper outlines a case study where a design team make the transition to a relational database type CAE system as they begin development a new product having migrated from the traditional wireframe CAD system dedicated to drafting. It deals with the further development effort in integrating CAD/CAM to deliver rapid prototyping and later to integrate the supply base world wide to realise improvements in time to market out to volume production. (For further study reference [1])

2 New ground rules to meet the Competitive Challenge.

Facing the challenge of having to develop a new generation of product in a shrinking development time window, with the flexibility to respond within hours to changing product requirements, it was decided to adapt a new approach to to the engineering design development process with a more progressive outlook. (Ref. 5) Aspects of the new approach included ;

- Team building with players selection based on ability.

- Use the most productive, state of the art applications and tools.

- Simplify administration process.

- Cost & performance were key considerations at every design step.

- Merge the skill sets of design and manufacturing engineers for faster development and better quality volume.

- Early purchasing organisation and supplier involvement. - identify process dependences for cost reduction and quality.

- Fast order completion processes for prototype materials

- Design engineering work directly with volume production plant from early in the development cycle - DFM

3 From CAD to CAE.

After a search and evaluation of the current offerings in the CAE space. EUCLID IS from Matra Datavision was the system selected as best at the time. All of the users had been experienced in using a 3D wireframe based system. The migration to s Solids modeler with a relational type, non file oriented database structure would be a challenge.

3.1 Euclid-IS Features.

EUCLID-IS is a mechanical CAD/CAM/CAE turnkey system. Primarily it is a Solids modeling system with layered applications. SOME its key strengths are perceived to be;

- Highly productive user interface - creating models was fast and simple by comparison with wireframe systems.

- DESIGN Applications such as sheet metal design offered true Computer Aided Engineering to the designer by automatically taking care of standard features.

- Relational Database structure - All parts in a database had a hierarchical relationship to one another. A Change to a top level part changed all other instances automatically.

- Project Management Access structure - Users accessed their work on a Project. Sub project, user basis. This enabled them to reference other users work without the risk of modifying it.

- Compact Database - the database was not file oriented and was highly compact requiring minimal mass storage along with past retrieval of objects.

- Visualisation was excellent at any level of assembly due to the efficiency of the hidden line removal algorithms, for both single components and large assemblies.

- Documentation supported postscript screen dumps executed quickly.

- True client serving of non display computes to a mainframe provided very fast response. The workstation only has to cope with display refresh.

- Batch processing meant that several activities could continue simultaneous with no impact to the users performance. Printing and plotting are simple examples.

3.1.1 Benefits

The system was installed on a cluster of 25 workstations and a VAX 6440 which with client serving accounted for the high performance even with 21 simultaneous users.

Productivity improvements of between 3 : 1 and 5 : 1 depending on the CAD task were observed..

The ratio of draughts people to designers was reversed from 3 : 1 to, 2 draughters to 5 designers. Further benefits will be outlined in the DESIGN PROCESS in section 4.

3.1.2 System development and support.

The suppliers of the system, Matra Datavision, worked very closely with the team. Customised applications and interfaces were developed as required and this some of effort is described here.

3.1.3 Downsides

It required a learning curve of between 6 to 9 months to develop proficiency. This was not a problem because it was planned for.

At that time the size of a single database was limited to 64000 links. This was exceeded on this project but that turned out to be a restriction rather than a gate. This is not a permanant feature of Euclid and the understanding is Matra Datavision plan to relieve this restriction in a future release.

Overall, the application is quite stable though it continues to be improved.

(For further study on Euclid, Ref. [7])

4 The Design Process.

From the macro product description in Euclid, the project is divided up into sub-projects with users assigned to each. That is how Euclid access is allocated. The mechanical packaging for this product, housing for a new midrange computer system, is mostly sheet metal with some plastic parts. The Euclid Sheet Metal Design application is a Computer Aided Design application in the real sense because it assists the designer with the development of the sheet metal part by taking care of the sheet metal features, such as bend and bend reliefs, dimples, perforations etc. automatically. It references a materials library with all of the material behavior characteristics defined there.

The sheet metal design process starts by selecting the material and stock thickness required for that part from the system library. Next the shape of the part is defined. Euclid automatically inserts the bends and bend reliefs. The display can instantly toggle between a flat pattern or a bent up or partially bent up part.

4.1 Prototyping.

When an assembly of parts is completely modeled the next step is usually to have them made and put them together. This is where quick turn around was most valuable because this tended to be an iterative process. It was clear that direct CAD/CAM interfaces with the machine shop could obviate the time and energy to prepare and manage drawings as well as provide other time saving and qualitative benefits.

Specialist expertise was assigned to work a solution in this area. The expectation was that a simple conversion to IGES was all that is necessary. But it emerged that the conversion to IGES for sheet metal data was not simple or at all possible initially. The task was pursued to a successful completion over an 18 month period and the solution is described here.

From the body of knowledge accumulated it was decided to take effort a step further and pursue the development and integration of the entire supply base world wide. While the main benefit of having fast turn around of parts was to design engineering it made a lot of sense to apply the incremental effort to extend those benefits to volume manufacturing as well. This will be described under the Supplier linkage project ,section 8. Rapid prototype turnaround technologies were also implemented for fast turn around model plastic parts and soft tooled plastic parts also and these are described in section 6.

5 CAD/CAM Interface Development.

A cloaer look at the CAD/CAM interface development effort will hopefully provide some insight into how to apply this knowledge in similar situations such as data driven processes that interpret geometric data. It is not possible to provide more than a cursory out line of the important considerations in the available.

5.1 Sheet metal programming.

The process for providing fast turn around sheet metal parts was to take the geometry from Euclid in IGES form and read it to sheet metal programming software at the machine shop or at an external suppliers. The difficulties with getting fast turn around sheet metal parts stemmed from the fact that the sheet metal models in Euclid were in a unique data format not compatible with surface or wireframe type geometric entities.

The Euclid CAD system is already described. SMP81 from Merry Mechanisation, is a sheet metal programming CAM application with a graphics front end which operates on almost any workstation. In this instance the sheet metal parts are modeled *from* the Euclid IGES. SMP81 was used as a base reference such that if it could read the IGES successfully, then other sheet metal CAD/CAM systems could do it. SMP81 was widely installed throughout the corporation so there was high availability of skilled expertise at the machine programming level.

The first problem was that Euclid sheet metal data was a unique type and not IGES writable. It could be converted to a solids model which could be displayed as wireframe. This solid could be translated to IGES but SMP81 could not read it, initially-Within the Euclid IGES preprocessor, parameters can be set to include or exclude certain types of geometry or attributes. These parameters were varied until it produced IGES that SMP81 could read. The key to this first step was not to process entities as copious data. ie, assemblies or sets of entities and secondly not to process opacity.

The part was successfully modeled in SMP81 from this IGES and fabricated at the machine shop. However the process was very unreliable and had potential for much improvement which would be essential to have in place by the time prototypes were being ordered for the full product assembly. Modeling the parts from early IGES required a lot of manual intervention. Goals were defined to provide IGES which ;

- Required minimal user intervention and the risk of errors .
- Could be tracked or modeled automatically in the reading CAD/CAM system.
- Was accurate and unambiguous ie, had good entity connectivity.
- Contained entity types that could readily be read as <u>features</u> by the CAM system.
- Had compact sized files to support Electronic Data Interchange.

5.2 USER APPLICATION DEVELOPMENT.

The wireframe generated from the solid model had no arc or circular entities, only linear figures and polylines. SMP81 can interpret a number of geometric shapes as the features they represent, for example an arc along an edge is interpreted as a bend. A circle, square rectangle or obround will be mapped as a hole where the center of the hole is important because this will be the center of the hit for the punch later. As such the Euclid features all had faceted edges and a lot of extra work involved in deriving center locations from the IGES if it is made up of polylines only. The geometry of the hole can readily changed once positioned in SMP81. Also, if entities do not match end to end (entity connectivity) or if they out of plane for the previous bend angle, modeling from IGES will fail. This was a regular occurrence initially.

Matra Datavision, the suppliers of Euclid, were involved to devise a solution to this problem. They (Mr. Peter Hajjar) produced a User Application in FORTRAN which could be invoked from the Euclid environment to replace all polylined arcs and circles with real arcs and circles. It was decided to work with a contoured representation of the sheet metal part. Thickness lines in various places represented the material side of the contour as well as the stock thickness value. This is perfectly adequate since all of the sheet metal CAM systems only require thickness defined as a constant value. The advantages of using the contour only were, it kept the IGES file size down and it was much easier to track or model from since wireframe fog was minimised. The User Application actually created a new representation of the whole part. and this is what was translated to IGES.

The subsequent Euclid -> IGES translation process also presented some problems. For example it still has difficulty processing large files. If the entity count is less than 100 there will be no problem. If it is between 100 and 180 then unprocessed graphics need to be reselected. If the count is greater than 180 then the entity count will have to be reduced by making assemblies of entities groups or types.

Another observation was that IGES would not represent geometry where multiple levels of assemblies were captured. For example if the lowest level is a single line, arc or circle, the next level is a polyline or string of lines. Next is a Linear Figure (LFIG) which would be a closed string of and entities. Next is a PERF which is an assembly of Polylines or LFIGs within a closed contour on a single plane. Above that is an assembly of everything previous and above the assembly is a hierarchical assembly of assemblies. IGES can capture Assemblies more successfully as as *copious data* entities. Recently many other CAD/CAM systems have added support for *copious data* entities. Even with the User Application, there are still about 10 steps required to prepare IGES of Sheet metal. I managed to get this down to three by writing a series of macro programs and a screen menu which are called within Euclid. (Further study on IGES, Ref. [2] [3])

6 APPLYING THE NEW PROCESS.

As it happened, necessity dictated the requirement for very fast turn around of prototypes with no time to prepare drawings. This was the ideal opportunity to put the new streamlined process to the test in a live situation. With the *object* required for fabrication recalled in a Euclid session, the process was as follows;

A. Run EUCLID macros to generate IGES for each part.

B. The IGES is written over the network to a location where the machine shop can read it from.

C. Read the IGES into the SMP81 IGES preprocessor for verification.

D. Iterate A and C if necessary until IGES is perfectly representative of the part.

E. The designer informs the machine programmers that the IGES files are ready and verified. The time to get to this step varies from about 3 to 25 minutes.

F. The machine programmers track the IGES into an SMP81 model.

  1. The time to generate the SMP model will vary but averages about 15 minutes.

  2. A fully dimensioned flat pattern with a table of tools requires takes about 10 seconds to generate in SMP81. This can be plotted on a HP pen plotter.

  3. The programmer invokes a post processor for the machine that will make the part and generate a Numerical Control file. This part of the application enabled the programmer to :

    a. Assign tooling,

    b. Select and position the sheet stock,

    c. Arrange the flat parts on this sheet.

    d. Optimise the layout,

    e. Animate the punching sequence,

    f. Optimise the punching sequence

    g. Generate the NC file

G. The NC file is down loaded to the machine controller and when the machine is set up the punching operation commences.

H. The flat parts are bent up on a break press. The backstop settings are called from the flat pattern drawing. More modern presses can take this data from SMP81 directly.

Without really pushing, the best performance achieved was a 3 hour turn around for an assembly of six parts, ie. 3 hours from the decision to have them made to having parts in hand from a facility about 6 mile away. Essentially we had same day turn around capability when required. Inspection of prototypes at the machine shop is not necessary. The parts are sent back to the Design engineering labs for assembly and testing.

6.1 Lead time comparison

Generally speaking the benefits from this process are derived in modeling, prototyping and volume production startup situations and introducing Engineering Change Orders. For prototyping the savings are :

• The time and resources it takes to do a drawing, print it and send it to the shop.

• The time savings in generating the flat pattern are in the order of 10 to 1 over working out the development on a drawing board.

- To generate a flat on SMP81 from a drawing takes about two to three times as long as it does from IGES increasing with the complexity of the part.. This is true only for in house programmers who are highly skilled with CAD/CAM and have access to fast systems and networks.

Different metrics apply to the external supply base where we have also established favorable lead times.

6.2 Quality

Throughout the development effort, the high quality of parts inspired the use of IGES for live engineering prototypes. One reason for this is that IGES represents geometry to an accuracy of five places of decimals. Drawings show dimensions to two places of decimals at best and these are input manually by the programmer who may further round off or make an error.

There are other advantages to the production environment where by the drawing and the IGES must correlate. Ambiguities or misrepresentations can be identified early and usually clarified without referring back to design engineering.

7 Fast Turn around for PLASTIC components.

The process by which plastic components designed in EUCLID were prototyped was easier to implement than the sheet metal process. During the investigation we looked at several processes in the area of Solids to Models and Models to solids. The models to solids process are not mature and were not essential to our needs The SOLIDS to models space is becoming well established through Stereolithography (SLA).

7.1 Rapid prototyping through Stereolithography.

The process from 3D Systems is well known and widely documented at this stage. (Ref. [3]). Briefly a shape is recreated by layering cross sections of the part one on top of another. This is achieved by having a mirror guided laser scan each cross section onto the surface of a liquid resin. This path cures solid. The solid piece is resting on a table which lowers through the liquid and the next cross section is drawn on top of that.

IMPLEMENTATION

This is a data driven process where the input is a slice file from a solids modeling system. Matra Datavision the suppliers of the Euclid system supplied us with a user application which generated the Slice File. The debug time to get this up and running and delivering working parts was just a couple of hours. Once a suitable resin is selected and shrinkage is allowed for the process is set up.

BENEFITS.

This process provides a 2 day turn around on parts that would have otherwise taken 3 weeks to procure from a model shop. It also costs about half the alternative.

DOWNSIDE.

The disadvantages are that size is restricted and only relatively small parts can be processed, though parts can be welded together. Larger machines are becomming available

There is usually some finishing required on the part but this is best taken care of before the part cures. (For further study refer to [3][6])

7.2 Tool development for Plastic injection moulding.

The lead time for tool delivery for Plastic injection moulded parts is usually between 12 and 16 weeks for our requirements, That is with drawings and no CAD data and for single impression steel soft tools. As described earlier, Euclid already offered productivity benefits in the time it took to design the part and subsequently in the time to to produce a drawing. The plastic part is described as a Solids object within the Euclid database. It can be translated into other data formats such as IGES or mesh models for analysis.

IGES for mould tool Manufacture.

IGES does not support SOLIDS data types directly. However the Euclid IGES translator will output a wireframe polyhedral of the solid model. Some CAE systems can read this polyhedral and automatically generate a solid from it which can be used immediately for tool design. Wireframe CAD systems will read it as a wireframe of line entities only or else may not interpret anything from the IGES. (For further study, Ref. [5])

Surfaces entities.

Surface entities are essential to *Machining Applications* on any CAD/CAM system. On the wireframe based systems the IGES wireframe provides a structure on which to pin surfaces. Resurfacing the part is unnecessary when IGES is generated from EUCLID surface models. However, because each tool maker had a different CAD/CAM system and because of the low number of plastic parts in the product, the simplest and most reliable process was to send wireframe geometry to everyone. On a SOLIDS based CAE system, IGES with surfaces can readily be turned into a SOLID model. These would normally be turnkey systems with all of the machining and CNC milling applications embedded. This Solid model can be taken through all of the applications with no further transformation other than material shrinkage compensation.

Plastic flow analysis for mould design.

A CAE analysis tool called MoldFlow, profiles the plastic material flow within a plastics injection mould. It depicts graphically thermal and pressure isobars as well as flow vectors. We use it to help the tool designer position gates, venting and cooling channels to provide good fill and uniform cooling in the mould. This analysis is done in-house and the results are passed onto the tooling designer for reference only.

MoldfFow is supplied by MOLDFLOW INC and it supports most workstations and PC. It does not read IGES directly so the model is created interactively in 2D. We are evaluating a Surface model IGES interpreter for MoldFlow currently.

8 Supplier CAD linkage project

8.1 Objectives.

The objective was to reduce time to market for new product introduction to volume manufacturing through the development of CAD/CAM and rapid prototyping capability at external suppliers of sheet metal, world widefor the corporation. The success of the rapid prototyping effort suggested that this was a good time to extend the development to the external supply base for sheet metal. At this point no supplier had and CAD/CAM expertise or equipment.

8.2 Scope

The scope of the project included;

A. Sell the concept and the potential benefits of the technology to volume manufacturing and the supply base.

B. Resource allocation and training in volume manufacturing facilities.

C. Consultation on selecting hardware and software - suppliers choice.

D. A program of developing and testing prototype fabrication from IGES

E. Introduction of Electronic Data Interchange technology

F. Position resources for ongoing support within manufacturing. facilities

8.3 Achievements

- A total of seven suppliers participated in the project All of them invested a CAD/CAM system for sheet metal programming. The choice of software varied between 5 different packages. The hardware consisted of PC based and high perform ance workstations.

- A generic form of IGES from Euclid sheet metal was determined.

- All of the suppliers invested in modems for Electronic Data Interchange.

- All suppliers developed the skill sets to program from IGES geometry received electronically.

8.4 Benefits

- Large quantities of unique parts can be set up for volume production in what averages out at one third less time .

- Higher quality production parts result from programming from the original geometry with no round off errors.

- More complex geometries can be designed and manufactured.

- Ambiguities or illegibles in drawings or written specifications can be identified and resolved thus reducing the potential for scrap material.

9 CONCLUSION

As experienced users of CAD/CAM will know the investment required to achieve customised functionality often exceeds initial projections, as it did in this case. When the window of opportunity expanded to integrate the external supply base the opportunity had to be grasped while the momentum was established and the resources in place. The primary justification for the investment comes from the return to Engineering. The lead time savings buy additional design time and help the project maintain aggressive delivery schedules. Furthermore these benefits can apply to any future new product development effort though the investment is recouped in a single development cycle.

The benefits of this level of CAD/CAM/CAE integration with volume manufacturing occur at new product start up and design change introduction. Apart from the time savings in programming and set up, the confidence in the data and the availability of a second source to verify drawing specifications is a significant advantage.

Acknowledgments

This effort came about and concluded successfully through the diligent insight and support of Charles R. Barker, Pauline B Nist, Steve Holmes, Steve Noyes, Paul Curran, Eugene McCabe of Digital Equipment Corporation and Peter Hajjar, Tim Illingworth , Gina Jackson of Matra Datavision.

REFERENCES

[1] Zimmers E. P., Groover M. P.: *CAD/CAM Computer Aided Design & Manufacturing*. Prentice Hall, New Jersey 1984.

[2] Browne Gerald, Herbstritt Eric, Pietrow Glen: Making design transfer work: CADkey to Computervision and back using IGES., *Mechanical Engineering Systems*, Vol. 1,No 5, pp.35-44, March/April 1991.

[3] Leonard,LaVerne: Will rapid prototyping be part of your future, *Plastics Design Forum*, volume no:.16/No. 1, pp 15-22, 1991.

[4] Taraman, Dr. Khalil S. editor: Society of Manufacturing Engineers, CASA., *CAD/CAM Integration and innovation.*, Chap.4,5,6, 1985

[5] Vesey Joseph T.: The New Competitors : They Think in terms of 'Speed-to-market', *IEEE Engineering Management Review*, Vol. 19 No.4 pp. 12-18 Winter 1991

[6] Nutt Kent,: Automated Prototyping techniques: A market report., *Mechanical Engineering Systems*, Vol. 1,No 5 pp.45-47 March/April 1991.

[7]McKinnis Craig, Brusch Richard,: Convair goes Concurrent, *Computer Aided Engineering* Vol 10 No 2 pp. 18-27, February,1991.

INTEGRATION OF CAE/CAD/CAM SYSTEMS: A CONCURRENT ENGINEERING STUDY

Marcos G. D. Bortoli

Embraco S.A., Joinville, S.C., Brazil

ABSTRACT

This work describes the use of an integrated CAE/CAD/CAM system to design and manufacture the housing of a hermetic compressor for household refrigeration. This product has restrictions in terms of noise level and the housing shape has a significant role in this sense. The process to manufacture this component is stamping, in which the main characteristic is the high tool investment, even during the prototype phase. Other important characteristic is the long time needed to build the stamping tool. Therefore, it is highly undesired to redesign the component, after the stamping tool has been manufactured.

The main topics related to this project, are: surface modeling due to the complexity of the component shape; geometric restrictions considered for designing the component; use of the finite element method for engineering analysis; tool design based on the components created during the design phase; automatic generation of NC programs; use of DNC (Direct Numerical Control) to download programs to CNC machines. The product development cycle was made in a concurrent engineering environment. This kind of approach and the integrated CAE/CAD/CAM system allowed to reach the project stated goal within schedule.

INTRODUCTION

Today's rapidly changing technology combined with fast communication is shortening the overall product life cycle. To keep pace with technology, the development phase of new products needs to be completed faster than ever before.

Until recently, the industries used the traditional serial engineering cycle for developing their products. The development followed a linear path, with each step beginning only after the latest has been completed. Engineers worked in isolated departments, and the designs were thrown over the wall that divided the product design from the manufacture. Changes often required comming back to conceptual design and increase in cost as time went on.

In order to shortening the product development cycle, and to be more cost effective, the industries are to changing from the traditional serial engineering cycle to the concurrent engineering. In this cycle, suitable for the 1990s life style [01], the team approach all the aspects of product development simultaneously contrary to the traditional serial method. Most changes come in the early stages when they are easily an unexpensively made. Fewer prototypes are needed, and the ones that are built often require only fine-tunning. The end result: a product that takes less time to develop, has higher quality, and less cost since expensive changes and prototypes are virtually eliminated [02].

The objective of this paper is to present the main topics related to developing the design of a new housing for a hermetic refrigeration compressor. The product development cycle was made in a concurrent engineering environment, and based on an integrated CAE/CAD/CAM system. This kind of approach allowed to reach the project stated goals within schedule.

CHARACTERISTIC OF THE PROJECT

The main goal of the project was to improve the role of the housing in the reduction of the global noise level radiated by the compressor [03]. To be effective the changes in the housing should decrease the noise radiation in the range of 1600-3000 Hz. This means that we had to shift the first resonance frequency of the housing to beyond 3000 Hz. The chosen way to reach the objective, was to change the housing shape increasing its stiffness [04].

In changing the shape of the housing, there were some geometric restrictions that had to be obeyed. The main restrictions were related with the external dimensions of the housing. The compressors are used in refrigerators, freezers, etc, where the space for its location is already defined. In this manner, to avoid troubles to the manufacturers, the external dimensions of the compressor should be kept, nearly unchanged at least for its height, length and the distance A-A' (A-A' is the smallest distance between the top of compressor and refrigerator bottom, showed at figure 1). Since only the housing would be modified, the changes should also be consistent with the other components of the compressor. Mainly for the parts that are also standard for other existing models, such as the base plate and the support plate of the eletrical starting device. The figure 2 shows the old housing and its main components.

The manufacturing process of the housing is stamping. The stamping tool is complex in its characteristics: size, shape and dimensional tolerances. Stamping requires a high tool investiment, and long time is needed to build the tool, even the prototype one. The changes in the product, after the tool has been manufactured, have a big impact over the economic and cronological development of the project. Therefore, the use of a relibility engineering analysis tool, during the conception phase, would be fundamental for the sucess of the project.

DEVELOPMENT OF THE PROJECT

The methodology of concurrent engineering was used to develop the project. Several sectors of the company (Research and Development, Product Engineering, Sales, Process Engineering, Tool Shop, Industrial Design, Quality and Procurement) were involved since the beginning of the project. In this way, more suggestions were analysed during the conception phase of the product. This procedure decreased the number of changes in the product after the conception phase.

All the development of the project was based on an integrated CAE/CAD/CAM system. Below, are presented details of the use of this system in the different phases of the project. These phases were done simultaneously, and with high exchange of information.

Design of the product

The solid model is the fundamental link for the integration of the CAE/CAD/CAM systems. The model created at the design phase, represents the basic input for the development of the other activities of the system, as for example, engineering analysis using the finite element method, automatic coding to CNC machines, or to build a sample on resin by a rapid prototyping systems [05,06]. This integration increases the design efficiency for the model will be built only once. The solid model is also, an excellent communication vehicle for the project team in the environment of concurrent engineering, because it makes easy to visualize and to understand the model, since the preliminary phases of conception [07].

The increase of the design quality is another important aspect associated with the use of solid models. With solid modeling a higher consistency of the design interpretation is achieved, wich otherwise would generaly be only identified at the manufacturing stage. Since the database contains only the information of a specific solid model, the uniqueness of the information is guaranteed. There is also the advantage of having a better control on the design changes that have to be introduced on the original model. Clearance analysis and assembly checking with other components is another advantage of the solid modeling technique.

From the solid model, other useful information can be obtained for the project, such as volumes, areas, weight, moments of inertia, and center of gravity

Detailing the design is a slow phase of the project, and sometimes it may take more time than the solid modeling phase itself. It is estimated that the time for the detailing phase is reduced by 50% if the detailing is made based on a solid model then if based on a CAD bi-dimensional drawing. From the solid modeling technique it is also much easier to get perspectives and exploded views. With an integrated system the detailing phase may be unnecessary for some phases, such as finite element analysis and automatic programming for CNC machines.

The housing of a hermetic compressor can only be represented by a set of complex surfaces, i.e.,a surface that cannot be represented by primitives such as straight lines, cylinders, blocks, spheres, or cones [08]. All the models created and analysed were obtained by using the Bezier surfaces. The methodology used throughout the project, is described below. The first step, was to generate curves which had to conciliate the geometrical restrictions of the product, and also had to result in a favorable shape that produced an increase of the shell stiffness. The second step was to build the paths on which ocurred the transitions of the curves. And the last step was to build the surface which contained the curves and the paths. The major difficulty was to get a smooth and coherent transition between the several surfaces that form the shell. Several times the shading process was used to detect possible failures on the surface, which would not have been identified by the conventional representation by lines. The change on a surface, even small modifications, implied in rebuilding the whole surface, making use only of the curves and paths. It is important to mention that the surface that form the shell, may be converted in solids, if desired.

Engineering analysis of the product

The use of engineering analysis tools, based on eletronic prototypes, are the most adequate for product development. These tools are flexible, and of fast response, speeding up the development and otimization of the product. Therefore, more options can be analysed and tested, with less cost and reduced time, when compared with the experimental analysis with prototypes. These tools are very important when building actual prototypes is expensive and time demanding.

The engineering tool generally used to calculate numerically the resonance frequency and vibration modes of the structures is the finite element method (FEM). The basic concept of FEM is the fact that any structure can be represented as an assembly of discret parts, which are all connected to each other by a finite number of points, called nodes. The set of all the node coordinates, and how they are connected to form the elements, is called mesh. Besides the mesh, the method needs information about the material properties of the structure and displacement constraints (boundary conditions). The accuracy of the results will depend mainly of two factors. The first is related with the accuracy of the finite element model created to represent the actual structure. The second, is the quality of the element formulation used to calculate the solution.

Among all the information needed to build the finite element model, the most troublesome, in order to get a rapid solution, is the generation of the mesh. Luckly, the capacity of the CAD systems to generate meshes over the models created in the system, is continuously increasing. Therefore, the mesh can be created faster by the CAD system if compared to manual mesh generation, or, using the limited mesh generator of the finite element solvers. Others advantages associated with CAD/FEM integration are: avoidance of rebuilting the model by some other way, that would be a duplication of work; and, the uniqueness of the information is guaranteed for the projected and analysed model.

In this study, the finite element mesh was built on models created by using surfaces of the CAD system. Although the mesh generator avalaible in the system is able to make an automatic mesh on the surface, this approach was avoided because it would be restricted to the use of triangular elements only. To get a good solution using this

kind of elements, a great mesh density is needed, i.e., a great number
of elements. The consequence of using great mesh density is to
increase the time processing of the computer to find the solution. The
solution was then, to use quadrangular elements where they were
possible. The consequence was that the patchs of the surface had to be
arranged to allow the use of the mesh generator. It is only possible
to generate the mesh with quadrangular elements when the patch has at
least, four edges (vide figure 3). Therefore, the mesh was created
patch by patch, supplying the quantity of divisions on each edge. To
reduce the computer processing time, the housing symmetry was used.
The figure 4 shows a mesh of a finite element model to represent a
housing shape. After the meshing process was completed, the model data
was stored in a file, that was the input data to the finite element
solver.

To calculate the resonance frequencies and shape modes of the
model, a in-house developed finite element solver was used [09,10].
This solver was previously tested, and validated experimentally. The
solver checked the mesh created on the CAD, and fixed it when
necessary. The solver also automatically applied the boundary
conditions. The results were analysed on line on graphic video
terminal, where it was possible to see the animation of the shape modes
and getting the values of the resonance frequencies. After the
analysis, if the results were not adequate, it was necessary to go back
to the conception phase. This procedure was repeated successively
until reaching the desired value for the frequencies, which fulfill the
goals of the project.

The tool design

The compressor housing is formed by the body and the cover. These
components are stamped in a progressive tool. The manufacturing
process and components assembly began to be studied during the
conception phase. During this phase, also a good solution of the
product technical function, and its manufacturing process, was
searched. Some examples are the study about the shape of the
connecting border between the body and the cover, and the maintenance
of the stamping tool. Also the sequence of the stamping process began
to be studied in the conception phase too.

After the product was defined, but before the detailed design of
the product was ready, were began the modeling of the main components
of the tool, punches and dies using the CAD system. These components
were modeled with the same technique used to develop the product. It
is important to say that the housing model was not used in the tool
desing. Not even in the tool calibration stage. This happens because
the design of the dies and punches have to account for deformation and
contractions of the process. The model of the product was used only as
a reference to be followed for the the tool models. Here it is
necessary to mention the importance of exchanging interdepartamental
experiences for the development of the models on the CAD system, to
increase the productivity.

Manufacturing of the stamping tool

The software to generate automatically the codes to define the tool path to CNC machines is based on surfaces. This software is able to simulate, on the screen of the workstation, the tool motion in near real time to further check the program. Simulation is especially important in complex machining because it helps programmers visualize the part a tool path. The advantage of using the automatic programming is emphasized when the shape of the piece is complex, and to write the program using the usual form depends on the skill of the programmer. Another advantage is to avoid manual adjustement to the tool, guaranteeing its shape when substitute pieces are needed. The codes automatically generated also assure a greater coherence between the design and real tool.

The dies and the punches of the housing stamping tool were manufactured by using CNC machines. The codes to the machines were by using software described before. It used the surfaces created during the design phase of the tools. The greatest difficult found was that sometimes, the surfaces were inadequate for making the path tool, what implied in rebuilding the surfaces. Because of the deficiency of the postprocessors, it was necessary to edit the programs to introduce information about comments and technological information. The use of the CAM reduced 2/3 of the programming time, when compared to the traditional way.

The programs were dowloaded to CNC machines by DNC (Direct Numerical Control), which are all connected on the same net. There is also a safe portable equipament, that is able to download the programs directly to CNC machine. It was used when there were any problems with the net.

Manufacturing of the compressor housing and practical results

In manufacturing the body, it was necessary to make two adjustments on the tool, to get a good body in terms of the dimensional tolerances. During the first test of the tool, there were problems in the second stamping stage, with excessive lateral deformation of the body. After making some changes in the tool, the deformation decreased, but some remained. Although the important dimensions, like lenght, height and width were within the specified tolerances. The problem was solved after the inclusion of a calibration die at the fifth stage.

The cover tool worked very well since the first test, and it was not necessary any adjustment.

A experimental modal analysis was made to evaluate the resonance frequency and the mode shapes of the new housing [11]. The experimental results agreed well with the numeric values predicted. The increase of the first resonance frequency was about 80% within the predicted value.

FINAL COMMENTS

The advantages that the integrated CAE/CAD/CAM system within a concurrent engineering environment offered for the development of the new compressor housing project, are shown in three aspects: quality, time and economical benifit.

The improvement of quality is directly related with an increase of the quantity of options and suggestions that were analysed to optimize the product. This optimization considered the aspects of technical function, manufacturing, and the assembly of the product. The integration of the system guaranteed a consistent interchange of information among all the departaments involved in the project. For example, the finite element analysis was done using the models created on the CAD, and the same happen when making the code to CNC machines.

To analyse better the time spent, the project should be divided in two parts. The first is related with the time that was reduced using a integrated CAE/CAD/CAM system. The best benefit was avoiding to make the work twice, and using the system to antecipate the beginnig of the tasks as soon as possible. For example, to create a mesh to finite element model, due the CAD/FEM integration, it was not necessary to wait for the detailed design. Imagine the time nedeed to make the detailed design of about 50 models that were analysed, and afterwards, to built the mesh using other finite element pre-processor. The same comments can be made about the CAD/CAM integration, where the NC codes were made over the models created in the tool design phase. The other face shown, was the reduction of the time using concurrent engineering. With all the departaments involved since the conception phase of the product, the changes on the product, after it has been defined at first time, decreased.

This method of development of a product reduced the cost, because it avoided great changes on the product, after it has been defined. It reduced costs of re-designing, changes in the tool, and prototypes manufacturing. The other economical aspect is due the overall decrease of the development cycle time. The product can be launched earlier in the market, keeping or increasing the market share.

ACKNOWLEDGMENT

We would like to thank all the personnel in this interdisciplinary project team that developed this work, wich is a real example of concurrent engineering.

REFERENCES

1. Mills R, Beckert B, Carrabine L: The future of product development. Computer-Aided Engineering, October 1991, p 38-46.

2. Kempfer L: CAD/CAM Management. Computer-Aided Engineering, December 1991, p 56.

3. Roys B, Soedel W: On the acoustics of small high-speed compressors: a review and discussion. Noise Control Engineering Journal, volume no.:32 1989, pp 25-29.

4. Lowery D C: An improved shape for hermetic compressor housing. International Compressor Engineering Conference at Purdue, 1984, pp 285-290.

5. Asheley S: Rapid prototyping systems. Mechanical Engineering, April 1991, pp 34-43.

6. Lindsay K F: Rapid prototyping shapes up as low-cost modeling alternative. Modern Plastics International, November 1990, pp 76-78.

7. Mills R, Beckert B: Software Tools. Computer Aided Engineering, October 1991, p 58.

8. Potter C: Nurbs vs. Bezier. Computer Graphics World, October 1990, pp 77-83.

9. Bortoli M G D, Driessen J L, Barbieri R: Elementos finitos: desenvolvimento e uso como ferramenta de projeto. 4º Simpósio Sobre CAE/CAD/CAM, São Paulo.

10. Bortoli M G D, Barbieri R: Desenvolvimento do método de elementos finitos na indústria. II Congresso de Engenharia Mecânica Norte-Nordeste, July 1992.

11. Sangói R, Ozelame A E: Análise dinâmica experimental da nova carcaça F. Internal Report Engineering, January 1992.

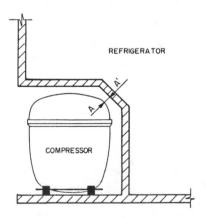

FIG. 1   Distance A-A'

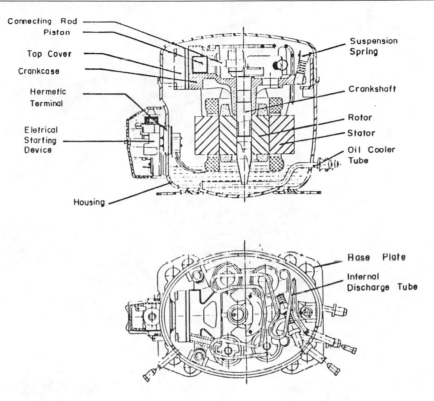

FIG. 2 The main components of the compressor

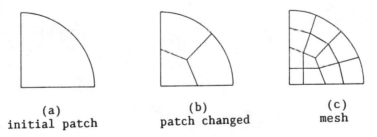

(a)
initial patch

(b)
patch changed

(c)
mesh

FIG. 3 Generation of quadrangular elements

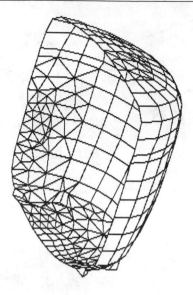

FIG. 4   A mesh of a finite element model

THE DEREGULATION OF NUMERICALLY CONTROLLED MACHINE TOOLS

Frederick J. Robinson and Mark A. Fugelso

Department of Industrial Engineering, University of Minnesota,
Duluth, Minnesota, U.S.A.

ABSTRACT

From the early days of numerically controlled machines, which were programmed from punched cards or paper tape, industry has progressed to the computer driven controller (CNC) which makes use of advances in microelectronics. During the past 20 years the computing power available in machine tool controllers has steadily increased but at the present time the capabilities of these controllers is considerably exceeded by relatively 'low cost' personal computers. Whereas modern CNC equipment can frequently receive data from an external source such as a PC, the machine operation is in fact directly controlled from the system supplied with the machine tool which generally operates from its own limited memory. In this paper it is argued that such arrangements constitute an unreasonable bottleneck and that it is now time to redesign machine tool controllers enabling them to interface directly with modern external computers from which all program commands would be derived. In this way the rapid advances being made in computer technology could be incorporated into machine tool control as they become available and at a relatively low cost and future advances such as the possibility for on line CAM or changes in the application of artificial intelligence (AI) techniques could be readily accommodated. Work in this area which is under study at UMD is discussed, together with suggestions concerning the design of the controller and interface unit that would be necessary for the system proposed.

BACKGROUND

Early numerically controlled (NC) machine tools introduced in the 1950s were based on World War II servomechanism technology, linked to the recently developed digital computer. Servomechanisms were employed to drive the slideways of conventionally designed machine tools and the accuracy achieved depended to a large extent on the type of feedback generator fitted. There was much variation in the design of these servos and special classifications, such as 'point-to-point', 'continuous path', 'absolute' and 'incremental positioning', became familiar terms to those involved with such equipment.

Communication with these early machines was achieved by using modified versions of existing telecommunications technology, both in terms of hardware and 'software'. Punched paper tape, with its associated tape reading and preparation equipment, was employed and formats such as the E.I.A. 8-hole tape code were adopted. In the early days a range of programming methods was tried which has reduced over the years to the generally accepted 'word address' system now widely in use.

This equipment gave good accuracy and excellent repeatability but it was extremely expensive. Whereas they were not dependent on the skill of a machinist to achieve good results, these machines were not that easy to program and the use of a separate computer, running a high level language such as APT, was frequently necessary for more complicated components. In addition, much of this early equipment developed a poor reputation for reliability, possibly the result of using vacuum tube and relay electronics in the unfriendly workshop environment.

As a result of the above mentioned problems, NC equipment took a long time to become accepted in many shops and those machines that were installed were generally owned by the larger specialist firms in shipbuilding and aerospace. In such companies computing facilities already existed and the special requirements of their products outweighed the problems associated with early NC machine tools.

By the late 1960s many of the problems referred to above had been somewhat alleviated. Reliability had increased considerably by the use of solid state electronics components and there had been a great deal of standardization in programming formats. In addition, servo design had settled down and some of the classifications referred to earlier ceased to have major significance. Punched paper or mylar tape became the preferred programming medium and word address format, with the familiar "G" and "M" codes, had become widely accepted. These machines still differed from each other considerably however, and following the production of a 'general' cutter path for a component profile, using a language like APT, it was necessary to generate the program using a 'post processor' specific to the machine tool to be used.

NC machines were still rather expensive and parts having complicated curved profiles had to be programmed using a separate digital computer. Such equipment was still rare at the end of the 1960s in a general machine shop and although these early machines indicated the promise of efficient production for small batches and a simplification of jigs and fixtures, such advantages had not been generally realized in industry.

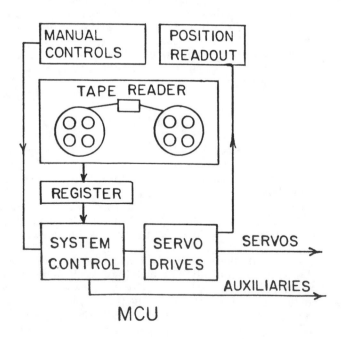

FIG. 1 *Block diagram of the machine control unit for an early tape controlled machine tool. In these systems the tape itself served as the memory but with limited access by the operator.*

FIG. 1 provides a block diagram of a machine control unit (MCU) for one of the early NC machine tools. In such systems the tape program may be regarded as a special kind of 'read only memory' (ROM) with limited addressing capability and the short term memory or 'register' could hold only one or two information blocks at a time. The MCU was normally located inside a separate cabinet which also contained the electrical power supplies and servo amplifiers, in addition to the tape reading and system management electronics. Manual override was usually provided for at the console of the MCU and position readout was achieved by displaying the signals from the position feedback generators. The tape program, or memory, could not be accessed by the operator except by sight reading the tape or running it through a teletype machine. Also, it was not possible to alter the instructions coded on the tape in order to accommodate changes in tool dimensions or a different machine tool.

THE ADVENT OF CNC

Arrival of the microprocessor and the memory chip toward the end of the 1960s made possible a major advance in the design of machine controllers for NC equipment. Relatively large and accessible memories could now be incorporated into the MCU, together with a microprocessor capable of carrying out calculations at high speed. MCU's now effectively became dedicated special purpose computers designed to operate the servos and spindles on machine tools. In addition to the control capability, these new MCU's could not only store large programs, such instructions could be modified to suit a change of cutter dimensions, for example. Control panels for the MCUs incorporated full keyboards with alphanumeric input capability and a visual display unit (VDU) output which enabled the operator to select data at will from memory or position feedback. These controllers became known as computer numerical control (CNC) units and in fact they are what we still have today.

CNC controllers operate by reading instructions from memory into which a program has been entered, either from the console keyboard, from a tape reader or from an external microcomputer. Program instructions may be viewed and edited by the operator as required, generally without interfering with the control of the machine tool. Also, specific program instructions, such as position commands, may be selectively changed as they are read to accommodate, for example, a cutter of slightly differing dimensions from that for which the original profile had been defined. This latter feature has now become refined into the system of stored cutter offset values which one normally finds on a typical CNC machining center. Program profiles may now be defined as the required finish machined shape and the necessary calculations to compute the constant difference curve required to achieve this objective for a given cutter are generated 'on the fly'.

This above mentioned feature has also provided an important simplification for machines equipped with tool changing capability. Prior to controllers having the ability to compute cutter compensation, machines that changed cutters had to be provided with standardized tooling which needed accurate calibration prior to loading into the magazine. This requirement frequently necessitated a separate workstation almost as complicated and expensive as a machine tool.

A further and most important advance that has resulted from the introduction of computing capability into the MCU relates to the location of workpieces on the machine tool, prior to the actual machining operation itself. As many will recall, locating the workpiece accurately in relation to the reference frame of an NC machine tool was essential and inaccurate location was a potential source of error in the finished part. The introduction of touch probes during the last 10 years has greatly facilitated such operations whereby the local reference frame of the workpiece blank is established which in turn enables the necessary rotation and translation calculations to be made by the controller, relating the part to the machine's reference frame.

In addition to the above features, CNC controllers incorporate various 'potted' routines, such as circular interpolation, in fast ROM 'firmware' which provides for extremely close tolerances to be held at reasonable feedrates. Finally, one should not ignore the advance made possible in positioning accuracy by calibration and software correction. This technique enables the calibrated accuracy of a system to be programmed into the controller itself and also affords the possibility of re-calibration to accommodate wear during the machine's life.

Clearly CNC has resulted in a number of important advances which have made NC equipment much more reliable and also more 'user friendly'. These changes have certainly resulted in much wider use of NC machinery which is now commonly found in most machine shops. However, the initial capital cost of a CNC machine tool is still large and represents a major investment for a small company.

The controller represents about 40% of the CNC machine tool cost but as a computer the MCU has a somewhat limited capability. Market needs have established the relatively limited range of features available on modern MCU's with the result that less frequently produced parts, such as those having contoured profiles, are less well provided for. Memory capacity in CNC systems is generally more than adequate for the majority of machining programs likely to be needed but when the longer programs associated with accurate contouring operations are required, such capacity is shown to be hopelessly inadequate. On computer workstations we now expect 'megabytes of memory', for about $40 per Mbyte; moreover, the larger memories now available on general purpose micro and mini computers [1] could not be accessed easily by the microprocessors normally supplied with a CNC MCU.

Whereas the machine tool itself is likely to have a useful working life between 10 and 30 years, depending on the application, the MCU becomes obsolete far more rapidly. Recent developments in microcomputers clearly show that obsolescence is reached in months rather than years in the micro-electronics field, with the result that we are generally left with a perfectly good machine tool linked to a hopelessly outdated special purpose computer.

If we consider the cost of the MCU we arrive at a similar result. At the time of purchase the MCU represents a major part of the cost; within months that value has depreciated because the 'computer' rapidly becomes obsolete. Retrofitting the machine with a new dedicated system is, however, no simple task; neither is it a low cost operation. Since the market for machine controllers is nowhere near the size of the general computer market, the cost benefits resulting from high production rates are not realized in machine tool control.

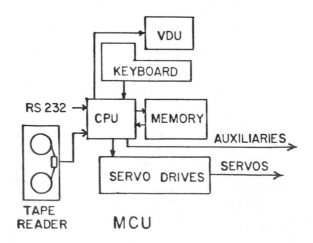

FIG. 2 *Block diagram of a CNC controller for a machine tool. Input is by means of a keyboard and a VDU is provided for visual monitoring. Programs are held in the memory and the processing unit is capable of modifying program instructions as they are supplied to the real time servo controller.*

It is clear that CNC MCUs will always lag behind general purpose computers in both performance and cost reductions. Moreover, this situation is likely to become more marked in the future. FIG. 2 shows a block diagram for a typical CNC controller. In this system a reasonable memory is provided, together with a processing center for calculations and a keyboard and VDU as input and output elements. Generally these controllers incorporate a conventional tape reader which is arranged to feed directly into the MCU memory.

THE DEREGULATED CNC CONTROLLER

Whereas the introduction of computing power to the MCU, resulting in CNC rather than tape controlled NC, represented a major step forward, this step introduced a bottleneck in the control of machine tools. The constraint is that the speed of operation of a CNC controller is limited by the power of the central processing unit (CPU) fitted. Similarly, the memory built into the controller will be limited by the CPU power provided. Whereas this computing power might well be adequate or even advanced at the time of installation, recent events have clearly demonstrated that the rapid developments in computer and communications technology will soon make any MCU dated, long before the machine tool itself has become obsolete. As already mentioned, the computing capability, in terms of speed and memory, associated with CNC controllers is easily surpassed by the capability of a modern PC, costing a fraction of the price paid for the special purpose unit. It is also apparent that this situation is in fact accelerating and the gap between the power and cost of a general purpose PC compared to a special purpose unit built to control a machine tool might well become greater in the future.

There are also technical reasons why we need to upgrade the capacity of CNC machine tool controllers. Whereas a wide range of typical machining operations can be carried out on current machines, contouring operations having long programs and many cutter paths rapidly fill up the available memory in a CNC controller. The authors discovered the truth of this latter point during some 3-axis contouring carried out in 1989 and 1990 [2,3]. In the course of this study components having compound curved surfaces were 'reverse engineered' on to a CAD database using a computer controlled coordinate measuring machine (CMM). Following any required changes performed at the CAD station, NC cutter paths were generated using CAM software, to enable the parts to be contoured on a CNC vertical machining center. Transferring the computer generated cutter path programs from the CAD workstation to the machine tool controller was carried out using a standard PC and the relative memory size between the PC and the CNC controller was all too apparent. In fact the CNC memory was completely filled by a series of cutter path programs required to machine a relatively small (3 in. cube) workpiece to a low degree of surface finish.

Other workers have identified similar limits to current CNC controller capability [4,5] and their views appear to coincide with the criticisms listed in this paper. The solutions proposed, however, are to write down all the foreseeable requirements for machine tool controllers into the future and from this advanced specification, attempt to design a new class of CNC controllers. Whereas a standard approach of this type will no doubt work, it will result in another design for a dedicated computer and one which we believe is likely to be extremely expensive. Furthermore, since this solution does not provide for simple retrofitting, any more than on current machines, CNC controllers of the new design would also become obsolete in time. It should also be remembered that such a solution would still be aimed at the relatively small machine tool control market, compared to the general purpose computer market, and would therefore be unlikely to provide similar cost benefits resulting from large batch production.

In the authors' view, CNC controllers should be split so that only circuitry and hardware associated with operation of the servos and spindles on the machine tool remain dedicated to that specific unit. These elements, forming a simplified MCU, should be fitted with a standardized interface unit, which would enable the MCU to be coupled to any general purpose computer workstation. The simplified controller, which is illustrated in

FIG. 3, would also include all necessary power supplies, manual controls and a position readout device. Furthermore, the servos of such a system should incorporate some degree of microprocessor control so that position feedback calibration could be incorporated. Such corrections could be programmed into erasable/programmable read only memory (EPROM) chips following calibration and fitted directly into the feedback generator circuitry.

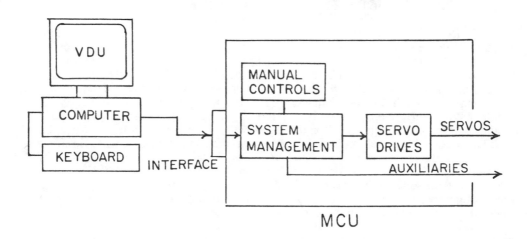

FIG. 3 *Schematic diagram showing a deregulated CNC system having a simplified MCU, which is coupled to a general purpose command computer by way of a standardized interface unit.*

It is also suggested that all other functions currently available on CNC machine controllers such as command programs, interpolation, special routines and workpiece location, should be handled by the separate general purpose computer. On this computer would be installed the necessary specialized software which would be coupled to the MCU by way of the interface unit, enabling the command computer to be changed as improved versions become available. The specialized software, capable of handling all current and possible future requirements, could be developed for such an application and this arrangement could also provide for compatibility with a range of programming languages.

In this way it could be ensured that CNC machine tools would always be controlled by the latest computing equipment. The power and cost benefits, likely to arise from advances in computer technology, would thereby become readily available to machine tool operators. In addition, advances in machine tool control, such as the introduction of 'on line CAM' and the incorporation of 'AI', both of which will require more powerful processors and greatly increased memories, could be accommodated into the field of CNC relatively easily and at minimal cost.

Retrofitting such a system, which at present is both complicated and costly as previously mentioned, would reduce to either changing software or increasing the power of the command computer used to control the MCU.

This proposal resembles in many ways the method by which computer controlled CMM units are operated. These machines are currently supplied with servo driven axes powered by dedicated electronics, to which the manufacturer adds a proprietary general purpose computer, on which is installed specialized software for the tasks to be undertaken. In such 'islands of automation', communications with systems outside the island are possible but generally require special translating software. Furthermore, at the present time, it is not a simple matter to change the type of computer supplied with the machine.

At UMD the authors are currently engaged in a project which involves the replacement of a proprietary MCU, fitted to a CNC lathe, with a general purpose PC and writing the necessary specialized software routines to enable operation of the lathe. From this work it is hoped to establish data on the relative signal traffic requirements for real time servo control, auxilliary functions and general programming commands, on a real system under operating conditions. In addition, Fugelso [6] has developed and successfully operated a CNC drill sharpening machine which follows the general principles of the deregulated controller outlined in this paper. On this machine, axis control is effected by servo motors driven by off-the-shelf electronics components and the system is commanded from a '286 PC'. This computer is equipped with the necessary specialized software, written in the 'C' language, which enables the machine to grind drills according to the operator's requirements. Attention has been given to rendering the system as 'user friendly' as possible; the operator simply responds to a menu driven program once the software has been engaged.

DISCUSSION

Clearly the design of the interface unit is critical to satisfactory deregulation, both in terms of its physical and electronics features. It is the authors' proposition that the controller supplied with the machine tool should be limited to the servo drive elements and power supplies as far as possible. Whereas these drive elements might well incorporate some microelectronics rendering them 'intelligent' and perhaps incorporating position calibration corrections in an EPROM element in the encoder that could be changed as the machine wears, as much of the programming and control as possible should be achieved by means of software installed on a general computer.

The ideal deregulated controller would have every hardware component such as feedback encoders and resolvers, servo amplifiers, coolant on/off relays, spindle speed controller, CNC computer and workpiece position measuring probe connected to a local area network (LAN) that could transmit and receive messages virtually instantaneously, or in less than 1 msec, in a standardized format. In such a case, 'retrofitting' could be achieved simply by replacing those elements that needed updating, the remaining system elements functioning as before. Development tools could be provided to adapt and alter the software interfacing the newly installed system elements, which is the technique commonly employed for microcomputer peripherals, such as printers, mice and monitors, at the present time. With standardized communications protocols, the hardware becomes a set of separate independent units, each one by itself relatively easy to cope with.

At present LAN technology cannot provide a means to transmit data fast enough to permit the closing of a servo loop over a network that also carries other, frequently quite long, messages; feedback signals from encoders cannot be ignored for more than 3 or 4 msec. One possible technique to deal with this problem is suggested in the schematic diagram illustrated in FIG. 4. In this suggestion it is proposed that real time operations, such as the feedback and feedforward messages controlling the servos, are provided with a special LAN which would be reserved for this traffic only. A second LAN, possibly similar to the first, would also be provided in parallel to carry other traffic. System elements would be coupled to the 'double LAN' according to their function.

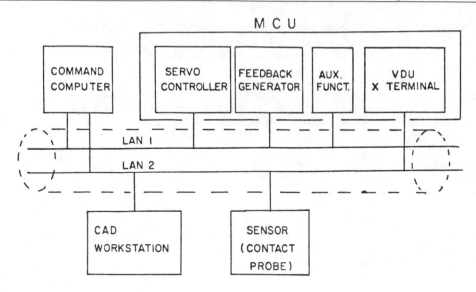

FIG. 4 *Schematic of the suggested method of handling communications in a deregulated machine tool controller using a double LAN. One network is reserved for real time servo operations and a second network is provided for all other traffic. System elements are coupled to these LANs according to their function in the controller.*

Another technique would be to use dedicated circuitry to close the machine servo loops as in the past but to arrange for the servo controllers to communicate with each other and the command computer over a single standard LAN. In this case it is still considered likely that communication delays of more than 10 or 20 msec. might result, which would be difficult to cope with. For example, the transmission of long 'paper tape images' could not be permitted to occupy the LAN while the machine slides were in motion.

In a deregulated controller, eliminating the use of proprietary microcomputer busses will be most important because no component that is incompatible with these busses can be introduced into the system. Each bus has to have its own computer board and its own memory, which means that each MCU has to have its own computer system. Each component cannot be located remotely from the bus because it will be connected by a multi-wired cable to its proprietary interface card plugged into the bus. Such arrangements are the opposite of the position being presented in this paper.

On a capable LAN, one CNC computer might be able to control several machine tools, depending on the complexity of the command programs being implemented. However, multi-axis machines might require the undivided attention of the command computer, whereas a relatively simple drilling machine performing point-to-point positioning operations, almost certainly would not. In a manufacturing cell having several machine tools controlled by one CNC computer, the MCU display for each machine could be a window on an 'X' terminal.

One of the issues that has tended to keep hardware in machine tool control systems 'tightly bundled' together with complicated circuitry, and one for which the authors offer no suggested solutions at the time of writing, arises from the way in which alarm and error signals are dealt with. Generally error signals are transmitted through hardware pathways to make certain that 'runaway' situations do not occur. This important failsafe feature will have to be addressed in a deregulated machine tool controller so that the revised system offers protection which is the same as or better than current equipment.

CONCLUSIONS

In order to take advantage of technical advances in computer technology, CNC machine tool controllers should be redesigned so that only those system elements concerned directly with the operation of the machine tool remain dedicated. Such a revised MCU should be provided with a standardized interface, so that modern powerful general purpose computers could be coupled to this interface to provide the necessary command function.

Features such as programming, changes of programming, cutter offset, special cutter path routines and workpiece location programs should be handled by specialized software installed on the command computer. This software is specialized in the sense that it relates to machine tool operations but it should be written in a high level language, such as 'C' and would therefore be portable.

In this way retrofitting machine tools as their MCUs become outdated would become a relatively simple procedure involving the changing of a command computer or the software installed on it. This technique would also enable advantages to be gained by the owner due to likely cost reductions resulting from the rapid development of general purpose computers. Furthermore, increased requirements, with respect to computing power and memory capacity, which are likely to be required with the introduction of 'on line CAM' and 'AI' could be readily accommodated.

REFERENCES

1. Puttré, M.: Engineering Workstations: Top Performance Makes for Tough Choices, *Mechanical Engineering,* A.S.M.E., November 1991.
2. Robinson, F.J. and Fugelso, M.A.: Automated Manufacturing Cell for Prototypes, F.A.I.M. Conference, Limerick, Ireland, March 1991.
3. Robinson, F.J. and Fugelso, M.A.: Computer Controlled Pattern Production, Manufacturing International Conference, Dallas, Texas, April 1992.
4. Cincinnati Milacron: Needs Analysis - Intelligent Machining Workstation Initiative, Report No. F33615-86-C-5038, Cincinnati, Ohio, May 1987.
5. Greenfield, I. and Wright, P.K.: A Generic User-Level Specification for Open-System Machine Controllers, ASME Winter Annual Meeting, San Francisco, California, December 1989.
6. Fugelso, M.A. and Vangsness, J.D.: A Computer Controlled Contouring Twist Drill Grinding Machine, Submitted to *International Journal of Machine Tools and Manufacturing,* Pergamon, UK, January 1992.

ACKNOWLEDGEMENT

The authors wish to thank Mrs. Jean Laundergan for her assistance in the preparation of this paper.

A FITTING TECHNIQUE FOR THE ESTIMATION OF FORM ERRORS

Kung C. Wu, Jason G. Song, and Thomas J. McLean

Department of Mechanical and Industrial Engineering, University of Texas, El Paso, Texas, U.S.A.

ABSTRACT

Evaluation of form errors, such as the straightness, flatness and cylindricity, etc., of workpieces is a primary application of the Coordinate Measuring Machines(CMMs) for precision manufacturing and automated production. Among the various techniques available for estimating the form errors from measured points, least-squares fit and min-max procedure are the most commonly used. The least-squares fit has a computational advantage over the other techniques but provides the least estimation confidence level. This paper presents an iterative procedures called the boundary least-squares fit. Experimental results have shown that the measuring confidence level of the boundary least-squares fit is significantly higher than that of the least-squares. It also provides a computational advantage comparable to the least-squares method. The boundary least-squares method is a significant advancement in estimation theory for form errors. It provides the manufacturing engineers an unique method for evaluating form deviations regardless of the manufacturing process producing the part.

The conventional least-squares method fits the mean surface from all the sample points. A form error is estimated by finding the maximum distance between the surfaces passing through each sample point and parallel to the mean surface. With the boundary least-squares method, the mean surface is used as a reference to select a much smaller set of "boundary" points from the sample data. A new surface is fitted from the boundary points. Iterations of boundary point selections and least-squares fit continue until the mean surfaces converge or a predefined number of iterations are completed. Experimental data reveals that the difference between the minimum zone form errors and that estimated by the boundary least-squares method have decreased from approximately 11.6% to 1.1% after three iterations. The boundary point selection strategies and other factors affecting the efficiency of the algorithm have been studied and are presented in this paper.

INTRODUCTION

Automation and high precision are the trends in modern manufacturing. Computer integrated manufacturing processes require in-line flexible inspection systems for verifying the quality, in terms of form errors, of the manufacturing process. Conventional form error is inspected by functional gauges or evaluated by fitting standard templates on measured surfaces. Accurate measurement of the form deviation is difficult using the hard gauges, especially for three dimensional forms. High precision processes demand accurate measurement systems for part inspection. Zero-defect production can only be achieved by active quality control system that utilizes fully automated in-line inspection system and numerical controllers for real-time error correction. Coordinate Measuring Machines (CMMs) are the ideal equipment designed to implement such inspection systems.

With the advent of CMMs, sample points of the surfaces, called features or forms, of interest are measured with a high degree of flexibility, speed, and accuracy. Form deviations are derived from the sample data using statistical or numerical procedures. One of the major deficiencies of CMM technology is the lack of standard algorithms for evaluating form errors (straightness, flatness, cylindricity, etc.). Porta and Waedele [1] concluded from their study that substitute geometries generated for given set of coordinate data vary among CMMs. These differences are directly attributed to the software implementation of data reduction algorithms. More research is necessary to characterize substitute geometry algorithms. Many algorithms for evaluating form deviations are now available. These algorithms can be grouped into two categories: the estimation procedures and the optimization procedures. Example of the estimation procedures is the least squares method. Min-max and complex hull are the examples of the optimization procedures. In general, the estimation procedures enjoy computation simplicities but suffer from uncertainties due to the estimation nature. The optimization procedures, on the other hand, provide exact form errors but require massive amount of computation time. A new technique that possesses the characteristics of both procedures without their shortcomings is thus needed for evaluating form errors.

This paper describes an iterative procedure called the boundary least-squares method. Experimental results have shown that measuring confidence level of the boundary least-squares method not only outperforms the conventional least-squares fit by as much as 40%, but also provides a computation advantage comparable to the least square method.

MEAN TRUE-LINE DEVIATION AND LEAST-SQUARES DEVIATION

In the early 40s, British standard (B.S. 863-1939) defined the form deviation as the distance between the highest and lowest points measured from the mean true-line. The evaluation method that could be used to generate mean true-lines from sample data, however, was not specified. In 1968, Scarr [2] proposed the use of least-squares error as the method to estimate mean true-lines. The ANSI (American National Standard Institute) standard B89.3.1-1972 recognized the least-squares as acceptable for determining circularity. Since then the least-squares method has become the standard procedure for evaluating form deviations in manufacturing industries. Statistically, least-squares method represents a point estimation procedure in which the m real parameters of a function, denoted by $f(x_j, j=1,2,3,...,m)$ or simply $f(x_m)$, are estimated from the observation data $y(i)$, $i=1,2,3,...,n$. The observation data represents a degraded version of the function by an additive noise. In vector form, the observation data is represented by the equation,

$$y_k(i) = f_k(x_m,i) + v_k(i) \quad i=1,2,3,...,n \tag{1}$$

where

$y_k(i)$ is a k-dimension vector represent the k components of the observation data,
x_m is a m-dimension vector represent the m parameters of the real function f,
$v(i)$ is the additive noise. Notice that both the function f and the noise v have the same dimension as the observation data y.

If the additive noise is a random noise of known distribution, the conditional probability density function of the observation data $p[y_k(i), i=1,2,3,...,n \mid x_m]$ can then be determined. The conditional probability density is also known as the "likelihood" function $l(x_m)$ of the estimation problem. The likelihood function is written as

$$l(x_m) \equiv p[y_k(i), i=1,2,3,...,n \mid x_m] \tag{2}$$

For example, if the additive noise $v_k(i)$ is a white noise as the gaussian noise, the likelihood function can be written as,

$$l(x_m) \equiv \prod_{i=1}^{n} p(y_k(i) \mid x_m)$$
$$= C(\sigma^2) \exp[-\sum_{i=1}^{n} (y_k(i) - f_k(x_m))'(y_k(i) - f_k(x_m))/\sigma^2] \tag{3}$$

where

$C(\sigma^2)$ is the normalization constant independent of x_m,
σ^2 is the variance of the zero-mean gaussian noise, i.e., $v_k(i) = N(0,\sigma^2 I)$.

The maximum likelihood estimates of the parameters as defined in Eq. 2 are the set of values that makes the observation data most likely to occur. In other words, considering the likelihood function as a real function of m parameters, the best estimates are the values of x_m so that the likelihood function is maximized. For additive white gaussian noise, i.e., Eq. 3, maximizing $l(x_m)$ lead to minimizing the quadratic function,

$$\min \sum_{i=1}^{n} (y_k(i) - f_k(x_m))'(y_k(i) - f_k(x_m)) \tag{4}$$

Eq. 4 is well known as the least-squares criteria. It can be shown [3,4,5] that the least-square estimation of parameters degraded by additive white noise, is unbiased and the variance of the estimation is of the form

$$\sigma^2_x = K(\sigma^2) / n \tag{5}$$

where $K(\sigma^2)$ is a real function and n is the number of sample points.

Two important observations of the least-squares estimation:

(a)   if the feature of interest, i.e., the form of the workpiece, is degraded by additive white gaussian noise, least-square fit estimates the true feature unbiasedly from the observation data. Furthermore, estimated feature converges to the true feature, i.e., the mean true-line, if the sample size n approaches infinity, i.e., $\sigma_x \to 0$ as $n \to \infty$. Form deviation evaluated from the least-squares reference thus converges to the mean true-line deviation. This relation is shown graphically in Fig. 1.

(b)   If the noise degrading the observation data is not an additive white noise, the convergence property of the least-squares deviation to the mean true-line deviation is not guaranteed. The statistical meaning of the least-squares fit as discussed above is no longer valid.   The least-squares fit in this case is simply a mathematical curve-fitting tool. The form errors derived from this reference deviate from the mean true-line form error by an unknown distance. This situation is also depicted in Fig. 1.

MINIMUM ZONE DEVIATION

In 1982, another definition of form deviation, the minimum zone deviation was recommended by American National Standards Institute (ANSI Y14.5M). Instead of using a reference line as the mean true-line, form deviation is defined as the normal distance between the maximum inscribing and the minimum circumscribing features that bound the sample points of the surface of interest. The minimum zone definition leads naturally to the optimization procedures for evaluating the form deviation. There are many optimization procedures available today for determining the minimum zone deviations. Examples are the min-max, convex-hull, simplex searching, and method of polynomials [6,7].   There are three problems common to all of the optimization procedures:

(a)   the optimization procedures stop the search for the true minimum value where a local minimum is encountered. This problem can only be solved if a global search is employed.

(b)   the sample size of the feature of interest must be large to cover the whole profile of the workpiece, i.e., analytical equation describing the feature can be derived from the discrete sample points. Fig. 2 depicts the relation of the "minimum zone" deviation determined by the optimization procedure and the actual minimum zone deviation as a function of the sample size [8].

(c)   optimization procedures are computational expensive. In addition, global search for the minimum zone deviation usually takes a large number of iterations that add more complexity to the already expensive computation.

Because of the computation complexity, optimization procedures have limited applications in practice. Furthermore, as shown in Fig. 2, form deviations evaluated by any optimization procedure will be under estimated if the sample size is too small to represent the actual profile of the workpiece. This implies that bad parts may be mistakenly judged as good ones. This type of error is highly unacceptable in precision manufacturing. To overcome this problem, large sample size must be used at the expense of increased computation time.

LEAST-SQUARES DEVIATION vs MINIMUM ZONE DEVIATION

On the manufacturing floor, computing efficiency is the key to the survival of software algorithms. Optimization procedures demand high computing power. Least-squares fitting, in contrast, enjoys computational simplicity and thus is the most popular technique for evaluating form deviations. In general, the least-squares deviation does not converge to the minimum zone deviation. Therefore, it is necessary to establish the relationship between the minimum zone and least-squares deviations in order to fulfill the tolerance requirements. It can be proven (beyond the scope of this paper) that the least-squares deviation converges to the minimum zone deviation if the feature noise, $v_k(i)$ as in Eq. 1, is additive gaussian white noise. This relation is illustrated graphically in Fig. 3.

In reality, form deviations are the combined effect of process characteristics, CMM measurement noise, and electronics noise of the NC controllers. The noise caused by the process itself is machine dependent. It is most likely a real function of unknown parameters, (a sinusoidal function of unknown amplitude and frequency, for example) rather than a random variable. The measurement and electronics noise can be random but not necessarily gaussian noise. Without knowing the probability density of the combined noise $v_k(i)$, the least-squares deviation cannot be related to the minimum zone deviation in a predicable manner. Empirically, the least-squares deviation for flatness, circularity, and cylindricity has been found to be 5 to 25 % larger than the minimum zone deviation[6]. Nevertheless, these figures can only be used as references. Reliable form deviations must be evaluated algorithmically not by multiplying a fixed percentage to least-squares deviations.

BOUNDARY LEAST-SQUARES METHOD

The basis of the boundary least-squares method is built upon the observation that the necessary condition to approximate minimum zone deviation by least-squares fit is the requirement that the feature noise $v_k(i)$ is an additive white noise. Assuming that the observation data can be grouped into two subsets: one subset of points contains only the data degraded by random white noise and the other subset contains all other points. The former subset must contain enough data points to preserve the geometric properties of the form to be evaluated. If the data points that satisfy the above conditions exist and can be identified from the whole sample points, the least-squares deviation calculated using this subset of data points will lead to the minimum zone deviation. The boundary least-squares method proposed in the paper is designed to identify and utilize this subset of data points.

In general, any given set of observation points provides two pieces of information: the geometry of the workpiece and the form deviation. Considering the example of evaluating the straightness as shown in Fig. 4, let O be the set of points falling within a narrow region adjacent to the maximum inscribing and minimum circumscribing lines defining the minimum zone deviation, denoted by $f_{max}(x)$ and $f_{min}(x)$, respectively. Let I be the rest of points. It is shown in Fig. 4 that the subset O contains mostly the form error information and I provides mainly the actual profile of the workpiece. Let f(x) be the mean of $f_{max}(x)$ and $f_{min}(x)$. clearly, the least-squares deviation will approximate the minimum zone deviation if f(x) can be estimated by the least-squares fit. Mathematically, the data points in O can be viewed as the observation data of the mean feature f(x) degraded by a bi-gaussian random noise. That is

$$y_k(i) = f_k(x_m) + v_k(i) \qquad\qquad i \in O \qquad\qquad (6)$$

where $v_k(i)$ is the zero-mean bi-gaussian random noise.

The term bi-gaussian is used here because the random noise $v_k(i)$ consists of two gaussian function equally spaced from its zero mean as shown in Fig. 4. The bi-gaussian distribution can be obtained mathematically by convoluting a zero-mean gaussian function with two unit-impulse functions, $\delta(t+t_0)$ and $\delta(t-t_0)$. Since gaussian is a random white noise and the linear convolution is a linear operator, the bi-gaussian noise is also a random white noise. The least-squares estimation of the mean feature $f(x_m)$ using Eq. 6 will be very close to the actual one. Let the term boundary least-squares deviation represent the form deviation calculated from this estimated mean feature. It is obvious then the boundary least-squares deviation approaches the minimum zone deviation if a set of boundary points satisfying Eq. 6 is used.

Strategies to Select Boundary Points

As indicated in Fig. 4, the selection of boundary points requires the establishment of $f_{min}(x_m)$ and $f_{max}(x_m)$. This can be accomplished by using the least-squares mean calculated from all the sample points. The boundary points are selected by taking 10-20% of the total points which have longer distance from the reference mean. Using these boundary points, a new least-squares mean is calculated and another set of boundary points is selected using this new reference. Iterations are repeated until the selected boundary points converge to a fixed set of points.

Software Implementation

The software implementation of the boundary least-square method is describes as follows:

step 1.   pre-process the sample data to eliminate the high-frequency low-amplitude measurement noise, i.e., process the input data with a low pass filter.

step 2.   find the least-squares mean using all the sample points and the least-squares deviation.

step 3. calculate the signed normal distance of each point to the reference mean. The points outside the mean have positive distance values and those points inside have negative distance values.

step 4. sort the data points according to their distance values from positive maximum to negative maximum.

step 5. select top 10% of upper points and bottom 10% of the lower points from the sorted data.

step 6. calculate the least-squares mean using the boundary points selected in step 5 and the least-squares deviation of the new reference mean. The least-squares deviation are evaluated using the whole sample points.

step 7. if the least-squares deviation determined in step 6 is smaller than that of step 2, i.e., the boundary points of the form as in Eq. 6 exist and are identified, repeat steps 3 to 6 for another iteration; otherwise go to step 8.

step 8. report the best evaluation of the form error.

EXAMPLES

To illustrate the effectiveness of the boundary least-squares method, several examples were analyzed. The results are compared to that of using conventional least-squares and optimization procedures. Table 1 to 3 summarize the results of six different examples. Raw data used in these examples is given in the appendix.

Examples 1 and 2 used computer simulated data for evaluating straightness and flatness. The equations used to generate these data are given as follows:

Straightness: $y(x) = 0.01[\sin(x) + v(x)], \ x = 0,1,2,\ldots,180$ (7)

Flatness: $z(x,y) = 0.01[\sin(x+y) + v(x,y)], \quad x, y = 0,1,2,\ldots,180$

where $y(x)$ and $z(x,y)$ are less than or equal to 0.01,
     $v(x)$ and $v(x,y)$ are random number between 0 and 0.5.

The above equations model the feature noise as a sinusoidal function plus a random value arbitrarily specified by the user rather than gaussian noise. As shown in Table 1, the least-squares deviation of the straightness and flatness are 150% and 42% larger than the minimum zone deviations, respectively. The boundary least-squares deviations, on the other hand, has improved the difference to 2.7% and 8.1% respectively, after three iterations.

Examples 3 to 6 used actual coordinate data measured by CMM machines. These data were analyzed by Murthy and Abdin[5] and Traband et al.[6] using various algorithms. Their experimental results as well as the boundary least-squares deviations are tabulated in Table 2 and 3.

Table 1. Evaluations of Straightness 1 and Flatness 2

Ex*	n*	LS* Method(1)	O*	BLS* Method(2)	Actual Value(3)	RE* (1-3)/3	RE (2-3)/3
1	181	2.5051E-2	20	1.0266E-2	1.0000E-2	150%	2.7%
2	361	2.8427E-2	36	2.1616E-2	2.0000E-2	42%	8.1%

*Ex: Example, LS: Least-Squares, BLS: Boundary Least Squares,
RE: Relative Error, n: total number of data points,
O: number of boundary points selected.

Table 2. Evaluation of Straightness

Ex	n	LS Method(1)	O	BLS Method(2)	CH* Method(3)	CMM	RE (1-3)/3	RE (2-3)/3
3	15	5.377E-3	8	5.203E-3	5.186E-3	7.3E-3	7.2%	0.3%
4	25	1.463E-3	6	1.326E-3	1.311E-3	1.5E-3	11.6%	1.1%

*CH: Convex Hull.

Table 3. Evaluation of Flatness

Ex	n	LS Method(1)	O	BLS Method(2)	CH Method(3)	CMM	RE (1-3)/3	RE (2-3)/3
5	20	4.381E-2	6	4.227E-2	4.185E-2	4.47E-2	4.7%	1%
6	25	3.033E-3*	24	2.693E-3	2.897E-3*	6.6E-3	-	-

* This number is found to be 2.709E-3 by using conventional least-squares fit.

Discussion

As illustrated in the examples, the boundary least-squares method described in this paper provides significant improvement on form error estimation than the conventional least-squares. The proposed method is an iterative least-squares procedures. To achieve computational efficiency, three iterations were used for all of the examples. In addition, the number of data points used in the second and the third iteration were only 10-20% of the total sample points. The computational complexity of the boundary least-squares method is comparable to that of conventional least-squares.

One important issue that was not addressed explicitly in this paper was the necessary sample size to support the boundary least-squares method. As indicated in the proceeding sections, the selected boundary points must preserve both the geometry and form deviation information. Empirically, it is found that 10% of the sample size of 100 to 500 points provide satisfactory results for evaluating straightness, flatness, and circularity. Additional study must be performed to determined the optimal sample sizes for various applications.

CONCLUSION

The boundary least-squares method described in this paper is proven to be superior to the conventional least-squares in performance. A maximum of 10% improvement using actual coordinate data has been observed. It is also superior to the optimization procedures in speed. In all of the examples, the boundary least-squares deviations are is 1%-10% larger than the minimum zone deviations. This illustrates the immunity of the boundary least-squares method to unknown process noise. The insensitivity to process noise is important because various manufacturing processes contribute different process deviations to the final forms of the parts. The boundary least-squares method is a significant advancement in estimation theory for form deviations. It provides the manufacturing engineers an unique method for evaluating form deviations regardless of the manufacturing process producing the parts.

REFERENCES

[1]   Porta,C., Waedele, F.: Testing of Three Coordinate Measuring Machine Evaluation Algorithms, Commission of European Communities, 1986

[2]   Scarr, A.J.: Use of the Least Squares Line and Plane in the Measurements of Straightness and Flatness, *Proceedings of the Institution of Mechanical Engineering,* p.531, 1968.

[3]   Cramer, H.: Mathematical Methods of Statistics, Princeton University Press, Princeton, N.J., 1946.

[4]   Nahi, N.E.: Estimation Theory and Applications, John Wiley and Sons, New York, 1969.

[5]   Acharya,P.K. and Henderson,T.C.: Parameter Estimation and Error Analysis of Range Data, *Proceeding of the International Conference on Pattern Recognition,* pp.1709-1714, 1988.

[6]   Murphy, T.S.R. and Abdin, S.Z.: Minimum Zone Evaluation of Surfaces, *International Journal of Machine Tool Design and Research,* VOl.20, pp.123-136, 1980.

[7]   Traband, M.T., Joshi, S. and Cavalier, T.M.: Evaluation of Straightness and Flatness Tolerances Using the Minimum Zone, *Manufacturing Review,* Vol. 2, Sep. 1989, pp.189-195.

[8]   Hsu, J.P., Hsu, T.W., Filliben, J., and Hopp, T.: Status Report on Statistical Inspection in Coordinate Measuring for ASA/NIST/NSF 1990-91ResearchFellowship Program, June 1991.

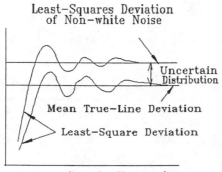

Fig.1 The relation of the mean true-line
deviation and the least-squares deviation

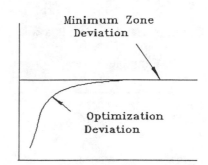

Fig.2 Form deviation obtained
by optimization approaches

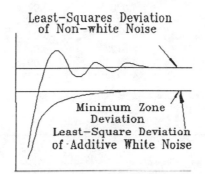

Fig.3 The relation of the minimum zone
deviation and the least-squares deviaton

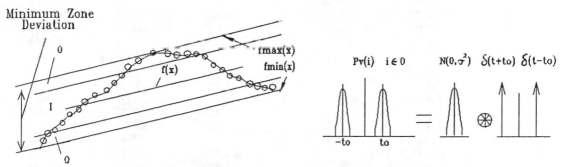

Fig.4 The relation of flatness and f(x)

APPENDIX

Example 1 and 2: data is available upon request.

Example 3.

x	y
0.3952	-0.0032
0.6953	-0.0016
0.9669	-0.0042
1.2762	-0.0028
1.5797	-0.0028
1.8593	-0.0037
2.1333	-0.0007
2.4197	-0.0010
2.6001	0.0007
2.8590	0.0007
3.0662	0.0017
3.2165	0.0025
3.2165	-0.0017
3.4217	0.0026
3.6179	0.0027
3.8185	0.0047

Example 4.

x	y	x	y
0.2845	-0.0034	5.3603	-0.0030
0.6600	-0.0032	5.6535	-0.0032
1.2041	-0.0030	5.9058	-0.0020
1.4994	-0.0035	6.0774	-0.0019
1.8494	-0.0036	6.2962	-0.0019
2.2261	-0.0025	6.5240	-0.0017
2.5724	-0.0028	6.7114	-0.0017
2.9076	-0.0026	6.9996	-0.0019
3.2548	-0.0031	7.2076	-0.0017
3.4142	-0.0031		
3.6307	-0.0029		
3.9237	-0.0029		
4.2647	-0.0028		
4.5122	-0.0028		
4.8150	-0.0027		
5.1334	-0.0027		

Example 5.

x	y	z
0.3846	0.2416	-0.0828
1.5008	0.2922	-0.0821
2.3107	0.3289	-0.0787
2.9817	0.3593	-0.0789
3.6964	0.3917	-0.0760
3.6743	0.8794	-0.0785
3.1195	0.8543	-0.0735
2.3552	0.8196	-0.0745
1.5875	0.7849	-0.0714
0.5573	0.7382	-0.0740
0.5413	1.0921	-0.0730
1.2205	1.1229	-0.0727
2.1673	1.1658	-0.0716
3.0881	1.2076	-0.0749
3.8459	1.2419	-0.0799
3.8305	1.5796	-0.0848
3.2057	1.5514	-0.0410
2.4230	1.5159	-0.0759
1.6710	1.4819	-0.0746
0.5263	1.4300	-0.0745

Example 6.

x	y	z
0.0255	0.2994	0.0005
1.4992	0.3371	0.0013
2.6656	0.3726	0.0000
3.5978	0.4009	0.0005
4.6241	0.4321	-0.0007
4.5989	1.2640	0.0001
3.4451	1.2289	0.0008
2.7096	1.2066	0.0004
1.6726	1.2968	0.0014
0.5273	1.2620	0.0009
0.1683	2.1413	-0.0002
0.9906	2.1663	0.0010
2.5485	2.1801	0.0008
3.4605	2.1369	1.1000
4.8632	2.1795	-0.0017
4.8401	2.9417	-0.0014
3.6557	2.9058	0.0012
2.4224	2.8683	0.0012
1.3839	2.8368	0.0011
0.4966	2.8098	-0.0002
0.4672	3.7751	-0.0008
1.6709	3.8116	0.0001
2.8864	3.8486	0.0006
3.7562	3.8750	0.0008
4.6746	3.9029	-0.0003

Part V
Flexible Manufacturing Systems

GROUP TECHNOLOGY APPLICATION FOR DESIGN OF MANUFACTURING CELLS

[1]Ali K. Kamrani and [2]Hamid R. Parsaei

[1]*University of Missouri, Rolla, Missouri, U.S.A.*
[2]*Industrial Engineering Department, University of Louisville,*
Louisville, Kentucky, U.S.A.

ABSTRACT

This paper presents the application of linear programming to develop a methodology which uses design and manufacturing attributes to form manufacturing cells. The methodology is implemented in two phases. In phase I, parts are grouped into part families based on their design and manufacturing dissimilarities. In phase II, the machines are grouped into manufacturing cells based on relevant operational costs and part families are assigned to the cells using an optimization technique.

INTRODUCTION

Group Technology

Group technology is one of the major concepts used in the design of flexible manufacturing systems and manufacturing cells. Group technology is a philosophy which identifies similar parts and groups them into part families. The group technology concept is based on grouping the parts according to their design attributes (Physical shape and size) and manufacturing attributes (Processing sequence). The task of grouping parts into families requires careful planning and consideration. Grouping may be done by visual inspection of parts [1]. This method lacks accuracy and sophistication. In the second method, Production Flow Analysis (PFA) [2], grouping is done based on the required operations of each part. This method relies on the information from route sheets. The third method of grouping is classification and coding [3,4,5]. In this method, each part is identified in terms of its design and processing features. Three types of coding methods are used in classification and coding: 1) Monocode, in which the meaning of each digit of the code is dependent upon the immediate predecessor digit. An advantage of this approach is the complete information and the disadvantage is the relationship between the digits, 2) Polycode, in which the meaning of the digits used in the code is independent of the other digits. Each digit is used to specify a particular feature of the part, and 3) Hybrid, which is the combination of the monocode and polycode. Most coding systems are of hybrid structure. This type of code consists of a series of small polycodes within a monocode structure.

Cellular Manufacturing

Cellular manufacturing, also known as family production, or flexible manufacturing, can be described as a manufacturing process which produces families of parts within a single line or cell of machines operated by machinists or robots that operate only within the line or cell. Some of the benefits of the cellular manufacturing systems are:

1. Elimination of or decrease in setup time and setup cost.
2. Greater manufacturing flexibility.
3. Less floor space around the machines.
4. Higher productivity.

Manufacturing cells can be categorized into two groups, 1) Manned cells which require a trained operator to perform different tasks and, 2) Unmanned cells

where cells are automated and consist of robots, material handling equipment, and programmable machines.

Design of Cellular Manufacturing Systems

Cellular Manufacturing is one of the major applications of group technology philosophy. Cellular manufacturing groups the required machines in a cell to produce a limited number of part families which have already been identified by some method of classification.

In recent years, the use of clustering techniques to form manufacturing cells have been considered by several researchers. This method requires the calculation of similarity or dissimilarity coefficients. After the coefficients are determined, the machines with closest measures will be grouped into one cell. Similarity and dissimilarity coefficients represent the relationship between parts. The values of these coefficients generally range from 0 to 1. If the value assigned to a coefficient is '0', this indicates that there exist no similarity and the value of '1' indicates strong similarity.

These coefficients can be used to group objects using clustering methods and algorithms. Two methods widely used for clustering are hierarchical clustering and non-hierarchical clustering.

The hierarchial method is the technique which is often used for clustering. This method results in a graph known as a dendrogram which groups a set of given information or data into small clusters based on their similarities. Small clusters are then grouped and ranked into larger clusters. The hierarchial method is accomplished in two forms, agglomerative and divisive. The Agglomerative method builds clusters by gathering the objects sequentially. The Divisive method performs consecutive division of a set of information or data into clusters, until no further division is possible.

The non-hierarchial method, also known as partition clustering, creates various clusters where the objects within a cluster are much similar to each other, and rather different from the objects of other clusters.

A PROPOSED DESIGN METHODOLOGY FOR FORMING MANUFACTURING CELLS

A two-phase methodology for the design of a cellular manufacturing system is proposed. Figure 1 illustrates the proposed model [6]. The phase I of the methodology identifies part families using a coding system, KAMKODE to determine the dissimilarity measures between parts. Phase II of the methodology determines the number of cells, assignment of the parts to the cells, the number of machine types, the number of tool types, and the number of fixture types, required to meet the production demand and other constraints.

KAMKODE is a classification and coding system developed for part identification. The code is an 18-digit number and the structure is illustrated in Figure 2. It includes information about both design and manufacturing. These attributes are represented by numerical values. The required information for development of the codes can be accessed from design and manufacturing databases.

Phase I : Forming Part Families

In this phase, the methodology selects parts which are similar in their design and manufacturing features, and groups them into families. The disagreement indices [7] method is used for measuring the degree of association between parts. For binary and nominal variables the disagreement index between parts i and j for attribute k is measured by:

$$d_{ijk} = \begin{cases} 1, & \text{if } R_{ik} <> R_{jk} \\ 0, & \text{otherwise} \end{cases} \qquad (1)$$

where: d_{ijk} = Disagreement index between parts i and j for attribute k.
R_{ik} = Rank of part i for attribute k.
R_{jk} = Rank of part j for attribute k.

The linear disagreement index for ordinal variables is measured using the following function:

$$d_{ijk} = |R_{ik} - R_{jk}| / (m - 1) \qquad (2)$$

where: m = Number of classes for attribute k.
m - 1 = Maximum rank difference between part i and j.

Each attributes may be weighted based on the degree of their importance for design using the following category:

Very Important	1.0
Important	0.75
Not Important	0.5
Necessary	0.25

The dissimilarity index for processing and end operation sequences is calculated using equation [8]:

$$d_{ijk} = \sum_{o} (q_{iok} * q_{jok}) / \sum_{o} (q_{iok} + q_{jok} - q_{iok} * q_{jok}) \qquad (3)$$

where:

$$q_{iok} = \begin{cases} 0, \text{Part i requires operation o for attribute k} \\ 1, \text{otherwise} \end{cases}$$

and the dissimilarity indices for tools and fixture is calculated using equation [9]:

$$d_{ijk} = (NUM_{ik} + NUM_{jk} - 2NUM_{ijk}) / (NUM_{ik} + NUM_{jk}) \qquad (4)$$

where: NUM_{jk} = Number of required tool/fixture (k) by part j.
NUM_{ijk} = Number of common tool/fixture (k) to both parts i and j.

The dissimilarity indices for processing and end operation machines for parts i and j is calculated using the Hamming Metric as below [10]:

$$d_{ijk} = \sum_{m} \delta(X_{imk}, X_{jmk}) \qquad (5)$$

where:

$$\delta(X_{imk}, X_{jmk}) = \begin{cases} 1, \text{if } X_{imk} <> X_{jmk} \\ 0, \text{otherwise} \end{cases} \qquad (6)$$

and, X_{imk} = Part i uses machine m for attribute k.

Finally, the weighted dissimilarity measure between parts i and j is measured as:

$$D_{ij} = \sum_k (w_k * d_{ijk}) / \sum_k w_k \qquad (7)$$

where: w_k = Weight assigned to attribute k

d_{ijk} = Disagreement index between parts i and j for attribute k

D_{ij} = Weighted dissimilarity measure between parts i and j

Part assignment to families can be done using the 0-1 integer-linear program formulated as follows:

Coefficients

p = Number of Parts

K = Required Number of Part Families

D_{ij} = Dissimilarity measure between Part i and j, $D_{ij} = D_{ji}$

Mathematical Model

MINIMIZE

$$\sum_{ij}\sum D_{ij} * x_{ij} \quad , i,j = 1,2,....p \qquad (8)$$

subject to

$$\sum_{j=1,...,p} x_{ij} = 1 \quad \text{for } i = 1,2,.....,p \qquad (9)$$

$$\sum_{j=1,...,p} x_{jj} = K \qquad (10)$$

$$x_{ij} \leq x_{jj} \quad \text{for all i and j} \qquad (11)$$

where:

$$x_{ij} = \begin{cases} 1, \text{ if part i belong to group j} \\ 0, \text{ otherwise} \end{cases}$$

The objective function of this model will minimize the total sum of dissimilarities. Each part is assigned to only one family (10), the number of families is selected by the operator (11), and parts are assigned to a family if and only if that family has been created (12).

Phase II : Manufacturing Cell Formation

In Phase II, the methodology takes into account several relevant operational costs. The formulation is described below:

Index Sets

$$o : \text{Operation} \qquad o = 1,2,...O$$
$$p : \text{Part} \qquad p = 1,2,...P$$
$$m : \text{Machine Type} \qquad m = 1,2,...M$$
$$g : \text{Part Group} \qquad g = 1,2,...G$$
$$c : \text{Cell}, C \le G \qquad c = 1,2,...C$$
$$t : \text{Tool Type} \qquad t = 1,2,...T$$
$$f : \text{Fixture Type} \qquad f = 1,2,...F$$

Decision Variables

$$X_{gc} = \begin{cases} 1, & \text{if group } g \text{ is assigned to cell } c \\ 0, & \text{otherwise} \end{cases}$$

$$Y_{pc} = \begin{cases} 1, & \text{if part } p \text{ is assigned to cell } c \\ 0, & \text{otherwise} \end{cases}$$

NM_{mc} = Number of type m machines assigned to cell c

NF_{fmc} = Number of type f fixtures assigned to machine type m in cell c

NT_{tmc} = Number of type t tools assigned to machine type m in cell c

Mathematical Cost Model

MINIMIZE

$$\sum_{mc} CM_m * NM_{mc} + \tag{12}$$

$$\sum_{fmc} CF_f * NF_{fmc} * \sigma_{fm} +$$

$$\sum_{pc} CT_t * NT_{tmc} * \sigma'_{tm} +$$

$$\sum_{tmc} (NMV_p - 1) * d_p * CH_{pc} * Y_{pc} +$$

$$\sum_{pmo} CI_{pm} * TI_{pm} * d_p * \mu_{omp} * NM_{mc} +$$

$$\sum_{mgc} TS_{mgs} * CS_{mgs} * \alpha'_{mg} * X_{gc} +$$

$$\sum_{pmc} [CP_{pm} * R_m + CRW_m * (1 - R_m)] * \beta_{pm} * TMD_{pm} * Y_{pc}$$

Constraints

$$\sum_m \sum_c CM_m * NM_{mc} \leq BM \tag{13}$$

$$\sum_f \sum_m \sum_c CF_f * NF_{fmc} * \sigma_{fm} \leq BF \tag{14}$$

$$\sum_t \sum_m \sum_c CT_t * NT_{tmc} * \sigma'_{tm} \leq BT \tag{15}$$

$$\sum_p \sum_m \sum_o CI_{pm} * TI_{pm} * d_p * \mu_{omp} * NM_{mc} \leq BI \tag{16}$$

$$\sum_p \sum_c (NMV_p - 1) * d_p * CH_{pc} * Y_{pc} \leq BMH \tag{17}$$

$$\sum_p \beta_{pm} * TMD_{pm} * Y_{pc} \leq TP_m * NM_{mc} \quad \text{for all } m,c \tag{18}$$

$$\sum_p \beta_{pm} * \beta'_{tp} * \sigma'_{tm} * TMD_{pm} * Y_{pc} \leq TL_t * NT_{tmc} \quad \text{for all } m,c,t \tag{19}$$

$$NF_{fmc} * \sigma_{fm} \geq NM_{mc} \quad \text{for all } f,m,c \tag{20}$$

$$\sum_d d_p * Y_{pc} \leq IC_c \quad \text{for all } c \tag{21}$$

$$\sum_c X_{gc} = 1 \text{ for all } g \tag{22}$$

$$\alpha_{pg} * Y_{pc} = X_{gc} \text{ for all } p,g,c \tag{23}$$
$$X_{gc} \in (0,1) \text{ for all } g,c \tag{24}$$
$$Y_{pc} \in (0,1) \text{ for all } p,c$$
$$NM_{mc} \geq 0 \text{ and Integer for all } m,c$$
$$NF_{fmc} \geq 0 \text{ and Integer for all } f,m,c$$
$$NT_{tmc} \geq 0 \text{ and Integer for all } t,m,c$$

The objective function, equation 12, minimizes the total sum of costs for machine investment, fixture investment, tool investment, material handling, inspection, setup, and machine operation including rework. Equations 13 through 17 limit the expenses based on the available budgets set for machine, fixture, tool, inspection and material handling by the firm. Equation 18 ensures that the capacities of machine types assigned to each cell are not violated. Equation 19 considers the tool life of each tool type. Equation 20 balances the required number of fixtures for machine types, since each duplicated machine requires a matching number of fixtures. The maximum number of parts allowed in a cell for the purpose of flexibility is determined using equation 21. Equation 22 ensures each part family is assigned to one cell, and assignment of all members of a part family to one cell is guaranteed using equation 23. Equation 24 ensures the integerality and binary results of the decision variables.

NUMERICAL EXAMPLE

The numerical example considers 5 parts, 4 process machine types, 2 end operation machine types, 4 process operations, 2 end operations, 5 tool types, and 4 fixture types. A sample of the characteristics and the codes used for part identification is illustrated in Table 1. The annual demand of these 5 part types are illustrated in Table 2. Tables 3,4,5,6, and 7 illustrate the results of phase I and phase II of the methodology. The developed model is solved using LINDO.

CONCLUSION

In this article, a two phase methodology is developed for forming machine cells using the dissimilarity measure between design and manufacturing attributes of parts. An 18-digit classification and coding system is also developed which can be easily constructed using the available data from the manufacturing and design database of a firm. The linear integer mathematical model for phase I and a mix-integer mathematical model for phase II of the methodology is used to solve a numerical design problem. This model can be further expanded by incorporating other cost factors which are involved in the manufacturing system design. These costs may include inventory and work-in-process.

Table 1. A Sample of Coding Structure

Attribute	Part Characteristics	Code Number
General Shape	Rotational	1
Material	Steel	3
Maximum Diameter	0.75" < D ≤ 1.5"	2
Overall Length	3" < L ≤ 7"	3
Inside Hole Diameter	No Hole	1
Product Type	Type b	2
Number of Process Operations	2	2
Process Operation Sequence	P. Operation #1 P. Operation #4	14
Number of Process Machines	1	1
Processing Machine Type	Process Machine Type 2	2
Number of Tools	2	2
Tool Type	Types 2,3	23
Number of Fixtures	1	1
Fixture Type	Type 2	2
Number of End Operations	2	2
End Operation Sequence	E. Operation #1 E. Operation #2	12
Number of End Operation Machines	2	2
End Operation Machine Type	E.O. Machine Types 1,2	12

Table 2. Annual demands of parts

Part	d_p
1	9000
2	7400
3	7700
4	6500
5	7800

Table 3. Phase I results

Family	Part 1 2 3 4 5
1	1 1 1
2	1 1

Table 4. Cell Configuration

Part Number	Family Number	Cell Number
1 2 5	1	1
3 4	2	2

Table 5. Number of machine types and their assignments

Machine Type	Cell Number	Number of Duplicates
1	1	5
2	1	6
4	1	2
5	1	4
6	1	5
-------	-------	-------
3	2	5
4	2	2
5	2	2
6	2	2

Table 6. Number of tool types and their assignments

Tool Type	Machine Type	Cell Number	Number of Duplicates
1	1	1	60
2	2	1	74
3	1	1	39
3	2	1	66
4	4	1	21
-------	-------	-------	--------
2	3	2	43
4	3	2	41
4	4	2	22
5	4	2	20

Table 7. Number of fixture types and their assignments

Fixture Type	Machine Type	Cell Number	Number of Duplicates
1	1	1	5
2	2	1	6
4	1	1	5
4	4	1	2
-------	-------	-------	--------
2	3	2	5
3	3	2	5
4	4	2	2

REFERENCES

1. Snead, C.S., "**Group Technology: Foundation for Competitive Manufacturing**," Van Nostrand Reinhold, New York, 1989.

2. Groover, M.P., "**Automation, Production Systems, and Computer-Integrated Manufacturing**," Prentice-Hall, Englewood Cliffs, New Jersey, 1987.

3. Allen, D.K. and P.R. Smith, "**Computer-Aide Process Planning**," Computer-Aided Manufacturing Laboratory, Brigham Young University, Provo, Utah, 1980.

4. Houtzeel, A., and B. Schilperoort, "**A Chain-Structure Part Classification System (MICLASS) and Group Technology**," Proceeding of 13th Annual Meeting and Technical Conference, Cincinnati, Ohio, pp.383-400, 1976.

5. Optiz, H. "**A Classification System to Describe Workpieces**," Pergamon Press, Elmsford, New York, 1970.

6. Kamrani, A. "**A Methodology for Forming Machine Cells in a Computer Integrated Manufacturing Environment**," Doctor of Philosophy Dissertation, Department of Industrial Engineering, University of Louisville, August 1991.

7. Anderberg, M.R., "**Cluster Analysis for Applications**," Academic Press, New York, 1973.

8. McAuley, J., "**Machine Grouping for Efficient Production**," Production Engineer, 1972.

9. Dutta, S.P., R.S. Lashkari, G.Nadoli and T. Ravi, "**A Heuristic Procedure for Determining Manufacturing Families Form Design-Based Grouping for Flexible Manufacturing Systems**," Computer and Industrial Engineering, Vol. 10, No. 3, pp. 193-201, 1986.

10. Lee, R. "**Clustering Analysis and its Applications**," Advances in Information System Science, Plenum Press, New York, 1981.

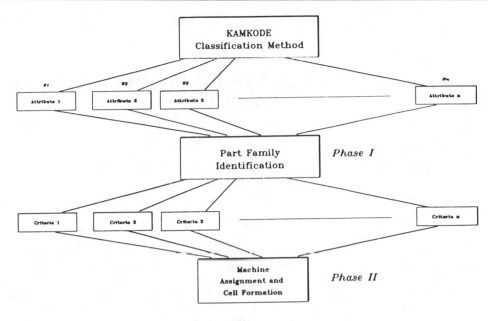

FIGURE 1. PROPOSED METHODOLOGY

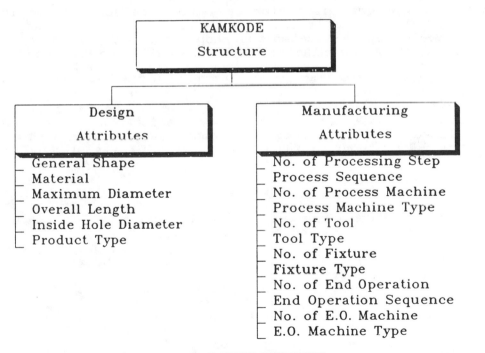

FIGURE 2. KAMKODE STRUCTURE

AN INTEGRATED CELLULAR MANUFACTURING SYSTEM APPROACH

Samir B. Billatos and Rishi Kumar

Department of Mechanical Engineering, University of Connecticut,
Storrs, Connecticut, U.S.A.

ABSTRACT

Cellular manufacturing systems (CMS) represent the logical arrangement of machines into groups or clusters of machines to process parts or families of parts by forming manufacturing cells. Recent research in CMS has predominantly dealt with an initial incidence matrix which consists of process plans formed for each part. The objectives of this paper are to design an integrated CMS (ICMS) that forms alternate process plans for each part, selects the optimum process plan out of these plans, forms the Group Technology (GT) cells, and allocates the shop floor devices needed for independent functioning of the formed cell. The process plans are formed using part and machine file information. The optimum plan is selected using the similarity coefficient method. Proposed CMS also delineates a method for process planning of a component, not part of existing database, and interacts with CAD software such as CADKEY. The newly developed ICMS that incorporates above ideas is applied to one of the parts existing in the database as a case study.

NOTATION

c_{mi}	Manufacturing cost of part i ($/min)
c_{si}	Subcontracting cost of part i ($/part)
d	Device used on shop floor (machine, tool, gripper, etc.)
MAX_z	Max. number of machines in machine cell MC-z
n	Number of parts
N_i	Set of alternate process plans for part i
P_f	Possible process plan for part i, f = i1, i2, i3,..
S_{jk}	Similarity coefficient between machines j and k
S_{P1P2}	Similarity coefficient between process plans P_1 & P_2
t_{ji}	Time used by machine j in processing part i
T_j	Max. processing time available on machine j
v_i	Production volume of part i
β_i	Profitability factor of part i
S_{ij}	Similarity coefficient between components i and j

s_{ij} Score between component i and j on attribute k.
X_{ik} Weight assigned to component i for attribute k
x_{jk} Weight assigned to component j for attribute k
R_k Range of attribute k taken over the population space
K Number of attributes

1. INTRODUCTION

Group Technology (GT) is an established manufacturing philosophy with new far reaching implications. The basic concept of GT is in identifying parts and products with similar processing requirements, (i.e., relevant attributes must be classified to form groups or families of related items) and taking advantage of the similarities between the members of each family to rationalize procedures during all stages of design, planning and manufacturing. For component parts, the attributes classified would include shape features, size, material, etc. For products and sub-assemblies, details of specification and performance would be relevant.

Part family grouping is an important step for successful GT applications. In grouping part families, scheduling for optimum sequencing and machine loading can be improved by considering the production data such as lot size, annual production plan, etc. The associated machine cells are formed such that one or more part-families can be fully processed within a single cell. This process is called "Machine Component Grouping". Similarities in the processing requirements of the parts within each cell would rationalize tooling, fixturing, etc. The flow of material within each cell can be facilitated by the use of conveyors, robots, etc. The main advantages of using a cellular layout are [1-3]:

1. Inventory reduction of 60%-85%.
2. Direct labor savings of 30%-50%.
3. Better utilization of available facilities (80%-90%).
4. Reduced floor space of 40%-50%.

2. PROBLEM DEFINITION

Advantages of CMS are met by forming cells considering proper scheduling, process routing, etc. Therefore, the problem is reduced to implementation of the concept of cellular manufacturing starting from the design of the part to the production of the complete end product.

The design of the part should lead to the development of the process plans for that part. A GT algorithm can then be used to select the optimum process plan and form the various cells. Indecision can arise as part of the solution. This must be examined and alternative solutions suggested.

Most of researchers such as Vakharia et al [5], Kusiak [6], Wu et al [7], and Rajmani [8] have either focussed on clustering algorithms or dealt with the selection of the process plan. For example, Kusiak [6] has dealt with Expert Systems and clustering algorithms along with other factors, in the formation of the cell. But the controversial case of bottleneck machines or parts, has not been considered.

Seifoddini [9] and Andenberg [10] focussed on the formation of the Similarity Coefficient matrix. A similarity coefficient (S) between two machines m and n, can be defined as follows:

$$S_{mn} = \frac{P_{bm}}{P_{em}}$$

(1)

where P_{bm} is the number of part types being processed on both machines, and P_{em} is the number of part types being processed on either machine. This coefficient was based on the Jaccard Similarity Coefficient [10] as explained below.

Consider the simple machine-part chart in Figure 1. This chart shows the process plan for each part. For example, part number 1 is machined on machine 2, shown by "1" (a non zero value) in position (2,1) of the matrix. Similarly, part number 2 is not being manufactured on either machine 1 or 2, shown by "0" in positions (1,2) and (2,2). Figure 1 can be organized to provide matches between attributes of machines 1 and 2 as shown in Figure 2.

	PART PLAN		
MACHINES	1	2	3
1	0	0	1
2	1	0	1

FIG. 1: MACHINE-PART CHART

	MACHINE 2	
MACHINE 1	1	0
1	a	b
0	c	d

FIG. 2: ATTRIBUTES

where a = number of (1,1) matches
      b = number of (0,1) matches
      c = number of (1,0) matches
      d = number of (0,0) matches

The similarity coefficient between machines 1 and 2 in terms of a, b, c and d will be as follows:

$$S_{12} = \frac{a}{a+b+c}$$

(2)

A major drawback to this similarity coefficient is that it is based on the number of part types being processed on the two machines rather than the actual number of operations being performed on these machines.

3. OBJECTIVES

The GT cell formation problem involves optimization of cell layout under constraints related to machine capacity, material handling system capability, machine cell dimensions, and technological requirements [12]. To design a cellular manufacturing system under these constraints, our paper is focused to achieve the following objectives:

1.   Develop an algorithm for alternate process planning.
2.   Develop a clustering algorithm for cell formation using an intelligent optimization procedure.

Process plans will be developed using the design data file (part file) where the part design will be used in the cell formation. It is likely that there may be more than one process plan for each part. All the plans together would form the initial incidence matrix. Using the Similarity Coefficient algorithm, the optimum process plan would be selected for each part to form the final incidence matrix. The clustering algorithm would then be used to form the GT cells. In case of controversies, suitable remedies will be suggested.

4. TECHNICAL APPROACH

Most of research in CMS was concerned with the formation of cells considering the initial incidence matrix as already available. Other research was concerned with different approaches for computerization of process planning. But the integration of computer aided process planning (CAPP) and CMS has not been fully explored. In this research a new program called Integrated Cellular Manufacturing System (ICMS) has been developed. The capabilities of this program can be outlined as follows:

♦   Providing alternate process plans of each part using part file information.
♦   Viewing any or all of the parts existing in the database.
♦   Forming the initial incidence matrix comprising of the myriad process plans possible for each part.
♦   Suggesting suitable process plan for a new part whose part information is not contained in the database. Using dendogram tables to substantiate above claims or to help in suggesting alternatives.
♦   Viewing new-part CDL files prepared using CADKEY (for new part to interact with ICMS).
♦   Running the clustering algorithm (the Fortran program) to form the final GT cells, and also viewing graphically this final layout.

The technical approach has been divided into two steps. The first step is the formation of alternate process plans for each part by forming the initial incidence matrix and applying the GT algorithm (which would ensure selection of an optimum process plan for each part) to form the final incidence matrix, and hence the GT cells. The second step is the selection of the process plan for a new part that is not a part of the part file. These steps are explained below.

4.1   Process Planning and Cell Formation

The   process   plans   are   generated   by   the   program considering data from the machine data files and the part files. The part file consists of design information in terms of line, ellipse (for circle) and arc.

The machine data file contains information about hole making machines, milling machines and turning machines. The tolerances are given in terms of a range (a maximum and a minimum). The values of the tolerances are in units of microns (10^{-6} m). The process planning of each part is developed based on the tolerances that are to be achieved and the incidence matrix is formed by considering process plans of all parts.

The tolerance data available in the part file is compared to the data from the machine file. Accordingly, machines capable of achieving required tolerances are selected for machining of that part. There may be different machines possible for processing the same part, and thus, there will be alternate process plans. These process plans are evaluated and the most optimum one recommended for use in forming the final incidence matrix. This procedure is developed using the GT algorithm which consists of three subroutines:

Subroutine SCP: Calculates the Similarity Coefficient between parts using equation (3):

$$S_{jk} = \frac{\sum_{i=1}^{n} v_i \, n_{jki} \, x_i}{\sum_{i=1}^{n} v_i \, n_{jki} \, y_i} \qquad (3)$$

IF $t_{ji} > 0$ AND $t_{ki} > 0$ then $x_i = 1$
ELSE $x_i = 0$

IF $t_{ji} > 0$ OR $t_{ki} > 0$ then $y_i = 1$
ELSE $y_i = 0$

Based on the SCP values it selects the process plan from different alternate process plans for each part, which are used to form the final incidence matrix.

Subroutine SCM: Calculates the Similarity Coefficient between machines. Once the most optimum process plan is selected for the part, forming machine cells and part families would result in a CMS. The procedure used here, modified from [11,12], selects the process plan by the algorithm and compares a permutation of all process plans (generated by an optimization module written in C) on the basis of the "weighted profitability factor" (α_k). This is defined as:

$$\alpha_k = \beta_i + \sum_{P_2 \in P^{(k-1)}} S_{P_1 P_2} \qquad for \ all \ P_1 \in N_i \qquad (4)$$

Where β_i is the profitability factor and S_{P1P2} is the similarity coefficient between two process plans, say P_1 and P_2. N_i is the set of all possible process plans for part i. N_1, N_2, ..., N_n is the sequence of process plan sets for parts 1, 2, ..., n. If $P^{(k)}$ = {P_1, P_2, ..., P_k} denotes the selected elements at stage k of the algorithm, then P_1, P_2, ..., P_k are chosen as they have the highest value of α in N_1, N_2, ..., N_k, respectively. The similarity coefficient S_{P1P2} is defined as,

$$S_{P_1P_2} = \frac{\sum_d w_d \, \delta_1 (t_{1d}, t_{2d})}{\sum_d \delta_2 (t_{1d}, t_{2d})} \tag{5}$$

$$\delta_1 (t_{1d}, t_{2d}) \quad = \quad 1 \qquad \text{if } t_{1d} \text{ AND } t_{2d} > 0$$
$$= \quad 0 \qquad \text{otherwise}$$

$$\delta_2 (t_{1d}, t_{2d}) \quad = \quad 1 \qquad \text{if } t_{1d} \text{ OR } t_{2d} > 0$$
$$= \quad 0 \qquad \text{otherwise}$$

W_d is the weight coefficient of the device (machine, tool, or auxiliary device). The weight coefficient indicates a relative importance of the device on the shop floor. The inclusion of the similarity coefficient between processes in the criterion for selection (weighted profitability factor) results in a selection process which reduces the number of different types of machines and auxiliary devices. This means a reduction in size of the manufacturing system, making the problem of scheduling, allocation, etc., easier to solve.

The Clustering Algorithm: Uses the final incidence matrix from SCP and forms the GT cells. The cells are formed to ensure against creation of any part bottlenecks. This subroutine is discussed in detail in section 5.

4.2   Process Planning of New Parts

For a part that does not exist in the database (new part) there are two possible alternatives:
♦   Add information about the part in the part file (in a fixed format).
♦   Develop an algorithm that uses the codes of the database parts and compares them with that of the new part to decide on the process plan.

The second alternative is preferred since parts manufactured regularly are only included in the part file. The decision is made using a grouping algorithm based on a similarity measure matrix between the parts [13]. But, instead of using part codes to form groups between the most similar parts as in [13], codes are used in the selection of process plans for a new part considering existing database process plans. The selected process plan may then be varied to suit the exact requirements of the new part. The flow chart in Figure 3 illustrates how the ICMS program is intelligent enough to react to these two different approaches.

Subroutine SM (Similarity Measure)

The main objective of this subroutine is to convert the part codes (for any pair of parts) into a form that can be readily utilized for forming desirable groups. This similarity measure is then used for the selection of the process plan for the new part by the grouping algorithm. The end result of this is the dendogram table that gives the exact match up of the similarity measure with different parts of the new part. The similarity coefficient between parts can be defined as follows [13]:

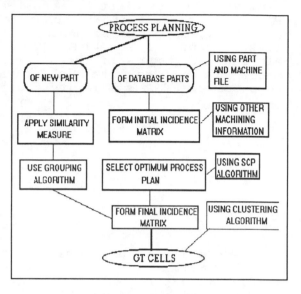

FIG. 3: FLOW CHART OF ICMS

$$S_{ij} = \frac{\sum\limits_{k=1}^{K} s_{ijk}}{\sum\limits_{k=1}^{K} \delta_{ijk}} \quad \textbf{(6)} \qquad\qquad S_{ij} = 1 - \frac{|x_{ik}-x_{jk}|}{R_k} \quad \textbf{(7)}$$

where, δ_{ijk} is 1 if comparison between components i and j is possible for attribute k, and 0 otherwise.

The Grouping Subroutine (Grouping Algorithm)

The steps are enumerated herewith:
1. Compute S_{ij} $\forall_{ij}$, $i \neq j$ and form matrix D.
2. Choose maximum S_{ij} (similar parts) and form a group G_1.
3. Recompute similarity coefficients and find maximum S_{ij}.
4. If S_{ij} is such that $i(j) \in G$, add $i(j)$ to G. Otherwise initiate another group G_r.
5. If $i \in G$ $\forall_i$ STOP. Else, go to step 3.

5. APPLICATION

Before introducing our applications, let us explore some of the features of ICMS. The main menu is shown in Figure 4. The user is given options of viewing the help file to get acquainted with the program capabilities as well as viewing the parts that are contained in the database. There are seven parts of the database as of now and are shown in Figure 5. The machining considered for these parts are drilling and finishing (boring, reaming, etc), turning and milling. Parts

#1 to #5 need milling and hole making operation. Part #6 needs turning and milling. Part #7 needs turning.

In addition, the user can edit the data files and query process plans for a new part.

5.1 Incidence Matrix

This is formed by the process plans. A sample incidence matrix is shown in Figure 6. The first row is the subcontracting cost (C_{si}) and second is job number. The first column is the cost of using the machine (C_{mi}). The second column is the time available every hour by the machine (T_j in min.). (Time is assumed lost in set-up). The third column is machine number. The other values represent the time in minutes (calculated as: pieces required per hour (V_i) multiplied by machining time per piece).

```
F1:  Help
F2:  Edit/View Data Files
F3:  Viewing database files
F4:  Available process plans
F5:  For New Process plans
F6:  Read from CDL file
F7:  Run GT algorithm
ESC: Quit/Exit ICMS
```

FIG. 4: THE MAIN MENU

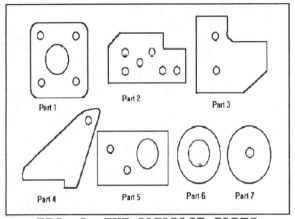

FIG. 5: THE DATABASE PARTS

5.2 Running the GT Algorithm

There are three different components of the GT algorithm:

- The Similarity Coefficient between machines.
- The Similarity Coefficient between parts.
- The actual clustering algorithm.

Using the incidence matrix, the Fortran module forms the GT cells. This program primarily aims to remove part bottleneck. This eliminates any form of interaction between the cells and thus tries to eliminate any form of travelling time and cost

			.6	.5	.5	.5	.4	.19	.2	.3	.4
			11	12	21	22	31	32	33	41	42
2 40	1		14		14		21		10	12	10
3 40	2	11	21	27	10	10		10	12	10	20
3 40	3		10	21	20			22	10	10	
4 59	4	21	10		10	11		29			
4 60	5	21			10			29	10		

FIG. 6: INCIDENCE MATRIX

involved. The output of the program appears in the following format for each cell:

♦   Machines and Parts which form the cell.
♦   Similarity coefficient between the process plans
    (used to select the process plan for each part).
♦   The machine similarity coefficient to resolve any
    controversies.
♦   Machine bottlenecks.
♦   Status of time-availability for each machine.
♦   Number of machines used in each cell in reference to
    MAX_z.
♦   Manufacturing cost and the profitability factor.
♦   The allocated shop floor devices d, in each cell.

In case the machine number limitation of each cell is
violated, then the user should consider similarity coefficient
between machines and accordingly drop the machine after
assessing values from the table and other factors.

5.3   Case Study

In this section, the actual usage of the program shall be
illustrated, by forming the process plan for a part, the
selection of the best process plan, and the formation of the
GT cells. These three major steps are possible by selecting
the respective menu options of ICMS. As seen in Figure 5, Part
1 has five holes that are to be machined.
Referring to the part file, it is found that the
tolerance for the peripheral holes of part 1 is 0.4 μm, while
the center hole needs a tolerance of 6.3 μm. Matching the
tolerance values from the machine data file (the tolerance
file), it is concluded that there are four different process
plans possible for this part. The milling machine has two
alternatives, whereas for the hole making process, machines #3
and #4 can be used. Machine #3 is a drilling machine and
machine #4 is a boring machine.
After forming the initial incidence matrix, from these
different plans for each individual part, the most optimum one
has to be selected (using the Similarity Coefficient Algorithm
(Equation 3)) to contribute to the final incidence matrix.
Using the SCP algorithm, the incidence matrix can be used
to form the similarity coefficient matrix as shown in Figure
7. Using these similarity coefficient values, the conclusion
is made regarding the process plan to be selected for each
individual part. The process plan number 1, of part 1 (process
plan #11) is to be selected, as part of the input to the final
incidence matrix.
Thus, applying the SCP algorithm, on an initial incidence
matrix, the final incidence matrix is formed. It was found
that process plans 11, 21, 32, 44, 52, 62 and 71 are selected.
Applying the clustering algorithm, the final cells are formed
as shown in Figure 8.
Some critical information analyzed by ICMS and the
suggestions offered thereupon are explained below.

♦   To resolve bottleneck, the suggestion is to use more than
    one machine. Otherwise, use the similarity coefficients
    or dendogram table between machines to make a decision.

```
   11 12 13 14  21   22   23   24   31   32   41   42   43   44   51   52   61   62   71  72
11.0.0 .0  .0 5.0 .40  .60  .14  2  1.3  .40  .60  .40  .60  .75  .75  .14  .29.17  .17
12.0.0.0.0 2.5 .13  .50  .0  1.3  1.0  .3  2.50 .33  1.0  1.3  .60  .33  .1   .14  .14
13.0.0.0.0 1.3 .14 1.3  .40  .75  5.0  .14  .29  1.0  1.3  .3  2.0  .0   .6   .17  .17
14.0.0.0.0 1.0 .00 1.0  .13  .60  2.5  .13  1.0  .75  2.5  .6  1.3  .13  .25  .14  .14
21.0.0.0 .0  .0  .00 .0   .0  4.0  2.0  .50  .75  .5  .75  1.0  1.0  .17  .13  .2   .2
22.0.0.0 .0  .0  .00 .0   .00 .20  .0   .25  .17  .00  .00  .2   .0   .25  .17  .0   .0
23.0.0.0 .0  .0  .00 .0   .0  1.0  2.0  .17  .13  1.5  .75  .4  1.0  .0   .33  .2   .0
24.0.0.0 .0  .0  .00 .0   .0   .0  .17  .00  .25  .17  .0   .2   .0   .50  .0   .0
31.0.0.0 .0  .0  .00 .0   .0   .0  .0  .67  .40  .67  .40 1.5   .5  .2   .14  .25  .0
32.0.0.0 .0  .0  .00 .0   .0   .0  .0  .17  .33  1.5  2.00 .40  4.  .0   .33  .20  .2
41.0.0.0 .0  .0  .0  .0   .0   .0  .0   .0   .0   .0   .0  .67  .2  .25  .17  .00  .0
42.0.0.0 .0  .0  .0  .0   .0   .0  .0   .0   .0   .0   .0  1.0  .4  .5   .0   .0   .2
43.0.0.0 .0  .0  .0  .0   .0   .0  .0   .0   .0   .0   .0   .2 .67  .0   .5   .33  .0
44.0.0.0 .0  .0  .0  .0   .0   .0  .0   .0   .0   .0   .0  .14 1.0  .0   .33  .20  .2
51.0.0.0 .0  .0  .0  .0   .0   .0  .0   .0   .0   .0   .0   .0  .0  .67  .0   .00  .0
52.0.0.0 .0  .0  .0  .0   .0   .0  .0   .0   .0   .0   .0   .0  .0   .0  .14  .0  .25
61.0.0.0 .0  .0  .0  .0   .0   .0  .0   .0   .0   .0   .0   .0  .0   .0  .00  .33  .0
62.0.0.0 .0  .0  .0  .0   .0   .0  .0   .0   .0   .0   .0   .0  .0   .0   .0  .67  .0
71.0.0.0 .0  .0  .0  .0   .0   .0  .0   .0   .0   .0   .0   .0  .0   .0   .0   .0  .0
72.0.0.0 .0  .0 .0  .0   .0   .0   .0  .0   .0   .0   .0   .0  .0   .0   .0   .0  .0
.6 .55.55 .5  .4  .19 .20 -.50 .1 .15  .6   .15  .35  .45  .4  .19  .45   .4  .19.4
```

FIG. 7: THE SIMILARITY COEFFICIENT MATRIX

- Total number of machines repeated after formation of each cell is described.
- If time is insufficient to complete machining, it suggests introducing more machines based on earlier information.
- It checks whether the machine number constraint is satisfied.
- It checks for the number of machines that could still be added to a cell.
- It handles single machine cells by merging them with another GT cell.

```
    | 11 21 | 32 44 52 62 71
 _  |_____|_____
 1  |  3  5 |
 3  |  4  5 |
 6  |  6  5 |
 _  |_____|_____
 2  |       | 3  4 10  6
 3  |       | 4  8  2
 4  |       |    6
 6  |       | 5     6
 8  |       |          10 12
```

FIG. 8: THE FINAL GT CELLS

5. CONCLUSIONS

In this research, an integrated cellular manufacturing system (ICMS) has been developed that can be used to form flexible manufacturing cells. The integration is in terms of design and production planning. For this purpose, a program, ICMS, has been developed that is user friendly and menu driven. ICMS incorporates three different operations: hole making, milling and turning. It is an intelligent system because it recommends alternate process plans for parts existing in database as well as new parts. It also has the capability of communicating with computer aided design systems that are commercially available such as CADKEY.

6. ACKNOWLEDGEMENTS

The authors wish to thank the University of Connecticut Research Foundation and the Precision Manufacturing Center for their Financial support.

REFERENCES

1. Gallagher CC, Knight WA: Group Technology, CAD/CAM and flexible automation. Proceedings of the IXth ICPR :pp 2340, 1987.
2. Snead CS: Group Technology - Foundation for Competitive manufacturing, Van Nostrand Reinhold, NY, 1989.
3. Gallagher CC, Knight WA: Group Technology Production Methods in Manufacturing, Ellis Harwood Ltd., 1986.
4. Luggen WW: FM Cells and systems (Prentice Hall) 1991.
5. Vakharia AJ, Wemmerlov U: A new design method for cellular manufacturing systems, Proceedings of the IXth ICPR, 1987, pp. 2357.
6. Kusiak A: Branching algorithms for solving the Group Technology Problem, J. Mfg Systems, V. 10, No. 4, 1991.
7. Wu HL, Venugopal R, Barash MM: Design of a Cellular Manufacturing system: A syntactic Pattern recognition approach". J. Mfg Systems, SME, Vol. 5, No. 2, 1986, pp. 81.
8. Rajamani D, Singh N, Aneja YP: Integrated design of cellular manufacturing systems in the presence of alternative process plans, Int. Journal of Production Research, Vol. 28, 1990, pp. 1541-1554.
9. Seifoddini H: Incorporation of the production volume in machine cells formation in group technology applications Proceedings, the IXth ICPR, 1987, pp. 2348
10. Andenberg MR: Cluster analysis for applications Academic Press, NY, 1973.
11. Kusiak A: Intelligent Manufacturing Systems, Prentice Hall International Series in Industrial and Systems Engineering, 1990.
12. Kusiak A, Chow WS: Efficient Solving of the GT Problem: J. Mfg Systems, Vol. 6, No. 2, 1987.
13. Offodile FO: Applications of Similarity Coefficient Method to Parts Coding and Classification Analysis in Group Technology, J. Mfg Systems, V. 10, No. 6, 1991, pp 442.
14. Vannelli A, Ravikumar K: A method for finding minimal bottle-neck cells for grouping part-machine families, IJPR 1985.

DESIGN OF MANUFACTURING CELLS VIA TABU SEARCH

[1]Rasaratnam Logendran, Parthasarathi Ramakrishna, and [2]Chelliah Sriskandarajah

[1]*Department of Industrial and Manufacturing Engineering, Oregon State Univesity, Corvallis, Oregon, U.S.A.*
[2]*Department of Industrial Engineering, University of Toronto, Toronto, Ontario, Canada*

ABSTRACT

The objective of this paper is to present an accurate and a more realistic approach to the selection of machines and a process plan for each part in the design of cellular manufacturing systems. The model is formulated as a binary linear programming problem over a planning horizon of one year. As the problem is proven NP-hard, we propose a higher level heuristic algorithm, based upon a concept known as tabu search. The algorithm is further extended into two methods - 1 and 2. Method 1 corresponds to saving all possible best configurations, created by perturbing each part about its annual operating cost, in the candidate list, while method 2 saves only the first best configuration, known as the best first strategy. A host of computer programs have been written in MICROSOFT-QUICKC, and the algorithm is implemented on a GATEWAY 2000-486 microcomputer. When the search is set to be terminated after evaluating the first ten local optima, the results obtained from solving an example problem show that method 1 determines a local (near) optimum that is within 3.3% of the optimal solution and method 2 the optimal solution.

INTRODUCTION

Over the past two decades or so significant improvement in manufacturing productivity has been realized by applying the concepts of cellular manufacturing (CM) to batch oriented small to large sized parts manufacturing companies. In the area of automated manufacturing, significant improvements have also been made by extending the concepts of CM to other manufacturing technologies such as robotics and flexible manufacturing systems. Reduction in set-up time, throughput time and work-in-process inventories, simplified flow of raw materials and parts, and improved human relations have been some of the advantages of CM systems over functional manufacturing systems [1-3]. Grouping parts that require similar processing needs into part families and machines that meet these needs into machine cells has been the focus of cellular manufacturing (CM) systems. Once the part and machine assignments to manufacturing cells are completed, each cell, consisting of similar part types and one or more dissimilar machine types, will function independently with minimal or no interaction with other cells.

PROBLEM STATEMENT

The problem of identifying parts and machines to each manufacturing cell, referred to as cell formation, has been investigated extensively and reported in the literature. All of these studies have assumed that each part has a unique process plan and that each operation of a part can only be performed on one machine [4-11]. In actuality, however, each part can have two or more processing plans and each operation associated with a part can be performed on alternative machines [12-14].

The model presented by Rajamani et al. [14] is an improvement over the others because it incorporates the limitations on budget and capacity of machines explicitly into the formulation when alternative process plans are present. The budgetary limitations, however, is placed on the operating cost, which is really the cost of producing all parts. Realistically, the parts manufacturing company should be producing all of what is demanded of each part as it is a known deterministic value. Thus, imposing a limit on operating cost leads to an unrealistic model formulation.

In this paper we present an accurate and a more realistic approach to the problem of selecting machines and a process plan for each part in the design of cellular manufacturing systems. The remaining sections of the paper are organized as follows. In the next section, we present a suitable mathematical model for the problem. The computational complexity of the problem is then stated. A higher level heuristic algorithm for solving the problem, based upon a concept known as tabu search, is presented next. The results obtained from solving an example problem using the proposed solution algorithm are described and the paper is concluded.

THE MODEL

The model is formulated over a planning horizon of one year as a binary linear programming problem. An amortized cost for each machine type based upon its useful life is evaluated and used in the model. It is assumed that the annual demand for each part type and the available capacity of each machine type are known. Additionally, for each process plan, the time and cost required to perform a specific operation of a part on a machine is known. A list of notations used in the formulation of the model as well as the model are presented below.

Notations

i	=	1, 2, ..., m machines
j	=	1, 2, ..., n parts
p	=	1, 2, ..., P_j process plans for part j
k	=	1, 2, ..., K(j,p) operations for (j,p) combination
X_i	=	number of machines of type i
Z_{jp}	=	$\begin{cases} 1 & \text{if part j is produced using plan p} \\ 0 & \text{otherwise} \end{cases}$

$$Y_{ik}(j,p) \quad = \quad \begin{cases} 1 & \text{if operation k for (j,p) combination is} \\ & \text{performed on machine i} \\ \\ 0 & \text{otherwise} \end{cases}$$

ac_i = amortized cost per machine of type i

d_j = annual demand for part j

at_i = available capacity on each machine of type i

$oc_{ik}(j,p)$ = operating cost for performing operation k on machine i for (j,p) combination

$pt_{ik}(j,p)$ = processing time to perform operation k on machine i for (j,p) combination

$$a_{ik} \quad = \quad \begin{cases} 1 & \text{if machine i can perform operation k} \\ 0 & \text{otherwise} \end{cases}$$

$$b_k(j,p) \quad = \quad \begin{cases} 1 & \text{if operation k has to be performed} \\ & \text{for the (j,p) combination} \\ \\ 0 & \text{otherwise} \end{cases}$$

In the above notations, X_i, Z_{jp}, $Y_{ik}(j,p)$ are decision variables, while a_{ik} and $b_k(j,p)$ are coefficients used in the formulation of the model.

Model

The total amortized cost of machines

$$= \sum_{i=1}^{m} ac_i \cdot X_i$$

The total annual operating cost for processing all parts

$$= \sum_{\substack{\text{over all} \\ j,p,i,k}} d_j \cdot Y_{ik}(j,p) \cdot oc_{ik}(j,p)$$

Thus, the model can be represented as:

Minimize

Total Annual Cost

$$= \sum_{i=1}^{m} ac_i \cdot X_i + \sum_{\substack{\text{over all} \\ j,p,i,k}} d_j \cdot Y_{ik}(j,p) \cdot oc_{ik}(j,p)$$

subject to:

$$\sum_p Z_{jp} = 1 ; \qquad j = 1, 2, ..., n \tag{1}$$

$$\sum_{i=1}^{m} a_{ik} \cdot Y_{ik}(j,p) = b_k(j,p) \cdot Z_{jp} \; ; \qquad \begin{array}{l} k = 1, 2, ..., K(j,P) \\ j = 1, 2, ..., n \\ p = 1, 2, ..., P_j \end{array} \qquad (2)$$

$$\sum_{\substack{\text{over all} \\ j,p,k}} d_j \cdot Y_{ik}(j,p) \cdot pt_{ik}(j,p) \le at_i \cdot X_i \; ; \qquad i = 1, 2, ..., m \qquad (3)$$

$$X_i \ge 0 \text{ and integer} \; ; \qquad i = 1, 2, ..., m$$

$$Z_{jp} = (0,1) \; ; \qquad \begin{array}{l} j = 1, 2, ..., n \\ p = 1, 2, ..., P_j \end{array}$$

$$Y_{ik}(j,p) = (0,1) \; ; \qquad \begin{array}{l} i = 1, 2, ..., m \\ k = 1, 2, ..., K(j,p) \\ j = 1, 2, ..., n \\ p = 1, 2, ..., P_j \end{array}$$

The objective function is formulated as minimization of the sum of the amortized cost of machines and the operating cost of producing all parts demanded over a period of one year. Three major sets of constraints are included in the model. First requires that only one process plan is selected for each part. Second requires that all of the operations on a part as per the process plan chosen are performed on one of the available machines. Finally, the third constraint set ensures that the machine capacity required for producing all parts does not exceed the available machine capacity.

COMPUTATIONAL COMPLEXITY OF THE PROBLEM

Binary integer programming problems belong to a class of NP-complete problems [15], and it turns out that our problem is no exception. The computational complexity of the problem was investigated, and it was shown NP-hard by considering a special case of our problem, denoted Q, and proving that a known NP-complete problem P (2-partition) is polynomially reducible to Q [16]. As the problem is NP-hard, while it is possible to develop an algorithm based upon a standard optimizing procedure such as a branch-and-bound technique, such an algorithm would turn out to be too time consuming even for a problem with moderate number of parts and process plans. Coupled to this is the fact that the proposed model is less restrictive than that by Rajamani et al. [14] as there is no restriction imposed on the budget for operating cost of producing all parts. It means that even for a small problem a large number of feasible combinations of solutions needs to be attempted on the tree before an optimal solution can be found. In the next section a higher level heuristic algorithm based upon a concept known as tabu search for solving large-scale practical problems is presented.

HEURISTIC ALGORITHM

Successful applications of tabu search has been reported by Glover [17], for obtaining optimal or near optimal solutions to a wide variety of classical and practical problems. It is a higher level heuristic for solving combinatorial optimization problems so designed to overcome the problem of being trapped in a local optimum, if one encountered, during the search. We propose a heuristic algorithm based upon tabu search for the selection of machines and a process plan for each part in the design of cellular manufacturing systems.

Background

To demonstrate the underlying concepts of the heuristic algorithm, an example problem, previously considered by Rajamani et al. [14], is considered. The problem has 4 parts and 3 machines. Table 1 presents the data on operations required of each part as per a process plan and the annual demand for each part. Data on capacity and amortized cost of each machine and the operations that can be performed on each machine are presented in Table 2. Table 3 presents the processing time and operating cost for performing an operation of a part on a machine as per a process plan. For instance, consider part 1 (P1) in Table 1 which has two process plans. First process plan requires that operations 1 and 2 be performed, while the second requires that operations 2 and 3 be performed. The data in Table 2 further indicate that operation 1 can be performed on either machine 1 (M1) or M3; operation 2 on either M2 or M3; and operation 3 on either M1 or M2. Thus, as per process plan 1 there are four different ways P1 can be processed. That is: operations 1 and 2 can be performed on M1 and M2, or M1 and M3, or M3 and M2, or M3 and M3, respectively. Similarly, there are four different ways P1 can be processed as per process plan 2. First the annual operating cost for each part is evaluated with every process plan. The annual operating cost can be evaluated as the product of the per unit operating cost and the annual demand. For P1 above there are eight such annual operating costs.

The annual operating costs evaluated for each part are then sorted in the non-increasing order. The initial feasible solution is determined as the one corresponding to the minimum annual operating cost on every part. The total annual cost for this solution is the sum of the minimum annual operating costs for all parts and the amortized cost of machines satisfying the capacity requirements. Initially the aspiration level (AL) is set equal to this total cost. There are two other lists created, namely the index list (IL) and the candidate list (CL). The IL contains the local optima evaluated as the search progresses. The CL consists of potential configurations chosen to perform future perturbations. A configuration here refers to a solution point. For instance, for this example problem with 4-parts, a configuration given by (1,1,1,1) corresponds to the initial solution, as 1 here denotes the smallest annual operating cost achievable on each part. There are both *forward* and *backward* perturbations performed on each part during the search. A forward perturbation performed on a part corresponds to the next adjacent incremental annual operating cost for that part. Thus, a forward perturbation performed on P1 using (1,1,1,1) as a seed would give us ($\underline{2}$,1,1,1). Notice that the element corresponding to P1 ($\underline{2}$) is underscored to indicate that it is now tabu, implying that it was the last part perturbed to create this configuration. Similarly, a backward perturbation performed on a part corresponds to the next adjacent decremental annual operating cost for that part. Evidently, no backward perturbation can be performed on the initial solution (configuration). However, if a backward perturbation is performed on P1 using ($\underline{2}$,1,1,1) as a seed, it would give us the initial solution (1,1,1,1), and it is the only part on which a backward perturbation can be performed for that seed. The initial feasible solution (i.e., 1,1,1,1) is assumed equal to the first local optimum, and is therefore inserted as the first configuration in the IL. As (1,1,1,1) is going to be used as a potential seed to perform forward perturbations it is also inserted as the first configuration in the CL.

TABLE 1

Data on $b_k(j,p)$ indicating operation k of part j to be performed as per process plan p; and the demand d_j for part j

Operation	Part j = 1 Process Plan		Part j = 2 Process Plan		Part j = 3 Process Plan			Part j = 4 Process Plan	
	p = 1	p = 2	p = 1	p = 2	p = 1	p = 2	p = 3	p = 1	p = 2
k = 1	1		1		1		1	1	1
k = 2	1	1	1	1	1	1	1	1	1
k = 3		1	1	1		1	1	1	
Demand	10		10		10			10	

TABLE 2

Data on a_{ik} indicating if operation k can be performed on machine i; capacity (at_i) on machine i; and the amortized cost (ac_i) of machine i

	Machine		
	i = 1	i = 2	i = 3
k = 1	1		1
k = 2		1	1
k = 3	1	1	
Capacity	100	100	100
Cost	100	250	300

Forward perturbations are performed using (1,1,1,1) as a seed. Four configurations are developed. They are: (2,1,1,1), (1,2,1,1), (1,1,2,1), (1,1,1,2). The total annual cost evaluated for each of these configurations is 1030, 1320, 1220, and 1330, respectively, and the total annual cost for (1,1,1,1) is 1320. As 1030 is the smallest and is less than the total cost for the seed (1320), (2,1,1,1) is chosen as the next configuration admitted to the CL. For the purpose of explanation, we assign *a star* to this configuration as it has the potential of becoming the next local optimum. It will become the next local optimum provided the configurations generated by performing both forward and backward perturbations using (2,1,1,1) as a seed gave us either a total cost of 1030 or higher. Indeed this is the case, and therefore, (2,1,1,1) becomes the next local optimum, and is assigned *two stars* (as it is already a local optimum) and admitted to the IL. The aspiration level is set equal to 1030, and P1 gets a *tabu* status in the configuration (2,1,1,1) with an underscore. This status is both permanent and temporary. It is permanent for backward perturbations as performing a backward perturbation on P1 would result in (1,1,1,1), which has already been considered as a seed. It is temporary for forward perturbations because the tabu ·status can be overridden if the configuration resulting from performing a forward perturbation on P1 (i.e., 3,1,1,1) gave a total annual cost lower than the current AL.

Four new configurations generated by performing forward perturbations on each part using (2,1,1,1) as a seed are: (3,1,1,1), (2,2,1,1), (2,1,2,1), and (2,1,1,2). The total annual cost associated with each configuration is 1290, 1030, 1030, and 1040, respectively. As 1030 is the smallest of the four and is equal to the total annual cost of the seed (i.e., 2,1,1,1), (2,1,1,1) is chosen as the next local optimum and is admitted to the IL as mentioned before. Notice that neither (2,2,1,1) nor (2,1,2,1) will receive *a star* as the total cost evaluated is not lower than that of the seed, and therefore have no potential for becoming a new local optimum in the future as the search progresses. Two different approaches are attempted with the heuristic algorithm. The first (method 1) is based upon all possible best configurations. As per this approach both (2,2,1,1) and (2,1,2,1) will be admitted to the CL in that order. The second approach (method 2) is based upon what is known as the *best first* strategy. According to this approach only (2,2,1,1) will be admitted next to the CL. Once the forward and backward perturbations are performed on a

TABLE 3

The processing time ($pt_{ik}(j,p)$) and operating cost ($oc_{ik}(j,p)$) required for machine i to perform operation k on part j using process plan p

		j = 1		j = 2		j = 3			j = 4	
		p = 1	p = 2	p = 1	p = 2	p = 1	p = 2	p = 3	p = 1	p = 2
k = 1	i = 1	5,3		3,4		2,2		8,1	1,2	9,7
	i = 3	7,2		4,3		2,2		9,2	2,1	8,9
k = 2	i = 2	3,5	9,8	7,8	3,3	3,3	1,2	5,9	2,3	9,8
	i = 3	4,3	7,9	7,7	2,3	4,4	2,4	3,10	2,4	10,9
k = 3	i = 1		8,8	10,9	6,5		11,7	7,4	3,5	
	i = 2		7,7	8,9	6,6		8,8	9,5	2,6	

configuration chosen as the seed, it is crossed out from the CL. The reason is that when a new configuration is about to be inserted into the CL it is checked against those configurations crossed out to ensure that there is no repetition, thus resulting in an inefficient search. The algorithm is terminated either after a fixed number of local optima is evaluated or when a prescribed CPU time lapses.

RESULTS

The main program for the proposed heuristic algorithm with each method is written in MICROSOFT QUICKC, and implemented on a GATEWAY 2000 - 486 based microcomputer with 33 MHz clock speed and 640K RAM. In order to compare the relative performances of both methods with respect to identifying the optimal solution for the example problem, a program, also written in MICROSOFT QUICKC, is used to determine explicitly the optimal solution. It required evaluating the total annual cost of 18,432 different combinations as P1, P2, P3, and P4 have 8, 12, 16, and 12 different operating costs as per process plans. The optimal solution is presented Table 4. The minimum total annual cost for the optimal solution, given by sum of the amortized cost of machines (550) and operating costs of all parts (350), is 900. The heuristic algorithm with both methods determined the configuration given by (4,1,5,2) as the local optimum corresponding to the optimal solution with a minimum total annual cost of 900. Figures 1 and 2 show the variation in total annual cost vs. local optimum number with method 1 and 2, respectively. The optimum solution is the 65th local optimum inserted into the IL with method 1 and the 9th local optimum with method 2. Correspondingly, they are the 328th and 37th entries into the CL. If the search is set to terminate after evaluating ten local optima, then method 1 would have determined a best near optimal solution with a total annual cost of 930 as the 7th entry into the IL and method 2 would have determined the optimal solution as the 9th entry into the IL. It means that method 1 determines a near optimal solution that is within 3.3% of the optimal before the search is terminated. The CPU times to determine the local optimum corresponding to the minimum total annual cost of 900 are 7 and 2.3 sec., with methods 1 and 2, respectively.

CONCLUSIONS

A realistic approach to the selection of machines and a unique process plan for each part in the design of cellular manufacturing systems is presented. As the problem is proven NP-hard, a higher level heuristic algorithm, based upon a concept known as tabu search, is developed. The algorithm is further extended into two methods (1 & 2). Method 1 corresponds to saving all possible best configurations into the CL while method 2 saves only the first best configuration, known as the best first strategy. Results obtained from solving an example problem show that method 1 determines a local (near) optimum that is within 3.3% of the optimal solution and method 2 the optimal solution when the search is set to terminate after evaluating the first ten local optima. Research is now continued to perform an extensive comparison of both methods using test problems corresponding to different problem structures in the design of cellular manufacturing systems.

TABLE 4
Optimal Solution

	Number of Machines
i = 1	3
i = 2	1
i = 3	0

Operation	Part			
	j = 1 p = 1	j = 2 p = 2	j = 3 p = 2	j = 4 p = 1
k = 1	i = 1			i = 1
k = 2	i = 2	i = 2	i = 2	i = 2
k = 3		i = 1	i = 1	i = 1
Operating Cost	80	80	90	100

ACKNOWLEDGEMENTS

This research is funded in part by the National Science Foundation (USA) Grant No. DDM-9108507, and by the Natural Sciences and Engineering Research Council of Canada Grant No. 0GP0104900 and a Connaught Grant from University of Toronto, Canada. Their support is gratefully acknowledged.

REFERENCES

1. Burbidge JL: *The Introduction to Group Technology.* New York, John Wiley, 1975.
2. Kusiak A, Heragu SS: Group technology. *Computers in Industry*, 9, pp 83-91, 1987.
3. Wemmerlov U, Hyer, NL: Research issues in cellular manufacturing. *International Journal of Production Research*, 25, pp 413-431, 1987.
4. King JR: Machine-component grouping in production flow analysis: an approach using rank order clustering algorithm. *International Journal of Production Research*, 18, pp 213-232, 1980.
5. Chan HM, Milner DA: Direct clustering algorithm for group formation in cellular manufacture. *Journal of Manufacturing Systems*, 1, pp 64-76, 1982.
6. Chandrasekaran MP, Rajagopalan R: ZODIAC - An algorithm for concurrent formation of part families and machine cells. *International Journal of Production Research*, 25, pp 835-850, 1987.
7. Tabucanon MT, Ojha R: ICRMA - A heuristic approach for intracell flow reduction in cellular manufacturing. *Material Flow*, 4, pp 189-197, 1987.

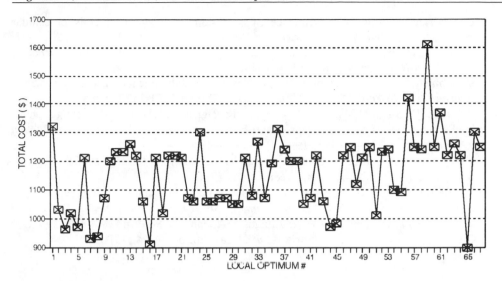

FIG. 1 Total Cost vs. Local Optimum for Heuristic Algorithm - Method 1.

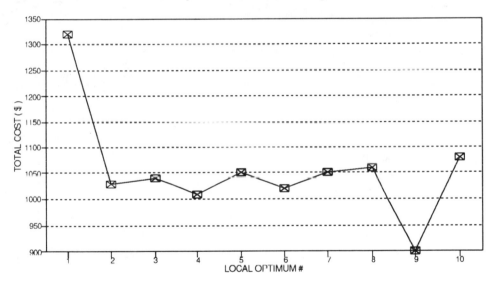

FIG. 2 Total Cost vs. Local Optimum for Heuristic Algorithm - Method 2.

8. Logendran R: A workload based model for minimizing total intercell and intracell moves in cellular manufacturing. *International Journal of Production Research*, 28, pp 913-925, 1990.

9. Harhalakis G, Nagi R, Proth JM: An efficient heuristic in manufacturing cell formation for group technology applications. *International Journal of Production Research*, 28, pp 185-198, 1990.

10. Askin RG, Chiu KS: A graph partitioning procedure for machine assignment and cell formation in group technology. *International Journal of Production Research*, 28, pp 1555-1572, 1990.

11. Logendran R: Impact of sequence of operations and layout of cells in cellular manufacturing. *International Journal of Production Research*, 29, pp 375-390, 1991.

12. Kusiak A: The generalized group technology concept. *International Journal of Production Research*, 25, pp 561-569, 1987.

13. Choobineh F: A framework for the design of cellular manufacturing systems. *International Journal of Production Research*, 26, pp 1161-1172, 1988.

14. Rajamani D, Singh N, Aneja YP: Integrated design of cellular manufacturing systems in the presence of alternative process plans. *International Journal of Production Research*, 28, pp 1541-1554, 1990.

15. Garey MR, Johnson DS: *Computers and Interactability: A Guide to the Theory of NP-Completeness*. New York, Freeman and Company, 1979.

16. Logendran R: Effect of alternative process plans in manufacturing cell design: complexity issues. *Proceedings 18th Annual NSF Grantees Conference on Design and Manufacturing Systems Research*, Atlanta, GA, January 8-10, pp 885-889, 1992.

17. Glover F: Tabu search: a tutorial. *Interfaces*, 20, pp 74-94, 1990.

ASPECTS OF INFORMATION MANAGEMENT IN CELLULAR MANUFACTURING SYSTEMS

[1]Chris O'Brien, Heiko Schäfer, and [2]John K. Bryant

[1]*Department of Manufacturing Engineering,*
University of Nottingham, Nottingham, U.K.
[2]*Dunlop Cox, Ltd., Nottingham, U.K.*

ABSTRACT

This paper illustrates the importance of effective information management in Cellular Manufacturing Systems, with emphasis on the particular situation of a European manufacturer of automotive components. This company's relationship with its suppliers and customers has been strongly affected by the current popularity of several established manufacturing philosophies, such as Simultaneous Engineering, Total Quality Management and Just-in-Time production.

The combination of these developments means that the entire manufacturing system used to meet market demands has to be totally reliable in terms of delivery and quality, in order to ensure the continuity of quality production and hence total customer satisfaction. In 1987, the company decided to respond to these requirements by relocating to a new, purpose-built factory and implementing a Cellular Manufacturing System based on the principles of Just-in-Time operation.

The discussion refers to the company's experiences with the development of new control and information structures to match its revised manufacturing strategy. These structures are continually being adapted to the new system, and the paper explores the particular impact of the cellular structure on a press tool management system currently under development.

INTRODUCTION

The Automotive Industry

Two forces have dominated developments in the European motor industry over the past two decades: on the one hand, increasingly fierce competition has made unremitting, radical improvements in productivity and quality essential for survival, whilst on the other hand this tough requirement has been matched by an upsurge in extremely effective manufacturing philosophies. The current economic climate has further increased the pressure on companies operating in this industry to strain their analytical and creative resources in a search for organisational strategies and tactics which comply with the nature of their markets.

Many of the successful manufacturing philosophies have affected or directly concerned the relationship between the manufacturers of chiefly assembled goods ("implosive manufacture": many sub-assemblies -> one product), such as cars, and their suppliers. The most significant of these are **Simultaneous Engineering** practices and **Just-in-Time** delivery. These developments have forced the suppliers to reconsider their approach to customer satisfaction, especially in terms of the speed of their response to customer requests (for goods, but also for change). The closer links to customers have also made Quality Assurance measures on the part of the supplier mandatory for continued corporate survival.

The Increasing Importance of Flexibility

All aspects of the suppliers' operations have been affected by these pervasive changes to organisational structure and philosophy. Measures of performance have been revised, and approaches such as World Class Manufacturing [1] and Lean Production [2] have introduced a new emphasis on true operational effectiveness. It is interesting to reflect that these dramatic improvements in the supplying organisations were often caused by increasing market pressures experienced by this particular industry. Indeed several of the modern manufacturing philosophies have their origin in the shift of manufacturing product profiles, from large volumes of mass-produced items to smaller quantities of customised goods. The objective of designing manufacturing systems is therefore no longer the construction of a system which can produce a given product as efficiently as possible. Instead, a structure has to be developed which can produce a variety of goods efficiently, while also being responsive and robust to changes in its market. This flexibility required of modern manufacturing systems to serve this changed demand is one of the key reasons for the current interest in Cellular Manufacturing, since the cellular structure, apart from the well documented operational benefits, creates a certain simplicity on the shop floor which makes it easier to adapt the system to a changing outside world. The relatively encapsulated nature of the cells also makes alterations in the product portfolio of the company much easier, since product are localised in their cells.

CELLULAR MANUFACTURING

Thus Cellular Manufacturing (CM) [3-5], as an application of the more general Group Technology philosophy [6-7], is a major device in the satisfaction of the stringent requirements of the car industry. The list of implementations shows that CM not only leads to operational benefits (such as reduced lead times and improved material flows) but can indeed secure a fundamental competitive advantage over more traditional production structures.

Applications of the Group Technology philosophy by actual car manufacturers have had varying degrees of success [7-9], especially as far as the "hard" benefits are concerned. This could be attributed to the category of market these organisations are serving, which may not be entirely suitable for the application of Cellular Manufacturing. The impact on industrial relations, however, has generally been positive [10]. On the other hand, organisations in the supply chain for car manufacturers tend to produce larger volumes of products, and hence there have been more applications of CM in this sector [3,11]. These companies have also often found it easier to realise the expected benefits.

From Hardware to Software

The shop floor developments of CM have of course affected the rest of the organisations concerned very strongly. The more immediate effects have involved a restructuring of the administration, with the cellular structures on the factory floor being reflected in almost every department (not just production control and design but also in the sales force, or even the organisation of the canteen). However, a more subtle and often overlooked change has been required in the control structures governing the operation of the system, i.e. the paperwork and information control operating "behind the scenes" of the manufacturing organisation. The impact of CM on shop floor information management is a particularly neglected aspect [12]. Modern Information Technology makes distributed intelligence at the site of data collection easily accessible, and the structure of CM reduces the complexity of these "data treatment at source" tasks even further.

However, too much emphasis is often placed on efficiency as opposed to effectiveness, and the different role and structure of information **management** under CM is hardly explored. Each cell in a Group Technology manufacturing system functions as a small business unit, with all its information needs of input and output, which are radically different from the needs of a functional department. Conversely, many categories of information which were previously only collected from or distributed to a single department now have to be circulated across the entire shop floor. This raises issues such as dangers to information integrity through communication between cells or data duplication. There is not generally an increased volume of information to be controlled, but the restructuring of the manufacturing system is certainly reflected in the information flow patterns supporting them.

Importance of IT for the Factory of the Future

It would be difficult to repudiate the importance of Information Technology for the factory of the future in converting the masses of readily available data into useful information [13]. There has always been plenty of data flying around in manufacturing organisations, but historically there was never enough computational power available to convert these into information/knowledge about the system. The quantum leaps in computational performance over the past few decades have virtually removed this bottleneck, but they have revealed a new one: descriptive models which allow a prediction of performance, and especially variations in this performance, are not available for the more complicated manufacturing processes. Thus many organisations still frequently run dry of knowledge about what is going on on the shop floor, in spite of the considerable computational power at their disposal. An information system merely presents the potential for converting data into information, but unless this process is carefully controlled the output may be an excess of information, which is just as useless as the previous excess of data.

Information Management in the Evolving Company

One long-term benefit of Cellular Manufacturing is the provision of an explicit path for growth of the system. Each machine cell is dedicated to the production of a family of related parts, and the composition of the cells can be altered as the product portfolio of the company changes. This is in stark contrast to traditional functional arrangements of facilities, where the decision to re-configure the layout is not necessarily in response to changes in the products. The foremen in charge of departments in the traditional scenario are more likely to see the arrangement of their machines as a function of changes in these facilities, and changes in part routings may not be reflected in the layout.

This path for evolution also applies to the numerical technologies used in the system. Information Technology is characterised by extremely rapid developments, which in a functionally arranged system will lead to a homogenous mixture of all ages of IT equipment (both hardware and software). Since a CM system will regularly introduce new cells and phase out older ones, the different generations of hardware and software will be localised. This has important implications for communication among these devices, and also beyond the cell boundaries, i.e. the integration of the cell's IT structure in the company networks (interfaces to each cell can be customised to the level of technology available there).

INFORMATION MANAGEMENT

A Case Study

The features of information management described above were experienced during the development of a tool management system for a European supplier of automotive components, whose manufacturing system is a very effective implementation of the Cellular Manufacturing principles in a Just-in-Time (JIT) environment. This company moved into new premises in 1987, as part of a reaction to the changing customer/supplier relationships in the industry. This relocation allowed it to implement a radical change in manufacturing strategy. The new factory consists of a number of self-contained manufacturing cells, operating JIT internally and externally, which can manufacture a product from raw material to despatch in a matter of hours.

As expected, this change in manufacturing strategy has affected most aspects of the company. The Department of Manufacturing Engineering and Operations Management of the University of Nottingham has been involved with the company under Teaching Company Programmes throughout its transition from more traditional manufacture to cellular production [14,15]. After the physical restructuring of the layout was complete, the control structure has had to be adapted, and a current project on tool management is part of this development [16].

The manufacturing processes in this system usually occur in the stages BLANKING - FORMING - FABRICATION - PAINTING - ASSEMBLY. The first two of these stages entail the use of fairly large, complex press tools (up to 5 t), which may consist of hundreds of components. The current project concerns the development of a customised maintenance management system for these press tools, of which there are nearly 1500 at the company.

The purpose of tool management is to ensure that tools are available for production when needed. For non-consumable tools, effective tool management extends beyond tool location/issue control, to efficiently planned and effectively integrated tool maintenance. The maintenance requirements and failure rates of the press tools used for blanking and forming operations are currently at a level which is not coherent with the company's stringent performance objectives. Since the company operates a JIT system, no part should be produced unless it is genuinely required by the next process stage. This means that all production is implicitly urgent; it also means that all unscheduled tool repairs impose unexpected delays on the production schedule - and often directly on the customer. The reliability of press tools is so important that some other manufacturers schedule their tools for inspection after every run, to ensure that tools stored and considered ready for use are not **disasters waiting to happen** as soon as the tool is required for its next run. The aim is to install a system which will pursue the company's philosophy of continuous improvement by

(a) identifying trends in press tool maintenance, reacting to these trends, and acting to reduce them, and

(b) identifying abnormal repairs and acting to eliminate them by analyzing possible failure causes.

The achievement of these goals hinges on two activities: recording an accurate and complete tool use and maintenance history, and converting the collected data into helpful information. This task is complicated and promising enough to warrant the use of a computerised information system, with the following three roles

• monitoring the use and maintenance of the tools,

• improving the timing of tool maintenance based on this tool history (such that tools are serviced **before** failure), and

• identifying and highlighting possible sources of recurring problems (to management and everybody involved directly).

The system will control very detailed information about tool use and maintenance, but all elements of this information will be related to potential tool failure causes. A key factor in the approach is that the system will not serve as a tool information library for general reference; it is designed with the achievement of a specific aim in mind. This approach is what is lacking from the commercially available tool management systems [17-19], which aim to improve the efficiency of existing procedures. To a certain extent, the company mainframe already provides a general reference service for the press tools.

Information flow paths to and from the system embrace the entire organisation. Initial communications are limited, since the system only requires detailed data on tool use and maintenance from the production cells and the tool room, while its output is sufficiently low in volume to be circulated as hard copy. Future developments, however, may also cover such areas as spares purchasing, personnel administration, subcontracted orders, or other tasks requiring communications across the site - or even beyond the company boundaries. This makes the issue of system integration in the company's "infostructure" crucial to a long-term success. Communication with tool component suppliers, subcontractors, tool designers (internal and external) and all Engineering departments may accelerate the development processes in the company, so the project also has implications for the company's drive towards Concurrent Engineering.

After the strategy and tactics of this project were relatively clear, the next issue addressed was how to develop and implement an information system which can achieve these objectives. A survey of available off-the-shelf tool management systems had revealed that these systems, although very powerful, would not address the specific difficulties experienced by the company and would therefore not solve the problems. Thus a new system is being developed for the project. This left the difficult decision of whether to host the system on the company's mainframe or to develop a localised system. The company mainframe covers purchasing, accounting, production control, part design records and production engineering. It would, in theory, be able to provide all necessary information and would certainly facilitate the distribution of output, but compared with a localised system the use of the mainframe would entail several significant drawbacks:

- information on tool use would have to be calculated from the mainframe's production schedules, incurring a significant time lag

- detailed information on tool use (which individual press, operator) would not be available to the mainframe system

- development would be more difficult, using an uncommon operating system and programming environment

- implementation would be less flexible; installation, debugging and corrections would involve disruptions of the entire system

The current focus of the project is the development of a prototype system, which will only monitor and manage the maintenance of a single cell's tools. This system is being developed as a stand-alone system on a micro-computer, with only limited communication links to other parts of the company's computer structure (via the mainframe and networks). Data on tool use and maintenance will be supplied by the cell leader and tool room manager respectively, and entered manually by the project leader on a daily basis. This manual data entry and output circulation is the major drawback of the choice of platform, since it is an option which is not feasible for control of all tools in the factory. The project should, however, benefit greatly from the flexibility and power of the programming environment. Greater control over the user interface and general appearance of the system will also be an advantage in this context. Once the prototype has been installed, tested and evaluated, the implementation of the final system will have to be carefully considered. Paperless operation of a system based on the prototype may be possible if microcomputers are installed in a few remaining cells.

The integration of such a system in the information infrastructure of the company is a very important issue (see above). Tool maintenance not only affects all aspects of production (including subsequent operations) but also depends for its effectiveness on a knowledge of true urgencies and bottlenecks. Likewise, the capacity of the tool room and the availability of the tool from the cell should be checked before booking a tool for a preventive service. The bulk of input and output of the system will take place in the tool room, although this is not necessarily the place where the benefit will be most evident. The manufacturing cells using the tools should find that their tools become more reliable, as preventive work replaces repair work - and the sum of these decreases in magnitude.

Impact of Cellular Manufacturing

A point which was noted during the project was the convenience of a cellular layout and distributed cell based control systems for this type of development, as described above. The levels of Information Technology available for data collection and results distribution vary significantly from cell to cell, and cell-specific interfaces will be required. The main benefit in this, apart from the facility for future development of the system, is that the interfaces will be customised. Also, much smaller numbers of tools are dealt with in each cell than if the presses were all located in a press shop, and manual control would be feasible for most cells. Another advantage important during development of the system is that individual cells can be isolated for analysis and prototype work.

Feedback from the production process into the design stages can be considered a final aim in this advancement of more effective tool control through more effective information management. The effects of tool failures can only be avoided by eliminating their causes, and this task will be made much easier by supplying structured information directly to the Engineering departments. This information should include experienced tooling problems, as well as possible failure causes (rather than just maintenance summaries etc.). Such feedback will aim to achieve an even more reliable system in the future by preventing experienced problems in the first place. The interesting issue of change management will also have to be addressed upon implementation of the final software.

CONCLUSION

The impact of Group Technology on information management in a manufacturing organisation is severe. On a surface level, it affects the interfaces for input/output collection and distribution, but the underlying structure of the support systems will also be affected very strongly by the choice of this type of manufacturing system.

These information management issues can also have strategic implications for a company adopting CM. The cell structure will affect the technological evolution of the company, in terms of Information Technology as well as production technology. The encapsulated nature of each cell provides a path for dynamic growth of the organisation without disrupting the remainder of the system.

Several current approaches to Open Systems Architectures describe the application of such architectures to the control of manufacturing systems (e.g. the CIM/OSA model developed under ESPRIT). These provide a reference for different levels of technological sophistication in manufacturing organisations. They can be used to illustrate the strategic impact of CM, and the different control needs which arise at different levels in the hierarchy. In a more traditional type of production system lower in the hierarchy, an ad hoc response to appearing problems may suffice. Higher up in the classification, however, total control is essential for achieving any form of management of the production process. This control has to be provided at the machine tool level, incorporating architectures, processes and communications. All of these have to be operational and reliable, otherwise a complex system will collapse sooner or later ("link or fail").

The move towards the higher levels of these hierarchies, which must rank as a principal strategic aim for the modern manufacturing organisation seeking to remain competitive, can be facilitated by the adoption of Cellular Manufacturing. The encapsulation of sub-production units will obviously reduce the risks and consequences of control failure. However, the inherently simpler flow through a Group Technology cell further improves the grounds for testing and improvement practices. It could be postulated that the company described here might not have reached its current level of control sophistication, and the resulting manufacturing performance, without the use of Cellular Manufacturing. This illustrates the importance of positive control at the top of the organisation for continued domination of the operational processes.

REFERENCES

[1] Schonberger RJ: World Class Manufacturing: The Lessons of Simplicity Applied. Free Press, NY, USA, 1986.

[2] Womack JP, Jones DT, Roos D: The Machine That Changed the World. Rawson, NY, USA, 1990.

[3] Hyer NL, Wemmerlöv U: Group Technology and Productivity. Harvard Bus Rev 4: pp 140-149, 1984.

[4] Ham I, Hitomi K, Yoshida T: Group Technology. Kluwer-Nijhoff, Boston, MA, USA, 1985.

[5] Gallagher, CC, Knight, WA: Group Technology Production Methods in Manufacture. Ellis Horwood, Chichester, UK, 1973/1986.

[6] Mitrofanov SP: Scientific Principles of Group Technology. (English Translation), National Lending Library for Science and Technology, Wetherby, UK, 1959.

[7] Burbidge JL: <u>Production Flow Analysis for Planning Group Technology</u>. Clarendon, Oxford, UK. (refers to Volvo plant in Kalmar, Sweden)

[8] Anon: Fiat - where they hope £50m will buy happy workers. <u>Sunday Times</u>, 8/7/1973. (cited in [10])

[9] Schlote S, Kowalewsky R: Flexible Fabrik (Flexible Factory). <u>Wirtschaftswoche</u> 38: pp 52-64, 1991.

[10] Fazakerley GM: Group Technology: Social Benefits and Social Problems. <u>Prod Eng</u> 53: pp 383-386, 1974.

[11] Kellock B: Supplier Spotlight. <u>MAPE</u> 148/3777: pp 119-124, 1990.

[12] Schäfer H: <u>Group Technology Information Management for Effective Cell Formation</u>. PhD thesis, University of Wales, College of Cardiff, UK, 1991.

[13] Bullinger H-J, Niemeier J, Fähnrich K-P: Information Technology: Strategic Weapon for the Factory of the Future. <u>Proc 10th ICPR</u>, Nottingham, UK, 1989.

[14] O'Brien C, Chalk S, Grey S, White AC, Wormell NC: An Application in the Automotive Components Industry, in Voss CA (ed): <u>Just-in-Time Manufacture</u>. IFS, Bedford, UK, 1987, pp 303-317.

[15] O'Brien C: Costs, Benefits and Traumas in Changing from Traditional Process to Cellular Manufacture in Automobile Component Fabrication. <u>Proc 5th Int Work Sem Prod Econ</u>, Innsbruck, Austria, 1988.

[16] O'Brien C, Schäfer H, Bryant JK: An Information Infrastructure for Effective Press Tool Management. <u>Proc Conf Techn Transfer & Implement '92</u>, London, UK, 1992.

[17] Zuin D: Tool Management. <u>Ind Comp</u> 7: pp 30-31, 1990.

[18] Granger C: (two articles) <u>MAPE</u> 148/3789: pp 32 33/41 48, 1990.

[19] Devaney W: Tool Management Systems. <u>Proc 2nd BIMTTC</u>, NMTBA, USA, pp 1.199-1.200, 1984.

A PRODUCTION CONTROL SYSTEM FOR A FLEXIBLE MANUFACTURING SYSTEM — DESCRIPTION AND EVALUATION

Jannes Slomp and Gerard J.C. Gaalman

University of Groningen, Groningen, The Netherlands

ABSTRACT

This paper discusses an hierarchical production control system for a flexible manufacturing system (FMS) consisting of two identical machining centres. The production control system is made up of four levels: (i) determination of job lists, (ii) assignment of jobs to each machining centre, (iii) scheduling of jobs on each machining centre and (iv) monitoring and control. The production control system is now functioning for more than two years. The planning department, the production manager and the operators of the FMS are contented with the system.

One year after implementation, the production control system is evaluated over a period of seven weeks. The evaluation is done on the basis of the scheduling results and the actual weekly output of the FMS. It appears that the production control system performs well in practice. However, the evaluation shows also some critical aspects. In the first place, the assignment of certain production control tasks to the foreman, instead of the operators, is criticised. In the second place, the frequency of the determination of job lists (=job release) is debatable. Finally, it is shown that by means of smaller batch sizes, the MRP lead-times can be reduced.

A general conclusion extracted from this research is that one should not accept satisfaction as an argument to refrain from evaluation. A careful evaluation may lead to further improvements of the production control system and, consequently, of the efficiency and effectivity of the FMS. A regular evaluation fits in a philosophy of continuous improvement.

1. INTRODUCTION

A Flexible Manufacturing System (FMS) is an integrated manufacturing system consisting of automated workstations linked by a material handling system capable of processing different jobs simultaneously. Its effective use in a particular situation is largely determined by its degree of *efficiency*, *flexibility* and *multifunctionality*.

Efficiency. In an FMS, all the preparatory work, like the clamping of parts on pallets/fixtures and the presetting of tools, can be done during machining time. This results in an almost complete absence of changeover times while processing different parts sequentially. An important aspect of efficiency furthermore concerns the capability of an FMS to operate during an unmanned night shift. However, the efficiency of an FMS is limited. Often, the buffer storage of pallets in the FMS is not large enough to support a fully unmanned night shift. Sufficient workload for each machining centre of an FMS is also an important condition for an efficient operation.

Flexibility. Due to the presence of all essential equipment in the system, like pallets, fixtures, tools and NC programs, and through the absence of changeover times, an FMS can easily change from the processing of one mix of jobs to another mix. However, the flexibility of the system is limited by the availability of equipment (e.g. a unique cutting tool cannot be used simultaneously on different machining centres). Furthermore, the introduction of new part types may cause significant preparation times (NC programs, clamping plans) and loss of production time (caused by testing activities).

Multifunctionality. An FMS is capable of performing several processing steps of a job which traditionally were executed on subsequent production units. This multifunctionality can be realized in two different ways: (i) through the use of multifunctional machining centres or (ii) through the integration of various types of work stations in one system. However, the multifunctionality of an FMS is limited. Certain finishing processes, such as tempering and painting, are seldom integrated in an FMS.

An effective use of an FMS requires the integration of the efficiency, flexibility and multifunctionality aspects of the system. A well designed production control system is a key condition to deal properly with the specific characteristics of the FMS. Such a system is often based on a hierarchical decision structure, see e.g. [1]. This paper presents an hierarchical production control system for a particular FMS.

In literature much attention is paid to the development of sophisticated solutions for specific production control problems of FMSs. Most articles are focused on the presentation of heuristics and algorithms (for an overview, see [2] or [3]). There is however less attention for the practicability of the proposed solutions. Only a few authors deal with real-life situations and if they do, for instance [4] and [5], there is no indication whether the proposed production control systems are used in practice or not. In this paper, a production control system developed for a particular FMS is presented and its use in practice is evaluated.

The paper proceeds as follows. Section 2 describes the FMS considered in this paper and the aspects which complicate the production control. Section 3 gives the hierarchical structure of the production control system. Section 4 describes briefly the interactive scheduling tool which supports the human scheduler. Section 5 presents an evaluation of the production control system over a period of seven weeks. Some concluding remarks are given in Section 6.

2. THE FMS AND ITS PRODUCTION CONTROL ASPECTS

2.1. Description of the FMS and its environment

Figure 1 gives a schematic representation of the particular FMS considered in this paper. The FMS consists of two identical machining centres (M1 and M2) linked together with an automatic material handling and transportation system. A parts pallet storage, which has a capacity of eight pallets (1-8), is integrated with the material handling system. Each of the two machines can house a maximum of 60 tools in the tool magazine (T1 and T2). Both the exchange of cutting tools in the tool magazines and the clamping and unclamping of the parts on the pallets is done manually. Two part clamping and unclamping stations (C/U) are integrated in the system.

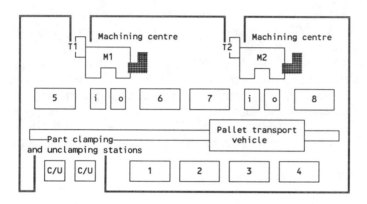

FIG. 1 Schematic representation of the particular FMS.

The FMS is able to manufacture different part-types in a random sequence without loss of production time (if it is operated properly, see below). At the moment of evaluation, the FMS could manufacture 17 different part-types. Tools, fixtures and machining programmes for these part-types were available. A part-type may require more than one operation. An operation corresponds with a visit to one of the machining centres. The processing times of operations range from 14 to 34 minutes. Generally, the next operation on a part requires another fixture/pallet.

At the beginning of each week, new jobs are released to the FMS. The batch sizes of the jobs vary from 10 to 200 parts. Before manufacturing can start, an operator has to (i) prepare the required pallet(s)/ fixture(s), (ii) load the tool magazines with the appropriate tools and (iii) instruct the FMS-computer. After this, one or more parts of a job can be clamped on a pallet/fixture. The FMS-computer controls the material flow in the system. The automated material handling system transports the palletised parts either to the parts pallet storage or directly to a machining centre. After machining, the automated material handling system transports the palletised parts either to the parts pallet storage or directly to a clamping/unclamping station. On this clamping/unclamping station the parts are replaced by new parts. The refilled pallet/fixture continues its way through the system. The parts on which an operation is performed, may wait for their next operation.

The FMS operates in a two manned shifts and a partial third unmanned shift. The size of the parts pallet storage (8 pallets) limits the realizable production time in the unmanned shift.

2.2. Production control aspects

Several aspects complicate the production control of this particular FMS. The main aspects, which are due to the configuration of this type of FMS, are listed below:

(1) Within the FMS, only one pallet/fixture is deployed to support an operation that has to be performed on the parts of a job. A reason for this restriction are the involved costs of preparing a second pallet/fixture for the same (type of) operation. The restriction has consequences for the throughput times on the FMS. To avoid idle time of the machining centres, caused by the time needed for unclamping/clamping and transport of a pallet, an operation on a machining centre is mixed with other operations; the so-called operation-mix. By this, throughput times become mix-dependent;

(2) To avoid tool changes during the processing of an operation-mix (see aspect 1) it is desirable that the number of cutting tools needed for this mix is less than or equal to the capacity of the tool magazine. Furthermore, it is desirable to minimize the number of tool changes when the operation-mix has to change;

(3) A unique cutting tool cannot be used simultaneously on both machining centres. Consequently, the operations on both machining centres should be geared for one another;

(4) A pallet/fixture can often be used for operations on several jobs c.q. part types. These operations cannot be processed simultaneously on the FMS.

(5) To utilize the unmanned shift as much as possible, it is desirable to have enough parts in the parts pallet storage at the start of the nightshift. There is room for maximal eight pallets, therefore operations with long processing times should be selected;

(6) Operations of the jobs should be scheduled in such a manner that the work is done within the due dates given by the MRP system of the firm.

To a certain extent, all these aspects can be controlled by human judgement and experience. However, even in the most simple FMS, computer aided tools would be helpful for people who are responsible for the production control of the FMS. Computerised interactive decision aids would enable the people to better use their judgement and experience in generating a realistic work schedule.

3. THE PRODUCTION CONTROL SYSTEM

In solving the complex production control problems for Flexible Manufacturing Systems, two basic approaches can be distinguished. In the first approach (called an off-line approach), several decisions are periodically taken prior to the start of actual production activities. These decisions concerns, for instance, batching, loading and scheduling (see [1] and [2]). In the second approach (called an on-line approach), the planning, scheduling and control decisions are dynamically taken based upon the actual status of the system at the time a machine completes a specific job in the shop.

In the specific situation described here, an off-line approach is used to deal with various production control aspects. The resulting decision making hierarchy is shown in Figure 2.

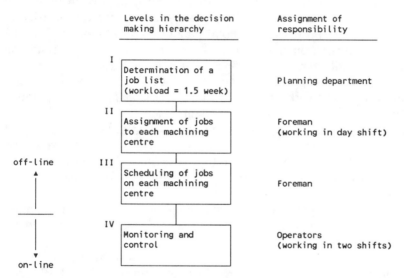

FIG. 2 The hierarchical structure of the production control system.

Level I. At the beginning of each week, an MRP print-out is generated that contains all future jobs of the FMS. The planning department selects a number of these jobs for immediate release to the FMS. This selection is based upon the due dates generated by the MRP system of the firm (aspect 6, Section 2.2) and the workload of the FMS. The total amount of work is limited to 1.5 week. The FMS is given two weeks to manufacture each job.

The job-list, created by the planning department, is formally given to the foreman in whose group the FMS is functioning. The foreman may accept the jobs or not. When the jobs in his opinion cannot be manufactured efficiently, for example because of a bad mixture with respect to the required fixtures (see aspect 4, Section 2.2), he can refuse. So there is some bargaining between the planning department and the foreman. As soon as the foreman has excepted the list, he is held to realize the lead times of the released jobs (= two weeks).

Level II. The foreman allocates the released jobs to each machining centre. Several aspects complicate this allocation activity. As far as possible, job allocation should be done such that each fixture and each unique cutting tool is needed on just one machining centre. By doing this, the machining centres can be considered independently of each other during the remainder of the procedure. In this way, aspect 3 (the aspect of unique cutting tools) of Section 2.2 cause no problems anymore.

Level III. The third level in the production control hierarchy deals with the scheduling of jobs on each machining centre. The human-scheduler (= foreman) has to deal with several aspects. The number of tools needed for a mix of operations should be less than or equal to the capacity of the tool-magazine of a machining centre (aspect 2, Section 2.2). Operations that need the same pallet/fixture cannot be placed in one mix (aspect 4, Section 2.2). Finally, the throughput time of a job should not exceed two weeks (see aspect 1, Section 2.2).

The interactive tool, presented in Section 4, can be seen as an helpful instrument for the foreman. Furthermore, the proposed (machine) schedules give useful information about the appropriate timing of preparation activities.

Level IV The (machine) schedules can be helpful in monitoring and controlling the system. Comparing with reality the FMS-schedule gives information whether production requirements and lead times are being met or not.

4. THE INTERACTIVE SCHEDULING TOOL

An extensive description of the interactive scheduling tool can be found in [6]. With help of the tool the user is able to generate workable schedules, in a menu-controlled way. This Section only describes the output, i.e. the representation of scheduling data.

Figure 3 shows two ways of presenting the same scheduling data. Figure 4a is a quite normal representation (a Gantt Chart) in which one can see in what sequence the operations a, b and c have to be done. Because of the batch sizes of jobs, each operation has to be performed several times. Figure 3b stresses the fact that operations are running in a mix during a certain period of time. At each moment the operation-mix can be seen. The 'relative load' in Figure 3b (the y-axis) shows the percentage of time needed for each of the operations in an operation-mix.

The alternative way of presenting scheduling data, see Figure 3b, was chosen as a starting-point for the development of an interactive scheduling tool. This way of presenting data has important advantages above the conventional one. First of all, it presents the scheduling data in a more concise form. In case of considerable batchsizes of jobs, the conventional way of presenting scheduling data would not provide a readable scheme. The readability of the alternative presentation is independent of the batchsizes. A second advantage of the alternative presentation relates to its 'operation-mix orientation': The major production control problems of the human scheduler (=foreman) are constraints with respect to the operation-mix (i.c. aspects 1, 2 and 4, see Section 2.2).

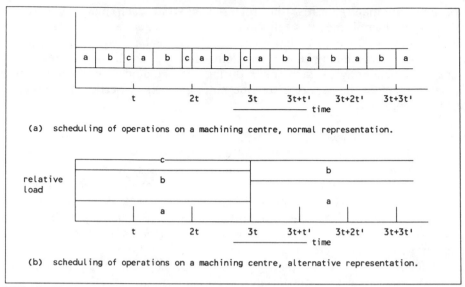

(a)   scheduling of operations on a machining centre, normal representation.

(b)   scheduling of operations on a machining centre, alternative representation.

FIG. 3   Two way of presenting the scheduling data of a machining centre.

The interactive tool is used in the particular case presented in this paper. Figure 4 shows an example of a schedule which was made for one machining centre at the beginning of a week.   Each week (=5 days) a new schedule is performed.   It must be noted that the presentation of scheduling data suggests continuous production.   In the particular situation presented in this paper, this is not the case: The   FMS is working in a two shifts situation.   This fact is integrated in Figure 4 by adjusting the time-axis: Each day represents the real available production time per day.

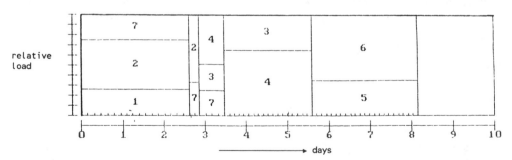

FIG. 4   A real life example of a scheduling result.

5 EVALUATION

Interviews with the people involved in the production control system showed a general contentment with the system. Nevertheless, a more detailed evaluation may lead to new ideas. One year after implementation, the production control system is evaluated over a period of seven weeks. The evaluation consists of three parts. Firstly, the use of the scheduling tool is evaluated. Secondly, the planned (scheduled) weekly output is compared with the real output of the FMS. Finally, the methods used at each level of the production control hierarchy are evaluated.

5.1. Allocation of the scheduling task

The interactive scheduling tool is used by the foreman in whose group the FMS functions. One of the reasons for the acceptation of the tool is undoubtedly the fact that the foreman has participated in the development of the tool. During the development the foreman was direct responsible for the effective and efficient use of the FMS. This situation has been changed recently: The production manager of the firm has upgraded the task of the foreman, he is now responsible for more groups of machines, and consequently, his involvement with the daily functioning of the FMS has been decreased. This fact might have negative consequences since there is no immediate feedback anymore between on-line and off-line activities (see figure 2). This aspect will be illustrated more explicitly in the remaining of this Section.

5.2. Planned (scheduled) versus realized weekly output of the FMS

At the end of each week the detailed output of the FMS is reported to the planning department through an automatic data communication link. While making a new job list the planning department takes into account the remaining workload. In Figure 5 the planned (scheduled) and real weekly output are compared by means of the so-called 'Committed Volume Performance (CVP)'. The 'Deviation-CVP' is defined as follows:

$$\text{Deviation-CVP} = \frac{\text{real production - planned production}}{\text{planned production}} \times 100\,\%$$

As can be seen, the Deviation-CVP per machining centre varies between -30% and 30%. This is caused by, for instance, the unpredictable output of the unmanned shift and the number of breakdowns. Due to the unforeseeable capacity of each machining centre, it is advisable to assign the jobs as late as possible. In the production control hierarchy (Figure 2), the assignment of jobs is done weekly, by the foreman. There is no immediate feedback of the FMS to the foreman (see Section 5.1.). A more on-line assignment of jobs can be realized by the allocation of the production control tasks of level II and III (see Figure 2) to the operators. The operators should have the authority for an on-line move of jobs from one machining centre to the other. The interactive scheduling tool could support the operators in evaluating the consequences of a reassignment of jobs.

In the production control system the planning department authorizes the FMS (i.c. foreman, operators) to perform the released jobs within two week. For machining centre 1 the due date performance is almost optimal: 98 % of the jobs assigned to machining centre 1 are finished within 2 weeks. The due date performance of machining centre 2 is less perfect: 19% of the assigned jobs are too late. To a certain extent this is caused by a major breakdown in week 2-3. The operators did not have the authority to move jobs from machining centre 2 to machining centre 1. This furthermore supports the idea of a reassignment of production control tasks, as mentioned before.

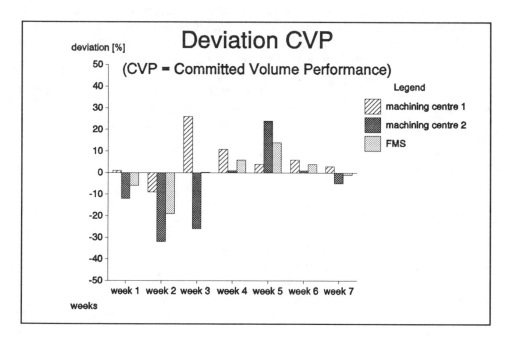

FIG. 5 Deviation between Real and Committed Volume Performance.

5.3. The methods used at each level in the production control hierarchy

The production control system consists of four levels (see Figure 2). Here, the methods used at each level will be evaluated.

Level I   Every week the planning department releases new jobs for the FMS. The selection of jobs is primarily based upon the due dates generated by the MRP system of the firm. The planning department is not allowed to establish priorities within the job-list. As mentioned before, the FMS is given two weeks to manufacture each of the released jobs. In the evaluated period 80% of the jobs were released too late with respect to the MRP due dates. It is observed that the planning department has some strong preferences concerning urgent jobs to be processed within one week (ergo, the next week). A possible way to deal with these preferences without changing the essence of the production control structure, is by doubling the frequency of work release and limiting the total amount of released work to 1 week (instead of 1.5 week). This, however, stresses the flexibility at levels II and III of the production control hierarchy.

Level II   At the second level of the production control system, the foreman assigns jobs to each of the machining centres. It appears that he has assigned the jobs such that during the week there was no simultaneous claims of the machining centres with respect to the fixtures and the cutting tools.

Level III   The scheduling of jobs on each machining centre is done at the third level of the production control system. In the evaluation the schedules of the machining centres for each week are analyzed. The main focus was on the 'Job Throughput Times' of the jobs. The Job Throughput Time is defined as follows:

Job Throughput Time = Finishing time - Starting time
                              of a job   of a job

(Weekends are excluded in the calculation of the Job Throughput Times).

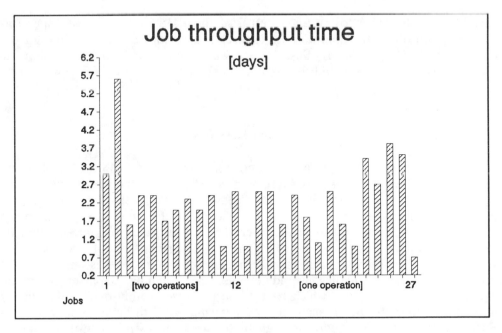

FIG. 6 Throughput times of jobs on machining centre 1.

Figure 6 shows the 'Job Throughput Times' of jobs on machining centre 1. The figure enables one to identify the essential parameters which determine the Job Throughput time. In the evaluation period 27 jobs were processed on machining centre 1. Each part of twelve jobs (1-12) had to be processed in two operations. The parts of the other jobs (13-27) were processed in one operation. As can be seen in Figure 6, the Job Throughput Times of the first twelve jobs do not significantly differ from the other jobs. This can be explained by the fact that the foreman usually schedules the different operations of one job in one operation-mix (Operations 3 and 4 in Figure 4, for instance, belong to the same job). Thus the Job Throughput Time is not dependent on the number of operations per part. Furthermore, it appears that the *individual* processing times per pallet play a minor role with respect to the Job Throughput Time. This can be explained by the fact that the presence of an operation-mix levels out the relative impact of the short and long processing times. The most important parameter that determines the Job Throughput Time appears to be the batch size of a job; more precisely, the number of pallets required for the parts of a job to perform an operation. Figure 6 supports this: The average number of pallets required for an operation of a job is 27. The required number of pallets for (an operation of) Job 2, see Figure 8, is 54. The number of pallets needed for jobs 23, 25 and 26 are respectively 45, 41 and 41.

A conclusion of the above analysis is that the MRP system, c.q. the planning department, can control the Job Throughput Times by means of the batch sizes of the jobs. In this particular 'production-on-stock' case, the MRP lead-time can be reduced by decreasing the batch sizes (for example, from two weeks to one week).

Level IV   The monitoring and control of the FMS is done by the operators. The foreman offers the operators every week a hard-copy of the schedule for that coming week (Figure 4). By this, the necessary preparatory activities, such as the exchange of cutting tools and the preparation of pallets/fixtures for new jobs, are directed.

6. CONCLUSIONS

In this paper a production control system for a particular FMS is described and evaluated. The people involved are very satisfied with the production control system. However, the evaluation shows some critical aspects and leads to some recommendations. In the first place, the allocation of certain production control tasks to the foreman, instead of the operators, is criticised. It is argued that the operators should have more flexibility in the assignment of jobs to machining centres. In the second place, the frequency of job release is discussed. A doubling of the frequency would enable the planning department to put preferences on urgent jobs which have to be processed within one week. Finally, it is shown that the batch sizes of jobs determine the throughput times on the FMS. Therefore, in case of smaller batch sizes, the MRP lead-times can be reduced.

A more general conclusion extracted from this research is that one should not accept satisfaction as an argument to refrain from evaluation. A careful evaluation may lead to further improvements of a production control system and, consequently, of the efficiency and effectivity of a FMS. A regular evaluation fits in a philosophy of continuous improvement.

REFERENCES

1. Stecke KE: Design, planning, scheduling and control problems of flexible manufacturing systems. *Annals of Operations Research* 3:3-12, 1985.
2. Van Looveren AJ, Gelders LF, Van Wassenhove LN: A Review of FMS Planning Models, in Kusiak A: *Modelling and design of flexible manufacturing systems*, Amsterdam, Elsevier Science Publishers B.V., 1986, pp 3 30.
3. Buzacott JA, Yao DD: Flexible manufacturing systems: a review of analytical models. *Management Science* 32:890-905, 1986.
4. Stecke KE: Algorithms for efficient planning and operation of a particular FMS. *International Journal of Flexible Manufacturing Systems* 1:287-324, 1989.
5. Aanen E: *Planning and scheduling in a flexible manufacturing system*, thesis. University of Twente, The Netherlands, 1988.
6. Slomp J, Gupta JND: An interactive tool for scheduling jobs in a flexible manufacturing environment. Accepted for publication in the *International Journal of Computer Integrated Manufacturing Systems*, Butterworth-Heinemann Ltd, 1992.

THE DESIGN AND IMPLEMENTATION OF A DATA COLLECTION AND CONTROL SYSTEM IN A FLEXIBLE MANUFACTURING ENVIRONMENT FOR APPLE COMPUTER

Noel Conaty

Apple Computer, Holly Hill, Cork, Ireland

ABSTRACT

In June 1989, a joint project was started between two of Apple's manufacturing sites in Singapore and Cork (Ireland). Its objective was to provide an integrated system for product serialisation, process data collection and the control of product localisation. The aim of this paper is to share with you the experience we had developing this project.

In Section I, I will briefly outline the complexity of manufacturing products for the international market in our Cork plant, then describe our previous problems with data collection, out of which most of the requirements for the project were drawn up. In Section II, I will focus on the design and implementation of this project. Finally I will conclude by giving an account of the results achieved and the possible developments emerging from this project.

SECTION I

MANUFACTURING PROFILE

The Apple Computer manufacturing plant in Cork Ireland is broken up into the following areas:
- Board Shop with ASRS for incoming parts,
- Systems Build and Test with ASRS (Automatic Storage & Retrieval System) for Run In,
- Distribution Carousel Warehouse for generic product,
- Product localisation and Packout,
- European Service Centre.

The Board Shop manufactures and tests a wide range of PCB's for System build and for direct shipping. Over sixty different product configurations are built in the Systems build and test area.

Product assembly and test is handled on FMS (Flexible Manufacturing System) moving tables and Macintosh controlled serial lines. The products are then Run in tested on an ASRS controlled by a Macintosh running A/UX (Apple's version of UNIX). After Run in the products are automatically transported to the Distribution Carousel Warehouse. The Cork site operates a replenishment program, where the generic products in the Distribution Carousel Warehouse are localised as required. Products are automatically pulled from the carousels in the correct order and ship pallet quantity, and are routed to one of three pack lanes. At the pack lanes the correct language and version of software is installed on the product's hard drive, then the keyboards, manuals, cables are added to the box. The complete contents of the box are verified against customer requirements, by scanning the bar code label on each item. The objectives of this program are to keep the localised product inventory low and the customer service levels high. Product can be shipped overland to any of our European markets within 36 hours. When one product is sold, then another one is manufactured. The stored products may be localised with software and accessories in as many as 26 languages. The Cork plant packs out in excess of 1500 different localised product types. The Product localisations are increasing as we provide the emerging markets of Eastern European countries, furthermore, we handle simultaneous product introductions. We are presently witnessing an increase in the demand for customised products, (for example: a customer may request certain specific applications to be installed on the hard drive).

HISTORICAL BACKGROUND TO DATA COLLECTION

From the onset of the project we encountered a lot of justified cynicism towards data collection. In Apple there existed many DCS's (Data Collection System), which were usually designed by some department outside the manufacturing plant. They were therefore not adapted to our specific needs, and often proved impractical in a rugged environment. The main problems with the existing DCS's were the following:

Since there was no integration with existing control systems, ensuring accurate data entry was difficult. When a product enters the workstation it is locked into position until the tests are completed. When this task is done, the operator may release the product to the next workstation without entering the test results to the DCS.

There was no integration with the Test equipment and most of the data entry was manual which meant that data verification was problematic. A fault code entered by one operator may be different than that entered by another operator.

At Test workstations there could be two Macintosh computers, one as test equipment, and the other for Data collection. Entering data on two systems: the DCWS (Data collection workstation) and the Test Macintosh caused extra work at the workstation.

Report generation was slow and suffered from data integrity because of the above problems. The DCWS for new products and processes were cumbersome to set up and required individual updating at each workstation .

The shopfloor networks were unreliable and required a lot of maintenance.

The DCS's were not designed to handle the increasing growth and changes in business.

Training was an issue as the DCWS's were difficult to use with many levels of screens for data entry and the operators frequently changed jobs on the shopfloor. In many cases untrained operators used the systems and this also resulted in data inaccuracy.

Installing a DCS was difficult as the process was not designed with data collection in mind. The Systems build process consisted of two FMS moving tables each with twenty two workstations to build the low volume high mix products. The Serial lines built the high volume low mix products. A workstation on the FMS moving tables would handle many product types and a product type could be built on either process FMS or serial, this proved difficult for the DCS to handle. The PCB shop consists of four complete surface mount cells, each PCB and system built has a bar code serial label attached which uniquely identifies the product. This is used to track the product through the build and pack out process and to track the product quality in the field.

SECTION II

REQUIREMENTS

For customer requirements, (Finance, Materials, Engineering, Quality, Human Resources, Production, Shipping and MIS) an in depth survey was carried out in both, Cork and Singapore. The purpose of the survey was twofold, firstly to ascertain the requirements for a shopfloor data collection system, secondly to list all the problems with the existing DCS's.

What we need to know about a product as it moves through the process is the following:

what workstation is it at ?

who is working on it and for how long ?

what work will be carried out ?

what will be the result of it ?

where will it go when finished ?

A requirements document was drawn up, which outlined that the system should provide the following: Timely and accurate reports on Product quality, Productivity and process downtime, and Paperless rework, paperless tracking and product routing verification.

At the DCWS, local reporting is required for Statistical Process Control; if the process goes out of control, then the DCWS will stop the line and alert the process engineer. The DCWS will provide a report on the failures that caused the out of control condition. Defect paretos and yields are also required at the DCWS .

To ensure that only trained operators work at a particular workstation and on the DCWS, then the DCS should provide the capability to lock out untrained operators from the DCWS. The operator training records will be maintained on the DCS database.

The DCS will be used as a central location for all the process and product information, and all shopfloor systems will be updated as required from the DCS database. We want to centralise the control of process information, for example, to ensure that testers have the latest version of test software or control equipment have the correct routing for a new product type. All this equipment would normally be updated individually and manually which leaves room for error.

The response time at the workstation is critical, the operator cannot afford to wait for the database to respond before the product can be released to the next workstation. For many workstations the product cycle time is less than a minute, so it is essential that data collection does not adversely affect this time. The database response at the workstation has to be instantaneous. Downtime at the workstation caused by DCS or networks is not acceptable either. We do not want the DCS to become a burden on production. To guarantee maximum uptime and optimum database response, each DCWS has to be able to operate in standalone mode, to put it in other words, the DCWS should have the capability to operate disconnected from the database.

To ensure data integrity, the keyboard entry should be kept to a minimum and should be verified whenever possible. The data collected will vary depending on the function of the workstation (Assembly, Test, Rework..). The data collection screens will vary also, but they should still maintain the same overall look and feel. The number of screens operated by a user should be kept to a minimum of two. Multiple screens are known to confuse the user and increase the training required for users transferring to different jobs.

At all workstations with test equipment, the operator interface for the tester will be through the DCWS. When the operator scans the product bar code, the serial number is passed to the tester to start the test. When the test is finished, the results are passed back to the DCWS. So all operator interface to any test or control equipment will be through the DCWS. This is to provide the operator with a common and simple interface and will also reduce the training required for stations with multiple equipment. Prior to this, an operator might have had to interface with three Macintosh computers at the workstation for running tests, line control and data collection.

SYSTEM DESIGN

The system was designed to support production for two eight hour shifts, five days a week. The reports are run at shift end, and system maintenance is carried out at the weekend. The sizing and choice of vendor for the database and platform was based on the current build schedules and projected growth over a five year period. The database selected was Ingrés, which is a relational database that supports SQL (Structured query language). The availability of connectivity products between the Vax and Macintosh was a major factor for deciding on a Vax platform. Off the shelf client server software was available for the database to the Macintosh. The database and platform were the preferred choice across Apple Corporate. This enabled most of the development time to be spent on the DCWS application.

4th Dimension a Macintosh based 4GL (Fourth generation language) was used to build data entry screens for database maintenance and for reporting stations. The user may run predefined queries at the reporting stations. All these queries are tested so as not to adversely affect the response of the database to the DCWS. User adhoc queries are not allowed for that reason. 4th Dimension was also used for prototyping DCWS's. It has a very short development time, which was necessary to test initial functionality and database response times. Silver Run, a Computer Aided Software Engineering tool, was used to design the logical data model for the manufacturing process. Silver Run then created the tables and the physical database was denormalised for performance. Rules are used extensively for database performance, this reduces the number of queries from the DCWS. One query will trigger a rule to run a procedure that updates the necessary tables.

Workstation Configuration

Information for each workstation is stored on the database; this information is called the workstation configuration and is made up of the following:
- first we define the workstation type: Assembly, Test, Rework, etc... There are up to 20 workstation types between the sites.
- next we need to know what equipment is connected to the workstation, it could be a selection of the following Warehouse Vax for product entry to the distribution warehouse, PLC (Programmable logic controller) for controlling the stops at the workstation to ensure data entry, Bar code printer to print the serial labels for the product housings , Inkjet printer

to print the serial bar code on the PCB's or ICT (incircuit tester). These are just some of the equipment we have connected to the DCWS.

- each workstation type will have it's own predefined list of downtime reasons for that particular workstation.

- each workstation will have a field for the version of the DCWS application. This is used for application version control and the phased implementation of new versions on the shopfloor.

We will now cover the product information stored on the database for a workstation.

We start with a list of the product types valid for the workstation. Some workstations handle in excess of 20 different product types. For each product type we need to know the following:

- the test software and version required to test the product.

- parameters for control equipment, each product type will have specific requirements for automatic connection to test equipment.

- for each board type a list of components by location and possible vendors is required to ensure accurate for data entry.

- a list of the defect symptoms, all data entry is by selection from predefined lists, this will make data entry simpler and ensure accuracy.

- product test parameters.

- SPC (Statistical Process Control)parameters, for example sample size, upper and lower control limits.

- product routing for passes and fails.

- localisation information Language of keyboard and product accessories.

- finally, the timestamp: if any workstation information on the database is modified, the timestamp will change. The workstation configuration is downloaded to the DCWS when the Macintosh is first turned on at the workstation, or if the timestamp has changed from the last download. The workstation configuration is saved on the hard drive of the Macintosh and may take up to 2 meg, this provides the DCWS with the capability of working in a standalone mode, that is, disconnected from the database. If the DCWS Macintosh is connected to Test Equipment or Control Equipment at the workstation then it will pass on any configuration information required by that equipment. As shown in Fig 1, the DCWS, depending on its configuration, will communicate to many shopfloor systems, printers, and testers.

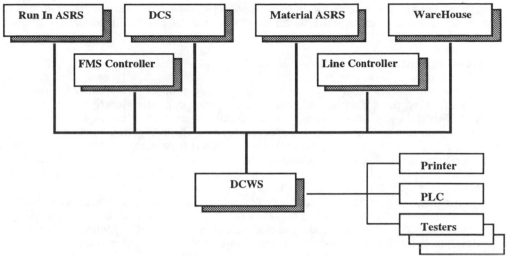

Fig 1 DCWS interfaces to Shopfloor systems

DCWS Application

The Macintosh SE was chosen for the DCWS because it is robust, compact and self contained with a built-in monitor and built-in networking and serial communications. With its small footprint the Macintosh takes up very little room at the workstation. The Macintosh graphical user interface is used to simplify data entry and reporting. There is very little keyboard entry, most manual data is entered with the mouse by selecting from predefined lists. With the varied functions of the workstation, as shown from the workstation configuration above, it was decided to build one DCWS application that would handle all the workstation types. The application is a collection of modules residing on a platform to handle their interaction. The development languages used were C and Pascal. The workstation configuration on the database will decide what application modules are active for that workstation. Some of the modules represent areas on the screen, which are called panes. The application screen will vary from workstation to workstation, depending on what panes are active. To develop a module, strict, but simple guidelines are laid down for interface and display. If any third party equipment is connected to the DCWS, the message formats and communication protocols are standardised. The above mentioned approach is used to provide a controlled development environment between the sites, while allowing the sites to develop their own application modules to suit site specific needs. With one application the look and feel of the user interface will remain the same over the various workstation types. The development time for new workstation types is a fraction of the time required to develop new applications. Source code management is in operation between the sites, so as to maintain only one version of the application.

Since there are 150 DCWS's at the site, it is difficult to maintain version control of the installed DCWS's. There is a field on the database for each workstation, which contains the version of the DCWS to be run at that workstation. This field is queried at regular intervals by the DCWS and if it differs with the version of the DCWS running, then the DCWS will shut down and read the new version of the DCWS from a predefined fileserver and start up again. When a new version of the DCWS is to be installed, the version field is changed on the database for all DCWS's. Each DCWS will quit and reload the new version. The field change is managed so that the DCWS's will not try to download the new application at the same time. This gives us application version control from the database and reduces maintenance time.

System operation

As mentioned above, the system was designed to support production over two eight hour shifts five days a week and was based on the projected growth over a five year period. This radically changed early on in the implementation. With the release of a wide range of products and the associated growth increase, the production work cycle changed to three eight hour shifts with weekend work. This left little time for reporting and system maintenance. With the DCWS's integrated to the workstation we had to ensure maximum DCWS uptime and optimum response times from the DCS. To maintain the DCS performance requirements the following steps were successfully taken:
- when running daily reports most DCWS's are disconnected automatically from the database, except the ones requiring maximum connection time. The disconnected DCWS's are to operate in standalone mode and will reconnect automatically when the reports are completed. Procedures will be run on the database to reconcile the saved data from the DCWS's.
- all the system and database management functions are automated and Saturday is reserved for these functions, when most DCWS's are disconnected. All Saturday's production data is saved locally and sent up to the database on Sunday.
- reports are run to monitor the performance of the system and to alert maintenance people of possible DCWS downtime.

Over a twenty four hour period the database activity was measured at 200,000 queries from the shopfloor. To maintain the high transaction speeds, there was a trade off on real time reporting. Reports are run at shift end, day end and weekend. We are currently investigating options to provide real time reporting.

DCWS operation

When the DCWS application is started for the first time at a new workstation, the operator is asked to enter workstation ID. The DCWS will query the database with the given ID and the workstation configuration is downloaded from the database. After the download, the DCWS will configure itself based on the download information for example Workstation type and functions. The configuration information is stored on the DCWS internal hard drive. The DCWS will only reread the configuration if the database is changed for that DCWS. The operator enters his/her ID number which is then verified. If the operator is untrained, the DCWS will report this back to the database and log the operator off. All product data entered is first saved to the internal hard drive and is then sent to the database. If the database confirms that the data was sent successfully, then the saved data is removed from the internal drive, otherwise the data will be sent later. This will ensure that no data is lost. If for any reason the transaction response time is greater than 5 seconds, depending on query type, then the DCWS will disconnect and reconnect later. All product information as it is collected, for example number of passes, fails and fail information is saved locally on the hard drive. In the event of a power down at the workstation, the application will rebuild its reports and screens from the saved data, so that the operator may carry on.

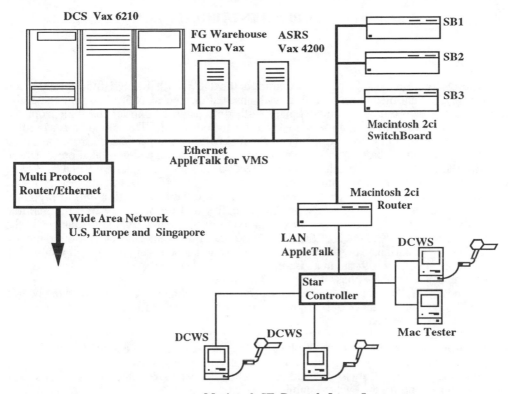

Fig 2. System Architecture

System Architecture

There are currently 150 DCWS connected to the Vax 6210. A DCWS is made up of an Apple Macintosh SE with 2 meg of ram and a 40 megabyte internal hard drive with a bar code laser scanner connected through the keyboard port. The bar code label used for tracking the product through the process is applied to the product housing. The old DCS's used light pens to read the bar code label. This was discontinued because the pens scratched the product housings and damaged the bar codes. A product may go through 10 workstations and the bar code label would have to be read at each workstation. The bar code label would be lucky to survive all this scanning.

Up to four workstations are connected to a port on the star controller and there are twelve ports for each star controller. All the network devices at the workstation are daisy chained. The Router connects LocalTalk and Ethernet segment and isolates traffic between them.The SwitchBoard (Macintosh 2cx) is a concentrator for the DCWS to reduce the number of connections to the database. Up to 30 DCWS may be connected to a SwitchBoard and there are eight Switchboards with two connections each to the database. The Switchboards may be configured for the number of connections to the Vax and the number of DCWS per SwitchBoard. Each DCWS is assigned to a particular SwitchBoard.

IMPLEMENTATION

Phase 1

Both Singapore and Cork have installed the DCS with limited functionality and reporting. This entailed no integration with any test or process control equipment. Daily reports are generated automatically using 4th dimension to extract the data from the database and present it in a simplified format for the customers. This raw data is stored on a fileserver for further engineering analysis. Weekly reports are generated from the stored data using Macintosh based applications for data manipulation and graphical presentation. Locally the DCWS report product yields defect paretos. For this phase 150 DCWS were installed at the site over a nine month period. During the installation, training was provided for application usage. User groups were setup to ease the introduction and to highlight problems, so that improvements on functionality and reports may be implemented promptly. During this stage many new station types evolved, mainly due to changes in the manufacturing process. Much of the time was spent fine tuning the DCS, both database and DCWS application, to ensure maximum uptime and automating the system monitoring techniques, especially with the increase in production schedules.

Our main problems in phase 1 were the following:
- since the DCWS's were not connected to the workstation controls, ensuring data entry was difficult. The numbers varied greatly at common workstations as it was up to the operator to enter the data.
- we had no links to the business system, so keeping the DCS updated with process change information was a major problem in phase 1. The process was in constant change with the increasing number of new products and country localisations. We had no process in place to make sure that the DCS was updated on time and this led to data integrity problems.
- it took some time for the DCS to stabilise, and during this period the system required a lot of monitoring to guarantee maximum uptime. With production running 24 hours a day the system performance suffered. Much of the system monitoring is automated now with the Vax running performance reports. If any problem occurs, the Vax using electronic mail will report it to the maintenance department.
- during this time it was difficult to get the DCS accepted as an integral part of the process. To the manufacturing engineer, data collection was always an afterthought when designing a manufacturing process and the DCWS Macintosh would usually end up in some obscure position at the workstation.

Phase 2

With the DCS installed and in a stable condition, we have started implementing phase 2. It consists of integrating the DCWS with any automation, control processes and test equipment at the workstation. This is a time consuming process as we interface to multivendor equipment for data acquisition. All product serialisation from PCB's to systems is controlled centrally from the DCS, in order to make sure that each product receives a unique serial number. We have integrated the DCWS's to the FMS moving tables and Serial lines and, as a result, the data accuracy has improved greatly . The products will not move to the next workstation until the data is entered. During this phase we are also concentrating on real time report generation using 4th Dimension (4GL). Procedural structures are being put in place so that all the shopfloor systems are updated timely with process change information and, where possible, this will be automated by integrating the DCS with the business system.

CONCLUSIONS

Future Developments

The DCS will be the foundation for other projects, for example we intend to develop a project using the information collected from the DCS to provide guided fault finding. The DCS collects all the product failure information from symptom to fix down to component and vendor.

The DCS will be expanded for tracking the quality of parts supplied by vendors and to collect information for machine preventative maintenance (PM) to make sure that the PM is carried out and on time. The DCWS currently collects process information for the DCS and for the Generic warehouse. In the future, the DCWS will also display scheduling information for the scheduling system, request material from the material ASRS and build instructions for the operator.

The conclusions to date are:

When we started the project, we had in mind data collection only, but as the system evolved we began to realise its true potential reaching far beyond our initial concepts. We have laid down the foundation which will provide us with a competitive edge in a customer demand environment throughout the 90s.

For this reason it has being important that the control of the design and development remain with the manufacturing sites, and is seen as an ongoing continuous improvement process, rather than a once off solution.

- credibility was always foremost in our minds, so a low profile implementation is important to slowly build up credibility and customer acceptance.
- in hindsight, it would have being better to limit phase 1 implementation to a confined area and build up the DCWS functions, instead we had sitewide implementation and a lot of time was spent fire fighting.
- our business is rapidly changing to customer demand manufacturing, so, the DCS has given production greater control on product tracking and localisation. This means that the customer receives what he ordered and on time.
- the central control provided by the DCS ensures that the correct procedures and product revisions are used in the process, and that only trained operators may work at the workstation. This was a factor in the achievement of ISO 9002 certification for the Cork site.
- with the implementation of the DCS and its integration with automated test and process control systems, the problem of insufficient technical maintenance was highlighted. The systems are gradually becoming more automated and dependent on each other. If one link in the chain breaks, then productivity is greatly affected. The maintenance requirements should be planned into the system requirements. We realised this during the implementation, when the DCS caused production downtime mostly due to inadequate maintenance. Operators should play a more active role in maintenance, as the workstation becomes more automated, especially in the area of preventative maintenance.
- the integrity of the reports is most important, so that engineers and management may

make informed decisions. It is essential that an information process exists and preferably automated, where all the shopfloor systems are updated on time, otherwise data integrity will suffer. Taking into consideration our experiences with distributed systems and the difficulty of keeping information consistent, we have opted for central control of process information residing on the DCS.

- in order to have accurate data entry, it is essential that the DCWS is an integral part of the process. For example, if data is not entered on the DCWS, then the process stops. In our experience if data collection can be bypassed at the workstation, it will. All data entered should be validated, in otherwords, the operator should select from predefined lists only.

- a comprehensive training program is required prior to implementation, for all the people involved with the DCS, from the users of the data to the people who enter the data. The training should not only cover the DCWS operation but also its objectives and highlight the benefits for product quality and productivity. Such training should alleviate people's natural apprehension towards data collection.

ACKNOWLEDGEMENTS

I would like to thank my colleagues in the CIM department of Apple Cork: Tommy O'Connell, John Brennan, Karl Grabe, Gerard Spillane, Joop Kaashoek and Ken FitzGerald Smith for their assistance at various stages in the completion of this paper.

THE APPLICATION OF A GENERIC CRITICAL RESOURCE SCHEDULER TO A MANUFACTURING SYSTEM

[1]John Driscoll and [2]Simon F. Hurley

[1]*Department of Mechanical Engineering, University of Surrey, Surrey, England*
[2]*ISE Department, Virginia Polytechnic Institute and State University,*
Blacksburg, Virginia, U.S.A.

ABSTRACT

Examining the flow of work through any batch manufacturing environment is an area of research identified early in the development of Industrial Engineering and is of relevance in todays competitive environment. This paper outlines the current status of a generic scheduler, including an introduction to the simulation software developed to support the model and an application in a Job Shop Environment.

The Generic Critical Resource scheduler has three major component parts; the Universal Transfer for representation of a work centre, methods used to identify the "Critical Resource" in any manufacturing system and the process of scheduling all work around the "Critical Resource". The three areas are commented on in this paper.

Extensive software, developed in a high level structured programming language, creates a user orientated environment which encapsulates the Critical Resource scheduler at the PC level. Thirty one computer programs access eighteen project files in order to undertake considerably detailed scheduling simulations. The structure of the computer suite will be outlined in the paper. The model and the simulation software are demonstrated through an example set in a simple Job Shop environment, with a theoretical programme of work selected to show the features of the software with respect to variable sequencing of work ordering models, applied over a finite time horizon.

INTRODUCTION

The job sequencing problem may be defined as:

> *"Given n jobs to be processed, each has a setup time, processing time, and a due date. In order to be completed, each job is to be processed at several machines. It is required to sequence these jobs on the machines in order to optimise certain performance criteria"*
> *Elsayed et al* [1].

With an industrial environment, where capacity sensitive sequencing is required, an approach of current interest is those procedures involving "bottleneck analysis".   The principle of "bottleneck scheduling" involves the identification of one or more critical work centres that dictate the overall rate of manufacturing throughput.   Once identified, work is sequenced for these "bottleneck" work centres and all dependent manufacturing work centres are then scheduled to the work sequence of the critical resource set.

Originally identified as a solution procedure by Baker [2], the most well known application of the approach is OPT (Optimised Production Technology) [4]. Both the comments of Baker and the OPT procedure are discussed later in this paper. One feature of OPT is the secrecy with which the operation of the supporting commercial package is surrounded and the proprietary nature of the software. This restriction has created the need for a more general "bottleneck scheduling" philosophy.   Within this paper one such philosophy, the MSA approach (Manufacturing Simulation and Analysis) is explained and shown in operation.

BOTTLENECK SCHEDULING

The first two modern references of significance to bottleneck scheduling can be attributed to a text by Baker [2] and a paper by Bellofatto [3]. Quoting from Baker (page 10) [2],

"For example, in multiple-operation processes there is often a bottleneck stage, and the treatment of the bottleneck with single-machine analysis may determine the properties of the entire schedule"

Baker then goes on to discuss relevant single machine scheduling approaches. A very similar comment was put forward by Bellofatto [3]. The "bottleneck scheduling" method unusually generated little interest until the later appearance of OPT. The potential of using this approach exists with nearly all manufacturing facilities, for unless perfectly balanced and equally loaded, bottlenecks must exist.

Bottleneck Analysis

Goldratt and Fox [4] defined a bottleneck as:

"any machine whose capacity is equal to or less than the demand placed upon it".

This definition is limited in scope in two ways. Firstly, it is only relevant to work centres within a manufacturing facility. Secondly, only resources which are fully loaded or overloaded are technically regarded as a bottleneck. Burbidge [5] sought to widen the definition from critical workstation to include critical management functions:

"the work centre or management function which limits the manufacturing output from a factory".

An over-lengthy procurement process, lengthening the order to delivery lead times, would be an example of Burbidge's thinking.   Acknowledging the existence of bottleneck resources or processes leads to the question of how quantitatively these critical resources are identified. In a notable article on the subject of capacity analysis and bottleneck identification Karamarkar et al [6] demonstrate that product throughput is limited by excessive queuing delays and puts the view that a a bottleneck is any work centre or machine with a substantial queue.

Identifying the real queue locations is virtually impossible by mathematical analysis in a dynamic manufacturing system. The most powerful method of identifying these "queue length" based critical resource positions is simulation, discussed later in this paper after examination of the most widely known critical resource package, OPT.

Optimised Production Technology

The concepts of Optimised Production Technology (OPT), the most well known capacity sensitive scheduling method was introduced by Goldratt and Fox in the popular management book The Goal [4]. Subsequently the OPT software and methodology have been reviewed widely by Everdell [7], Jacobs [8], [9], Manning [10], Meleton [11] and Plenert and Best [12]. No paper has, however, reviewed the software methods employed to identify the bottleneck resources, to calculate lead times, to calculate due dates and to create schedules.

The Methodology Claimed by Goldratt and Fox to be a radical approach to manufacturing, OPT is based upon nine non-quantitative tenants which schedule workload around identified bottleneck machines. Non-bottleneck machines are considered "servants: of key critical manufacturing resources and the point is made that a loss of production from key resources is a loss to the whole manufacturing system.

OPT software has been widely applied in batch manufacturing environments. Operating with sales data, resource information, product routes, bills of materials and work patterns, production is modelled for existing factories.

The OPT use of scheduling around the bottleneck is an example of a dominant global heuristic, where the ordering assigned to the bottleneck is applied to all connected machines. A further development within the OPT software is the availability of variable lead times and batch sizes. The mathematical supporting proof of variability is provided independently by David and LeFevre [13].

The assessment of schedule "quality" with OPT is based upon three criteria: throughput, inventory level and operating expense. Throughput is defined as the rate at which the system generates money through sales. The Inventory criteria is all the money the system invests in purchasing items the system intends to sell and Operating Expense is the money the system spends in turning inventory into throughput.

The OPT philosophy is viewed as having a lack of credibility within the academic community because of the secrecy surrounding the mathematical scheduling algorithms, where the lack of explicit formulae significantly reduces the ability to test the validity of OPT computer software results. The fictional style of The Goal only serves to increase the lack of credibility. In the less pragmatic and more applied world of industry, OPT is expected to prove itself by supporting the control of manufacture. The list of original industrial users is extensive including Ford, General Electric, Caterpillar and Perkins Diesels. Whether the software remains in use long term has yet to be seen.

The authors supports Jacob's opinion [8], [9] that the OPT approach is actually an amalgamation of unreferenced earlier work. For example, Goldratt stresses the fundamental importance to the scheduling approach of the bottleneck or critical resource. A point identified by earlier authors Baker [2] and Bellofatto [3] in 1974 as discussed earlier in the paper. Vollman [14] identifies an "A" class MRP systems implementation with finite loading would create a package superior to OPT.

SIMULATION IN PRODUCTION CONTROL

The role simulation can play in production sequencing and scheduling is given by:

"In its broadest sense computer simulation is the process of designing a mathematical-logical model of a real system and experimenting with this model on a computer". [15]

Manufacturing systems may be viewed as queuing networks. Discrete event simulation can be applied extensively to analyse the queuing properties of manufacturing systems. Resources (work centres) are modelled as servers or transformation processes. Customers (parts to be processed) enter a transaction at some predetermined time (arrival event) and leave (departure event) at some predetermined point in time in the future. This dynamic discrete process when modeled accurately, allowed detailed technical evaluation of production planning and control.

The original role of simulation was to support the justification of new manufacturing facilities and evaluation of resource requirements or equipment needs. A newer, and now more important role for simulation, as a discrete modeler in manufacturing, is set in the shorter term on-going scenario, fulfilling the role of aid to the decision making areas of equipment scheduling, shop order release, and work order scheduling.

From the review of "bottleneck scheduling" approaches and the general role of simulation the case has been made for a new "open" modeler in the critical resource modeling area. The remainder of this paper outlines the modeler (MSA) capable of working with critical resource scheduling and an illustrating test case set of simulations is given for the procedure.

This newer role for simulation permits multiple sequencing methods to be modeled quickly, fairly, and inexpensively without disturbing actual operations. The multiple priority rules available under simulation control lead to a more productive and rapid response to production changes. This is in contrast to analytical solutions to scheduling problems in job shops, which have been of limited use.

THE MSA SIMULATION AND SEQUENCING METHODOLOGY

The MSA (Manufacturing Simulation and Analysis) procedures form a widely applicable generic and intuitive simulation and sequencing methodology. Building upon the advantages of simulation and the existence of a bottleneck resource in all but perfectly balanced job shops, MSA uses this critical resource to expertly aid production sequencing.

An introduction to five key aspects to MSA are described in this paper.

1. The Generic Modeler
2. The Simulation Driver
3. The Simulation Database
4. Sequencing Methodology
5. The Software

The Generic Modeler

The MSA modeler uses one entity type a "Universal Transfer" (UT) to model all processes. A UT is a resource with two input queues, a work centre and one output queue. The MSA modeler does not maintain an explicit representation of a facility network. Each UT is defined in isolation and each job, via the Sequencing Database, "self-carries" its route through the facility.

Having no defined network has several advantages. If a new resource is to be added or one deleted a network model does not have to be accessed and modified. Only the Sequencing Database needs to be referenced to determine if any deleted resource is used by any job about to be scheduled. To edit the routes which jobs take through a facility requires only the BoM file be edited, the Sequencing Database is updated automatically.

No specific gateway resource need be identified. When a job is launched the initial components are identified and loaded into the relevant queues. Jobs can be launched at any work centre, visit any as many times as is required and exit the system from any. The use of the universal UT approach can therefore be seen to be highly versatile and speedy in model construction.

The Simulation Driver

The simulation driver is designed to operate in a highly versatile manner by having jobs "self-carry" all related processing data from the UT to UT. Each defined UT contains the logic variables which dictate how jobs passing through will be processed. The simulation driver is basically a discrete event simulator which uses two primary databases: a Sequencing Database and a Resources Database both of which are described in the next section.

The majority of manufacturing events occur at discrete points in time. For this reason the MSA simulator module utilises the "Discrete Simulation" *world view* to drive the simulation clock. Two non-conditional (time-dependent) events are defined:

1. set up complete; and
2. processing complete;

the other events being conditional on the state of the model:

3. transfer of jobs between UT's;
4. launch jobs from the Work Pending File; and
5. load job onto an idle process.

Events 1 and 2 define event times. At each event time the conditions for 3, 4 and 5 are checked to see if they are ready to fire.

The Simulation Database

The *Sequencing Database* contains a record for all production stages for every defined end item, sub-assembly and component. Locating all the information in one central location speeds up the execution of the program. Two versions of the Sequencing Database are maintained: a master and a run-time version. The master version is created immediately after a BoM definition has been input into the software. Five fields in the master version are blank to be initialised when the run-time version is created as a job is about to be launched to the facility. Each record is twenty one fields split into four groups: Simulation Control, Product Description, Batch Sizing and Succeeding Product Information.

The *Resource Database* has one record per process defined. Each record has thirteen fields split into the groups; Information, Logic and Operational. Each record fully defines a UT, for example, the queue length, resource based set up (if required), queuing discipline, how many jobs may be worked on simultaneously. Each UT may be configured individually or all globally set to defaults and edited later. A job residing in an input queue is represented by a ten digit number. The first five digits are the Sequencing Database record number for the particular part on the specific work centre. The second five digits is the record number of the part that it will become when it has finished processing at the current work centre.

The Sequencing Methodology

The sequencing methodology uses the designated Critical Resource as the focus for assigning priorities to jobs within the system.

There are five methods by which the Critical Resource can be identified and selected. The resource can be identified manually when the significance of particular work centres is under investigation. The Critical Resource can be selected randomly, frequently used to generate results that can be compared to the results from other selection procedures.

There are three structured methods built into the software for computer selection of the Critical Resource. The first method, LM1 static loading model, utilises highest expected loading as generated by the Work file. The second method, LM2 dynamic loading model, simulates the period of interest to determine loading on each resource and selects highest simulated loading to identify the Critical Resource. The third method, QM1 dynamic queuing model, again simulates the period of interest but in this case determines the Critical Resource from queuing statistics. The two statistical results used are time average queue length and time average processing time.

Jobs are released to the shop floor in the form of a weekly work plan, representative of real manufacturing practice and a departure from the more common distribution based job arrivals. The weekly work plan is updated to include jobs that have become required. Job requirement is calculated by back tracking from job due date by the total work content of the critical path akin to a Just-In-Time procedure.

From the Work Pending List a Criticality List is generated. The list contains all the parts which are to be processed by the designated Critical Resource. The list is ordered with respect to the currently active dispatching rule. There are four dispatching rules or procedures built into the MSA philosophy; random, first come-first served (FCFS), shortest processing time (SPT) and earliest due date (EDD). The random selection again allows the generation of test results for efficiency comparison. The three remaining rules are well established priority rules, with FCFS being known for simplicity and SPT and EDD being known for high quality results.

The Work Pending file is ordered with respect to the Criticality List and the jobs are launched following the strict order of this file. If there is insufficient room in any of the input queues for a job about to be launched is given the status "blocked" and no component of it is launched. The next job is then selected for launching.

The selection of which job to load from an input queue onto an idle work centre is a two step process. Firstly, a list of jobs which are *eligible* for launching is generated from those currently in the input queue. To be eligible all the components must be at the work centre at least in sufficient quantity to produce one transfer batch, the size of which is determined by the scheduler. The list of eligible jobs is compared against the Criticality List. The job which is highest on the Criticality List has the highest priority and is loaded onto the resource and processing begins. If none of the eligible jobs are to be processed by the Critical Resource (hence not on the Criticality List) the currently active dispatching rule is used to set priority.

The third stage to the total Sequencing Heuristic is to remove jobs from the Criticality List once they have been processed by the Critical Resource. This ensures that any Critical Resource bound job has the highest priority.

The Software

The MSA program suite is the software implementation of the research detailed above. The programs are in four functional *groups*: Control, Systems Management, Project Management, Simulation. Within each group three *types* of programs are used: Primary, Support, Library. In total there are thirty one linked compiled computer programs supporting the current version of MSA as illustrated in Figure 1. Approximately thirteen thousand lines of code are built into the thirty one computer programs An important feature of the language is the portability of the programs. The compiled code is designed to be portable executable on any IBM or IBM compatible computer with a hard disk running DOS 2.0 or later.

The supporting database consists of Software Control files (to direct simulation activity) and Project Files (to represent individual manufacturing projects). A total of fifteen individual projects may be defined with a maximum of 999 simulation runs per project.

Each project consists of a root file set and the individual file sets for the simulation runs known as *versions*. It is possible to run a simulation directly from the root thus creating a new version or alternatively to load a previously created version and continue that simulation run. This feature allows the loading of any project-version-period combination and creation of a new project from this start point. Any version of any project can itself be transferred to become the root version of a whole new project. The ability to rapidly construct and model manufacturing problems allows the scheduler to try several experimental simulation runs and then copy the "best" results to be the master copy and actual issued manufacturing sequence for production.

The software is completely menu driven in order that an end user does not have to be familiar with which program to use or file to load to achieve a certain goal. The user stays within the menu environment for as long as the program is active. A goal of the software was to create a powerful simulation tool which was user friendly and this has been achieved by the use of more than thirty menus. Wherever possible the same menu design and logic has been adhered to.

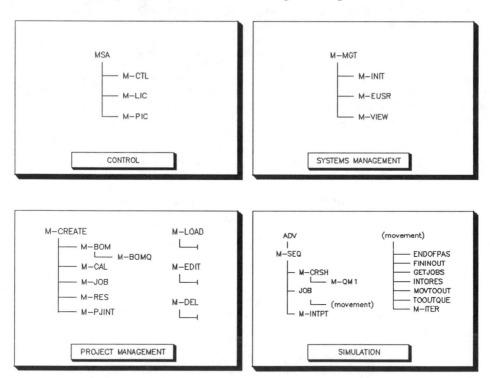

Figure 1 : The MSA Software Structure

AN ILLUSTRATING TEST CASE

To demonstrate the potential of the MSA approach an illustrating test case showing manufacturing simulation for a 12 work centre 16 job per time period scenario is given at this point. The problem is one of a series of theoretical cases developed to test the software and to examine two initial questions in modeling manufacturing systems: determining the start-up to steady state boundary time period and decisions on stable and unstable manufacture.

A frequent problem with simulating production is that many simulations start with empty and idle production resources. Work is then fed into this resource until the normal level of work loading is established, referred to as steady state operation. MSA models this progressive build up and records on a period by period basis the level of work and Manufacturing Error (ME) where ME is defined as actual flowtime minus theoretical flowtime.

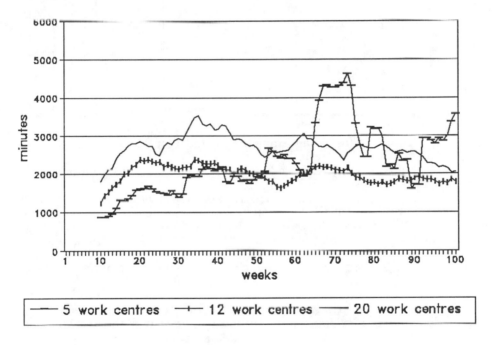

Figure 2 : ME Moving Average for 100 Periods with 16 Jobs

Figure 2 shows the ME for three different factory layouts using the sixteen product manufacturing program. Statistical tests on the ME (t-test) identify the start of steady state production at period 70. The start of steady state is found by calculating a 95% ME mean confidence interval and using this band to detect the point at which the mean consistently falls within the interval.

Even in the steady state production time period (period 70 to 100) the week by week variability in performance may identify an "unstable" manufacturing situation. The 5 work centre factory shown in Figure 2 demonstrates this instability with ME performance varying considerably. Where instability is evident the results

	average mean flow time			percentage of jobs tardy		
Load	70%	100%	200%	70%	100%	200%
	value rank	value rank	value rank	value rank	value rank	value rank
CRITICAL RESOURCE IDENTIFICATION METHOD						
Random	2201 3	6846 3	10090 3	46 1=	41 3	81 4
LM1	2183 1	7175 4	10019 1	47 2	49 4	77 3
LM2	2194 2	6719 2	10066 2	48 3	40 2	56 1
QM1	2248 4	6679 1	10329 4	46 1=	34 1	65 2
DISPATCHING RULE						
Random	2224 4	6984 4	10597 4	47 2=	41 1=	70 3
FCFS	2208 2	6858 3	10169 3	47 2=	42 2	69 2=
SPT	2138 1	6705 2	9907 2	42 1	41 1=	64 1
EDD	2257 4	6872 1	9831 1	47 2=	44 3	69 2=

Table 1: Illustrating Example Selected Test Results

from all production assignment procedures should be treated with caution.

For the 12 work centre 16 product scenario the simulation, based on a random selection of Critical Resource and a random sequencing rule is "frozen" at time period 100 to form the basis of substantial further experimentation on Critical Resource selection and sequencing. By varying the batch size quantities for the 16 products the effect of loading at many levels is simulated for combinations of CR selection and sequencing rule. An extract of the results is shown in Table 1.

The most consistent Critical Resource selection method for under-loaded, evenly loaded and overloaded manufacturing scenarios was the LM2 model. The procedure was ranked second in all three scenarios for throughput time, showing its versatility and was ranked equal first for lowest percentage job tardiness (customer satisfaction) in heavy and overload situations.

On the question of dispatching rule SPT, with its reputation for minimising the mean flow time, performed extremely well being first ranked in throughput time and consistently first in customer satisfaction.

The results refer to simulation over the time interval period 100 to 130. This again demonstrates the power of the MSA approach where from any selected start point extensive testing of CR identification methods and dispatching rule can be carried out to determine the best manufacturing programme. The full test results collect some 35 relevant parameters for analysis of manufacturing effectiveness and customer satisfaction.

COMMENTS AND CONCLUSIONS

Within this paper a new approach to scheduling work for the job shop environment has been explained and demonstrated. The most significant comments and conclusions from the overall work would include:

The MSA philosophy and software provides a viable scheduling approach for medium sized realistic manufacturing scenarios.

The concept of using a Critical Resource as the basis of scheduling has been shown to work. Further experimentation will allow comparison with conventional scheduling work from both the view of schedule quality and computational efficiency.

Within the heuristics available for Critical Resource Identification and job ordering two heuristics have shown the most favourable results; LM2 and SPT respectively.

The modular nature of software development means that additional work is possible on this micro computer based planning approach.

REFERENCES

1 Elsayed EA, Boucher TO: *Analysis and Control of Production Systems*. New Jersey, Prentice-Hall, 1985.

2 Baker KR: *Introduction To Sequencing and Scheduling*. New York, Wiley, 1974.

3 Bellofatto WR: Lead Time vs The Production Control System. *Prod & Inv Mgt*, 2:14-22, 1974.

4 Goldratt EM, Fox RE: *The Goal : Excellence In Manufacturing*. North River Press, 1984.

5 Burbidge JL: *IFIP Glossary Of Terms Used In Production Control*, North Holland, 1987.

6 Karmarkar US, Kekre S, et al: Capacity Analysis of a Manufacturing Cell. J Of Manuf'g Systems 6:3, 165-175, 1988.

7 Everdell R: MRPII, JIT and OPT : Not A Choice But A Synergy. *APICS Readings in Zero Inventory*, 144-149, 1981.

8 Jacobs FR: OPT Uncovered : Many Production Planning and Scheduling Concepts Can Be Applied With or Without The Software. Industrial Engineering, 16:10 32-41, 1984.

9 Jacobs FR: The OPT Scheduling System : A Review Of A New Production Scheduling System. Prod & Inv Mgt, 24:3, 7-51, 1983.

10   Manning EJ: OPT : A Production Technique Case Study. *BPICS Control* 11:3, 17-21, 1985

11   Meleton MP: OPT - Fantasy Or Breakthrough?. Prod & Inv Mgt 27:2 13-21, 1986.

12   Plenert G, Best TD: MRP, JIT and OPT : What's Best?. *Prod & Inv Mgt* 27:2, 22-29, 1986.

13   David HI, LeFevre DC: Finite Capacity Scheduling With OPT. *APICS Conference Proceedings 1981*, 178-181, 1981.

14   Vollmann TE: OPT As An Enhancement To MRP II. *Prod & Inv Mgt* 27:2 38-47 1986.

15   Pritsker AAB: Compilation of Definitions of Simulation. *Simulation* 33:61-63, 1979

SCHEDULING A GROUP TECHNOLOGY MANUFACTURING CELL USING A HYBRID NEURAL NETWORK

[1]John F. C. Khaw, Beng S. Lim, and [2]Lennie E. N. Lim

[1]*Gintic Institute of CIM, Singapore*
[2]*Nanyang Technological University, Singapore*

ABSTRACT

Group Technology (GT) is an innovative and powerful manufacturing strategy to increase efficiency and productivity of batch production by grouping similar parts together in families, and arranging machines required to process these families in manufacturing cells. Benefits from GT implementation include setup-time reduction, simplification of materials flow, increase of capacity, and improvement in production planning and control. However currently there are still very few tools available for GT scheduling. Conventional MRP II systems are not designed for GT scheduling, nor are some commercial finite capacity scheduling systems. This paper describes the use of a hybrid neural network for scheduling in a real GT cell. The proposed hybrid neural network has a three dimensional structure which is designed to capture the temporal relationship of time dependent scheduling data and variables. Two algorithms, namely the recurrent backpropagation and interactive activation and competitive functions are used in the scheduling neural network. Knowledge representation is based on distributed approach where each scheduling object is represented by a pattern of microfeatures. The procedure for developing the proposed system involves three main phases, namely data generation, network learning, and scheduling application phases. Preliminary results have shown that the proposed system is able to generate satisfactory schedules for the GT cell when compared to schedules generated by a simulation model.

INTRODUCTION

Group Technology (GT) is a manufacturing strategy whose main idea is to capitalize on similar and recurrent activities. Major applications of GT in a manufacturing company include design standardization, process planning, and cellular manufacturing. In design standardization, GT concepts have been used to develop a parts design database that is classified and coded by significant part attributes. The engineer can then retrieve designs that are similar to the codes of the part he or she is designing. Using parts classification and coding and GT principles as a foundation, process planning systems have been used to group parts into families based on their manufacturing characteristics. For each part family, a standard process plan is established. In the shop floor, the concepts of GT have been applied to cellular manufacturing by grouping the necessary machines in a cell to fabricate families of parts according to the process plan.

Scheduling in a GT manufacturing cell is still a relatively new area of research in cellular manufacturing [1]. Typically a scheduling problem involves assigning start and end times and machines to each operation of a work order which is subjected to a set of constraints (eg.

resource constraints, quality requirements, priority of customer, etc.). For the past three decades, numerous methods have been developed to tackle scheduling problem in the shop floor environment. Most of these methods however, are not designed for scheduling in GT manufacturing cells which are based on a production flow line concept designed to minimize the entire cycle time. Manufacturing resource planning (MRP II) systems for example, are the most popular production planning tools for multi-product, small lot size job shop production environment. Likewise, GT is useful in solving problems associated with the same environment. In spite of this apparent common application characteristic, the two concepts are in conflict in their operational characteristics.

When GT is applied to the manufacturing of parts with multi-level product structure, the emphasis is on combining the parts in terms of their common manufacturing characteristics regardless of the level of the product structure they are in or the timing of when they should be manufactured. MRP II, however, strictly considers the position of the part in the product structure hierarchy and the timing to meet the part's need date. The conflict, then, is that GT wants similar parts manufactured at the same time regardless of the schedule impact whereas MRP II wants the parts manufactured at the correct scheduled time regardless of the manufacturing impact. Although attempts to integrate GT with MRP II have been done, such systems are still not widely available [1].

Other approaches to GT scheduling include finite capacity scheduling (eg. OPT) and analytical tools (such as linear programming). These approaches tend to be too rigid and too complicated to be applicable in a real-life GT scheduling environment. Linear programming, for example, requires accurate problem formulation to make a sharp distinction between constraints (which must be satisfied) and costs. A solution which achieves a very low cost but violates one or two of the constraints is simply not allowed. In GT scheduling, it is usually preferable to schedule work orders with the same part number together so that the number of machine setups can be reduced. But this is not an absolute constraint. It does not apply when a work order has a different due date priority. These types of "soft" constraints are inherent in GT scheduling and the optimal solution to a GT scheduling problem is the one which minimizes the total violation of soft constraints.

Until recently, neural network technology has received an enormous amount of attention from both the academics and practitioners. Although neural network technology is still at its infancy stage, it has demonstrated its potential and capability in a wide area of applications such as pattern recognition, speech understanding, forecasting [2], optimization, and robotics control. In the area of optimization, neural network technology has been used to solve the traveling salesman and task allocation problems. The next section gives an introduction to neural network technology and describes some of its properties. The complexities of GT scheduling in a real GT production environment are then discussed. The paper then describes the use of a neural network approach for GT scheduling. The neural network model has a three dimensional structure together with two algorithms. One of which is used to train the neural network for capturing the GT scheduling logic. The other algorithm is used to perform on-line dynamic scheduling. Finally some experimental results will also be presented and discussed.

NEURAL NETWORK TECHNOLOGY

In essence neural network models can be described as mathematical models of the biological brain. They consist of powerful learning algorithms which allow the application systems to learn from past examples. They also possess many unique and useful properties which offer a completely different computational paradigm as compared to conventional approaches. These properties are briefly described in the following paragraphs.

In general, a neural network consists of two major parts: the individual processing elements (PEs) and the network of interconnections among the PEs. A neural network may have many PEs in multiple layers topology. Basically each PE, as illustrated in Figure 1, performs three functions.

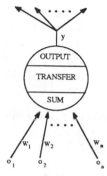

Figure 1: A Typical Processing Element

First it accepts incoming signals from the outputs o_j of other PEs. The output values are modified by a weighing factor w_{ij}, representing the strength of connection i to connection j. A net function takes the weighted sum of all impinging signals, ie.

$$net_j = \Sigma \, w_{ij}o_j \qquad (1)$$

Often excitatory and inhibitory biases are fed to the inputs of the PEs to enforce certain desired behavior and relationships. Second this net input value is used to perform simple operations to calculate an activation value. The activation value for a PE i at time t, $a_i(t)$, represents the current state of the unit. The set of activation values may be continuous or discrete. A new state of activation function, F, for $a_i(t)$ can be defined as

$$a_i(t+1) = F[a_i(t), net_{ij}] \qquad (2)$$

F is usually some kind of threshold function with no activation until the net input exceeds a certain value. Third an output function are used to determine the final output value $y_j(t)$. An output function $f[a_i(t)]$ maps the current state of activation $a_i(t)$ to an output value $y_i(t)$, ie.

$$f[a_i(t+1)] = y_j(t+1) \qquad (3)$$

A major consideration of neural network processing is stability. A neural network of many PEs must be constructed so that it will converge to some stable state regardless of the starting state. Many neural network models are capable of learning by dynamically modifying their interconnection weights (strengths). There are a number of learning schemes that adapts the connection weights to recognize new patterns.

Neural networks use many, very simple PEs as computational units. These PEs are connected into a multi-layered architecture to form a network. Each PE can operate in parallel with the rest and thus massive parallelism of the system can be achieved. The network as a whole will try to satisfy as many constraints as possible when it reaches a solution. An important point is that each PE performs a nonlinear transformation which allows the network to approximate any complex function.

Artificial neural networks exhibit many interesting characteristics such as memory association in which part of the contents in the memory of a PE is used as a key to retrieve or recall the rest of the contents. This important characteristic not only enables neural networks to process incomplete data, but to handle noisy data as well. Another characteristic of neural networks is related to conflict resolution. The networks behave in such a way that all PEs cooperate together to arrive at a solution which satisfies the largest number of constraints. Sometimes, they also compete among themselves. This is the magnification of the interaction between excitatory and inhibitory forces within each PE.

Knowledge in neural networks is stored in connection weights and biases. Thus knowledge is represented in a totally distributed fashion. To solve a problem using a neural network, every PE and every connection plays a part, albeit a small part. The neural network as a whole collectively computes a solution using the three equations discussed above. There are two advantages to this distributed representation of knowledge. Firstly this approach is fail-save. If some PEs or connections are damaged, the performance of neural networks will degrade gracefully, rather than cease to operate completely. Secondly, PEs and layers can be added or deleted without having to reconstruct the new network. Instead one need only to invoke the learning algorithm to adjust the weights and biases to absorb new information.

Learning is an indispensable feature in neural network modelling. There are a variety of learning algorithms for different neural network models. In general learning algorithms can be categorized into two kinds; ie. supervised learning and unsupervised learning. In supervised learning, desired output of the neural network is used to compare with the actual output. The learning algorithm is to adjust the connection weights to minimize total mean-squared error. Whereas in unsupervised learning, the learning algorithm usually requires PEs to compete for input. The weights of the winning PE are tuned closer to the input. This positive reinforcement gradually organizes the PEs of the network into specific patterns.

THE COMPLEXITIES OF GROUP TECHNOLOGY SCHEDULING

In this section, a real-life GT scheduling problem will be described in details. The company in this case study is a make-to-order manufacturer of electrical submergible pump for the petroleum industry. The plant facility incorporates all phases of manufacturing processes which include a foundry, a machine shop, and an assembly area. The foundry operation produces high quality castings. From the foundry, raw castings are sent to the machine shop where components are machined to specifications. The machine shop is organized in multi-tool GT manufacturing cells for greater flexibilities. Component parts are inspected before the assembly process begins. After a finished component leaves the assembly area, it is sent to the test well. Here all the finished components (ie. protectors, motors, and pumps) are subjected to rigorous testing and inspection.

The company has a full-fledged MRP II system installed with Kanban system working within the MRP II system. Although the company has long been a Class A MRP II user, scheduling in the machine shop and assembly area is mainly performed manually by planners. The main reason was that MRP II makes an invalid assumption that capacity is always infinite. Another weakness of MRP II lies in the fact that it is inappropriate for handling GT scheduling. The role of MRP II system in the company is reduced to planning materials availability, controlling the material plan, and maintaining material and process information. Detailed capacity planning and work order (WO) scheduling is done manually by individual planner in charge of a particular manufacturing cell. The planning horizon is typically in the range of 4 to 8 weeks.

Machines and operators in the machine shop and assembly area are arranged in a number of manufacturing or work cells. At this moment, there are 4 assembly area work cells and 8 machine shop work cells. Each work cell consists of a number of different machines capable of producing different families of parts. These machines may be grouped into work centers within the work cell. Tied with each work center, one can define the normal manning level. The sum of the normal manning level for all work centers under a work cell is equal to the aggregate manning defined for that work cell. Manning in the company is divided into 3 shifts. The work cell is called the Stage cell which consists of five work centers. Work center 2083 consists of 3 non-convertible impeller machines, while 2084 has 3 non-convertible diffuser machines. Machines in 2062 and 2063 are convertible (ie. from impeller to diffuser, and vice versa). Due to environmental constraint (such as air conditioning requirement), a particular part can only be processed on certain machines.

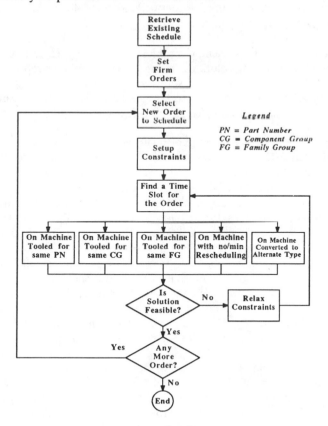

Figure 2: Exisiting Scheduling Logic

Shop floor scheduling in the company is carried out whenever a new customer order is received or whenever there is a constraint change in existing WOs, such as machine breakdown or changes in customer's order. The scheduling logic for the machine shop work cells is shown in Figure 2. From the work order list, a WO with the highest score (which is calculated based on order duedate, priority, and earliest start date) is selected for scheduling. While satisfying the need date, the WO is scheduled in one of the following ways: Using the aggregate manpower available for the work cell, the work order is first scheduled on the machine which has been setup for the same part number. If this fails, then the work order is scheduled on the machine that has been tooled for the same component group. If this fails, then it is scheduled on the machine which has been setup for the same family group. If this fails too, then the work order is scheduled on the machine which will provide the earliest completion time for the work order. If all these fail, the planner may consider converting convertible machine to meet the constraints. Other constraint-relaxation rules include increasing the manpower level at the work cell and/or changing the need date and suggesting alternative dates for the WO.

A 3-DIMENSIONAL NEURAL NETWORK MODEL FOR GT SCHEDULING

Neural networks have shown good promises for solving combinatorial optimization and constraints satisfaction problems such as shop floor scheduling. Foo *et. al.* [3] and Zhou *et. al.* [4] have applied stochastic Hopfield network to solve 4-job 3-machine and 10-job 10-machine job shop scheduling problems respectively. These two networks, however, frequently produce constraint violating solutions. To improve the reliability of the networks, Van Hulle [5] reformulated the 4-job 3-machine problem as a goal programming problem which guarantee constraint satisfying solutions. Using Foo's knowledge representation approach, Khaw *et. al.* [6] have used Kohonen's self-organizing map to solve the 4-job 3-machine scheduling problem. Other approaches to scheduling problem include simulated annealing and Boltzmann machine, both of which have been investigated by Laarhoven *et. al.* [7].

Due to a large number of variables involved in generating a feasible schedule, it can be prohibitively difficult to formulate a mathematical model to reflect the functional relationship between the scheduling variables and the scheduling objectives to be achieved. The measurable quantities may be "the number of on-time deliveries", "machine utilization", "manufacturing cost", etc. This represents a collection of some unknown functions and variables as customer need dates, family part group, and available capacity. The proposed 3-dimensional scheduling neural network however, does not require the scheduling variables to be explicitly defined for achieving the objective function. It is based on the backpropagation learning algorithm to train a set of existing scheduling data. When the network converges to a solution, an internal model of the scheduling problem is constructed. The 3-D network is designed to accept several sets of scheduling input data together. Thus it is able to capture the temporal information contained in the sets. The output of the 3-D network is the machine selected for the work order with the start and end date. Figure 3 shows the network representation of the GT scheduling problem.

An essential consideration when designing neural network models is concerned with knowledge representation of the problem domain. There are two main approaches to knowledge representation in neural network models [8]. One is distributed, where each concept or object is represented by a pattern of microfeatures. The other is localist where single nodes are associated with concepts, and the casual relationship between concepts is represented by the strength of connections between them. Knowledge representation of the

GT scheduling problem is based on distributed approach. There is an input layer which represents scheduling input data such as work order data, capacity availability, performance measure desired, etc.. The output layer has a 3-D structure which consists of a number of sub-layers. Each sub-layer represents a time unit. The PEs in each sub-layer represents machines and setup times.

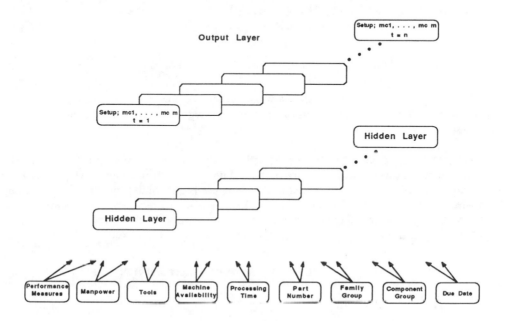

Figure 3: Network Representation of the GT Scheduling Problem

The learning algorithm used to train the 3-D network is based on back-propagation which is an iterative gradient descent algorithm designed to minimize the mean square error between the desired and the actual outputs. The main activation (transfer) function used in each processing element of the neural network is hyperbolic tangent function, although a sigmoid function can be used as well. The hyperbolic tangent function is defined as:

$$f(z) = \frac{e^z - e^{-z}}{e^z + e^{-z}} \tag{4}$$

where z, the summation function is defined as:

$$z_j = \sum_{i=0}^{n} w_{ji} x_i \tag{5}$$

where x_i is the current output state of jth processing element and w_{ji} is the connection weight joining ith processing element to jth processing element.

The 3-D scheduling neural network is trained by applying an input vector, computing the output vector, comparing the output vector to the desired output vector, and adjusting the weights to minimize the difference. The weight adjustment or error function is calculated using the backpropagation algorithm which is defined as:

$$w_{ji,s}(n+1) = w_{ji,s}(n) + \alpha * \varepsilon_{j,s} * x_{i,s-1} \qquad (6)$$

where $w_{ji,s}$ is the weight from neuron j to neuron i and s indicates the layer number. $\varepsilon_{j,s}$ is the scaled local error of neuron j in layer s. $x_{i,s-1}$ is the current output value of ith neuron in layer s-1. α is the learning coefficient. After a large number of presentation of the input vectors, the network will converge to a solution that minimizes the difference between the desired and measured network outputs.

For the scheduling process, the interactive activation and competitive neural network model is used instead. In this network, PEs are organized into a number of competitive pools. There are excitatory connections among PEs in different pools and inhibitory connections among PEs within the same pool. The output layer of the 3-D network consists of sublayers which implement the interactive activation and competitive network. Each PE receives an external input from the connected units. A combined input to PE i is calculated as follows:

$$net_i = \sum_j w_{ij} a_j + extinput_i \qquad (7)$$

where a_j is the activation of PE j and has value a_j for all $a_j > 0$; otherwise its value is 0.

The activation values are updated according to the following equations:

If $net_i > 0$, $\qquad a_i = a_i + net_i (max - a_i) - decay (a_i - rest)$ $\qquad (8)$

otherwise, $\qquad a_i = a_i + net_i (a_i - min) - decay (a_i - rest)$ $\qquad (9)$

where max = 1, min $\leq$ rest $\leq$ 0, and 0 < decay < 1

The decay rate determines how fast the stable condition is reached. In our simulation of the 3-D network, max = 1, min = -0.2, rest = 0, and decay = 0.07.

PROCEDURES FOR DEVELOPING THE PROPOSED SYSTEM

In this section, the procedures for developing the proposed hybrid neural network for GT scheduling is discussed in detailed. The procedures entail three main phases, namely a data generation phase, a network learning phase, and a scheduling application phase. The data generation phase involves several simulation studies of the Stage work cell. The purpose of the simulation studies is to generate pairs of input/output data for training the scheduling neural network. The simulation models developed are also intended for use in comparing output of scheduling neural network with results generated from the simulation models. For the simulation studies, SIMAN simulation language has been used to model the existing scheduling logic in the work cell. The input to the simulation model is scheduling data such as part number, family group, runtime, due date, priority, etc. and scheduling rules such as earliest due date first, etc. Machine schedule and performance measures are collected at the end of each simulation run.

For the network learning phase, pairs of input/ouput data are fed to the input and output layers of the 3-D network. The input portion of the data is a vector of scheduling data, scheduling rules, and performance measures. The output portion is a vector of machine schedule. The 3-D network is trained simply by being fed with the input-output pairs. The purpose of this phase is to capture the scheduling heuristics and constraint knowledge specific to the Stage work cell. The learning algorithm of the 3-D network is the backpropagation algorithm as described above.

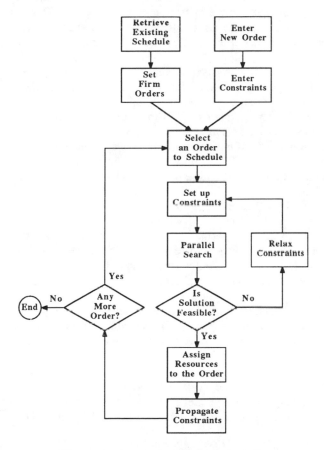

Figure 4: Dynamics of the Scheduling Process

In the scheduling application phase, the planner will use the proposed system when a new customer order is received or when a rescheduling is required. Through a graphical user interface, the planner will enter new work order data or enter new constraints (in case of rescheduling) into the system. The planner may also set firm planned orders as desired.

Work orders are scheduled, in the order of most-constrained first, sequentially by the trained 3-D network, which parallelly searches and settles for the most appropriate machine/start-date combination for the work order under simultaneous interaction of complex constraints. The dynamics of the scheduling process is shown in Figure 4. After the work order is assigned to a machine, the constraints imposed by the assignment are backpropagated by appropriate weight adjustments before the next work order is selected for assignment. This process is repeated until all the work orders are scheduled.

IMPLEMENTATION

The proposed system is currently being implemented using NeuralWorks Professional II Plus from NeuralWare, Inc. on a 486-33 PC. The user interface is created using the Microsoft Excel version 3.0. The user interface module consists of a reports/gantt-chart sub-module and a data maintenance sub-module. The data maintenance module is a relational database which stores customer order, resources, part, and transaction data. The 3-D scheduling neural network module will be linked to the user interface module through the dynamic data exchange which is a message protocol for data exchange between Windows programs.

The 3-D scheduling neural network consists of an input layer with 60 PEs, a 3-D hidden layer with 420 PEs, and a 3-D output layer of 210 PEs. The number of PEs in the input layer are determined from the types and number of scheduling input parameters currently used in the Stage work cell under study. The number of PEs in the output layer, on the other hand, represents a 7-day, as opposed to the original 8-week planning horizon used in the Stage cell. A shorter planning horizon was used in this initial stage because a smaller network is easier to debug and to experiment with different parameters. The number of PEs in the hidden layer is not constrained in a definite way by the problem at hand but is determined arbitrarily in this experimental study.

The learning time for the network to converge was well over 8 hours on the 486-33 PC. The preliminary results from the initial experiments were encouraging. After the network has converged, tests were performed to measure the accuracy of the network. It was found that the trained network is able to assign different work orders to the correct machines and to the correct numbers of time units according their process plans both at 95% accuracy. This shows that the trained 3-D network has captured the scheduling patterns for the Stage cell and has developed an internal model of the planner's scheduling heuristics since it is able to replicate the past scheduling decisions with high accuracy. The scheduling process of the 3-D network as described above, however, is still under development at the time of writing this paper. So it is still not available for further evaluation.

The efficacy of the resultant 3-D network will be evaluated based on a series of computational tests using both randomly generated data as well as real data from the Stage cell. A good measure of the quality of the solution generated by the 3-D network is the performance ratio, R:

$$R = \frac{T_{3-D}}{T_{siman}} \tag{10}$$

where T_{3-D} is the makespan of the schedule found by the 3-D neural network, and T_{siman} is the makespan of the schedule found by the SIMAN simulation model. A negative ratio will mean the 3-D network performs better than the SIMAN model, and vice versa.

CONCLUSION

We have outlined an innovative approach for GT scheduling using a 3-dimensional scheduling neural network. The proposed 3-D network is different from other scheduling neural networks in that it consists of an additional dimension to capture the temporal elements of scheduling. This is essential since scheduling problems deal with a lot of time dependent data and variables. In the near future we believe that the proposed 3-D network is capable of generating more accurate and more optimised schedules for the GT work cell concerned. This, however, will be at the expense of greater computing resources and time needed.

ACKNOWLEDGEMENT

This project is supported in part by Singapore Computer Systems Pte. Ltd. under contract STTG/90/0012. The authors would like to thank Dr. Francis Wong and Mr. Lim Joo Hwee of the Institute of Systems Science for providing ideas and comments related to this work.

REFERENCES

[1]   Snead, Charles S., Group Technology: Foundation for Competitive Manufacturing, Van Nostrand Reinhold, New York, 1989.

[2]   Wong, Francis, "Time Series Forecasting Using Backpropagation Neural Networks", *Neurocomputing*, Elsevier, 1991.

[3]   Foo, Y. P. and Y. Takefuji, "Stochastic Neural Networks for Solving Job Shop Scheduling: Part 1. Problem Formulation and Part 2. Architecture and Simulations", Proceedings of the IEEE 2nd International Conference on Neural Networks, San Diego, June 1988.

[4]   Zhou, D.N., V. Cherkassky, T.R. Baldwin, and D.W. Hong, "Scaling Neural Network for Job-Shop Scheduling", Proceedings of the International Joint Conference on Neural Networks, 1990, Vol. 3, pp. 889-894.

[5]   Van Hulle, M. M., "A Goal Programming Network for Mixed Integer Linear Programming: A Case Study for the Jobshop Scheduling Problem", International Journal of Neural Systems, Vol. 2 No. 3, 1991.

[6]   Khaw, F.C. John, Lim Beng Siong and Lennie Lim, "A New Approach to Shop Floor Scheduling in Advanced Manufacturing Systems", *Information Technology - Journal of the Singapore Computer Society*, December 1990, Vol. 3 No. 4., pp. 61-68.

[7]   Laarhoven, Peter Van, Emile Aarts and Jan Karel Lenstra, "Solving Job Shop Scheduling Problems by Simulated Annealing", Working Paper, Philips Research Laboratories, Eindhoven, 1988.

[8]   Zeidenberg, Matthew, *Neural Networks in Artificial Intelligence*, Ellis Horwood, 1990, Chapter 4: Knowledge Representation.

AN ARTIFICAL INTELLIGENCE APPROACH TO JOB SCHEDULING IN A FLEXIBLE MANUFACTURING SYSTEM

Robert A. Hanson and Osama K. Eyada

ISE Department, Virginia Polytechnic Institute and State University, Blacksburg, Virginia, U.S.A.

ABSTRACT

The dynamic environment of Flexible Manufacturing Systems (FMSs) makes real-time job scheduling difficult. Researchers have tried FMS job scheduling using techniques such as dispatching rules, heuristics, knowledge based systems, and expert systems. Dispatching rules and heuristics have been widely used with varying degrees of success, while knowledge based and expert systems implementing dispatching rules or heuristics tend to generate better schedules than those generated by dispatching rules or heuristics alone. However, these systems are considerably slower and cannot always generate the schedules in real-time.

Neural networks are now widely used by researchers to solve problems that were previously solved by heuristics or expert systems. Neural networks can solve these problems faster because they do not have to perform large numbers of enumerations nor do they need to evaluate rule conditions and then chain through a series of rules. Our approach is to create a system that uses both neural networks and expert systems to schedule jobs in an FMS. The expert systems will perform the job scheduling using dispatching rules as well as heuristics while the neural networks will select the appropriate expert system to implement. This system will enable us to generate good job schedules for the FMS which could be implemented in real-time applications.

INTRODUCTION

Flexible manufacturing systems are computerized batch production systems that consist of numerical control machines, automated material handling systems, and control systems. Their inherent flexibility enables them to efficiently produce different products in dynamic environments. However, this flexibility makes scheduling jobs on machines in FMSs a very difficult task. FMS scheduling problems belong to the NP-hard problem class. Problems that belong in the NP-hard class cannot be solved by a computer in polynomial time; the time required to obtain a solution grows exponentially as the problem size increases.

Simple dispatching rules as well as operations research techniques, such as integer programming and branch and bound method, have been implemented to solve FMS scheduling problems. However, simple dispatching rules are ineffective at generating good schedules because they are single-pass methods that cannot readily accommodate the arrival of new jobs nor machine breakdowns in dynamic FMS environments. On the other hand, integer programming, which involves formulating the scheduling problem in terms of linear constraint equations, and branch and bound, which is a tree-structured procedure that entails computing lower bounds to limit the search space, are computationally ineffective at solving realistic FMS scheduling problems.

Expert systems have been employed to schedule FMSs [1] in order to overcome the weaknesses of the previously mentioned methods. Advantages of employing expert systems for FMS scheduling include their ability to provide explanations as to how the final schedule was generated, to accommodate new job arrivals, and to consistently produce feasible schedules. The two main disadvantages of expert systems are that they become slow as the size of the system increases and they cannot learn.

Neural networks have also been employed for FMS scheduling [2]. They can quickly solve large scale problems, they can generalize and abstract, and they can work with incomplete information. Their drawbacks are that they cannot provide explanations as to how the solution was obtained, and they can take a long time to train.

The purpose of this paper is to describe the framework for a neural network-based expert system for FMS job scheduling. This system would combine the benefits from both expert systems and neural networks; it would quickly generate good schedules, it would be able to generalize, abstract and learn, and it would be able to provide explanations as to how the schedule was generated. This paper will first provide some background to FMS scheduling, expert systems, and neural networks. It will then describe the neural network-based expert system scheduler's framework, followed by concluding remarks.

RELATED WORK

FMS Scheduling

Scheduling involves the allocation of jobs to machines and the determination of their starting times. Jobs can consist of single or multiple operations that may or may not require more than one machine. The number of operations, their respective precedence constraints, the job routings, and machine and storage constraints are factors that make FMS job scheduling difficult. Typical assumptions made in order to simplify real-life simulation problems and to facilitate simulation studies include [1,3]:

No groups of similar machines
Machines never breakdown
No job delays due to lack of resources
Machines can only perform one operation at a time
Negligible transportation times between machines
Deterministic machine processing times
Fixed job routings
Jobs can only be processed on one machine at a time
No preemption
No job order cancellations
Fixed due dates.

There are many dispatching rules, heuristics, and algorithms in the literature that can be applied to scheduling in FMS environments [4,5,6,7]. Most of this literature deals with scheduling in order to minimize or maximize a single performance measure. [8,9,10] have shown that job schedules to optimize one measure subject to another can be generated in both one and two machine cases, however, there are currently no published applications of these bicriteria scheduling problems in FMS environments. Recent research includes scheduling FMSs using expert systems [1,11], Hopfield neural networks [2], and a combination of expert systems and backpropagation trained neural networks [3].

Expert Systems

Expert systems are intelligent computerized systems designed to solve complex problems that would normally require a human expert. An expert system consists of a knowledge acquisition system, a knowledge base, an inference engine, an explanation system, and a user interface, as shown in Figure 1.

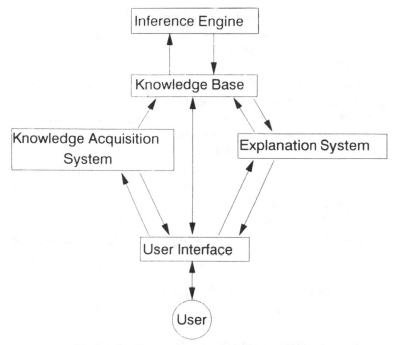

Figure 1. Components of an Expert System

The Knowledge Acquisition System. This component serves as the tool for providing the expert system with the necessary knowledge. The knowledge acquisition process usually involves expert(s) in a given field and a knowledge engineer who computerizes the knowledge provided by the expert(s). He/she then incorporates the knowledge extracted from the expert(s) in the knowledge base in a structured form.

The Knowledge Base. The knowledge obtained from the expert(s) is stored in the knowledge base. There are three main knowledge representation schemes [12]: semantic net-based, rule-based, and frame-based. Semantic net-based representation schemes are rarely used as they are complicated to create, they can be ambiguous, and they cannot easily deal with exceptions. Rule-based representation schemes are the most widely used because they are easy to construct and their production rules are similar to the "If-Then" statement in conventional programming languages. Frame-based representation schemes are a combination of the above two schemes. They possess the semantic net's inheritance property, which allows a frame to inherit characteristics and information from its parent frame, and they utilize production rules to infer new knowledge. Each frame consists of slots which can be filled with values, production rules, inferred knowledge, or left empty. The main drawbacks of frame-based systems are that they require a large area of memory and they are time-consuming to initialize.

The Inference Engine.   This is the expert system's "thinking" component, consisting of the interpreter and the scheduler. The interpreter decides on how the rules will be applied while the scheduler prioritizes the rules in an appropriate order [13]. The inference engine infers "new" knowledge from existing facts according to the inferencing procedure. This knowledge is then posted in an area of the working memory, usually referred to as the Blackboard.

The two inferencing procedures are forward and backward chaining. Forward chaining, or "data" driven, procedures check the premises of rules and, if they are satisfied, executes them. Backward chaining, or "goal" driven, procedures work in the opposite direction; they look for the conclusions that support the desired goal(s), and then check to see if the corresponding premises can be satisfied. Forward chaining obtains all possible inferred knowledge from the knowledge base, and is therefore inefficient for all but the smallest of knowledge bases, whereas backward chaining only infers the necessary knowledge to satisfy the required goal(s).

The Explanation System.   This component enables the user to ask the expert system why a particular rule was fired and how the result was reached.

The User Interface.   This component enables the user to input the information needed to solve the problem. The user can directly input the information, or select a choice (or choices) from pop-up menus.

Neural Networks

A neural network, also referred to as an artificial neural network or a neural net, consists of densely interconnected parallel distributed processing elements, referred to as neurons, with a structure similar to that of the human nervous system [14]. The network's neurons are arranged in layers, typically two to four in number. The first layer is known as the input layer, while the last is known as the output layer. The other layers, if there are more than two, are referred to as the hidden layers. Neurons are connected to the neurons in both of the adjacent layers, but they are not interconnected within a layer as shown in Figure 2.

A neural network operates in the following manner. Outputs from neurons in the previous layer, or inputs in the case of the input layer, are multiplied by weights and then summed for each neuron. A transfer function, if utilized by the network, is then applied to the summation of the weighted inputs in order to obtain the neuron's output. Commonly used transfer functions include linear functions, step functions, hyperbolic tangent functions, and sigmoid functions.

A neural network solves problems by responding in parallel to inputs at the neural network's input layer, and presenting the problem's solution in the output state of each of the neurons in the neural network's output layer. The network must first be trained in order to present a problem's correct solution. Training a network involves computing the errors, which are the differences between the actual and desired output states, and then following a procedure, which usually depends upon the network type, to minimize these errors.

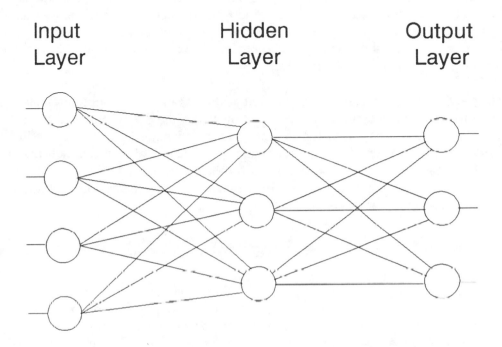

Input Layer   Hidden Layer   Output Layer

Figure 2. Structure of a Neural Network

      Three commonly used network types are backpropagation nets, counterpropagation nets, and Hopfield nets. The type of network to be used depends upon the type of problem to be addressed. The number of layers and their corresponding sizes can then be determined through experimentation.

Backpropagation Nets.   These are the most commonly used network type due to their versatility. They are trained by being shown many pairs of inputs and the corresponding desired outputs. In training, the error is multiplied by the derivative of the transfer function, and then propagated through the network starting at the output layer, back to the input layer; hence the name backpropagation.

One of the main disadvantages of backpropagation nets is the large amount of time required for training. The training process can involve thousands of pairs of inputs and outputs, which are each shown to the network over and over again. Other disadvantages include the paralysis effect during training, and the poor likelihood of locating the global minimum [14].

Counterpropagation Nets.   These networks consist of three layers: the input layer, the Kohonen layer, and the Grossberg layer.   Neurons in the Kohonen layer function in a winner takes all fashion. The Kohonen neuron with the largest input signal wins and outputs a one; all other Kohonen neurons output zeros.   Each neuron in the Grossberg layer simply outputs the weight that connects it with the "winning" neuron in the Kohonen layer.

Hopfield Nets.   These networks, unlike backpropagation and counterpropagation nets, are recurrent networks. In recurrent networks, the output obtained from the initial input is fed back into the network as a new input. This process continues until a steady-state set of outputs is obtained. The ability of recurrent networks to reach a steady state is an important factor. It has been shown [15] that a sufficient condition for reaching steady state is for the matrix of the weights to be symmetrical with zeros on the main diagonal.

PROPOSED FRAMEWORK STRUCTURE

The assumptions made about the FMS environment will be similar to those in other studies [1,3] with the exceptions that machines will be permitted to breakdown, and new jobs will be able to arrive at any time. This will result in the need to generate a new schedule at every occurrence of one of these events. It will be assumed that the operation being performed when a machine breaks down, if any, will be lost. However, all previous operations on the part will be unaffected.

The rescheduling needs to be performed in real time for an FMS to be efficient. Therefore the scheduling method employed must be highly interactive and responsive to the state of the FMS at all times. Following is the framework of the proposed Neural Network-Based Expert System Scheduler.

The system will be composed of expert systems and neural networks as shown in Figure 3. In short, the Neural Network Selector is an expert system that will determine which neural network to use. The selected neural network Expert System Scheduler Selector will identify which scheduling method to use and will select the appropriate Expert System Scheduler. The FMS Monitor is an expert system that will continuously monitor the state of the FMS.

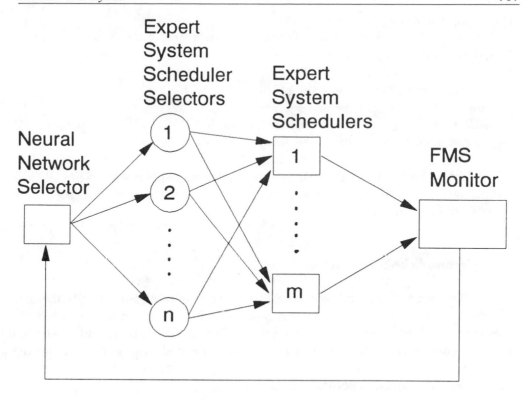

Figure 3. Architecture of the Proposed Scheduling System

Neural Network Selector

This expert system will select the appropriate neural network to implement based upon the desired scheduling objectives and the job data. The objectives in the study will include:

Minimize the mean flow time
Minimize the number of jobs tardy
Minimize the mean tardiness
Minimize the mean flow time subject to minimum number of jobs tardy
Minimize the mean tardiness subject to minimum number of jobs tardy.

The job data will consist of:

Number of jobs to be scheduled
Total number of operations
Minimum number of operations per job
Maximum number of operations per job
Number of jobs, if any, requiring only one operation
Number of operations that can be performed on each machine.

It is anticipated that the neural network selector will be written in C++. It will primarily be a rule-based expert system, although there may be some limited use of frame-like structures.

Expert System Scheduler Selectors

The network selected by the neural network selector will identify the best scheduling method to meet the desired objective, given the job data, and select the Expert System Scheduler to be implemented. The neural networks will be created using a commercially available software package and will either be of the backpropagation or counterpropagation types, depending upon each type's performance during the experimentation phase.

Expert System Schedulers

The selected expert system will schedule the jobs on the machines according to its scheduling method. Each Expert System Scheduler will implement a single method, such as a simple dispatching rule, a heuristic, or an algorithm. It is anticipated that these expert systems will be written in C++ and will solely be rule-based systems.

FMS Monitor

The FMS monitor will be an expert system who's sole purpose is to continuously monitor the state of the FMS. When changes occur, such as a machine breakdown or a new job arrival, the FMS Monitor will update the perceived state of the FMS in order to reschedule the system. This expert system will be written in C++ and is expected to utilize both production rules and frame-like structures.

CONCLUDING REMARKS

There are now new technologies available to help schedule flexible manufacturing systems. Two of these technologies, expert systems and neural networks, will be used in the described FMS scheduler system's architecture. This architecture will take advantage of the speed and learning ability of neural networks along with the explanation facilities provided by expert systems. The Neural Network-Based Expert System Scheduler is expected to provide good schedules for real time FMS applications. The quality of the schedules and the time required to generate them will be evaluated during the testing and validation phases of the research.

REFERENCES

1. Wu D: *An expert system approach for the control and scheduling of flexible manufacturing cells,* dissertation, Pennsylvania State University, 1987.

2. Foo Y-PS, Takefuji Y: Stochastic neural networks for solving job-shop scheduling: Parts 1 and 2, in <u>Proceedings of the SPIE Conference on Applications of Artificial Neural Networks</u>, Vol. 1469, Orlando, Florida, 1991, pp 121-128.

3. Rabelo LC: *A hybrid artificial neural networks and knowledge-based expert systems approach to flexible manufacturing system scheduling,* dissertation, University of Missouri-Rolla, 1990.

4. French S: <u>Sequencing and Scheduling: An Introduction to the Job-Shop,</u> Chichester, England, Ellis Horwood Limited, 1982.

5. Panwalker SS, Iskander W: A survey of scheduling rules, <u>Operations Research</u>, Vol. 25, No. 1: pp 45-61, 1977.

6. Blackstone JH, et al.: A state-of-the-art survey of dispatching rules for manufacturing job shop operations, <u>International Journal of Production Research</u>, Vol. 20, No. 1: pp 27-45, 1982.

7. Hutchison J, et al.: Scheduling for random job shop flexible manufacturing systems, in <u>Proceedings of the Third ORSA/TIMS Conference on Flexible Manufacturing Systems: Operations Research Models and Applications,</u> Amsterdam, The Netherlands, 1989, pp 161-166.

8. Bianco L, Ricciardelli S: Scheduling of a single machine to minimize total weighted completion time subject to release dates, <u>Naval Research Logistics Quaterly</u>, Vol. 29, No. 1, 1982, pp 151-167.

9. Sarin S, Hariharan R: A two machine bicriteria scheduling problem, Virginia Polytechnic Institute and State University, Blacksburg, Virginia 24061-0118.

10. Emmons H: One machine sequencing to minimize the weighted sum of completion times with secondary criteria, <u>Naval Research Logistics Quaterly,</u> Vol. 22, No. 1: pp 585-592, 1975.

11. Kusiak A: Scheduling automated manufacturing systems: a knowledge-based approach, in <u>Proceedings of the Third ORSA/TIMS Conference on Flexible Manufacturing Systems: Operations Research Models and Applications,</u> Amsterdam, The Netherlands, 1989, pp 377-382.

12. Rolston DW: <u>Artificial Intelligence and Expert Systems Development,</u> McGraw-Hill Co., New York, 1988.

13. Turban E: <u>Decision Support and Expert Systems,</u> Macmillan Publishing Co., New York, 1988.

14. Wasserman PD: <u>Neural Computing Theory and Practice,</u> Van Nostrand Reinhold, New York, 1989.

15. Cohen MA, Grossberg S: Absolute stability of global pattern formation and parallel memory storage by competitive neural networks, <u>IEEE Transactions on Systems, Man, and Cybernetics,</u> Vol. 13: pp 815-826, 1983.

THE APPLICATION OF MULTI-ATTRIBUTE TECHNIQUES IN THE DEVELOPMENT OF PERFORMANCE MEASUREMENT AND EVALUATION MODELS FOR CELLULAR MANUFACTURING

Hampton R. Liggett and Jaime Trevino

*Department of Industrial Engineering, North Carolina State University,
Raleigh, North Carolina, U.S.A.*

ABSTRACT

The reported benefits of cellular manufacturing include inventory reduction, cycle time reduction, space reductions (both in the manufacturing and warehousing areas), reduced material handling, improved quality, and increased worker satisfaction. For firms adopting the cellular manufacturing concept, the management of performance is becoming an increasingly important issue. The responsibility for decision making has filtered down from the operations manager to the shop floor supervisor to the individual work cell, resulting in an increased need for tools to assist in performance management.

Many models have been developed for assessing the overall performance of a manufacturing firm. While these techniques provide invaluable information for strategic (long-term) decision making, they provide little assistance for the operational decisions that must be made on a daily basis at the cell level. With respect to operational decision making, the existing performance measurement models are lacking in one or more of these critical areas: (1) the models lack the necessary detail to assist in operational decisions, (2) the models do not provide timely information for operational decision making, and (3) most of the models assume that performance indicators are independent of one another.

This paper proposes a framework for developing models which provide the necessary detail and timeliness for performance measurement and evaluation at the cell level. Multi-attribute techniques which specifically address strong interdependencies that may be present among the performance indicators can be incorporated in this framework. A method for the development of feedback mechanisms to assist in operational decision making is presented, and an ongoing application of the modelling framework is reported. Finally, a natural extension of this research, the development of cell-level decision support systems, is discussed.

INTRODUCTION: THE RESEARCH PROBLEM

As American manufacturers have begun to answer the challenge of international competition in a global market, the corporate mission has been drastically altered. Historically, the "goal of the firm" in the U.S. has been to maximize the future wealth of the stock-holders. Today's goal is multi-dimensional -- achieve increased market share, provide the best quality (both product and service), get the product to the customer with the minimum possible lead time, provide the "least cost" product to the consumer, and respond rapidly to the changing demands of a dynamic market. To accomplish this goal, manufacturers have implemented both improved technology (e.g., computer-aided design, robotics, CNC machine tools, etc.) and new approaches and philosophies to the control of the shop floor (e.g., just-in-time production, cellular manufacturing, statistical process control, etc.).

The "new multi-dimensional goal" mentioned above is evident in the areas of primary emphasis for today's leading manufacturers: (1) improved quality, (2) reduced inventories, (3) increased flexibility, (4) organization along product lines, (5) increased automation, and (6) more effective use of information [1]. The first four areas are at least partially achieved through the implementation of cellular manufacturing, the benefits of which are reported to include reductions in inventory, set-up time, cycle time, material handling costs, and floor space, as well as improvements in flexibility, overall quality, and worker satisfaction.

Greene and Sadowski define *cellular manufacturing* as "the physical division of the functional job shop's manufacturing machinery into production cells ... designed to produce ... a group of parts requiring similar machinery, machine operations, and/or jigs and fixtures" [2]. The concept was developed to deal with the explosive growth in product diversity and complexity which characterizes modern manufacturing.

Many firms implementing cellular manufacturing have, unfortunately, only realized a small portion of the potential benefits. One cause of this is a lack of effective tools to assist in cell management. Manufacturing decision making was delegated to the operations manager in the traditional plant environment. Today that responsibility has filtered down to the shop floor supervisor, and ultimately, to the individual workcell. This expanded responsibility results in a need for tools to assist in performance management at all levels of the manufacturing hierarchy.

In order to effectively manage, one must first evaluate. In order to evaluate, one must first measure. Thus the initial step in managing a manufacturing cell is the development of meaningful, comprehensive, measures of cell performance. Historically, the principal focus of performance measurement has been direct labor. This focus on labor and labor utilization, which is echoed by most financial reporting systems, has led to problems including costly increases in work-in-process (WIP) inventory and lack of attention to quality.

A great deal of research has been undertaken to determine the proper tool(s) for assessing performance at the organizational level, resulting in the development of several models which provide invaluable information to assist with strategic and tactical (long term) decision making. For example, Sink's seven measures of organizational system performance (effectiveness, efficiency, quality, profitability, productivity, quality of work life, and innovation) [3] capture the critical information from the perspective of monitoring and managing the overall performance of the firm. Such broad measures, however, provide little assistance for the operational decisions (e.g. changing component routings due to a machine failure) that must be made daily on the shop floor. Only by developing comprehensive measures of performance at the machine and/or cell levels can the necessary level of detail be attained.

With respect to operational decision making, existing techniques are lacking in one or more of the three critical areas listed below:

(1) They track only gross aggregate, often subjective, variables which lack the necessary detail to assist in operational decisions

(2) They do not provide timely information for operational decisions; they are intended to be applied once per reporting period (monthly or quarterly)

(3) They assume that the attributes of performance are independent of one another.

This paper illustrates a framework for developing comprehensive, theoretically robust performance measurement and evaluation models, at the manufacturing cell level. The modelling approach is first presented, along with a method for developing feedback mechanisms to assist in operational decision making. Next, an ongoing application of the modelling framework is reported. Finally, a natural extension of this research, the development of cell-level decision support systems, is discussed.

THE RESEARCH PLAN

The research plan is divided into three distinct phases, each of which consists of a series of tasks. The objective of Phase I is to develop a "generic" cell performance measurement and evaluation model. In Phase II, this model is customized to meet the needs of specific applications. Finally, in Phase III, the model is applied in the cells for which it has been customized. An overview of the research methodology is included in TABLE 1. Each of the phases and their associated tasks are described in the following sections.

TABLE 1
Overview of Research Methodology

PHASE I. Develop Generic Performance Measurement and Evaluation Model
Task A. Develop list of performance indicators and cell performance survey
Task B. Identify potential client firms and conduct survey
Task C. Analyze survey responses
PHASE II. Customize Performance Measurement and Evaluation Model
Task A. Conduct preliminary group meeting to modify model
Task B. Determine indicator weights and identify relationships among indicators
Task C. Determine method of measurement for each indicator
Task D. Develop "value function" for each indicator
PHASE III. Apply Customized Performance Measurement Model
Task A. Apply customized performance measurement model in "pilot study"
Task B. Analyze results of pilot study and adjust model as necessary
Task C. Apply finalized model and analyze results
Task D. Develop feedback mechanisms

Phase I. Develop "Generic" Performance Measurement and Evaluation Model

The "generic" performance measurement and evaluation model for manufacturing cells that was developed in Phase I is not intended to give an adequate representation of performance for any particular cell -- rather, it is intended to provide a starting point for the more specific modelling in Phase II.

Task A. Develop List of Performance Indicators and Cell Performance Survey   A list of forty manufacturing performance indicators was developed as the result of a review of over fifty journal articles related to manufacturing performance and in-depth interviews with three engineers familiar with cellular manufacturing. The performance indicators are listed in TABLE 2.

A survey which includes the performance indicators in TABLE 2 was developed as a part of this task. It consists of three sections, the first of which elicits descriptive information on a specific cell within the participating firm. The second section asks respondents which of the forty performance indicators are crucial to the performance of that cell, and the third section requires these crucial indicators to be grouped into the following seven higher-level categories: cell maintenance, flexibility, personnel issues, productivity, profitability, quality, and reliability.

TABLE 2
Cell-Level Indicators of Manufacturing Performance

Absenteeism	Machine Downtime	Rework
Corrective Maintenance	Machine Utilization	Scrap
Customer Returns	Material Flow Distance	Setup Time
Cycle Time	Material Handling Time	Storage Time
Direct Labor Cost	Overhead Cost	Throughput
Employee Suggestions	Preventive Maintenance	Transfer Batch Size (prod)
Finished Goods Inventory	Process Time	Transfer Batch Size (in)
Indirect Labor Cost	Product Variety	Value-Added Time
Inspection Time	Production Leadtime	Warranty Claims
Inventory Obsolescence	Production Lot Size	WIP Inventory
Inventory Turns	Queue Time	% on Time Orders (#)
Job Satisfaction	Raw Material Availability	% on Time Orders ($ value)
Lost Time Accidents	Raw Material Inventory	% Within Conformance
Lost Time Person-Hrs		

Task B. Identify Potential Client Firms and Conduct Survey   Fifteen firms that have implemented manufacturing cells were initially identified and contacted. Twelve agreed to participate in Phase I. Rather than attempting to include a larger number of companies in the study, a decision was made to limit the sample size, while ensuring the quality of the responses by conducting follow-up site visits. These site visits allowed first-hand information on cellular manufacturing in the client firms to be attained and provided a chance to clarify survey responses. A description of the participating cells is included in TABLE 3. Because some of the participants viewed portions of the information provided as confidential, the names of the firms are not included.

Task C. Analyze Survey Responses   In analyzing the survey responses, the frequency with which each performance indicator was selected as "crucial" and the number of instances that each "crucial" indicator was associated with each of the seven higher-level categories were determined. Any performance indicator associated with a particular category by 40% or more of the respondents is included in the generic cell performance measurement model that was developed on the basis of this survey. The resulting hierarchy is shown in TABLE 4. (None of the indicators were listed under the **reliability** category by 40% of the respondents; thus, it is not included in the model.)

TABLE 3
Description of Survey Participants *

Cell	Products **	Processes	Automation	Mach	Oper	Lot
A	tractor front axle (2)	milling, drilling, tapping, boring	some NC	6	1	40
B	solenoid air valve (15)	machining, welding, assembly, testing	NC; manual handling	8	7	200
C	fiberglass filters (3)	machining, gluing, assembly, testing	primarily manual	7	4	10
D	housings for electronic equipment (14)	blanking, forming, welding	NC; manual handling	11	7	40
E	clipper machine components (≈300)	machining, stamping, sanding, inspection	NC; manual handling	7	2	16
F	gasoline engine oil pan (3)	milling, boring, tapping, drilling	CNC; fully automated	3	2	120
G	quick-change sockets for PC boards (≈ ∞)	assembly, screen printing, inspection	some CNC, mostly manual	3	1	9
H	screw compressor rotors (6)	precision machining	CNC; manual handling	10	5	50
I	motor starters and contacts (300)	molding, grinding, assembly, testing	CNC; manual handling	10	8-9	20
J	switchgear housings (3000 +)	sheet metal punching, blanking, shearing	fully automated	5	1	25
K	personal computer CPU (1)	packaging	manual (except matl handling)	2	2-3	1
L	color monitor (5)	adjustment, inspection	mostly manual	6	1	1

* Mach --> number machines; Oper --> number operators; Lot --> average lot size
** *(number)* indicates the number of different products/components processed in cell

Phase II: Customize Performance Measurement and Evaluation Model

Because the "generic" performance measurement and evaluation model is based on input concerning a wide variety of cells (see TABLE 3), it must be customized for specific applications during Phase II. Five of the twelve firms from Phase I participated in this process.

Task A. Conduct Preliminary Group Meeting to Modify Model The "generic" model is customized through the use of a structured group participation format, the nominal group technique (NGT) [4],[5]. Participants in the NGT sessions (manufacturing engineers, managers, cell operators and supervisors) are asked to determine which (if any) of the forty performance indicators *not included* in the generic hierarchy should be added to the model and which of the indicators that *are included* in the generic model should be deleted.

TABLE 4

"Generic" Performance Measurement and Evaluation Model (categories are in **bold** type)

Cell Maintenance	**Flexibility**	**Personnel Issues**
Machine Downtime	Production Lot Size	Absenteeism
Corrective Maintenance	Setup Time	Employee Suggestions
Preventive Maintenance	Transfer Batch Size	Job Satisfaction
Productivity	**Profitability**	**Quality**
Absenteeism	Direct Labor Cost	Customer Returns
Cycle Time	Indirect Labor Cost	Inspection Time
Machine Downtime	Inventory Turns	Rework
Material Handling Time	Overhead Cost	Scrap
Process Time	Throughput	
Throughput		

Task B. Determine Indicator Weights and Identify Relationships Among Indicators The teams that participated in the NGT sessions also assist in determining weights and identifying relationships. The weights are developed as follows: the most important indicator under each category is assigned 100 "points", and the other indicators in that category are assigned points according to their importance relative to the top indicator. (Weights are developed in the same manner for the higher-level performance categories.) The nature of the relationships among each pair of indicators within each category are identified according to the following scale:

> +2 = Strong positive (direct) relationship
> +1 = Positive (direct) relationship
> 0 = No relationship
> -1 = Negative (inverse) relationship
> -2 = Strong negative (inverse) relationship.

Task C. Determine Method of Measurement for each Indicator For some performance indicators (e.g., absenteeism, customer returns), the proper method of measurement is obvious. The measurement of other indicators, including value-added time, may not be as straight-forward. Some method of tracking such indicators had to be developed in each of the participating firms. In addition, tables, charts, and graphs to facilitate data collection and analysis were designed for some of the participants.

Task D. Develop "Value Function" for each Indicator Because different units of measure are used for the various performance indicators (production lot size --> number of units, cycle time --> seconds/minutes/hours, etc.), some common unit of measure must be developed, or the indicators cannot be aggregated into an overall "score." This common unit of measure can be achieved with "value functions". According to von Winterfeldt and Edwards, "the 'natural' scale ... is converted to a value scale by means of judgments of relative value of preference strength." [6]. The process for eliciting a value function for a particular performance indicator consists of first identifying the feasible range of outcomes on the natural scale for that indicator by determining the "best case" (maximum attainable) and the "worst case" (minimum acceptable) levels. For some indicators, these levels may be clearly defined (e.g., for percent within conformance, 100% is obviously the "best case" outcome). In other instances, the endpoints of the feasible range must be defined by management policy. The best case outcome for each indicator is assigned a value of **100** points, and the worst case is assigned **0** points. The midpoint on the natural scale (value of **50** points) is then elicited by asking the following: "If *best case* is worth **$100.00** and *worst case* is worth **$0.00**, for what level of outcome on the natural scale would you be willing to pay **$50.00**?". (Phrasing the question in terms of money proved more meaningful to most of the participants.) On the basis of these three carefully assessed points, a reasonably representative value function can be constructed [6, p.236].

Phase III. Apply Customized Performance Measurement Model

The final phase in the research methodology is the application of the customized performance measurement and evaluation model developed in Phase II. At the time of this writing, two companies are involved in *Task B* of Phase III, and a third participant is preparing to begin *Task A*.

Task A. Apply Customized Performance Measurement Model in "Pilot Study"   The models developed in Phase II are applied in a one-month "pilot study." During this time, the levels of the indicators are recorded at the end of each shift. (Data should be collected more frequently if automated collection devices are available. In the absence of such devices, however, more frequent data collection is too time consuming.) A log of all non-routine occurrences must be maintained during the pilot study to provide a basis for validating the model.

Task B. Analyze Results of Pilot Study and Adjust Model as Necessary   An attempt to validate the performance measurement and evaluation hierarchy empirically is made at the completion of the pilot study. The data for each shift are aggregated into an overall cell performance index (**PI**) through the use of the following equation:

$$PI = \Sigma_j \left[\Sigma_i \left(v_i * w_{ij} \right) * W_{j} \right], \text{ where}$$

v_i = value function equivalent of performance indicator **i**

w_{ij} = weight of performance indicator **i** in category **j**

W_j = weight of category **j**

PI = overall cell performance index

The resulting **PI**'s for each shift are compared against the log maintained during the pilot study to determine whether or not the performance measurement hierarchy has reacted appropriately to any "hiccups" (non-routine occurrences) during that shift. Failure to react properly to a recorded event results in a need to "fine-tune" the model by:

(1) adding additional indicators; (2) adjusting the weighting scheme; (3) deleting redundant indicators; and/or (4) incorporating additional multiplicative or multilinear scaling constants in the calculations to explicitly consider interdependencies among the performance indicators in each category (see [6] and [7] for a thorough discussion of both multiplicative and multilinear scaling constants). Strong interdependencies among performance indicators are evidenced by either of two sources: large positive or negative correlations among two or more indicators included in a particular category, or the "strong" relationships identified in Phase II, *Task B*.

Task C. Apply Finalized Model and Analyze Results   Following any adjustments resulting from the analysis of the pilot study, the performance measurement and evaluation model is re-applied for a period of one month. The log described in *Task A* is again maintained.

    The same set of equations described in *Task B* (with the possible addition of the scaling constants mentioned above), are used to calculate the **PI** for each shift. Once again, the performance indices for each shift are compared against events recorded in the log to validate the model.

Task D. Develop Feedback Mechanisms   The intent of the framework described above is to develop models to assist in decision making at the cell level. In order for this to be accomplished, appropriate feedback mechanisms must be identified. Perhaps the most important aspect to be considered is the nature of the relationships among performance indicators (see Phase II, *Task B*). For example, the level of one indicator might suggest the need for a particular corrective action. Because this indicator has a strong inverse relationship with a second indicator (i.e., an improvement in *A* will be accompanied by a degradation in *B*) that is more heavily weighted, however, the corrective action suggested by the first indicator could lead to an overall degradation in cell performance. To further clarify the relationships among performance indicators, the chart shown in TABLE 5 is used. (This is actually a variation of failure-mode and effects analysis, or FMEA, described in [8].) In addition to explicitly identifying causal relationships among performance indicators, the chart in TABLE 5 converts input data from the performance measurement and evaluation model into appropriate information for assisting in cell management. (The chart provides a convenient format for converting this information into a simple rule-based system, as will be discussed in a later section.)

TABLE 5
Feedback Mechanism for Performance Measurement and Evaluation Model

Failure type	Indicator(s) associated with failure type	Probable cause(s)	Indicator(s) associated with cause(s)	Potential effect(s)	Indicator(s) associated with effect(s)

CASE STUDY

The following case study reports the ongoing implementation of a cell-level performance measurement and evaluation model in *Company F* (see TABLE 3). At the present time, the results of the pilot study are being analyzed (Phase III, *Task B*). Certain adjustments to the model have already been made on the basis of this analysis.

Phase II Results

During the first task of Phase II, several changes were made to the "generic" model. The participants in the preliminary group meeting (the cell supervisor, the supervisor of the "customer" assembly cell, and a manufacturing engineer) added five performance indicators to the generic model, while deleting four that were deemed to be inappropriate or redundant. The hierarchy resulting from Phase II, *Task A* is shown in TABLE 6.

TABLE 6
Company F's Hierarchy Following Phase II, *Task A*

Cell Maintenance	Flexibility	Personnel Issues
Corrective Maintenance	Production Lead Time	Absenteeism
Machine Downtime	Production Lot Size	Job Satisfaction
Preventive Maintenance	**Productivity**	**Quality**
Profitability	Absenteeism	Customer Returns
Customer Returns	Cycle Time	Inspection Time
Direct Labor Cost	Machine Downtime	Rework
Indirect Labor Cost	Machine Utilization	Scrap
Inventory Turns	Process Time	Straight Acceptance
Machine Downtime	Scrap	Warranty Claims
Overhead Cost	Straight Acceptance	
Value-Added Time	Throughput	
	Value-Added Time	

Following the establishment of weights for the performance indicators and the identification of relationships among the indicators, methods for measuring each indicator were determined. During this task, it was discovered that the following modifications to the model were also necessary: (1) the category **personnel issues** was deleted because *job satisfaction* could not be adequately measured on a daily basis and the only remaining indicator in that category, *absenteeism,* was also included under **productivity**; (2) *warranty claims, customer returns,* and *rework* were all eliminated (no *rework* is carried out in the cell; no *warranty claims* or *customer returns* due to workmanship in the cell have occurred since the cell was established); (3) *process time* was deleted because it is essentially the same as *cycle time* (the cell is machine-paced). The resulting model, complete with adjusted indicator and category weights, is shown in TABLE 7. (Categories are in **bold** type.)

TABLE 7
Company F Hierarchy at the Completion of Phase II, *Tasks B* and *C*

	Weight		Weight
Cell Maintenance	**18.2**	**Flexibility**	**17.4**
Machine Downtime	40.4	Production Lead Time	66.7
Preventive Maintenance	33.8	Production Lot Size	33.3
Corrective Maintenance	25.7	**Profitability**	**20.5**
Productivity	**21.2**	Direct Labor Cost	20.1
Absenteeism	12.3	Indirect Labor Cost	18.2
Cycle Time	11.2	Inventory Level	11.2
Machine Downtime	13.1	Machine Downtime	16.4
Machine Utilization	12.8	Overhead Cost	18.6
Scrap	13.6	Value-Added Time	15.6
Straight Acceptance	16.0	**Quality**	**22.7**
Throughput	10.4	Inspection Time	33.6
Value-Added Time	10.4	Scrap	27.7
		Straight Acceptance	38.7

Value functions have been constructed for each of the indicators in the performance measurement hierarchy. Examples of two of these value functions are included below.

Straight acceptance: The maximum attainable level is obviously 100% (zero defects). This was assigned a value of **100**. The minimum acceptable level for this indicator, 70%, was set by management policy and assigned a value of **0**. The participants from *Company F* determined that a *straight acceptance* level of 88% is worth one-half the value of zero defects (i.e., value of **50**).

Machine downtime: The best case outcome, 0% (no downtime) was assigned a value of **100**. The worst case outcome was determined to be 15% and assigned a value of **0**. The midpoint on the value scale, with a value of **50**, was set at 6%.

Note that it is entirely possible for a worst case level to be "violated" (e.g., *machine downtime* could exceed 15%). This results in an unacceptable condition, and a method to deal with this condition is incorporated in the model -- any violation of a worst case level will be assigned a value of **-1,000,000**. Thus, the **PI** will be negative, reflecting the unacceptable condition.

Phase III: Current Status of Implementation

At present, the results of the pilot study conducted with *Company F* are being analyzed and evaluated. Preliminary signs, based on both the comparison of the overall performance indices from each shift with recorded events and the examination of correlations among indicators, point to a need for a few minor modifications to the model: (1) the three financial indicators under **profitability** (*direct labor cost, indirect labor cost*, and *overhead cost*) can be combined into a new performance indicator, *operating costs* (they are all measured in terms of dollars and are roughly proportional to one another); (2) a new performance indicator, *raw material availability,* should be added to the model under the category **productivity** (there were a few instances during the study period when the cell had to be shut down due to a lack of incoming parts); and (3) *scrap* and *straight acceptance* should be replaced by two new indicators, *defective materials* and *poor workmanship,* which provide better insight into quality problems. (These changes will necessitate an adjustment of the weighting scheme.) The updated model, following the above-mentioned modifications, is shown in TABLE 8.

TABLE 8
Final Performance Measurement Hierarchy for *Company F*

Cell Maintenance	Productivity	Profitability
Machine Downtime	Absenteeism	Inventory Level
Corrective Maintenance	Cycle Time	Machine Downtime
Preventive Maintenance	Defective Materials	Operating Costs
	Machine Downtime	Value-Added Time
Flexibility	Machine Utilization	
Production Lead Time	Poor Workmanship	**Quality**
Production Lot Size	Raw Material Availability	Defective Materials
	Throughput	Inspection Time
	Value-Added Time	Poor Workmanship

In addition to the previously mentioned changes to *Company F's* performance hierarchy, possible interdependencies among two sets of indicators are being explored to determine whether of not multiplicative and/or multilinear scaling coefficients (see earlier discussion under Phase III, *Task B*) should be incorporated in the model. These sets of indicators are: (1) *machine downtime, machine utilization,* and *value-added time;* and (2) *inspection time* and *defective materials.*

CONCLUSION

The modelling framework for measuring and evaluating cell performance improves upon three critical flaws with respect to operational decision making, some (or all) of which are present in the published research on performance measurement. The necessary detail to facilitate operational decisions is obtained by tracking measurable performance indicators at the cell level. Secondly, timely feedback is developed by measuring and evaluating performance on a daily (or more frequent) basis. Finally, the modelling framework provides three mechanisms to address interrelationships among performance indicators: (1) they are identified by participants during the modelling process, (2) they can be incorporated into the calculations (see Phase III, *Task B*), and (3) the charting technique illustrated in TABLE 5 assists in the identification of causal relationships among the performance indicators.

Proper feedback from the performance measurement and evaluation model provides information for determining the ideal number of kanbans (under a pull system), determining proper batch sizes, reconfiguring the layout of a cell, and improving overall cell management. This overall improvement of cell operations would, in turn, provide invaluable information for the design and development of future cells.

Decision Support System Development

A logical extension of the methodology described above is the development of a decision support system (DSS) for managing manufacturing performance at the cell level. The DSS could take the form of a rule-base system or a simulation model, but the ideal format, a hybrid model, combines the advantages of simulation and rule-base systems -- the rule-base portion helps to identify potential corrective actions when problems occur, while the simulation portion allows the analyst to determine which course of corrective action is most suitable.

The necessary information for the development of the hybrid DSS described above can be obtained through the performance measurement and evaluation framework. A thorough analysis of conceivable types of "failures," their underlying causes, and their potential effects (following the format provided in TABLE 5) helps to identify causal relationships among performance indicators and provides the foundation for the rule-base portion of the DSS. In addition to information on causal relationships, the simulation portion of the model requires distributional information on the performance indicators, which can be statistically developed after the measurement and evaluation model has been operational for a sufficient period of time.

REFERENCES

1. R.A. Howell & S.R. Soucy: The new manufacturing environment: major trends for management accountants. *Management Accounting*, July, 1987, pp. 21-27.
2. Greene, T.J. and R.P. Sadowski: A review of cellular manufacturing assumptions, advantages, and design techniques. *Journal of Operations Management*, Vol. 4, No. 2, pp. 85-97.
3. Sink, D.S.: *Productivity Management: Planning, Measurement and Evaluation, Control, and Improvement*, New York, NY: John Wiley and Sons, 1985, pp. 40-46.
4. Sink, D.S.: Using the Nominal Group Technique effectively. *National Productivity Review*, Spring, 1983, 173-184.
5. Delbecq, A.L., A.H. Van de Ven, and D.A. Gustafson: *Group Techniques for Program Planning: A Guide to Nominal Group and Delphi Processes*, Glenview, IL: Scott, Foresman, 1975.
6. von Winterfeldt, D. & W. Edwards: *Decision Analysis and Behavioral Research*, New York, NY: Cambridge University Press, 1986, pp. 205-241.
7. Keeney, R.L. and H. Raiffa: *Decisions with Multiple Objectives: Preferences and Value Tradeoffs*, New York, NY: John Wiley, 1976.
8. Plsek, P.E.: FMEA for process quality planning. *Annual Quality Congress Transactions, Vol. 43, May 8-10, 1989, Toronto, Ontario, Canada*, Milwaukee, WI: ASQC, pp. 484-489.

PREDICTING THE PERFORMANCE OF FMS USING LEARNING CURVE THEORY

Patrick C. Walsh and Huw J. Lewis

University of Limerick, Limerick, Ireland

ABSTRACT

In recent times the methods and techniques used in the manufacturing industry have changed dramatically. Companies and organisations are moving away from the labour intensive factories of the past and have moved or are moving to machine orientated methods of manufacture.

Situations thus arise in the conversion from conventional to "New technology", where semi/ unskilled labour is reduced and replaced by skilled labour to maintain and programme the system. Traditionally the method used in a labour intensive system to rate and predict the performance of new operators is the learning curve. This has now been applied to the startup situation of flexible manufacturing systems (FMS) based on historical data. Thus fulfilling a role of rating FMS immediately after it has been implemented, even though labour now plays a diminished role with respect to systems performance.

Due to the investment involved in the application of FMS, it is in the interest of the company that the system produces at maximum capacity as soon as possible, after commissioning. The startup of the system must therefore be carefully monitored, and its efficiency rated. Thus performance of future FMS can be compared with the initial application. Further to this it is necessary to predict the likely performance of the system during the startup phase, based on its initial performance. From these predictions the production schedule, delivery response, product price, budgeting and cost control can be defined.

Thus several methods of applying the time constant learning curve as a performance predictor, based on aggregate hours producing for FMS startup have been investigated with varying success.

INTRODUCTION

In introducing FMS the problem associated with individual aspects of the system, could be compounded. It is therefore necessary to forecast the likely performance of the system during the startup stage. This can then be utilised by management to define the likely output of the system, plan capacity, estimate delivery dates and lead times, and define costs.

The factors that could affect the performance of the system during the startup and subsequent production scenario, can be defined graphically (FIG. 1). The diagram indicates the effect of product mix, production requirements, and stock levels on

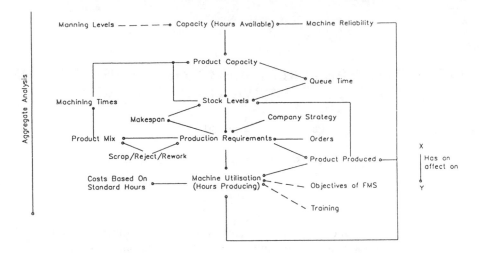

FIG. 1 Relationship of FMS startup factors

machine utilisation. These in turn are affected by other factors, such as, orders, queue times, and scrap/ rework/ reject levels. The product mix however can be constrained by the available stock, the backlog of work in the form of queue times, the level of rework, and the companies order books. The product mix and the completion of product, in line with the production requirements (schedule), will define the efficiency of the system. This is related to the capacity of the system , via machining times and can thus define the machine utilisation. However it is possible to have poor utilisation of the machine, but still have high efficiency due to the required product being produced at the correct time. This highlights one of the philosophies of FMS, in its economy of scope rather than its economy of scale, i.e. its ability to produce numerous small batches, with an optimal size of one component.

However investigation of machine utilisation is still valid, although it defines capacity and not efficiency. From FIG.1 it can be seen that the machine utilisation is affected by two main factors, the breakdown hours and the production requirements. It is therefore possible to investigate the performance of a system using aggregate data in the form of hours producing. If this is coupled with the breakdown hours, then the difference between the sum of the breakdown hours and the hours producing, when compared with the maximum available hours, will give the lost hours due to the production system, (e.g. idle time etc). Further to this, large fluctuations in the aggregate data can be investigated by stepping upwards through FIG. 1, from the machine utilisation to stock levels. By analyzing each of these factors their effect on machine utilisation can be determined, and the efficiency of the system defined in the form of product output against product requirement. This however could define unique complex models for individual companies, depending on the availability of data from the company, resulting in the comparison of performances of companies and for individual companies becoming difficult.

The data used in this analysis will be the aggregate hours producing, as used by Lewis [10] in his analysis of FMS startup. Here a series of time constant learning curves were successively applied to the startup scenario's of two FMS. It is therefore possible using the data to define the better method of predicting performance using learning curve theory. Two approaches have been utilised. The first predicts the final parameters of the learning curve, and the second uses an iterative method based on the Taylor expansion.

METHODS USED TO PREDICT THE OUTPUT OF A SYSTEM FROM LIMITED DATA.

Introduction:

The performance of a flexible manufacturing system, or any manufacturing based system is difficult to define. Factors affecting performance include the product mix, production requirements, stock levels, etc (FIG.1). However as Buzzacutt [1] points out;
    "It is impossible for analytical models to capture all the details of the system. It is necessary to decide how much detail to include- Too much makes the model impossible of solution and too little makes the model unrealistic. There has to be a compromise between the adequacy of representation and ease of solution."
    Thus due to the limitations of available data and in order that a general method of predicting performance could be developed, the performance criteria utilised for modelling was "hours producing". Thus any major variance away from the defined models could be investigated via the factors affecting the hours producing.
    There are many methods available to forecast changes in output. The majority of these methods utilize the trend of historical data and include:
    (a)   Average of all past data.
    (b)   Moving point average (based on a preset time slot)
    (c)   Exponential smoothing.
These methods (FIG.2) provide forecasts by establishing the trend of the actual data by smoothing. However, smoothing has the tendency of hiding possible reductions in capacity and adapts to poor conditions. Thus these methods are suitable only if short term forecasting is required and not applicable in the case of implementing FMS, where forecasted output is required over a long period. The merits of these long term forecasts are improved scheduling, improved allocation of resources and more accurate prediction of future production costs [2].
    Thus in order to facilitate the requirement of long term forecasting, learning curve modelling will be investigated as a performance predicting method.

Learning Curves

    There are many learning curve models available, however, to be effective a model must treat lines, discontinuities, false plateaux and excessive scatter as information carriers to trigger management induced investigative and corrective action [3]. Hackett [4] undertook a study of 18 different learning curve models, eg, DeJong model [5], Wiltshire model [6], Time Constant model (Bevis equation) [7], Mathematical model, etc. Of the models studied it was concluded that the Time Constant (Equation 1) is a good choice for general use; as it was shown to be robust (reliable under a wide range of behaviour patterns for the observed data) and simple to use. Thus it will be used here.

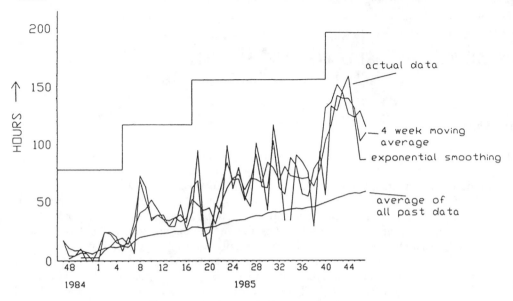

FIG.2 Forecasting techniques used on period A

$$Y_m = Y_c + Y_f (1 - e^{-t/\tau}) \qquad \{1\}$$

where

Y_m = Output rate at time t
Y_c = Initial output rate.
Y_f = Maximum achievable rate - Y_c.
t   = Time.
τ = Time constant.

The problem of forecasting future performance via learning curves is compounded by the inevitable "scatter" in industrial data [8]. The extent of the scatter contained in the startup is a function of the effectiveness of the startup. Hence a high degree of scatter could imply a bad startup, which makes prediction of future performance much more difficult. However, in a production situation, prediction of the future performance of a system has to be made from the performance of the system to date. This may be achieved by fitting a learning curve to the available data such that the residual sum of squares ($\Sigma(Y-Y_m)^2$) is minimized. This is done in the hope that the initial trend of the data, to date, also represents the future trend of the data.

Thus the individual parameters of the time constant learning curve need to be defined, if it is to be used as a prediction model. The initial output rate Y_c is well defined, and can be taken as the initial weeks output. The Y_f value and the time constant (τ) are therefore the two defining parameters of the prediction model which have to be established.

Determining Values for τ and Y_f .

The time constant τ has traditionally been found by finding the time it takes for the data to reach $0.63Y_f$ (FIG.3). However as the data is often fluctuating, the initial slope of the data is taken. This should then reach the Y_f value, at a time equal to the time constant.

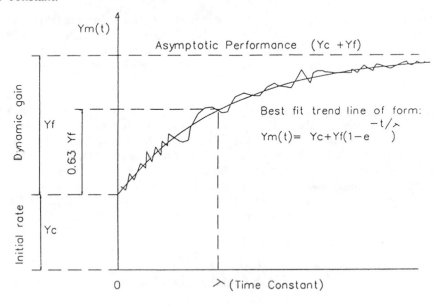

FIG.3 Time Constant Learning Curve Model

It thus arises that good estimates of the Y_f value are required for performance prediction, hence defining τ and the overall model. The methods of defining the parameters used in this case are:

(1) *Initial slope of the data.*   The initial slope may be defined simply by eye, by the gradient of two defined points in the data or by linear regression. It should be noted this method is not essentially a prediction technique, as it utilises the initial slope of the data to define τ. However, it has been used by Walsh [13] to test the suitability of the learning curve to fit the data and is thus included here.

(2) *Power curve method.*   This is a natural progression to using the initial slope. In this case the initial data is used to define a regression curve, in the form

$$Y = At^b \qquad\qquad \{2\}$$

where

Y = Output rate at time t.
A = Constant.
t = Time.
b = Reduction factor.

Having defined the regression equation, the time taken for the data to reach $0.63Y_f$ can be substituted as Y in {2} and calculating τ as t.

(3) *Iterative technique.*   An iterative procedure involving the use of a digital computer to establish the time constant (τ) such that the residual sum of squares is minimized. The assumption here is that good estimates of Y_c and Y_f are available.

(4) *Taylor series technique.*   A Taylor series technique for predicting ultimate parameters of the exponential learning law [9]. This is a two parameter technique (though it can be modified to incorporate a third parameter, ie Y_c) and involves the use of a computer algorithm to establish the best values of Y_f and τ given an initial estimate of both.

(5) *Estimating τ and Y_c using the prolate cycloid.*   The prolate cycloid curves lie within the plane curve family and are an alternative to using the power curve which is in the same family but describes a hyperbola. They have been shown not to follow the "learning pattern"[10]. However they do tend to plateaux at a maximum value. Therefore by defining the parameters of the cyclic curve using the initial output rate and the maximum hours available for production, an estimate of the Y_f value can be made, as the maximum value attained by the prolate cycloid. τ can then be found by dividing the time to reach $0.63Y_f$ by the time required to reach the prescribed Y_c value.

APPLYING THE PREDICTION TECHNIQUES TO INDUSTRIAL DATA

The data utilised in this analysis has been obtained from a company applying an FMS system [11]. The data shows signs of specific changes (FIG.4). These changes have been highlighted in previous work [10,11,12], and include, the addition of extra capacity, and new startups after stoppage. The effect of these stoppages on modelling using all the collected data have been noted, as follows [11].

(1)   A sequential curve occurs if there is an increase in capacity without production stoppage.

(2)   A new curve with a Y_c value below the end point of the previous curve, occurs if there is no increase in capacity, but there is a break in production.

Before analyzing the prediction efficiency of each method it is first necessary to establish the accuracy of each curve fitting technique based on the same historical data. This is vital because if the method being used to fit a curve to all the data for a startup period, is a bad fit, then it is no use using that method as a means of predicting future performance. Since the objective of each technique is to reduce the residual sum of squares to a minimum, it is therefore logical that this parameter be chosen as the criterion used to assess the efficiency of each method. Other methods that are used are the Chi-Squared goodness of fit test and a comparison of the actual accumulated versus predicted accumulated outputs.

For all the methods with the exception of the prolate cycloid (method 4) the predictability of the data was tested by each method in the following manner. All the

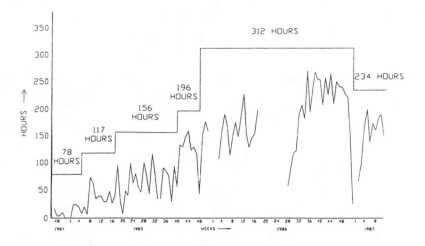

FIG.4 Hours producing for FMS

data points, for the period in question, were used to obtain a curve which best represents the performance of the system for that period. Then, less data points were progressively used until a point was reached where the curve obtained differed considerably from the 'best' curve.

This method of prediction is not aimed at trying to predict the immediate performance for the next week/ interval given the performance of the system to data. Rather, the objective is to be able to predict the output from the system at any time in the future given the minimum number of data points.

In order to test the prediction methods, the resulting models were tested against the known data. Thus for each defined period the data was increased in steps (four weeks), with new predictive learning curves being produced at each addition. This method was suitable for the linear and power regression models. For the iterative and taylors technique the opposite was carried out, with a steady reduction in data, until the resulting curves became indefinable. The prolate cycloid method utilised the maximum available hours and the initial weeks output to define Y_f and τ. Thus indicating a "one off" predictive curve based on initial data.

In all cases the Y_e value was taken as the initial weeks output. The resulting Y_f and hence time constant, was defined by the predicting method, resulting in differing τ and Y_f values.

The predictive curves for period A (Table 1) showed some similarity, with a range of time constants between 37 - 53 weeks, with the exclusion of method 5. This resulted in estimates of accumulated hours producing lying between -18% to 10% of the actual accumulated values (again with the exception of method 5) and with chi squared fits of above 80%.

The other periods did not show the levels of similarity between the time constants and Y_f values. Hence, varying degrees of success in predicting the likely outcomes were recorded. This is highlighted by the range of chi squared test values for each method and the fluctuations of estimated accumulated values around the actual

TABLE 1
Startup Analysis of an FMS

Period	Method Used	Best Curve for Period	R.S.S	Min Weeks Reqd. To Predict Curve(% of total)	Estimate of Total accumulated output (% difference)	Chi² test
A	1	$120(1-e^{-t/37})$	40082	46 (87)	2401 (-18)	80
	2	$156(1-e^{-t/45})$	41747	45 (85)	3248 (10)	90
	3	$154(1-e^{-t/45})$	37999	46 (87)	3260 (11)	90
	4	$154(1-e^{-t/53})$	34750	39 (74)	2905 (1.2)	97
	5	$4+112(1-e^{-t/4.84})$	195079	1 (2)	5490 (7)	--
B	1	$43+57(1-e^{-t/8})$	31458	--	3224 (10)	50
	2	$43+269(1-e^{-t/15.9})$	47863	12 (54)	3349 (14)	30
	3	$43+296(1-e^{-t/18})$	49722	18 (82)	3699 (26)	30
	4	$43+122(1-e^{-t/1.8})$	20178	16 (73)	2811 (4)	99
	5	$43+180(1-e^{-t/4.59})$	47980	1 (4)	3509 (20)	75
C	1	$55+185(1-e^{-t/5})$	71610	--	3761 (24)	50
	2	$55+179(1-e^{-t/30})$	74355	8 (33)	5315 (3.5)	20
	3	$55+257(1-e^{-t/12})$	117945	25 (100)	5100 (3)	--
	4	$55+184(1-e^{-t/4.6})$	69277	18 (72)	5135 (3)	50
	5	$33+165(1-e^{-t/4.48})$	69600	1 (4)	4842 (0.3)	50
D	1	$65+110(1-e^{-t/2.5})$	6568	--	1480 (2.3)	99
	2	$65+108(1-e^{-t/1.1})$	10808	4 (40)	1608 (6)	90
	3	$65+169(1-e^{-t/5.4})$	10041	10 (100)	1713 (13)	95
	4	$65+107.4(1-e^{-t/1.4})$	4469	6 (60)	1573 (4)	98
	5	$65+94(1-e^{-t/4.31})$	14267	1 (10)	1224 (19)	50

Method 1 - Initial Slope of the Data
Method 2 - Power Curve
Method 3 - Iterative Technique
Method 4 - Taylor Series
Method 5 - Prolate Cycloid

accumulated values.

However method 4 shows a consistently better residual sum of squares value than the other methods, coupled with good chi² test and an error in estimated accumulated hours producing ranging between 1.2 - 4% over the four periods. The values of the parameters of the Bevis equation, obtained using this method, represent the optimum, as regards reducing the residual sum of squares to a minimum. However, this method is very sensitive to changes in the data, for example, it was found that for period A, if the number of data points used in the analysis exceeded forty five the value of τ and Y_f obtained, became negative. This implies that the best time constant curve for the data has the form of a positive exponential curve. The reason for this is clear by analysising the raw data (FIG.4), ie, beyond week 45 two extra machines were being commissioned into the system causing an increase in output for the remainder of that period. This increase in output at the end of the period is the reason why a positive exponential curve is a better fit to the data than the conventional time constant curve. As mentioned previously , all startup periods are taken as being independent of each other (Y_c = first data point of period). This is used as the forgetting factor or slippage (loss of productivity between the end of one period and the start of another [8]) can be taken into consideration. From a prediction perspective this method is the most consistent of the methods discussed requiring less data (70% of total, on average) to accurately predict the final curve for a period.

The iterative technique (method 3) produces curves which are in all cases of lower quality. The reason for this is due to the fact that this method is constrained by the value of Y_f, which is set at the maximum capable capacity of the system. This therefore reduces the robustness of the time constant model since the only variable now remaining is the time constant. Because of this, method 3 proved to be the least efficient as regards predicting the final curve, requiring on average 92% of the data to make a good prediction. The assumption being made with this method is that the system will eventually be producing at its maximum capacity at some time in the future. However, the duration of the startup period in question may not be sufficiently long enough for this to occur, hence this method loses its credibility.

By using a power regression analysis on the data, to define the parameters of the learning curve the amount of data required to attain a suitable curve appears to have been reduced. However, even with three of the four periods only requiring 33-40% of the available data to describe the predicting curves, the chi^2 values were erratic, falling to below 30% in most cases. Thus indicating that the curves were not of good quality.

Method 5 being a method of predicting curves based on the initial output value, and the maximum available hours thus utilises only one weeks data in its prediction. The resulting curves could only produce a chi^2 test value of 50% or more in three of the four periods investigated. However estimates within 0.3-19% of the actual accumulated hours producing were made with this method.

DISCUSSION

By analyzing the productivity data associated with an FMS, it is most often practical to define the 'global' startup in terms of several discrete startups. These discrete startups can be further analyzed in turn by fitting learning curve models to them. This paper is concerned solely with the intricacies of the time constant model from a prediction perspective. Production data from a company involved in the implementation of an FMS has been analyzed and the various startup periods defined. Different techniques were then used to establish the parameters of the time constant learning curve model. The simplest technique used was that of establishing the initial slope of the data trend for each period (by eye or by regression analysis) and finding the point where the line obtained and the asymptote line ($Y_c + Y_f$) intersect. This intersection point corresponds to the time constant. It was discovered that the best curves were obtained if the value of Y_c was set equal to the first data point for the period in question. The reason for this is that usually a time period has elapsed between the end of one period and the start of another or some technological difficulties have arisen which cause a slippage in performance. This method does not give the optimal learning curve, but a good suboptimal solution may be obtained by trial and error, ie, by changing the number of data points used in the estimation of the initial slope different τ values are obtained.

A better method of establishing the time constant is by using a simple iterative algorithm (method 3). For this technique to be effective, it is vital that good estimates of Y_f are available for each period. Hence, this method may be used for modelling historical data where the value of Y_f can be accurately estimated (by smoothing techniques), however in a prediction situation the results show (Table 1) that this method is practically useless. This is due to the fact that accurate estimates of Y_f are not available.

The most accurate method of deriving the parameters of the time constant model was found to be a two parameter estimation technique in which a Taylor series expansion of the exponential control law, in conjunction with partial differentiation, is used to establish Y_f and τ such that the least squares error is minimised. An initial guess of Y_f and τ is entered and incrementally increased or decreased until the optimal is obtained, ie, when $\delta Y_f/Y_f < 0.0001$ and $\delta 1/\tau /1/\tau < 0.01$ (or otherwise specified). The initial guess is independent of the final values, however, they must be within reasonable limits for the programme to run. For prediction purposes this technique is the most consistent and requires the least amount of data (73%) to accurately predict future performance. The quantity of data required for accurate prediction is dependent on two factors:

(a) The extent of scatter in each startup period.

(b) The difference between Yc and Yf, i.e. the smaller this difference the better the results.

The above methods rely on the estimation of the Y_f value. This seems to be the crux of the estimation methods. Having defined this value and knowing Y_c, then the time constant and the overall curve can be found.

Method 5 is the only method that attempts to define a method of estimating the Y_f value and utilises the minimum amount of data; to describe the predicting curves. Thus, although the method gives poor chi^2 test fits and estimates of accumulated hours producing ranging from 7 - 20% of the actual values, it could be utilised as a rule of thumb, until, enough data has been gathered to utilise one of the other four methods.

CONCLUSION

In conclusion several methods have been looked at to predict the likely performance of the startup of FMS. Although good curve fits have been attained in utilising method 4, the taylor series expansion, at least 70% of available data was required, before a learning curve that predicted within 4% of actual accumulated hours producing was formed. This to some extent may be due to the use of aggregate data to define the curves, which will be influenced highly by other constraints on the system such as product mix, queues, etc.

An alternative method utilising the initial weeks output, and maximum available hours, produced estimates of accumulated output within 20% of actual values. These however had poor chi^2 test results. However it is felt as a rule of thumb that they would be useful as an initial prediction mechanism.

In all cases the definition of the Y_f value is critical. It is this value, which defines the time constant and hence the shape and outcome of the predicting curves.

REFERENCES

1. Buzacott JA: Modelling manufacturing systems. *Robotics and computer integrated manufacturing*, Vol 2, No 1, pp 25-32, 1985.
2. Globerson S: The deviation of actual performance around learning curve models. *IJPR*, 1984, Vol 22, No 1, pp 51-59.
3. Towill DR et al: Dynamics of capacity planning for flexible manufacturing system startup. *Engineering costs and production economics*, No.17, 1989, pp 55-60.
4. Hackett EA: Application of a set of learning curve models to repetitive tasks. *IERE Int. conf. on learning curves and progress functions*, London IERE conf Pub No. 52,pp 32-52, 1981.
5. DeJong JR: The effect of increasing skill on cycle times and its consequences for time standards. *Ergonomics*, 1, No.1, pp 51-60, 1957.
6. Wiltshire HC: The variation of cycle times with repetition for manual tasks. *Ergonomics*, Vol. 10, No.3, pp 331-347, 1967.
7. Towill DR: Systems engineering for startup management. *Euro 11 conf of International Federation & Operations Research Studies*, Stockholm, Sweden, 1976, pp 6-15.
8. Towill DR: The use of learning curve models for the prediction of batch production performance. *Int. Journal of Operations & Production Management,* Vol.15, No 2, 1985, pp 13-24
9. Bevis FW, et al: Prediction of operator performance during learning of repetitive tasks: *Int. J. Prod. Res.*, 1970, Vol 8, No.4, pp 293-305.
10. Lewis HJ: *Sequential learning curve models for AMT introduction using Taxonomic selection procedures.* Phd Thesis, University of Whales, 1991.
11. Lewis HJ et al: The introduction of AMT- Triumph or Trauma. *IMC6*, Dublin city university, pp 874-890, August '89.
12. Walsh P: *The application of learning curves to the startup phase of a flexible manufacturing system.* BEng thesis, University of Limerick, 1991.

THE CONSIDERATION OF HUMAN FACTORS IN THE DESIGN OF GROUP TECHNOLOGY MANUFACTURING SYSTEMS

Heiko Schäfer, Neil H. Darracott, Chris O'Brien, and John R. Wilson

Department of Manufacturing Engineering, University of Nottingham, Nottingham, U.K.

ABSTRACT

Group Technology (GT) has been considered a powerful manufacturing philosophy for many years. In response to a growing demand for aids to effective implementation, numerous sophisticated analytical techniques have been developed by a variety of academic researchers and industrial practitioners. However, in their search for the best configuration of machines these methodologies frequently neglect an important aspect of GT systems design: their analytical sophistication is often not matched by an equally careful design of the human factors of the system in question. In fact, most techniques for the design of GT manufacturing systems do not consider any Ergonomics aspects in their analysis.

This paper outlines the impact of Cellular Manufacture (the implementation of the GT philosophy) on the human factors of the manufacturing system. The most significant of these is the impact on the role of the worker (as opposed to Ergonomics details such as workplace design). The paper advocates the consideration of these aspects in the systems design process from the earliest stages and examines how this might be achieved. The work presented is a preliminary result from on-going research at the University of Nottingham, which also investigates the human factors impact of several other currently popular developments in Manufacturing Systems Engineering, such as Just-in-Time methods and Optimised Production Technology.

INTRODUCTION

Group Technology (GT) has recently been heralded as a major agent in the continuing drive for manufacturing competitiveness [1-4], since it can provide production systems with the structural simplicity which is considered essential for the modern manufacturing organisation [5,6]. Industrial implementations of the GT philosophy [7] can thus secure a strategic advantage which may be more significant than the established tactical and operational benefits of GT, such as reduced lead times, improved material flow etc.

In order for a GT system to achieve these goals, the configuration and design of each group has to be carefully considered. This task is far more difficult than might be expected; analytical complications usually make it impossible to evaluate all possible configurations [9], and a profusion of techniques have been developed to support the search for the "best" systems design (many of these reviewed in [10]). These techniques are often very sophisticated, since they have to overcome algorithmic obstacles while satisfying conflicting objective functions (e.g. simplicity of material flow vs. simplicity of part family).

However, there is more to a group than parts and machines: its effectiveness may depend very significantly on the work environment imposed on its human members - an aspect which has been neglected in almost all analytical research into the field. This situation may have been caused by the analytical complexity of GT systems design: the majority of algorithms available suffer from some form of factorial effect, and any consideration of human factors would imply the inclusion of much additional data in the analysis.

This paper examines the influence of Group Technology structures on the human factors of a manufacturing system and vice versa (Fig.1), advocating the consideration of this influence during systems design. Describing the general neglect of an important factor in a field of research is itself desirable, but the contribution increases in value when suggestions for improvement are made. An investigation of **how** a consideration of human factors might be incorporated in the design of GT systems therefore concludes the paper.

Figure 1: Relationship between HF and GT

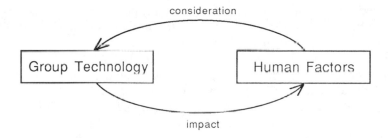

Definition

> **Group Technology (GT):** a manufacturing philosophy based on the underlying principle that the efficiencies of flow line manufacture can be achieved in a batch production environment - usually implemented as a system of highly autonomous manufacturing cells, the configuration of which can be based on various criteria [7,8,11-13].

Origins of the GT Philosophy

Group Technology was developed in response to a degradation of production control in increasingly large manufacturing systems producing small batches of output [14-16]. Due to combinatorial effects, the task of scheduling production in batches simply became too complicated as manufacturing systems evolved. Isolated implementations of what would now be considered Group Production were first described in the first half of this century [17,18]. These pioneering implementations in an industrial environment illustrated the potential of GT techniques for particular Engineering tasks, but showed no appreciation of the holistic nature of GT as a manufacturing philosophy. This underlying philosophy of GT was not formalised until 1959 [7]. Since then, GT has experienced three periods of particular research activity and popularity in industry:

1959-1970:	• formulation of the GT philosophy by Mitrofanov [7] and subsequent research [19-21]
1970-1980:	• increased research activity during and after major industrial successes, esp. in the UK [8,12,18,22]
recently:	• renewed interest through associations with the "simplicity" culture [1-6,23,24]

The Nature of Group Technology Research

It was during the second of these three phases that a rift seems to have developed between academic researchers and industrial practitioners of GT. The former delved into the development of increasingly powerful algorithms considering reduced volumes of data [9,10,13], whereas the designers of actual manufacturing systems, unaware of any useful techniques, often turned to more abstract frameworks and approaches for assistance in the design of their systems [12,19]. This separation of theory and practice has contributed to the neglect of human factors in the design of GT systems, since no tools were available for the incorporation of such aspects in the underlying analysis. Most research on social factors in GT also dates from this period [25-28]. Some academic researchers [8,15,17,18,29] have, from time to time, commented on this situation - emphasizing the broad impact of GT and advocating systems design based on more than just machine/part data, e.g.:

> "A cellular system is not just a simple arrangement of similar components, tooling and grouping of machines, but is a total manufacturing system which involves the complex interaction of market, management and men. It is a new approach to batch production affecting not only the production aspect of the company but the entire organisation of management." [18]

However, their warnings have largely been ignored. Over the years, the many industrial applications of the GT strategy have certainly been accompanied by a matching academic effort directed at improving the systems design methodologies (see [30] or [31] for a useful survey of this research). However, the aim of much of this research has been to overcome the algorithmic obstacles which almost inevitably appear during GT analysis, such as the combinatorial complexity dilemma, which can only be resolved using unconventional analytical methodologies

[9,24,32]. Only very few analytical techniques aim to incorporate a greater variety of aspects in their design of the manufacturing system (e.g. [33,34]), and not many of these achieve a true consideration of human factors.

The systems design decisions considered here are those affecting the overall layout of the manufacturing system, i.e. not the Ergonomics of individual work stations. These can also be affected by the choice of a cellular structure, e.g. due to the redesign of methods for stock transport and storage, but the organisational impact of GT is generally more severe, since it affects the roles of the workers, their social environment and many other aspects of the job ambience [28].

HUMAN FACTORS

Impact

The decision to use Group Technology in a manufacturing system will always have both beneficial and detrimental implications for intra-organizational cooperation and management relations. These positive and negative effects will also affect employees who do not work on the shop floor. This will happen as a result of the change in organizational structure diffusing upwards from the production level, as the remainder of the organization adapts to the structure of the manufacturing operations - a process which is often neglected by industrial practitioners. However, the impact of GT on aspects of human factors is of course most severe on the factory floor, including operators and front line management. These effects have been outlined in several sources [8,25-28,34]:

Group Structure　　The principal factor on the shop floor is the team structure of the workforce. This generally can improve communication between workers, since GT groups tend to be smaller than functional departments, which also simplifies the coordination of the activities required for the manufacture of a part. This is because the cell works as a small business unit and does not interact much with the rest of the factory. Working as a team should also affect the measurement of worker performance: the contribution of a team member to the cell's operation should be much more obvious and this should be used for assessment. Feedback from management, especially from senior management, can be much more direct for the same reason. However, feedback in the opposite direction, from workers to management, should also be more constructive, due to the smoother and simpler production flow through the cell, naturally highlighting bottlenecks and problem sources.

Some of the consequences of using GT can be very difficult to measure and trace to their causes. One of these is a frequently documented improvement of product quality which may arise from the particular social relations, attitudes and motivation in a group. The cohesive nature of a manufacturing group and the peer pressure which will result if workers identify and highlight the quality of each others' work can lead to substantial output improvements, in terms of quality and productivity. This depends to a large extent on the detailed work design and relations with management (as well as corporate identity etc.), but GT may facilitate the adoption of such mechanisms.

The most frequently described social benefit of GT is increased job satisfaction. Fazakerley [27] lists the contributions to job satisfaction to be expected from the use of GT as:

- worker involvement in decision making
- personalized work relationships
- variety in tasks
- freedom to determine methods and workplace layout
- meaningful pay incentives

Since a GT group typically employs fewer people than machines, each member of a GT manufacturing team will tend to possess far more skills than an operator in a corresponding functional department, where the members of the workforce essentially all perform the same function. Opportunities for training and job enrichment through job rotation are also far less remote in a GT system, as workers can rotate within a cell instead of moving to a distant and alien department.

Disadvantages   There are also some human factors disadvantages to working in a GT group. Some researchers [35] claim that in a GT system the functional isolation of traditional manufacture can be replaced by a similar group isolation, which may be more severe. A system adhering to the GT principles should aim to minimise all interactions between groups, and this can indeed lead to social isolation among sectors of the workforce. Also, expansion of a GT system will typically be through the creation of new cells, and the individual groups may not be as organic as functional departments (this can be seen as an advantage or disadvantage). GT groups are certainly less flexible than the departments in a traditional system, since their production system is tailored to the production of a particular part family. The work passing through the cell could therefore be considered more monotonous, but this aspect could be attacked (if desired) through job enrichment measures such as job rotation within a cell or other enriched forms of (semi-)autonomous group working. A conclusion from these observations is that the nature of the work environment under GT will depend heavily on management decisions (at all levels).

Indirect Consequences   Since the underlying principle of GT is achieving flow line behaviour in a batch production environment, flow through a GT cell is generally simpler, smoother, less chaotic and faster than the corresponding flow through a larger, functionally segregated system. Further effects on the work environment in the enterprise result from the matching change in the nature of the company's operational characteristics, as production delays, delivery times, throughput times, and levels of work-in-progress are reduced. That workers see components completed in the group is also often said to lay the basis for pride in achievement and resulting increases in morale and performance. It is also seen as encouraging group staff to spot and solve technical problems when and where they arise and transform the role of their leader from traditional foreman/trouble shooter to a true manager. Other indirect benefits claimed for GT include group cohesiveness (which can result from the separation of the work force) and increases in job satisfaction (due to the task variety and diversity of skills which almost inevitably accompany GT).

Resistance to Change   Implementations of GT can be classified into two categories by distinguishing between the case of an initial design incorporating the concepts of Group Technology and the (more common) re-configuration of existing facilities.

The problems facing the management of an organization considering a **transformation** to Group Technology, usually from an existing functional layout of facilities, are firstly those expected for any fundamental change to the activities of a work force. Employee resistance to the change could be quite substantial, although it certainly is a function of the quality of worker-management relations prior to the alterations. The misconception of GT as a management gimmick can obviously exacerbate a hostile situation. As an understanding of the reasons for a major change is crucial to the acceptance and support of the work force, such problems can only be overcome by management demonstrating their initiative and a willingness to communicate and train. A misapplication or failed implementation of GT will certainly amplify any industrial relations problems the company had prior to the change; the diffusion of awareness to worker level should therefore be a primary objective of management activity **before** the re-configuration takes place.

Even when the use of Group Technology is considered for the **initial design** of a plant layout, and difficulties arising from employee resistance to change are less pertinent, there are several important implications for management arising from the "change" to GT. This is because workers recruited for a newly designed system will also have a cultural heritage of their own, which may affect their response to the peculiarities of working under GT. If no appreciation of human factors is evident in the design of the system, the benefits of GT associated with increased job satisfaction may be foregone.

Effect of Neglect

It is a sad fact that human factors are often one of the last concerns of the designers of manufacturing systems [36,37]. The design of GT systems is no exception. The methodologies used to determine the layout of the system can be extremely sophisticated, but no technique seems to explicitly incorporate any Ergonomics aspects in its analysis. It is generally agreed, however, that the use of GT can have a positive effect on job satisfaction, participation and general industrial relations, which can lead to considerable improvements in labour productivity and product quality. Thus the neglect of human factors in the design of GT manufacturing systems may have an indirect impact on the commercial performance of such systems.

An important effect usually neglected is the larger number of roles for an operator in a GT system [38]. These roles may overlap and lead to a lack of role clarity, jeopardizing the positive effects of group cohesiveness etc. outlined above. This overlap of operator roles is depicted in Fig.2.

Figure 2: Model of Internal Groups in a GT Cell

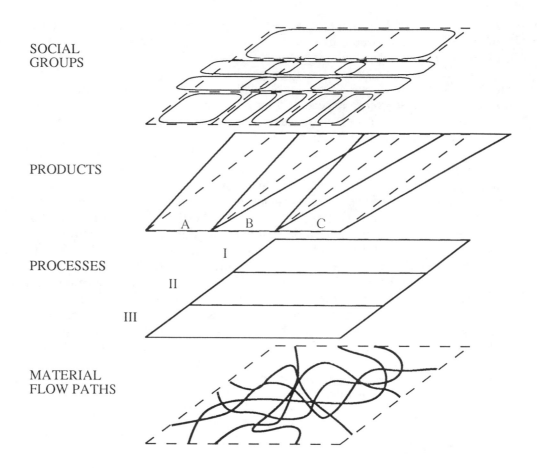

SOCIAL
GROUPS

PRODUCTS

PROCESSES

MATERIAL
FLOW PATHS

In a traditional company with a functional layout, machines are clustered according to function. This gives economies of scale and makes scheduling of jobs on the machines easier. In such a system, workers operating on similar **processes** will associate. In some cases, this may be the only role the worker is to enact ("I am a press operator"). In a GT system, on the other hand, machines are clustered in an arrangement to manufacture a particular range of parts, which will be related in some way. Therefore, workers operating on similar **products** will associate. Within the group, however, operators will assume other roles as well, and this can lead to a loss of the desirable GT group spirit.

Unless job rotation is extreme, operators in a cell will often be linked to particular processes for one reason or another (usually expertise). Associations may also develop to other operators on the same material flow path (who, in the simpler flow system, become visible and identifiable). Social groups may be established within a group based on the immediate physical or psychological work environment (visibility constraints, operator proximity, common interests, similar backgrounds).

Thus a loss of role clarity can easily develop, particularly in groups larger than the generally accepted optimum size of 10-20 members [8]. Further roles can arise, according to job function, job grade, or other parameters.

The final effect of a neglect of human factors in the design of GT manufacturing systems will be a reduced performance of that system in achieving its corporate objectives. The result of role overload will not be limited to reduced job satisfaction etc. but will eventually filter down to the bottom line and affect the financial returns generated by the organization.

The Integration of Human Factors in GT Systems Design

A consideration of Human Factors in the design of a GT manufacturing system can only be achieved through a careful consideration of the implications of the cellular structure for the roles of the workers in the system. The design of the management structure in particular has to be geared to ensuring that the job roles to be fulfilled by all group members are clear, straightforward and consistent.

The effective integration of Human Factors in the design of GT systems is therefore mostly a **qualitative** design task [39]. Areas requiring particular attention are the number of management grades, and the precise role of the cell leaders. This has to be carefully considered to ensure that these front-line managers are genuine team leaders, as opposed to functional foremen or departmental managers. It should be aimed to integrate their role into the group as far as possible. The aim is also to run a system with as few operator grades as possible, in order to ensure maximum clarity of role and thus reinforce the team-working spirit of the group, rather than superimposing the group member role onto a traditional job function. Another key concern for GT job design is the skill level of the workforce. In order to achieve a team-working environment, it will be necessary to train the workforce to a level where each member of a group is trained for several different jobs in his group. Such a multi-skilled workforce may be more expensive per member, but the cohesiveness and smoother operation of the group should justify the outlay.

The administration and technical support for the manufacturing team members is equally important. Here, measures such as group based pay, continuous improvement programs, and quality based assessment should be used to stress the team-working aspect of the operations for which the employees are being rewarded. Customer awareness and identification with the product can similarly be encouraged by these measures.

There is also a strong **quantitative** aspect to GT system design (particularly at the part families formation stage), and the systems designer should strive to incorporate a consideration of Human Factors in this branch of the analysis. Here, recent developments in the research environment are important, as they allow the inclusion of different types of data in the analysis [32-34]. It is not feasible to incorporate a consideration of human factors in GT systems at the design stage when conventional, algorithmic design techniques are used, but the emerging trend towards more holistic systems design methodologies makes this a real possibility.

A final point which must be emphasised is the importance of **timing** for the integration of Human Factors in GT systems design. For this integration to be effective, it needs to be formally included in the **initial design** of the manufacturing system, rather than attempting to resolve these aspects as they arise upon implementation of the system. Management also needs to have a clear view of their human resources policies during any recruitment phases and other actions prior to implementation which will influence the composition of the workforce. This may include promotion/demotion, or activities such as the design of payment and incentive schemes, etc.

CONCLUSION

Group Technology is often seen as an important device in the never-ending struggle of manufacturing organisations to improve the performance of their production systems. However, the achievement of the potential benefits depends on the provision of the right work environment for its underlying concepts to be implemented in practice. This is only possible through a proper design of the hierarchical and physical structure of the organisation, of employees' roles as well as plant layout. After some "lights-out-factory" enthusiasm in the past, the importance of humans in the factory of the future is now being reemphasised [40,41], and GT within CIM is playing an important part in these developments [23]. Autonomous, flexible work teams are often seen as a key element of the factory of the future [5,6], and the link to Group Technology structures is obvious.

Machine layouts and component design will obviously affect the human factors in a manufacturing system and thus always merit consideration. However, these are details which will not affect the survival of the entire organisation, whereas this may be at stake if the decision makers in the company fail to consider the nature of their workers' roles. Senior management need to demonstrate not only commitment to GT but also to the necessary organisational changes, which will often involve a complete restructuring of the workforce. Management of the company will also have to demonstrate commitment to training and experimentation (pilot cells etc.), preferably with worker participation. Some of the dramatic failures of GT in the past can be attributed to attempts to impose a GT layout on workers enacting functional, departmentalised roles. These roles of the employees have to be changed to achieve the true GT benefits.

> "Application of the cell concept has so far been limited mainly to grouping machines and machining families of components with virtually no effect on company organisation and management. Consequently, the full possible economic and other benefits have not been achieved." [18]

A more pessimistic observer could remark that implementations of Group Technology have often failed to achieve even the proposed Engineering benefits of GT - and rarely accomplished the full potential of Sinha's **"Total GT System"**. The fact that GT has nevertheless succeeded as a manufacturing strategy and remains a viable option illustrates that very potential.

REFERENCES

[1] Hollingum, J: Group Technology is Alive and Well. <u>FMS Mag</u> 7/4: pp 189-191, 1989.

[2] Ingersoll: <u>Competitive Manufacturing - The Quiet Revolution</u>. Ingersoll Engineers, Rugby, UK, 1990.

[3] McManus J: All Join Hands. <u>Manuf Engineer</u> 70/1: pp 36-38, 1991.

[4] Baran JJ: Tips on Tackling GT-Based Cells. <u>Manuf Engineering</u> 2: 46-49, 1991.

[5] Schonberger RJ: Frugal Manufacturing. <u>Harvard Bus Rev</u> 5: pp 95-100, 1987.

[6] Drucker PF: The Emerging Theory of Manufacturing. <u>Harvard Bus Rev</u> 3: pp 94-102, 1990.

[7] Mitrofanov SP: <u>Scientific Principles of Group Technology.</u> (English Translation), National Lending Library for Science and Technology, Wetherby, UK, 1959.

[8] Burbidge JL: <u>The Introduction of Group Technology</u>. Heinemann, London, UK, 1975.

[9] Brandon JA, Schäfer H, Huang GQ: An Approach to Production Flow Analysis Using Concepts from Information Theory, <u>Proc Conf Expert Planning Systems,</u> Brighton, UK, 1990, pp 124-129.

[10] Kusiak A: <u>Intelligent Manufacturing Systems</u> (Ch.8). Prentice-Hall, Englewood Cliffs, NJ, USA, 1990, pp 206-246.

[11] Edwards GAB: <u>Readings in Group Technology</u>. Machinery, Brighton, UK, 1971.

[12] Gallagher, CC, Knight, WA: <u>Group Technology Production Methods in Manufacture</u>. Ellis Horwood, Chichester, UK, 1973/1986.

[13] Ham I, Hitomi K, Yoshida T: <u>Group Technology</u>. Kluwer-Nijhoff, Boston, MA, USA, 1985.

[14] Schäfer H, Brandon JA: Effective Information Management for the Design, Implementation, Operation and Control of Group Technology Systems. <u>Proc SAMT '92</u>, Sunderland, UK, 1992.

[15] Williamson DTN: The Pattern of Batch Manufacture and its Influence on Machine Tool Design. <u>Proc IMechE</u> 182/1: pp 870-895, 1968.

[16] Gallagher CC: The History of Batch Production and Functional Factory Layout. <u>Char Mech Eng</u> 4: pp 73-76, 1980.

[17] Flanders RE: Design, Manufacture, and Production Control of a Standard Machine. <u>Trans ASME</u> 46: pp 691-738, 1925.

[18] Sinha RK, Hollier RH, Grayson TJ: Cellular Manufacturing Systems. <u>Char Mech Eng</u> 12: pp 37-41, 1980.

[19] Burbidge JL: Production Flow Analysis. <u>Prod Eng</u> 42: pp 742-762, 1963.

[20] Petrov VA: <u>Flowline Group Production Planning</u>. Business Publ, London, UK, 1966 (translated 1968).

[21] Solaja VB, Urosevic SM: Optimization of Group Technology Lines by Methods Developed in IAMA, Beograd. <u>Proc 1st ILO Sem Group Tech</u>, Turin, Italy, 1969, pp 157-176.

[22] Willey PCT, Dale BG: Group Technology - some factors which influence its success. <u>Char Mech Eng</u> 9: pp 76-80, 1979.

[23] Kurimoto A, So K: CIM - Manufacturing Strategy. <u>Proc Conf Factory 2001</u>, Cambridge, UK, 1990, pp 6-10.

[24] Brandon JA, Schäfer H: Strategies for Exploiting Emerging Programming Philosophies to Re-invigorate Group Technology. <u>Proc 29th MaTaDoR Conf</u>, Manchester, UK, 1992.

[25] Burbidge JL: <u>Group Production Methods and Humanisation of Work: The Evidence in Industrialised Countries</u>. IILS, Geneva, Switzerland, 1976.

[26] Naumova NF: Social Factors in the Emotional Attitudes Towards Work, in Osipov GV (ed): <u>Industry and Labour in the USSR</u>. Tavistock, London, UK 1970 (translated 1976).

[27] Fazakerley GM: Group Technology: Social Benefits and Social Problems. <u>Prod Eng</u> 53: pp 383-386, 1974.

[28] Whitaker D: Effect of Group Technology on the Shop Floor. <u>Prod Eng</u> 53: pp 193-194, 1974.

[29] Burbidge JL: Whatever Happened to GT? <u>Management Today</u> 78/9: pp 87-90, 1978.

[30] King JR, Nakornchai V: Machine-Component Group Formation in Group Technology: Review and Extension. <u>IJPR</u> 20: pp 117-133, 1982.

[31] Waghodekar PH, Sahu S: Group Technology: A Research Bibliography. <u>Opsearch</u> 20: pp 225-249, 1983.

[32] Schäfer H: <u>Group Technology Information Management for Effective Cell Formation</u>. PhD thesis, University of Wales, College of Cardiff, UK, 1991.

[33] Askin RG, Subramanian SP: A Cost-Based Heuristic for Group Technology Configuration. <u>IJPR</u> 25: pp 101-113, 1987.

[34] Nagai Y, Tenda S, Shingu T: Determination of Similar Task Types by Use of the Multidimensional Classification Method: Towards Improving Quality of Work Life and Job Satisfaction. <u>IJPR</u> 18: pp 307-332, 1980.

[35] Leonard R, Rathmill K: The Group Technology Myths. <u>Management Today</u> 77/1: pp 66-69, 1977.

[36] Trought B: Manufacturing Systems Design: Incorporating the Human Dimension, in Chandler J (ed): <u>Advances in Manufacturing Technology</u> 4 (Proc 5th NCPR). Kogan Page, London, UK: pp 148-152, 1989.

[37] Daniellou F, Garrigou A: Human Factors in Design: Sociotechnics of Ergonomics, to appear in Helander M (ed): <u>Human Factors in Design for Manufacturability and Process Planning</u>. Taylor and Francis, London, UK, 1991.

[38] Darracott N, Wilson JR: Tacit Skills and Knowledge of Production Workers: An Overview. <u>Internal IOE Report</u>, Inst Occ Ergon, University of Nottingham, UK, 1991.

[39] O'Brien C: Costs, Benefits and Traumas in Changing from Traditional Process to Cellular Manufacture in Automobile Component Fabrication. <u>Proc 5th Int Work Sem Prod Econ</u>, Innsbruck, Austria, 1988.

[40] Sictmann R: CIM braucht die Schnittstelle zum Menschen (CIM needs the human interface). <u>VDI-N</u> 44/18: p 33, 1990 (in German).

[41] Salvendy G: Human Aspects in Integrated Manufacturing Systems (ICOMS), in Gerstenfeld A, Bullinger H-J, Warnecke H-J (eds): <u>Manufacturing Research: Organizational and Institutional Issues</u>. Elsevier, Amsterdam, Netherlands: pp 99-116, 1986.

REAL-TIME DECISION MAKING USING NEURAL NETS, SIMULATION, AND GENETIC ALGORITHMS

[1]Albert T. Jones and [2] Luis C. Rabelo

[1]*National Institute of Standards and Technology, Gaithersburg, Maryland, U.S.A.*
[2]*Industrial and Systems Engineering Department,*
Ohio University, Athens, Ohio, U.S.A.

ABSTRACT

In this paper, we outline a real-time decision making approach which integrates artificial neural networks, real-time simulation, and genetic algorithms. Artificial neural nets are used to generate a small set of attractive plans and schedules. Real-time simulation predicts the impact of each on the future evolution of the manufacturing system as quantified in one or more performance measures. Genetic algorithms are used to combine attractive alternatives into a single "best" decision. This approach forms the heart of the analysis tools to be incorporated into a generic architecture for an intelligent manufacturing controller. This controller can be used at any level within a shop floor control hierarchy for a computer integrated manufacturing system. For each assigned job, this controller decides which process plan will be used and the start and finish times for all of the tasks in that plan. In addition, the controller monitors its subordinates' execution of those tasks and recovers from reported errors.

INTRODUCTION

A wide variety of controllers are used in manufacturing today. There is general agreement that it is desirable to make these controllers more "intelligent". There is, however, no agreement in what constitutes an intelligent controller. In our view, an intelligent controller does four functions: planning, scheduling, monitoring, and recovering from problems. The planning function is responsible for deciding what operations are required to complete an assigned task. The scheduling function decides when each of these operations will start and end. The monitoring function compares the actual execution of operations with the desired schedule. Whenever a problem occurs, the controller must, if possible, recover and get back on schedule, generate another schedule, or find a completely new plan.

In this paper we briefly review the generic architecture for intelligent controllers proposed in [1]. We then describe an approach for carrying out the optimization procedures contained within that architecture. Those procedures determine the plan to be followed in completing each job and the sequence of start and finish times for each of the tasks in that plan. We propose an approach which integrates artificial neural networks, real-time Monte Carlo simulation, and genetic algorithms. While we do not preclude the use of this approach for equipment controllers, we limit our discussion in this paper to higher level controllers which coordinate the activities of many pieces of equipment.

THE GENERIC CONTROL ARCHITECTURE

The generic architecture for the intelligent controller proposed in [1] is shown in Figure 1. It performs four major functions which expand and generalize those developed in [2,3]: assessment, optimization, execution, and monitoring. These functions are described in the following sections.

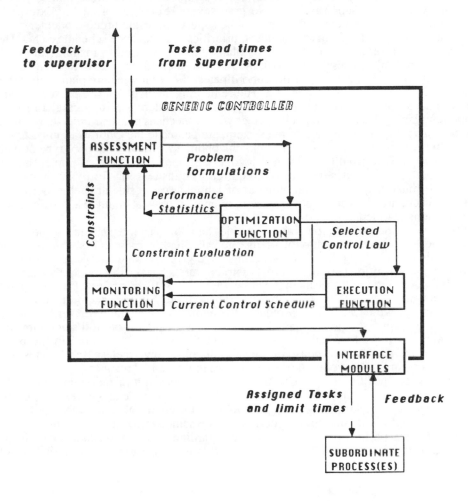

FIGURE 1. Generic Control Architecture

Assessment Function

The Assessment Function (AF) formulates all "planning and scheduling" problems based on the current state of its subordinates and input from its supervisor. Three sets of inputs are used in this problem formulation task. The first is a list of assigned tasks with due dates and other constraints which are specified by the controller's supervisor. The second is the feedback information from the Monitoring Function describing 1) the current state of the system, and 2) the projected system response under the current control law . The third comes from the Optimization Function which summarizes its current success in finding an optimal solution to the problems it has been given to solve. In formulating these optimization problems, the AF first specifies the constraints. There are two types: hard and soft. All of these constraints must be chosen so that the due dates imposed by its supervisor are not violated.

Hard constraints are those that cannot be violated either by the optimization and execution functions in the same controller or by subordinate controllers. These constraints are typically of three types: external, internal, and those related to the physical dynamics of subordinates. We give several examples of each. Externally imposed hard constraints come from either the supervisor or the process planner. The former are usually in the form of due dates, priorities, maintenance schedules, etc. The latter results in a set of candidate process plans that can be used to manufacture each product. If multiple process plans exist, the AF will screen them to eliminate infeasible plans based on the current state of the subordinates. For example, any plan requiring a "downed"-machine must be eliminated. The remaining alternatives are passed as hard constraints to the Optimization Function which will determine the actual run-time production plan. This run-time plan together with collections of scheduling alternatives, which are also based on the state of the system, are passed to the Optimization Function as hard constraints to be used in determining the optimal schedule. The third type of hard constraint is derived from the physical layout of subordinates. For example, the AF must account for the times that will be required to transfer a job from one process to another. It must also consider the number of jobs that can reside at each subordinate location. There are also certain physical assignments of jobs that could result in deadlock which must be avoided.

The specification of soft constraints provides an additional mechanism for the AF to control the evolution and behavior of its subordinates. In most cases, these can be violated without violating the hard constraints. We give several examples. While the due date for a given job may be a hard constraint, at each layer there may be start and finish times for some of the tasks that can be viewed as soft constraints. Even though a subordinate may be capable of storing several jobs, the AF may decide to impose a Kanban strategy which significantly reduces the number of jobs that will ever be queued before a given process. The AF may also impose soft constraints on utilization rates for subordinates. Unlike hard constraints, soft constraints are not essential in the basic definition of the projected response of the subordinates. Nevertheless, their imposition will further constrain the evolution of subordinates in the desired manner.

The AF also specifies the performance criteria for each optimization problem to be considered by the Optimization Function. These criteria can be related to the "performance" of individual subordinates or jobs, or the entire collection of subordinates or jobs. Examples for subordinate performance include utilization, and throughput. Examples for job performance include lateness, tardiness, and makespan. The AF can also assign priorities to individual jobs and weights to the performance criteria. All of these can be changed to reflect the current state of the system.

Optimization Function (OF)

The Optimization Function (OF) has two major responsibilities. First, it must solve the planning and scheduling (optimization) problems posed by the AF by selecting a run-time production plan and scheduling rule from those passed down by the AF. Second, it must restore feasibility (optimality) whenever 1) the Execution Function cannot restore feasibility using the selected plan and rule, (its hard constraint is violated) and 2) the AF has determined that the current projected system response is unacceptable based on either the performance criteria or soft constraints. Strategies for carrying out this type of error recovery are discussed in [1].

As noted above, these optimization problems include a set of performance criteria, hard constraints and soft constraints. Sub-gradient optimization techniques [4] are commonly used to solve this type of problem when everything is assumed deterministic.

In such solution techniques the soft constraints can be included in the objective function as a penalty function. We are extending this notion to stochastic systems utilizing real-time simulation analysis (see below), in which processing durations for individual tasks are sampled from probability distributions. These distributions are derived from historical data and are updated as the tasks are processed. In performing this analysis, the notion of dominance from multi-criteria decision-making is fundamental. Like other decision-making methodologies, the techniques for multi-criteria decision-making are most developed for static, deterministic decision-making. However, some work, notably on stochastic dominance, has also been done for the stochastic case [5,6].

For each assigned job, the output from the OF consists of a process plan, estimated processing durations (pessimistic, optimistic, and average) and a scheduling rule. These are all used by the Execution Function to determine the actual start and finish times for the tasks in that plan.

Execution Function (EF)

The Execution Function (EF) implements the alternatives selected by the Optimization Function. In particular, it addresses the following two tasks: it uses the scheduling rule selected by the Optimization Function to compute limit times for each subordinate task in the selected production plan, and it restores feasibility using that rule when violations of those limit times arise from the actions of the subordinate subsystems.

The primary focus of the EF is to generate the limit times (start and finish) for each task based on the selected production plan and scheduling rule. For a new job, this is done using a single-pass real-time simulation initialized to the current state of the system. In this single pass simulation, the EF must select a unique duration for each task from those passed down by the Optimization Function. The selection of this unique duration is strongly influenced by the variance and shape of the probability distribution used by the Optimization Function. This, in turn, is completely determined by the amount of historical data that exist for both the job and the subordinate. Whenever the variability is high, then the pessimistic estimate should be chosen. Whenever, the variability is low the average should be chosen. Under special circumstances- such as a new tool has just arrived, or better than average recent performance - the optimistic value can be chosen.

There are of course tradeoffs to be considered in making this selection. If the pessimistic value is selected, which has higher probability of being met, then fewer deviations can be expected between the planned and realized system evolutions. However, selecting a duration with a high probability of completion also implies a higher probability that the subordinate will finish early and be idle for a longer than expected time. On the other hand, the selection of a optimistic duration increases the probability that the task will not be completed as planned. This means that more conflicts will arise between planned and realized system performance.

Once the durations are chosen, the single-pass simulation is run to obtain the limit times. Once these limit times are determined they are passed, along with the tasks, to 1) subordinates where exact trajectories are computed, and 2) the Monitoring Function to be evaluated against the constraints specified by the Assessment Function.

Monitoring Function (MF)

The Monitoring Function has two major responsibilities: updating the system state, and evaluating proposed system responses against the hard and soft constraints set by the Assessment function.

The system state summarizes the operational status, buffer status, and job status of subordinates. This expands the concept of "world model" espoused in [2]. The exact definition changes as one moves from one layer to another. The Monitoring function generates this state information by aggregating the detailed feedback data from the various subordinates. The Monitoring function does not communicate directly with subordinates. It communicates through an Interface module. This module performs all of the functions necessary to make meaningful communication possible between a supervisor and its subordinates. This updated system state is then broadcasted to all of the other functions, because it is the starting point for all of the real-time analysis performed by the generic controller.

The Monitoring function also evaluates each decision against the current set of constraints. Recall that these constraints are of two types: hard (which cannot be violated) and soft (which are desirable goals). Hard constraints are imposed by the supervisor on assessment, by assessment on optimization, by optimization on execution, and by execution on subordinates. As we have seen, a series of refinements takes place in the sense that the hard constraints imposed by the Assessment function on the Optimization function are a subset of those imposed on the Assessment function by the supervisor. The same is true for the Optimization and Execution functions. Each time one of these functions makes a decision, the monitoring function will determine if any of these hard constraints have been violated. Each time a violation occurs, the Monitoring function will identify the constraint(s) and inform the appropriate function.

For the soft constraints, the situation is a little different. These are tracked continually and transmitted to the assessment function where several things can happen. Consider a soft constraint on System Utilization (SU > 90%). The assessment function can choose to ignore the violation when it is small (SU = 85%). On the other hand, the Assessment Function could simply request the Optimization Function to generate a new schedule or pick a new plan, whenever this violation is outside some threshold. We allow the threshold to a two sided limit (perhaps 70% < SU < 95%). This allows the Assessment Function to react when the utilization is too small (ie machines sitting idle) or too large (ie bottlenecks may be occurring).

ANALYSIS TOOLS

As noted above, the Optimization Function is responsible for solving the decision-making (optimization) problems posed by the Assessment Function. There are two types: planning and scheduling. If we examine the structure of these problems, we find that they are very similar. In each case, a series of alternatives must be evaluated against a collection of performance criteria. The goal is to select the "best" alternative, relative to those performance measures, which meets all the hard constraints. We propose an integration of three techniques: artificial neural networks, real-time simulation, and genetic algorithms. Our research, to date, has been restricted to scheduling, but we are hopeful that this same approach can be used for planning.

Artificial Neural Networks - The First Step.

The first step is to select a small list of attractive scheduling rules from a larger list of such rules. This is referred to as "Candidate rule selection" in Figure 2. For example, we might want to find the best five dispatching rules from the list of all known dispatching rules. These five could be chosen so that each one maximizes (or minimizes) at least one of the performance measures, with no regard to the others. In addition, we would, whenever a problem occurs, like to use to select those rules most likely to result in a quick recovery. To carry out this part of the analysis, we propose to use a neural network approach [7,8]. Neural networks have shown good promises for solving manufacturing scheduling problems. Their applications have particularly been dedicated to NP-complete problems, and in specific to the Traveling Salesman and Job Shop Scheduling problems. (For a good survey, see [9].) However, the direct application of neural networks for scheduling problems has not produced consistently optimal, real-time solutions for realistic problems. There are two reasons: limitations in hardware and the lack of algorithms. Because of the distributed, hierarchical nature of the controllers proposed in [1] we do not anticipate this problem in our research.

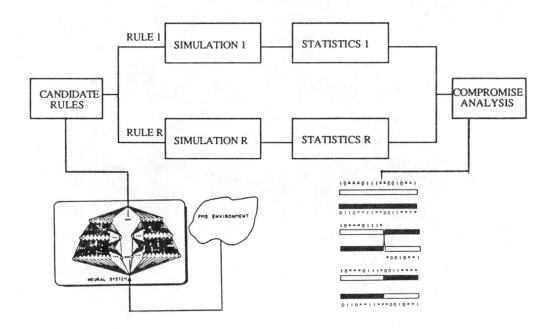

FIGURE 2. The Integrated Decision-Making Framework

Network Paradigms for Candidate Rule Selection. The optimal neural network paradigm for "candidate rule selection" should be established by analytical and experimental procedures. Supervised learning (because solutions from previous experience or simulation are known) should be emphasized.

The paradigms should be selected due to their general acceptance, mapping capabilities, and suitability to be automated. Integration of several of them to solve the problem should also be considered. The paradigms should be assessed on the following factors:

(1) Trainability: The impact of the initialization and learning parameters in the probability of the network learning training sets for candidate rule selection.
(2) Generalization: The ability of the network to model the behavior of the specific FMS relationship. This is tested utilizing unseen cases.
(3) Computational characteristics: Storage needed for programs and data files, training time, and execution time.
(4) Re-Trainability: Ability to be retrained when the training set has been incremented or modified.

Several paradigms are possible candidates for "Candidate Rule Selection". In this research, we will focus our initial efforts on Backpropagation and Fuzzy ART. These paradigms have been selected due to their supervisory learning mode, non-linear and fuzzy modeling capabilities, generalization potential, and suitability of the algorithms to be automated.

Backpropagation [10] is a supervised learning paradigm that learns adequate internal representations using deterministic units to provide a mapping from input to output. The backpropagation artificial neural network structure is a hierarchical design consisting of fully interconnected layers or rows of processing units. Each unit itself comprised of several layers or rows of processing elements (see Figure 3). In this network, the value

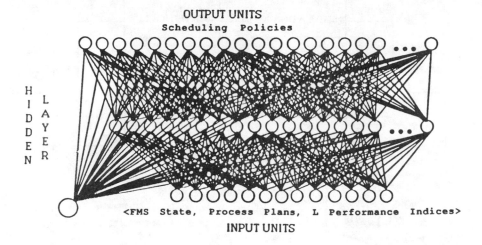

FIGURE 3. Back Propagation Neural Network for Candidate Selection

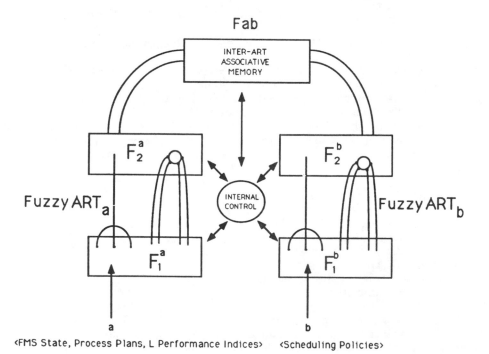

FIGURE 4. Fuzzy ART Neural Network for Candidate Selection

from the input might represent the state of the FMS, the process plans, and the L performance indices. The output might represent the available scheduling policies from which a quantitative evaluation is going to be obtained.

On the other hand, Fuzzy ARTMAP[11] is a supervised learning of recognition categories in response to arbitrary sequences of analog or binary inputs. It incorporates two Fuzzy ART modules (networks which utilize the fuzzy set theory conjunction in the learning rules in order to develop stable categories) that are interlinked through a mapping field Fab (see Figure 4). The inputs to the Fuzzy ARTa Module (also outputs for the entire neural structure if Fuzzy ARTb is considered the input Module) could be the state of the FMS, the process plans, and L performance indices. The inputs to the Fuzzy ARTb Module (also outputs for the entire neural structure if Fuzzy ARTa is considered the input Module) might represent the available scheduling policies.

Input Feature Space. In order to design a system for the "Candidate Rule Selection" process based on neural networks an appropriate representation of the "state" of the FMS, should be developed. The state contains aggregate information about all in-process jobs, processes, and the active schedule. This aggregate state information is derived from data for each job includes job ID, current location (buffer, transporter, or process), due date, expected completion time (if it has previously been scheduled), shop floor release time, current active routing and expected/actual start and finish time at each process in that routing. The state of the FMS processes should indicate according to their type the jobs, batch size, the planned start and delivery time of the jobs, and information about the transporters. The current schedule and its variables. The process plans information is provided by the processing times and routings. The L performance indices such as Average Tardiness, Job Flow Time, Job Productivity, and Process Utilization will be required.

The state of the FMS, process plans, and the L performance indices should be translated into a suitable representation for the neural networks to provide the desired evaluation of the different scheduling rules available in the system. Previous studies have indicated guidelines for this task [9,12].

Neural Networks Knowledge Acquisition Process. The results of past scheduling cases should be presented to the neural structure in order to develop the desired behavior (i.e. gives good candidates). If information on past behavior is not available, simulation could be used as part of the knowledge acquisition process. The simulator will solve the problem for each scheduling rule, and the results will be recorded. The neural network training process will continue even after the system is put in place, continuously updating knowledge from actual cases. This will be critical whenever rescheduling is required to offset delays that result from system bottlenecks and breakdowns.

Real-Time Simulation

After these candidates have been determined, each of them will be evaluated (see Figure 2) using an analysis technique termed real-time Monte Carlo simulation [13]. The outputs from these simulation trials yield the projected schedule of events under each decision alternative. These schedules are then used to compute the values of the various performance measures and constraints imposed by the Assessment Function. Since R alternatives must be considered, we plan to run the R real-time simulations concurrently to avoid unacceptable timing delays in the analysis. At this time, we are using a network of Personal Computers, with each one simulating a single alternative. We hope to move on to a parallel processing machine in the near future. The length of each simulation trial for a given simulation engine is determined by the length of the planning horizon of the controller.

This form of Monte Carlo simulation differs considerably from traditional discrete-event simulation in two ways. First, each simulation trial is initialized to the current system state as updated in real-time by the Monitoring Function (see above). This results in the second important distinction - the simulations are neither terminating nor steady state [14]. They are not terminating because the initial conditions may change from one trial to the next. Furthermore, they are not steady state because we are specifically interested in analyzing the transient phenomena associated with the near-term system response. The question is, from a statistical perspective, does this really matter. Our early experiments indicate that the answer is a definite maybe. That is, the inclusion or exclusion of new events corresponding to changes in the initial state can bias the statistical estimates of certain, but not all, performance measures. Another issue is related to how many trials and how much information is needed to calculate the estimates of these measures. Currently, we use the most recent 100 simulation trials for statistical analysis. That is, as each new simulation trial is generated by a given engine, the oldest simulation trial is removed.

Given the last 100 simulation trials for each of the R alternatives, a series of statistical calculations are made for the given performance criteria for each individual trial: the empirical probability density function associated with each criterion, means, variances, and the correlations between each pair of the *L* performance measures. We have found cases where the correlation is positive, zero, and negative [15]. These correlations are useful as they provide the statistical tradeoffs that exist among the considered performance criteria for each decision alternative. For example, if the correlations between two criteria are nearly one, then we can replace two distinct criteria with a single criteria for that decision alternative. However, a positive correlation under one decision alternative may not necessarily imply that the same two criteria will also be positively correlated under another decision alternative.

The computed statistics and associated empirical probability density functions are passed to the compromise analysis element where the decision alternative which currently provides the best statistical compromise among the performance criteria will be selected. To date, most of this development has been formulated to address the maximum expected value of a utility function that would be achieved if a given decision alternative was adopted. In the next section, we discuss a new approach to compromise analysis, genetic algorithms.

Genetic Algorithms

No matter how the utility function described above is constructed, only one rule from the candidate list can be selected. This causes an undesirable situation whenever there are negatively correlated performance measures, because no one rule can maximize (or minimize) all objectives simultaneously. Conceptually, one would like to create a new "rule" which 1) combines the best features of the most attractive rules, 2) eliminates the worst features of those rules, and 3) simultaneously achieves satisfactory levels of performance for all objectives. Our approach does not deal with the rules themselves, but rather the actual schedules that result from applying those rules. Consequently, we seek to generate a new schedule from these candidate schedules. To do this, we will use a genetic algorithm approach.

The basic execution cycle for a genetic algorithm is depicted in Figure 5 and explained below:

1. The population is initialized using the specific procedures. This initialization process will result in a set of chromosomes.

2. Each member of the population is evaluated, using the objective function.

3. Select candidates according to strength - relative ranking according to the objective function. Candidates with the highest fitness rating have a greater probability of reproduction.

4. The population undergoes reproduction until a stopping criteria is met. Reproduction consists of a number of iterations of the following three steps:

(**a**)   One or more parents are chosen to reproduce. Selection is stochastic.

(**b**)   Genetic operators are applied to the parents to produce the offsprings.

(**c**)   The performance of the new population is evaluated.

5. Iterate the process.

FIGURE 5. Execution Cycle for a Genetic Algorithm

When a genetic algorithm is run using a representation that encodes solutions to a problem such as scheduling and operators that can generate better offsprings, the methodology can produce populations of better individuals (e.g., schedules), converging finally on results close to a global optimum. Knowledge of the the domain can often be exploited to improve the genetic algorithm's performance through the incorporation of new operators or the utilization of integrated reasoning architectures. Examples of this approach can be found in [17,18].

The compromise analysis (see Figure 2) process carried out by a genetic algorithm can be thought of as a complex hierarchical generate and test process. The generator produces building blocks which are combined into schedules. At various points in the procedure, tests are made that help weed out poor building blocks and promote the use of good ones. To reduce and support uncertainty management of the search space, the previous two steps (candidate rules selection and parallel simulation with statistical analysis) provide partial solutions to the problem of compromise analysis. Reducing uncertainty makes the search process more effective, with the complexity of the scheduling problem becoming more manageable in the process (see Figure 6).

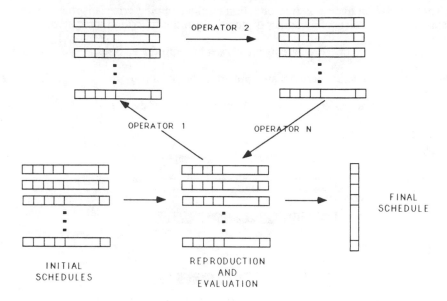

FIGURE 6. Genetic Algorithm for Compromise Analysis

SUMMARY

In this paper, we have specified an generic architecture for a controller which provides the building block for an integrated shop floor control system. This controller contains a generic set of functions that provides a high degree of intelligence and the ability to respond to disturbances in real-time. Those functions are Assessment, Optimization, Execution, and Monitoring. They effectively integrate both decision-making and control. The decisions are formulated by the assessment function and solved by the optimization function. The execution and monitoring functions provide the most of the control. The paper also discussed a new integrated framework for addressing the real-time analysis a problems essential to successful implementation of that architecture. It consist of neural networks to provide candidate solutions, real-time Monte Carlo simulation to carry detailed performance analysis of each of those candidates, and genetic algorithms to complete the compromise analysis.

REFERENCES

1. Davis W, Jones A, Saleh A: A generic architecture for intelligent control systems, *Computer Integrated Manufacturing Systems*, to appear, 1991.

2. Albus J, Quintero R, Lumia R, Herman M, Kilmer R, Goodwin K: Concept for a reference model architecture for real-time intelligent control, *NISTIR-1277*, National Institute of Standards and Technology, Gaithersburg, Md., April, 1990.

3. Jones A, Saleh A: A multilevel/multilayer architecture for intelligent shop floor control, *International Journal of Computer Integrated Manufacturing special issue on Intelligent Control,* Vol. 3, No. 1, 60-70, 1990.

4. Geoffrion A: Elements of large-scale mathematical programming, *Management Science,* Vol. 16, No. 11, 652-691, 1970.

5. Fishburn PC: Stochastic dominance without transitive preferences, *Management Science*, Vol. 24, No. 12, 1268-1277,1978.

6. White CC, Sage AP: Second order stochastic dominance for multiple criteria evaluation and decision support, *Proc. of International Conference on Cybernetics and Society*, Sponsored by IEEE Systems, Man and Cybernetics, Atlanta, Georgia, 572-576, 1981.

7. Hopfield J, Tank D, Neural computation of decisions in optimization problems, *Biological Cybernetics,* Vol. 52, 141-152, 1985.

8. Kohonen T: An introduction to neural computing, *Neural Networks*, Vol. 1, No.1, 3-16, 1988.

9. Rabelo L: A hybrid artificial neural network and expert system approach to flexible manufacturing system scheduling, PhD Thesis, University of Missouri-Rolla, 1990.

10. Rumelhart D and the PDP Research Group: *Parallel Distributed Processing: Explorations in the Microstructure of Cognition*, Vol. 1: Foundations, Cambridge, MA: MIT Press/Bradford Books, 1988.

11. Carpenter G, Grossberg S, Rosen D: FUZZY ART: Fast stable learning and categorization of analog patterns by an adaptive resonance system, CAS/CNS-TR-91-015, Boston University, 1991.

12. Rabelo L, Alptekin S: Synergy of neural networks and expert systems for FMS scheduling, *Proceedings of the Third ORSA/TIMS Conference on Flexible Manufacturing Systems: Operations Research Models and Applications*, Cambridge, Massachusetts, Elsevier Science Publishers B. V., 361-366, 1989.

13. Davis W, Jones A: Issues in real-time simulation for flexible manufacturing systems, *Proceedings of the European Simulation Multiconference*, Rome, Italy, June 7-9, 1989.

14. Law A M, Kelton WD: *Simulation, Modeling and Analysis*, McGraw-Hill, New York, 1982.

15. Davis W, Wang H, Hsieh C: Experimental studies in real-time Monte Carlo simulation, *IEEE Transactions on Systems, Man and Cybernetics*, Vol. 21, No. 4, 802-814, 1991.

16. Goldberg D: *Genetic Algorithms in Machine Learning*, Addison-Wesley, Menlo Park, California, 1988.

17. Biegel J, Davern J: Genetic Algorithms and Job Shop Scheduling, *Computers and Industrial Engineering*, Vol. 19, No. 1, 81-91, 1990.

18. Parunak H: Characterizing the Manufacturing Scheduling Problem, *Journal of Manufacturing Systems*, Vol. 10, No. 3, 241-259, 1991.

DESIGN AND ANALYSIS OF FLEXIBLE MANUFACTURING SYSTEMS USING ARTIFICIAL INTELLIGENCE AND SIMULATION

[1]Luis C. Rabelo, [1]Girish Bakshi, and [2]Anil Malkani

[1]*Industrial and Systems Engineering Department, Ohio State University, Athens, Ohio, U.S.A.*
[2]*Electrical and Computer Engineering Department, Ohio State University, Athens, Ohio, U.S.A.*

ABSTRACT

Simulation has been extensively utilized as a technique for the design, dynamic control, and scheduling of flexible manufacturing systems (FMSs). For the design, this has resulted in long and iterative processes. For dynamic control and scheduling, "on line" simulation has only been able to evaluate a small number of alternatives due to timeliness requirements imposed by the short time horizon of the decision-making process. Artificial neural networks could generalize from a small set of simulation runs in order to model the behavior of the systems. Therefore artificial neural networks could be utilized to optimize the simulation process. Examples and comparisons will be provided to illustrate the possible benefits of the integration of artificial neural networks and simulation in order to have higher levels of performance in FMSs.

INTRODUCTION

Simulation modeling is a powerful experimental technique available for the analysis and study of flexible manufacturing systems whose complexity has precluded the use of a solo effective analytical solving technique. Simulation modeling has been utilized as a method to determine desirable variable-values, and inter-relationships between the variables of the FMS under analysis. Simulation may be integrated with artificial neural networks in order to speed up and improve the design and analysis processes of FMSs.

Artificial neural networks might help the step-wise refinement of an abstract conceptual model of a real-world system to obtain a detailed functional model by supporting the analysis of structural and behavioral properties of the manufacturing system under consideration. Traditionally, mathematical and formal models have been utilized for this refinement process. However, these methodologies have the tendency to simplify complex models (i.e., complicated analytic tractability) and therefore omissions of important concepts are abundant. This process may be either qualitative or quantitative. Artificial neural networks handle both categorizations. Artificial neural networks could process experimental data for finding functional and more realistic relationships beyond the common approach of finding a function which in many cases (specifically if new data or changes on-line are produced) do not represent at all times the concrete facts of the modeled system.

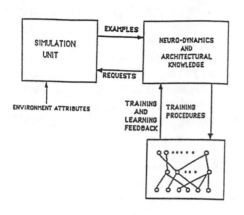

FIG. 1 Integration of Artificial Neural Networks and simulation in Design, Scheduling, and Dynamic control of FMS.

Qualitative tools are used if the number of parameters under consideration is not very large and the diversity of the world or object under investigation is not too great. Artificial neural networks scale in general with the problem, and their computational features avoid complicated preprocessing making them a viable alternative for traditional quantitative tools such as mathematical statistics, integral equations, and simulation. Their fuzzy modeling capabilities and hierarchical structures allow them to support the concretization of qualitative models into quantitative ones.

Simulation techniques have been extensively used for FMS areas such as design, process control scheduling, quality control, inventory management. In manufacturing process control, real-time simulation may be used to simulate future expected event variables (such as quality of a product) based on variable-values at current time (such as machine wear and tear), and use this analysis to adjust the process to obtain desired future variable-values. Control of manufacturing processes and FMS workcells needs the collection of massive amount of data. Adjustments to control parameters and strategies based on the values of the data collected are necessary in real time. These adjustments are made to return the processes to an operating state or to avoid "out of control parameters" or re-scheduling of operations in order to meet shop floor performance criteria. Such control needs to be performed on a real-time basis, and using traditional simulation methods are often time consuming and therefore not suitable for this purpose.

In this paper, the utility of artificial neural networks to simulation in FMSs, specifically the use of artificial neural networks to speed up iterative design processes and the responsiveness of on-line simulation will be discussed (see Figure 1). Two illustrations using artificial neural networks and simulation for the design and dynamic control of FMS, and real-time FMS scheduling will be presented.

ARTIFICIAL NEURAL NETWORKS:

The primary purpose of all neural systems is centralized control of various biological functions. In higher animals the major capacity of the central nervous system is connected with behavior, that is, control of the state of the organism with respect to the environment. However, in neural computing we conceptualize in idealized form, the sensory, motor functions and thinking of the physical neural systems.

Artificial Neural Networks may be defined as :
"Artificial Neural Networks" are massively parallel interconnected networks of simple (usually adaptive) elements and their hierarchical organizations which are intended to

interact with the objects of the real world in the same way as biological nervous systems do.[1]

Artificial neural networks are information/processing systems composed of large number of interconnected and fundamentally simple processing elements. By virtue of their computational structure, artificial neural networks feature attractive characteristics such as graceful degradation, robust recall with fragmented and noisy data, parallel distributed processing, generalization to patterns outside of the training set, nonlinear modeling capabilities, and learning.   The various learning mechanisms can be broadly classified as:

Supervised Learning	-	learning with a teacher.
Unsupervised Learning	-	self organize the categories.
Reinforcement Learning	-	add feedback to unsupervised learning to evaluate the pattern classification process.

The spectrum of different artificial neural networks paradigms is quite extensive. For example, the network architecture range from simplistic perceptrons to the hierarchical neocognitron. In addition, there are a large number of algorithms for modification of the adaptive coefficients. The various paradigms developed have their limitations and strengths, hence one must identify the suitable application areas for which they lend themselves. Our research is mainly developed on the standard backpropagation and fuzzy ARTMAP paradigms of artificial neural networks.

Backpropagation Artificial Neural Network Paradigm

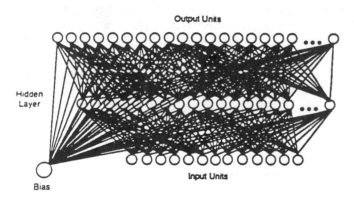

FIG 2. A Backpropagation network

The artificial neural network utilized in this research, developed by Rumelhart [2] learns adequate internal representations using deterministic units to provide a mapping from input to output. The backpropagation artificial neural network structure is a hierarchical design consisting of fully interconnected layers or rows of processing units (with each unit itself comprised of several layers or rows of processing elements, see Fig 2). In this network, the value from the output layer, which is obtained by propagating forward the input vector, is compared with the targeted one.

The error is measured based on the following objective function :

$$E_p = \frac{1}{2}\sum_i (t_i - O_i)^2$$

where

p = represent each input vector/target output vector.

t_i = target output for the ith output node

O_i = respond\se obtained from the ith node.

The weight of the interconnections are iteratively modified in order to reduce the measured error. Each weight update is determined by the partial derivative of the error with respect to each weight (W)

$$\Delta W_{ij} \sim \frac{\partial E_p}{\partial W_{ij}}$$

The partial derivative of the error in this hierarchical neural network structure is facilitated by the utilization of the sigmoid functions (non decreasing and derivative transfer functions) to generate the output processing element.

The training process using backpropagation is a difficult problem[3]. It is needed to find an appropriate architecture (e.g number of hidden units, number of hidden layers etc.), adequate size and quality of training data, satisfactory initialization(e.g initial weights), learning parameter values (e.g learning rates), and to avoid over-training effects (performance degradation due to prolonged training).

In this research the following approaches were utilized:

1. To help find an appropriate architecture (i.e number of hidden units) an interactive addition of nodes was performed as proposed by the dynamic node creation (DNC). DNC is a methodology of adding nodes to the hidden layer(s) of the network during training. A new node is added if the root mean square (RMS) error curve has flattened out to an unacceptable level. This process is stopped when the desired performance has been achieved.[4]

2. To speed convergence behavior, the selection of parameters such as learning rates and the utilization of a momentum factor were utilized. The learning rule utilized consisted of a weight update using momentum (β) with the exception that each weight had its own "adaptive" learning rate parameter (μ) . The "adaptive" learning rate strategy increments the μ(s) by a small constant if the current partial derivative of the objective function with respect to the weight (W) and the exponential average of the previous derivatives has the same sign, Otherwise μ will be decremented by proportion to its value. The updating equation of the weight is defined by using W_{ij} as the weight value located between nodes i and j, t is the present iteration, Δw is the weight increment which equals to the product of the μ and the partial derived of the objective function with respect to the weight ($\delta E/\delta w_{ij}$),

$$W_{ij}(t) = W_{ij}(t-1) + \Delta W_{ij}(t) + \beta \Delta W_{ij}(t-1)$$

The β changes dynamically, because each problem has a range of optimal β values to avoid oscillations.

Fuzzy ARTMAP

Fuzzy ART[5] incorporates the basic architecture and neuro-dynamics of ART systems. Fuzzy ART is designed as generalization of ART1. However, the set theory intersection operator ($\cap$) of ART1 is replaced by the fuzzy set theory conjunction(Λ). This fuzzy operator makes Fuzzy ART capable of handling both analog and binary data. Nevertheless, Fuzzy ART also has other features such as :

    a) Fast-commit slow-recode   : It combines fast learning with a forgetting rule that buffers system memory against noise.

    b) Input preprocessing
        1) Normalization   : Prevents category proliferation.
        2) Complement Coding   : Preserves individual feature amplitude.

For each input (I) presented to the network of dimension M (2M in complement coding) in the interval [0,1], the net value at the output is calculated as

$$net_j(I) = \frac{|I \wedge W_j|}{\alpha + |W_j|}$$

Where $\alpha = 0.00001$
        I is the input vector
        W_j is adaptive weights or LTM traces (initialized to 1)
        Λ is the fuzzy And operator
The max of $net_j(I)$ is chosen. For this node J, it is checked if the vigilance (ρ) criteria is met, i.e.

$$\frac{|I \wedge W_j|}{|I|} \geq \rho$$

resulting in node J being committed, else reset occurs at the selected output node and is inhibited for further competition for input I. The above process is repeated till the vigilance criteria equation is satisfied.

If the output neuron satisfies the vigilance criteria then learning takes place using

$$W_j^{NEW} = (I \wedge W_j^{OLD})$$

Fuzzy ARTMAP [6] is a supervised learning of recognition categories in response to arbitrary sequences of analog or binary inputs. It incorporates two Fuzzy ART modules i.e Fuzzy ARTa and Fuzzy ARTb that are interlinked through a mapping field F^{ab}. The inputs to the Fuzzy ART modules are presented in complement form, i.e A =(a, a^c) (see Figure 3).

During training, at the start of each input presentation the Fuzzy ARTa vigilance factor equals the baseline vigilance of F_a, and the map field vigilance parameter is set to 1.0 . When a prediction by Fuzzy ARTa is disconfirmed at Fuzzy ARTb i.e F^{ab} output vector $X_{ab} = 0$, match tracking is induced. Match tracking rule raises the module A vigilance to

$$|Xa|/|A| + 0.0001 \quad \text{(in our implementation)},$$

Where, $|X^a| = |A \wedge W_j^a|$.
A is the input, i.e A $=(a, a^c)$,
X_a is the output vector of Fuzzy ARTa module,
$|X|$ is summation of all the components of X, and
W_j^a is the weight vector from the jth output node of Fuzzy ART 'A' to F_{ab}.

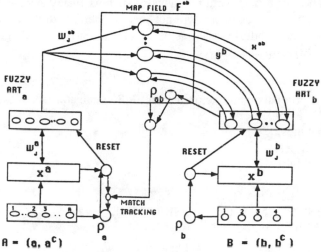

When Match tracking occurs, Fuzzy ARTa search leads to activation of another output neuron. If the prediction by Fuzzy ARTa is confirmed at Fuzzy ARTb (F^{ab} output vector $X_{ab} = 1$), Map field learning takes place. J learns to predict the Fuzzy ARTb category K, and sets $W_{JK}^{ab} = 1$.

FIG 3. Fuzzy ARTMAP.

INTEGRATION OF ARTIFICIAL NEURAL NETWORKS AND SIMULATION FOR DESIGN AND DYNAMIC CONTROL OF FMS

Simulation has been utilized as a methodology to support the "FMS design" task which includes issues such as machine availability and compatibility, layout, part families, and control unit(s). Simulation has been utilized due to the difficulty of setting realistic mathematical programming models for this task. However, using simulation for the design task involves the utilization of a long and iterative process. Chryssolouris et al [7,8], proposed the utilization of artificial neural networks in the design stages to expedite and improve the design process of manufacturing systems by using the inverse of the simulation function. We consider that the knowledge developed during the design stages of the manufacturing system should also be utilized for dynamic control of the FMS system by using artificial neural networks which emphasize bi-directional information flows (i.e., the inputs could be utilized as the outputs and vice versa).

The argument developed above is supported by several observations. As an example, different philosophies may be employed in developing FMS workcells [9]. Several strategies call for applying a FMS integration strategy that allows for the exchange of units as needed in order to modify the functions of the FMS (i.e., high flexibility). In addition, during the production run of a FMS its performance might drop considerable. A dynamic control systems could play sensitivity analysis with some exchange of units or control/scheduling policies in order to regain the productivity level. Therefore, the knowledge developed by using some simulation runs to train the networks to help to approximate a good solution to the design problem could be also utilized to change configurations or control/scheduling policies in real-time in order to re-establish the desired performance level.   In addition, artificial neural networks due to their inherent speed, provide a structure which could be updated easily to incorporate shop floor events as they happen in real-time and therefore more responsive than "on-line" simulation.   It is appropriate to remember that FMS are systems which evolve through nonlinear interactions. This limits the effective utilization of "on-line" simulation.

In addition, the utilization of a bi-directional network such as Fuzzy ARTMAP will support the development of a strategies which will not be so vulnerable to the many-to-one problem in neural networks paradigms such as backpropagation (i.e., averaging outputs when several answers exist for the same input).

Case Study

There are three types of products manufactured at an FMS which consists of three cells. The possible configurations are:
CELL 1   Milling (1 to 3 machines);
CELL 2   Drilling (1 to 3 machines);
CELL 3   Turning (1 to 3 machines);

The routing for each product type are:

PRODUCT TYPE	PROCESS PLAN	PROCESSING TIME
1	Milling	10
	Drilling	3
	Turning	15
2	Drilling	8
	Milling	7
3	Turning	10
	Milling	3

The above setup was simulated using SIMAN language. Statistics on performance measures such as Flowtime, Tardiness, Completion-time, Machine Utilization and Work-in-process were collected. There are 3^3 possible configurations (i.e. 1-3 milling 1-3 drilling, 1-3 turning in each work center). Further each dispatching rule adds an independent dimension and all the configurations have to be simulated for the respective rule. Two dispatching rules, EDD (Earliest Due Date) and SPT (Shortest processing Time) are used in this study.

Training Phase

Eighteen training pairs or input-output vectors are generated using simulation. The specific training pair is selected using a one-third fractional factorial design since each of the independent resource variables can exist at three levels. The training data is given in Table I. All the input-output elements are normalized between 0 and 1 to train it using Fuzzy ARTMAP. The training is Supervised i.e the training sample has both input and output pairs. Training is performed with the following adjustable control parameters till optimal results are obtained :

  i) fa vigilance factor [0,1].

  ii) fb vigilance factor [0,1].

iii) fab vigilance factor [0,1], in our implementation it was set to 1.

 iv) fa and fb choice factor, in our implementation it was set to 0.0001.

  v) beta the learning rate is set to 1.

The network was able to generalize the training data using only one epoch (i.e presenting data samples only once). The training was done using Fuzzy ARTMAP simulator on the SPARC workstation.

TABLE I

Training Samples and Performance Parameters

INPUT				OUTPUT			
M	D	T	R	F	TT	MT	C
1	1	3	SPT	41.500	105.000	35.000	88.00
1	2	1	SPT	45.700	120.000	40.000	98.00
1	3	2	SPT	42.100	105.000	35.000	88.00
2	1	1	SPT	42.300	96.000	32.000	90.00
2	2	2	SPT	29.200	22.000	22.000	58.00
2	3	3	SPT	27.700	18.000	18.000	54.00
3	1	2	SPT	30.400	16.000	16.000	52.00
3	2	3	SPT	24.200	7.000	7.000	43.00
3	3	1	SPT	39.700	66.000	32.000	90.00
1	1	3	EDD	47.100	91.000	13.000	88.00
1	2	1	EDD	52.300	122.003	17.429	90.00
1	3	2	EDD	46.500	91.000	13.000	88.00
2	1	1	EDD	44.300	99.000	24.750	90.00
2	2	2	EDD	30.100	8.000	4.000	55.00
2	3	3	EDD	28.800	6.000	3.000	55.00
3	1	2	EDD	30.100	13.000	13.000	52.00
3	2	3	EDD	25.200	7.000	7.000	43.00
3	3	1	EDD	41.400	99.000	24.750	90.00

M=milling D=drilling T=turning R=dispatching rule
F=flowtime TT=total tardiness, M=mean tardiness
C=completion time.

Testing Phase

    In the testing phase the trained neural network is fed the input pattern, the output of the network is compared with the simulation outputs to determine the generalization capability of the artificial neural network used.

    The testing mode of Fuzzy ARTMAP simulator was used to generate the networks response to the testing samples presented. The output generated by the network was compared to the actual output generated using simulation, for comparison and statistics purposes. (Table II).

TABLE II
Results Obtained Using Fuzzy ARTMAP ($\rho_a = [0.5, 0.6]$)

Performance measures	# of samples	# correct responses	% correct
FLOW TIME	36	36	100
TOTAL TARDINESS	36	33	91.67
MEAN TARDINESS	36	29	80.55
COMPLETION TIME	36	36	100

INTEGRATION OF ARTIFICIAL NEURAL NETWORKS AND SIMULATION FOR FMS SCHEDULING.

The FMS scheduling problem has been defined as an open deterministic dynamic job shop scheduling model. FMS scheduling problems belong to the class "NP-hard". Several techniques such as mathematical programming and analytical models, dispatching rules and simulation, look ahead algorithms, and other unique approaches such as simulated annealing, space filling curves, genetic algorithms, and artificial neural networks (based on relaxation models such as Hopfield networks) have been applied to FMS scheduling problems [10,11]. Most of these existing FMS scheduling methodologies are not effective for real/time tasks. In addition, the knowledge of scheduling of FMS is mostly system-specific, not well developed and highly correlated with shop floor status, resulting in the absence of recognized sources of expertise [12]. This gives rise to problems on effective knowledge acquisition and knowledge representation schemes, and limits the development of generic knowledge-based expert systems for FMS scheduling. Consequently , an FMS scheduling system should have numerous problem-solving representation schemes. Each of these problem solving strategies should be selected according to the situation imposed by the scheduling environment. Consequently, a predictor mechanism can provide the flexibility of using a single scheduling policy to meet the required performance criterion, for a given manufacturing environment and the scheduling problem Artificial neural networks (based on learning models such as backpropagation networks) are strong candidates for this task.

This prediction mechanism should take into consideration several factors which define the scheduling environment while selecting the best scheduling policy for a specific case. Simulation would support the knowledge acquisition process by generating several cases in order to induce the behavioral rules of that FMS.

The utilization of artificial neural networks as predictor mechanisms for FMS scheduling differs from previous approaches to FMS scheduling using artificial neural networks. Previous approaches emphasized the utilization of artificial neural networks based on relaxation models (i.e., pre-assembled systems which relaxes from input to output along a predefined energy contour) and their direct application to the scheduling problem (i.e., They solve the scheduling). These models have not produced consistently optimal/real time solutions due to the limitations in hardware and the algorithms. Our approach utilizes artificial neural networks based on systems which emphasize learning by mapping adjustments (i.e., backpropagation) which are more developed and have been applied successfully in several domains.

Assumptions, Performance Criterion, and Scheduling Policies Used

The assumptions for this investigation are common to those utilized in simulation studies performed by other researchers [13,14,15,16]. These assumptions include the following:
a) Processing times are deterministic
b) Due dates are fixed
c) There are no groups of similar machines
d) The required resources to carry out the jobs have been identified.
e) The job routings are fixed and unique
f) Pre-emption is not allowed
h) Transportation times are negligible.

For these studies the performance criterion which will be used is tardiness. Tardiness has been selected as the optimization criterion due to the importance of meeting in industrial problems.

The following dispatching rules were selected due to their utilization in previous scheduling studies involving due dates measures: SPT (Shortest Processing Time), CR (Critical Ratio), LWR (Least Work Remaining), EDD (Earliest Due Date), SLACK (Minimum Slack Time), S/OPN (Slack per Operation).

Input Feature Space and Knowledge Acquisition Process

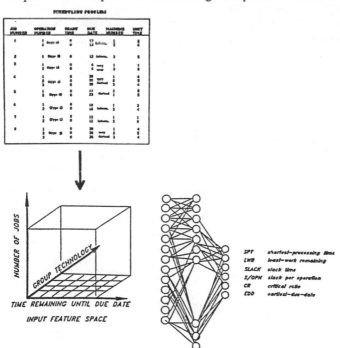

FIG 4. FMS Scheduling using ANN

In order to teach to an artificial neural network concepts about FMS scheduling and Tardiness, an effective input feature space should be developed (see Figure 4). This input feature space should contain:
1) FMS scheduling problem characteristics such as types and the number of jobs to be scheduled, along with their respective routings and processing times.
2) Due-date information
3) Manufacturing environment information such as number of machines, capacity.

Training examples should be generated in order to train a network to provide the correct characterization of the manufacturing environments suitable for various scheduling rules and the chosen performance criterion. In order to generate training examples, a performance evaluation using simulation of SPT, EDD, LWR, S/OPN, SLACK, and CR for the same set of manufacturing situations was carried out. The number of situations runs needed to provide a good training set for the artificial neural network was established by using the artificial neural network performance to determine if new data sets were needed.

Experiments

As an example, the case of a 10-job 4-machine scheduling problem will be presented. For this case an artificial neural network using the standard backpropagation paradigm and the techniques previously mentioned was developed. This artificial neural network developed, had a training set of 200 examples. The training session took 35 hours on an IBM PS/2 Model 70. The artificial neural network selected had 15 input units (defining the job database in terms of the machining processes and due date information) corresponding to the input feature vector size, 6 output units corresponding to each specific dispatching rule, and 68 hidden units in one hidden layer. The network was tested for 100 "new" examples and the performance of the network was found to be consistent (See Table III). These results were superior to those obtained using other dispatching rules. The execution time of the artificial neural network (being trained) was around 50 milliseconds on the same computer without neuromorphic hardware. This response time is fast enough to be considered adequate for real time and dynamic scheduling problems.

Another experiment involved the case of a 8-job 5-machine scheduling problem. The artificial neural network developed using backpropagation had 400 training examples. The resulting architecture has 15 input units, 6 output units, and 85 units in one hidden layer. This network was tested with 100 "new" examples and it yielded better results than the other dispatching rules (see Table IV).

TABLE III. 10-Job 4-Machine problems   TABLE IV 8-Job 5-Machine problems

LOWEST TARDINESS FREQUENCY - 100 problems

SPT	LWR	SLACK	S/OPN	CR	EDD	ANN
12	11	60	62	53	57	85

LOWEST TARDINESS FREQUENCY - 100 problems

SPT	LWR	SLACK	S/OPN	CR	EDD	ANN
24	57	3	6	5	28	74

AVERAGE TOTAL TARDINESS - 100 problems

SPT	LWR	SLACK	S/OPN	CR	EDD	DRC	ANN
82	83	72.9	72.6	73	76.3	69.6	70.6

AVERAGE TOTAL TARDINESS - 100 problems

SPT	LWR	SLACK	S/OPN	CR	EDD	DRC	ANN
221.1	208.3	299	257	272	222.8	204.7	207.2

DRC Dispatching rules combined   DRC Dispatching rules combined

CONCLUSION

Artificial neural networks have capabilities to learn, to perform massively parallel processing, and to adapt to complex environmental changes. These capabilities are especially significant in FMS for providing systems with abilities of learning and self-organization, and efficient real-time operation. In this paper we have demonstrated how the integration of neural networks and simulation has the potential to improve the design, dynamic control, and scheduling of FMS.

Currently other factors such as layout, correlation of performance measures etc, which affect an FMS are being added to our initial models. This would lead to a more generic framework and determine the degree of industrial strength that could be achieved with further research.

REFERENCES

1   Kohonen T: An Introduction to Neural Computing, Neural Networks, Vol. 1, No. 1,p. 3-16, 1988.
2.   Rumelhart D, McClelland, and the PDP Research Group: Parallel Distributed Processing: Explorations in the Microstructure of Cognition, Vol. 1: Foundations. Cambridge, MA: MIT Press/Bradford Books, 1988.
3.   Chauvin Y: Dynamic Behavior of Constrained Back-Propagation Networks, Advances in Neural Information Processing Systems 2, edited by David Touretzky, Morgan Kaufmann Publishers, p.642,1990.
4.   Ash T: Dynamic Node Creation, Technical Report, ICS 8901, University of California, San-Diego,1989.
5.   Carpenter G, Grossberg, Rosen D: FUZZY ART: Fast stable learning and categorization of analog patterns by an adaptive resonance system. CAS/CNS-TR-91-015, Boston University, 1991.
6.   Carpenter G, Grossberg S, Reynolds J, Rosen D: Fuzzy ARTMAP: A neural network architecture for supervised learning of analog multidimensional maps. CAS/CNS Technical Report, CAS/CNS-91 016, Boston University 1991.
7.   Chryssolouris G, Lee M, Pierce J, Domroese M: Use of neural networks for the design of manufacturing systems, Manufacturing Review, vol. 3, no. 3, p. 187-194, Sept.-1990.
8.   Chryssolouris G, Lee M, Domroese M: The use of neural networks in determining operational policies for manufacturing systems, Journal of Manufacturing Systems, vol. 10, no. 2, p.166-175, 1991.
9.   Mitchell F. H. Jr: CIM Systems, An introduction to Computer Integrated Manufacturing.
10.   Foo Y, Takefuji Y: Stochastic Neural Networks for solving job shop Scheduling: part 2, Architecture and Simulations, Proceedings of the IEEE international Conference on Neural Networks, published by IEEE TAB, p II283-II290, 1988.
11.   Foo Y, Takefuji Y: Integer Linear Programming Neural Networks for Job Shop Scheduling:, Proceedings of the IEEE international Conference on Neural Networks, published by IEEE TAB, p II341-II348, 1988.
12.   Wu S. D: An Expert System Approach for the Control and Scheduling of Flexible Manufacturing Cells, Ph.D. Thesis, The Pennsylvania State University, 1987.
13.   Blackstone J, Phillips D, Hogg G: A state-of-the-art survey of dispatching rules for manufacturing job shop operations, International Journal of Production Research, vol. 20, no.1, p. 27-45, 1982.
14.   Kiran A, Altpekin S: A Tardiness Heuristic For Scheduling Flexible Manufacturing Systems, 15th Conference on Production Research and Technology: Advances in Manufacturing Systems Integration and Processes, University of California at Berkeley, Berkeley, California,p.559-564, Jan 9-13, 1989.
15.   Rabelo L: A Hybrid Artificial Neural Networks and Knowledge-Based Expert System Approach to Flexible Manufacturing System Scheduling. Ph.D. Thesis. University of Missouri-Rolla, 1990.
16.   Rabelo L, Alptekin S: Synergy of Neural Networks and Expert Systems for FMS Scheduling, Proceedings of the Third ORSA/TIMS Conference on Flexible Manufacturing Systems: Operations Research Models and Applications, Cambridge, Massachusetts, Elsevier Science Publishers B. V.,1989, p. 361-366.

A SIMULATION APPROACH TO FMS DIMENSIONING INCLUDING ECONOMIC EVALUATIONS

Roberto Mosca, Pietro Giribone, and Alfredo Drago

Institute of Technology and Mechanical Plants, University of Genoa, Genoa, Italy

ABSTRACT

"We don't need factories of the future, but we need factories with a future". This sentence, from the first FAIM '91, underlines the extreme significance of preliminary plant evaluations in the development of factories which optimize not only production but also economic return.

This paper decribes such an integrated analysis which was conducted in collaboration with a producer of F.M.S. and automated work centers. The scope of the study was to define an optimal production system for a given annual plant output. To this end, a detailed model was developed, incorporating both a high degree of realism and flexibility to simulate the operating scenario.

The problem of plant optimization is often complex, in this case including selection of staffing levels, machine type, number and configuration, and number of input/output stations. Initial estimates of the possible ranges of these variables were made based on the anticipated production levels. A pre-processor was then used in conjunction with the simulator to vary individual parameters, examining their effect on both production and cost. From this simulator data a sensitivity analysis was performed to determine the relative importance of the various parameters, identifying which were more important for production and which had a greater effect on cost. The results of this analysis were used to define the most cost effective plant configuration.

It is important to remember that the purpose of the study was to arrive at the optimum plant design from a standpoint of management and cost effectivness, and not necessarily the maximum utilization of work centers. By following the process described, an optimum plant layout was determined which satisfied the criteria of both economy and productivity. So we can say that this project attained a double goal: on one hand, a simulation tecnique has been utilized to solve a plant dimensioning problem, and on the other, it highlights the fact that plant optimization must include a full examination of economic as well as production issues.

INTRODUCTION

This work describes a study conducted by a leading Italian company in the field of F.M.S. and the Institute of Technology and Mechanical Plants of the Faculty of Mechanical Engineering of Genoa.

This paper discusses a technical-economic feasibility study to upgrade a traditional plant with an F.M.S. line that has been specifically developed for a vast production range (production mix). The methods currently used by the the work centers to handle product diversity are completely inadequate and the plant must be shut down for extended periods of time in order to complete tool set-up procedures.

Thus, the paper develops the following main aspects of this problem:

- -evaluation of the real benefits of using a flexible line in a rapidly changing manufacturing plant;
- -determination of the optimum configuration of the new line, quantifying the type and number of work centers which are required to complete the production schedules without overdimensioning the plant;
- -selection of the proper management strategies to meet the production requirements relative to a "job" based set-up;
- -selection of the transport system to convey semi-finished products from the machines to the input/output stations.

This system might consist of a rail shuttle or a rotating multipallet magazine.

In this case, a deterministic and static analysis did not completely satisfy the proposed objectives and, above all, was unable to consider the random nature of the plant, such as the arrival of orders which are used to program the weekly production schedule. The most effective instrument to resolve these problems is the time and event-based discrete stochastic simulation. This approach resolves the dimension and management problems simultaneously, while analyzing their economical benefits.

THE PLANT UNDER EXAMINATION

The plant being studied presently consists of 4 traditional machine tools which are designed to produce almost one hundred different spare parts for hydraulic machines. This plant requires about 800 different types of tools almost all of which must be reproduced to satisfy current production requirements. The innovative proposal for this plant includes 2 or 3 (the simulation will indicate the exact number) in-line work centers which are installed, on the loading side, in front of a variable-capacity rotating multipallet system which is interfaced at 1 of the 2 ends with 2 load-unload stations where line operators prepare and handle the pieces. The tool control area is located on the machine side, opposite the multipallet system. This area is run by tool-room operators who must remove and replace worn tools. Each machine is equipped with a double-chain tool magazine with a capacity of 120 to 200 pockets (figure 1).

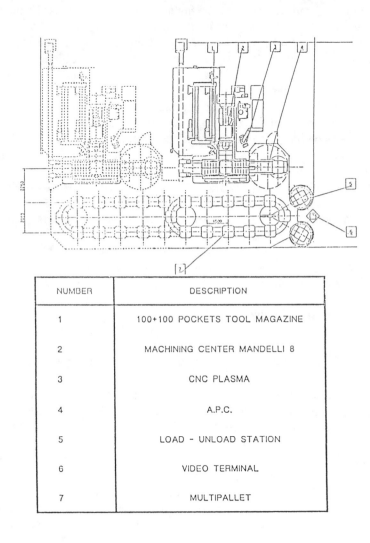

NUMBER	DESCRIPTION
1	100+100 POCKETS TOOL MAGAZINE
2	MACHINING CENTER MANDELLI 8
3	CNC PLASMA
4	A.P.C.
5	LOAD - UNLOAD STATION
6	VIDEO TERMINAL
7	MULTIPALLET

FIG. 1 Plant layout.

Though constant in a year, the production mix varies considerably each week: it consists of 83 different types of pieces supported by 62 types of equipment with a total of 140 different machining operations and 577 different tools involved in various part-programs. This manufacturing complex is also affected by another problem caused by a 4-hour unmanned shift in every 24 hours of production. During this shift the plant must continue to operate and manufacture parts automatically. Thus, a flexible production plant that operates independently during unmanned shifts is a very appropriate solution. A quality-quantity based evaluation of the advantages of such a proposal was performed using a discrete and stochastic detail simulation model.

THE SIMULATION PROGRAM

As mentioned above, due to the complex nature of the topics to be faced, it was very difficult to develop the simulation program, the only instrument capable of solving such a wide range of problems. In fact, some of the objectives established include: determination of the minimum-cost plant configuration and thus the minimum number of work centers, line and tool-room operators and the minimum multipallet capacity. In addition, an attempt was made to reduce the redundancy of circulating tools, i.e. the number of tool duplications required for production, through well-planned distribution of work center machining operations. Finally, studies concentrated on piece handling schemes, the management of the production mix and its distribution over the work centers as well as the reorganization of production in case of malfunctions to one of the two work centers. A stochastic event-based detail simulator, written in Fortran 77 according to Italsider techniques, i.e. with a time advancement mechanism scanned by alternating scan and rescan phases, was used to satisfy these objectives, while also considering the randomness of the events which may affect line operation.

Implementing what is by now a well-established simulator construction procedure, the first phase of the study concentrated on collecting plant data and analyzing the management policies that are currently used to operate the line. In particular, the analysis focused on all the critical elements of the line which were translated in terms of system variables. This preliminary phase indicated that in addition to the number of multipallet load positions, it was also necessary to test the number of load and unload stations, the number of operators, the amount of available equipment and the capacity of the tool magazines. Finally, an initial hypothesis considered a line with two work centers, a hypothesis which was confirmed since it more than guaranteed compliance with the assigned production level. In fact, it should not be forgotten that the analysis conducted indicated that the economic result could be used as an operative check and therefore, wherever possible, sets a limit on usable resources.

A block diagram of the model developed is illustrated in figure 2.

In addition to the simulation variable initialization phase, which we will discuss below, this diagram also illustrates the three blocks which manage the program time phases: the scan, rescan and transport management logic groups.

Besides the usual machine status control phases, the scan management logic

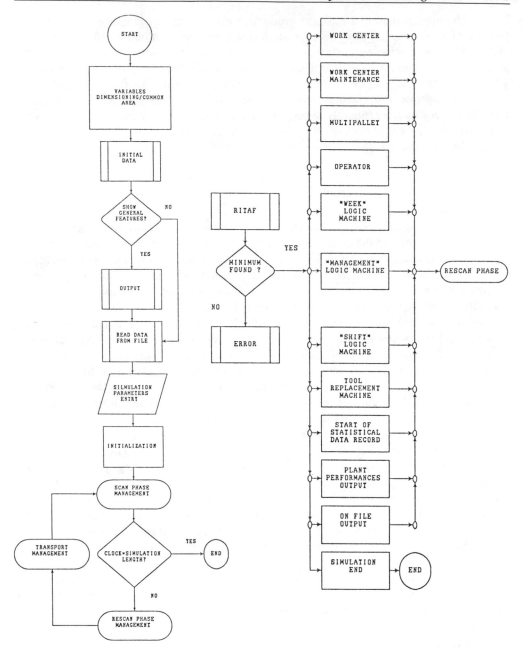

FIG. 2 Model block diagram   FIG. 3 Scan phase block diagram

group also includes the week logic machines and management logic machine (figure 3). The first one processes the weekly production schedule and allocates machining operations to the work centers. This task is executed using two priority criteria: saturation of the total hours of each machine so that it works for the entire week and maximum tool sharing criteria to limit their use to an absolute minimum. Each time a machining operation must be assigned to a machine, the number and type of tools which are required for such a machining operation are calculated by the part-program and the pieces to be machined are sent to the machine that already has the greatest number of tools on its tool chains. This minimizes the machine set-up phases at the beginning of each week and rationalizes tool stores, thus reducing redundancy to a minimum.

Instead, the "machine management" group handles the unmanned shift management problem, creating a so-called preparation phase. This phase is used so that when the unmanned shift begins, the multipallet contains only those pallets which keep the machine in service for the entire automatic operation phase and do not require assistance by personnel to replace tools. This result was obtained by conducting several experiments to identify the optimum length of the preparation phase and the temporary rules required to manage the line when operators are not present.

In the rescan phase in which all "conditioned" operations are started (figure 4), special efforts were focused on the retooling chain, which is used to replace worn tools. In particular, depending on the urgency of the replacement, the retooling index is increased to minimize tool turnaround times.

The transport management module was completely reviewed with respect to the concepts adopted in FMS line and rigid transfer line simulation applications. The final outcome of this evaluation is a reliable model which perfectly reproduces the behavior of the real system in terms of specific outputs: for example, each transport is completed while minimizing the rotation time of the tool mandrel by selecting the proper rotation direction (figure 5).

Therefore, the simulator was equipped with an initial input block in which a pre-processor shifts the plant configuration within the variability ranges assigned by the handler. This makes it possible to update the data required to evaluate the investment choices or management rules in terms of plant performances. As an additional SW product option, the results obtained by the simulator may be processed by an economic model which estimates the average plant operating cost by evaluating the average cost of the tools, the number and utilization of work centers, the cost of line personnel and machine depreciation.

As pointed out in the following paragraph, the system operating cost is a function of the different interactive variables which affect the overall cost in relation to their various weights. On the other hand, this work does not consider production profits which may be analyzed by concentrating on a realistic market survey. In any event, this may be a valid reason for further development of this model (figure 6). The stochastic simulator was validated analogically and statistically before being subjected to further processing. The analog test focused on the exact reproduction of the reality of the model, while the statistical part checked on the correct implementation of the stochastic

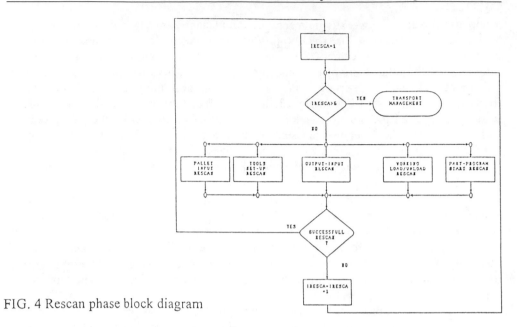

FIG. 4 Rescan phase block diagram

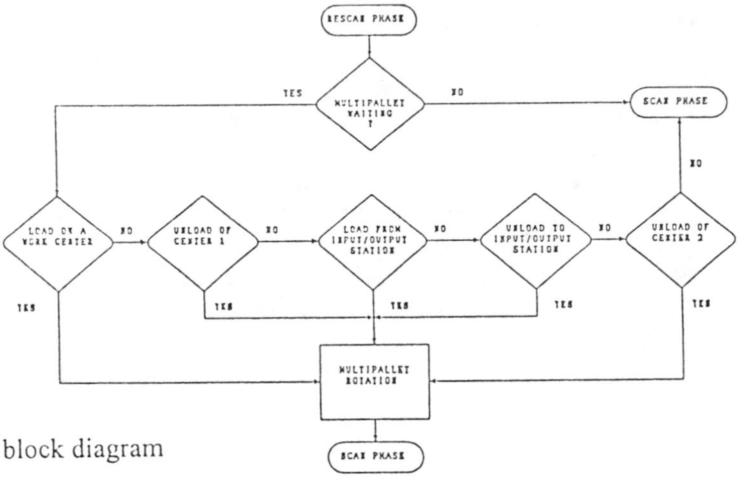

FIG. 5 Transport phase block diagram

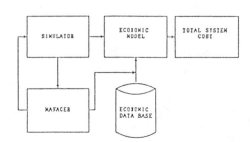

FIG. 6 Economic model block diagram

nature of the model.

For this purpose, the results of the simulation runs are tested evaluating the time evolution function of the M.S.P.E. [$y_i(t)$], where y_i = simulation objectives with i=1,...m, i.e. the variation of the experimental error for each simulator output. The same analysis identifies the optimum Run time, i.e. that simulation time value beyond which the results supplied are statistically stable and for which an extension of the run time does not correspond to a significant gain in terms of output reliability (figure 7).

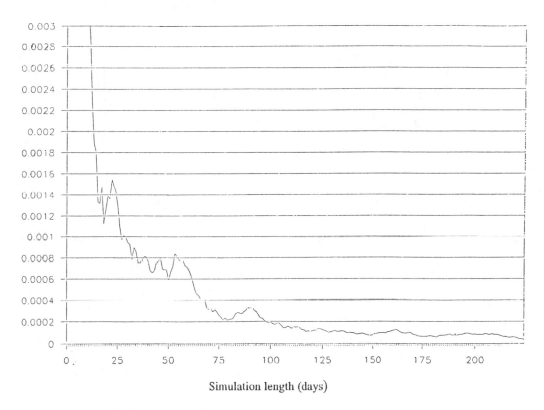

Simulation length (days)

FIG. 7 Mean Square Pure Error, work center 1.

EXPERIMENTATION PHASE AND ANALYSIS OF RESULTS

A designed experiment was used and involves the specific variables which are included in the relative variability ranges. System responses were analyzed in the various possible configurations. For this purpose, the analysis adopted a central compound design with a fractionized factorial core of 2^{5-1}. The independent variables together with the respective variability ranges are indicated in table 1.

TABLE 1
Variables Table

Independent Variables	Ranges
A) Multipallet capacity	8 – 16
B) Number of operators	2 – 4
C) Number of load-unload stations	1 – 3
D) Total number of tools	872 – 972
E) Tools magazine capacity	120 – 200

The m objective functions considered include: work center use coefficients, the production advancement coefficient - expressed as the ratio between the production at time t and the expected rated production at the same time - and the total annual cost. The rest of the analysis is conducted only on the last two objectives listed above.

The factorial analysis was used to evaluate the influence of the single independent variables on the response of the system together with their possible combinations. Tables 2 and 3 show that the variables which affect cost are A, B, C and D, while production is unaffected by B, confirming a well-known and basic concept, i.e. that a high level of automation greatly releases production from the human component. Returning to the main objectives, it is worthwhile analyzing the "production" and "total cost" objectives comparing them to the truly significant independent variables. The results of the simulation were then used to build polynomial type surfaces fitting the approximations of the "real" response surfaces of the system. Obviously, the response, which is represented by the set of equations obtained, is only valid within the range of variation of each individual variable.

TABLE 2
Cost Influence Factors

Variation source	Square sums	Degrees of freedom	Square averages	Fo
A	5.65E+03	1	5.65E+03	1280 *
B	5.55E+05	1	5.55E+05	1.26E+05 *
C	1.35E+03	1	1.35E+03	306.1 *
D	5.55E+03	1	5.55E+03	1261.4 *
AB	8.12	1	8.12	1.85
AC	9.92	1	9.92	2.26
AD	3.83	1	3.83	0.86
BC	6.40	1	6.40	0.15
BD	1.54	1	1.54	3.46
CD	1.92	1	1.92	0.45
ERROR	36	8	4.4	

TABLE 3
Production Influence Factors

Variation source	Square sums	Degrees of freedom	Square averages	Fo
A	5.50E-03	1	5.50E-03	18.55 *
B	6.20E-06	1	6.26E-06	0.02
C	2.60E-03	1	2.60E-03	8.67 *
D	2.20E-03	1	2.20E-03	7.52 *
AB	3.00F-04	1	3.00E-04	1.02
AC	2.52E-05	1	2.52E-05	0.08
AD	1.50E-04	1	1.50E-04	0.52
BC	3.61E-05	1	3.61E-05	0.12
BD	1.20E-05	1	1.23E-05	0.04
CD	4.41E-04	1	4.41E-04	1.47
ERROR	2.40E-03	8	3.30E-03	

The meta-models obtained and checked by a lack-of-fit test are:

$$P = 1.304 + 0.01882X_1 + 0.0128X_3 + 0.0117X_4 + 0.00314X_1^2 - 0.0207X_3^2 +$$

$$- 0.0192X_4^2 + 0.0052X_3X^4$$

$$C = 2136 + 19.08X_1 + 186.41X_2 + 8.94X_3 + 56.6X_4 - 4.44X_2^2 - 6.04X_4^2$$

These identify the surfaces fitting the response of the system whose maximum and minimum points represent possible system configurations. The analysis identified (five-two) parameter- based surfaces in the experiment domain, which are generally increasing monotonic functions, which, from the graphic point of view, describe $P(x_i)$ and $C(x_i)$ i = 1,...4. It can be easily deduced that the optimum point related to production corresponds to the maximum level of all the variables, while for the cost function this point corresponds to the lower limits of the range of all variables (two of the possible examples are shown in figures 8 and 9).

PRODUCTION   TOTAL COST

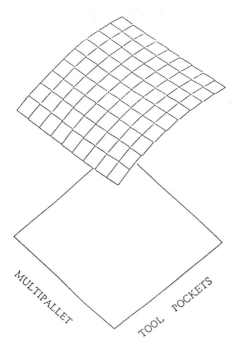

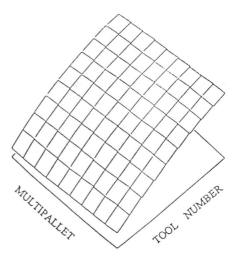

FIG. 8 Production curve   FIG. 9 Cost curve

Therefore, by setting all the variables to the proper values, it is possible to increase production compared to the current situation. However, it is also necessary to evaluate the added costs of such an increase, which can be obtained in the function $C(x_i)$.

Without getting into greater details regarding dimensions and quantities, which are obviously restricted by company security policies, the configuration which the simulator considers to be "excellent" and is currently in the plant installation phase is the one indicated in table 4.

TABLE 4
Optimum Configuration

Pieces	Description
2	Work centers
1	Load-unload stations
2	Operators
1	Tool-room operators
8	Places on multipallet
2	Tool magazines
120	Pockets on each tool magazine
872	Total number of tools

This is used to infer the design detail level achieved by the simulator.

CONCLUSIONS

Based on the previous discussion, we can conclude that the main objectives of this study have been achieved. It was possible to identify the minimum-cost plant configuration while complying with the annual production requirement. The overdimensioning solutions were rejected since they would have emerged from an a priori estimate of design parameters.

The design parameters with the greatest influence on the value of the responses considered were identified and finally it was shown how the proper simulative analysis can solve dimensioning, management and cost-reduction problems.

As previously mentioned, development of this study could be extended from a cost analysis to an analysis of the profits deriving from a particular production campaign set up at a specific market competition level.

REFERENCES

R. Mosca, P. Giribone, A. Drago, M. Schenone: An expert controller in F.M.S. management. *Proc. Faim '91* ,Limerick (EIRE)

R. Mosca, P. Giribone, A. Drago, M. Schenone: A simulation approach to a bi-shuttle F.M.S. tool-room management. *Proc IMACS '91*, Dublino (EIRE)

W. Hardeck: Simulation on-line off-line: simulation techniques. Integrated automation , january 1987

G. S. Fishman: Principles of discrete event simulation. John Wiley & Sons, N.Y., 1978

R. Mosca, P. Giribone, G. Guglielmone: A mathematical method for evaluating the importance of the input variables in simulation experiments. *II I.A.S.T.E.D. Int. Sym. Modelling, identification and controll*, Davos, march 2-5, 1982

G.E.P. Box, J.S. Hunter: The 2^{k-p} fractional factorial design. Technometrics, III, N° 3 and 4, 1961

H. Scheffe: The analisys of variance.John Wiley & Sons,N.Y.,1959

SIMULATION OF AUTOMATED PRODUCTION OF A BONDED GEAR BOX

[1]Mahmoud A. Younis and [2]Medhat M. Sadek

[1]*Engineering Department, American University, Cairo, Egypt*
[2]*Department of Mechanical Engineering, Kuwait University, Safat, Kuwait*

ABSTRACT

In this paper, a simulation program written in the BASIC language for a flexible manufacturing system (FMS), is developed for the fabrication of a bonded gear box casing. The package is used for an appropriate layout of the machining cells around a closed loop conveyor system. The routing of the working parts within the FMS system is also simulated. Parts are assigned to the appropriate machining cells by matching the part machining requirements (based on group technology) within the machine capabilities. A novel code system dependent on flowchart was used. This system is suitable for both prismatic and cylindrical parts as well as automatic assembly of the bonded gear box casing with robots. Finally the utilization rate of the production resources is predicted by the package using a search and fit program to minimize the idle time of the machine by optimizing the assignment of components to the machining cells.

INTRODUCTION

The whole philosophy of production automation is changing rapidly due to the emergence of versatile programmable industrial resources such as robots, CNC machines and material transfer devices. With such highly automated systems, long lead times and heavy work in progress inventories cannot be easily tolerated. One of the most promising solutions is the introduction of Flexible Manufacturing Systems (FMS) with full computer control. In such systems small batches can be machined automatically from first process to completion. A variety of workparts is introduced randomly to the system. This is made possible by applying a group technology to code workparts. Coding systems such as AUTOCAP [1] and ICAPP [2] dealt only with cylindrical and prismatic components respectively. OPITZ, MICLASS merely codes parts without consideration for process planning; and none of the mentioned systems encountered the assembly process. The initial capital costs of these systems are rather high. The use of a computer simulation model to predict the performance of the system before it is built is now widely used. Furthermore, it is seen as a tool for defining a suitable system configuration at the initial design stage. Again simulation in dynamic form can represent the interactions between the system components and provide a detailed prediction of its performance. Furthermore the model could also be used as a means of developing the software to control the real system. It also facilitates the optimization of all the independent parameters, identifies the possible bottle necks and examines the impact of introducing novel operating procedures and/or variations in batch size and scheduling policies. This computer model enables the analysis of complex interactive processes to be performed.

Once the model is established, tests can be applied to investigate the effect of batch size, cycle time variation, set-up time, lead time, work in process, machine utilization etc. This will result in the optimization of the plant design, reduce the installation time, maximize resource utilization, reduce lead time and stocks.

Examples are given in [3] to show how manufacturing simulation can be used to analyze materials handling system. One of these examples is the Bosh type accumulation conveyor system where two products are manufactured on the line [4]. The model is used to give information on the effect of batch size and plant load. The model is written so that other variables such as maximum allowable queue lengths into each operation, robot operational state and operational times, could be changed before a simulation run is started or during the run.

A recent paper [5] described how Fortran sub-routines could be used to look at the results of Simon simulation program. There are many investigators who dealt with the use of expert systems in simulation to help with the many aspects of the operation and control of FMS [6,7,8,9].

The present work illustrates a method of simulation modelling for a production plant, consisting of automated processing such as CNC machines and a workpart transfer system, such as a conveyor and robots. Here the system is tested in the manufacturing of parts of a sub-assembly of a gear box, which has to be assembled using the novel technology of adhesive bonding. A friendly user-active dialogue issues codes for both cylindrical and prismatic workparts taking into consideration type of material, design parameters, manufacturing processes, sequence of manufacturing and assembly.

The model can determine the system capacity to balance production and manage inventories, to relieve bottlenecks and to establish a possible optimization of the system resources. In this work the simulation technique is designed to use the computer graphics with animation to generate and manipulate images on the screen in order to accomplish various types of interaction between the user and the system. Through a process of trial and error, a near optimum solution to the production system has been obtained.

SYSTEM CONFIGURATIONS

Figure 1 shows a view of the bonded gear box casing which is to be manufactured and assembled. A system consisting of three CNC center lathes, three CNC milling machines and a closed loop conveyer has been used to manufacture the gear box casing components: corner plates, side plates and bushes. Three robots have been used to load and unload the components to the machines. Parts are washed, collected and palletized at the conveyer unloading station, then delivered to the robotic assembly station. Two specially designed jigs shown in Fig. 2 are used to facilitate the robot's bonded assembly process.

The first of these jigs is for the assembly of the corner joints, side joints and the base plate, while the second one is for assembly of the bushes with the plates. The jigs are designed to hold the parts in the required position and to maintain a uniform bond line thickness. A central computer is used to control the part-flow within the whole production line, using specially written software.

Flowchart of the code system

The flowchart of the system [9] is presented in Fig. 3. It consists mainly of three parts. In the first part, the user is asked a series of questions about the main characteristics of the workparts and the structure. The first question is concerned with the number of the available machines, which is going to be used in optimizing the machine schedules. The next question is about the number of parts in structure, which will be used in the organization of the assembly process. After this the user is asked about the workpart and the number of other parts connected to it. This is to be used in determination of the priority of workparts. The program offers the choice from three types of material and copper. In addition, the material type is used and depth of cut was optimized [10].

Later the speed, feed, machining time and the power consumption were calculated. Then the user is asked about the workpart class, if it is cylindrical or prismatic. The answer to this question constitutes the third digit of the issued code and will lead to the next part of the flowchart. In the second part of the program, the user is asked to describe the processes to be carried out. For cylindrical workparts, it is known that there are two opposite faces. So, the user has to enter the number of processes on face 1 and then starts specifying them, the same will be repeated for face 2. The processes that can be carried out on a cylindrical workpart which are: external and internal threading, external and internal chamfering, knurling, forming, recessing, parting off and issued automatically as soon as the process is entered. Fig. 4 shows the detailed flowchart of this part. As for prismatic workparts which have six faces, for each face, the number of processes is to be entered, then the processes to be carried out are specified. The processes that can be specified for prismatic parts are: end milling, face milling, side milling, slotting, drilling and threading. Each of these has a specified code that is issued as the process is entered. Fig. 5 shows the detailed flowchart of this part.

When any process is specified, the relevant sub-routine is called to give the necessary parameters related to that process. These parameters specify the part geometrical dimensions, location of the process and tool type. The third part of the flowchart uses the previously given information to calculate and print speed, feed, machining time and the power consumed. Then workpart code is issued.

Simulation model

At any processing station, the simulation model is constructed out of three interdependent sections:
A. The overall system model : This model analyzed each operation at each machine. It calculates the setting time, the machining time, the tool handling time, etc. The model then shows the cycle time for each operation, which consists of two main parts. One part corresponds to the periods of the resource activity and the other represents the idle times. Furthermore, the active times are subdivided to productive and non-productive periods [11], allowing the evaluation of the resource optimization levels
B. The machining cell model shown in Fig. 6, may be active or passive

depending on the availability of the required machine. The active machining cell model records the resource's idle and active periods by registering the time elapsed between the start and end of each period. The system updates these period times by receiving the outputs from all active process models of the production station. These periods are cross referenced by the system model to seek an optimum resource utilization.

The passive machining cell models collect information about the queuing times of the workpart when the resource is not available. It calculates the overall waiting time of the workpart by registering the outputs of all passive machining cell models.

C. The workpart routing between the production resources. All parts in the system are grouped into families with similar production routes in association with machine tools. Machining cells of the system are then formed according to Group Technology. All parts are introduced to the system at the same time and later on a part that completes the production phase, generates a request for the same type of part. Parts to be machined can be assigned to the appropriate machining requirements with the cell production capabilities.

The above collected information is then employed to achieve a maximum utilization of the resources. A search and fit program is used to schedule the next operation of the resource at any given time during simulation in order to eliminate or minimize its waiting times or passive periods.

The machining times are deterministic. They are calculated for each machining process using a special program to optimize the cutting parameters, depth of cut, feed and cutting speeds for minimum total machining costs [17]. Moreover the program continually compares the information received from all machinery cell models, the system models and the part routing models and determines the most suitable part to be released to the machining cell in order to minimize the idle time at each machine. In this way the desired solution for maximizing cell efficiency can be achieved.

RESULTS AND DISCUSSION

Fig. 7 and 8 show how the search and fit program can enhance the efficiency of one of the machinery cells consisting to a CNC center lathe, CNC milling machine, buffer storage for each machine and a robot serving the cell. Fig. 7 shows the sequential cycle of the robot as well as the center lathe and milling machine for first in, first out FIFO approach. It gave a mean utilization of 30, 65, 62 per cent for the robot, center lathe and milling machine respectively. These low utilization figures are mainly attributed to the fact that parts released to the conveyor at the loading station were seen to form a complete gear box casing, otherwise a shortage, or starving, for one part, will hinder or delay the assembly process. To overcome this a buffer storage for each finished workpart of the gear box casing was created at the assembly cell. A large reserve was needed to account for any machine failure at any of the machining cells. When a search and fit program was used with random release of the working parts, the utilization percentage for each machine was improved to 60, 99 and 95 for the robot, center lathe and milling machine respectively. The program searches all the machining time for all parts released to the system and fits the routing of the parts within the machining cell to fill the idle gaps a each machine, as shown in Fig. 8. For example in Fig. 7 when machining part E after part A on the center lathe, the idle time between part B and part E is eliminated. For the milling machine if part D is machined before part C the idle time is reduced to a minimum value.

CONCLUSION

This paper presents a computer package for the simulation of an FMS of a fabricated bonded structure consisting of NC-lathes, NC-milling machines, handling system and robots. This package is written in the BASIC language for its simplicity, flexibility and popularity. Computer graphics are used to clarify various movements of entities of the manufacturing system taking into consideration their synchronization. A new coding system that depends on flowcharts is presented . The concept of the coding depends on group technology and is issued automatically through the user - interactive program. The system distributes the workparts on the available machines and the code instructs the machine what process to be done in what sequence and at the optimum cutting parameters. A search and fit program is added to enhance the efficiency of the system.

REFERENCES

1. El-Midany T.T.,Davies B.J.: AUTOCAP - A dialogue system for planning sequence of operations for turning components. Int.J.MTRD , 1981.

2. Eskicioglu H.,Davies B.J. : An interactive process planning system for prismatic parts (ICAPP). Int. J. MTRD. 1981.

3. Novels M.D.. Hackwell G.B. :Graphical simulation of material handling system Proc. 4th European Conference Automated Manufacturing 50-7-528 May I.S.S.1987.

4. Haddock J. : An expert system framework based on a simulation generator.Simulation 48.No.2 Feb 45-53, 1987.

5. Beb - Ariek D.: A knowledge based system for based on simulation and control of F.M.S. Generator. Simulation - Applications in manufacturing. IFS Publications Ltd. U.K. 1986.

6. Shivnan J., Browne J.: AI - Based simulation of advanced manufacturing system Simulation - Applications in manufacturing. IFS Publications Ltd.UK.1986.

7. O'Keefe R. : Simulation and expert systems - A taxonomy and some examples. Simulations 46 No.1 Jan. pp 10-16. 1986.

8. Shannon M., and Adelsberger D.: Expert systems and simulation. Simulation 44 No. 6 June 275-284 ,1985.

9. Sadek M.M.. Younis M.A. Darweash S. &others: Unpublished Report from Kuwait University funded project ME 023., 1988.

10. Deprereux W.R. : Determining of optimal cutting parameters for economical application of NC machines Ph.D. Dissertation T.H. Aachen 1969.

11. Younis M.A. ,Sadek,M.M. : Simulation for cutting parameters optimization for an automated machining process. Unpublished report Kuwait Univ.1990.

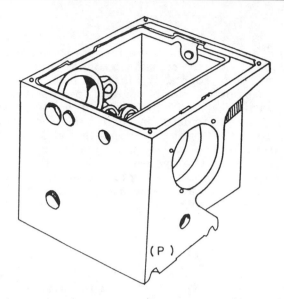

FIG. 1a THE M300 HARRISON CAST IRON LATHE GEAR BOX

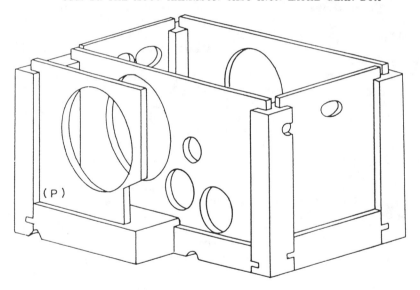

FIG 1b   THE PROPOSED BONDED GEAR BOX

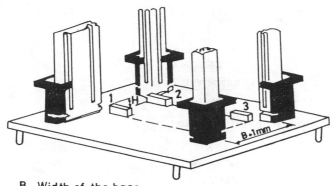

B_ Width of the base
H_ Height of the base ₊0·5 mm

FIG **2**a THE JIG USED FOR ASSEMBLY OF THE CORNER
SIDE JOINTS AND THE BASE PLATE

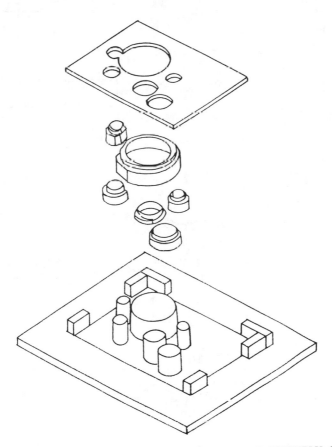

FIG. 2b A SAMPLE JIG USED FOR THE ASSEMBLY OF THE
BASE PLATE ALONG WITH ITS BUSHES

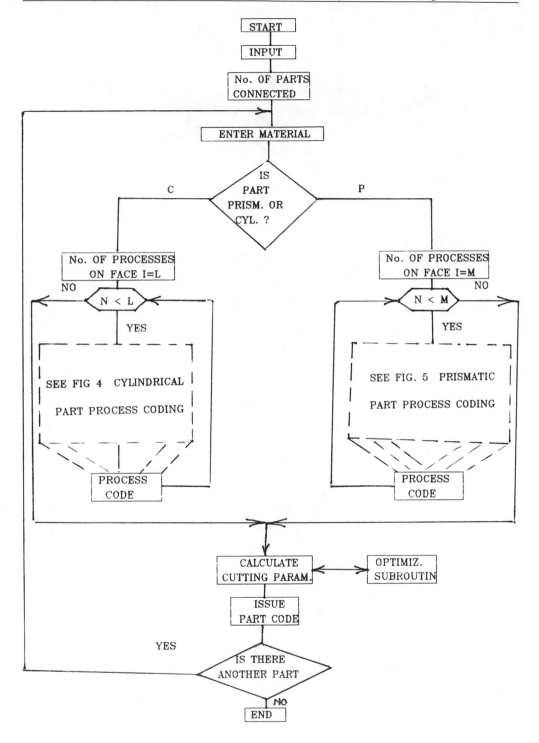

Fig. 3   FLOW CHART OF CODE SYSTEM

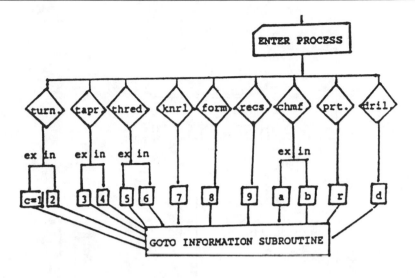

Fig. 4 PROCESS CODING FOR CYLINDRICAL PARTS

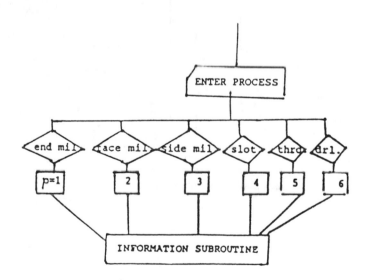

FIG. 5 PROCESS CODING FO PRISMATIC PARTS

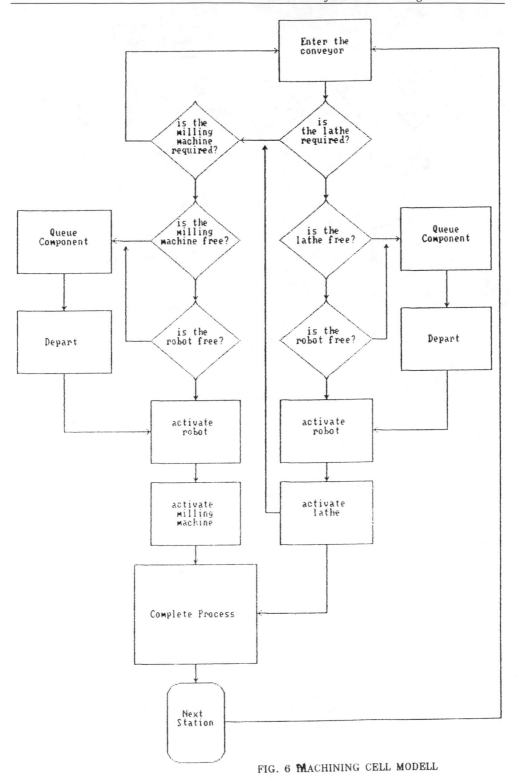

FIG. 6 MACHINING CELL MODELL

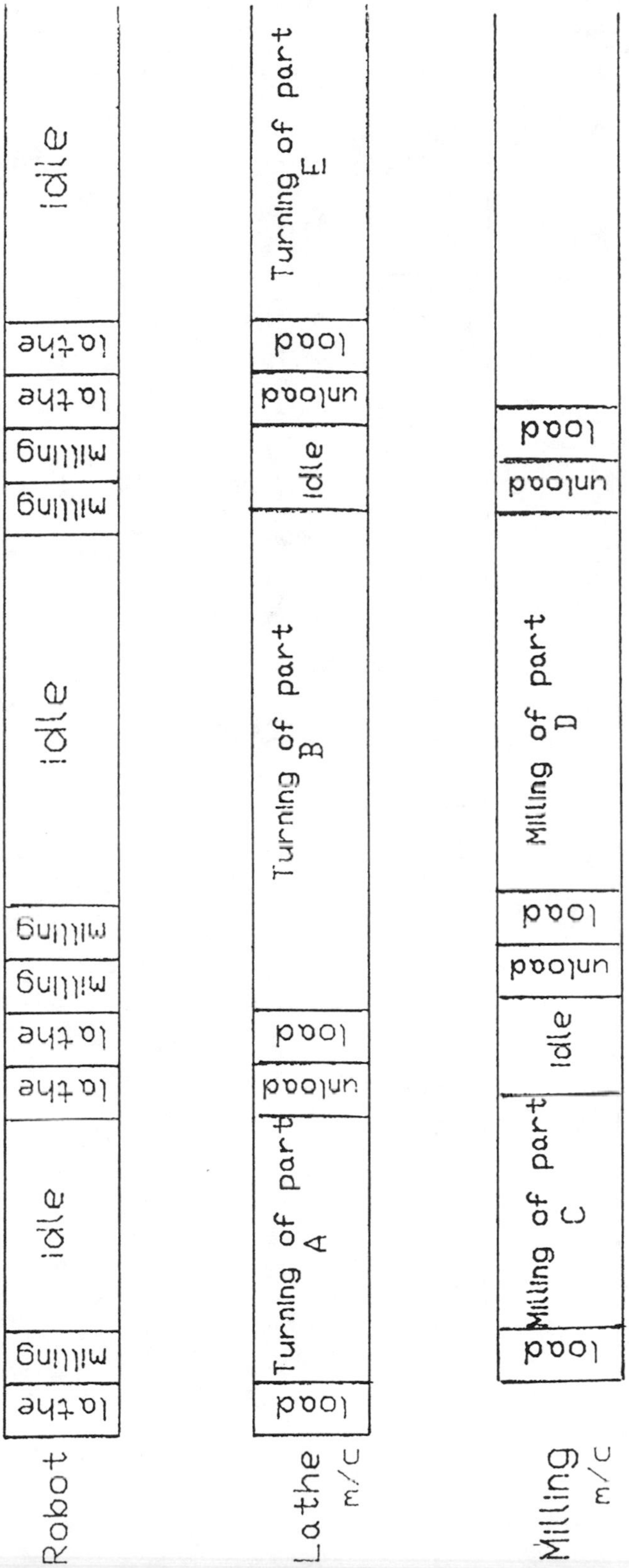

Fig. 7. Machining cell sequential cycles of operations using fifo approach

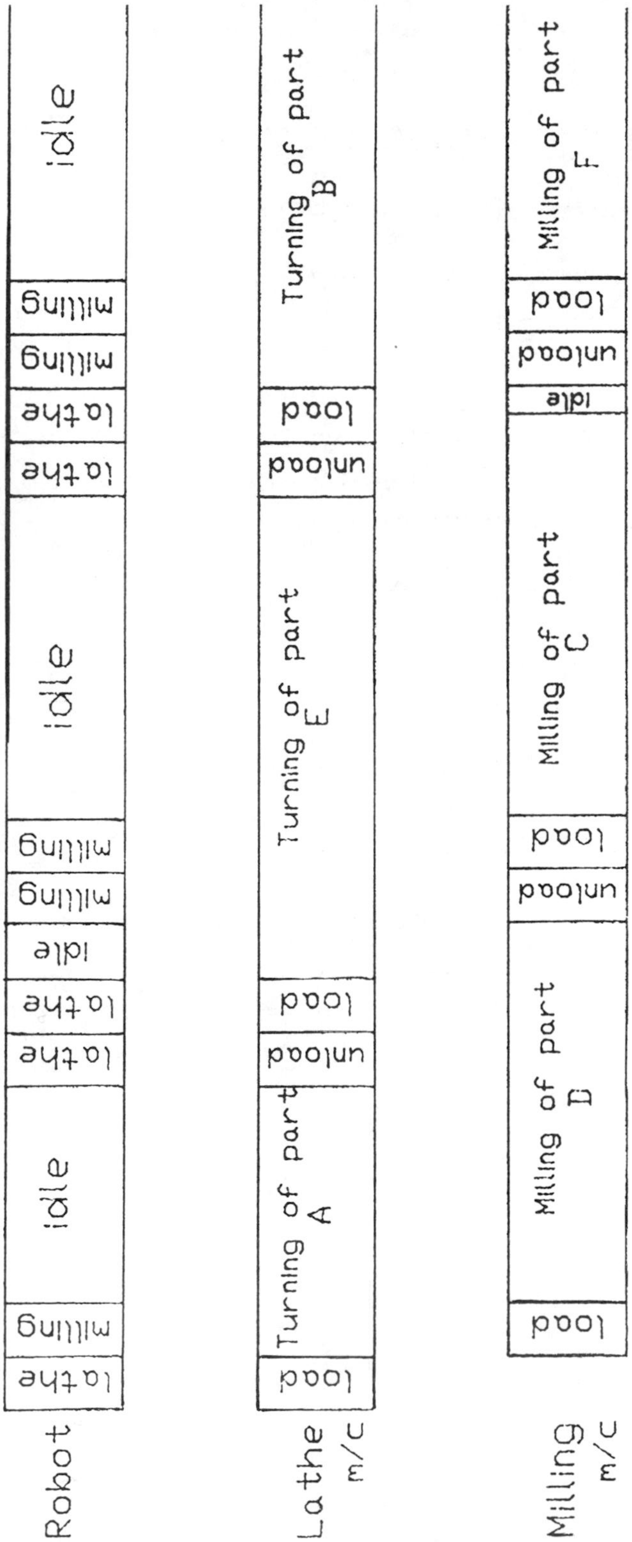

Fig 8.   Machining cell sequential cycle of operations using random release of parts and search-fit program.

VEHICLE SCHEDULING BY SIMULATED ANNEALING

Abdur Rahman and Frances Kerrigan

University of Limerick, Limerick, Ireland

Abstract

In this paper we consider the application of the 'Simulated Annealing' technique to the Vehicle Scheduling problem. A number of practical constrainsts are handled in the formulation including pick-ups, time windows, multiple depots and mixed fleet sizes. We present a selection of results obtained from problems with varying objectives.

1. Introduction

The Vehicle Scheduling Problem (**VSP**) is a challenging logistics management problem, of which many variations arise. The basic components of the problem are a fleet of vehicles with fixed capacities and a set of demands for transporting certain commodities between specified depots and customer sites.

The routing decision involves determining which of the demands will be satisfied by each vehicle and the route each vehicle will follow in servicing its assigned demand. In making decisions the total cost of operating the vehicle fleet must be minimized, where the cost includes fuel, personnel and vehicle depreciation. Generally, a variety of constraints accompany VSP problems, including varying capacity vehicles, backhauling, time windows and multiple commodities. The inclusion of all these factors transforms the VSP into a very complex and time consuming problem, but one with many applications in the real world.

Many variations of the VSP exist with perhaps the most important generalization being the time constrained VSP (TCVSP). Almost all approaches suffer from the limitation of overlooking of this aspect but some recent work has been done in particular by Christofides, Mingozzi & Toth[3], Desrosiers,Soumis & Desrochers[6], Desrosiers, Suave & Soumis [7], Kolen,Rinnooy Kan and Trienkens[10] and Solomon[14], Tinkemer and Gavish[16]. All methods to date have had only limited success working with just a single depot and ignoring other constraints.

The issue of backhauling or incorporating pickups is very important from a commercial point of view. It is a natural method of enhancing profits as the "deadhead" trip back to the depot is replaced with profitable activity e.g. pickups. Surprisingly, very little work has been done in this area and where backhauls are considered they are treated only at the end of a route. To generalise the optimization process pickups need to be incorporated from the first delivery leading to the obvious stage of dynamic routing.

The number of algorithms available for solving the VSP are numerous but are all constrained to solve only a particular aspect of any practical application. Algorithms are not flexible enough to incorporate additional constraints other than those for which they were originally designed. The recent introduction of Simulated Annealing as a heuristic for solving combinatorial optimization problems has meant that multiple constraints on the VSP can be incorporated easily and efficiently.

This paper formulates the problem, introduces the Simulated Annealing algorithm, highlights the work done on the constrained VSP using this technique, discusses the user interface and provides a graphical output of some results obtained from this method.

2. Mathematical Formulation

The VSP is fundamentally concerned with minimizing the distance travelled by a number of vehicles operating from a central depot, subject to vehicle capacity. Vehicles are stocked at the depot with the desired commodities for the customers concerned. Capacity constraints are violated if a vehicle must return to the depot for reloading in order to service all locations. Pick-ups are not allowed at any point.

This paper deals with the VSP on a much higher level. There are multiple constraints in designing the schedules to maximize profits for the distributer while simultaneously satisfying customer requirements.

The basic objective remains the same i.e. the total distance travelled by all vehicles should be minimized. However, to generalize the scheduling the following constraints are added :

        1. Single or multiple depots

        2. Independent specification of depots

        3. One or more vehicles operating from all depots

4. Various sizes and capacities of vehicles

5. Speed specification for each individual vehicle

6. Separate departure times for all vehicles

7. Equalization of route lengths for all vehicles

8. A delivery requirement for each customer

9. An optional pick-up from any customer

10. A time window for each customer specifying the latest acceptable time of service.

The constrained VSP is thus concerned with finding roughly equal routes for a set of vehicles, operating from multiple depots, to minimize the total distance travelled. The capacity constraints on each vehicle should be adhered to in connection with both deliveries and pick-ups and customer service deadlines should be met as far as possible.

The constrained VSP can be mathematically formulated using the following notation to provide a precise statement of the problem.

Constants

K = No. Vehicles

M = No. Depots

N = No. of customers which must be serviced

V_k = Capacity (Volume) of vehicle k

S_k = Speed of vehicle k

d_i = Delivery requirement of customer i

p_i = Pick-up requirement of customer i

tw_i = Time window for customer i

td_k = Departure time of vehicle k

c_{ij} = Direct distance between customers i and j

Formulation of the problem

Minimize $\quad \Sigma\ c_{ij}$

such that $\quad \Sigma d_i < V_k \qquad\qquad k = 1,.... K$

$\qquad\qquad \Sigma\ p_i < V_k \qquad\qquad k = 1,.... K$

$\qquad\qquad \Sigma\ td_k + c_{ij} / S_k \ < tw_i$

To deal effectively with this problem a number of weighting factors are introduced. These can be altered when tradeoffs are neccessary, i.e When the successful implementation of all constraints cannot be achieved. The user, therefore, has the capability to place more emphasis on meeting customer deadlines than incorporating all pick-ups or vice versa. The route equalization factor can be reduced if route lengths do not neccessarily have to be the same. Similarly the other factors can be altered to emphasize / deemphasize a particular scheduling requirement.

The cost function of the system to be optimized is given as :

cost function = α * maximum route distance

+ β * total distance / no. vehicles

+ γ * overcapacity along routes

+ σ * pick-up violations along routes

+ ε * time window violations along routes

By selecting different values of the parameters α , β, γ, σ and ε various solutions are possible. The main purpose of the α and β factors are in route distribution analysis and are usually chosen to make $\alpha + \beta = 1$. The best experimental values have been found as $\alpha=0.8$, $\beta = 0.2$. γ is used as a penalty factor when vehicle capacity has been exceeded with deliveries and its value changes as annealing proceeds. We found that values from 0 to 10 give the best results with large values of γ giving importance to operating within vehicle capacity bounds. The σ and ε factors operate in a similar manner to the γ factor. σ is used to penalize pickups that violate the capacity of the vehicles and σ is the penalty element when delivery deadlines are not met. These latter two penalty factors can be altered throughout the annealing process or can remain fixed. For a balanced scheduling technique we found that values of 1 for each factor gave the best results. If γ, σ and ε remain fixed at zero then infinite vehicle capacity and deadlines are assumed and the problem is transformed to that of route minimization.

Because the user has control over these 5 parameters the desired type of scheduling required can be achieved for individual distribution centres.

3. Simulated Annealing

Simulated annealing is now an well established iterative optimization technique [1,13,15] with the distinguishing characteristic that deteriorations in the system during iterations are conditionally accepted. This gives the algorithm a 'hill climbing ability' and thus a way of avioding local minima traps commonly associated with iterative improvement techniques. A number of parameters are associated with the technique which can be altered to suit the users requirement. The algorithm introduces a pseudo temperature, in the same units as the cost function, as a control parameter. This adheres to the analogy in physical systems while allowing greater freedom as to the type of optimization invoked. The results can thus be as good as desired rendering this method versatile, robust and very general.

A number of optimization problems can be solved using this technique as it is quite straightforward and easily extended to new areas. Despite its recent introduction it has already received a lot of attention. Kirkpatrick et al. [9] have applied it to the travelling salesman problem (TSP) and to block placement with some good quality results although the computational time was high. Aarts and Laarhoven[1] have found the algorithm to be a very powerful tool when applied to the TSP, the linear arrangement problem and the macro placement problem. In their view the main advantages were the generality and ease of implementation while once again the computational effort was a deterrent. Vecchi and Kirkpatrick[15] have tested the algorithm in a global wiring situation with promising results and many others have tested and improved on the basic structure. It has been thought of as an intriguing instance of "artificial intelligence" where the computer has arrived almost unaided at a solution that might have been thought to require the intervention of human intelligence.

4.2 Moveset Generation

Moveset generation forms the core of the simulated annealing algorithm and thus needs to be implemented at great speed. To facilitate this virtual swops are introduced when testing the feasibility of exchanging two nodes. If the exchange is not accepted then nothing needs to be done as the old configuration is kept. If, however, the exchange is accepted then a physical exchange of links must be made. The virtual swop mechanism prevents the swopping and swopping back of pointers when a move is rejected.

Using the array of pointers a virtual swop is easily implemented and the change in cost function can be found with relative ease by consideration of only the nodes adjoining the generated random nodes. There is no need to traverse entire routes as would be neccessary in a linked list formulation. An exchange can be achieved by simply swopping two pointers on the array as opposed to the saving and copying of data in an array of records type structure. Thus, the virtual swop and the quick exchange saves a lot of computational effort in this area.

The basic simulated annealing algorithm uses a direct exchange mechanism only when solving optimization problems. To complement this we have introduced a migration type swop where a node may be omitted from any designated route and included in another. This form of exchange enables modification of the initial assignment rendering it of minor importance to the quality of final solution. A virtual swop is also possible for this type of move and the exchange is similar to previously. A few more pointers need modification with this approach.

A loop exchange technique is carried out on each loop in an attempt to unravel any loop in the route which would make it suboptimal. However, in the constrained VSP the elimination of all loop may not be feasible because of pickup and time window violations. When two random numbers are generated belonging to the same route the reversing of both the inner and outer nodes is tried to see if any improvement is possible. If so the relevant pointers must be exchanged.

The migration and loop exchange type swops are more complicated than the direct swop but their inclusion in the optimization process has shown some good results. We have implemented the direct and migration type exchanges alternatively throughout the program. When a solution has been found the loop exchange mechanism is invoked to check optimality of routes.

5. Results

The simulated annealing approach to constrained vehicle scheduling was tried out on many problems and was found to be consistent and reliable in all cases. The runtimes were moderate but it must be remembered that execution was on an IBM PC using Turbo PASCAL. In some of the relevant literature faster execution times are presented but these programs were run on more powerful machines. Our system also includes a graphical interface and storage of all relevant data to facilitate user requirements which contributes additional time.

The results obtained from 3 problems, with varying constraints, are presented in this section. Each problem was run ten times with different parameter values in order to get a general overview of the system and to test for any correlation among parameters. The results are presented in tabular form giving the best and worst solutions in each case. The average and standard deviation of both the cost function and the computational time are also given to verify the consistency of the algorithm.

It is often very difficult to verify that the solution obtained is indeed optimal, especially when there are a number of constraints to be satisfied. For this reason problems 1 has been fabricated where the optimal solution is known. The results obtained with simulated annealing is the global minimum. Problem 2 has been generated randomly and it is almost impossible to verify optimality. However, in the solution presented there are no violations of either load, pick-ups or deadlines and the reduction in distance has been almost 60%. This is the best solution found for this problem but others which were very similar but having a slightly higher cost function were generated many times. In our opinion then, all solutions presented here are optimal.

5.1 Problem 1

This problem consists of 20 customers with delivery requirements, pick-up requirements and a latest acceptable time of service. There are two depots with one vehicle at each having a specified capacity, speed and departure time

Customer	X_co	Y_co	Del.	Pick-up	Deadline
1		60	100	DEPOT	
2	320	122	0	5	11.00am
3	160	72	8	6	08.35am
4	239	116	8	9	12.50pm
5	276	129	8	6	11.50am
6	213	58	7	8	09.50am
7	213	74	6	6	10.10am
8	212	98	5	5	0.40am
9	176	106	3	1	1.30am
10		246	106	DEPOT	
11	260	100	8	9	09.20am
12	285	100	5	3	09.45am
13	314	103	7	4	10.20am
14	160	87	4	4	08.20am
15	324	111	8	6	10.40am
16	296	129	9	6	11.30am
17	258	125	7	4	12.15pm
18	213	90	7	4	10.30am
19	189	51	4	5	09.20am
20	192	100	9	4	11.10am
21	239	123	6	5	12.40pm
22	160	59	6	2	08.50am

Fig. 1(a) **Input Customer File**

Depot	Site	No.Vehs.	Capacity	Speed	Dep.Time
1	1	1	60	60	08.00am
2	10	1	70	70	09.00am

Fig. 1(b) **Input Depot File**

PARAMETER	VALUE
No. Customers	20
No. Depots	2
No. Vehicles	2
Initial Cost Function	2082
Final Cost Function (best)	444 / 106 secs.
Final Cost Function (worst)	512 / 86 secs.
Average Cost Function	462 / 108 secs.
Standard Deviation	24 / 21 secs.
Load Violations	0
Pick-up Violations	0
Time Violations	0
Vehicle Loads	59 66
Optimal Solution ?	Yes

Table 1. Results for Problem 1.

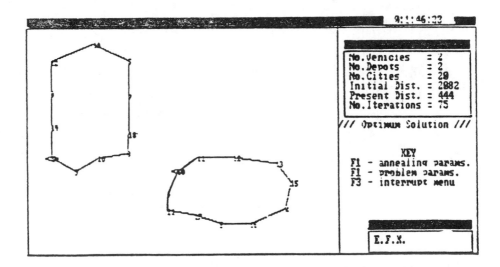

Fig.2 Graphical display of results of Problem 1.

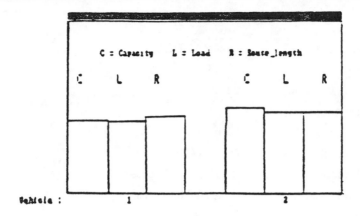

Fig.3 Comparison of route distribution (Problem 1)

5.2 Problem 2

This is a randomly generated problem with 50 customers serviced by 5 vehicles from two depots. All vehicles are identical and are synchronized to depart at the same time.

Input Depot File for problem 2

Depot	Site	No.Vehs.	Capacity	Speed	Dep.Time
1	6	2	60	60	08.00am
			60	60	08.00am
2	30	3	60	60	08.00am
			60	60	08.00am
			60	60	08.00am

PARAMETER	VALUE
No. Customers	50
No. Depots	2
No. Vehicles	5
Initial Cost Function	638
Final Cost Function (best)	270 / 698 secs.
Final Cost Function (worst)	412 / 562 secs.
Average Cost Function	324 / 680 secs.
Standard Deviation	38 / 27 secs.
Load Violations	0
Pick-up Violations	0
Time Violations	0
Vehicle Loads	60 57 58 43 50
Optimal Solution ?	Unknown

Table 2. Results for Problem 2.

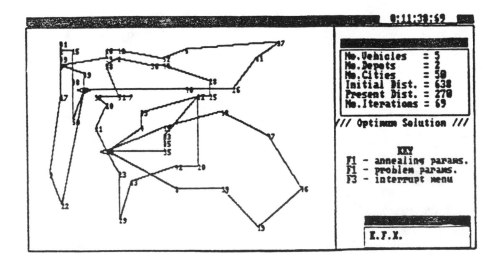

Fig.5 Graphical display of results of Problem 2.

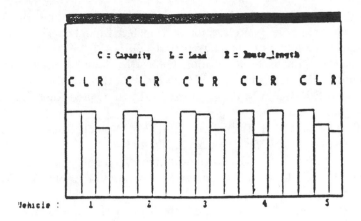

Fig.6 Comparison of route distribution (Problem 2)

The results obtained in all cases have been optimal or near optimal and produced in moderate run times. The consideration of the extra constraints generalizes the VSP and enforces practicality on it. In using simulated annealing the user can interrupt the optimization at any stage and take the results at that point. They are guaranteed to be more optimal than the original solution, thus ensuring savings from the distribution end and resulting in a larger proportion of satisfied customers.

References

1. Aarts E.H.L.,Laarhoven P.J.M. van, "Optimization by Statistical Cooling", Cave Workshop on CAD for VLSI,11 13 Dec. 1984.
2. Bodin L.,Golden A.,Assad A.,Ball M., "Routing and Scheduling of Vehicles and Crews :The State of the Art",Comput.Opns.Res. **10**, 62-212, 1983.
3. Christofides N.,Mingozzi A.,Toth P. "An Algorithm for the Time Constrained Travelling Salesman Problem", Technical Report, Imperial College, London, 1981.
4. Christofides N.,Mingozzi A.,Toth P. "State Space Relaxation Procedures for the Computation of Bounds to Routing Problems", Networks **11**, 145-164, 1981.
5. Clarke G., Wright J.W. "Scheduling of Vehicles from a Central Depot to a Number of Delivery Points", Opns.Res.**11**, 568-581, 1964.
6. Desrosiers J.,Soumis F.,Desrochers M. "Routing with Time Windows by Column Generation", Networks **14**,545-565,1984.

7. Desrosiers J.,Soumis F.,Desrochers M.,Suave M.,"Routing and Scheduling with Time Windows Solved by Network   Relaxation & Branch & Bound on Time Variables", Computer Scheduling of Public Transport,Vol II, 451-469 1985.

8. Geman S. and Geman D., "Stochastic Relaxation, Gibbs Distribution and the Bayesian Restoration of Images",IEEE Trans.Patt.Anal.Mach.Int., PAMI-6, 6, 721- 741,1984.

9. Kirkpatrick S., Gelatt C.D., Vecchi M.P. "Optimization by Simulated Annealing", Science, Vol. 220, 1983.

10.Kolen A.W.J.,Rinnooy Kan A.G.H.,Trienekens H.W.J.M., "Vehicle Routing with Time Windows", Opns.Res., Vol.35,   No 2, 266-273, 1987.

11.Laporte, Nobert G.Y., Arpin D. "An Exact Algorithm for Solving Capacitated Location Routing Problems", Location Decisions; Methodology & Applns.Vol.20, 246-257, 1986.

12.Magnanti T., "Combinatorial Optimization and Vehicle Fleet Planning : Perspectives and Prospects"Networks 11, 179-214, 1981.

13.Metropolis N.,Rosenbluth A.,Teller A.,Teller E.,"Equations of State Calculations by Fast Computing Machines", Jr. Chem. Phys., Vol 21, 1087, 1953.

14.Solomon M.M., "Algorithms for the Vehicle Routing and Scheduling Problems with Time Window Constraints",Opns.Res.Vol.35, No. 2, 254-265, 1987.

15.Vecchi M.P.,Kirkpatrick S., "Global Wiring by Simulated Annealing", IEEE Transactions on CAD, CAD 2, No.4 1983.

16.Tinkemer,K. & Gavish, B.," Parallel Savings Based Heuristic For The Delivery Problem", Operations Research, Vol. 39, No. 3, May-June, 1991.

SIMULATION OF AN UNPREDICTED FAILURE — FLEXIBLE MANUFACTURING SYSTEM (FMS)

[1]Mahmoud A. Younis and [2]Magdi S. Mahmoud

[1]*Engineering Department, American University, Cairo, Egypt*
[2]*Department of Electrical Engineering, Kuwait University, Safat, Kuwait*

ABSTRACT

The problem of dynamic part routing in an automated manufacturing system is described. The system consists of work stations capable of performing a number of different operations on a family of parts. In view of the unreliability of machines, the production control system should manage the production schedule given the random pattern of machine failure/repair. A three-level hierarchical control structure is suggested and simulated using stochastic generation of events and part movements. The obtained results show that production profile derived under the developed controller is feasible

1. INTRODUCTION

Modern manufacturing facilities vary more and more from traditional batch processing units which have always had inherent limitations. The fundamental concept behind modern manufacturing is flexibility which results from the constraints imposed upon the manufacturing environment by the market. The need for a short response time to change in demand, increasing variability in products, all together with the need for high production rate created the need for Flexible Manufacturing Systems (FMS), as a concept.

The FMS performance is normally derived from simulation of appropriate models because actual experimentation on existing system is not feasible. Comprehensive survey of the development of analytical models of FMS is presented in [1] and partially in [2]. It has been pointed out that analytical models, sometimes referred to as generative models, are quite useful for systems with relatively few operating parameters. The effects of machine failures, demand uncertainties and concurrent routing are difficult to cast. Other approaches include evaluative modelling [3], perturbation analysis [4] and network of queues [5-8].

Most manufacturing systems are large and complex. Therefore, it is natural to divide the control or management into a hierarchical multilevel structure. Each level is distinguished by: i) a length of the planning horizon, and ii) the kind of data required for the decision-making process.

Lower levels of the hierarchy typically have short horizons and use detailed information, while higher levels have longer hohizons and use highly aggregated data. It is to be noted that the nature of uncertainties at each level of control also varies. Planning and control activities usually involve managerial decision-making that can be implemented into a three-level structure . Typically, the operational level is concerned with the short-time production profile and hence it focuses on the joint determination of production rates of members of the part family while taken into account the effects of the demand, the level of the downstream buffer levels, and the reliability of workstations.

This paper examines the problem of dynamic part routing in FMS through computer simulation and using stochastic generation of events and part movements. Parts are loaded into the system in a way so that the production goals are met. A three-level hierarchical control structure is suggested. A computational algorithm is described for a multiproduct - multiworkstation production system. A simulation example is presented to study the performance of this control policy.

2. THE PROBLEM OF DYNAMIC PART ROUTING

Here we focus on the operational level and consider the flow control of determining the production rates for the part family. The horizon is set in relation to the master production plan.

Model Development

The FMS consists of M workstations producing N different part types. Work station m(m = 1,2,..,M) has L_m identical machines. Each part of type i requires Ki operations for its completion. A particular operation can be done at one or more different workstations. The time to complete operation k on a type i part at workstation m is a random variable with mean τ^k_{im} The flow rate of type i parts to station m for operation k is defined as $y^k_{im}(t)$.

Let u_i be the production rate of type i parts. Since no material is accumulated within the system, then conservation of flow implies that :

$$\sum_{m=1}^{M} y^K_{im} (t) = u_i(t) \qquad k =1,\ldots,K_i \ \& \ i =1,\ldots,N \quad (1)$$

Define $\underline{x}(t)=[x_1(t)\ldots x_N(t)]$ as the buffer state which measures the cumulative difference between production and demand for the parts. Thus, we have :

$$\frac{d\underline{x}(t)}{dt} = \underline{u}_i(t) - \underline{d}(t) \qquad x_i(o)=x_{io} \qquad (2)$$

where $\underline{u}_i=[u_1(t)\ldots u_N(t)]$ is the vector of production rates for the part family and $\underline{d}_i=[d_1(t)\ldots d_N(t)]$ is the vector of downstream (terminal or intermediate) demand rates, which is known over the interval $(0,t_f)$ where tf specifies the end of the production plan. The $\underline{d}(t)$ represents the production requirements set forth by the tactical level. Finished parts are stored in downstream buffers from where the downstream demand is satisfied. When $x_i(t) < 0$ backlogged demand occurs for part i whereas $x_i(t) > 0$ gives the size of the inventory stored in the downstream buffers. The state of the workstations is called the machine state and is denoted by the integer vector $\underline{a}(t)=[a_1 \ldots a_m(t)]$ with the component $a_m(t)$ represents the number of operational machines at station m. Recall that the production rate at any instant is limited by the capacity of the currently operational machines. We introduce $\Omega[\alpha(t)]$ to represent the capacity of the system and is defined as the set of all production rates u(t) such that there exist feasible flow rates $y^k_{im}(t)$ satisfying (1) and

$$\sum_{i=1}^{N} \sum_{k=1}^{K_i} y^k_{im} \tau^k_{im} \leq a_m \qquad m=1,\ldots\ldots M \qquad (3)$$

It is important to note that the product $y^k_{im} \tau^k_{im}$ constitutes the proportion of each unit time interval used by one or more operational machine at station m to perform operation k on type i parts. From (3), the left hand side is thus the total input work to station m per unit time due to the part flow rate y^k_{im} The inequality reflects limited capacity.

Given that a machine at station m is operational, the probability of a failure in an δt interval is $(p_m \delta t)$. On the other hand, the probability that a failed machine is repaired during δt interval is given by $(r_m \delta t)$ where p_m and r_m are the failure and repair rates for the machines at station m.

Considering the structure of the FMS, an appropriate dynamical model of the machine state, where P[E] representing the probability of occurrence of event E, is described by the following conditional probabilities.

$$P[a_m(t+\delta t) = \sigma + 1 \mid a_m(t) = \sigma] = \begin{cases} (L_m-1)\ ((r_m\ \delta t) & 0 \leq \sigma < L_m \\ 0 \end{cases} \qquad (4)$$

otherwise :

$$P[a_m(t+\delta t) = \sigma - 1 \mid a_m(t) = \sigma] = \begin{cases} \sigma\ (p_m\ \delta t) & 0 \leq \sigma < L_m \\ 0 \end{cases} \qquad (5)$$

It readily follows from (4) and (5) that :

$$P[a_m(t+\delta t) = \beta \mid a_m(t) = \Delta] = 0$$

if $|\beta - \Delta| > 1$ and for two different machines states j and n, the quantity :

$$\lambda_{jn}\ \delta t = P[\underline{a}(t+\delta t) = n \mid \underline{a}(t) = j]$$

defines the machine state transition probability which has the property that :

$$\lambda_{jj} = - \sum_{n} \lambda_{jn}$$

The above discussion suggests modelling the times between failures (TBF) and the times to repair (TTR) by exponentially distributed random variables with mean time between failure (MTBF) and mean time between repair (MTTR), where $(MTBF=1/p_m)$ and $(MTTR=1/r_m)$; respectively. Note from (4) and (5) that state transitions are due to the failure or repair of a single machine.

For convenience, we consider that failure and repair rates are independent of production rates and the number of operational machines.

Problem Statement

The dynamic part routing problem can now be stated. Given the buffer and machine states at arbitrary time; $\underline{x}(0)$ and $\underline{a}(0)$ along with the dynamic model (1)-(5), it is desired to determine the pattern of production over the interval $(0, t_f)$ such that the performance index.

$$J(\underline{x}, \underline{a}, o) = E \left\{ \int_{o}^{t_f} g[\underline{x}(t)] \ dt \ \middle| \ \underline{x}(o) = \underline{x}_0 \ , \ \underline{a}(o) = \underline{a}_0 \right\} \ (6)$$

is minimized where $E[,]$ is the expectation operator. The functional $g[\underline{x}(t)]$ penalizes the production profile for failing to meet the demand and for keeping an inventory of parts in the down-stream buffers. In this regard, $J(\underline{x}, \underline{a}, o)$ expresses the total penalty incurred by the producer in the interval $(0, t_f)$. For convenience, we select $g[\underline{x}(t)]$ to be component-wise monotone and convex; that is :

$$g[\underline{x}(t)] = \sum_{j} g_j [x_j(t)]$$

$$g_i(o) = o$$

$$\lim_{|x_i| \to \infty} g_i(x_i) = \infty \qquad\qquad (7)$$

$$g_{i-1}[x_{i-1}(t_{i-1})] < g_i[x_i(t_i)] < g_{i+1}[x_{i+1}(t_{i+1})] \quad \text{for all } i$$

Ideally, the optimal policy would be to produce at the demand rate which means that the buffer state is kept at zero level. This ideal case is far from reality in view of the random failure of machines.

The Optimal Production Policy

To characterize the optimal production policy, we consider the 'cost to go' when the applied policy is $u[x, a, t]$ as :

$$J^u(\underline{x}, \underline{a}, t) = E \left\{ \int_{t}^{t_f} g[\underline{x}(s)] \ ds \ \middle| \ \underline{x}(t) = \underline{x} \ , \ \underline{a}(t) = \underline{a} \right\} \ (8)$$

Following the development of Rishel [6], it can be shown that the optimal u* is determined by :

$$\min_{u \in \Omega(\underline{\alpha})} \quad [\, dJ^u / d\underline{x} \,] \, \underline{u} = R \qquad (9)$$

A systematic procedure to satisfy (9) subject to (1)-(7) result in a computational algorithm that will be described in the next section.

3. A COMPUTATIONAL ALGORITHM

A general purpose computational algorithm has been developed for a multiproduct, multistation flexible manufacturing system. The algorithm is developed in such a manner that a randomly chosen step function is used to model the time varying demand rate for each part type at a randomly generated intervals within the time horizon of the production plan.
The times between failure and to repair of the production resources are again modeled by exponentially distributed random variables. The means $(MTBF=1/p_m)$ and $(MTTR=1/r_m)$ are then computed from the randomly generated machines states $\alpha_m(t)$, where a reliability for each machine has been considered. The system is penalized equally for being ahead or behind demand requirements by a chosen cost function satisfying (Eq. 7):

$$g[x(t)] = \sum_{i=1}^{N} \alpha_i x_i^2 \qquad (10)$$

The buffer state $x_i(t)$ can be calculated by integration equation 2, thus:

$$x_i(t) = \int_{t}^{t_f} [\, u_i(t) - d_i(t) \,] \, dt \qquad (11)$$

From equations 8 and 9 and by using equations 10 and 11 one can get :

$$J^u(\underline{x},\underline{\alpha},t) = E \left\{ \int_{t}^{t_f} \sum_{j=1}^{N} \alpha_j \int_{o}^{s} \{ \, u_j(\tau) - d_j(\tau) \, \}^2 \, d\tau \right\} \Bigg|_{\substack{\underline{x}(t)=\underline{x} \\ \underline{\alpha}(t)=\underline{\alpha}}} \qquad (12)$$

and hence we obtain :

$$R = \min_{u \in \Omega(\underline{\alpha})} \quad [\, dJ^u / d\underline{x} \,] \, \underline{u} \qquad (13)$$

It should be emphasized that algebraic manipulations of (11)-(13) are performed by numerical iterations.
A flow chart for the computational algorithm is given in Fig. 1.

Following is a brief description of the various routines in the computer program:

CONTROL   is the main program which organizes the various coputational steps of the algorithm. It also reads the title and all input data such as: time length of the production plan,workstation M, number of part type N, etc.

DEMAND   Generates the time varying demand rate for each part by using a random number generator. The computer code is based on :

> For part i to part N
> TIME = RANDOM (1)
> DEM = RANDOM (TIME)

STEPS   The production plan can be monitored by performing the calculations at the end of a previously determined time interval. The time horizon TF of the production plan is divided into F intervals. The computer code is of the form:

> TF = 300 HR
> ST = 30 MIN
> REPEAT FOR Time Interval = 0
> TILL F = (TF * 60/ST)+1

ALFA   All possible machine states are generated randomly and the averages $MTBF=1/p_m$ and $MTTR=1/r_m$ are then evaluated

PART RELEASE   Generates all possible combination of parts entering the production line $us(1 \leq s \leq N)$ and $ur(1 \leq r \leq N)$.

PRODRATIO   The production rate at each instant is limited by capacity of the currently operational machines. At time t, the production rate must lie in a set $\Omega[\alpha(t)]$ which depends on the machine state. The program generates the production constrains sets for each part type combination at each machine state. The production vectors are then calculated.

PRODRATES   Calculate the production rate for each type u_i by assuming that parts production rates is simply oportional to the demand rates $(u_id_i=u_jd_j)$ and solving with the previously obtained production vectors. The values obtained are then normalized and stored in an array.

GAMMAA   curve fitting is applied to the production rates calculated at each time interval and a set of a time function production rates for part types within the time horizon of the production plan is computed.

PENALTY   The system is penalized for backlogged demand or inventory of a part type by applying the cost function Eq. 10 and computing the result R from Eq. 12. The calculation is repeated NSIM times and the minimum value of R is considered as an optimum value.

OUTPUT   This program organized the output results of the algorithm. It prints out the demand $d_i(t)$ and production rates $u_i(t)$ as well as the buffer state $x_i(t)$. The program gives the statistical data for the workstations. It computes the MTBF, MTTR as well as the availability and utilization of the available time at each workstation.

4. SIMULATION EXAMPLE

To demonstrate the application of the developed algorithm, we consider a typical case study; that is the manufacturing of windows, door and curtain walls made out of aluminum profile. Fig. 2 shows the configuration the FMS plant. Seven part types are manufactured on this FMS. They are divided into three groups by making use of the group technology, production systems. The first group consists of three different sizes of windows, and require five operations A, B, C, D and E. The second group consists of two sizes of doors and required four operations A, B, C and E, while the third group consists of two sizes of curtain walls and need only three operations A, B, and C.

The average processing times for, each part type is given in Table 1. Initially, the system has all machines operating and therefore the buffer state $x(t_o) = 0$. The demand ratio is taken to represent 80% of the system production capacity.

5. SIMULATION RESULTS

The system of the plant given in Fig. 2 was considered in computer simulation of the three-level hierarchical control structure based on stochastic generation of events and randomized part movements. The computational algorithm of Fig. 1 managed the production schedule considering the random pattern of machine failure/repair. The simulation model was run for an equivalent of 300hr. The flow control level implemented in the simulation computes the production rate $u(t)$, the demand rate $d(t)$ and the buffer state $\underline{x}(t)$ at 30 minutes intervals. It should be pointed out that the buffer state $x(t)$ refers to the accumulated difference between the actual production level and the desired demand and can therefore take on negative values, indicating backlogs. Throughout the computer simulation, the algorithm produces the following items per each workstation:

1) NOF : number of failures
2) MTBF : mean time between failures
3) MTTR : mean time to repair
4) ULT : average utilization time for full-load working
5) AVL : availability of the machines

Figure 3 give the production and demand rates as well as the actual buffer state for part type 3. Similar results are obtained for the other part types. Results show how the algorithm was able to control the production rate and let it track or follow the demand rate and thus keeping a small number of pieces in the downstream buffer. However, it can be seen from both figures that the accumulated buffer states oscillates around the zero level. The maximum value of the buffer state represents about ± 4% of the production rate. Here the cost function penalizes the controller equally for excess production and for backlogged demand. It is usually preferred to keep the buffer close to zero when all machines operate with no failure and to clear the backlog (negative value of buffer state) in case when failure occurs rather than to maintain high values of inventory to compensate for backlogs result from future failures.

The statistical data for each workstation is given in Table 2. The simulated random number of failures for the machines within a workstation is given as F. It is readily seen that stations 4 and 5 are more reliable and have the least number of failure of 96 and 94 respectively. Consequently both stations have the largest average meantime between failure (MTBF) 184.4 and 185.7 minutes respectively. No significant difference can be observed between the mean time to repair (MTTR) as well as the availability for all five stations. However, the controller has been able to attain a utilization of about 82% for each of work station 1,2, and 3. On the other hand stations 4 and 5 are lightly loaded with only 73% and 77% of the available time being used.

6. CONCLUSION

A problem of dynamic part routing in a flexible manufacturing system is described. The machines are unreliable. The production control system was able to meet a random production requirements while the machine fail and are repaired at random times. A three level hierarchical control structure is suggested. A general computational algorithm is described to fit into existing factory management structure. The algorithm is structured to handle a multiproduct – multiworking station production system. The example for a plant manufacturing door, windows, and wall curtains made out of aluminium profile is described. The example and the results show that it is possible to accurately track demand requirements while maintaining a low inventory level. The control policy is applied to the whole FMS and not merely to the first machine that the part type encounters. This eliminates the buildup of parts inside the system, reducing the scheduling problems while considering repair and failure information.

REFERENCES

1. Buzacott J.A and Yao.D.D: Flexible manufacturing systems: A review of analytical models. Mgmt. Sc., 32, 890-905 (1986).
2. Buzacott J.A.and Shanthikumar J.G.: Models for understanding flexible manufacturing systems, AIIE Trans. 12, 339-350 (1980).
3. Iwata K., Murotsu Y, Oba F., and Yasuda K: Production scheduling of flexible manufacturing systems, CIRP Annals. 31, 319-333 (1982).
4. Suri R., and Dille W. :On line optimization of flexible manufacturing systems using perturbation analysis. First ORSA/TIMS Conf. on Flexible Manufacturing Systems, Michigan (1984).
5. Kimemia J. and Gershwin S.B.: Flow optimization in flexible manufacturing systems, Int. J. Prod. Res., 23, 81-96 (1985).
6. Rishel R.: "Dynamic Programming and Minimum Principles for Systems with Jumps Markov Disturbances", SLAM I. Control, Vol. 13, 338-371.
7. Gershwin S.B.: "Representation and Analysis of Transfer lines with Machines that have Different Processing Rates," Annals of Operations Research, Vol. 9, 511-530 (1987).
8. Kamath M., Suri R., and Sanders J.L.: "Analytical Performance Models for Closed-loop Flexible Assembly Systems", Int. J. Flexible Manufacturing Systems, Vol. 1, 51-84 (1988).

Table 1. Processing Time For The Parts In Minutes

Part No.	Part Name	Working Time (min)					TOTAL TIME
		A	B	C	D	E	
1	Window (size 1)	2.0	3.1	2.5	1.5	2.1	11.2
2	Window (size 2)	1.6	2.9	2.4	1.5	1.9	10.3
3	Window (size 3)	1.4	2.8	2.2	1.4	1.8	09.6
4	Door (size 1)	3.1	3.9	4.2	N.A*	2.5	13.7
5	Door (size 2)	2.9	3.5	3.5	N.A	2.3	12.2
6	Curtain wall (size 1)	2.2	4.5	3.5	N.A	N.A	10.2
7	Curtain wall (size 2)	2.0	4.1	2.5	N.A	N.A	08.6

* N.A stands for Not Available

A --> Profile Cutting

B --> Assembly

C --> Fastening and Fitting

D --> Mounting Accessories

E --> Glazing and Sealing

Table 2. Statistical Data for Workstations

Station	Number of Failure (F)	MTBF* (min)	MTTR+ (min)	Availability (%)	Utilization (%)
1	123	145.4	30.2	83.7	82.6
2	114	155.3	30.3	89.3	82.5
3	100	177.0	33.3	85.2	82.7
4	96	184.4	34.1	85.3	73.3
5	94	185.7	32.9	86.3	77.9

* MTBF : Mean Time Between Failure

+ MTTR : Mean Time To Repaire

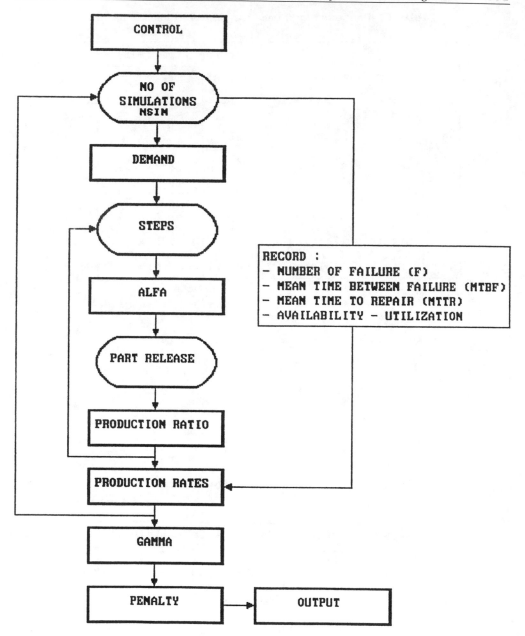

Fig. 1.   Flow chart for the computational algorithm.

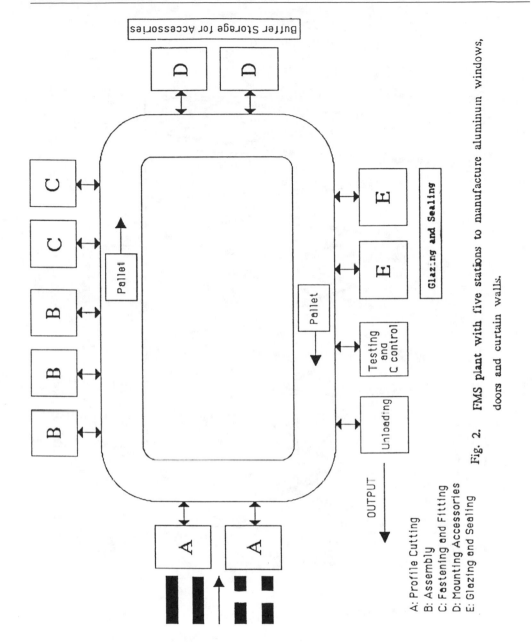

A: Profile Cutting
B: Assembly
C: Fastening and Fitting
D: Mounting Accessories
E: Glazing and Sealing

Fig. 2. FMS plant with five stations to manufacture aluminum windows, doors and curtain walls.

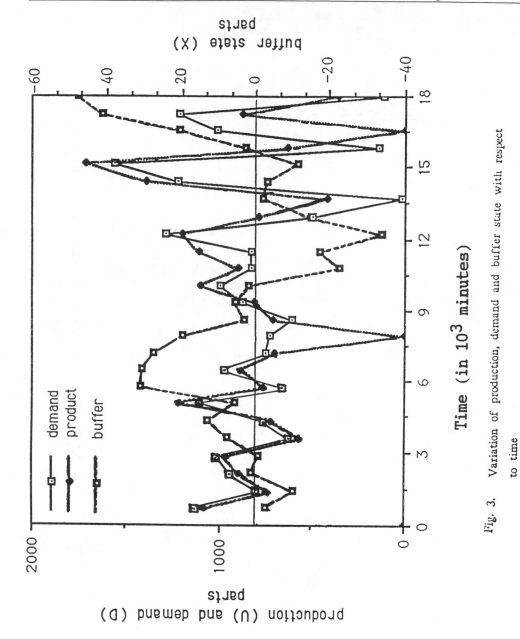

Fig. 3.   Variation of production, demand and buffer state with respect to time

Part VI
Increasing Competitiveness
Through Technology

FUZZY LOGIC BASED DIGITAL IMAGE PROCESSING

Jun Yan and Michael Ryan

School of Computer Applications, Dublin City University, Dublin, Ireland

ABSTRACT

The theory of fuzzy sets was first developed by Zadeh in 1965 as an extension to traditional set theory, along with the fuzzy logic to manipulate the fuzzy sets. Fuzzy logic has been used to carry out image processing functions such as gray level thresholding, edge detection and image enhancement. The image is usually considered as an array of NxN fuzzy singletons, each with a membership denoting the property of the image at that point. In this paper, the concept of a fuzzy measure is introduced. The index of fuzziness is used as a tool to extract the fuzzy property plane for a given image. The relationship between the extracted fuzzy property plane and the window size of the S-function is studied. An automatic procedure to select the optimum thresholds and construct the fuzzy property planes for multimodal images is proposed, where the human intuitive descriptions of the properties of the multimodal image are used to specify the multiple gray level regions. Based on the selected thresholds and the constructed fuzzy property planes, operations using the fuzzy operators MIN, MAX and INT for image thresholding, edge detection and image enhancement are examined and demonstrated.

INTRODUCTION

Fuzzy set theory was first developed in 1965 by Lotfi Zadeh as an extension to traditional set theory, along with the fuzzy logic to manipulate the fuzzy sets [1]. A fuzzy set allows for degrees of membership in a set. A membership function defines the grade of membership in a fuzzy set for all the possible members, and is typically expressed as a mathematical function or a set of discrete digital mumbers. This allows human observation, expressions and expert knowledge to be more closely modelled. Combining aspects of multivalued logic, probability theory, artificial intelligent (AI) and neural networks, fuzzy logic has been successfully established as an alternative approach to reasoning under uncertainties. Fuzzy logic makes tough problems much easier to solve by allowing a more natural representation of the situation being dealt with.

A fuzzy set A in an universe of discourse U is characterized by a membership function μ_A which takes values in the interval [0,1], namely, $\mu_A:U\to[0,1]$. A fuzzy set A in a discrete universe of discourse U with $u_1,u_2,...,u_n$ elements can be represented as:

$$A = \sum \mu_A(u_i)/u_i \qquad (1)$$

The element u in U at which $\mu_A(u) = 0.5$ is called the crossover point. A fuzzy set whose support ($\mu_A(u_i) > 0$) is a single point in U is referred to as fuzzy singleton. Depending on the applications, a membership function $\mu_A(u)$ can be defined by using the standard S-function, π-function, triangle shape, etc..

An image X of $M \times N$ dimension with L gray levels can be considered as an array of $M \times N$ fuzzy singletons, each with a membership denoting the property of the image at that point [4]. It has been found that low level digital image processing bears some fuzziness in nature due to the factors such as:

    a. the lack of the quantitative measurement of image quality,
    b. the loss of information due to many-to-one features in image grabbing,
    c. the inherent distortion or noise in the image being processed,
    d. the subjective selection of the membership functions.

Fuzzy sets have been applied to low level gray tone image processing operations such as gray level thresholding, edge detection, image skeleton, image enhancement and segmentation [4-11]. Among them, the gray level thresholding plays an important role. For example, in enhancing the contrast of an image, proper threshold levels must be selected so that some suitable non-linear transformation can highlight a desirable set of pixel intensities compared to others. Similarly, in image segmentation one needs a proper histogram threshold to partition the image space into meaningful regions.

Bimodal gray level thresholding based on fuzzy measures has been demonstrated by Pal and Rosenfeld [9]. The main feature of fuzzy measure based thresholding is the extraction of a fuzzy property plane from a gray tone image by minimizing the fuzziness between the gray tone image and its nearest two tone version. Several fuzzy measures have been employed to perform the measurements of fuzziness both in the spatial domain and in the gray level domain [6, 9 & 10]. Fuzzy geometry measures such as compactness have been used to evaluate the spatial fuzziness of an image, while the concepts of index of fuzziness and entropy have been employed to measure the gray level fuzziness of an image.

The effectiveness of the methods proposed in [6, 9 & 10] to perform thresholding using fuzzy measures in the gray level domain is limited by the following factors:

    a. the lack of the perceptions of the image properties.
    b. the lack of an automatic procedure to construct the fuzzy property plane and select the threshold, especially the lack of an automatic procedure to adjust the the window size of the S-function during the thresholding process.
    d. the failure to find the multithresholds for a multimodal image.

It has been long recognized that the human being can provide the quickest description of an input image. Such intuitive descriptions may provide rich information such as the properties of the noise distributions, the contrast and the gray level regions of the image. For a bimodal image with only two gray level regions denoting the objects and the background, only one threshold is require across the whole gray level range. Thus, the whole image properties can be expressed by only one fuzzy property plane. However, for a multimodal image with several gray level regions denoting the objects and background, it is necessary to select a threshold and construct a fuzzy property plane for each gray level region.

In this paper, the concept of fuzzy measure is introduced. The index of fuzziness is used as a tool to extract the fuzzy property plane for a given image. The relationship between the extracted fuzzy property plane and the window size of the S-function is studied. An automatic procedure to select the optimum thresholds and construct the fuzzy property planes for multimodal images is proposed, where the human intuitive descriptions of the properties of the multimodal image are used to specify the multiple gray level regions. Based on the selected thresholds and the constructed fuzzy property planes, operations using the fuzzy operators MIN, MAX and INT for image shresholding, edge detection and image enhancement are examined and demonstrated.

FUZZY MEASURES

Basic concepts of fuzzy measures

Fuzzy measure indicates the fuzziness of a set. In general, a measure of fuzziness is a function $f:P(X) \rightarrow R$, where $P(X)$ denotes the set of all fuzzy subsets of X. That is, the function f assigns a value $f(A)$ to each subset A of X that characterizes the degree of fuzziness of A. In order to qualify as a meaningful measure of fuzziness, f must satisfy the following requirements:

    a. $f(A) = 0$ if and only if A is a crisp set.

    b. If $A \langle B$, then $f(A) \leq f(B)$. Here $A \langle B$ denotes that A is less fuzzy than B (or sharper than B). The sharpness relation $A \langle B$ is defined by
$\mu_A(x) \leq \mu_B(x)$ for $\mu_B(x) \leq 0.5$ and $\mu_A(x) \geq \mu_B(x)$ for $\mu_B(x) \geq 0.5$.

    c. $f(A)$ reaches maximum if and only if A reaches maximal fuzzy $\mu_A(x) = 0.5$.

Among the measures of fuzziness developed so far, the index of fuzziness, which is defined in terms of a metric distance between a fuzzy set A and its nearest crisp set C, has been widely used in image processing. The nearest crisp set C of the fuzzy set A is denoted by $\mu_C(x) = 0$ if $\mu_A(x) \leq 0.5$ and $\mu_C(x) = 1$ if $\mu_A(x) \geq 0.5$. When the Hamming distance is used, the index of fuzziness is expressed by the function

$$f(A) = \Sigma \mid \mu_A(x) - \mu_C(x) \mid$$
$$\text{or} \quad f(A) = \Sigma \min\{ \mu_A(x), 1 - \mu_A(x) \} \tag{2}$$

Fuzzy measures in gray level domain

Fuzzy measures in the gray level domain of an image are provided using the "index of fuzziness". For an image X of size MxN with L gray level levels, $X:\{x_{mn}\}$, the fuzzy image plane can be denoted by $\mu_X:\{ \mu_X(x_{mn})/x_{mn} \}$, where $\mu_X(x_{mn})$ denotes the grade of possessing some fuzzy property (e.g. brightness) by the (m,n)th pixel intensity x_{mn}. Let $\mu_C(x_{mn})$ denote the property of the nearest two tone version of $X:\{x_{mn}\}$, the index of fuzziness can be defined as follows:

$$V(X) = [2/(MN)] \, \Sigma\Sigma \mid \mu_X(x_{mn}) - \mu_C(x_{mn}) \mid \tag{3}$$
$$\text{or} \quad V(X) = [2/(MN)] \, \Sigma\Sigma \min\{\mu_X(x_{mn}), 1 - \mu_X(x_{mn})\} \tag{4}$$

It can be proven that $V(X)=0$ for $\mu_X(x_{mn})=1$ or 0 and $V(X)=1$ for $\mu_X(x_{mn})=0.5$. For a gray tone image, the index of fuzziness reflects the average amount of fuzziness present in the image by measuring the distance between the extracted fuzzy property plane and its nearest crisp two tone version. Once the threshold is decided for an image, the nearest two tone image can be formulated and the fuzzy property plane can be constructed corresponding to each gray level. Since the value for each fuzzy singleton in the image is usually computed by using a user defined S-function, modification of the crossover point and the window size of the S-function results in different segmented images with varying index of fuzziness. For the fixed window size, proper selection of the crossover point will result in a minimum index of fuzziness. Thus, the optimum threshold can be selected as the gray level where the index of fuzziness reaches a minimum. The fuzzy property plane can be constructed with the selected threshold and window size.

FUZZY PROPERTY PLANE AND S-FUNCTION

For a fixed window size of the S-function, the gray level thresholding algorithm proposed by Pal and Rosenfeld [9] can be outlined as follows:

step 1. Construct the membership $\mu_X(x_{mn})$ for the fuzzy property plane using a user defined S-function, that is, $\mu_X(x_{mn}) = S(x_{mn}, a, b, c)$, where $b=l_i$ (gray level).

step 2. Compute the amount of fuzziness in μ_X corresponding to $b=l_i$ with $V(X)$.

step 3. Vary l_i from l_{min} to l_{max} and select $l_i=l_t$ for which $V(X)$ is a minimum.

The S-function used here is defined as

$$S(x_{mn}, a, b, c) = \begin{cases} 0 & x_{mn} < a \\ 2[(x_{mn}- a)/(c-a)]^2 & a \leq x_{mn} \leq b \\ 1 - 2[(x_{mn}- c)/(c-a)]^2 & b \leq x_{mn} \leq c \\ 1 & x_{mn} > c \end{cases} \quad (5)$$

The cross over point is given by $b = (a + c)/2$. The window size of the S-function is given by $w = c - a$. Thus, the a and b can be denoted by

$$\begin{cases} a = b - w/2 \\ c = b + w/2 \end{cases} \quad (6)$$

It is apparent that the value of the fuzzy singleton, $\mu_X(x_{mn})$, is affected not only by the selection of b but also by the selction of the window size w. An inappropriate window size will lead to a wrong selection of the threshold. Even for the same window size, different gray level search regions will result in different "optimum" thresholding results. This relationship, however, has not been emphasized or analysed in the literature to date. Here we examine the relationship in an example. For the simplicity, let us assume an image X of size 12x12 with 256 gray levels which is shown in APPENDIX A. Depending on the perceptions of the human operators, the input image can be interpreted in two ways according to its gray levels:

1). The image consists of two objects, each with different gray level. Thus, the first object has gray level about 110, the second one has gray level about 255 and the background has gray level lower than 20.

2). The image contains only one object with the gray level above either 110 or 200. The rest belong to the background.

According to the interpretations of whether the image is multimodal or biomodal, the gray level regions can be specified. In the case of the bimodal image, the global gray level region has been defined as a range from the minimum gray level to the maximum gray level of the image, that is from 10 to 255. In the case of the three-modal image, the local gray level regions have to be defined according to the properties of the object gray levels. Here, the first local gray level region is from 10 to 110 and second one is from 110 to 255. Hence, the selection of the optimum threshold has been carried out against the minimum index of fuzziness by searching each gray level region. TABLE 1 shows the selected thresholds with different window sizes in each gray level region.

TABLE 1. THRESHOLD vs. WINDOW SIZE & SEARCH RANGE

w	V(X) (*¹)	l_t (*¹)	V(X) (*²)	l_t (*²)	V(X) (*³)	l_t (*³)
5	0.0	22 - 247	0.0	22 - 107	0.0	113 - 247
20	0.0	29 - 240	0.0	29 - 100	0.0	120 - 240
50	0.0	44 - 85	0.0	44 - 85	0.001033	134 (**)
80	0.0	59 - 70	0.0	59 - 70	0.007704	145
100	0.000886	64 (**)	0.000886	64 (**)	0.012811	156
120	0.010260	69	0.010260	69	0.020995	154
150	0.035072	80	0.035072	80	0.040370	145
180	0.059708	138	0.061497	91	0.059708	138
200	0.071627	133	0.078558	99	0.071627	133
250	0.100645	131	0.127811	109	0.100645	131

(*¹) global search in gray level region [010, 255].
(*²) local search 1 in gray level region [010, 110].
(*³) local scarch 2 in gray level region [110, 255].
(**) optimum thresholds selected in each region.

From the above computations, it can be seen that:

a. The selection of the threshold l_t is heavily dependent on the window size and the gray level search range.

b. Even for the same image, different thresholds can be computed for different sizes of windows in a gray level region. In the case of a fixed search range, the value of the index of fuzziness V(X) provides one way to automatically select the optimum threshold and the window size: the window sizes and the thresholds can be selected as the optimum where V(X) reaches a minimum except for V(X) = 0.

c. It is hard to tell which computed threshold is optimum in the case of global search due to lack of the understanding of the object properties. The selected threshold in TABLE 1 through global search is 64 which does make sense if the gray level of the object in the image is assummed to be over 110. However, it fails for the object with gray level above 150, where the pixels with gray level under 150 should be considered as background.

d. Fuzzy measure based on local gray level search regions provides reliable threshold and window size selections. However, threshold selection based on local gray level search needs the assistance of the intutitive or human observed knowledge of the image.

FUZZY LOGIC BASED THRESHOLDING

It has been found that a reliable and meaningful threshold is difficult to select without the full understanding and perception of the image properties. A blind search in the global gray level range for a multimodal image will result in unreliable thresholds being selected. Hence, for a multimodal image with multiple gray level regions it is better to perform the thresholding in each gray level region of the image. In what follows we will use human peceptions and linguistic descriptions about the properties of the image to specify the gray level region. Then an automatic procedure to decide the window size of the S-function and to select the threshold in each gray level region will be proposed.

The determination of the gray level regions for a multimodal image

There are several methods to determine the gray level regions for a multimodal image. The gray level histogram mode analysis is one of the techniques being used to partition the gray levels of an image. However, with the increase of the histogram modes, especially when the image is corrupted by the noise, it is very difficulty to isolate the regions of interest. Moreover, whatever techniques are used, human intervention and judgement are always required to reach a final decision from the computed results. On the other hand, for a given image, the human operator can provide the quickest and most reliable descriptions of the properties of the image such as the constrast and the gray level regions in linguistic terms. Using the human descriptions to determine the gray level regions for a multimodal image is pursued here.

For the simplicity, let us consider the case of three-modal images. For an input image displayed on the monitor or computer screen, the human operator may describe the gray level regions of the objects and the background of the image as: "The objects in the image are located in the *two* gray level regions. The first gray level region is *very bright* and the second region is *less bright*. The background of the image is *very dark* ". Here the linguistic terms such as *very bright, less bright, very dark*, etc. provide rich information about the properties of the gray levels of the objects and the background in the image. FIG. 1 gives the definitions of the fuzzy set "BRIGHTNESS" of the image gray levels in the universe of discourse of l_{min} to l_{max}. The fuzzy set "BRIGHTNESS" can be expressed as:

$$\text{"BRIGHTNESS"} = \text{"}VD\text{"} + \text{"}D\text{"} + \text{"}LD\text{"} + \text{"}LB\text{"} + \text{"}B\text{"} + \text{"}VB\text{"}$$

Where   VD - very dark,   D - dark,   LD - less dark,
             LB - less bright, B - bright, VB - very bright

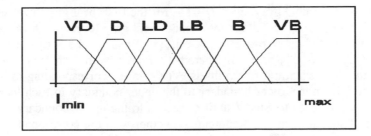

FIG. 1 Brightness fuzzy set & membership function

For most images to be analysed, the gray level regions for the objects can be expressed by fuzzy terms [*less bright, bright, very bright*] with respect to the background of the image. Let us define GL_1 to be the gray level of the first region of the objects, GL_2 to be the gray level of the second region of the objects, TS_1 to be the upper boundary for the first gray level region and TS_2 to be the upper boundary for the second gray level region. Thus, by incorporating the expertise knowledge of the image processing, a fuzzy logic rule base can be formulated as follows:

Rule 1: IF GL_1 is LB and GL_2 is VB   THEN TS_1 is LB and TS_2 is VB
Rule 2: IF GL_1 is LB and GL_2 is B   THEN TS_1 is LB and TS_2 is B
Rule 3: IF GL_1 is B and GL_2 is VB   THEN TS_1 is B   and TS_2 is VB
Rule 4: IF GL_1 is B and GL_2 is LB   THEN TS_1 is B   and TS_2 is LB
Rule 5: IF GL_1 is VB and GL_2 is LB   THEN TS_1 is VB and TS_2 is LB
Rule 6: IF GL_1 is VB and GL_2 is B   THEN TS_1 is VB and TS_2 is D

By manipulating the defined fuzzy rule base and the fuzzy membership function, the gray level boundaries for the regions, TS_1 and TS_2, can be inferred. The crisp values for the boundaries can be obtained either by using a look-up table method or by using the on-line defuzzifcation method. Thus, the first gray level region for the objects is of a range from l_{min} to [min(TS_1, TS_2)], while the second gray level region for the objects is of a range from [min(TS_1, TS_2)] to [max(TS_1, TS_2)].

An automatic thresholding procedure

The reliability of the threshold selections can be greatly enhanced by calculating the index of fuzziness in the local gray level regions. In a specified gray level region, the fuzzy property plane and the index of fuzziness is closely related to the window size of the S-function used. The selection of the window size is a typical process of trial and error. The adjustment of the window size is a time consuming task and is best carried out automatically. Thus, an automatic thresholding procedure can be formulated as follows:

Step 1. Describe the gray level regions of the image in linguistic terms and find the region boundaries TS_1 and TS_2

Step 2. Determine the local gray level search ranges: [l_{min}, min(TS_1, TS_2)] and [min(TS_1,TS_2), max(TS_1, TS_2)].

Step 3. Choose the initial window size w starting from small numbers (say w=5)

Step 4. Specify the initial gray level. $l_1 = l_{min}$ for the first local gray level search range, l_1 = min(TS_1, TS_2) for the second local gray level search range.

Step 5. Construct the fuzzy property plane $\mu_X(x_{mn})$ using a the S-function, that is, $\mu_X(x_{mn})$ = S(x_{mn}, a, b, c), where $b = l_i$ represents the ith gray level.

Step 6. Compute the amount of fuzziness, V(X), in μ_X corresponding to $b = l_i$.

Step 7. Vary l_i from the lower boundary to the upper boundary in each local gray level region, go back to Step 4 until l_i reaches to the upper boundary.

Step 8. Find the minimum V(X) computed together with the corresponding thresholds.

Step 9. In the case that at least more than one threshold are corresponding to the minimum V(X)=0, the selection of the window size failed. If the number of the selected thresholds l_t is *big*, increase the window size w in a *big* step; If the number of the selected thresholds l_t is *small*, increase the window size w in a *small* step. Then go back to Step 4 to repeat.

Step 10. In the case that the minimum V(X) is not equal to 0 and there is only one threshold corresponding to it, select $l_i = l_t$ for which V(X) is a minimum.
If the selected threshold is not *too close* the lower and the upper boundary of the local gray level search range, go to Step 11. Otherwise, decrease the window size *a little bit*, then go back to Step 4 to repeat.

Step 11. Try a few nearby window sizes (give the window size some disturbance) and choose the threshold l_t where V(X) is a minimum.

In the above procedure, the threshold is automatically selected by considering the index of the fuzziness under a given window size. The window size will be changed in the cases that either the computed minimum index of fuzziness is unreasonable or the selected threshold is too close to the boundaries of the local search range.

Using the above procedure, we have processed an image called "ORACLE" shown in FIG. 2.a in which a spacecraft, a few bright stars and different color clouds are present. If the blue clouds are taken as the background of the image, two gray level regions can be specified for the spacecraft, the stars and the gray clouds: the spacecraft and the stars are described as "*very bright*" and the gray clouds are described as "*less bright*". The first threshold is computed as $l_t = 87$ in the local search range from 50 to 130 for the gray clouds, and the second threshold is computed as $l_t = 189$ in the local search range from 130 to 230 for the spacecraft and the stars. The window sizes are computed as 30 for the first gray level region and 60 for the second gray level region.

FUZZY OPERATORS IN IMAGE PROCESSING

Several fuzzy operators have been used in low level image processing, which include: MAX, MIN, and INT. The fuzzy contrast intensification (INT) operator is taken as a tool for image thresholding and enhancements in the fuzzy property domain, while the fuzzy operators MIN and MAX have been combined to perform an enormous variety of digital image processing tasks such as edge detection, image skeltonization and smoothing.

INT and Image thresholding/enhancement

Recall that the fuzzy property plane of an image $X:\{x_{mn}\}$ is denoted by

$\mu_X:\{\mu_X(x_{mn})/x_{mn}\}$, where $\mu_X(x_{mn})$ denotes the grade of possessing some fuzzy property (e.g. brightness) by the (m,n)th pixel intensity x_{mn}. The fuzzy property plane for a given image can be constructed with the selected threshold and membership function. The image contrast can be enhanced by using the fuzzy operator INT on the fuzzy property plane. For the fuzzy set X, the contrast intensification operator (INT) generates another fuzzy set $X' = INT(X)$ with the membership function denoted by:

$$\mu_{X'}(x_{mn}) = \mu_{INT(X)}(x_{mn}) = \begin{cases} 2[\mu_X(x_{mn})]^2 & 0.0 \le \mu_X(x_{mn}) \le 0.5 \\ 1 - 2(1- \mu_X(x_{mn}))^2 & 0.5 \le \mu_X(x_{mn}) \le 1.0 \end{cases} \tag{7}$$

This operation reduces the fuzziness of the fuzzy set X by increasing the values of $\mu_X(x_{mn})$ which are above 0.5 and decreasing those which are below it. For the multimodal images with multithresholds and fuzzy property planes, the INT operation can be carried out on each fuzzy property plane which is defined in each local gray level region. Referring to eq.(5), eq.(6) and eq.(7), the pixel intensity x_{mn} after an INT operation can be denoted by:

$$x_{mn} = \begin{cases} (b - w/2) + w\ [\mu_X(x_{mn})/2]^{0.5} & 0.0 \le \mu_X(x_{mn}) \le 0.5 \\ (b + w/2) + w\ [(1 - \mu_X(x_{mn}))/2]^{0.5} & 0.5 \le \mu_X(x_{mn}) \le 1.0 \end{cases} \tag{8}$$

Using eq.(8), the INT operation can be easily implemented. The fuzzy thresholded image, which is also an enhanced version, for the "ORACLE" using the INT(X) operator based on the extracted fuzzy property planes is shown in FIG. 2.c, while the thresholded version using a threshold of 130 is shown in FIG. 2.b. It can be seen that the fuzzy thresholded version gives better thresholding and enhancement of the image without losing too much information.

The MIN and MAX operators and their opeartions

The fuzzy operators of MIN and MAX can be classified into point-wise functions (MIN_P, MAX_P) and local functions (MIN_L, MAX_L). The point-wise functions MIN_P and MAX_P, known as the fuzzy intersection and union, perform the operations at each point on the image. The local functions MIN_L and MAX_L, known as the minimum and maximum operation, change the gray level of the center pixel inside the specified mask which is moving on the image, that is, the MIN_L (MAX_L) replaces the digital value of the center pixel by the minimum (maximum) gray value of its neighbours and itself inside the specified window (say 3x3) in its immediate vicinity.

The MIN_L and MAX_L operators resemble shrinking and expanding operators. The MIN_L and MAX_L operations can be carried out either on the digital gray level image plane or on the fuzzy property planes of the image.

The edge detection of the gray level image can be achieved by taking the difference of the original image X and its shrunk version $MIN_L(X)$, that is:

$$FUZZY\ EDGES = X - MIN_L(X) \tag{9}$$

The strength of the detected edge at any point is proportional to the largest local digital value at that point in the original image. Using a 3x3 mask, the fuzzy edge detection has been carried out on "ORACLE" and the detected result is shown in FIG. 2.f, while the edge detections using Roberts and Sobel operators are shown in FIG. 2.d and FIG. 2.e. It can be found that the fuzzy approach provides a reliable edge detection for a multithreshold image since no edge has been lost due to weak strength. A visual comparision of the results with those obtained using Sobel or Roberts algorithm showed that the fuzzy approach performed better than Roberts and equally well as Sobel.

Image smoothing is often implemented to blur the image by attenuating the high spatial frequency components associated with edges and other abrupt changes in gray levels. This can be performed by employing the shrink operator followed by the expanding operator, that is,

$$\text{FUZZY SMOOTHING} = [MAX_L]^q [MIN_L]^q (X) \tag{10}$$

Where $q = 1,2,3,...N$. The higher the q, the greater is the degree of blurring.

SUMMARY

Gray level thresholding plays an important role in image processing. The use of thresholding as a tool in image enhancement and segmentation has remained an intensive research topic for a long time. A lot of effort has been made to find the "best" thresholds for distinguishing the desired features from the rest of the image, especially multithresholding which has attracted much more attention recently. The index of fuzziness has been used as a tool to select the thresholds and the window size of the S-function for an given image. The fuzzy property plane has been extracted by using the defined S-function. The extracted fuzzy property plane is closely related to the window size of the S-function and the gray level region in which the threshold is selected. However, identification of the gray level regions of an image is not an easy task. The proposed procedure to automatically select the optimum thresholds and adjust the window size of the S-function has been proven to be an effective tool to process the multimodal images, where the human intuitive descriptions of the properties of multimodal image are used to specify the gray level regions. Based on the selected thresholds and the constructed fuzzy property planes, image thresholding using the fuzzy operator INT has been performed. Experiments showed that the fuzzy thresholded version gives better thresholding and enhancement of the image without losing too much information.

Operations using the fuzzy operators MIN and MAX for edge detection and image smoothing have been examined. And the fuzzy edge detection for "ORACLE" using the operator MIN has been demonstrated. It has been found that the fuzzy approach provides a reliable edge detection for a multithreshold image since no edge has been lost due to weak strength. A visual comparision of the results with those obtained using Sobel or Roberts algorithm showed that the fuzzy approach performed better than Roberts and equally well as Sobel. The experimental results show that fuzzy operators provide an appealing approach for low level image processing. Further work is being carried out to investigate the computation efficiency of the proposed approach and the effectiveness of other fuzzy measures such as fuzzy entropy and fuzzy compactness which are not discussed in this paper.

a. original image

b. thresholded image

c. fuzzy thresholded image

d. roberts edge detection

e. sobel edge detection

f. fuzzy edge detection

FIG. 2   Image processing results for the ORACLE

APPENDIX A. THE INPUT IMAGE

```
012 013 014 015 016 017 017 018 019 012 015 016
014 015 018 019 018 013 014 010 010 010 210 110
010 010 010 210 210 210 210 210 210 010 010 010
010 010 010 255 210 210 220 210 210 010 010 010
010 010 010 210 250 150 220 170 210 110 010 010
010 010 010 210 210 190 250 210 210 110 010 010
010 010 010 210 230 210 250 210 210 110 010 010
010 010 010 210 210 180 210 210 210 110 010 010
010 010 010 190 010 010 010 010 110 110 010 010
010 210 011 012 010 010 010 010 110 110 010 010
010 110 010 010 010 010 010 010 110 110 010 010
010 010 010 010 010 010 010 010 010 010 010 010
```

This image can be interpreted as either bimodal image or three-modal image.

A). In the case of three modal, the gray level regions of the image can be classified as [10, 110] and [110, 255].

B). In the case of bimodal, the gray level region of the image is [10, 255] but the gray level of the object can be set as either 110 or 200.

REFERENCES

1. Zadeh LA: Fuzzy sets, Inform. & Contr. vol 8, 1965, pp338-353.
2. Zadeh LA: Outline of a new approach to the analysis complex systems and decision processes. IEEE Trans. Syst. Man & Cybern., vol SMC-3, 1973, pp28-44.
3. Self K: Design with fuzzy logic. IEEE Spectrum, 1990, pp42-44.
4. Pal SK, King RA: Image enhancement using smoothing with fuzzy sets. IEEE Trans. Syst. Man & Cybern., vol SMC-11, No.7, 1981, pp494-501.
5. Goetcherian V: From binary to gray tone image processing using fuzzy logic concepts. Pattern Recogn. vol 12, 1980, pp7-15.
6. Murphy CA, Pal SK: Fuzzy thresholding: mathematical framework, bound functions and weighted moving average technique. Pattern Recog. Lett., vol 11, 1990, pp197-206.
7. Nakagawa Y, Rosenfeld A: A note on the use of local min amd max operations in digital picture processing. IEEE Trans. Syst. Man & Cybern, vol SMC-8, No.8, 1978, pp632-635.
8. Pal SK: fuzzy skeletonization of an image. Pattern Recog. Lett., vol 10, 1989, pp17-23.
9. Pal SK, Rosenfeld A: Image enhancement and thresholding by optimization of fuzzy compactness. Pattern Recog. Lett., vol 7, 1988, pp 77-86.
10. Pal AK, Ghosh A: Index of are coverage of fuzzy image subsets and object extraction. Patt. Recog. Lett., vol 11, 1990, pp 831-841.
11. Li H, Yang HS: Fast and reliable image enhancement using fuzzy relaxation technique. Proc. of the 4th Int. Conf. Patt. Recog., U.K., 1988, pp 577-586.
12. Gonzalez RC, Wintz P: Digital image processing, Addison-Wesley, 1987

REAL-TIME IMAGE PROCESSING FOR OPTICAL INSPECTION OF ELECTROPLATED CONTACT SURFACES

[1]Paul Healy, [1]Kenneth Dawson, [2]Apostolos Kyritsis, and [2]Christos Kassiouras

[1]*Department of Computer Science, Trinity College, Dublin, Ireland*
[2]*Zenon, SA, Athens, Greece*

Abstract

A real time inspection system[1] has been developed to improve the yield of surface treated mechanical relay contacts by detecting defects at the micrometre level.

Relay contacts should have low electrical resistance, no adhesion and should not be impaired by electrical arcing. Meeting these specifications means that each contact surface must be inspected visually for defects. This visual control is expensive and subjective.

The inspection system uses a laser scanner to image the contact surfaces. Data streams from three detectors are fed into a multi-processor transputer system where real-time defect classification takes place. This process involves data reduction and both 1D and 2D signal correlation. The process engineer is presented with a real time display of defect statistics. Off-line visualization of defect information is also provided by the system.

1 OVERVIEW

1.1 Manufacturing Background

A mechanical relay contact production line manufactures up to 10 contacts per second of the type pictured in figure 1. Relay contacts are carried on a belt through an alloy coating process.

Full manual inspection of such output for defects is not feasible. Instead, currently a statistical off-line process is the best that can be used to monitor output quality. Since this can take place a number of hours after production, there is little scope for correcting the manufacturing process quickly.

---

[1]The research described in this paper has been supported by the Commission of the European Communities under Brite-EuRam Project 3076.

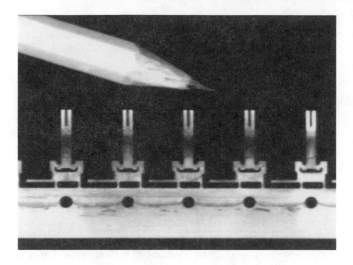

Figure 1: Photograph of relay contacts

A real time inspection system would allow continuous in-line monitoring of the quality of the plating system, and pave the way for the development of a closed loop control system. Towards this end a laser scanning system was combined with a fast image processor. Preliminary results are presented in this paper.

1.2   Image Acquisition

Three photodiodes collect light scattered from the beam of a laser which is scanned over the relay contacts in a horizontal direction. The belt carrying the contacts provides movement in the vertical plane.

1.3   Classification Rules

One dimensional correlation of the three input data streams gives defect location and classification. Peaks on signals can be used to identify edges, pits and protrusions. Discolourations can be identified by an intensity drops.

1.4   Principals of Image Processing

Image data is made available to the image processor in the form of digitized scanlines (one for each channel). In order to meet the resolution criteria each contact is digitized to a resolution of 1024 pixels per scan line (there are approximately 256 scan lines per contact).

One dimensional noise removal and data reduction is followed by a correlation process, which identifies edges, pits, protrusions, discolourations and burrs.

The task of processing incoming data from the optical system and classifying defects must be performed in real-time. This requirement has a significant impact on the design of both the hardware and software of the image processor.

In particular the throughput requirements mean that the task is significantly I/O bound at the interface to the optical system. This means the overall throughput of the inspection system will primarily depend on how fast raw data can be passed into the image processor, and once there, how quickly significant features can be segmented out of this data.

However once features have been identified and classification of defects begins, the system becomes CPU bound (i.e. the performance of the system at this stage will depend on how fast the chosen processor is).

Since there are three streams of data to be processed simultaneously for each side of the contact, there is significant scope to introduce parallelism into the processing of the incoming data.

In summary therefore the criteria for image processor hardware are:

† High bandwidth I/O capability.

† High performance processing.

† Support for parallel processing of multiple input streams.

1.5 Transputer Technology

The transputer is a 32 bit processor with both RISC and CISC features. Top of the range devices come with an on-chip FPU, and 4k of on-chip static memory [2].

Each transputer has four 20 Mbits/sec bidirectional DMA driven link engines which can be used to construct Multiple Instruction Multiple Data (MIMD) parallel processing networks.

The transputer has a well defined model for on-chip and off-chip parallel processing. A network of concurrent processes can be distributed over a network of processors in order to efficiently distribute I/O and processing requirements. The system can be programmed in a high level language, for example C [3].

Processes can be used to simulate the presence of hardware I/O devices. In this manner the different parts of a system can be built and tested in a modular manner.

The current generation of transputers (e.g. the INMOS T801) exhibit the following I/O characteristics:

† 60Mbytes/sec data transfer rate to and from external memory.

† 4 bidirectional 20Mbits/sec serial links, with a bidirectional data rate of 2.4Mbytes/sec per link.

† 630ns response to interrupts.

A 30MHz device can deliver 30MIPS (peak-transputer instructions), and 4.3Mflops peak. Use of internal (in-chip) static memory can significantly improve performance.

2 IMAGE ACQUISITION

The output from three analogue to digital converters are fed to transputer processors responsible for data reduction. Feature data is transferred to a classifier process, which is the next stage in the processing pipeline.

Raw image data from the laser scanner is shown in figures 2 and 3.

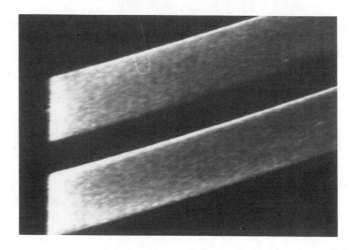

Figure 2: Raw data from a detector positioned to the left side of the scanning laser beam.

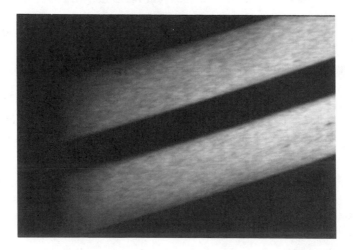

Figure 3: Raw data from a detector in the line of the laser beam.

3 IMAGE PROCESSING

Transputer technology was chosen as the processing platform in order that the module can be readily expanded to meet the throughput requirements of the overall system. For example if at any stage the data bandwidth exceeds the computational capacity of the processor on which it is being executed, the workload can be subdivided among a number of processors.

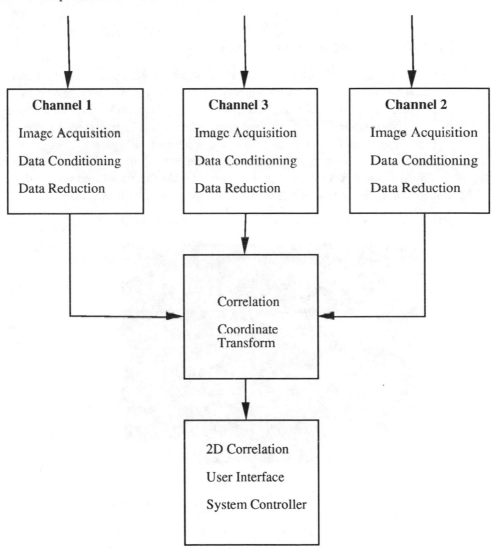

Figure 4: Outline of the prototype inspection system.

3.1 Data Reduction

The data reduction process is responsible for the extraction of the maximum intensity out of a conditioned data set of intensities, and for the evaluation of the actual position of the defect.

One data reduction process is run for each of the three input data streams. These three input streams of features are the input to the next stage of the pipeline (the classification process).

3.2 Feature Classification Procedures

3.2.1 Rule-Based Approach Classification is managed through the correlation of the three input data streams (fed from the three data reduction processes). Features are classified into categories such as edges and defects (e.g. pits, protrusions). Defect (pits, protrusions, discolourations and burrs) and edge information is then sent to the next stages of the process pipeline. Typical results are illustrated in Figure 5.

Edge information is sent to a process responsible for calculating edge orientation, and defect information is sent to a process responsible for transforming the defect coordinate information from a scan line reference frame to a contact reference frame.

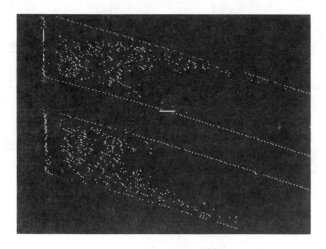

Figure 5: Typical results from the rule based correlation process.

3.2.2 Back-Propagation Neural Network The current classification process is quite restrictive from the point of view of ease of modification. Work is currently underway with the intention of building a back-propagation neural network classifier to replace it. This data driven approach will allow a more flexible approach to classifying defects.

3.3 Correlation in 2 Dimensions

The output of the feature classifier is a set of data units, where each unit corresponds to a suspected defect and contains:

† Contact identification number.

† Location of the defect point on the contact.

† Class of the suspected error.

† Characteristics such as length and intensity information.

3.3.1 Sorting of Input Stream Data units from the input stream arrive in such a way that information relating to one contact can be interlaced with data from another contact, i.e. data corresponding to neighbouring contacts can be mixed. This results from the fact that a single scanline can extend over two contacts.

The incoming data stream is therefore split and data units are sorted and buffered to separate queues, each queue corresponding to a single contact. The 2D correlation process can start when all data units corresponding to one contact are gathered together.

3.3.2 2D Correlation All data units corresponding to points of the same suspected defect are grouped together by this process. Each resulting group of points is identified as an area of a single potential defect.

Each group is then examined separately to decide if a defect is in fact present. In the case of a defect being found, its class is decided upon and its characteristics such as position and size are determined.

3.3.3 Analysis Stage The statistical distribution of defect size and position is computed. Various features of the contact can be examined; e.g. total number of defects, maximum (or average) defect size, most frequent defect class and areas of spatial concentration of defects.

3.3.4 Result Visualization Visualization of the results of contact analysis is essential for the human operator. The system provides an off-line tool for graphical representation of contacts where size, class and spatial distribution of defects are displayed.

The user interface also presents to the process engineer a real-time graphical representation of the temporal evolution of defect statistics.

4 SECOND GENERATION TRANSPUTERS

The T9000 is a second generation transputer featuring a pipelined superscaler micro-architecture. The CPU performance is expected to reach 200 MIPS peak and 25 MFLOPS peak. Link performance will increased to a bidirectional rate of 16Mbytes/sec. Other features will include enhanced support for resource sharing [4].

This development holds out the possibility of reducing hardware complexity, by reducing the number of component processors in the system, while at the same time it seems probable that increased throughput/performance will be available.

5 CURRENT STATUS

It has proven possible to identify defects using real data gathered in the laboratory. The next step will be to measure and improve the reliability of defect classification algorithms and to build on the current slow-speed prototype to produce a system capable of processing data at real rates.

With this in mind the second generation transputer device from INMOS (the T9000 when it becomes available) will hopefully be substituted for the current architecture for an expected I/O and processing gain of about 5-10 times.

6 SUMMARY

A slow speed prototype is currently working using real data. Current research is concerned with improving the quality of defect identifications and the overall speed of the system.

The inspection project described in this paper has involved a fusion of technical skills across a broad spectrum running from experiments with laser scanning apparatus through parallel processing systems and conventional image processing techniques.

The expected benefits include the production of a real time inspection system, with benefits forecast in the area of increased yield on the production line. Manufacturing engineers will be given a tool that allows real time display of the defects appearing on the production line.

Although outside the remit of the current project, progression to a closed-loop inspection/automated manufacturing process is possible.

7 Acknowledgements

The partners in project BE 3076 are: Siemens AG - Corporate Production and Logistics, responsible for image acquisition and signal conditioning; Trinity College Dublin, responsible for image preprocessing; and ZENON SA responsible for image postprocessing and the user interface.

The initial idea for the inspection system described in this paper came from Dr. R. Schneider of Siemens AG. Subsequently the overall system design is also due to Dr. R. Schneider (who leads the BE3076 Project), D. Spriegel (who was also responsible for providing raw data measured in the laboratory setup) of Siemens AG, Munich, and Dr. D. Vernon of Trinity College, Dublin.

8   References

[1] 'BE 3076 Edited Progress Report 1', September 1991.

[2] 'The Transputer Databook', INMOS (1989).

[3] 'Parallel C User Guide', 3L Ltd (1989).

[4] 'IMS T9000 Preliminary Information', INMOS (April 1991).

CAD MODEL GENERATION FOR PLANNER PARTS

Chien-Nan Huang and Saeid Motavalli

Department of Industrial Engineering, Wichita State University, Wichita, Kansas, U.S.A.

ABSTRACT

Creating a CAD database for a product is an essential part of the design process. In the automated manufacturing environment the CAD database is used to create process plan, production plan, and to analyze the design. In some design cases the CAD model of the product does not exist, such as redesigning an old part, or designing the product by building a clay model.

This paper presents a machine vision system capable of creating a contour CAD drawing for various flat sheet metal components typical in aerospace industry. The process of image acquisition is accomplished by the integration of a CNC machine and a CCD line scan camera. The camera is setup to acquire an image of the part with the field of view of the camera covering only a small portion of the part surface. The complete surface of the part is scanned by combining several images. An algorithm is developed to create a line drawing of the whole surface of the scanned part. The developed algorithms and the system setup are discussed and sample results are detailed.

INTRODUCTION

Engineering drawing is the basis for the design of a product, the creation of the process plan, and engineering analysis of the product. The creation of an engineering drawing is a fundamental task in the product design process. An engineering drawing is created either from scratch or from a prototype based on the functional requirements for the product. Therefore, the engineering drawing must be carefully evaluated and modified to meet both functional and manufacturing requirements of the product. The task of refining the design of a product is a repetitive and crucial procedure which can have a large impact on the production cost.

In the current engineering environment, engineering drawings are created and modified in the Computer Aided Design (CAD) systems, and the CAD database is a direct input to the CAPP (Computer Aided Process Planning) systems or to the numerically-controlled tool path generation softwares. Often, it is necessary to create a CAD model for an existing part, and refabricate the part from the created CAD drawing i.e., where a conceptual prototype is physically built before the drawing is created, or when the design of an existing part must be modified for which a CAD drawing does not exist. Extensive effort is usually required to create a CAD model in order to accomplish this reverse engineering process.

The problem is more obvious in the aerospace industry. Thousands of sheet metal parts are used to assemble an airplane, and they often need to be reproduced or modified. In the cases where an older airplane has to be redesigned, the CAD drawings from the existing parts are often created by electronic digitizers or CMM (Coordinate Measurement Machine), following the contours of the parts. The procedure of using digitizing tablets to trace the contour of the part is time consuming and is a potential source of error. In addition, the followed path is not retractable, thus the digitizing procedure should be completed in one shot. Coordinate Measuring Machines can be used to digitize the part surfaces, but this procedure has the drawback of high cost and long digitizing times. Both of these procedures require an operator to follow the surface contour manually which is not acceptable in an automated manufacturing environment.

Recent advances in machine vision technology have triggered a number of new applications for this technology in the manufacturing environment. The main applications include: inspection and measurement, location analysis, part recognition, and robot guidance. Recently, machine vision has been successfully applied to the creation of CAD model for existing parts in the form of an engineering drawing [1]. This approach has opened up a new alternative to part digitization and the generation of CAD models for existing parts.

This paper describes the developments of a vision system integrated with a CNC machine for the creation of CAD models for a flat part in a reverse engineering environment. The developed algorithms for CAD model generation is presented and sample results are illustrated. The developed system has the capability of creating CAD model for different sizes and shapes of planner parts with acceptable accuracy.

BACKGROUND

The developed CAD model generation system uses machine vision technology. The process involves the extraction of geometric data from an image of the part, the interpretation of the extracted data and the representation of the created image in an existing CAD system. The procedure consists of three basic steps; thresholding, image feature description, and representation (curve fitting).

The first step, thresholding, is used to separate the object from the background. Thresholding is a simple technique for part recognition or object boundary extraction. Various automated threshold selection techniques have been introduced in the literature. Most thresholding techniques involve the analysis of the image histogram, which is the representation of the number of pixels with different gray levels [2-5]. In these techniques, the histogram of the image is constructed. If the objects in the image have a different gray level from an uniform background, the histogram of the gray levels would be bimodal, with two peaks and a valley. A threshold value selected at the gray level at the valley of the histogram, which separates the two peaks, can separate the object from the background.

Another method for the selection of a threshold level is associated with the use of statistical analysis of the image gray levels [6-8]. These threshold selection methods involve the statistical analysis of the distribution of the grey levels in the image without interpretation of the histogram.

The next step in the process of CAD model generation is the extraction and description of the features of the part. Various feature description methods have been suggested. Two common methods are boundary following and chain codes [9].

Boundary following techniques consist of the tracing the points on the boundaries of the object. The boundary points are defined as a series of connected points on the object edges in the binary image. Two kinds of boundary points are defined: four-connected and eight-connected. An eight connected boundary point is defined as a point connected to at least one of its eight pixel in its 3x3 neighborhood where at least one of the neighboring pixels belongs to the background.

Chain codes is also used to represent the boundary by a connected sequence of points. In chain codes, the boundary points are defined with four or eight connectivity. In order to reduce the number of codes and avoid noise or imperfect segmentation, a larger grid space than the space between neighboring pixels is usually used for the coding process. Therefore, only a starting point is chosen as its original location; the other points on the boundary are replaced by the predefined successive displacement grid points along the boundary.

The last procedure involved in the CAD model development is the image representation. A sequence of boundary points are represented using straight lines or curves. The boundary points extracted from the image are not sometimes the true positions of the object boundary. Noise involved in the image processing operation such as the effect of lighting conditions creates inaccuracies in the process. Parametric curves are usually used to represent a sequence of points extracted from the image. The advantage of these curves is that they smooth the fluctuation in the acquired data. The most often used polynomial curves are cubic splines [10].

SYSTEM CONFIGURATION

The workstation used in the system includes a microcomputer and a CNC milling machine with its associated control system and software. The microcomputer is equipped with a high-resolution frame grabber, a video camera, and two high-resolution monitors: a monochrome video monitor and a color VGA monitor. The solid state video camera is a Charge Coupled Device (CCD) with a macro len. The PC controls the camera which is mounted on the CNC

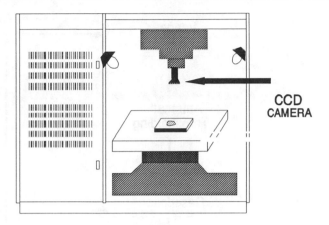

FIG. 1 The position of the camera and lights

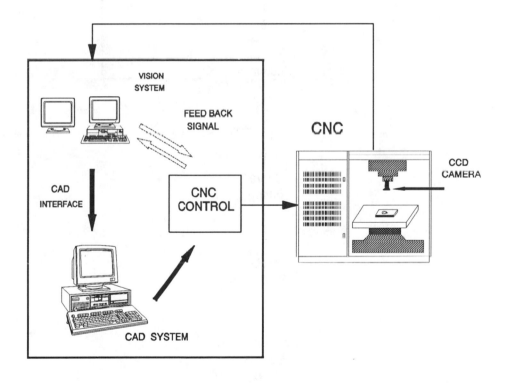

FIG. 2 The setup of the system

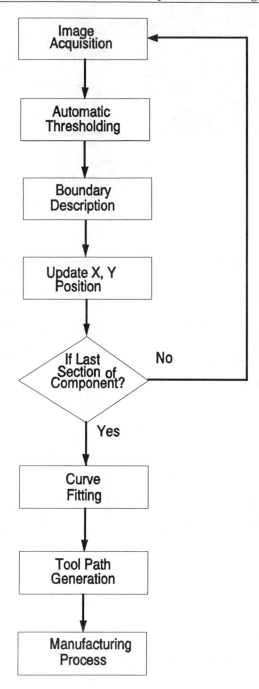

FIG. 3 The operation steps in the system

spindle. The camera is mounted on the CNC spindle to provide accurate camera positioning and to enable the scanning of larger-size parts. The CNC machine is programmed such that it relocates the part after each image capture and the complete part surface is scanned with a sequence of image captures. A light box with uniform lighting is mounted on the CNC work table, and the object is fixed on the light box.

In addition to backlighting, two incandescent light sources were also used. The positioning of the lights and the camera are shown in Figure 1, and the system setup is illustrated in Figure 2. An appropriate period of dwell time between each table movement is predefined to allow enough time for image acquisition and processing. The size of the field of view and the number of movements is dependent on the size of the part, and the required accuracy. The part surface is scanned in a row by row format from left to right.

ALGORITHM DEVELOPMENT

Figure 3 illustrates the steps involved in the operation of the system. The four major steps in developing the CAD model generation algorithm are: image acquisition, automated thresholding, boundary description, and curve fitting. They are described in the following sections:

Image Acquisition

An image of a flat object is acquired using a CCD camera. The analog image grabbed by the camera is digitized and stored in the frame buffer. The object is lighted with an evenly illuminated light box and two side lights. The field of view of the camera is selected relatively small (2x2") to obtain greater accuracy. Therefore, for larger objects each image covers only a small portion of the object boundary. The CNC machine is then used to move the camera sequentially to cover the whole surface of the part.

Image Thresholding

The thresholding technique used is based on the work by Tsai [12]. The method is based on the moment-preserving principle which assumes the moments of the original image are equal to the moments of the thresholded image. Moments are computed from the histogram of the image, using the following relation:

$$m_i = \sum_j p_j (z_j)^i$$

where n_j is the total number of the pixels in the image with grey value z_j, and p_j = n_j/n, and n is the total number of pixels in the image. The pixels of the image are then classified into two categories: above-threshold and below-threshold, and two categories are represented by z_0 and z_1. The z_j is replaced by z_0 or z_1 and the moments of these two categories are set equal as follow:

$$m_i = p_0 (z_0)^i + p_1 (z_1)^i$$

where i = 1,2,3, and $p_0 + p_1$ = 1. After solving the above equations, we can choose a threshold value closest to the p_0-tale of the histogram of the image. This thresholding technique can be extended for the situation that the background and the objects are not easily distinguishable.

Boundary Description

The boundary description methods, such as boundary following and chain codes techniques can be only successful in tracing along the close contours. These methods are not capable of reliably extracting the points on a portion of the object boundary. For large objects, each image covers only a portion of the object boundary, thus we are not dealing with closed boundaries.

The developed boundary description algorithm extracts the boundary points based on the examination of the four-neighborhood pixels of any point on the binary image. After the image is thresholded, the boundary of the object in the binary image is considered as the connection of the black pixels (object points) with at least one white pixel in their four connected neighborhood. The boundary points are scanned from left to right and top to bottom. If a boundary point is found, its X and Y coordinates are written to a data file. The coordinates are given in terms of the position of the points with respect to the world coordinate system. The translation of coordinates from pixels to inches are performed using a calibration process which utilizes a grid.

The appropriate table displacements are assigned as input to the software to update the X, Y coordinates with respect to the world coordinate system. This displacement is a factor of the chosen field of view which is selected with respect to the required accuracy. The X, Y coordinates acquired from the consecutive images are then appended into the same data file.

Curve Fitting

The extracted boundary coordinate data is used as input to a curve fitting process to be used for the generation of NC tool path. We use parametric cubic splines to describe the boundary of the object. The advantage of fitting cubic splines is that they tend to smooth the input data by approximating the boundary by a piecewise smooth curve. Thus they are capable of removing noisy input data. The boundary coordinates acquired in the previous step are stored in a data file. An interface is developed which converts the X,Y coordinates to be used as input to MASTERCAM, a CAD/CAM software capable of creating NC codes and also fitting spline curves to the input data. The input data to the MASTERCAM is presented as a cloud of points. The MASTERCAM is then used to fit spline curves to the acquired boundary point. The selected contours are automatically closed after the first point, the last point and the contour direction is specified. The contour is then processed by the MASTERCAM and the NC tool path program is generated.

MANUFACTURING PROCESS

Once the contours have been established, tool path generation module of the MASTERCAM is used to generate the tool path for the contours. Based on the generated tool path and machining parameters, input by the user, the sequence of NC program code is created and downloaded to a CNC milling machine.

To synchronize the image capture, a predefined time interval is allocated between the consecutive image captures. This time interval is determined by the required time for the NC machine to move the part and the image analysis time. The system can be equipped with a feed back control to send a signal to the vision system to capture the next image when the movement of the part has been completed. The final step in the process is to reproduce the part based on the created NC program.

RESULTS

A sample sheet metal part with circular shape and three internal holes was selected to demonstrate the performance of the system. For this sample part, it was determined that three movement and four image acquisition have to be performed to cover the whole surface of the part with reasonable accuracy. The number of movements is dependent on the size of the part and the required accuracy.

The four images acquired in the previous step were thresholded to binary images by the moment preserving method. The threshold value for this particular application was selected as 157 (0 for black and 255 for white). Based on the developed algorithm, the boundary points were extracted by determining four connected neighborhoods. The coordinate data are stored in a file in ASCII format. Along the sequence of image capture and processing, the boundary points extracted from each binary image were appended to the created data file. The stored data were then retrieved by MASTERCAM and displayed on the monitor. There were four contours on the test part, therefore four spline curves were passed through the corresponding contour points. Figure 4 displays the splines passing through the cloud of boundary points.

SUMMARY

The results obtained from the developed machine vision system, clearly demonstrate that machine vision can be used as a CAD model generation tool for implementation in the manufacturing environment. Especially, the developed system is very helpful in the design modification stage. The system is capable of scanning large complex shape parts. The use of indexing table would be more appropriate for the positioning of the part. For Large flat components such as in aerospace industry, this system can be used as an alternative for the creation of CAD models.

With the aid of structured lighting, the 3-D capability can be added to the system. The accuracy of the system depends mainly on the resolution of the camera, the chosen field of view, and the appropriate illumination. Even a small region of shadows in the image may cause errors in the digitization process. Therefore, controlled lighting is essential to the accuracy of the acquired results.

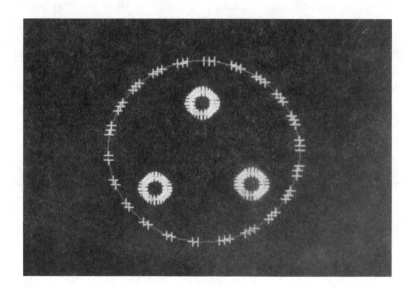

FIG 4. The CAD model of the planner part

REFERENCES

1.　Motavalli S, Bidanda B: A part image reconstruction system for reverse engineering of design modifications. *Journal of Manufacturing Systems*, 10:

pp 383-395, 1991.

2. Weszka J: A survey of threshold selection techniques. *Comput. Vision, Graphics, and Image Process.*, 7: pp 259-265, 1979.

3. Otsu N: A threshold selection method from gray-level histograms. IEEE Trans. Systems, Man and Cybern., SMC-9.: pp 62-66, 1979.

4. Weszka J, Rosenfeld A: Histogram modification for threshold selection. *IEEE Trans. Systems, Man and Cybern., SMC-9.*: pp 38-52, 1979.

5. Pun T: A new method for grey level picture thresholding using the entropy of the histogram. *Signal Process.*, 2: pp 223-237, 1980.

6. Deravi F, Pal SK: Grey level thresholding using second-order statistics. *Pattern Recognition Lett.*, 1: pp 417-422, 1983.

7. Kittler J et al.: Threshold selection based on a simple image statistic. *Comput. Vision, Graphics, and Image Process.*, 30: pp 125-147, 1985.

8. Rosenfeld A, Smith R: Thresholding using relaxation. *IEEE Trans. Pattern Anal. Mach. Intell.*, PAMI-3.: pp 598-606, 1981.

9. Ballard, DH, Brown, CM: Computer Vision, Prentice-Hall, Englewood Cliffs, New Jersey, 1982.

10. Shirai Y: Three-Dimensional Computer Vision, Springer-Verlag, New York, 1987.

11. Parvin, B, Medione G: Adaptive multiscale feature extraction from range data. *Comput. Vision, Graphics, and Image Process.*, 45: pp 346-356, 1989.

12. Tsai WH: Moment-preserving thresholding: a new approach. *Comput. Vision, Graphics, and Image Process.*, 29: pp 377-393, 1985.

13. Nagao, M, Matsuyama T: Edge preserving smoothing. *Comput. Graphics and Image Process.*, 9: pp 394-407, 1979.

OPTIMIZATION OF VAPOR PHASE REFLOW PARAMETERS DURING THE MANUFACTURE OF FILTER PIN COMPONENTS

Olagunju Oyeleye and El-Amine Lehtihet

Department of Industrial Engineering, Pennsylvania State University, University Park, Pennsylvania, U.S.A.

ABSTRACT

This paper describes the optimization of vapor phase soldering parameters during the manufacture of filter pin components. Defective parts can be produced due to improper soldering conditions or a drift in vapor temperature over time. These defects lead to the rejection of components and ultimately an increase in the final cost of assembled components. It is therefore important to optimize vapor phase soldering conditions by minimizing the occurrence of defects. Experiments were performed to select optimal soldering conditions for Sn-96 solder preform and obtain a vapor phase fluid/blending replacement strategy to control vapor temperature drift over time. Optimal soldering conditions were determined in two stages. A pilot factorial experiment was used to select important soldering parameters. Additional experiments were performed on several levels of soldering temperature and time and three different optimization techniques (minimax, average sum of defects, and graphical technique) were used to select the optimal vapor phase soldering condition. To maintain peak vapor fluid temperature, a fluid blending strategy was developed by obtaining the relationship between blending conditions, soldering temperature, and the drift of temperature over time. Results show that soldering temperature and time are the only two parameters that significantly affect the occurrence of defects. In addition, the three optimization techniques produce an identical optimal soldering condition. Furthermore, a linear relationship between the percentage of blended fluids and desired vapor temperature is obtained.

1. INTRODUCTION

Filter pins consist of a suitably plated ceramic element soldered to a pin and are used in circuits for electromagnetic interference control. During their manufacture, the soldering process must be tightly controlled in order to provide a high reliability part. In addition, components must have an adequate solder fillet to provide proper frequency response for the type of filtering provided, eliminate any solder creepage up the leads which would interfere with the mating interface of the components, and be produced via a mass production technique (such as the vapor phase process) to supply components at a competitive price. In using the vapor phase soldering process, defective parts can be produced if soldering conditions are inadequate or if a drift in peak temperature of the vapor occurs. It is therefore important to find optimal strategies to address both of these problems so that the occurrence of defective components is minimized. By optimizing soldering conditions the cost of manufacturing these components will be minimized.

In this paper, optimal vapor phase soldering conditions are determined by minimizing the occurrence of defects during filter pin manufacture. Possible defect causes were identified from the literature and experiments designed to isolate responsible variables. Optimization techniques were then utilized in obtaining soldering conditions. Finally, an optimal fluid blending/replacement strategy was developed.

2. DESCRIPTION OF DEFECTS

A variety of defects can be produced during the manufacture of ceramic filter pins. These defects include:

1. Voids around leads. Voids greater than 25% are unacceptable.
2. Solder bonded to leads and capacitors. All solder joints must be soldered a minimum of 270° around.
3. Excessive solder wicking up leads. Maximum wicking allowed 0.075".

Figure 1. provides an illustration of each of these defects.

3. DETERMINATION OF OPTIMAL REFLOW PARAMETERS

Optimal Vapor phase soldering parameters were determined in three stages. These stages involved:

1. The identification of factors that can cause defect variation during soldering.
2. The use of statistical experimental design techniques in obtaining important reflow parameters.
3. The use of optimization techniques to obtain optimal reflow parameters.

3.1 Factors that Contribute to the Occurrence of Soldering Defects

Several factors affect the quality of a soldered joint. These factors include:
1. The nature of the substrate surfaces involved.

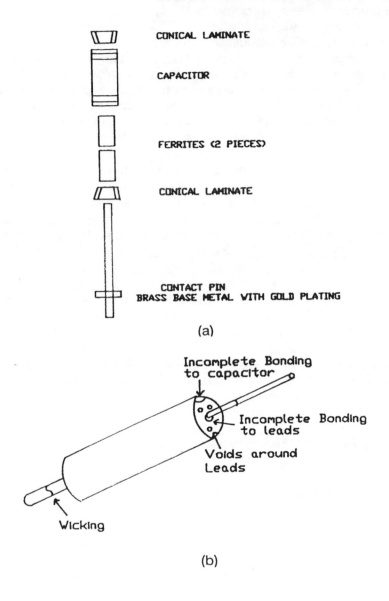

Figure 1 (a) Filter pin assembly   (b) Filter pin defects

2. Composition and purity of the solder alloy.
3. Activity of the flux used [1,2,3].
4. Fixturing of the components [4].
5. Temperature profile of the filter pin [5,6,7].

Presolder cleaning of the substrate (contact pin surface) should be done to avoid the problem of performing experiments on substrates of unknown quality. Contaminants reduce solderability and can exist on the surface to be soldered in the form of oxidation, grease, oil, paint, pencil markings, atmospheric dirt, and some other forms. An ultrasonic cleaner was used to clean all substrates prior to soldering.

Composition and purity of the solder alloy can also affect the quality of the joint produced. If the composition of the solder changes, properties such as solder melting point are affected. This can influence the formation of defects such as wicking (temperature too high), or cold joints (temperature too low). In addition, the purity of the solder can affect parameters such as void formation, and the bonding to the leads and/or capacitor terminations. Since no control could be excersized over this parameter, its effect was randomized in the experiment.

Flux activity also plays an important role during the soldering of filter pins. If the flux is not in sufficient quantity at the soldered joint, or the flux is not active enough, this could affect the quality of the soldered joint. Since the flux and solder were received prepackaged in the solder preform, there was no direct way to influence the activity of the flux.

The type of fixture used also plays an important role in the final quality of the soldered joint. Some criteria used in designing the fixture include the following: required joint spacings and positioning must be maintained throughout a soldering cycle, fixture should not act as a heat sink, it should not be easily wetted, it should not oxidize significantly or change dimensions over long periods of use, it should not interfere with the formation of the soldered joint, it should be easy to load and unload, and it should not allow vibration of the component.

The temperature profile of the filter pin also affects the quality of the soldered joint. The profile of a filter pin in a vapor reflow oven consists of:
1. The preheat temperature.
2. The preheat time.
3. The soldering temperature.
4. The soldering time.

3.2 Experimental Design for the Determination of Important Reflow Parameters

The basic philosophy behind the selection of an optimal set of reflow parameters is to select a set of parameters, (i.e. any combination of soldering temperature, soldering time, preheat temperature, and preheat time) such that the occurrence of defects within a soldered joint will be minimized.

Testing various combinations of each reflow parameter at many levels is time consuming and very expensive. If any of the reflow parameters does not influence the occurrence of defects then it is unnecessary to vary that parameter. Consequently a pilot study, using a 3^3 factorial design was done to determine the important reflow parameters [8]. The design consisted of 3 levels of preheat temperature (232, 329, & 426°C), preheat time (20, 60, & 100s), and soldering time (15, 30, & 45s). This study investigated the influence of preheat temperature, preheat time, and soldering time on defect occurrence. Soldering temperature was not varied at this stage because experimentation with the infrared reflow process indicated that it was important [9] and due to the high costs involved in varying this parameter.

Results from the initial study indicate that only soldering time and temperature affect the occurrence of solder defects.

3.3. Optimal Reflow Parameters

3.3.1 Experimental Setup for the Determination of Optimal Reflow Parameters

The vapor phase reflow machine has the capability to control reflow time from 0-50s [10]. For the Sn-96, Ag-4 solder preform, with a melting point of 221°C [1], two fluids with boiling temperatures of 215°C & 240°C were blended together in various proportions to achieve varying boiling temperatures.

Due to the cost of the vapor, the highest blended temperature attained was 234°C. The minimum amount of time to successfully solder a joint at this temperature was 10s. Using the above information, a 2 factor experiment with 9 levels of time (10 - 50s at 5s intervals) and 6 levels of temperature (224-234°C at 2°C intervals) was developed. A summary of results obtained is provided in Table 1.

TABLE 1
Summary of Results for Effect of optimal Variables on Defect Occurrence

Defect	Cause	Maximum Allowed	Maximum Value Obtained
Wicking	Soldering time	0.075"	0.0506"
Voids around leads	Unexplained from experimental parameters used	25%	6.05%
Solder not bonded to leads	Unexplained from experimental parameters used	90°	360°
Solder not bonded to capacitor	Fixture	90°	360°

TABLE 1
Continued

Defect	Minimum value obtained	Mean value obtained	Contribution
Wicking	0.00"	0.02645"	Increase in soldering temperature and time increases wicking
Voids around leads	0%	0.0292%	Voids occurred randomly with respect to parameters
Solder not bonded to leads	0°	37.14°	Increase in soldering temperature and time reduces unbonded portion to leads
Solder not bonded to capacitor	0°	37.85°	Increase in soldering temperature and time reduces unbonded portion to capacitor

3.3.2 Determination of Optimum Vapor Phase Reflow Parameters   Optimum vapor phase reflow parameters are determined by selecting a set of parameters that produce the minimum amount of defects. In selecting the optimal point, the following assumptions were made:

1. Each defect is equally important.
2. Any point at which a defect exceeds its maximum value is eliminated. This criteria translates to:

    0" < wicking <= 0.075"
    unbonded lead or capacitor <= 90°
    voids around leads <= 25%.

3. The ideal soldered joint should have:

    360° bonding of solder to leads and capacitor
    0% voids around leads
    and 0.0375" (0.075"/2) of wicking on the joint.

4. In order to compare defects, the same measurement scale should be used for all defects. Consequently, all defect measurements were converted to percentages:

    Bonding to leads and capacitors = (unbonded portion)/ 360°
    Percentage void area was used as determined
    Wicking = ABS[(wicking - 0.0375")/0.0375"].

Using the above assumptions, 2 different optimization techniques were used. The first method involved minimizing the average set of defects at each level of temperature and time. The second approach involved selecting the maximum value of each defect at each level of soldering temperature and time and then finding the minimum of the entire set [11].

These strategies are formulated below.

Given:

    L = Percentage of solder bonded to leads
    C = Percentage of solder bonded to capacitor
    V = Percentage of voids formed around leads
    W = Absolute value of (Wicking - 1/2 of maximum allowed wicking) / 1/2 of
        maximum allowed wicking.
    N = Number of defect measurements = 4

        Minimization of Average sum of defects   (1)

Minimize $\{(L + C + V + W)/4\}$

Subject to:
        L <= 25%
        C <= 25%
        V <= 25%
        W <= 100%

        Minimization of Maximum value of defects   (2)

Minimize (Maximum [L, C, V, W])

Subject to:
        L <= 25%
        C <= 25%
        V <= 25%
        W <= 100%

Analysis of the results shows optimum soldering parameters to be 232°C and 40s. Histograms of these results are also shown in figure 2. These histograms indicate that using both the average sum and minimax procedures, the optimum soldering conditions are at 232°C and 40s.

4. VAPOR FLUID BLENDING/REPLACEMENT STRATEGY

The vapor phase process utilizes specially formulated fluids to produce a vapor phase necessary for component reflow. Occasionally, two different vapor fluids are blended together to achieve a boiling temperature that cannot be achieved by a single vapor fluid. Over a period of time, the more volatile fluid tends to progressively evaporate. This results in a progressive drift or change in peak

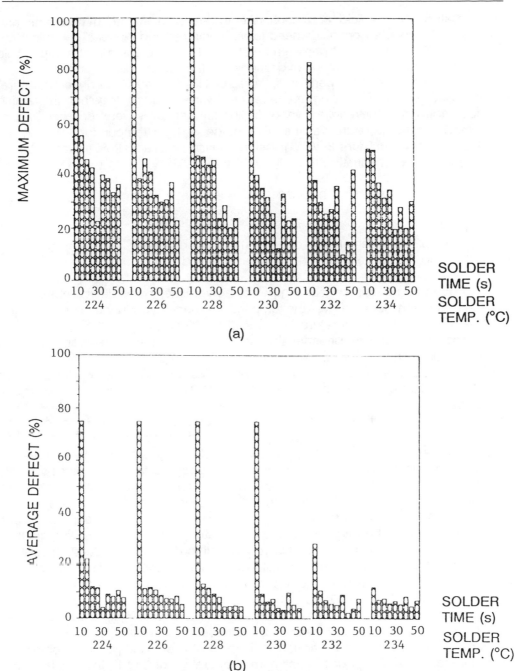

Figure 2 (a) Histogram showing the variation of maximum defect value at each
soldering point at different soldering temperatures and time.
(b) Histogram showing the variation of average defect value at each
soldering point at different soldering temperatures and time.

temperature of the fluid and consequently leads to soldering problems and defects. For this reason, fluids need to be monitored and blended periodically to maintain a constant, ideal peak temperature. The development of a vapor fluid blending/replacement strategy is described in the rest of this section.

To achieve various temperatures, two fluids HS240 (240°C boiling temperature) and FC5312 (219°C boiling temperature) were blended together in varying percentages, and the resultant vapor boiling temperatures were recorded. A plot showing the temperature vs. the percentage of high temperature fluid in the mixture is shown in figure 3. A regression equation with a 96.96% fit was obtained and is provided in equation 3.

$$\%HS240 = Temp*4.5 - 966.333 \qquad (3)$$

where:
   %HS240 = percentage of 240°C fluid in the mixture
   temp = fluid temperature between 219 and 240°C.

To observe the drift in peak temperature over time, it was decided to observe variations in temperature over the temperature range used in the experiment (i.e. 224-234°C) in two temperature intervals (224-228°C and 230-234°C). This was necessary because to observe the drift in fluid temperature over the entire temperature range would have required 18 liters of fluid. This quantity of fluid would have come in contact with the workpiece. By using the two temperature ranges mentioned above, a starting amount of 9.5 liters of fluid was required in both cases.

To obtain the drift over time, the optimum vapor soldering conditions obtained in the last section were used. These conditions included a soldering time of 40s, soldering temperature of 232°C, preheat time of 60s, and preheat temperature of 625°F. In addition, because it took 15 minutes for the fixture to return to room temperature, a time interval of 15 minutes between each vapor cycle was used. Figure 4 shows the drift of temperature over time in hours using the temperature ranges 224-228°C and 230-234°C. These plots indicate, as expected, that the drift in temperature at the lower end of the temperature range is a lot more rapid than at higher temperatures. This plot is very useful to determine a blending strategy since it indicates that, for example, fluids blended at 224°C would have to be replenished every 1.5 hrs., those blended at 227°C every 9.5 hrs, and those blended at 232°C every 17 hrs.

Using figure 3 and its corresponding regression equation, it can be calculated that in order to maintain a constant vapor temperature, 4.5% by volume of the low temperature fluid should be added at the end of every drift period to obtain the original temperature. For the optimum temperature of 232°C, this implies that if the starting fluid volume was 8 liters, at the end of 17 hrs, 4.5% of 8 liters (i.e. 0.36 liters) by volume of the low temperature FC5312 fluid should be added.

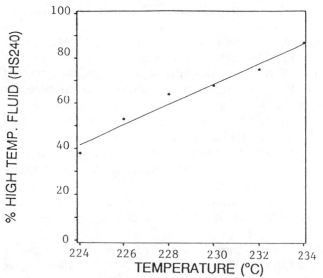

Figure 3 Percentage of 240°C boiling point fluid in 240/219°C fluid mix plotted against the boiling temperature of the vapor.

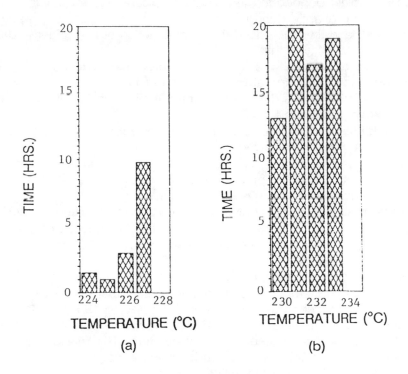

Figure 4 Drift of temperature over time
(a) 224-228°C range   (b) 230-234°C range

9. CONCLUSION

This paper describes the optimization of vapor phase soldering parameters during the manufacture of filter pin components.

Preliminary results indicated that of all the vapor phase parameters that can be varied, i.e. (soldering time, soldering temperature, preheat time, and preheat temperature), only the soldering time and temperature had an impact on the occurrence of defects. In addition, close scrutiny of the soldered joints revealed that if the joints were not properly held in place by the fixture, there is a possibility of defects occurring due to component movement during soldering.

An optimal set of vapor phase parameters cannot be determined if the parameters used have no significant effect on the quantity of defects formed. Consequently, the soldering temperature and time were used in finding an optimum set of parameters. The effect of each of these parameters on defectssi discussed below:

   a. Void formation around leads occurred randomly across the experimental range. However, the maximum amount of void area formed was 6.05% and this defect occurred in only 2 of the 108 samples made.

   b. As the soldering temperature and time are varied, the amount of solder unbonded to the capacitor reduced.

   c. Results of solder bonded to the capacitor indicated that even if the solder was displaced slightly during its manufacture, an increase in soldering temperature and time increased the wetting action of the solder and thus reduced the quantity of this defect.

   d. The amount of wicking was found to be proportional to an increase in either the soldering temperature or soldering time.

   e. Defects soldered at 10s did not produce any bonding for all temperature values except at 232 and 234°C.

   f. Joints soldered at 224°C had a rough and grainy appearance and did not appear to be properly reflowed.

Using a minimax, average sum of defects, and graphical approach, the optimal vapor phase parameters at 625°F preheat temperature and 60s preheat time was 232°C soldering temperature and 40s soldering time.

Vapor fluids that are blended together to achieve a different blending temperature from either of the two primary fluids tend to have a drift in peak soldering temperature of the fluid over time. Results from the blending of a 240°C and 219°C fluid indicated that:

   a. As the peak temperature of the blended product is increased the amount of time for the fluid temperatures to drift for this temperature is significantly reduced. These times vary from 1.5 hrs. at 224°C to 19 hrs. at 232°C.

   b. A relationship was obtained between the percentage of high temperature (HS240) fluid and the peak soldering temperature. This relationship is only valid within the 219-240°C range and is given by
   $$\%HS240 = TEMP*4.5 - 966.333.$$

   c. The time taken for the peak temperature to drift in hours was obtained over a temperature range of 224-234°C. Also the optimal amount of low temperature (219°C) fluid that is required to be blended periodically with the fluid to determine temperature drifting was determined to be 4.5% of the original fluid volume. In the case of the optimal vapor temperature this quantity of fluid would be added every 17 hrs.

REFERENCES

1. Manko, HH: *Solders and Soldering*, 2nd Edition, New York, Mcgraw Hill Book Company, 1979.
2. Manko, HH: *Soldering Handbook for Printed Circuits and Surface Mounting*, New York, Van Nostrand Reinhold Company, 1986.
3. Woodgate, RW: *The Handbook of Machine Soldering*, 2nd Edition, New York, John Wiley and Sons, 1988.
4. Thwaites CJ: *Capillary Joining-Brazing and Soft Soldering*, England, Research Studies Press Division of John Wiley and Sons, 1982.
5. Hinch, SW: *Handbook of Surface Mount Technology*, New York, Longman Scientific & Technical 1988.
6. Markstein, HW: SMT Reflow: IR or Vapor Phase ?, *Electronic Packaging and Production*, Vol. 27 #1, Jan 1987, pp. 60-63.
7. Hutchins, CL: Optimizing the Vapor Phase and IR Reflow Process, *Electronic Packaging and Production*, Vol 28 # 2, February 1988 pp. 106-108.
8. Montgomery, DC: *Design and Analysis of Experiments*, 2nd Edition, New York, John Wiley & Sons Inc. 1984.
9. Oyeleye OO & Lehtihet EA: *Midyear Report on the Manufacture of Filter Pin Components*, March 7th 1989.
10. *Centech VP-1000 Operations Manual*, 1989.
11. Gabler, S: *Minimax Solutions in Sampling from Finite Populations*, Germany Springer-Verlag, 1990, pp 13.

A VISUAL INSPECTION SENSOR PLATFORM — VISP

James A. Mahon and Sean M. O'Neill

*Department of Computer Science, Vision and Sensor Research Unit,
Trinity College, Dublin, Ireland*

ABSTRACT

This paper describes a PCB inspection system called VISP (Visual Inspection Sensor Platform). This is a low cost, high speed system for inspecting surface mount printed circuit boards and other objects. It is highly flexible using off the shelf components to minimize cost and design effort. The system consists of a core mechanical, electronic and software kernel to which special purpose hardware and software can be added to perform specific inspection tasks. It is possible to maintain 80% commonality between all systems at this level, with less than 20% applications code per system. The system can currently perform chip capacitor inspection and solder paste area measurement with solder bond verification under development. The system is based on transputer controlled frame grabbers and can use an additional 1-3 T800 transputers for image analysis if a large increase in processing power is required. It can currently operate at 5 square inches per second for components as small as 0805 resistors and capacitors.

INTRODUCTION

The placement of surface mounted chip components on surface mounted boards is an imperfect process, especially when combined with a glue dispensing system. Components can be incorrectly placed in terms of X,Y and may be skewed. Additionally, if the glue by which they are affixed to the board is not present in sufficient quantities, the devices may lose their position entirely. It is difficult to test for the devices electronically as they are often in parallel with IC devices of larger capacity, whose tolerances (±5, 10%) swamp the value of the chip component. As has been described [1] human inspections are not ideal as the task is exacting and tedious. Even well motivated operators are only 70% effective while fatigued or distracted operators can be as low as 50% effective. Additionally, as the technology moves on and devices sizes shrink, manual inspection becomes even more difficult. Clearly there is a demand for automatic visual inspection in this field.

Review of previous work

PCB inspection can trace its roots back to the pioneering work of Ejiri et. al. [2] at Hitachi Inc. for bare board inspection, but it was not until the mid 1980's that inspection was performed for SMD components with the work of Buffa [3]. Systems exist today from a number of sources using a number of technologies: mainly 2D and 3D vision. SVS has produced a system based on a dense range map which is described in [4]. This uses a dense range mapped based on a synchronized laser scanner similar to that described in [5]. This system can produce synchronized range and intensity images of flat surfaces such as PCBs. Two tasks are implemented; solder paste inspection and component placement inspection. Each uses different resolution and runs at different speeds. The component placement system model 7620 uses 0.002 inch (50 micron) pixel size and can inspect at 1-2 square inches/second while the solder paste system uses 0.001 inch pixel size and can inspect .5 to 1 square inches / second.

2D systems can be characterised by the Cimflex 5515/6/7 series from Control Automation [6]. These use four CCD cameras each viewing the PCB at an angle of approximately 30º to the vertical, 90º to each other. Illumination is by a dome of LEDs angled around the view area under software control. The 5515 can operate at up to 12 square inches per second for component placement inspection, with 0.002 inch pixel size and 1 inch square field of view. They cost in the order of $200K to $250K depending on options. The 2D system can be seen to be much faster than the range imaging system, while is is argued by Trail in [7] that 3D data is inherently more reliable that 2D grey scale intensity data. Here it is argued that factors such as specularity and variance in component colour from batch to batch of components make it difficult for a 2D system to function well. While 2D systems may have difficulty with more demanding tasks such as solder joint inspection [1], it seems likely that they still have much to offer in tasks such as component placement. The number of Cimflex systems sold (in excess of 200 units) would [8] seem to confirm this point.

The Inspection task

The system was to inspect bottom-side PCBs for presence/absence of components and component placement and skew. It was also to inspect for presence and location of glue spots and to run fully in line in the production facility at a speed no less than the line speed. This allowed 40 seconds to inspect a PCB 12 x 11 inches in size. The false accept rate was planned to be less than 1 in 10,000, and the false reject rate was planned to be less than 1 in 10,000. Of these, the false accept rate is the most critical as the boards are manually reworked, and a false reject can be ignored. The system was to be insensitive to normal manufacturing variabilities including variable PCB substrate colour and variable chip colour and appearance from batch (manufacturer) to batch. The devices were surrounded by white box markings which had been for manual inspection and which could not be immediately removed on changeover to an automatic system. One problem of the boxes was that they were often broken by vias on the PCB, badly printed or merged into one another if the device spacing was very close. Some of these issues can be seen in Fig. 1.

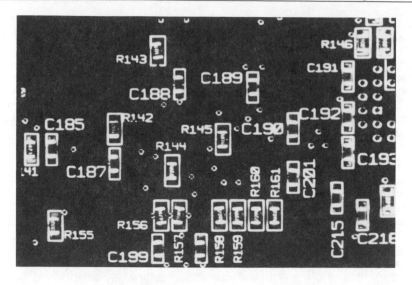

FIG. 1. Appearance of the PCB under Low level Illumination

The components differed in two ways: they could have different coloured bodies, and their ends could have different shapes and finishes. The substrates were generally green in colour but had conductors crossing them and copper hatching on their surfaces.

Thus we may summarise the scene variabilities:

Substrate:	Any colour, may contain conductors.
Bounding boxes:	Broken print, merging with boxes and characters.
Components:	Different coloured bodies, different shaped and finished ends.

System Hardware

The system was hosted in an 80386 based PC clone. It consisted of an Aerotech Unidex-14 X-Y table and with DC servo drives and a Digithurst TM monochrome transputer controlled frame grabber. This could grab images at 720x512 resolution and had 1 MByte of local program memory and a 25 Mhz T800 transputer processor, running at 5 wait states for local memory access. The system had a 24 line Opto22 digital I/O controller. A Sony XC77CERR camera was used with a 35 mm Pentax lens, giving a field of view of approximately 1.6 x 1.2 inches, with a pixel size of 0.0022 inches. Lighting was provided by a compact square of 4 x 20 W Philips PLC high frequency fluorescent bulbs, each of which could be driven separately under software control. These were placed approximately 10 mm above the PCB to ensure low level illumination of the devices for best image segmentation. See Fig.2 for details.

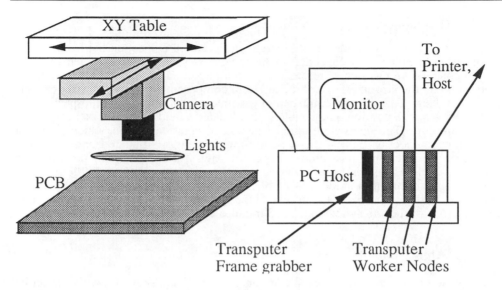

Fig. 2. VISP Hardware architecture.

Software

The system consisted of one main program running on the transputer framestore. This was coded in 3L parallel C, although only one processor was deemed necessary. The framestore communicated with the PC host using a version of the Inmos Alien file server protocol to perform such tasks as screen and disk I/O and X-Y table control.

Motor control

Control of the table movement is a critical issue. There are two basic approaches: a non-stop raster-scanning type motion such as that used by Cimflex, or stop/start motion such as that used by IRT [9] and Mahon et. al. [10,11,12] in earlier systems. Each has its advantages: a scanning system allows a much higher number (at least 6) of images per second to be acquired and processed, as there is no need to constantly accelerate and decelerate the camera head. On the other hand, much more processing power is required to process the data as it is coming in much faster and must be processed in the time available, or the system will not be ready to acquire the next image. This usually requires several image processing modules and a parallel architecture, all of which increase the cost, programming difficulty and development effort. A point to point system, though slower, is much simpler. Once the image is acquired, the next move can be initiated and the image processing can begin. The move and processing can continue in parallel until both have terminated, at which stage the next image can be acquired and the cycle can continue. In most cases, the processing will be faster than the move, but if this is not so (in the case of densely packed regions on the board) the system simply waits for the processing to complete. No image frames need be lost. The system can be designed around a single processor working at an average, rather than a worst case speed without a significant decrease in overall processing time, while the programming difficulty and effort are greatly reduced. In the final system, the overall inspection time was within specification, so it was decided to leave the system configured as a point to point system. Another benefit of this approach is reduced X-Y table wear due to shorter distance travelled.

CAD data

The system could automatically generate an optimised inspection list from a PCB layout file. The PCB layout file contained the locations, types, names and orientations of the components, and was cross referenced with a part description database containing information on the size and shape of the different component types and the sizes of their bounding boxes. These can be combined to generate a set of views for an inspection list where a view was a position on the table which located a field of view for a camera at a given resolution. The views were initially generated in raster fashion, which yielded a sub-optimal inspection path. This was optimised using a simulated annealing[13] algorithm to generate an almost optimal inspection path. The entire training process takes about five minutes, assuming no new devices have to be added to the component database, and all the information is available. Once the view files have been generated, they can be loaded at any time, and in general the system could be changed over from one product to one of the same width in about 3 minutes by a technician.

Imaging Algorithms

The vision task was to locate each component in X, Y and skew (Ø) with respect its bounding box in a short time (< 50 msec) as accurately as possible (See Fig. 3 for details). The problems of the inspection task have been discussed, but to these must be added the problem of the component end covering and thus merging with the bounding box. Thus a simple thresholding and blob tracker based approach such as SRI connectivity [14] could not be used.

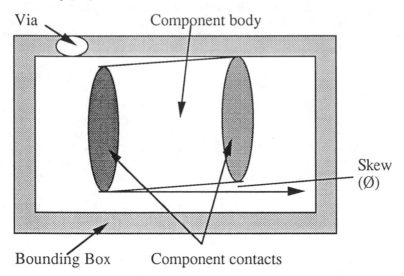

FIG. 3. Appearance of a Component and bounding Box

The first task is to locate the bounding box, which, as mentioned can be fragmented due to vias (in one case by 5 vias) and can also merge with other boxes. This calls for a template matching or cross correlation based algorithm as these have good resistance to topological damage, and the size and orientation of the boxes are known in advance.

However unless hardware is available, such as the Cognex system [15], 2D cross correlation is extremely slow. Nonetheless, the task is suitable for 1D correlation as the boxes are reasonably symmetric and regular, and thus the box can be found first along the X axis, and then along the Y axis. This was performed to approximately 0.5 thou. repeatability using sub-pixel resolution techniques.

Once the box had be located, the task remained to locate the component within it in X, Y and Ø. Again, this was done one dimension at a time. The system was lit so as to emphasise the specular contacts on the ends of the components, and the image analysis task was to locate these accurately, or decide that they were not present. The component is first located along its body. The colour of the centre of this is not known at run time (as it can change from component batch to batch), and thus is a "don't care" in the correlation template mask. The inside of the box is summed to a 1D vector, and correlated with the mask. If the maximum of the result is below a threshold for that component type, the component is deemed to be absent, else it is deemed to be present, and located at the position of the maximum point in the output trace. See Fig. 4.

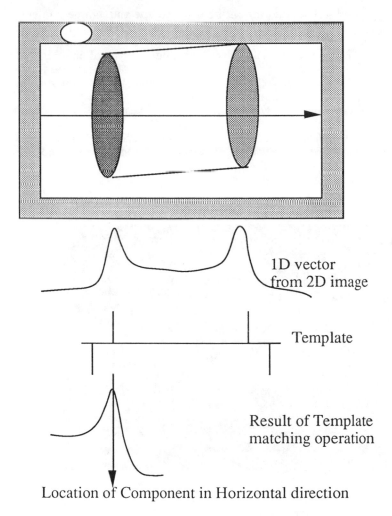

1D vector from 2D image

Template

Result of Template matching operation

Location of Component in Horizontal direction

FIG. 4. Location of Component in Horizontal direction

The system now searches for the vertical position of each of the chip contacts using a different 1D technique. The vector is generated, and then thresholded to two types: contact or box, and pad or substrate. It was not possible to separate component contact from bounding box. Once thresholded, the centre of the box can be located simply whether or not it is occluding the box or not. The decision to use thresholding was taken only after it was finally deemed impossible to obtain robust results from correlation techniques due to the occlusion problems. The correlation techniques were measured to be accurate to 1/4 pixel, while the thresholding techniques were only accurate to 1 pixel but dealt with occlusion properly. After each end has been located, a skew measurement can be generated, and the component can be classified against a set of offset thresholds for each device type, or against special individual thresholds for exceptional devices.

The results can be sent to a printer, disk, a host data collection system, or can be displayed on the video monitor for manual verification.

Installation

The systems was moved to a laboratory on the beta test site and a large number of PCBs were passed through it. These highlighted several problems mainly in skew and vertical offset measurement. These were solved by the threshold based vertical location algorithms as described. The system was initially inspecting one type of board in under 25 seconds per board while the line cycle time was in the order of 55 seconds for that product. It was thus decided to decrease the field of view from 2.4 x 1.8 inches to 1.6 x 1.2 inches to increase accuracy and robustness. This increased the cycle time to 40 seconds, which was still acceptable as it was less than the line "heartbeat" of 55 seconds. It generally increased confidence in the system, especially with the smaller 0805 components. The system was then moved out to the production line and, after running alongside it for a week, was moved in-line and performed well from day one, improving yield downstream and finding a high level of acceptance from line supervisors and operators alike.

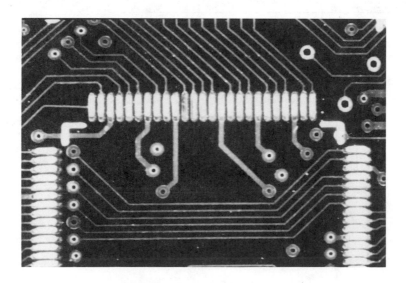

FIG 5.   Image of Solder Paste Deposits

Results

The system was installed, and was able to inspect 12" x 11" boards containing 360 components in 50 seconds and 200 components in 39 seconds, averaging about 2-6 ->3.4 square inches per second, or about 7 components per second, after the field of view had been decreased. Previously it had been about 5 square inches and 12 components per second, but this speed was unnecessary. The system can measure to about $1/4$ pixel accuracy (3 sigma repeatability) under good conditions reducing to about 1 pixel worst case. This yields about one false accept / day, a huge improvement on the manual inspection which it replaced.

A solder paste inspection system was also developed using similar lighting techniques. It performed well for 2D area measurement using binary thresholding and boundary tracking, but when results from 2D area classifications were correlated against solder joint failures at final test, there was only a 20% correspondence. Thus the system was not further developed, although results were promising (see Fig. 5. for an image of solder paste). However, the system was modified to perform solder paste inspection in 1 week, which implies that one of the design considerations, that of rapid algorithm development, was achieved.

Conclusions and Future Work

An automatic chip component inspection system was designed, developed and is now being used in-line as in an SMT assembly line. The success of the project was greatly helped by a thorough approach to testing by the use of repeatability testing and the use of a large enough (30 board) 10,000 component set of test cases during development. Even so, there was considerable work to be done when the system was brought to the plant for off-line testing as a much greater variability of scene was encountered than had been previously seen. The multi-level (macro / micro) tests and the use of software test points and graphics enabled this beta test to proceed rapidly. In an ideal situation the variability, pad surface and component contact appearance caused more trouble than expected, showing the shortcomings of grey scale analysis of specular objects or objects with surface colour variability. An approach more like the SVS 3D scanner [4] would overcome this, but would have proved too expensive to implement and perhaps too slow in operation. The 2D approach eventually proved adequate for the task, but required a lot of extra development time that a better (3D) sensor could have eliminated.

Future directions include the development of a 4/5 camera solder joint inspection system, and a low cost single axis system for inspecting much smaller PCBs. The single axis system could be much smaller and cheaper to produce (< $40K it is hoped), and could open up new inspection areas currently too price sensitive to afford the technology. At a hardware level, it is hoped to port the system to an EISA based PC clone with a colour frame grabber which is currently under development. This should further reduce costs while availing of the dynamism of the mainstream PC marketplace. The pace of transputer development was found to be too slow, which is why this technology is being abandoned. It was not deemed necessary to use a second transputer processor as the system was mostly bottlenecked by XY table movement rather than the image analysis, although this might have changed had a solder joint inspection system been developed. The EISA bus should allow images to be transferred at 33 MBytes per second which is faster than that achievable (2 MBytes / second / link) with transputers at present. It is hoped to either transfer images to the main I486 processor, or to other slave processors on the EISA bus. As long as these are programmable in C, porting the code to them should be a realistic proposition.

The system was successful, using off the shelf technology to solve a tricky inspection problem and do so within specification in terms of speed and accuracy, and at a low cost.

Acknowledgments

This work was funded by Eolas, the Irish Science and Technology Agency, and Apple Computer Ltd., Cork, Ireland.

References

[1] Iversen, W. R: "Automated board inspection gets more important", Electronic World News, 19-24, Oct 21, 1991.

[2] Ejiri, M. Uno, T. Michihiro, M. and Ikeda, S: "A process for detecting defects in complicated patterns", Computer Graphics and Image Processing: 326-339, 1973.

[3] Buffa, M: "Process Control for Automated Component Assembly of Surface Mounted Devices Utilising a machine Vision Controller", SME Proc. Vision 85, 5-180 - 5-200, 1985.

[4] Kelley, R. Collins, M. Jakimcius, A. Svetkoff, D. "Automatic Inspection of SMD Printed Circuit Boards Using 3-D Triangulation", SME proc. Vision 88, 7-33 - 7-47.

[5] Rioux M, et. al: "Design of a large depth of field 3D camera for robot vision", Opt. Eng. 26(12) 1245-1250 1987.

[6] Control Automation, General product Literature 1989.

[7] Trail D., Jakimcius, A. "Inspection Surface mounted PCBs with 3D Laser Scanners", Surface Mount Technology, April 1988.

[8] Zuech, N. "Automating Solder-Joint Inspection", Test & Measurement World, pp. 68-76, Oct 1989.

[9] Juha, M., "The economics of automated X-ray inspection for solder quality", IRT Corp. Publ., San Diego, CA, Feb 1986.

[10] Mahon J, Harris, N. & Vernon D., "Automatic Visual Inspection of Solder Paste Deposition on Surface Mount Technology PCBs", Computers in Industry 12, pp 31-42 1989.

[11] Mahon, J: "Automatic 3-D inspection of solder paste on surface mount printed circuit boards", Journal of Materials Processing Technology, 26 (1991) 245-256.

[12] Mahon, J: "Automatic Inspection of Solder Joints on Surface Mount PCBs", Proc. 7th conf. of the Irish Manufacturing Committee on AMT. Sep. 1990.

[13] Kirkpartick, C: "Optimisation by Simulated Annealing", Science, pp 671-680, May 1983.

[14] Agin, G: "Computer Vision Systems for Industrial Inspection and Assembly". Computer, pp 11-20, May 1980.

[15] Silver, W: "Alignment and Gaging Using Normalized Correlation Search", Proc. SME Vision 87, pp 5-33 - 5-56, 1987.

EXPERT SYSTEM FOR PACKAGE INTERCONNECTION SYSTEM SELECTION

Frank A. Stam

Digital Equipment International B.V., Galway, Ireland

ABSTRACT

The miniaturizing trend in the electronic industry involves developments in semiconductor technology, circuit design, and packaging of electronic components. Development in packaging is not advancing as quickly as the other development areas and consequently the overall development is impeded by packaging.

Over the years a myriad of packaging alternatives were introduced, none of them prevailing as a recognized solution. As a result the process of selecting an appropriate packaging or interconnection system is rather complicated. The objective of this paper is to shed some light on the various aspects playing a role in this packaging selection process and to indicate how an Expert System application can be used as a design and decision making tool for that.

In order to obtain a manageable system and because most radical changes in technology appear here, the problem domain is limited to single package interconnections. The disadvantage of this limitation is that real products (populated boards or substrates) can not be reasoned upon. The system is linked to several databases and a front and an end user interface which makes it an easy to use, powerful selection tool.

The system starts its reasoning based on some user inputs about user requirements and Integrated Circuit information, and then tries to find a packaging solution by matching a package type, a substrate, an interconnect technology, a solder joint and an interconnection method. A suitable solution complies with mechanical, thermal, and electrical rules.

Validity checking of the Expert System application functionality and extendibility of the knowledge base system can be done by studying and experimenting with new and existing interconnection systems. Work done so far indicated that more technical details should be included to make the system more accurate.

INTRODUCTION

The driving force in the electronic industry is mankind's desire for information processing power. The work described in this paper focuses on high performance chip packaging, as this is where packaging has failed to keep up with the IC (Integrated Circuit) technology developments. Usually IC innovations require packing more gates in a tighter space. The increased packaging density leads to higher power dissipation, more heat generation, the retention of signal integrity, less reliability, other substrate requirements, other bonding techniques etc. Dealing with these technological problems is difficult, especially when a reduction in size and manufacturing costs is also sought after. Packaging is in a continuous state of catching up with IC technology, in the meanwhile creating opportunities for new products such as smart cards or shirt pocket computers. Migration of performance and features from large into smaller existing systems can also be seen, e.g. personal computers becoming able to take on super computer functions.

Interconnection systems where performance is not the main factor (e.g. the heart pacemaker requires reliability and small size) usually don't require advanced interconnect techniques, but should still be incorporated in the Expert System application.

Increasing Performance

The performance [1] of a product is its ability to process information. One way of expressing a product's performance is the rate at which integer MIPS (Millions of Instructions Per Second) are processed by the CPU (Central Processing Unit). Theoretically the MIPS rate is equal to the clock frequency divided by the number of cycles per instruction. Thus the performance of a system can be improved by increasing the clock frequency or decreasing the number of cycles per instruction. The clock frequency is dependent on the IC technology (different propagation delays per gate) used, the wiring distance from package-to-package, the type of interconnections, and memory topology. The number of cycles per instruction is dependent on the logic architecture, e.g. vector, array, or co- processing. With more gates in the CPU the number of cycles per instruction will drop and the performance could increase.

A Knowledge Based Tool

The packaging design process usually takes place in an uncontrolled, unsystematic, and inefficient manner as so many parameter interactions are involved. A knowledge based package was proposed to optimize this design process. It was decided that an Expert System Shell would be used, to make it possible for people without knowledge of Artificial Intelligence languages, to develop an E.S. (Expert System) application. It would also reduce the lead time for development, as the programming burden would not be there. Necessary requirements were: a modular development environment; good portability to peripheral facilities such as databases, user interfaces, and user sub-routines; easy extendibility of the knowledge base system; and an Object Oriented approach integrated in a rule based system. Other aspects which played a role in the package selection were the availability of various AI paradigms, hardware platform support, cost, the maximum application size, and the corporate usage of the package. In [2] some E.S. Shells are compared. Eventually NEXPERT OBJECT™, from NEURON DATA Inc., for an IBM compatible PC running a DOS operating system is chosen.

Similar Work Done

A literature search showed that not many E.S. applications were developed in this field of technology. There are other tools around such as IC simulators and thermal modelling packages, but only a few E.S. applications (most diagnostic systems for welding processes) exist. One E.S. application [3] is developed for Integrated Circuit package design. Some overlap with the E.S. application proposed in this paper occurs, but their objectives are completely different. This package tries to provide a complete packaging solution, where the other one just concentrates on one aspect of it. Judging from the development status, its functionality seems to be impeded by hooking up a CAD system, a Finite Element Package, and various databases.

The structure of this paper is as follows: first a brief overview is given of the problem domain and the relevant aspects to be implemented in the E.S. application, next the implementation of this information in the E.S. application is discussed, and finally the validation of the functionality of the E.S. application will be discussed.

PROBLEM DOMAIN

Before starting the design process the users packaging requirements should be defined. This can be done by giving them a priority value (between 1 and 10). The following user requirements are distinguished: cost, size, technical risk, performance, reliability, and environment. This qualitative information can be used in cases where insufficient information is available or where qualitative comparisons are required.

IC Technology

A logical starting point for package-to-substrate interconnection system design is the package. One of the first things to decide on is what *IC technology* or what *package type* to use. Table 1 shows some IC technology criteria which could be used when ASIC's (Application Specific Integrated Circuits) are considered.

TABLE 1
IC Technology Criteria

Technology	Gates/cm²	Cost/Gate	Gate Delay	Pow.Dis./Gate
ECL	< 40,000	(4-10)*X	0.2 - 3 ns	3 - 80 mW
CMOS	< 60,000	X	1 - 8 ns	0.02 - 0.9 mW
GaAs	< 30,000	(6-12)*X	0.01 - 0.2 ns	0.05 - 1.2 mW

The cost/gate isn't quantified because it is too much of a changing factor. To make an effective selection, some idea about the gate density and the gate propagation delay or clock frequency ($gate\_delay \approx 1 / (20 * F_{clock})$) should be present.

Package Type

If no ASIC's are considered usually a package is selected straight away. Prerequisite are the number of package terminals, the clock frequency, and the die dimensions. The number of package terminals or number of I/O (Input/Output) ports can be estimated with Rent's rule supposing that the package complexity is known, i.e. microprocessor, gate array, DRAM etc.. In the equation below **k** and **p** are constants dependent on the package its function.

$$N_{i/o} = k * N^p_{gates}$$

The die dimensions can be chosen arbitrarily, but might require adjustment since the lead or pin pitch is dependent on it, which on its turn affects the potential interconnect technologies. There is also a thermal consideration, a small die size is good from a wiring delay point of view, but extra cooling or a heat sink might be required.

It is optional having the I/O ports on the perimeter of the die or at the bottom of the die (Pin or Land Grid Array). Assuming that a perimeter chip has leads on 4 sides and a square die with cavity up, then the pad/pin pitch equations are as follows:

$$P_{pitch, perimeter} = \frac{4 * L_{die\_side}}{N_{i/o}} \quad ; \quad P_{pitch, grid\_array} = \sqrt{\frac{L^2_{die\_side}}{N_{i/o}}}$$

Figure 1 illustrates the above equations, and shows that for the higher lead counts, Pin Grid Array (PGA) is much more beneficial when size is an issue.

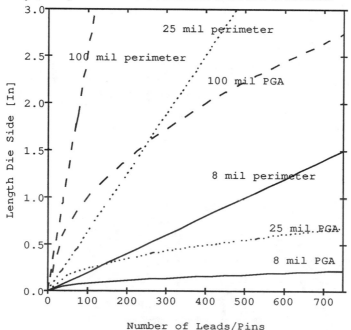

Fig. 1 Leads on die perimeter vs. pins on die area

After the lead/pin count and pitch is determined, a package selection has to be made. A matrix consisting of all the available package types and selection criteria [7],[8] could be used for selection. Table 2 shows a part of this matrix with a few package types and criteria. When comparing packages, different types of encapsulation should be avoided. Two package types are left out on purpose, namely COB (Chip On Board)and leadless package types. For COB the encapsulating of the chip takes place after wirebonding or Tape Automated Bonding (TAB) the chip to the board which makes it difficult to compare to other package types. The leadless packages are omitted as lead calculations can't be applied.

Some criteria (reliability and cost) are expressed relative to each other, as accurate data isn't available. The data should give a general idea how packages compare. For instance the cost of DIP (Dual-In-line Package) vs. PLCC (Plastic Leaded Chip Carrier) can not be expressed with a constant factor, as it is dependent on the lead count. DIP is actual cheaper for a lead count < 30.

TABLE 2
Package Type Selection Matrix

	DIP-plastic	PLCC	QFP	PGA	TAB
Lead Count	≤ 68	≤ 124	≤ 600	≤ 500	≤ 800
Lead Pitch mls	≥ 100	≥ 50	≥ 12.5	≥ 50	≥ 3
Reliability	Y	1.2*Y	1.2*Y	1.3*Y	0.9*Y
Cost	1.1*X	X	1.1*X	1.5*X	1.25*X

QFP = Quad Flat Package

Interconnect Technology

After a *package type* is selected, the *interconnect technology,* the *substrate type*, and the *solder joint* will be looked at. There is no fixed order, sometimes the package type requires a certain substrate type which might require a specific interconnect technology, other times a package type requires a certain interconnect technology etc.. Many publications are available which compare or discuss the different interconnect technologies, e.g. [4],[5],[6]. At the moment 5 major technologies are considered (Area TAB, BIP™, GE overlay are not well established yet):

1. **PTH** (Plated Through Hole)

2. **SMT** (Surface Mount Technology) mass reflow

3. **C4** (Controlled Collapse Chip Connection) or Flip-Chip technology

4. **TAB** (Tape Automated Bonding)

5. **Wirebonding**

In the context of this paper it is not appropriate to list all the pros and cons of these technologies, but to give an idea a few key features are given. PTH is not used very much anymore, but the low lead count DIP is well known. For mass reflow the minimum lead pitch is about 20 mil, this means that for lead pitches smaller than 20 mil C4, TAB or wirebonding has to be used. C4 has low bonding inductance which makes it very suitable for CMOS devices. Also C4 has an excellent electrical performance and with an area array of 10 mil pitch which allows for lead count up to 1600. TAB can be done for lead pitches down to 3 mil and up to 1000 leads. TAB is very suitable for high volumes (e.g. watches), and is a low cost interconnect option. TAB's clean bonding interface gives it an advantage over wirebonding. Wirebonding is a relatively slow process, can handle 6 mil pitch and can easily adjust for package or board changes. The advantage of wirebonding over C4 and TAB is the low startup capital required and the presence of an infrastructure for a well understood process. For both C4 and TAB complex metallurgical aspects play an important role with respect to reliability issues.

Substrate Type

The type of substrate/ Printed Wiring Board/ Multi Chip Module to use is more or less directed by the package type and closely related to the chosen interconnect technology. E.g. if C4 is the desired interconnect technology, only a thin-film multilayer substrate can be used to match the fine surface pitch of the pads and to provide the high density routing for the array escapes.

Basically there are four types of substrates:

        a. Organic-based printed circuit boards

        b. Thick-film ceramic

        c. Thin-film single or multilayer

        d. Multi Chip Modules

One of the main considerations when choosing a substrate is the TCE (Thermal Coefficient of Expansion). If the substrate material doesn't have a matching TCE with the solder joint, mechanical stresses can be induced. Dielectric properties and thermal conductivity of the substrate materials also play an important role.

Solder Joint Design

As the solder joint forms the interconnection medium between package and substrate, the solder joint characteristics play a crucial role in the overall performance of the interconnection system The solder joint has to comply with mechanical, electrical, thermal, and chemical rules. The design sequence starts with material selection for base metal, flux, and solder, after that geometrical aspects of the joint are assessed [9].

Base Metal   Selection is done with the help of a material table showing all the metals and physical characteristics such as electromotive potential, solderability, electrical resistivity, melting point, TCE, modulus of elasticity in tension, Brinell hardness etc.. Electromotive potential has importance in terms of corrosion. The solderability will have an effect on the flux choice. The electrical resistivity could cause over heating the joint.

The melting point is important in combination with the solder. The significance of the TCE is already indicated.

Flux   The flux selection consists of two steps. Step 1: based on the solderability of the base metal an inorganic, a non-rosin organic, or a rosin organic (water white or activated) will be chosen. Step 2: picking a flux from the group chosen in step 1 suitable for the assembly.

Solder   To make a solder selection two data tables are required. One with the solder alloy composition, melting temperature, tensile strength, shear strength, TCE, etc.. And another table showing the intermetallic compounds between base metals and solders. First a temperature range should be chosen, than the mechanical strength, the electrical resistivity, and the TCE mismatch between base metal and solder should be checked. When intermetallic compounds are formed which have proved detrimental to the solder joint, metal plating of the base metal could be an option.

Joint geometry   Only two types of joints are considered, the so called lap joint as seen in TAB and the butt joint as seen in PGA. With simple mechanical calculations the strength of a solder joint can be predicted, the same for the electrical resistivity. An example of calculated joint strength vs. a measured joint strength follows later.

Interconnection Method

A wide variety of soldering or bonding equipment is available today. Methods such as adhesive bonding, mechanical joining or welding are not considered in this context, because of the absence of a solder joint. Wirebonding is another method that does not require a solder joint either, but it has proven to be a viable joining method.

A few important selection considerations are:
- Type of heat transfer: conduction, convection, or radiation;
- Local or mass heat application;
- The time-temperature relationship (e.g. flux at specific temperature for certain time)
- No heat application, but induction, resistance, or ultrasonics soldering/ bonding;

Discussion of the various methods can be found in [10] and [11]. A final comparison of the different methods should be based on criteria such as cost, speed, reliability, technical risk, safety, special tooling required, product flexibility, process control, automation level, infrastructure, capital investment etc. (e.g. Laser bonding vs. Single point parallel gap resistance reflow soldering).

In figure 2 the packaging design process is represented graphically. In reality this process is much more complicated with many iterative loops and other interactions, but in order to develop a manageable E.S. application a simplified model is used. The dashed line shows a link which is not implemented in the E.S. application, namely the effect of other devices to be connected to a substrate. This means that a realistic product situation can't be studied, which is a considerable weakness of the system. A possible solution could be the development of an independent knowledge base system which incorporates substrate design and contains a component database.

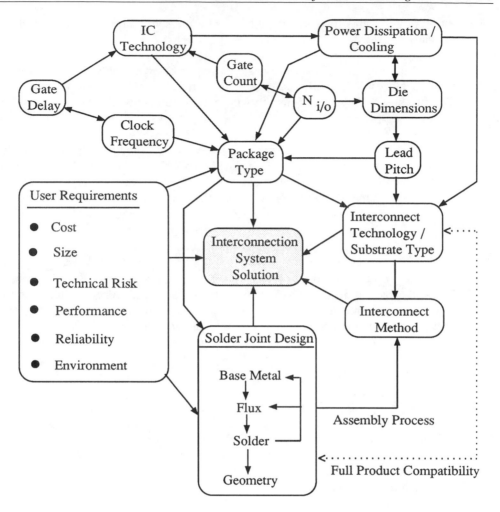

FIG. 2 Design steps and interactions for interconnect packaging

EXPERT SYSTEM DEVELOPMENT

Functionality Nexpert Object

Nexpert Object can capture the facts and procedural information in rules, but it can also capture descriptive information in classes and objects. Various AI paradigms such as *"Order of Sources", "If-Change demons", "Pattern Matching"*, integrated *"Forward and Backward Chaining", "Interpretations", "Dynamic behavior"* (rules can create and delete objects), and *"Inheritance"* are supported. A special facility called *"Knowledge Islands"* is incorporated to enable modular development. It also provides good portability to various external systems.

Domain Implementation

Architecture   Figure 3 shows the architecture of the E.S. application. Development of the E.S. application is done in a Microsoft™ Windows/ Mouse environment, but the finished E.S. application is delivered on a different platform with a specific form oriented front/ end user interface. The user can enter relevant data on a screen form via single choice lists, popup lists, multiple choice lists, selection tables, or input tables. After data entry the interface stores this data in the appropriate knowledge base property slots. The end interface is used as a report facility, it is a combination of static text and dynamic information taken from the processed E.S. application. The report should show the reasoning process and the made selections.

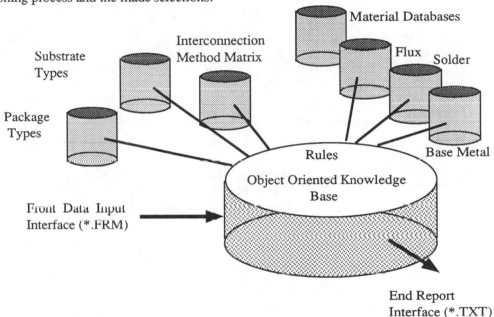

FIG. 3   Architecture for Interconnection Expert System Application

Object Oriented Knowledge Base   In Nexpert the whole application domain can be described in Classes, Subclasses, Objects and Subobjects. A "class" is a generalization of a set of objects, while an "object" would be a thing, idea, person, or place in the application domain. A "subclass" represents a subset of another class, which means that it has all the characteristics of that other class. A "subobject" represents a relationship of the type **"is a part of"**. Classes, Objects, and Subobjects can be described by properties. Below some examples are given from the class "Package". Other classes in this application

Class	Subclass	Properties	Object	Subobject	Properties
Package	Pin Grid	Terminal Pitch	PGA	Pin	Pin Diameter
	Perimeter	Gates / cm^2	TAB	Tape	Tape Material
	DRAM	Power dis./ gate	QFP	Lead	Lead Thickness
		Cost / gate	PLCC		Lead Form

are: substrate, solder joint, interconnect technologies, interconnection methods, flux etc..

Inference Engine　　The objective of the inference engine or reasoning mechanism of the E.S. application is investigating a hypothesis which represents a packaging solution. The suggested hypothesis consists of 4 hypotheses or subgoals itself, and all 4 of them have to evaluate to "true" to make the suggested hypothesis true. The 4 subgoals represent selecting a Package Type, an Interconnection Technology, a Solder Joint, and an Interconnection Method. There is no specific order of achieving these subgoals since the **if change** meta-slot can trigger rule re-evaluation when rule parameter values change. However, a specified order of firing the subgoal rules can be achieved by setting **rule priorities**. As the subgoals consist of hypotheses themselves, a tree like reasoning structure exists on which **backward chaining** is applied.

Example of a Rule　　An example of a "rule" in Nexpert Object is given which shows how the interaction with a database takes place. A rule consists of one hypothesis, left-hand side (LHS) conditions and right-hand side (RHS) actions. If all LHS conditions are evaluated to "true", the hypothesis (boolean slot) is set to "true" and the RHS actions are carried out. The following rule makes a solder selection from a Soft-Solder Alloys properties table [9]. The table is stored in a so called Nexpert flat-file database or NXPDB file (Solders.NXP), in ASCII format. The column headings are: Sn, Pb, Sb, Cd, Ag, Tmelt, Tens_str, Shear_str, Elect_res, TCE, etc.

Rule name:　　Solder-Selection　　　　Hypothesis: SS
LHS Conditions:　　FS = TRUE　　　　　　　　BS = TRUE
　　　　　　　　　　Solder.tmelt_low > 0　　　Solder.tmelt_high > 0

RHS Actions:　　　DO Basemet.TCE - 2 = Solder.TCE_low
　　　　　　　　　DO Basemet.TCE + 2 = Solder.TCE_high
　　　　　　　　　RETRIEVE "Solders.NXP"

QUERY: (Tmelt IN Solder.tmelt_low : Solder.tmelt_high) AND
　　　　(TCE BETWEEN Solder.TCE_low AND Solder.TCE_high)

CURSOR: Solder_retrieved

DATABASE FIELDS	NEXPERT PROPERTIES.
Tmelt	Solder.Tmelt
TCE	Solder.TCE
Tens_str	Solder.Tens_str

FS = hypothesis for Flux Selection; **BS** = hypothesis for Base Metal Selection;
Tens_str = Tensile strength; **Elec_res** = Electrical Resistivity;
Tmelt = Melting Temperature Liquidus; **Solder** and **Basemet** are objects;
tmelt_low, tmelt_high, TCE_low, TCE_high, Tmelt, TCE, and **Tens_str** are properties of **Solder**;
TCE is a property of **Basemet.**

First the rule tests the LHS conditions, i.e. if a flux and a base metal are selected, and if a temperature range for the solder is defined. If so, SS becomes "true" and the RHS actions are started. In order to get a matching TCE for the base metal and the solder, a TCE range is specified for the solder to be selected. Next a sequential retrieve of one record from the database Solders.NXP is requested. With the Nexpert Query Language it becomes possible to filter one specific data record out. In this case a solder is selected that falls in a certain melting temperature range and which has a TCE matching with the earlier selected base metal. If a solder is available that fulfills the requirements, the list of indicated database fields will be retrieved and the corresponding values will be copied into the indicated property slots of the object **Solder**. At the same time the value of the variable Solder_retrieved (initial value 0) is increased by 1, which can be used for firing the "intermetallic compounds" rule (the required solder alloy elements percentages can be obtained from the solders database). In case no suitable solder is found, the variable Solder_retrieved is set to -1. The hypothesis SS is set to "false" and the rule will be fired again when the ranges are changed.

VALIDATING THE E.S. APPLICATION

To validate the functionality of the E.S. application, designed interconnection systems should be tested in practice. The problem is, that designed interconnection systems not often materialize. And if so, the designed system often forms a part of a bigger system and can't be tested standalone. However those obstacles can be overcome by studying existing interconnection systems for which assembly results can be predicted theoretically. Simple solder joint calculations can be done to predict the mechanical strength, the electrical resistance, the reliability, and the heat transfer of the solder joint. Comparing the theoretical and practical results gives some information on the validity of the E.S. application.

The following example shows a mechanical strength calculation of a 20 mil TAB lead which is compared with the pull strength obtained from a gang bonded 20 mil TAB lead. Figure 4 shows the geometrical configuration of this interconnection system.

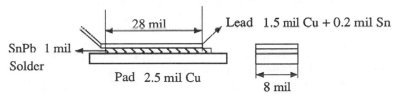

FIG. 4   20 mil TAB Lead Bonded to Pad

There are three different strengths to consider: tensile strength of solder, shear strength of solder, and tensile strength of copper. The general equation to calculate the strength is:

$$F = \sigma * A$$

whereby F is the strength, σ is the materials tensile or shear tension as can be found in handbooks, and A is the area to which the force is applied. First the constants are given:

$$\sigma_{s,j} = \quad 394 \text{ kg/cm}^2 \quad (\text{sj} = \text{shear joint})$$
$$\sigma_{t,j} = \quad 534 \text{ kg/cm}^2 \quad (\text{tj} = \text{tensile joint})$$
$$\sigma_{t,l} = 5992 \text{ kg/cm}^2 \quad (\text{tl} = \text{tensile lead})$$

$A_{s,j} = A_{t,j} = 28 * 8 \text{ mil}^2 = 1.45\text{E-}3 \text{ cm}^2$ Area is the full copper / solder interface.

$A_{t,l} = \quad 8 * 1.5 \text{ mil}^2 = 7.74\text{E-}5 \text{ cm}^2$ Area is the cross section of the lead

The strengths are:
$$F_{shear,solder} = 571 \text{ gram}$$
$$F_{tensile,solder} = 774 \text{ gram}$$
$$F_{tensile,copper} = 464 \text{ gram}$$

From above strengths can be concluded that the the solder joint is stronger than the copper lead. This is confirmed by the "broken lead" pull tests. However, the measured strength ($\approx$ 130 gram) is far from the predicted 464 gram. Even a 20% safety factor does not close the gap. Experiments showed that the best bonding results were obtained for bonding temperatures approximately 350°C, while the optimum bonding temperature was expected to be about 250°C. The reason for this is the very thin 1 mil solder layer. The bonding temperature rises hyperbolic for such thin solder layers, while the potential bond strength is reduced significantly. A solder thickness of about 3 mil, leads to optimum results. This validating exercise demonstrated that the E.S. application isn't very accurate yet and needs expansion with "solder thickness" functionality.

Similar calculations can be carried out with electrical, thermal, and reliability equations. As in the above example, this could result in better assessment of the validity of the functionality of the E.S. application and expansion of the E.S. application..

In the above validating exercise it is assumed that an optimized assembly process is in place. As characterization and optimization of the assembly process form a part of packaging design, the E.S. application supports the assembly optimization process by assessing initial parameter settings for a selected interconnection method. The updating of the knowledge base for this interconnection method parameter assessment facility is largely based on experimental work.

CONCLUSIONS

- The subject of packaging design is considered to be a black art, as no systematic design process is available. Because of the many inconsistencies and inefficiencies related to packaging design there is a definite need for an E.S. application. However, the problem domain is very large and unmanageable as a whole. Limiting the E.S. application to single package (IC's only) assembly, gives more control but makes realistic modelling impossible. A solution could be, developing of separate knowledge base systems for: silicon design, substrate design, and package assembly.

- At the moment the E.S. application consists of a backward chaining rules framework and an object oriented knowledge base able to find satisfying solutions for a Package Type, an Interconnect Technology, a Solder Joint, and an Interconnection Method A weakness of the system is its lack of detail and unsuitable boundary conditions, which prevents it from being a realistic tool. Strong points are its powerful database analyzing capabilities and easy to use front and end interfaces.

- Although Nexpert Object has taken away most of the programming burden, it still requires quite a long learning curve before the system features successfully can be employed

- Validation of the functionality of the E.S. application is difficult as not many test cases are available. However, validation information can also be gained from exercises with existing interconnection systems. The discrepancy between theoretical calculations and practice gives an indication about the accuracy of the system.

REFERENCES:

1. Bogatin E.: What is Performance, Skinner R. (ed.): *High Performance Packaging Solutions*, Scottsdale, Integrated Circuit Engineering Corporation, 1991, pp 2-1 - 2-16.

2. Payne E.C., McArthur R.C.: Appendix A - Generic shell programming and rule operators: *Developing Expert Systems,* New York, John Wiley & Sons Inc. , 1990, pp 357-159.

3. Decker D.R., Hillman D.J., Voros R.S.: A Knowledge-Based System For Integrated Circuit Package Design: *Proceedings of the Technical Program NEPCON WEST'89,* Anaheim, California, 1989, pp 420-425.

4. Mahalingan M., Andrews J.: TAB vs. Wire Bond - Relative Thermal Performance. *Connection Technology*: pp 37 - 42, January 1987.

5. Feindel D.: TAB and COB Compared. *Electronic Production:* pp 11- 15, October 1990.

6. Munns A.: Flip-Chip Solder Bonding for Microelectronic Applications. *Metals and Materials*: pp 22 - 25, January 1989.

7. Sinnadurai F.N.: High Density Packaging of Chips and Subcircuits, *Handbook of Microelectronics*, Ayr, Electrochemical Publications Ltd., 1985, pp 92 -133.

8. Ohsaki T.: Electronic Packaging in the 90s - A Perspective from Asia. *Proceedings of the IEEE IRPS,* pp 1 - 33, 1990.

9. Manko H.H.: Designing the Solder Joint: *Solders and Soldering,* New York, McGraw-Hill Book Company, 1979, pp 124 - 168.

10. Heilmann N.: A Comparison of Vaporphase, Infrared and Hotgas Soldering: *Proceedings of the IEEE IRPS,* pp 70 - 72, 1988.

11. Stockham N.R.: Joining Techniques for Fine Pitch Surface Mount Devices: *Proceedings of High Performance Packaging Conference*, London, pp 58 - 67, 1990.

MACHINE LEARNING MODEL FOR APPLICATION IN MACHINING PROCESS PLANNING

B. Gopalakrishnan and Jianfeng Song

*Department of Industrial Engineering, West Virginia University,
Morgantown, West Virginia, U.S.A.*

ABSTRACT

This paper deals with the development of machine learning models which can be used in process planning for machining. Particularly, the operations of a PUMA industrial size CNC lathe are used as a basis to validate the models which are used in process planning for the manufacture of axisymmetric parts. Several machine learning techniques are examined for their suitability for application in this domain, and the reasons for adopting explanation based learning is explained. The development of two models, namely the preliminary model and the final learning model are examined. The part descriptions provide input to the system, a preliminary model which determines a "rough" process plan is developed using factual information and previous information on the planning of similar parts. The final learning model is refined from the preliminary model by heuristics and criteria about process planning. The knowledge gained from developing a process plan in this manner is generalized as a macro operation and stored in a frame based data environment. The aspects relating to knowledge base development, criteria used for evaluating process plans, data flow, and other pertinent issues are described. The paper concludes with the description of implementation issues.

1. INTRODUCTION

Computer-Aided Process Planning (CAPP) has been a focus of study for last two decades and many CAPP systems have been developed [1]. Basically, the existing process planning systems can be distinguished as variant and generative systems. A variant system utilizes the part similarity in shape and properties to retrieve existing standard process plans for a part. Process plans are exhaustively listed in a retrieval system. A generative system is embodied with some type of decision logic inside to synthesize process information and create a process plan for a part. The decision logic such as decision table or decision tree controls the decision making for the selection of detailed processes based on a part description. Recently, AI techniques are introduced in generative systems. TIPPS [1], GARI, EXCAP, Intelligent Process Planner [2], Turbo-CAPP [3], and QTC [4] are some examples which adopt the expert system approach. Those systems represent planning knowledge in rules. The part being planned is described using a special notation of attributes, which represent shape, tolerance, surface finish, etc. Search methods are used to find the goal of planning by given values of attributes.

A traditional process planning system has its serious limitation in terms of system development. Its planning knowledge is almost static and its planning ability is mostly determined in its design stage. It suffers from the progress of manufacturing technology, the update of manufacturing facility, and the change of manufacturing environment. A variant system needs plans for all part families, and a generative system requires completely developed databases or production rules, which would lead to high cost and long time for system development and update. For instance, CPPP [1] took an average of 4 person-months' effort to develop a process model. An intelligent process planning system with self-learning ability can be very helpful from the view of system development, maintenance, and update.

Inductive machine learning technique provides an alternative to utilizing knowledge base techniques. Machine learning is related to system self-improvement, adaptation to new or different circumstances, behavior modification, or concept formulation and refinement [5]. Inductive machine learning can be defined as any automatic acquisition of explicit and implicit domain knowledge for a computing program or system so that the program or system performance in domain problem-solving is improved over time.

Similarity based learning and explanation based learning are two major approaches of inductive learning. Similarity-based learning needs plenty examples to form training sets, which is preferred for learning knowledge from a data-intensive domain [6]. The example size must be large enough to cover domain theory. The feature set of the examples must be large enough to discriminate positive and negative examples. Similarity-based learning also has the inductive leap problem [7]. All the features of the examples are equally treated without preferring one discriminating feature over the others, which is not always true in real world.

As an alternative, explanation-based learning deals with learning knowledge from a single example. The system knows the complete theory of a domain and a definition of a concept. The goal of learning is to learn a better definition of the concept. An explanation of how a particular example accomplishes the concept is constructed using domain knowledge. The features of the example which are identified by this explanation are converted to a more general definition of the concept. The learning method is preferred for knowledge enhancement in knowledge-intensive domains [8].

The concepts of inductive machine learning have been applied in engineering and manufacturing areas recently. The application scope covers rule learning for parameter selection in turning process planning [9], learning diagnosis knowledge of engine faults [10, 11] and power plant faults [12], learning control rules from data for process real-time monitoring [13], and learning patterns of applying scheduling policies in real time scheduling for a flexible manufacturing system [14]. All these applications adopt the approach of similarity based learning. However, we haven't seen a report of learning based CAPP systems.

Process planning is the course to construct a manufacturing routine for a part. For the purpose of learning in terms of constructing process plans, explanation-based learning is an appropriate approach. There exist strong domain theory and heuristics for problem-solving in machining process planning. The task for a computer-aided process planning system is how to learn more effective planning knowledge from its problem solving. The study presented here is focused on the development of an explanation-based learning method for the task of process planning for machining. A typical PUMA6 CNC turning machine which is a single turret engine lathe is used in the development and validation of the learning methodology for machining process planning.

In the following sections, the characteristics of axisymmetric parts are discussed. Then the scheme of an explanation based learning system for machining process planning is proposed. Knowledge representation and knowledge organization in the system are explained. An example is given to explain the information flow in the system.

2. CHARACTERISTICS OF AXISYMMETRIC PARTS

Most axisymmetric parts are produced in lathes in a manufacturing site. Normally a lathe provides operations of turning, boring, facing, grooving, chamfering and threading, and end drilling. Process planning for lathes refers to selecting operations with corresponding cutting conditions and tools, and establishing their execution order and tool path. The requirements of tolerance and surface roughness, geometry shape and dimensions, material and property of a part all contribute to the selection of operations, cutting conditions and tools.

2.1. Features of Axisymmetric Parts

Axisymmetric parts have simple features and are produced on lathes mostly. For the purpose of this study, feature is a kind of geometry which can be formed in one type of tool movement with associated dimension and quality requirement. Features can be further divided as geometry feature, dimension feature, and quality feature. Figure 1 lists the elemental geometric features which can be processed in a turning machine. The features include cylinder, taper, arc, end face, groove, chamfer, and thread on both external and internal surfaces. The dark lines mark the surfaces created by one tool movement. We further define cylinder, taper, and end face as base features since they construct the basic shape of an axisymmetric part. Dimension features are the necessary parameters to describe a geometry feature. Quality features are the quality requirement of a geometry feature. The quality requirements include tolerance and surface roughness. For example the dimension feature of a cylinder is its diameter and length. The tolerance of the diameter and the surface roughness are the associated quality features.

2.2. Activities in Process Planning for a CNC Lathe

The activities in process planning for a CNC lathe include stock selection, operation selection and sequencing, machining parameter selection, machine and tool selection, tool path planning, coolant selection, and code generation. These activities are standard for other types of machines as well. For the purpose of developing a learning system for machining process planning, the focused activities are operation selection and sequencing.

Operation selection is the activity of selecting proper operations to change the geometry of stock to a required shape. The operations provided by the PUMA6 CNC lathe include turning and boring, grooving and necking, threading and chamfering, facing and

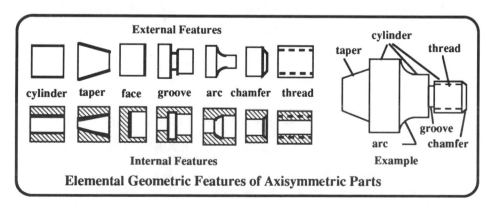

Figure 1. Elemental Geometric Features of Axisymmetric Parts

partition, and end drilling. Each type of operations can produce a certain type of surfaces with certain quality. A selected operation is only a feasible step to generate an elemental geometry. For a complex geometry, the execution order of selected operations becomes important. The geometry of a part is the main clue in operation selection as well as operation sequencing. They are subjected to operation constraints, tooling constraints, and quality constraints as well.

Tool selection is an activity of choosing proper tools for a type of cutting operation. It has a tight relation with selected operations. The CNC engine lathe is equipped with single-point tools. Tool geometry and tool material are the big considerations in tool selection. Machining parameters are cutting speed, tool feed rate, and depth of cut for one path. These parameters are critical for machining processes Rough cut and finish cut are distinguished by the value sets of parameters. Tool life, power consumption rate, production rate, cost, and surface quality all contribute to the selection of parameters.

3. EBL FOR MACHINING PROCESS PLANNING: SYSTEM SCHEME

An explanation-based learning system for machining process planning is presented here. The learning system focuses on learning for better problem-solving. It increases its planning knowledge and improves its planning ability from its problem-solving for presented problems. The model of a process plan is constructed using existing operations. A macro operation is generalized from the model. The developed macro operation can be used to construct plans for similar parts.

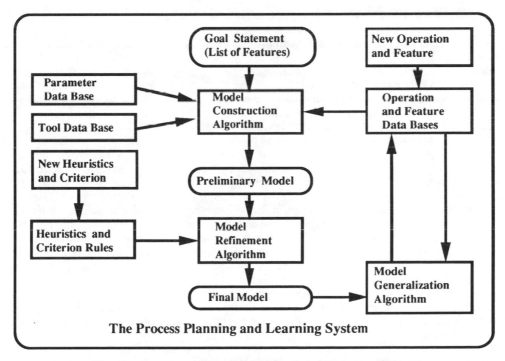

Figure 2. The Process Planning and Learning System

The process planning learning system is illustrated in Figure 2. It consists of three algorithms and supported data bases and rule bases. The three algorithms perform preliminary model construction, model refinement, and model generalization functions. They carry out the process planning activities of operation selection and sequencing, tool selection, and machining parameter selection. And more importantly, they conduct the learning scheme.

We assume that a part is cut from a cylindrical stock material which gives the minimum scrap. The planing task is to construct a plan which can effectively transfer an object from its initial state (stock) to its final state (designed part). The goal statement is a list of features which describe the final state of an object. It includes three types of feature description which are geometry description, dimension description, and quality description of features. The geometry description defines the type of a feature. The dimension description specifies the values of geometry parameters of the feature and the quality description specifies the quality requirement for the feature. The geometry description is called as the symbolic goal specification also, and the others construct the numerical goal specification.

The model construction algorithm carries out the preliminary selection of operations, tools, and machining parameters. The preliminary model is refined according to heuristics and criterion rules. The refinement results a final model which is a better selection of operations and tools, and a better operation sequence.

The final model is the source for generation of macro operations. It is transferred into a temporary operation frame. Then the temporary frame is generalized against existing macro operations. A new frame is created or a more general frame is refined into the knowledge base, which represents the increase of the system planning ability.

4. PROCESS PLANNING KNOWLEDGE REPRESENTATION

The process planning background knowledge is the part of domain theory in the learning system. The domain theory is the knowledge to describe the operation and features, and their relations. The process planning background knowledge is represented in frame here. There exist three types of knowledge frames: the operation routine frame which describes an essential operation in detail, the elemental feature frame which establishes the relation between the feature and operation frames, and the macro operation frame which holds a sequence of operations and the corresponding compound feature.

The operation routine frame is the representation of operation knowledge which shows the routine of a single operation, including corresponding tool type, cutting path, necessary input, parameters to be determined, and so on. The operations include rough turning and finish turning, threading, grooving, chamfering, and end drilling. Each type except drilling can further divided into internal and external operations, right cut, left cut, and neutral cut operations. An operation frame has a *command* slot which is the name of the operation in which the operation can be referred. The *input* slot indicates the necessary data to determine parameters. The *parameter* slot lists the functions of determining tool, machining parameters, clearance of path, and allowance to finish. The *constraint* slot lists the range of machining parameters, the suitable size of feature, and the quality range. And the *action* slot gives the tool path or movement.

The elemental feature frame contains a decision table which establishes the connection of the feature to appropriate operations and tool types. Taking necessary input from a goal (feature), A selection of proper operation and tool type is reached from the decision table.

A macro operation is a generalized operation sequence for a type of combined feature. A macro operation frame contains at least three slots. The *command* slot specifies the access of the macro operation. The *compound feature* slot gives the combined feature which is processed by the macro operation. The slot of *actions* contains a series of operation commands and tool pointers in sequence. The order they are listed in the slot is the sequence of the operations.

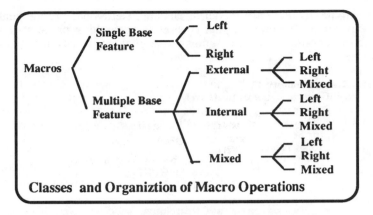

Classes and Organiztion of Macro Operations

Figure 3. The Classes of Macro Operations

Macro operations arc grouped as a class in terms of the properties of compound features. The classes are nested in the knowledge base, as illustrated in Figure 3. The macros with single base feature are established with the system. However, multiple base feature macros are expected to be learned by the system. Under the class of multiple base features, there are classes of External, Internal, and Mixed. Under each of the three classes, there are three other classes of Left, Right, and Mixed. A macro, with its compound feature consisting external and right direction features only, is in the class of External-Right. The increase of macro operation frames in the system represents the self-improvement of process planning knowledge and problem-solving ability.

5. MODEL CONSTRUCTION AND GENERALIZATION

The learning system acquires planning knowledge from its own problem solving. At first, it constructs a preliminary model using existing factual information and domain theory. The preliminary model is refined according to heuristics and planning criteria. A final model which satisfies the criteria is obtained and generalized into a macro frame which is saved into the knowledge base and facilitates planning of similar parts. The knowledge increase in this manner represents the improvement of system planning ability.

5.1. Preliminary Model Construction

One of the tasks of the learning system is to construct preliminary models for the input features of a part or the goal specification of a given problem. A preliminary model is the initial explanation for the goal specification. It is the list of the operations each of which is related to an elemental or compound feature.

A preliminary model is constructed using macros and operation routines step by step. Figure 4 illustrates the algorithm of preliminary model construction. The goal statement is copied to the first element of goal set GOAL[1]. The current goal pointer GP and the number of elements of goal set are set to 1 at beginning. The GP pointed goal becomes the current goal. A macro whose compound feature has a same class as the current goal is read in. The macro whose compound feature covers the current goal is the one being searched for. It is copied to the operation list array OP[GP]. Its features and actions are modified to fit the current goal. Next goal becomes the current goal and the searching starts again. If no macro in the class is found to cover it, the current goal is divided into a set of sub-goals using heuristics. Then the sub-goals are inserted back to the goal set from the point of GP pointed. The current goal is reset and searching process starts again. The searching process continues until all goals in the goal set are covered by macros. At the end, the operation array is the preliminary model.

```
Begin Construction
    Copy Goal Statement to GOAL[1]
    Set GP=1, K=1
    CURRENT_GOAL = GOAL[GP]
        Begin Searching
            Read in a macro
                If CURRENT_GOAL covered by macro.feature
                    Copy macro to OP[GP]
                    Modify OP[GP].feature and OP[GP].action
                    GP ++
                    Jump out Searching
            Until no one left in the class
            Divide CURRENT_GOAL to SUBGOAL[1..p]
            Shift GOAL[GP+1..K] to GOAL{GP+p..K+p-1}
            Insert SUBGOAL[1..p] to GOAL[GP..GP+p-1]
            K=K+p-1
        End Searching
    Until GOAL[GP] is empty
End Construction
```

Figure 4. The Algorithm of Preliminary Model Construction

If the goal set formation fails, it suggests the incomplete of domain theory or factual knowledge. New frames may need to complement the background knowledge. The output of the algorithm is the preliminary model of the plan which is the list of preliminarily selected operations, tools, and parameters for all goals in the sequence of listed features.

5.2. Model Refinement for Satisfaction

For each goal, the selected operations and tools provide a feasible solution to it. The preliminary models may provide the feasible solution to the goal specification as well. However, the solution may not be acceptable due to its redundancy in operation and tool selection, inefficiency in operation sequence, or contradictions among the operations and tools. The preliminary model needs modification to become acceptable.

The model refinement is the procedure of using planning heuristics and planning criteria. They are represented in production rules. The model refinement is a reasoning process of the production rules. The operations in preliminary model are generalized and sequenced according to specified criteria and heuristic sequences. The associated tool selection is modified as well to accomplish the operation generalization and sequencing. After the refinement, the final model is obtained. The resulting model is a better solution for the same problem.

5.3. Model Generalization for Learning

The learning issue in the system is to generalize the final model and develop macro operations. The developed macro operations represent the increase of planning knowledge of the system. The planning ability of the system is improved by adding the developed macro operations.

The generalization algorithm which is given in Figure 5 is the core of the learning scheme. First the generalization algorithm takes the goal specification, the preliminary model, and the final model as its input. The number of macros in the model is counted and the classes of the macros are checked to determine the case of generalization for the model. A model constructed by directly applying one macro operation without any expansion is the first case. No generalization is required in this case. A model constructed by one macro operation with certain expansion is case 2, in which the macro is modified so that it includes the expansion. If more than one macros are used in the model construction and all the macros are come from the same class, a new macro is generated from the final model. The new macro is then modified to cover all original macros which construct the final model. The new macro frame is added to the class and the original macro frames are removed from the class. This is the third case. If the macros in the model belong to more than one classes, this is the case 4. The generalization is divided into two steps. First, macros with same class in the model are grouped together. A group of macros in preliminary and final models, combining with their associated goals, construct one sub-task of generalization. The sub-level generalization is in class generalization as case 1, case 2, or case 3. For group i, the input is P[i] the corresponding portion of macros in preliminary model, F[i] in final model, and G[i] the corresponding portion of goal sets. After first step in class generalization, a new macro is generated to combine macros belonging to different classes. The new macro is saved into the class of Mixed.

```
Begin  Generalization
    Read in the preliminary model, the final model, and the goal statement
    Begin  Case  Generalizing
        Check  the  models' composition
        Switch  to  a  corresponding  case
            CASE  1
                End  Case  Switching
            CASE  2
                Modify  the  macro  with  the  expended  features
            CASE  3
                Generate  a  new  macro  frame  with  the  final  model
                Modify  the  macro  to  fit  its  original  macros
                Remove  the  original  macros  from  the  class
            CASE  4
                Generalization  within  class
                    Group  same  class  macros  in  preliminary  model
                    Form  P  set,  F  set,  and  G  set  respectively
                    Call  Case  Generalizing  using  P[i],  F[i],  and  G[i]  for  each
                    group
                Generalization  in  new  class
                    Generate  a  new  macro  belonging  to  a  new  (mixed)  class  with
                    the  final  model
                    Check  the  coverage  of  the  macro  in  the  class
                    Modify  the  new  macro  to  fit  all  covered  macros
                    Remove  the  covered  macros  from  the  class
    End  Case  Generalizing
End  Generalization
```

Figure 5. The Algorithm of Model Generalization

6. AN EXAMPLE

In this section, an example is presented to illustrate how the learning system works. The engineering drawing of an axisymmetric part is shown in Figure 6. The major features of the part are cylinders and a thread. The part is a positive example of domain. The problem solving task is to create a process plan for the given part.

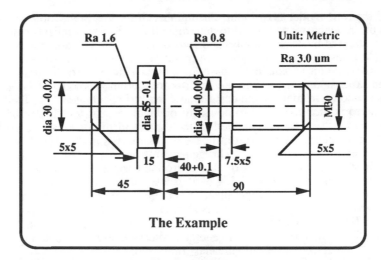

Figure 6. The Engineering Drawing of The Part

The feature file of the part is derived from the drawing and shown in Table 1. There are total of nine elemental geometry features in this part. Each geometry feature is listed with associated dimensional feature and quality requirements. The important features are the cylinder and steps which have high requirements of tolerance and surface roughness.

An appropriate material stock, say with diameter 60 mm is selected, which is based on the criterion of minimum scrap. The problem becomes how to construct an operator to connect the initial state with the goal state, or how to make a process plan to convert the material stock to the final part. Formally, with the decision of initial state (material stock), the problem can be restated as to generate an operator (process plan) which transfers the initial state (material stock) to the goal state (part).

Table 1. The Feature File and the Goal Set of the Part

The Feature File and the Goal Set of the Given Example				
Goal	F_No	Geometry	Dimension	Quality
1	1	right Face and	D30	Ra 3.0
2	2	Chamfer on	5x5	Ra 3.0
	3	right Thread with	M30, L50	Ra 3.0
	4	Groove and	D20, L7.5	Ra 3.0
3	5	right Cylinder and	D40, L40	Ra 0.8, DT -0.005, DL 0.1
4	6	Open Cylinder and	D55, L15	Ra 3.0, DT -0.1
5	7	left Cylinder with	D30, L40	Ra 1.6, DT -0.02
	8	Chamfer and	5x5	Ra 3.0
6	9	left Face	D30	Ra 3.0

The model construction algorithm forms a goal set by dividing the goal step by step. At beginning, the operation knowledge base contains only the operation routines and single base feature macro operations. The goal set is formed with one base feature golas, which is shown in Table 1 as well.

A preliminary model is constructed simultaneously. A set of operations and corresponding tools are selected. Those operations and tools are listed in the order of goals as shown in Table 2. The operation column records the operations for each goal. Each operation is given by its name and cutting direction. The word following the dash represents the direction of tool movement. **Right** indicates tool moving from right to left and **Left** means tool moving from left to right. **In** represents tool moving into the part in radius and **Out** indicates tool moving away the part in radius. Tools are given by their identification name too. **OD** and **ID** are the abbreviations of outside diameter and inside diameter and stand for external and internal cut respectively. **RH**, **LH** and **N** indicate the tool type as right hand tool, left hand tool, and neutral tool respectively. The following number indicates the nose angle for neutral tools and cutting edge angle for right hand and left hand tools. Then follows the series number of tools.

The model is inefficient in operation and tool selections, and operation sequence. Redundant operations exist in the model. The Chamfering-Left operation conflicts with the CutOff-In operation since the former needs the later to open the left end for it. Model modification is required. The refinement of model is relied on the model evaluation criteria and the expertise of operation sequencing. Table 3 presents the final model which lists the operations, tools, processed features, and machining parameters respectively.

Table 2. The Preliminary Model

\multicolumn{5}{c}{The Preliminary Model of the Example}
Goal	Macro	Operation	Tool	Feature
1	Ext-R-Cylinder	RoughFacing-In	OD-RH-75-1	right Face
		FinishFacing-In	OD-RH-75-1	
2	Ext-R-Cylinder	RoughTurning-Right	OD-RH-90-1	right Cylinder
		FinishTurning-Right	OD-RH-90-2	
		Grooving-In	OD-N-180-1	Groove
		Threading-Right	OD-N-60-1	
		Chamfering-Right	OD-N-90-1	Chamfer
3	Ext-R-Cylinder	RoughTurning-Right	OD-RH-90-1	right Cylinder
		SemiTurning-Right	OD-RH-90-2	
		FinishTurning-Right	OD-RH-90-2	
4	Ext-R-Cylinder	RoughTurning-Right	OD-RH-90-1	Open Cylinder
		FinishTurning-Right	OD-RH-90-2	
5	Ext-L-Cylinder	RoughTurning-Left	OD-LH-90-1	left Cylinder
		FinishTurning-Left	OD-LH-90-2	
		Chamfering-Left	OD-N-90-1	Chamfer
6	Ext-L-Face	CutOff-In	OD-N-180-2	left Face

The final model removes redundancy, inefficiency, and contradiction from the preliminary model. Comparing with the preliminary model, the final model requires fewer tools and operations. All RoughTurning-Right operations are unified to one operation. All FinishTurning-Right operations except the one for the right cylinder are unified to one operation too. The Chamfering-Right is absorbed by FinishTurning-Right and the Chamfering-Left is absorbed by CuttOff-In. The FinishTurning-Right for the right step remains unchanged since it is necessary step to create a close tolerance and a high surface finish. Redundant operations are removed in the final model. The contradiction of Chamfering-Left and CutOff-In is solved too.

The final model is the resource for the generalization of macro operations. First the goal specification and its final model are filed in the temperate macro operation frame. Then the generalization algorithm checks the model construction. Since multiple macro frames are used in the model and the compound feature belongs to the class of External-Mixed, it is generalization case 4. The generalization includes two steps. At the first step, all contingent right macros are generalized into a new macro which is a multiple right cylinder macro in the class of External-Right. The new macro is directly saved into the knowledge

Table 3 The Final Model

#	Operation	Tool	Features	Parameters
	The Final Model of the Given Example			

#	Operation	Tool	Features	Parameters
1	RoughTurning-Right	OD-RH-90-1	right Cylinder, open Cylinder, left Cylinder	DC=2mm/path; CS=200m/min; FR=0.5mm/rev.
2	RoughTurning-Left	OD-LH-90-1	left Cylinder	DC=2mm/path; CS=200m/min; FR=0.5mm/rev.
3	FinishTurning-Left	OD-LH-90-2	left Cylinder	DC=0.1mm/path; CS=350m/min; FR=0.2mm/rev.
4	FinishTurning-Right	OD-RH-90-2	right Face, Chamfer, right Cylinder, open Cylinder	DC=0.1mm/path; CS=350m/min; FR=0.2mm/rev.
5	FinishTurning-Right	OD-RH-90-2	right Cylinder	DC=0.005mm/path; CS=400m/min; FR=0.1mm/rev.
6	Grooving-In	OD-N-180-1	Groove	FR=0.2mm/rev; CS=150m/min
7	Threading-Right	OD-N-60-1	Thread	DC=0.2mm/path; DCF=0.024mm; CS=150 rev/min; FD=3.5mm/rev
8	CutOff-In	OD-N-180-2	Chamfer, left Face	CS=150m/min; FR=0.2mm/rev

base since no multiple base feature macros exist at its class. Since only one left macro is used, no generalization is necessary. At the second step, A new macro is generated to combine the right macro and left macro. The new macro belongs to the class of External-Mixed.The abstract of the second generated macro is illustrated in Figure 7. Both generalized macros can be used for similar part planning.

Macro Operation Frame

Command: Ext-Mix-Cylinder

Compound_Feature: 1. right Cylinder (1) Thread (2) Groove
                        2. open Cylinder
Actions: 3. left Cylinder

Operations:	*Tools:*	*Feature*
RoughTurning	OD-RH-90	1, 2
RoughTurning	OD-LH-90	3
FinishTurning	OD-RH-90	1,2
FinishTurning	OD-LH-90	3
Grooveing	OD-N-180	1(2)
Threading	OD-N-60	1(1)

Figure 7. One Generated Macro Operation Frame

7. CONCLUSION

Manufacturing process planning is the key activity to connect a product design with manufacturing processes. Machine learning is a promising approach for the automation of process planning system construction. The study presented focuses on developing an explanation-based learning methodology to solve the problem of machining process planning and to improve its planning ability from its problem solving. The system learns process planning knowledge by solving positive problems using known background knowledge, then generalizing the solution. The generalized knowledge can be used beneficially for planning similar parts. The process planning system developed in this manner has the properties of robust, adaptiveness, and self-maintenance.

REFERENCES

1　Chang T.C. and Wysk R.A., *An Introduction to Automated Process Planning Systems,* Prentice-Hall, Inc., 1985

2　Gupta T. and Ghosh B.K., A survey of expert systems in manufacturing and process planning, *Computers in Industry,* v11, pp. 195-204, 1988

3　Wang H.P. and Wysk R. A., Intelligent reasoning for process planning, *Computers in Industry,* v8, pp. 293-309, 1987

4　Chang T.C., *Expert Process Planning for Manufacturing,* Addison-Wesley Publishing Company, 1990

5　Schakoff R.J., *Artificial Intelligence: An Engineering Approach,* McGraw-Hill Publishing Company, 1990

6　Michalski R.S., A theory and methodology of inductive learning, *Artificial Intelligence,* v20, pp. 111-161, 1983

7　Segre A.M., *Machine Learning of Robot Assembly Plans,* Kluwer Academic Publishers, 1988

8　Mitchell T.M., Keller R.M., and Kedar-Cabelli S.T., Explanation-based generalization: a unifying view, *Machine Learning,* v1, pp. 47-80, 1986

9　Lu S.C. and Zhang G., A combined inductive learning and experimental design approach to manufacturing operation planning, *Journal of Manufacturing Systems,* v9, n2, pp. 103-115, 1990

10　Gupta U.K. and Ali M., Hierarchical representation and machine learning from faulty jet engine behavioral examples to detect real time abnormal conditions, *Proceedings of The 1st International Conference on Industrial and Engineering Applications of Artificial Intelligence and Expert Systems,* pp. 710-720, 1988

11　Ke M.and Ali M., MLS, a machine learning system for engine fault diagnosis, *Proceedings of The 1st International Conference on Industrial and Engineering Applications of Artificial Intelligence and Expert Systems,* pp. 721-727, 1988

12　Subramanian C. and Ali M., Detecting abnormal situations from real time power plant data using machine learning, *Proceedings of The 1st International Conference on Industrial and Engineering Applications of Artificial Intelligence and Expert Systems,* pp. 738-746, 1988

13　Walburn D.H. and Powner E.T., Automated acquisition of knowledge for an expert system for process control, *IEE Proceedings,* part E, v136, n6, pp. 548-556, 1989

14　Park S.C., Piramuthu S., Raman N., and Shaw M.J., Integrating inductive learning and simulations in rule-based scheduling, in Gottlob, G. and W. Nejdl (eds.): *Expert Systems in Engineering Principle and Applications,* Springer-Verlag, 1990, pp 152-167

RESTRAINT ANALYSIS OF ROBOT GRIPPERS

Edward C. De Meter

Pennsylvania State University, State College, Pennsylvania, U.S.A.

ABSTRACT

Previous works have applied restraint analysis to robot grippers. However they have been limited to point contact models. This paper presents models of planar, spherical, cylindrical, and frictional planar contact. It presents two algorithms for determining whether a gripper provides total restraint. Finally it presents a general algorithm for evaluating whether gripper actuator intensities are sufficient to insure part restraint throughout transport. An illustrative example is provided.

INTRODUCTION

The primary function of a robot gripper is restrain a part during transport. Two questions regarding the design of a robot gripper are: 1) does it provide total restraint, and 2) are gripper actuator intensities sufficient to hold the part during transport. These questions can be answered through the application of restraint analysis.

Restraint analysis combines screw theory and algebra to describe the true instantaneous freedom of a rigid body restrained by contact. True freedom reflects the ability of an object to move relative to multiple contact surfaces. This is in contrast to kinematic freedom which reflects the ability of an object to move relative to multiple contact surfaces and still remain in contact. Restraint analysis formulations have been based on static equilibrium or virtual power. Statics equilibrium analysis is used to determine whether the forces provided by contact regions are sufficient to resist all external forces acting on an object. Using a static equilibrium formulation, Lakshminarayana[1] showed that a minimum of seven points of frictionless contact are needed to provide total restraint. Salisbury and Roth[2] showed that for a set of point contacts to provide total restraint, the null space of a matrix containing their respective wrenches must contain at least one strictly positive vector. Kerr and Roth[3] formulated a linear program to determine the optimum set of intensities for frictional point contacts to resist an external load without slippage. Li and Sastry[4] developed three quality measures for evaluating the stability of a grasp. Chou et al[5] developed a synthesis methodology for determining the location of point contacts in three orthogonal planes. Chou[6] developed a methodology for evaluating the ability of frictionless point contacts to resist simple sequences of forces. Markenscoff et al[7] proved that a minimum of four frictional point contacts are needed for total restraint.

Ohwovoriole and Roth[8] introduced the concepts of repelling and contrary screws. They introduced the concept of true freedom as a means of describing all possible ways in which an object may move in space. They also showed that if a twist is reciprocal/repelling to the wrenches supplied by the vertices of a convex planar region, then it is reciprocal/repelling to all wrenches supplied by the region. Ohwovoriole[9] was the first to use virtual power to analyze rigid body restraint. He showed that a set of point contacts totally restrain a body if there are no screws about which the body may twist without causing negative virtual power. Nguyen[10] used virtual power to examine the geometric conditions for which a set of point contacts provide total restraint. Assada and Kitawaga[11] developed a linear program for determining whether a set of frictionless point contacts provide total restraint. Chou et al[12] formulated a virtual power model to determine whether a machining fixture provides total restraint. Bausch and Youcef-Toumi[13] developed the concept of a motion stop, which measures the virtual work at a frictionless point contact with respect to a single twist. Other related works include Sturges[14], Montana[15], and Assada and Andre[16].

Robot grippers often rely on frictional surface contact. Unfortunately the papers described thus far do not address this area. This paper presents finite models of frictionless and frictional surface contact. It presents linear programming formulations based on virtual power and static equilibrium which can be used to determine whether the contact surfaces of a gripper provide total restraint. Finally it presents a general formulation for determining the sufficiency of gripper actuator intensities during part transport.

TWISTS, WRENCHES, AND RATE OF VIRTUAL WORK

The instantaneous velocity of a rigid body is minimally described by the rotational velocity, $\bar{\Omega}$, of the body about a point and the translational velocity, $\bar{v}$, of the point. Associated with a body in motion is a unique, instantaneous screw axis, Ψ. For each point on Ψ, $\bar{\Omega}$ and $\bar{v}$ share the same direction, $\bar{d}$, as Ψ. Defining ω and ς as the intensities (signed magnitudes) of $\bar{\Omega}$ and $\bar{v}$ respectively, the ratio, κ: $\kappa = \varsigma/\omega$ is called the pitch of Ψ. In combination, $\bar{\Omega}$ and $\bar{v}$ define a twist, $\bar{t}$.

Analogously, a system of forces and couples acting on a rigid body can be resolved into a single force, $\bar{f}$, and couple, $\bar{c}$, directed along a unique screw axis, Ψ. Defining λ and ρ to be the intensities of $\bar{f}$ and $\bar{c}$ respectively, $\kappa = \rho/\lambda$. In combination, $\bar{f}$ and $\bar{c}$ define a wrench, $\bar{w}$. $\bar{t}$ and $\bar{w}$ may be represented as dual vectors in R^6 such that

$$\bar{t} = [\bar{\Omega}^T, \bar{v}_o^T]^T \tag{1}$$

$$\bar{w} = [\bar{f}^T, \bar{m}_o^T]^T \tag{2}$$

with:
$$\bar{v}_o = (\bar{r} \times \bar{\Omega}) + \bar{v} \tag{3}$$

$$\bar{m}_o = (\bar{r} \times \bar{f}) + \bar{c} \tag{4}$$

where: $\bar{r}$ = an arbitrary point on Ψ,

$\bar{v}_o$ = translational velocity of the reference origin, and

$\bar{m}_o$ = moment about the reference origin.

The virtual power, W, of a body undergoing a twist, $\bar{t}$, while being acted upon by a wrench, $\bar{w}$, is the permutated dot product, *, between the two vectors such that:

$$W = \bar{f} \cdot \bar{v}_o + \bar{m}_o \cdot \bar{\Omega} \tag{5}$$

If $W = 0$, then $\bar{t}$ and $\bar{w}$ are reciprocal[17]. If $W > 0$, then $\bar{t}$ and $\bar{w}$ are repelling. If $W < 0$, then $\bar{t}$ and $\bar{w}$ are contrary[8].

FINITE MODELS OF SURFACE CONTACT

When a body is restrained by a contact region, S, all points within that region are capable of exerting an infinite number of reactive forces (wrenches of zero pitch) on the body. If contact is frictionless, each point, $\bar{r}$, can exert an infinite number of forces which pass through the point and toward the body along the surface normal. The intensity of each force within this set is nonnegative and is dependent upon body dynamics as well as other wrenches acting on the body. If contact is characterized by columbic friction, each point can exert an infinite number of forces which exist within a friction cone. The apex of the cone is the contact point and its axis is directed along the surface normal. As with the frictionless case, the intensity of each force within this set is nonnegative.

To analyze the restraint of a body in contact with multiple regions, it is necessary to model the forces at each point within the contact regions. This is a formidable task unless the relationship between these forces is understood. Fortunately virtual power analysis provides insight into this relationship.

It is well known that if a body is restrained by a finite set of wrenches $\{\bar{w}_1, \ldots, \bar{w}_n\}$, its true freedom is described by all twists which are reciprocal or repelling to all wrenches in this set[9]. Mathematically this is expressed as:

$$A\bar{t} \geq 0 \tag{6}$$

where: $A = \begin{vmatrix} \overline{m}_{o_1}, & \overline{f}_1 \\ . & . \\ . & . \\ . & . \\ \overline{m}_{o_n}, & \overline{f}_n \end{vmatrix}$

All twists which satisfy [6] exist within a closed polyhedral cone. In addition, all twists lie within the nonnegative linear hull of $\{\overline{t}_1, \ldots, \overline{t}_q\}$, where $\overline{t}_i$ is an extreme direction of the cone and $q \geq n$. However all twists which satisfy [6] are also reciprocal or repelling to all wrenches within the nonnegative linear hull of $\{\overline{w}_1, \ldots, \overline{w}_n\}$. As a consequence, if the set of wrenches, w, provided by a contact region exist within a nonnegative linear hull, only the extreme directions of the set need to be modeled.

Ohwovoriole and Roth[8] proved that for frictionless, convex, planar contact regions, w is the nonnegative linear hull of a finite set of wrenches which act through the bounding vertices of the contact region. De Meter[18] expanded this work to prove that the same is true for restricted cases of frictionless spherical, and cylindrical contact. For a spherical region, the restriction is that it is bounded by n vertices and n circular arcs and that the included angle between any two surface normals is less than π. Circular contact is considered a degenerate case. For a cylindrical region, the restriction is that it is bounded by 4 vertices, 2 circular arcs, and 2 lines and that the included angle between any two surface normals is less than π. Examples of these regions and their spanning wrenches are provided in Figure 1.

For frictional point contact, an excellent approximation is achieved if w is modeled by the extreme directions of a polyhedral cone inscribed within the friction cone. Likewise for frictional convex, planar contact, excellent approximations are achieved if w is modeled with the extreme directions of polyhedral cones inscribed within the friction cones at the bounding vertices. Examples of these contact regions and their spanning wrenches are provided in Figure 2. Marginal approximations are achieved if a similar technique is applied to restricted spherical, cylindrical, and circular contact regions.

If w consists of the wrenches exerted by multiple contact regions, then w is spanned by the union of the extreme directions of the individual contact regions. Thus entire planar, spherical, and cylindrical surfaces can be modeled as composites of the restricted regions described above.

ANALYSIS OF TOTAL RESTRAINT

While held by a gripper, a part is often acted upon by small, random, external forces. To determine if the gripper can resist these external forces, total restraint analysis is applied. Assume that a gripper contacts a part over multiple regions such that w is spanned by the wrenches $\{\overline{w}_1, \ldots \overline{w}_n\}$. To determine whether w completely restrains the workpiece, virtual power analysis or statics analysis is applied. Using virtual power, we determine whether the workpiece can experience a twist

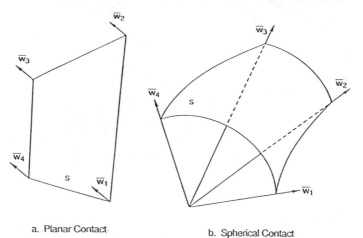

a. Planar Contact   b. Spherical Contact

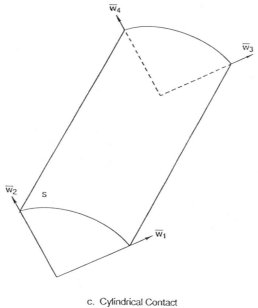

c. Cylindrical Contact

Figure 1: Frictionless Contact

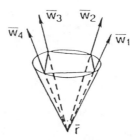

b. Planar Contact

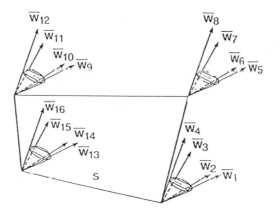

a. Point Contact

Figure 2: Frictional Contact

which will result in either zero or positive virtual power at each contact region. Such a twist exists if it satisfies [6]. If no nonzero solution to [6] exists, then the workpiece is completely restrained[9].

A linear program may be formulated to determine whether a nonzero solution to [6] exists. To do so, [6] is transformed into standard format by replacing $\bar{t}$ with $\bar{t}_1 - \bar{t}_2$, where $\bar{t}_1$ and $\bar{t}_2$ are strictly nonnegative vectors. In addition, a nonnegative slack vector, $\bar{t}_s$, is subtracted to change the virtual power constraints to equalities. This results in the following:

$$A\bar{t}_1 - A\bar{t}_2 - \bar{t}_s = [0] \qquad\qquad [7]$$

$$\bar{t}_1 \geq [0], \ \bar{t}_2 \geq [0], \ \bar{t}_s \geq [0]$$

To search for a solution, Assada and Kitawaga[11] advocate sequentially constraining each element within the resultant of $\overline{t}_1 - \overline{t}_2$ to be a positive or negative constant. However an alternative approach is to sequentially maximize and minimize each element within $\overline{t}_1 - \overline{t}_2$. The resulting formulation is:

For i = 1 ... 6,   [8]

    Maximize $t_{1_i} - t_{2_i}$ subject to [7]

    Minimize $t_{1_i} - t_{2_i}$ subject to [7]

If a nonzero solution to a subproblem of [8] is found, it will be unbounded, since all twists which satisfy [7] exist within a cone. In addition, it will also be an extreme direction of the set. If no nonzero solutions are found, then the gripper completely restrains the body. Note that both [8] and Assada and Kitawaga's formulation require 12 iterations.

Using static equilibrium analysis, it is necessary to determine whether a wrench exists within w which can resist an arbitrary external wrench, $\overline{w}_e$, acting on the body. This is equivalent to determining whether $W = R^6$. Since w is the nonnegative linear hull of $\{\overline{w}_1, \ldots \overline{w}_n\}$, this is represented as:

$$B\overline{\lambda} = -\overline{w}_e \qquad\qquad\qquad [9]$$

$$\overline{\lambda} \geq [0]$$

where: $B = [\overline{w}_1, \ldots, \overline{w}_n]$ and

$$\overline{\lambda} \in R^n$$

[9] will always have a solution if B is full rank, and if a strictly positive vector, $\overline{\lambda}_h$, exists within the null space of B. From linear algebra it is known that the null space of B is minimally represented as the linear hull of $\{\overline{\vartheta}_1 \ldots \overline{\vartheta}_r\}$, where r = n- rank (B) and $\overline{\vartheta}_i \in R^n$. Let C = $[\overline{\vartheta}_1 \ldots \overline{\vartheta}_r]$. Then $\overline{\lambda}_h = C\overline{\sigma}$ where $\overline{\sigma} \in R^r$. Salisbury and Roth[2] have shown that if w completely restrains the body, a solution must exist for:

$$C\overline{\sigma} > [0] \qquad\qquad\qquad [10]$$

All solutions to [10] exist within an open polyhedral cone. Since an open polyhedral cone and its closed counterpart share the same extreme directions, [10] can only have a solution if:

$$C\overline{\sigma} \geq [0] \qquad\qquad\qquad [11]$$

has a nonzero solution. This may be determined using a formulation similar to [8]. Unfortunately a solution to [11] may only satisfy $C\overline{\sigma} = 0$, and thus must be checked. A more direct approach is to formulate a linear program whose objective function is to force $C\overline{\sigma}$ to be strictly positive. This is done by changing [11] into an equality by subtracting a strictly positive vector, $\bar{\sigma}_s$, from $C\bar{\sigma}$, and creating an objective function which attempts to drive each element of $\bar{\sigma}_s$ positive. In standard form this is:

Maximize e [12]

subject to: $C\bar{\sigma}_1 - C\bar{\sigma}_2 - \bar{\sigma}_s = [0]$

$e \leq \sigma_{s_i} \forall i$

$\bar{\sigma}_1 \geq [0], \bar{\sigma}_2 \geq [0], \bar{\sigma}_s \geq [0]$

If a positive solution, Ξ, is found to [12], then a strictly positive vector exists within the null space of B, and the part is completely restrained.

GRIPPER ACTUATOR INTENSITY ANALYSIS

Grippers are mechanisms which rely on force and torque actuation to exert contact forces. During part transport, it is critical that gripper actuator intensities are sufficient for restraint. The following model is derived to determine actuator intensity sufficiency.

Assume that a gripper consists of M fingers which contact a part along M regions. Fingers 1, ..., L are actuated. Assume that finger j consists of ι_j actuators and that $\bar{\alpha}_j$ is comprised of their intensities. Assume that finger j exerts a wrench, $\bar{w}_j$. If finger mass and moment of inertia are small relative to the part, then the effects of gravity and inertia forces on the finger can be ignored. In this case, a direct linear relationship exists be between α_j and $\bar{w}_j$. This relationship is expressed as:

$\bar{\alpha}_j = F_j \bar{w}_j$ [13]

$\bar{\alpha}_j \geq [0]$

where: F_j = a $\iota_j \times 6$ actuator intensity constraint matrix

Assume that the wrenches exerted by contact region j on the part are spanned by $\bar{w}_{j_1}, \ldots, \bar{w}_{j_{\tau_j}}$, where τ_j equals the number of spanning wrenches. Then $\bar{w}_j$ must also satisfy:

$\bar{w}_j = W_j \bar{\lambda}_j$ [14]

$\bar{\lambda}_j \geq [0]$

where: $W_j = [\bar{w}_{j_1}, \ldots, \bar{w}_{j_{\tau_j}}]$, and

$\bar{\lambda}_j$ = the intensity vector for $\{\bar{w}_{j_1}, \ldots, \bar{w}_{j_{\tau_j}}\}$

Combining [13] and [14] the following constraints are derived for finger j:

$$\overline{\alpha}_j = F_j W_j \overline{\lambda}_j \qquad\qquad [15]$$

$$\overline{\lambda}_j \geq [0]$$

During transport, the part is subject to gravitational forces, inertia forces, and contact forces. Assume that the travel time is $t^{\cdot}$ seconds. Let $\overline{\xi}(t)$ and $\overline{X}(t)$ be the instantaneous inertia force and inertia moment acting on the part. Let $\overline{a}(t)$ be the acceleration of the center of mass of the part, and $\overline{0}(t)$ be the angular acceleration of the part. If $\overline{\xi}(t)$, $\overline{X}(t)$, $\overline{a}(t)$, and $\overline{0}(t)$ are defined relative to a frame attached to the part center of mass, then the following relationships exist:

$$\overline{\xi}(t) = -m\overline{a}(t) \qquad\qquad [16]$$

$$\overline{X}(t) = -\Xi\overline{0}(t)$$

$$0 \leq t \leq t^{\cdot}$$

where: m = the part mass, and

Ξ = the part inertial matrix

Assuming that the contact wrenches are defined with respect to the reference frame at the part center of mass, part motion is subject to the following dynamic equilibrium constraints:

$$\sum_{j=1}^{M} W_j \lambda_j + \left| \begin{array}{c} \overline{\xi}(t) \\ \overline{X}(t) \end{array} \right| + \overline{g} = [0] \qquad\qquad [17]$$

$$\overline{\lambda}_j > = [0] \ \forall \ j, \ 0 \leq t \leq t^{\cdot}$$

where: $\overline{g}$ = gravitational wrench acting on the part

[15] and [17] must be satisfied if the part is not to displace relative to the gripper. Let $\overline{\alpha} = [\overline{\alpha}_1^T, \ldots, \overline{\alpha}_L^T]^T$ and $\overline{\zeta} = [\overline{\lambda}_1^T, \ldots, \overline{\lambda}_M^T]^T$. Then [15] and [17] maybe combined to form:

$$Q\overline{\zeta} = \overline{P}(t) \qquad\qquad [18]$$

$$\overline{\zeta} \geq [0], \ 0 \leq t \leq t^{\cdot}$$

where: $\overline{P}(t) = \left| \begin{array}{c} \overline{\alpha} \\ -\left| \begin{array}{c} \overline{\xi}(t) \\ \overline{X}(t) \end{array} \right| - \overline{g} \end{array} \right|$, and

$$Q = \begin{vmatrix} F_1 W_1 & & [0] \\ & \cdot & \\ [0] & F_i W_i & [0] \\ & \cdot & \\ [0] & & F_L W_L \\ W_1, & \dots, & W_M \end{vmatrix}$$

[18] will always be satisfied if $\overline{P}(t)$ exists within the nonnegative column space of Q for $0 \leq t \leq t^*$. In general $\overline{P}(t)$ is nonlinear and nonconvex. As such it makes a sufficiency check difficult. However this problem is overcome if $\overline{P}(t)$ is replaced by a polygonal approximation[19]. For example, assume that $\overline{P}(t)$ is replaced by a polygon with η vertices such that $\overline{P}_1 = \overline{P}(t_1), \dots, \overline{P}_\eta = \overline{P}(t_\eta)$. If $\overline{P}_1, \dots, \overline{P}_\eta$ exist within the nonnegative column space of Q, then each point within the polygon is guaranteed to exist since it can be represented as a convex combination of any two vertices.

Using this approximation, $\overline{\alpha}$ is guaranteed to be sufficient if the following subproblems have solutions:

for $j = 1, \dots, \eta$ [19]

$$Q \overline{\zeta}_j = \overline{P}_j$$

$$\overline{\zeta}_j \geq [0]$$

Solutions may be found using a phase 1 linear program such as the ones described previously.

ROBOT GRIPPER EXAMPLE

Consider the transport of the part shown in Figure 3. The gripper consists of two fingers, one of which is actuated with a revolute joint. The intensity of this joint is defined as α. The contact regions between the fingers and the part are illustrated in Figure 4. Wrench data is provided in Table 1. Since one actuator controls the displacement of both fingers, derivation of Q yields:

$$Q = \begin{vmatrix} F_1 W_1 & F_2 W_2 \\ W_1 & W_2 \end{vmatrix}$$ [20]

with: $F_1 = [2\ 0\ -4\ 0\ -1\ 0]$,

$F_2 = [2\ 0\ 4\ 0\ 1\ 0]$,

$W_1 = [\overline{w}_1, \dots, \overline{w}_{16}]$, and

$W_2 = [\overline{w}_{17}, \ldots, \overline{w}_{32}]$

The gripper is attached to the two link manipulator shown in Figure 5. Its link parameters are defined in Table 2. The gripper reference frame is defined to be coincident with the part reference frame. As a result, $\overline{a}(t)$ and $\overline{O}(t)$ are equal to the linear and angular acceleration of the gripper reference frame. $t^{\bullet}$ equals one second, and the trajectories of $\theta_1(t)$ and $\theta_2(t)$ are defined by the functions:

$$\theta_1(t) = \theta_2(t) = 3.142 t^2 - 2.094 t^3 \qquad [21]$$

$$\dot{\theta}_1(t) = \dot{\theta}_2(t) = 6.283 t - 6.283 t^2$$

$$\ddot{\theta}_1(t) = \ddot{\theta}_2(t) = 6.283 - 12.566 t$$

$0 \leq t \leq 1$ sec.

Consequently $\overline{a}_i(t)$ and $\overline{O}_i(t)$ are defined by:

$$\overline{a}(\theta_1(t),\ \theta_2(t)) = \begin{vmatrix} \ddot{\theta}_1 \sin\theta_2 - \dot{\theta}_2^2 \cos\theta_2 - (\dot{\theta}_1 + \dot{\theta}_2)^2 \\ \ddot{\theta}_1 \cos\theta_2 + \dot{\theta}_1^2 \sin\theta_2 + (\ddot{\theta}_1 + \ddot{\theta}_2) \\ 0 \end{vmatrix} \qquad [22]$$

$$\overline{O}(\theta_1(t),\ \theta_2(t)) = \begin{vmatrix} 0 \\ 0 \\ \ddot{\theta}_1 + \ddot{\theta}_2 \end{vmatrix}$$

$0 \leq t \leq 1$ sec.

The inertial matrix of the part is:

$$\Xi = \begin{vmatrix} I_{xx} & 0 & 0 \\ 0 & I_{yy} & 0 \\ 0 & 0 & I_{zz} \end{vmatrix}$$

However due to the zero terms in $\overline{O}_i(t)$, only I_{zz} appears in the application of [16]. Values for m and I_{zz} are 13.584 lb_m and 58.86 $lb_m\text{-}in^2$ respectively. Values for $\overline{\xi}(t)$ and $\overline{X}(t)$ at t = 0, .2, .4, .6, .8, and 1 second are presented in Table 3. Evaluating $\overline{P}_1, \ldots, \overline{P}_6$ at these time intervals and applying [19] leads to the conclusion that $\alpha = 100$ lb_f is sufficient for restraint while $\alpha = 80$ lb_f is not.

CONCLUSIONS

Restraint analysis provides valuable information with respect to the design of robot grippers. It is used to evaluate total restraint and the sufficiency of gripper actuator intensities to restrain part motion during

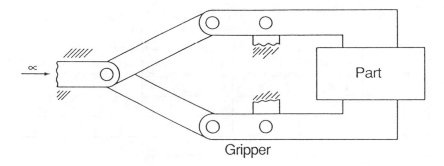

Figure 3: Gripper and Part

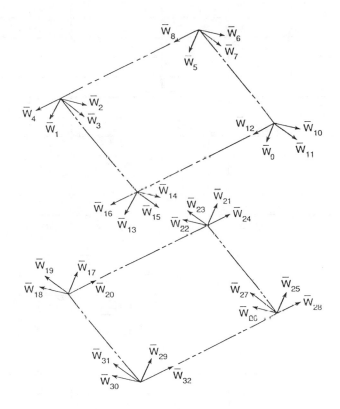

Figure 4: Contact Regions

TABLE 1
Spanning Wrench Data

$\overline{w}$	f_x (lb_f)	f_y (lb_f)	f_z (lb_f)	m_{o_x} (lb_f-in)	m_{o_y} (lb_f-in)	m_{o_z} (lb_f-in
1	.577	0	-1	1	1.577	.577
2	-.577	0	-1	1	.423	-.577
3	0	.577	-1	.423	1	.577
4	0	-.577	-1	1.577	1	-.577
5	.577	0	-1	1	-.423	.577
6	-.577	0	-1	1	-1.577	-.577
7	0	.577	-1	.423	-1	-.577
8	0	-.577	-1	1.577	-1	.577
9	.577	0	-1	-1	-.423	-.577
10	-.577	0	-1	-1	-1.577	.577
11	0	.577	-1	-1.577	-1	-.577
12	0	-.577	-1	-.423	-1	.577
13	.577	0	-1	-1	1.577	-.577
14	-.577	0	-1	-1	.423	.577
15	0	.577	-1	-1.577	1	.577
16	0	-.577	-1	-.423	1	-.577
17	.577	0	1	-1	-1.577	.577
18	-.577	0	1	-1	-.423	-.577
19	0	.577	1	-.423	-1	.577
20	0	-.577	1	-1.577	-1	-.577
21	.577	0	1	-1	.423	.577
22	-.577	0	1	-1	1.577	-.577
23	0	.577	1	-.423	1	-.577
24	0	-.577	1	-1.577	1	.577
25	.577	0	1	1	.423	-.577
26	-.577	0	1	1	1.577	.577
27	0	.577	1	1.577	1	-.577
28	0	-.577	1	.423	1	.577
29	.577	0	1	1	-1.577	-.577
30	-.577	0	1	1	-.423	.577
31	0	.577	1	1.577	-1	.577
32	0	-.577	1	.423	-1	-.577

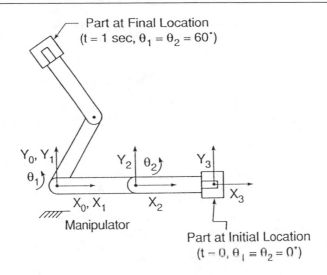

Figure 5: Two Link Manipulator

TABLE 2
Robot Link Parameters

Link i	γ_{i-1} (rad)	$a_{i-1}(in)$	b_i (in)	θ_i (rad)
1	0	0	0	$\theta_1(t)$
2	0	20	0	$\theta_2(t)$
3	0	20	0	0

TABLE 3
Inertia Force Data

t (sec)	ξ_x (lb_f)	ξ_y (lb_f)	ξ_z (lb_f)	X_x (lb_f-in)	X_y (lb_f-in)	X_z (lb_f-in)
0	0	13.253	0	0	0	1.914
.2	-3.260	8.013	0	0	0	1.149
.4	-7.568	3.167	0	0	0	.383
.6	-8.194	-1.452	0	0	0	-.383
.8	-5.400	-6.296	0	0	0	-1.149
1.0	-3.826	-11.045	0	0	0	-1.914

transport.   Future work will investigate formulations and solution techniques for the optimal design of grippers.

REFERENCES

1. Lakshminarayana, K., "Mechanics of Form Closure," <u>ASME Technical Paper, 78-DET-32</u>, 1979.

2. Salisbury, J. K., and B. Roth, "Kinematic and Force Analysis of Articulated Mechanical Hands," <u>ASME Journal of Mechanisms, Transmissions, and Automation in Design</u>, Vol. 105, 1983, pp. 35-41.

3. Kerr, J., and B. Roth, 1986, "Analysis of Multi-fingered Hands," <u>International Journal of Robotics Research</u>, Vol. 4, No. 4, 1986, pp. 3-17.

4. Li, Z., and S. Sastry, "Task-Oriented Optimal Grasping by Multi-fingered Robot Hands," <u>IEEE Journal of Robotics and Automation</u>, Vol. 4, No. 1, 1988, pp. 32-44.

5. Chou, Y. C., Chandru, V., and M. M. Barash, "A Mathematical Approach to Automatic Configuration of Machining Fixtures: Analysis and Synthesis," ASME <u>Journal of Engineering for Industry</u>, Vol. 111, 1989, pp. 299-306.

6. Chou, Y. C., "A Methodology for Automatic Layout of Fixture Elements Based on Machining Forces Considerations," <u>Proceedings of the ASME Winter Annual Meeting</u>, 1990, pp. 181-189.

7. Markenscoff, X., Ni, L., and C. H. Papadimitriou, "The Geometry of Grasping," <u>International Journal of Robotics Research</u>, Vol. 9, No. 1, 1990, pp. 61 - 74.

8. Ohwovoriole, E. N. and B. Roth, 1981, "An Extension of Screw Theory," <u>Journal of Mechanical Design</u>, Vol. 103, 1981, pp. 725-735.

9. Ohwovoriole, E. N., 1987, "Kinematics and Friction in Grasping by Robotic Hands," <u>Journal of Mechanisms, Transmissions, and Automation in Design</u>, Vol. 109, 1987, pp. 398-404.

10. Nguyen, V., "Constructing Force-Closure Grasps," <u>International Journal of Robotics Research</u>, Vol. 7, No. 3, 1988, pp. 3-16.

11. Assada, H., and M. Kitagawa, "Kinematic Analysis and Planning for Form Closure Grasps by Robotic Hands," <u>Robotics and Computer Integrated Manufacturing</u>, Vol. 5, No. 4, 1989, pp. 293-299.

12. Chou, Y. C., Chandru, V., and M. M. Barash, "A Mathematical Approach to Automatic Design of Machining Fixtures: Analysis and Synthesis," <u>Proceedings of the ASME Winter Annual Meeting</u>, 1987, pp. 11-27.

13. Bausch, J. J, and K. Youcef-Toumi, 1990, "Computer Planning Methods for Automated Fixture Layout Synthesis," <u>Proceedings of the Manufacturing International 90 Conference</u>, Vol 1., 1990, pp. 225-232.

14. Sturges, R. H., "A Three-Dimensional Assembly Task Quantification with Application to Machine Dexterity," <u>International Journal of Robotics Research</u>, Vol. 7, No. 4, 1988, pp. 34-78.

15. Montana, D. J., "The Kinematics of Contact and Grasp," <u>International Journal of Robotics Research</u>, Vol. 7, 1988, No. 3, pp. 17-32.

16. Assada, H., and B. Andre, 1985, "Kinematic Analysis of Workpart Fixturing for Flexible Assembly with Automatically Reconfigurable Fixtures," <u>IEEE Journal of Robotics and Automation</u>, Vol. RA-1, No. 2, 1985, pp. 86-94.

17. Ball, R. S., <u>A Treatise on the Theory of Screws</u>, Cambridge University Press, Cambridge, England, 1900.

18. De Meter, E. C., "Restraint Analysis of Rigid Bodies Constrained by Surface Contact," <u>IMSE Working Paper 91-133</u>, July 1991, pp. 1-33.

19. De Meter, E. C., "Restraint Analysis of Assembly Work Carriers," <u>IMSE Working Paper 92-104</u>, February 20, 1992, pp. 1-24.

ECONOMIC DESIGN OF MATERIAL HANDLING SYSTEMS

[1]James S. Noble and [2]J.M.A. Tanchoco

[1]*Industrial Engineering Program, University of Washington,*
Seattle, Washington, U.S.A.
[2]*School of Industrial Engineering, Purdue University,*
West Lafayette, Indiana, U.S.A.

ABSTRACT

Material handling systems perform a key integrating function in today's dynamic and integrated manufacturing environment. However, the economics of material handling systems have tended to be considered only in design evaluation, not design generation. The objective of this paper is to explore the integration of economic objectives in design generation. The result of this integration is that robust material handling systems, capable of supporting dynamic, product life-cycle induced manufacturing system objectives are designed. This is achieved through the development of a framework for the selection and specification of an integrated material handling system.

The framework utilizes a design procedure based on a modified branch and bound concept which is guided by a marginal analysis based modification strategy. The implementation of the design framework included the automatic generation of discrete optimization, capacity, queueing, and simulation models, all integrated through a knowledge-based environment. The application of the design procedure results in the development of a material handling system that is justifiable based on designer specified performance and economic criteria.

1.0 INTRODUCTION

Material handling systems perform a key integrating function in today's dynamic and integrated manufacturing environment. Failure to fully integrate material handling design within the larger manufacturing system design problem results in designs that are unable to complete successfully in today's markets. However, typical approaches for material handling system (MHS) design have tended to isolate the analysis with respect to specific design objectives and material handling technologies. Figure 1 shows the integrated nature of MHS design. This paper describes the development of a framework for integrated economic design of MHSs that support product life-cycle induced manufacturing system objectives.

First, the background of the economic design problem is discussed. Second, the MHS design framework and procedural approach is described. Third, an illustrative example of the design approach is presented.

2.0 BACKGROUND OF THE ECONOMIC DESIGN APPROACH

Different design approaches that incorporate economic considerations have been developed for both specific material handling components and for a complete MHS. The following is a review of design approaches that both explicitly and implicitly incorporate

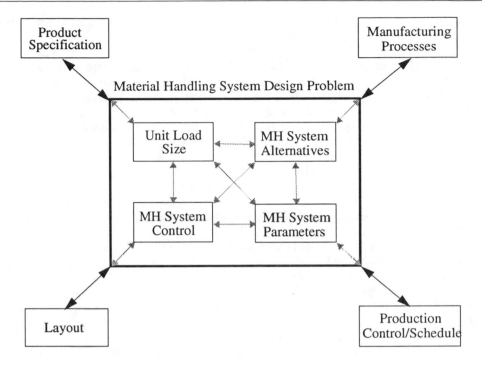

Figure 1 Material Handling System Design

economic criteria within the MHS design process. A more extensive review of economic design in manufacturing can be found in Noble [1].

An economic approach to the belt conveyor specification problem was developed by Tanchoco et al. [2]. They formulated the problem to minimize total capital cost such that the operating constraints are satisfied. Solutions are obtained using a decomposition and implicit enumeration procedure. Scott et al. [3] developed a combined performance and cost model design methodology for unit load conveyors. Their model combined the energy and maintenance requirements to generate a capital and operating cost estimate. The cost estimate was then used in searching for a design solution, as well as assisting in the development of operating conditions. Ziai and Sule [4] developed an equipment selection methodology for selecting between conveyors and forklift trucks that used a mixed integer program to obtain an initial solution. Upon arriving at the initial solution, a cost driven improvement heuristic was used to improve the solutions equipment utilization.

The complete material handling selection problem was first addressed using an economic approach by Webster and Reed [5]. The problem of minimizing capital, operating, and unit load changing cost with respect to material flow requirements and transporter capacity, was modeled as an integer program and solved using a heuristic solution procedure. Hassan, et al. [6] reformulated Webster and Reed's model and then developed a construction algorithm based upon the unique characteristics of the material handling equipment selection model. The construction algorithm sought first to minimize cost, while also minimizing overall equipment variation and maximize equipment utilization.

Multi-attribute or knowledge-based methodologies are approaches to economic design that include economic issues in an implicit manner. Shelton and Jones [7] used twenty three attributes (including cost) in an additive form of the multi-attribute value function for the selection of automated guided vehicles. Liang, et al. [8] developed a multi-attribute approach to the specification of material handling equipment that utilizes the analytic hierarchy process combined with a database. The database consisted of material handling equipment defined by 24 characteristics and the following attributes: layout flexibility, route flexibility, space utilization, safety, and operating costs. A knowledge based multi-attribute design approach was proposed by Gabbert and Brown [9] for specifying the types of material handling equipment to use in a given manufacturing scenario. Their approach allowed for decision maker attribute preferences by updating attribute weights. An expert system for the selection of material handling equipment was developed by Fisher, et al. [10]. Economic factors related to the selection of material handling equipment were implicitly embedded in the selection rules in the knowledge base.

A variety of different cost models have also been developed to analyze material handling systems at different levels. An overall material handling system cost framework was discussed by Apple [11]. Müller [12] explored cost relationships for a variety of material handling equipment. Dahlstrom [13] explored the applicability of automated guided vehicles (AGVs), fork-lift trucks, and fixed roller conveyors based on general cost models of each type of equipment. General cost models for AGVs have been developed by Daum [14] and for conveyors by Shultz [15] and Wiltse [16].

Each material handling design approach reviewed considers economic factors in some manner. However, none explore the ability to project the effect of design decisions nor do they provide an explicitly integrated economic and performance basis for MHS design and justification. Therefore, this paper builds upon the *design justification* concept proposed by Noble and Tanchoco [17, 18] to provide these abilities.

Design justification refers to an economic design procedure where the economic ramifications of design decisions are considered concurrently with design development. The goal of *design justification* is to guide the designer to a design that is justifiable from both a performance and economic perspective. The underlying conceptual approach of *design justification* utilizes a design approach that requires economic justification at each design decision to ensure a functionally and economically viable system design. The concurrent design approach in *design justification* is in contrast to the traditional sequential design and justification process.

There are four characteristic elements of *design justification*: 1) integrated engineering and economic design of the system so that both occur simultaneously; 2) goal directed design so that the goals/objectives of the design are achieved through a branching and bounding procedure that guides the designer throughout the design process; 3) decision maker centered design to provide an information processing ability and presentation mode that enables a designer to make good decisions; 4) knowledge guided design to provide a design procedure where critical points in the redesign or modification of the current design have a modification strategy which will guide the design efficiently. The following sections will discuss the development of a framework for the design justification of material handling systems.

3.0 COMPONENTS OF THE MHS DESIGN JUSTIFICATION FRAMEWORK

The focus of this paper is the development of a framework from which material handling systems can be selected and then specified using an economic design approach termed *design justification*. The MHS design justification framework has six major components: system designer, design interface, design inference model, model generator, rule base, and database.

The system designer maintains the controlling role in the design process, experimenting with, and assimilating the design information that is generated within the design environment. The design interface provides the system designer with a medium through which to interact with the design system. The design inference model performs two primary functions. The first function is to combine the knowledge of the system designer, rule base, database and intermediate design results to aid in guiding the design process. The second function is to distribute information to the system designer, the model generator and the database so that the analysis can continue.

The model generator uses the rule base and database to generate appropriate model abstractions of the MHS being analyzed. The rule base contains knowledge of the ranking of material handling equipment alternatives, the determination of the types of MHSs to be explored, the general analysis approach, the level of model abstraction, and the determination of critical parameters. The database serves to store, retrieve and update information on process plans, product characteristics, operation equipment, material handling equipment, physical layout, material flow path layout, and intermediate design results.

The following section will discuss and illustrate the components of the material handling design justification framework and how they interrelate within the *design justification* procedure.

4.0 MHS DESIGN JUSTIFICATION PROCEDURE

There are three major phases within the material handling *design justification* procedure: 1) the development of manufacturing system data; 2) the development of MHS alternatives; 3) the analysis and modification of system alternatives, resulting in a final selection and specification of a MHS. Figure 2 highlights the procedural approach embedded in the design justification framework. The following describes each phase in the procedure.

4.1 Manufacturing System Data

The first phase of the *design justification* procedure is the importation and preparation of manufacturing system data. The MHS *design justification* procedure requires an instance of the: product description (size, weight, quantity); process plan (operation types, sequence and times; number of machines at each operation); physical layout (departmental assignments, spatial relationships); and unit load description (size, weight, quantity). A data structure for product, process and unit load was developed so that a consistent representation exists for the various functions within the design procedure. The data for the physical layout is developed using CAD software. CAD data files are imported into a graphical interface for the analysis of MHSs developed by Rembold and Tanchoco [19].

4.2 MHS Alternative Generation

The second stage of the *design justification* procedure is a two step procedure to develop MHS alternatives. As is shown in figure 2, this stage is a solution branching phase where the MHS alternatives that merit analysis are generated. The two steps work together as the first step reduces the number of equipment options for each material flow so that the second step can configure a complete system efficiently.

The first step of the alternative generation process ranks equipment alternatives for each material flowpath in the system using a knowledge-based procedure. A material handling equipment knowledge-based was developed that is a variation of MATHES [10]. The

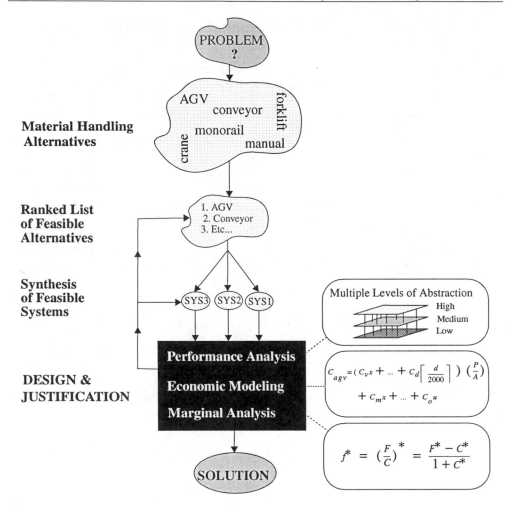

Figure 2 Economic Design Approach for Material Handling System

knowledge-base uses four parameters as the basis for ranking equipment: path type, path distance, flow rate, and unit load size. The ranking procedure utilizes a subset of the EMY-CIN certainty factor calculus to obtain a measure of how well a given equipment type matches the material flow. The results of the equipment ranking are passed to the equipment selection/system generator step of the solution branching stage. The second step of the alternative generation process utilizes an integer programming formulation of the equipment selection problem that is solved using a modification of the construction algorithm developed by Hassan et al. [6].

Several issues must be addressed concerning the use of the generation procedure. First, if the rules in the rule base are appropriate, then obtaining the highest level of certainty would be the goal. However, based on the variety of material flows that can occur in a given system this is not always desirable. Therefore, experimentation with different certainty factor cutoffs for each system is required. Second, the formulation of the model requires an aggregate fixed cost for capital expenditures and an aggregate variable cost for operating expenses. The result is that the cost values associated with a generated system do not always accurately reflect the actual system cost. Rather, it tends to be a good indicator of the relative comparison between different alternatives. Third, since the results of the formulation are based primarily on capacity requirements the systems dynamic nature is not captured. This effect is accounted for as the model typically over specifies the number of each equipment type, reflecting the fact that the formulation does not account for material flow routing.

4.3 MHS Analysis and Modification - Solution Bounding

The objective of *design justification* is to enable the designer to proceed through the design process in such a manner that each design decision made is justifiable based on the objectives of the design problem. Therefore, the crux of the material handling design justification framework occurs in the analysis and modification stage of the design procedure. Where the role of system analysis is to provide both economic and performance based solution bounds.

The analysis and modification stage is comprised of four functions: system classification, system performance modelling, system cost and flexibility modelling, and combined performance and economic analysis. The following describes each of these four functions individually and discusses how they are implemented within the material handling design justification framework.

MHS Classification The design and analysis of a MHS is unique based on the equipment composition of the system. As a result a MHS classification scheme was developed based upon each equipment's share of the total system material flow. The classification scheme developed only considers horizontal material handling equipment of four general types: manual, truck, conveyor, and automated guided vehicle.

MHS Performance Modeling The MHS classification scheme provides the basis from which to model material handling systems at different levels of model abstraction. This allows for systems to be modelled at the appropriate level to obtain useful information on system performance and reduces the complexity and time associated with model development. The designer is guided quickly to the appropriate abstraction of the system model to obtain results that reflect the level of confidence required. Two functions performed in the design procedure are related to the abstraction of the system model. First, models are generated for the initial abstraction of the MHS based upon the initial alternative system selection and specification by the designer. Second, models are generated to provide lower levels of abstraction of the MHS to allow the designer to automatically go to the next appropriate level of abstraction if it is merited, or to allow the designer to change the MHS specification and have the model regenerated to represent the altered system.

Three levels of system abstraction were implemented: deterministic capacity analysis, queueing analysis, and simulation. In each level an automatic model generator was developed. The deterministic capacity analysis calculates both operation capacity and material handling capacity. The closed queueing network program, CAN-Q [20] was used as the basis of the queueing network generator. The SLAM simulation language with the material handling extension was used as the basis of the simulation generator [21]. The simulation model generated is a mixed network-discrete event model with built-in statistical significance testing.

MHS Cost and Flexibility Two other measures of MHS performance are used within the material handling design justification framework. The first is a MHS flexibility metric which was developed to measure the ability of a system to adapt to change. The second is an extensive cost modelling structure for each type of material handling equipment which was developed to measure the economic impact of design decisions.

Because the material handling function is a primary integrator within a manufacturing system, the ease with which a facility can be adapted is critically linked to MHS design. MHS flexibility has been defined as the ability to move different parts types efficiently for proper positioning and processing through the manufacturing facility it serves [22]. The development of different metrics to measure MHS flexibility has been limited, therefore, a metric was developed to measure the combined ability of a material handling system to reconfigure, so as to handle new material flows, and handle additional material flow. The metric of MHS flexibility is defined as:

$$F_{mhs} = \sum_{i=1}^{N} x_i u_i \beta_i \qquad (1)$$

where, x_i is the number of each type of equipment, u_i is a measure of capacity based upon unit load, equipment speed, and flow path distance, and β_i is a measure of the relative rerouting cost based upon the inverse of the approximate installation cost.

Detailed cost models were developed so that an accurate assessment of the economic performance of the system could be obtained. The goal of modelling the cost of the different types of equipment is to capture not just capital costs, but also installation, operating, and maintenance costs. It was found in the modelling process that there are three major categories of models based upon the commonality shared between model structure. The three categories are: automated guided vehicles, trucks/manual, conveyors.

MHS Combined Performance and Economic Analysis   The mechanism within the design justification framework that enables it to meet the combined performance and economic justification design objective is a combination of system sensitivity analysis and an interactive marginal analysis directed modification strategy.

There are two main areas of sensitivity analysis that are supported in the material handling design justification framework: capacity and design objective. Capacity sensitivity is explored through the development of a unit cost curve that provides the designer with a considerable amount of information on cost performance. The unit cost curves provide a visual perception of the system cost flexibility with respect to material flow rate and the cost of changing system capacity requirements. Sensitivity analysis on the design objective entails analyzing the system with respect to minimum job makespan or maximum MHS flexibility. These two types of analysis are facilitated through a variety of decision graphics that are provided to the designer.

Determining where to alter a system design so that the net effect will satisfy the design objectives is a key component of the MHS design justification framework. The implementation of a multi-level modification strategy builds upon, and utilizes, all of the components discussed. The first level modification addresses the alteration of individual parameters of a system and instantiates the analysis of a MHS. The second level modification pertains to the alteration of the analysis tool used to set solution bounds. This alteration requires returning to the first level to restart the analysis procedure. The third level modification explores altering the equipment selection. This alteration requires returning to the first level to restart the analysis procedure once the equipment changes are determined. The fourth level modification of a system design re-specifies the material flow paths within the system. This alteration first requires returning to the third level to obtain a new system design, which is then followed by returning to the first level to restart the analysis procedure.

At the first, third and fourth modification levels there are performance and economic trade-offs that must be considered. The mechanism within the framework for this analysis is a marginal analysis network that shows the marginal or relative incremental effect of any absolute or incremental percentage alteration to the system on the performance measures of throughput, total MHS cost, material handling unit cost, MHS flexibility, and unit flexibility cost. The marginal analysis used in the modification strategy utilizes the incremental calculus developed by Eilon [23]. The marginal analysis was implemented using a Lotus 1-2-3 spreadsheet to provide for ease in conducting sensitivity analysis and presenting graphical information. The designer can utilize generalized marginal trade-off graphs to assist in making modifications. The effects of exploring different parameters on the overall analysis measures are displayed to the designer.

5.0 ECONOMIC DESIGN OF A MHS

The design justification framework for economic design of material handling systems is highlighted in the following example. The first stage of design procedure begins by generating MHS alternatives through the two step procedure described previously. The generation of system alternatives was limited for this example to combinations of three types of material handling equipment: unit load automated guided vehicles (AUL), powered roller conveyors (CPR) and/or forklift trucks (TFL). Table 1 gives the results of generating different alternatives ranked in descending cost order.

Table 1
Generated Material Handling System Alternatives

System Alternative	Total Cost	Total Flexibility	Certainty Factor
1. 31 CPR	$103,789	816.7	33.4
2. 3 TFL, 19 CPR	195,504	1207.4	72.2
3. 6 AUL, 19 CPR	218,005	944.9	61.3
4. 11 AUL, 12 CPR	290,177	1226.0	74.2
5. 28 AUL	445,035	1575.0	82.7
6. 10 TFL, 4 AUL, 3 CPR	450,976	2258.8	88.9
7. 13 TFL	488,240	2600.0	88.4

The second stage of the economic design procedure begins by requiring that the designer determine the objectives of the design process. This maintains the designer as the key decision maker throughout the design process. The nature of the procedure requires that the designer re-evaluate the design objectives and conduct an analysis on the effect of changing the design objectives. The justification of each design decision provides a quantitative analysis of the reasoning supporting each design decision which can then be used as the basis of an overall justification argument. In this example the design objective is determined to be the development of a MHS that gives a wide range of economical operating conditions, while providing at least a "moderate" level of system flexibility. Based on this objective the design procedure starts bounding the solution using system alternative #2.

The analysis of a MHS is conducted utilizing the three-level design analysis framework. Based upon the composition of system alternative #2 (3 TFL and 19 CPR), a deterministic capacity analysis is pursued first, followed by a stochastic simulation. The capacity analysis results show that a maximum capacity due to processing ability is 143 units/hour, and the maximum capacity due to material handling constraints of 76.6 units/hour. Therefore, the 3 TFL are the bottleneck in the system. Prior to analyzing the incremental effect of altering the number of TFL in the system, the next level of modelling abstraction is explored to obtain more accurate measures of system performance based on system dynamics. A four level simulation is automatically generated varying the number of TFL within each shop loading. The MHS utilizing the minimum number of TFL at a statistically equivalent maximum level of throughput was selected. Based on the simulation results, a detailed unit cost curve is developed and the marginal analysis within the TFL-CPR system alternative begins.

The marginal analysis is conducted by incrementally adding or deleting one unit of TFL from the 4 TFL-CPR system. From this, the relative incremental effect on the performance measures are noted. The interpretation of the results of the marginal analysis reveals several interesting characteristics of the TFL-CPR based system. First, the number of TFL, total system cost, and total system flexibility all follow a linear relationship. Hence, there is no increasing or decreasing marginal benefit associated with these changes. Second, the relative incremental unit cost values display a decreasing, followed by a negative, marginal benefit from increasing the number of TFL. Third, unit flexibility exhibits a decreasing marginal benefit behavior. The designer concludes the following with respect to changing the number of TFL in the system: 1) the system displays acceptable economic and performance characteristics in the 55 to 115 throughput/hour range, 2) it would be desirable to obtain a higher marginal flexibility to cost ratio, and 3) the system expands easily within the 55 to 115 capacity range. A second area of analysis within the TFL-CPR system concerns the benefit that could be derived from changing conveyor speed. Simulations are conducted comparing the throughput of the TFL-CPR system operating at 60 fpm and 120 fpm. A 90% Welch confidence interval was constructed on the difference of the throughput values, which reveals that conveyor speed changes do not result in statistically better throughput values for this system, therefore, all marginal relationships give a negative benefit.

The marginal analysis conducted so far has served to inform the designer of the performance ranges of the MHS alternative TFL-CPR and the type of system changes which reflect positive marginal benefit. The next stage of the marginal analysis explores possible equipment changes that would better meet the design objective.

The marginal analysis between alternatives begins using a single production level from which complete simulation and statistical significance test results for the systems comprised of AUL-CPR, CPR, AUL, and TFL were obtained. The marginal analysis now focuses on the effect of changing from the TFL-CPR system to another system. The ratio of the relative incremental total flexibility to the relative incremental total cost (TF/TC) is as the overall measure of the change. The TF/TC ratio for changing to the AUL-CPR is found to be - 8.99. This is implies that there is a highly negative marginal benefit for this system compared to TFL-CPR (i.e. the AUL-CPR system has a higher total cost and a lower system flexibility). The same result is found for the AUL system, except the TF/TC ratio reflects an even higher negative marginal benefit. The CPR and TFL system both have a positive TF/TC ratio, yet for different reasons. In the CPR system there is a decrease in both total cost and total flexibility, with the decrease occurring in approximately the same proportions. For the TFL system there is an increase in both total cost and total flexibility, with the cost increasing at a greater rate than the flexibility.

The designer has now looked at some of the marginal effects of changes within the system and between systems. The decision remains as to which alternative best satisfies the design objectives. The initial system, TFL-CPR, provides a compromise solution based upon the objective. It might be valuable for the analysis to continue further by selecting either the CPR or TFL systems for more detailed analysis. The analysis could also be extended to explore the marginal effect of changing from the TFL-CPR system to one of the other MHS system alternatives at a different shop loading level. To aid in making these decisions a detailed material handling unit cost relationships for each alternative is developed that supports the cost sensitivity analysis of each alternative with respect to the production level. Whichever analysis path the designer may pursue, a quantified decision path of the performance and economic trade-offs will exist to justify the final design.

6.0 SUMMARY

The economic design of a MHS has been illustrated through the application of the *design justification* procedure. The example given shows how the concurrent design and justification occurs. Specifically, the marginal analysis within a given alternative gave the designer the ability to determine whether changing the performance ranges of the system alternative was justifiable. The marginal analysis between different system alternatives gave the designer the ability to see what would be gained or lost, and at what relative rate the changes would occur. The relative incremental values between systems perform the concurrent design and justification function by allowing the designer to show why a given system is preferable to another. Hence, the goal of *design justification*, which is to enable the system designer to be able to quantify and project the trade-offs between different system capabilities and even design objectives, is obtained.

The material handling design justification framework allows for an integrated systems approach to the MHS selection and specification problem. The framework maintains the system designer as the key decision maker, while enhancing the ability to make decisions that are justifiable with respect to design objectives. Branching on system alternatives is accomplished using a combined knowledge-based system and integer programming model. Bounding system design solutions occurs using automatically generated models within a marginal analysis directed modification strategy. The combined effect of the branching and bounding scheme enables the designer to develop a MHS that provides the material handling capabilities that are justifiable from both a performance and economic perspective.

REFERENCES

1. Noble, J.S., "A Framework for the Design Justification of Material Handling Systems," Unpublished Ph.D. Dissertation, Purdue University, West Lafayette, IN, December 1991.

2. Tanchoco, J.M.A, R.P. Davis, R.A. Wysk, and M.H. Agee, "Component Selection and Operating Specifications for Belt Conveyor Systems," *Proceedings, 1979 Spring Industrial Engineering Conference*, pp. 112-119, San Francisco, CA, May, 1979.

3. Scott, H.A., J.M.A. Tanchoco, and M.H. Agee, "Unit Load (belt) Conveyor Operating Design and Cost," *International Journal of Production Research,* vol. 21, no. 1, pp. 95-107, 1983.

4. Ziai, M.R., and D.R. Sule, "Computerized Materials Handling and Facility Layout Design," *Computers and Industrial Engineering*, vol. 17, no. 1-4, pp. 55-60, 1989.

5. Webster, D.B. and R. Reed, Jr., "A Material Handling System Selection Model," *AIIE Transactions*, vol. 3, no. 1, pp. 13- 21, 1971.

6. Hassan, M.M.D., G.L. Hogg, and D.R. Smith, "A Construction Algorithm for the Selection and Assignment of Materials Handling Equipment," *International Journal of Production Research*, vol. 23, no. 2, pp. 381-392, 1985.

7. Shelton, D. and M.S. Jones, "A Selection Method for Automated Guided Vehicles," *Material Flow*, vol. 4, no. 1, pp. 97-107, 1987.

8. Liang, M., S.P. Dutta, and G. Abdou, "A New Approach to Material Handling Equipment Selection in a Manufacturing Environment," *Proceedings, 1989 International Industrial Engineering Conference*, pp. 225-230, Toronto, Canada, May 14-17, 1989.

9. Gabbert, P.S and D.E. Brown, "A Knowledge-Based Approach to Materials Handling System Design in Manufacturing Facilities," *Proceedings, 1987 International Industrial Engineering Conference*, pp. 445-451, Washington, D.C., May 17-20, 1987.

10. Fisher, E.L., J.B. Farber, and M.G. Kay, "MATHES: An Expert System for Material Handling Equipment Selection," *Engineering Cost and Production Economics*, vol. 14, no. 4, pp. 297-310, 1988.

11. Apple, J.M., *Material Handling Systems Design*, Ronald Press, New York, 1972.

12. Müller, T., "Comparison of Operating Costs Between Different Transport Systems," *Proceedings, 1st International Conference on Automated Guide Vehicle Systems*, ed. R-H. Hollier, pp. 145-155, Stratford-upon-Avon, UK, June 2-4, 1981.

13. Dahlstrom, K.J., "Where to Use AGV Systems, Manual Fork Lifts, Traditional Fixed Roller Conveyor Systems Respectively," *Proceedings, 1st International Conference on Automated Guided Vehicle Systems*, ed. R-H. Hollier, pp. 173-182, Stratford-upon-Avon, UK, June 2-4, 1981.

14. Daum, M., "Operating Costs of AGV Systems," *Proceedings, 5th International Conference on Automated Guided Vehicle Systems*, ed. T. Takahashi, pp. 311-328, Tokyo, Japan, October 6-8, 1987.

15. Schultz, G.A., "Estimating Total Cost of a Package-Conveyor System," *Plant Engineering*, vol. 32, no. 8, pp. 149-152, April 13, 1978.

16. Wiltse, J., "Eight Simple Steps to an Accurate Forecast of Conveyor System Costs," *Modern Materials Handling*, vol. 34, no. 2, pp. 60-63, 1979.

17. Noble, J.S. and J.M.A. Tanchoco, "Cost Modeling for Design Justification," in *Cost Analysis Applications of Economics and Operations Research, Lecture Notes in Economics and Mathematical Systems*, ed. T.R. Gulledge, Jr., L.A. Litteral, pp. 197-213, Springer-Verlag, New York, 1989.

18. Noble, J.S. and J.M.A Tanchoco, "Concurrent Design and Economic Justification in Developing a Product," *International Journal of Production Research*, vol. 28, no. 7, pp. 1225-1238, 1990.

19. Rembold, B. and J.M.A. Tanchoco, "Network Editor for Material Handling System Design", Technical Report, School of Industrial Engineering, Purdue University, 1991.

20. Solberg, J.J., "A Mathematical Model of Computerized Manufacturing Systems," *Proceedings, 4th International Conference on Production Research*, pp. 22-30, Tokyo, Japan, August 1977.

21. Pritsker, A.A.B., *Introduction to Simulation and SLAM II*, Halsted Press, John Wiley and Sons, New York, NY, Third Edition, 1986.

22. Sethi, A.K. and S.P. Sethi, "Flexibility in Manufacturing: A Survey," *International Journal of Flexible Manufacturing Systems*, vol. 2, no. 4, pp. 289-328, 1990.

23. Eilon, S., *The Art of Reckoning - Analysis of Performance Criteria*, Academic Press, London, 1984.

DESIGN AND SELECTION OF MATERIAL HANDLING SYSTEMS

Samir B. Billatos and Anand V. Thummalapalli

Department of Mechanical Engineering, University of Connecticut, Storrs, Connecticut, U.S.A.

ABSTRACT

Design and Selection of material handling systems (MHS) require extensive knowledge of facts, relationships, and rules that are very specific if used within a manufacturing system operating in a changing environment. The selection process may have to satisfy multiple objectives and to make efficient use of the extensive knowledge that characterizes the MHS selection. The design process may have to be carried out after facilities layout, although many aspects of MHS clearly influence the choice of layout of manufacturing systems. This is one of many interdependent problems that creates difficulties for the designer. In this paper, a knowledge based Expert System for Material Handling (ESMAH) is proposed to accommodate these complexities. The knowledge base incorporates five factors relating system performance, design criteria, material handling equipment specifications, operational environment and cell layout. The expert system recommends the best design and can help the manufacturing facility to operate intelligently.

1. INTRODUCTION

One of the most important aspects in the newly developed automated manufacturing systems is the design of material handling systems. Material handling embraces all of the basic operations involved in the movement of bulk, packaged, and individual products in a semisolid or a solid state by means of machinery, and within the limits of a place of business. It thus includes movement of raw material to workstations, semi-finished products between workstations, and removal of the finished products to their storage locations [1].

Material handling, however, is different from transportation, where material is moved from suppliers to places of business or from places of business to customers. Material handling systems can impact building requirements and department arrangements, as well as affect ongoing operations, such as the time needed to produce a part. The cost associated with material handling can sometimes account for 30-55 percent of the total manufacturing costs [2,3]. Effective design and management of material handling systems can reduce the manufacturing cost by as much as 15-30 percent [1-3], and the integration of these systems into the manufacturing environment has significant potential savings.

The design of MHS is a very complicated process due to many factors as [1-3]:

1. Multiple, conflicting, and noncommensurate design criteria, i.e., designer must consider the tradeoffs between the performance of the system on multiple criteria to minimize cost, maximize reliability, etc.).
2. Changing design specifications, either from the client or from the design process itself as new information becomes available.
3. Rapidly changing commercial products as evidenced by the growing number of MHS types and vendors, etc.
4. Uncertainty in the operational environment. This implies that design aids in this process must be capable of supporting reasoning under uncertainty with imprecisely articulated and changing specifications.

In this research, a knowledge based **E**xpert **S**ystem for **MA**terial **H**andling (ESMAH) is proposed to accommodate these complexities. The knowledge base incorporates the five factors listed in the Abstract through the use of the following procedure: (1) identify the task factors and input values for each criterion based on user's requirements, (2) generate alternative MH equipment, (3) generate the criteria matrix, and (4) compare the total difference and the total accession of each alternative from the user's criteria.

With such a structure of knowledge base, the proposed research would advance the knowledge of MHS within manufacturing systems operating in changing manufacturing environments such as flexible manufacturing systems (FMS), Just-In-Time (JIT), etc.

2. PROBLEM FORMULATION AND BACKGROUND

A good material handling system should possess most or all of the following characteristics [1]:

*   Handling combined with processing whenever possible.
*   Automation whenever possible.
*   Minimum manual handling.
*   Protection of material provided.
*   Minimum congestion or delay.
*   Minimum variation in equipment types.
*   Economical.

With all of these requirements the problem faced by many designers is the selection and design of an efficient MHS. Optimum and near-optimal design has been addressed in literature [5-8]. General design steps for MHS are:

▶   State the intended function of the MHS whether it is

for a warehouse, a manufacturing system, or assemblies.
▸ Determine the material, its characteristics and quantities involved.
▸ Identify the moves, their origin and destination, their path and their length.
▸ Determine basic handling system and the degree of mechanization desired.
▸ Perform an initial screening of suitable equipment and select a set of them to be evaluated on the basis of such measures as cost and utilization.

This design procedure is constrained by the following factors:

o Cost of the equipment and availability of funds.
o Physical characteristics of the building and the available space.
o Management attitude towards safety and employee welfare.
o Degree of involvement between handling and processing.

These issues and others will be addressed in this research.

3. OBJECTIVES

The objective of this research is to develop an intelligent MHS to be integrated in changing manufacturing environments such as FMS, JIT, etc. This is achieved as follows:

1. Designing the initial framework of database that includes existing MH equipment.
2. Developing a knowledge base that includes specifications and information concerning system selection, equipment justification, conflict identification, and layout.
3. Developing a knowledge based expert system (ESMAH) that includes a variety of design rules and quantitative rules that can recommend the best design and can help the system to operate intelligently.
4. Integrating ESMAH within the manufacturing system. This will enhance understanding of the problem and expedite solution.

The above objectives are specifically designed to address common problems in the design and operation of MHS including efficiency, effectiveness, responsiveness, etc. These issues have a direct impact on the benefits obtained from modernizing the factory floor, and affect decisions of great practical value and generic importance. Unfortunately, these issues have not been satisfactorily answered in literature nor realized in practice.

4. TECHNICAL APPROACH

The expert system of this research has a main program

which interfaces with algorithms and other programs. External programs were developed as needed to make the system intelligent. The tasks involved in development are:

Task 1: Obtain the specifications and information concerning system selection, equipment justification, conflict identification, and layout. To achieve this task, the following data are needed:

1. Number of machines in the system.
2. Type of layout.
3. Level of automation of the manufacturing system.
4. Degree of production flexibility (layout, routing, etc.).
5. Attributes of handling tasks for the expert system to select the appropriate MH equipment (e.g., weight/load, frequency, handling percentage, space utilization, etc.).

Most of these data were obtained from literature survey [1-8]. To avoid any conflict in data obtained and/or retrieved, data were classified into two different classes (A and B as detailed in section 5). For example, selection of an AGV requires a certain degree of automation. In addition, information solicited from users includes the range of weight/load different systems can handle, ideal type of material properties, the level of automation required, etc.

Task 2: Design an initial database framework that includes existing MH equipment and the corresponding factors and criteria.

An application of the selection factors and criteria are illustrated in [5]. The factors are considered as the indispensable functional characteristics to perform MH tasks and the basis to select alternative equipment categories. The criteria, which represents the degree of conformity to design, application and specification, serve to judge the most preferable category.

The database should be easily updated and/or modified, be efficient enough to do other conventional jobs as query and take short time to successfully answer the query.

Task 3: Develop a knowledge based expert system that includes a variety of design rules and quantitative rules that can recommend the best design and can help the system to operate intelligently.

The structure of the proposed expert system, ESMAH, is shown in Figure 1. It consists of: (i) a knowledge base built with algorithms, scheduling and expert decision rules; (ii) an input-output module, and (iii) an inference engine.

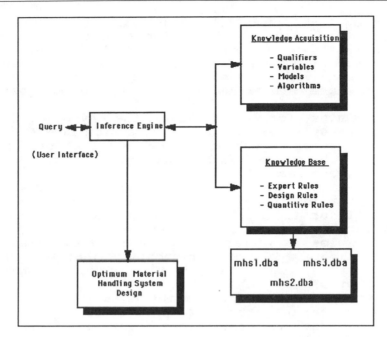

Figure 1 : Structure of ESMAH

4.1 Knowledge Representation

Effective representation of knowledge is one of the key issues in knowledge-based systems. Domain knowledge has many forms, including descriptive definitions of domain-specific terms, descriptions of individual objects, classes of objects and their inter-relationships, and criteria for making decisions. The most frequently used knowledge representation schemes are: First order logic, Horn clause subset of first order logic, Production rules, Model-based reasoning, Frames, and Semantic networks [4].

The knowledge base in model-based systems is ideal for MHS selection and design since in this design process there are number of models to be considered and, for a particular case of MH requirement, specific systems could easily be identified. If, for example, there are number of MH systems then each system can be represented as follows:

    Object   : Material handling system (e.g., AGV)
    Attributes: Volume of flow;
                Path;
                Size of load;
                Distance between facilities;
                        ....

The knowledge base (KB) of this study contains information on manufacturing processes, machines within the cell, material handling equipment, and retrieval of information needed by logic rules from an existing database. This KB could be limited to component's functions, preference, and reusability which are all stored in a heuristic manner. More information such as replacing one component with an alternative can simply be added to the existing KB. The KB will consist of a number of rules, some of which have been studied and collected from literature [5,6]. They are categorized into the following sets:

SET 1:   Rules for determining the type of layout. Determining the degree of system flexibility and type of layout given the varieties of parts and machines used within the system (e.g., Cellular, Group Technology, etc.)

SET 2:   Rules for determining layout arrangement. Determining the layout configuration based on floor dimensions such as the number of rows for the multi-row linear layout.

SET 3:   Rules for selecting material handling equipment (require information such as weight/load, type of speed, space utilization, floor conditions, etc.).

Examples of these rules could be as follows:

IF: number of parts in the system is >5, and production rate (pieces/hour) is >30

THEN: cellular or GT layout.

IF: number of parts in the system is >16, and production rate (pieces/hour) is <2

THEN: process layout, and linear single-row.

The KB essentially contains the database part which has been discussed earlier. The database files are saved separately so that these files can be changed or replaced without any modification to the sub-programs, thus making the KB more efficient and flexible. If these files have to change, i.e., to be used for different purposes, then the following limitations have to be observed:

o   The arrangement of a new database or changing the existing one is limited to the number of models in the model base (it can be avoided if used within the limit of the programs handling capabilities).

o   The files which contain the knowledge should be limited to the number of existing files (it can be avoided if additional files are loaded when consulting the database).

o   The graphical representation of the layout are presently limited to four basic types (it can be modified to include additional types).

Now, to integrate the ESMAH within the manufacturing system, the following additional sets of rules need to be implemented:

SET 4:   Rules that define various possibilities of linking basic machining operations with MHS in the system.

SET 5:   Rules for combining operation sequences that

define which operation sequences should be adopted for the system considering the new layout and the new MHS.

SET 6: Rules for local interface that monitor machine/MHS status and issue task commands and programs. It initiates part processing and checks rules 4 and 5 above.

These rules have to be implemented into the system for further expansion. The rules are usually customized based on the usage of the KB and also to the limit of the models it contains. Presently, KBESMA is in the prototype stage and has not been tested in an actual industrial environment.

5. APPLICATION

The methodology used in this study is based on one of the Tandem knowledge-based systems which is known as the model based system. The objects for the model based system are the four basic types of MHS defined as Class A as shown in Figure 2. The subsystems in this figure represent the Class B models.

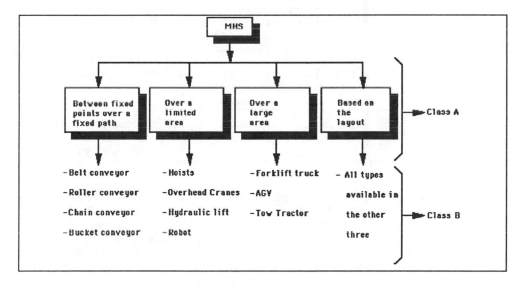

Figure 2: Tree Diagram of Material Handling Systems

5.1 Algorithm for the Selection Process

As stated earlier, our goal is to select the appropriate material handling system. To achieve this goal, the different models available need to be closely examined. This requires a number of assumptions that could be based on the previous knowledge available in books, research papers and/or experience. Consider the following four basic types of material handling systems [7]: (i) Between fixed points over a fixed path, (ii) Over limited area, (iii) Over large area, and (iv) Based on layout.

These four types can be considered as the four basic models, (denote them as class A models). Now, decide on the various attributes to be considered for the selection process in class A. These could be as follows [8]: (i) Type of layout, (ii) Type of area being considered, and (iii) Type of path. These are also called the main parameters in this system. To select one of these, there are some fixed constraints and some facts that have to be considered as shown in Table 1.

TABLE 1

Models	Constraints		
	Type of area	Type of path	Type of layout
Between fixed points over a fixed path	fixed distance	fixed	none
Over a limited area	specific area	variable in specific area	none
Over a large area	complete layout	variable in complete layout	none
Based on layout	none	none	fixed position, process or product-flow

If type of area and path is not given then input the type of layout. If the type of layout is other than those mentioned above then select one of the other class A models as shown in Table 2. Table 2 did not include the first model in class A (Table 1). Its selection is a difficult decision and requires extensive user interaction. Table 2 did not also include the fourth model shown in Table 1. It is selected only when the type of layout is given.

TABLE 2

Models	Type of layout
Over a limited area	linear single row, circular row arrangement
Over a large area	multi-row

The question remaining is how to select one of the class B models. To do this, some input based on the class A model selected (called minor parameters) is required. The proper model could be found through model based reasoning (MBR). MBR uses certain rules and facts to identify the proper model by matching the requirements of a particular model. The constraints for selecting one of the models of class B, if the first model in class A (between fixed points over a fixed path) is selected, are summarized in Table 3.

TABLE 3

| Class B | Constraints | | |
	Type of material	Wt. of material	Frequency of movement
Belt	boxes, bags and raw material	light	high
Roller	even base or to assemble parts on line	medium	low
Chain	boxes, pallets or to pull bulk material	medium	medium
Bucket	moving raw or liquid type materials	medium	medium
Overhead monorail	frames for painting or objects for baking	heavy	low

The constraints for selecting one of the models of class B, if the second model in class A (over a limited area) is selected, are summarized in Table 4 along with some of the facts or rules. In Table 4, if the type of material is raw and it is something like Iron ore or coal then the expert system has to select the first model in class A even though it is not for limited area because it is more convenient to operate one of the conveyors here. The constraints for selecting one of the models of class B, if the third model in class A (over large area) is selected, are summarized in Table 5 along with some of the facts or rules. For example, if a high degree of automation is to be implemented and the space available is small, then a robot in class B should be selected.

The constraint to select one of the class B models, if the fourth model in class A (Based on layout) is selected, are summarized in Table 6 along with some of the facts or rules. Once all of these are decided it is necessary to make a note of the values for the levels of various constraints, i.e., weight of material, type of area, frequency of movement and other constraints where the levels are low, medium or high. This depends on the application but it can be generalized using the procedure illustrated in Figure 3.

TABLE 4

Class B	Constraints			
	Wt. of material	Type of material	Required lift	Space available
Hoist	medium	boxes, frames, pallets	high	medium
Overhead Crane	heavy	boxes, frames, pallets	high	large
Hydraulic lift	heavy	anything with a proper base	medium	small
Robot	light	where a proper hold is possible	low	small

TABLE 5

Class B	Constraints			
	Flexibility in usage	Degree of automation	Wt. of material	Space available
Forklift truck	high	low	light	small
Tractor-trailer train	low	low	heavy	large
AGV	medium	high	medium	medium

TABLE 6

Class B	Constraints		
	Type of material	Type of production	Degree auto'n
Fixed position layout	subassembly, heavy material	One or two produced	low
In process layout	pallets, in process material	batch production	medium
Product flow layout	standard and identical type	flow type mass production	high

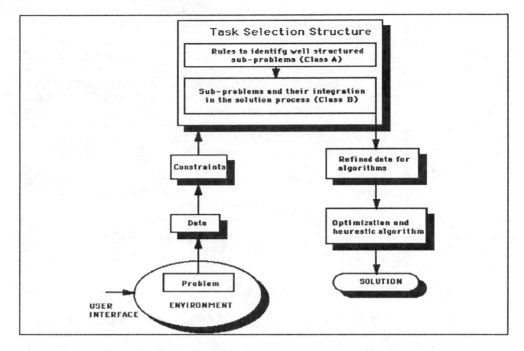

Figure 3: Selection Process within ESMAH

5.2 Illustrative Example

ESMAH is a stand alone program which runs in the Prolog shell. The program is user friendly as it is menu driven and prompts the user to input the data based on a help screen which gives summary about the possible answers and explanation. The main menu which pops up in the beginning gives various options to query, append, insert, and delete a rule or fact. An example query is as follows: Selection from the main menu was to query the system. The input for the first three screens of the query where: process type layout, large area, and path is variable. With these inputs the system had selected class A model and also loaded the required database automatically. This selection invoked another menu asking questions related to: heavy material, medium flexibility, less automation, and low space utilization. For the system this data was enough to suggest FORKLIFT as the MHS for a similar situation.

The above suggestion came out without many questions because the user had answered most of the questions which lead to a MHS selection. For example, if the user had not input the layout then the system would pop more questions, e.g., the number of parts in the system, production rate, etc. The graphics representation is the help tool for the layout which would also pop up along with the suggestion.

6. CONCLUSIONS AND FUTURE RESEARCH

The methodology presented in this research uses model based reasoning where model representation provides a better understanding of the selection process. In this scheme the input data will be minimized which means that the user does not have to provide excessive data not required by the expert system. The system developed is therefore a user friendly and capable of including a new model or changing an existing model without difficulty. The methodology is flexible enough to provide the right MHS system regardless of the layout.

The additional databases can be developed based on the facility requirement. Improvement depends on the need of the facility and the usage of it as the database is dynamic in nature and has the capability to be updated when necessary. The database can also be shared by the facility planning department because of its flexibility.

REFERENCES

1. Sule DR: <u>Manufacturing Facilities Location, Planning and Design</u>, PWS-KENT Publishing Company, Boston, 1988.
2. Elsayed EA: Algorithms for optimal Material Handling in Automatic Warehousing Systems, <u>International Journal of Production Research</u>, Vol. 19, No. 5, 1981, pp. 525-535.
3. Gabbert P, Brown DE: Knowledge-Based Computer-Aided Design of Material Handling Systems, <u>IEEE Trans. on Systems, Man and Cybernetics</u>, Vol. 19, No. 2, 1989, pp. 188-196.
4. Kusiak A: <u>Artificial Intelligence Implications for CIM</u>, IFS (Publications) Ltd, UK, 1988, pp. 484-503.
5. Abdou GH, Dutta SP: An integrated Approach to Facilities Layout Using Expert Systems, <u>Int. J. Prod. Research</u>, Vol. 28, No. 4, 1990, pp. 685-708.
6. Billatos SB: Design of Intelligent Manufacturing Cells, <u>Proceedings of the SME AutoFACT Conference</u>, Detroit, MI, November 1990.
7. Kulwier RA: <u>Materials Handling Handbook</u>, John Wiley & Sons, NY, 1985.
8. Apple JM: <u>Plant Layout and Materials Handling</u>, Ronald Press, NY, 1950.

RELIABILITY OF CUTTING FORCES IN MACHINABILITY EVALUATION

Osama K. Eyada

Department of Industrial and Systems Engineering, Virginia Polytechnic Institute and State University, Blacksburg, Virginia, U.S.A.

ABSTRACT

This paper describes a cutting forces test utilized on free-cutting steels; to justify previous conclusions that cutting forces do not correlate with the machinability ratings of automatic screw/bar machine tests; to investigate the sensitivity and repeatability of the resultant force components (cutting, feed and radial forces) in differentiating between the machinability of materials; and to determine whether the machinability ranking is independent of cutting conditions (speed, feed and depth of cut). The test results correlated with those of an automatic screw/bar machine test. The machinability ranking was independent of cutting conditions. The feed force was the most sensitive resultant force component to machinability variations.

INTRODUCTION

With the trend towards automated manufacturing, the machinability of work materials is becoming important. The consistency of work materials, cutting tools and cutting fluids are vital factors to the reliability of modern manufacturing systems. The high demand for accuracy and consistency in the input factors of machining operations is intensifying pressure on the manufacturers of work materials, cutting tools and cutting fluids. This work is concerned with materials manufacturers whose problem is not only how to design new materials that would provide better material properties and/or machinability, but also how to control the improved machinability in daily production (i.e., within and among batches).

Machinability is one of the most important concepts in metal cutting. Yet, scientists and investigators do not agree on how it should be defined or tested. Machinability is sometimes considered as the ease with which materials are cut using appropriate cutting tools, and at other times as the material resistance to machining. Either way, terms such as ease, resistance and appropriate are themselves subject to individuals' definitions. Ease or resistance can be expressed in terms of obtained part qualities (e.g., surface finish, surface integrity or dimensional accuracy) and/or material effects on tools (e.g., tool wear, cutting forces or cutting temperature). What is appropriate for a machining operation is not appropriate for another. For example, a single tool geometry for the optimum machining of a wide range of materials does not exist. Additionally, machinability data are usually presented in the form of a single index (where a standard material is assigned a 100 percent and the others are ranked relatively) based on the speed for a 60 minute tool life or any other indicator [1]. The machinability ratings of materials can be dependent on the type of test as well as the machinability indicator. Questions regarding how machinability should be defined and tested, whether machinability results should be provided in absolute or relative values, etc. are addressed in detail in reference [2].

Over two hundred references were retrieved when the literature was searched for machinability tests conducted within only a five-year period [2]. The majority of these tests can be utilized to investigate the performance of work materials, cutting tools or cutting fluids. Machinability tests that closely simulate production conditions, such as the automatic screw/bar machine tests, are commonly regarded as the most reliable ones. Although various automatic screw machine tests are described in the literature [3], also known as Production Performance Tests (PPT) or simulated production tests, their testing procedures are roughly the same. Variations are usually in terms of parts produced and/or testing domain, where feed, speed or both are varied. Typically, the actual production of a certain part is simulated under controlled conditions on a single or multiple spindle machine. The objective is to determine the maximum production rate (pcs/hr) for each material through searching the feed, speed or both the feed and speed domain by trial-and-error and under some constraints. These are usually an upper surface roughness limit and an average tool life of 6-8 hours of actual running time (cutting and indexing times). Machinability ratings based on these tests results could be dependent on the part produced, number of spindles and the employed search procedure. A standard method to overcome some of these problems is currently underdevelopment [3].

Free-cutting steels are usually machined on automatic screw machines and, hence, their machinability space is usually determined by automatic screw/bar machine tests. However, such tests are expensive and time consuming (usually, 5 weeks are needed to evaluate the machinability of one material) and, thus, only few materials can be investigated. Also, they do not utilize any replications and this makes it difficult to assess the real variables that affect machinability. A short term test that can provide results in agreement with those of the PPT is needed. Such a test can be used to screen a wider range of materials than the PPT and, in turn, qualify only those materials with high machinability potential for final screening by the PPT. In addition, it can be utilized in the steel mill for machinability assurance. This paper describes one of three short-term tests utilized on some free-cutting steels with machinability ratings based on a PPT to meet the above objectives [2].

Although the results of many cutting forces tests ([4] to [18]) did not correlate with industrial ratings (PPT results), a conventional cutting forces test was still investigated. The cutting forces are the sources for the heat generated during metal cutting. This heat raises the cutting temperature which has been shown to correlate well with commercial ratings [19]. It was, hence, questionable why the cutting forces do not correlate well with the PPT results while the cutting temperature does. This may be attributed to the high scatter in the cutting forces readings. Tests conducted by Redford [20] to investigate on-line tool wear assessment revealed that cutting forces readings gave more scatter than the cutting temperature measurements, especially when the tool was worn. However, the effect of data scatter can be reduced through replications. Despite the added costs of replications, cutting forces tests would still be advantages over cutting temperature tests: (1) the setup of cutting forces tests is simpler, and (2) the machining time required for the cutting forces to reach the steady state condition is less than that of the cutting temperature. The conversion of mechanical energy into heat which raises the cutting temperature takes time. The steady state temperature was reached after 25 seconds of cutting in a work by Redford [20], while 20 seconds was required in another work by Mills and Akhtar [21].

On the other hand, the work of Fersing and Smith [6] and De Filippi [12] indicated that the feed force increases faster than the cutting force with tool wear. Colwell, Mazur and De Vries [22] concluded that the feed force is a better indicator of the beginning of the accelerated tool wear region than the cutting force. Otherwise stated, the feed force seems more sensitive to variations in the machining process than the cutting force. Perhaps because of that and the small magnitude of the radial force, the majority of the indicators utilized in cutting forces tests are related to the feed force. However, previous studies did not address the relative sensitivity of the cutting forces components (cutting, feed and radial force) in differentiating among the materials machinability. Particularly, the radial force is often neglected due to its smaller magnitude relative to the other two components, while it is the one that directly affects the tool nose wear and, thus, the dimensional accuracy of the machined part. Cook [23] mentioned that the degree of tolerance degradation is associated with the wear of the tool nose radius and flank, and that it generally correlates well with the tool flank wear. Finally, the machinability ranking of materials was independent of feed pressure in a sawing test [2] and feed in a slotting test [3]. It was, then, questionable whether the machinability ranking is independent of other cutting parameters, such as the cutting speed and depth of cut.

A Cutting Forces Test (CFT) was, therefore, utilized to (1) justify previous conclusions that cutting forces do not correlate with the PPT results, (2) to investigate the sensitivity and repeatability of the resultant force components (cutting, feed, and radial forces) in differentiating between the machinability of materials, and (3) to determine whether the machinability ranking of materials is independent of cutting conditions.

A CUTTING FORCES TEST

Cutting forces are almost measured in every machining study. In addition to their importance for the design of machine tools, cutting tools and fixtures, they can be used either directly or indirectly for machinability evaluation. The majority of cutting forces tests performed on a lathe can be classified into two groups: conventional and non-conventional. The first group would include the tests that directly utilize cutting forces measured on a conventional setup in their analyses. The second group would include those tests that require an unconventional setup (i.e., turning at constant feed force tests [10] and turning at increasing speed tests [18]). The main drawback to non-conventional cutting forces tests is the special setup required. Such setups are not available in most laboratories.

Conventional cutting forces tests can be further classified into orthogonal and normal turning. To achieve orthogonality on a lathe, some workpiece preparations, certain workpiece geometries, or a special tool geometry would be required. For normal turning, typical cutting forces indicators include: forces obtained under certain conditions [5, 24, 25], feed for a given feed force value or the feed force resulted at a certain feed [4], time to reach a certain feed force value [26], and horsepower calculated from measured forces [27].

The utilized CFT is of the conventional type where cutting forces are measured under certain conditions. The work materials were thirteen different designs for the 1213 and 12L14 grades of the 12XX series of free-cutting steels [2]. All tests were performed on a LeBlond 20 inch (508 mm) heavy duty lathe. The lathe was equipped with a Kistler 9257A dynamometer, 9403 tool holder, and 5008 charge amplifier. The data acquisition and display system consisted of a smartscope, a monitor, and an oscilloscope. In order to avoid possible errors in re-sharpening HSS cutting tools (commonly used for machining free-cutting steels), indexable carbide inserts (WC) were utilized. The utilized inserts were Kennametal TNG322, a triangle insert of a 2/64 inch (.8 mm) nose radius, of the K45 type. A Kennametal

KTBR-643 INS TN-32 tool holder (i.e., a tool geometry of -5, -5, 5, 5, 15, 18 degrees) was utilized. A negative tool holder was chosen, since a negative insert provides more cutting edges than a positive one. The above tool geometry was employed due to the small diameter of the bars, 1-inch (25.4 mm), and the fact that the larger the side cutting edge angle, the larger the radial force and, hence, vibration.

RESULTS AND DISCUSSION

Preliminary tests, performed on a random sample of three materials, were aimed at: (1) investigating the sensitivity of the resultant force components in differentiating between the machinability of materials, (2) determining whether the machinability ranking of materials is independent of cutting conditions, and (3) locating a cutting condition at which screening among the materials is statistically most powerful. Figure 1 shows the CFT as an input/output system and Table 1 presents the utilized five testing conditions which reflect the speed (v), feed (f), depth of cut (d), and the interaction term (fd) effects from the conditions at which the cutting parameters are low. The interaction term (fd) represented a rough cutting condition. Three replications were taken for each material under each of these cutting conditions. Although these cutting conditions are very moderate for free-cutting steels, because the machine was discovered to lose power above a certain cutting condition, they still can reflect the effects of speed, feed, depth of cut, and a rough cutting condition on the machinability ranking of the materials.

3 Materials
WC Insert
5 Cutting Conditions   ]   MACHINABILITY - Cutting Forces
Dry   ]   (Turning)

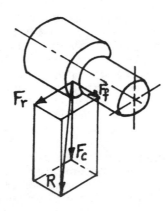

FIG. 1 The CFT as Input/Output System

TABLE 1
Testing Conditions

(1)	(d)	(f)
v1 = 400 sfpm (120 m/min) f1 = .009 ipr (.225 mm/r) d1 = .040 inch (1 mm)	d2 = .080 inch (2 mm)	f2 = .018 ipr (.457 mm/r)
(v)	(fd)	
v2 = 600 sfpm (180 m/min)	f2 = .018 ipr (.457 mm/r) d2 = .080 inch (2 mm)	

The bars were held between a collet and a live center with around one foot (304 mm) exposure to avoid vibration. For each test, a new cutting edge was utilized to first remove a thin outer-skin, .005 inch (.13 mm) depth of cut, and then cut for 13 seconds under the test conditions. The removal of the outer-skin provides a uniform depth of cut and, thus, eliminates a possible source of noise. The first three seconds were regarded as the tool break-in period. The smartscope was set to record 500 readings per 10 seconds and to display the slope of a fitted line to these measurements. The slopes were examined to insure that all the readings were taken after the tool break-in period, and that they were not significantly different between materials and/or testing conditions. The average value of the 500 readings was, then, recorded for each cutting force component.

Table 2 provides the averages of the three replications and the total averages under the different testing conditions. The results indicate that the effects of the different cutting conditions were in agreement to the expected general trends; the higher the speed, the lower the forces, and the higher the feed, depth of cut, or both, the higher the forces. Table 3 summarizes the CFT preliminary results in terms of F-ratios for among (Fa) and within (Fw) materials based on an analysis of variance, and Duncan ranking results [28]. In table 3, the materials were ranked first, second and third based on their cutting forces results for ease of display. The presence of a bar under the three materials ranking indicates that there is no statistical significance between the materials results based on the Duncan test. From the results, it can be concluded that:

(1) The cutting force did not differentiate between the materials, while both the feed and radial forces did.

(2) The ranking of the materials based on either the feed or the radial force was, in general, independent of the cutting conditions (speed, feed and depth of cut).

(3) The ranking of the materials was also independent of the type of cut; rough (fd) or finish (v) turning.

(4) Under the cutting condition of the feed effect (f), there was no overlapping in the Duncan ranking for both the feed and radial forces. Also, the feed force under the (fd) cutting condition effect, did not show overlapping. However, for the radial force (Fr) at the feed-effect cutting condition (f), screening among the materials was statistically most powerful. The F-ratio for among materials (Fa) was the highest given that the F-ratio for within materials (Fw) was insignificant.

TABLE 2
Averages of CFT Preliminary Results (Newton)

Force	Material	(1)	(v)	(f)	(d)	(fd)
Fc	A	320.4	299.4	540.5	648.7	1094.6
	B	339.5	307.7	581.0	688.3	1129.0
	I	329.1	308.5	560.1	609.2	1092.2
	Average	329.7	305.2	560.5	648.7	1105.3
Ff	A	306.0	274.5	428.8	702.5	976.4
	B	324.1	284.4	465.6	786.3	1064.3
	I	338.6	322.8	514.7	796.2	1116.2
	Average	322.9	293.9	469.7	761.7	1052.3
Fr	A	244.3	217.5	417.2	398.1	636.0
	B	264.1	226.5	479.3	453.0	696.4
	I	278.6	256.8	501.4	424.0	704.9
	Average	262.3	233.6	466.0	425.0	679.1

TABLE 3
Summary of CFT Preliminary Results

	Cutting Force (Fc)			Feed Force (Ff)			Radial Force (Fr)		
	Fa	Fw	Rank	Fa	Fw	Rank	Fa	Fw	Rank
(1)	5.8	2.3	1 3 2	17.8	5.1	1 2 3	8.0	.15	1 2 3
(v)	.4	.1	1 2 3	55.8	3.2	1 2 3	13.9	.30	1 2 3
(f)	5.9	7.6	1 3 2	314.2	42.5	1 2 3	97.2	.90	1 2 3
(d)	2.2	1.6	3 1 2	4.8	.1	1 2 3	3.4	.01	1 3 2
(fd)	2.8	1.4	3 1 2	34.1	1.1	1 2 3	15.5	.70	1 2 3

Fa : F-ratio among materials
Fw : F-ratio within materials (replications)

The correlation coefficient between the radial force results at the feed-effect cutting condition (f) and the PPT results was a 100% (i.e., positive correlation). Normally, the higher the cutting forces, the lower the machinability. However, the following data indicates that the radial force results did also correlate well with the values of the hardness (but negative correlation). Since a negative tool holder was used, apparently more forces were needed to bring the materials to a state where they can be sheared under compression stresses.

Material	HBN	PPT (pcs/hr)	Fr (Newton)
A	192	336	417.4
B	170	383	479.4
I	156	538	501.4

From the preliminary results, it can be concluded that the CFT does differentiate between the work materials, the cutting force is not a reliable machinability indicator, both the radial and feed forces are sensitive to machinability variations, and the machinability ranking of materials is cutting condition independent. Also, a cutting condition at which screening among the materials is statistically most powerful was located, the cutting condition of the feed effect (f) in Table 1.

The remaining ten of the thirteen work materials were randomly machined under the feed-effect cutting condition. A summary of the F-ratios in terms of among and within materials for the three forces is given below. The F-ratios confirm the preliminary testing conclusion that the cutting force, which has an insignificant ratio, is not a reliable indicator for evaluating the machinability of materials. Although both the feed and radial forces have shown significant differences among materials, variations within materials (replications) were significant for the radial force. Perhaps a larger sample size than three replications (i.e., more degrees of freedom) could result in insignificant variations within materials for the radial force. Unfortunately, due to stock limit, this was not investigated further. Nonetheless, it can be concluded that the feed force is the most reliable indicator among the components of the resultant force for the machinability evaluation of free-cutting steels. This is followed by the radial force.

Force	F-ratio Among		F-ratio Within	
F_c	3.8	(Insignificant)	34.3	(Significant)
F_f	100.9	(Significant)	2.7	(Insignificant)
F_r	90.9	(Significant)	19.0	(Significant)

Table 4 provides the production rates of an automatic screw machine test (PPT) and the feed force results determined after the Duncan analysis, and the absolute difference between the two tests' ranking. The biggest disagreements among the two tests' results were reflected in steels A, B and C. The CFT does not show these materials as unmachinable as the screw machine test. Examination of the CFT results in relation to a sawing test [2] and a slotting test [3] results pointed to two materials that had very high variations within their replications, steels A and G. They were, hence, considered nonhomogeneous materials (or belong to a different population). Also, the results of a simple linear regression analyses on the remaining data indicated that steel H is an influential point on the regression model

[29]. After disregarding these three steels, the two tests results had a 70% correlation, a typical value when correlating machinability data.

CONCLUSIONS

The results of the cutting forces test (CFT) on free-cutting steels indicated that the machinability ranking is independent of cutting conditions (speed, feed and depth of cut). The feed force was the most sensitive resultant force component to variations in the materials machinability. This was followed by the radial force. The cutting force measurements gave the highest scatter and, in turn, it is not a reliable machinability indicator. The CFT results did correlate with those of a PPT and, thus, it can be utilized to qualify potential materials for the production performance test.

ACKNOWLEDGMENTS

The author would like to thank Bethlehem Steel Corporation for their cooperation and support throughout this project.

TABLE 4
Production Rates and CFT Results

Material	HBN	PPT pcs/hr	PPT Rank	CFT Ff	CFT Rank	Absolute Difference in Ranking
A	192	336	8	422.5	2	6
B	170	383	7	463.5	3	4
C	192	388	6	422.5	2	4
D	170	440	5	516.6	4	1
E	167	440	5	463.5	3	2
F	187	440	5	463.5	3	2
G	207	572	2	422.5	2	0
H	207	572	2	403.9	1	1
I	156	538	3	516.6	4	1
J	153	620	1	516.6	4	3
K	156	620	1	516.6	4	3
L	156	502	4	532.3	5	1
M	159	502	4	532.3	5	1

REFERENCES

[1]   SAE, "Estimated Mechanical Properties and Machinability of Hot Rolled and Cold Drawn Carbon Steel Bars," *SAE Standard J414a, SAE Handbook*, 1981, pp. 3.14-3.17.

[2]   Eyada, O.K., "Short-Term Tests For Evaluating The Machinability of Free-Cutting Steel," *Ph.D. Dissertation, Lehigh University, 1987. (U.M.I. number 8729532)*

[3]   Eyada, O.K. and Kane, G.E., "An End Milling-Slotting Test for Evaluating The Machinability of Free-Cutting Steel," *Transactions, NAMRI/SME*, May 1989, pp. 103-107.

[4]   Murphy, D.W. and Aylward, P.T., "Machinability of Steel," *Homer Research Laboratories, Bethlehem Steel Corporation*, Bethlehem, Pennsylvania, 1964.

[5]   Litwa, R., "Development of Substitute For Leaded Steels," Unpublished Report, File Ref. 1204-2CQz-50-81022, December 1983, *Homer Research Laboratory, Bethlehem Steel Corporation*, Bethlehem, Pennsylvania.

[6]   Fersing, L. and Smith, D.N., "Machinability Research with J & L Tool Dynamometer on Titanium 150A," *ASME Paper No. 53-A-207*, 1953.

[7]   Mayer, Jr., J.E. and Stauffer, D.J., "Effects of Tool Edge Hone and Chamfer on Wear Life," *Transactions of SME*, Vol. III, 1975, pp. 5-15.

[8]   Schmidt, A.O., Ham, I. and Wilson, G.F., "Comparative Performance Study of Different Tool Geometries," *Proceedings, 7th IMTDR Conference*, 1966, pp. 435-462.

[9]   Loria, E.A. and Walker, D.R., "Force and Temperature Measurements are A Rapid Guide to Tool Life Evaluation," *Iron Age*, Vol. 174, 1954, pp. 156-158.

[10]   Boulger, F.W., Shaw, H.L. and Johnson, H.E., "Constant-Pressure Lathe Test For Measuring The Machinability of Free-Cutting Steels," *Transactions of ASME*, Vol. 71, 1949, pp. 431-446.

[11]   Troup, G.B., "Machinability Tests: A Critique," *Am. Mach.*, Vol. 110, 1966, pp. 127-131.

[12]   De Filippi, A., "Analysis and Proposal For Short-Time Machinability Tests," *Influence of Metallurgy on Machinability*, ASM, 1975, pp. 396-412.

[13]   Shaw, M.C., Smith, P.A., Loewen, E.G. and Cook, N.H., "The Influence of Lead on Metal-Cutting Forces and Temperatures," *Transactions of ASME*, Vol. 79, 1957, pp. 1143-1154.

[14]   Milovic, R. and Wallbank, J., "The Machining of Low Carbon Free Cutting Steels with HSS Tools," *The Machinability of Engineering Materials*, ASM, 1983, pp. 23-41.

[15]   Araki T. and Yamamoto, S., "An Evaluation of Machinability of Low Alloy Steel Materials With or Without Heat Treatment," *Machinability Testing and Utilization of Machining Data*, ASM, 1979, pp. 117-131.

[16]   Nead, J.H., Sims, C.E. and Harder, O.E., "Properties of Some Free-Machining Lead-Bearing Steels, Part 2," *Metals and Alloys*, Vol. 10, 1939, pp. 109-114.

[17]   Schlesinger, G., "Machinability of Metals," *The Tool Engineer*, Vol. 17, 1947, pp. 18-27.

[18]   Heritier, B., Duet, R., Thivellier, D. and Maitrepierre, P., "Turning at Increasing Speed: A Convenient Method to Assess Machinability," *The Machinability of Engineering Materials*, ASM, 1983, pp. 490-511.

[19]   Kane, G.E. and Groover, M.P., "The Use of Cutting Temperature as a Measure of The Machinability of Steels," *ASME Technical Paper No. MR67-199*, 1967.

[20]   Redford, A.H., "On-line Assessment of Tool Wear Using Wear-Temperature Gradient Relationships," *Machinability Testing and Utilization of Machining Data*, ASM, 1979, pp. 367-379.

[21] Mills ,B. and Akhtar, S., "A Metallurgical and Machining Study of Free Machining Low Carbon Steels," *Influence of Metallurgy on Machinability,* ASM, 1975, pp. 73-88.

[22] Colwell, L.V., Mazur, J.C. and De Vries, W.R., "Analytical Strategies For Automatic Tracking of Tool Wear," *17th IMTDR Conference, 1976,* pp. 276-282.

[23] Cook, N.H., "Tool Wear and Tool Life," J. Engr. Ind., November 1973, pp. 931-938.

[24] Boston, O.W., "Machinability of Metals," *Transactions of American Society For Steel Treating,* Vol. 13, 1928, pp. 49-61.

[25] Boulger, F.W., "Influence of Metallurgical Properties on Metal-Cutting Operations," *American Society of Tool Engineers,* paper number 80, 1958.

[26] Colwell, L.V. and McKee, R.E., "Evaluation of Bandsaw Performance," *Transactions of ASME,* Vol. 76, 1954, pp. 951-960.

[27] Boston, O.W., "Methods of Tests For Determining The Machinability of Metals in General, with Results," *Transactions of American Society For Steel Treating,* Vol. 16, 1929, pp. 659-710.

[28] Miller, I. and Freund, J.E., *Probability and Statistics For Engineers,* Prentice-Hall, Inc., 1965.

[29] Cook, R.D., "Detection of Influential Observation in Linear Regression," *Technometric,* Vol. 19, 1977, pp. 15-18.

ULTRA-HIGH SPEED MACHINING — THE PROBLEMS ASSOCIATED WITH SPINDLE DESIGNS, TOOLING AND MACHINED PARTS WHEN MILLING

Graham T. Smith

Engineering Division, Southampton Institute, Southampton, U.K.

ABSTRACT

Once a rotating spindle is revolved above 20 000 RPM, considerable limitations exist in spindle design, cutter balance and controller processing speed. Recently several machine tool builders have manufactured equipment with special purpose bearings, although often high-speed spindle cartridges are retrofitted to conventional machines.

Cutter balance is crucial for any rotating tooling at high speed, as vibration, surface finish and cutter life are severely reduced if this is not undertaken. Cutters require balancing in two planes, but when this has been resolved, the spindle nose can still swell as centrifugal force reduces the taper contact and modifies the spindle/holder relationship.

The workpiece can be subjected to a range of unusual effects resulting from ultra-high speed milling, chip clogging - poor geometry and evacuation of swarf can scratch and gouge the machined surface. "Workpiece abuse" on aluminium alloys can be present when cutters are "ramped-down" into a pocket to be milled using area clearance cycles, and this surface degradation is difficult to initially identify. However, major benefits accrue from using ultra-high speed milling operations and as such their implementation will increase for the foreseeable future.

INTRODUCTION

Initially one might ask the question: "Why do we need to rotate cutters at ultra-high speeds?" There are a number of advantages that occur from adopting such a progressive machining strategy (FIG 1) and they can be succinctly summarised as follows:

- o   direct benefits   – improved machining efficiency,
  -       – reduction in cutting tool variety,
  -       – reduction in distortion of workpiece,
  -       – effectiveness of swarf removal,
- o   indirect benefits – quality of finish improved,
  -       – increased cutter life,
  -       – capability of machining thin walls/ section,
  -       – reduced changes in material properties.

These production improvements are by no means all, with the component having a superior finish which usually does not require deburring, whilst fixturing can be of simple design [1]. If these are the advantages, then one has to expect there to be problems associated with rotating tooling at ultra-high speeds [2]. However, prior to a discussion about these disadvantages which must be addressed, it is worth considering the "milestones" that have lead to the rapidly expanding change in rotating tooling philosophy of late.

HISTORICAL PERSPECTIVE

Possibly the first work on ultra-high speed machining (UHSM) was that of Salomon, who ran a series of experiments from 1924-'31 [3]. The patent was founded upon a series of cutting speed curves plotted against cutting temperatures for a range of materials (FIG 2).   In these tests, Salomon achieved peripheral cutter speeds of 16,500 smm, using helical milling cutters when machining aluminium. He [3] contended that the cutting temperature peaked at a specific cutting speed – the "supercritical speed" – and as this speed increased still further the temperature decreased.

Furthermore, either side of this "supercritical speed" zone were suggested unworkable regimes where cutters could not withstand the severe forces and temperatures generated [4]. As the cutting speed increased beyond the "supercritical speed" the temperatures once again reduced to those expected by normal cutting conditions, permitting practical cutting operations.   This meant that the same cutting temperature existed in the normal speed range as well as being established at a higher range of speeds. During the Second World War much of the supporting experimental information on Salomon's work was lost, with none of the participants in this research being alive to comment on the data obtained and speculation on how this information was collated.

The practical data by Salomon was based upon experiments on non-ferrous alloys in particular the soft aluminium and red cast brass – which is unmachinable with High-speed steel cutters between 60 to 330 smm – with the curves for bronze and ferrous alloys being obtained by extrapolation and were not verified in the laboratory [4].   Any theoretical rationale is not available and the experimental procedures unclear, but Salomon can be considered the founder of ultra-high speed machining – beyond those considered by the Taylor equations.

Four periods of advancement in the field of UHSM can be established [4], with the first being during the 1920-'50's, each being separated by a significant event.

Obviously the instigator being Salomon in the 1920's, with the first major funded research project being commissioned by the United States Air Force (USAF) from 1958 to 1961.   During the previous decades little interest occurred apart from that by Vaughn [5], who became aware of the Salomon patent – acquiring limited information through the United States Consul in Berlin.   Vaughn's group were familiar with much of the terminology, due to their experiences from oil well perforation at high speeds – oil well drilling utilises the practice of explosive perforation cutters being used to perforate casings on oil wells.   Such background work meant that Vaughn of Lockheed set the scene for the second development period.

The USAF Materials Laboratory Study – alluded to previously – awarded Lockheed the contract to evaluate the response of a selected number of high-strength materials to cutting speeds up to 152,400 smm. The principal objective of this work was to increase producibility, whilst improving quality and efficiency of the fabrication of aircraft/missile components.   The experimental apparatus included cannons on sleds to obtain the desired exit velocity for cutting speeds, but unfortunately the results did not show how such speeds might be incorporated into a production application, moreover an analytical model of the high-rate cutting phenomenon was not developed.

During the 1960's activity occurred from some notable work by Arndt [6] in Australia; Recht [7] in the United States; Fenton [8] in the United Kingdom; Okushima [9] and Tanaka [10] in Japan. Although this increase in research was at this stage both without practical application and disparate in nature.

The third period of development was instigated by the United States Navy who contracted a series of studies in the early 1970's, in conjunction with Lockheed. Here, the objective was to determine the feasibility of utilising high–speed machining in a production environment – principally for aluminium machining and later, with nickel–aluminium–bronze. King's team [11] demonstrated that it was economically feasible to introduce high–speed–machining procedures into a production environment to realise significant improvements in productivity,. Such work [11] promoted significant activity and

interest in both experimental and applied research and it soon became clear that attention needed to be focussed for all of these small and diverse research studies.

Yet again in 1979, the USAF awarded a contract to the General Electric Company in this instance, to provide a scientific basis for faster metal removal via high–speed and laser–assisted machining. A further contract was given to General Electric to evaluate the production implications of the first, with a consortium of industrial companies and universities – which instigated the fourth period of development – whereby machine tool and tooling companies developed specific products for the market.

MACHINE TOOL & TOOLING – implications when UHSM

In order to gain significant benefits from milling operations when UHSM, several problems not associated with conventional rotational speeds must be overcome [12]:

- o   machine tool's accuracy and rigidity,
- o   cutting spindle's design and performance,
- o   tool holder balance and cutter design,
- o   axis drive control capability,
- o   controller processing speed.

MACHINE TOOL'S ACCURACY AND RIGIDITY

Using UHSM technology the higher feedrates associated with these cutting practices, means that faster acceleration/deceleration requirements are demanded so that optimum machine performance can be realised. Considerably greater stresses are imposed on the machine tool's structure – not by the cutting process which reduces the loading, but by the greater velocities induced by higher feedrates – which follows that it must be designed around a more rigid structure. However, this is contrary to what is demanded by an UHSM strategy, in that lighter and more responsive moving elements are desirable with machine tools being carefully designed to take account of excess weight. If the weight is reduced too much this will influence the machine's ability to absorb vibrations, as a good damping capacity is essential on any machine to minimise the influence of chatter which can have disastrous effects on tool life [13 & 14]

CUTTING SPINDLE'S DESIGN & PERFORMANCE

One of the bi-products of such high stock removal rates, is the excess volume of hot swarf which must be removed speedily and efficiently from the vicinity of the machine – which is easier to achieve in a horizontal configuration than vertical. Heat in general on any machine tool becomes a problem [15] particularly as many milling spindles utilise direct – drives, with the motor being mounted in-situ with the spindle. Motors create heat [15] and it is usual to refrigerate the spindle to minimise the problem of compensating for thermal drift when in use. Drive spindles are usually of three types

either:
- o   magnetic "active" spindles,
- o   pneumatic spindles,
- o   hybrid spindles

The reason why conventional ball bearing spindles are not specified at around 20,000 RPM, is that with an upper rotational velocity of 80 m/s the balls lose contact with the journal walls and rapidly wear out due to a combination of factors; centrifugal force, frictional effects and roundness modifications [12]. A typical magnetic spindle on the one hand, has a speed range from 4,000 to 40,000 RPM delivery 40kW of continuous power with a peak of 51.2 kW. These "active" magnetic spindles can maintain 1 μm maximum run-out by digital control of the current to the magnets – initiated by radial and axial sensors – continuously monitoring position 10,000 times per second. Further refinements to the spindle occur and these temperature – controlled milling spindles, maintain dynamic balance regardless of the cutting loads, an important criteria when attempting to reduce vibrations.

Pneumatic spindles have existed for sometime [12] with aerostatic bearings equalising the forces exerted whilst cutting and remaining centralised within its housing, whilst achieving dynamic balance. Such spindles can be rotated at exceptionally high speeds, by virtue of the "ideal" condition of minimal metal-to-metal contact, although the low power – torque – delivery is somewhat lacking. To answer the criticism of pneumatic drive spindles, the hybrid design was developed, which incorporates an aerodynamic thrust bearing with spiral grooves which can withstand up to three times the static loads of the conventional aerostatic bearings. A typical hybrid (aerodynamic) bearing can achieve rotational speeds of between 20,000 to 30,000 RPM with 15.5kW at peak speed.

Several problems faced by all these spindle drives, is the fact that the spindle nose grows non-uniformly with increasing peripheral speed and that momentum force can be a significant factor during machining operations [16]. The centrifugal force's influence on spindle nose growth is proportional to the square of the rotational speed and the taper exhibits a "bell-mouthing", which in turn leads to a loss of register of the tool holder's spindle taper coupling. In order to overcome this problem of loss of contact at high rotational speeds, shoulder-contacting toolholder designs which marginally elastically deform the spindle taper, giving secure taper-to-face contact [17].

TOOLHOLDER & CUTTER DESIGN

As one might expect when tooling is rotated at UHSM the problem of balance becomes of some importance. Many companies attempt to utilise unbalanced tooling and this might be acceptable at speeds up to 10,000 RPM, however when this is carried forward to higher rotational speeds then out- of-balance becomes a major problem [18]. Unbalance of a rotating body can be defined as the condition existing when the principal mass – "axis of inertia" – does not coincide with its rotational axis (FIG 3). Such an undesirable state of affairs can be illustrated by considering a Φ50mm face mill rotating at 15000 RPM having a peripheral speed approaching 150 miles per hour, which may prove disastrous if unbalanced!

Basically, there are types of unbalance [18] of rotating tooling that can exist:
- o   static unbalance – single plane,
- o   couple unbalance,
- o   dynamic unbalance – dual plane.

Static unbalance – single plane

This unbalance occurs when the mass does not coincide with the rotational axis, but is parallel to it and the force created by such unbalancing, is equal in magnitude at both ends of the rotating body. If some relief – metal removal – on the holder body equal to the out-of- balance mass occurs, then a nominal static balance is achieved.

Couple unbalance

Under these circumstances the cutter/holder – mass axis – does not coincide with the rotational axis, but intersects it at the centre of gravity of the cutter's body. Under such circumstances the force vectors equalise, but are $180°$ apart.

Dynamic unbalance – dual plane.

Such a condition of tool/holder arises when the axis does not coincide with the rotational axis and is not either parallel to, or intersecting this axis (FIG 3).

For any rotating tooling estimating the cutter unbalance [18] using the following variables is possible:

"M" = cutter/holder mass,
"S" = mass centre,
"e" = displacement of mass centre,
"r" = distance from centre of tooling, to the centre of gravity of mass (m),
"ω" = angular velocity,
"m" = mass unbalance
"U" = rotor unbalance.

Determining the relative unbalance (U) of a rotating tool/holder can be found by:

$$U = M.e \quad \text{or, alternatively } U = m.r \quad ......(i)$$

It is usual to express unbalance in terms of the product of the mass times distance, typically using the units "g-mm"

Finding the magnitude of centrifugal force produced by the rotating tooling with a given unbalance, can be established as follows:

$$F = U.\omega^2 \quad ......(ii)$$

with "ω" being the angular velocity in radians/sec.

The formula to find "ω" is expressed as:

$$\omega = \frac{2\pi.rpm}{60} \quad ......(iii)$$

Therefore by combining formulae (i) and (iii) in (ii), we obtain:

$$F = m.r. \left(\frac{2\pi.rpm}{60}\right)^2 \quad ......(iv)$$

As shown in equation (iv), the centrifugal force caused by the tooling unbalance increased with the square of the speed, in a similar manner to the spindle nose taper growth discussed earlier. However, assuming that the toolholder has initially low unbalance, this becomes a problem if rotational speeds are increased beyond 10,000 RPM. By way of illustration – with most toolholders exhibiting single-plane unbalance – research has shown [18] that the initial unbalance of typical tooling is around 250 g-mm. When such tooling is rotated at 15,000 RPM, this 250 g-mm generates a continuous radial force of 642.6N.

Unbalanced tooling can introduce considerable detrimental effects on not only the machine tool – high centrifugal force causing internal stress leading to premature spindle failure – but on the cutter's life and workpiece's surface finish. This unbalance can be traced back to several sources:

o   toolholders of the V-flange type might have different depth drive/slots, this being part of the inherent design,
o   holders for some endmills having set screws for locking the

cutter, causing unbalance,
o   out-of-balance caused by the unground V-flange base,
o   the collet & collet nut are the primary recurring source of unbalance in high-speed toolholders.

N.B. Most of these problems can be eliminated by modifying the tooling design.

As can be seen in (FIG 3), the marginally eccentric adjustable balance rings can be adjusted to give a degree of single-plane balancing and several manufacturers offer a range of differing adjustment methods for high- speed milling/drilling applications.

Finally, consideration needs to be made of the level of balance – quality required and as a milling cutter is expected to withstand both high rotational speeds and cutting forces, it can be considered as a "Rigid Rotating Body". This allows one to use the ANSI S.19-1975 standard for achieving balance (TABLES 1 & 2), which defines the permissible residual unbalance for a rotating body relative to its maximum speed. This standard and its equivalents (ISO1940), assigns different balance – quality grades – "G"-numbers – related to groupings of rotating bodies (TABLE 1) based on the experiences gained with a variety of sizes, speeds and types. The balance – quality grade "G" equals the specific unbalance "e" times the rotation speed "ω":

balance – quality G = e.ω
with the units being in mm/sec
Furthermore, as described earlier:

$$e = \frac{U}{M} \quad ......(i)$$

.˙. solving for "U", we obtain:

$$U = \frac{9.5.M.G}{rpm} \quad .....(v)$$

If one looks at TABLE 1 it shows that the balance – quality for machine tool drives is G2.5, although in many instances the value should approach G1.0 – the specification for grinding machine drives – as speeds are compatible today. However, if G2.5 is used, then the following example illustrates the balance-quality necessary using a toolholder weighing 3kg, rotating at 25,000 RPM:

$$U \text{ (higher)} = \frac{9.5.3000g. \ 2.5 \text{ mm/sec}}{25,000}$$

.˙. U higher = 2.85 g-mm

As alluded to above, this is the "worst case" and the tooling should approach that of G1.0 this produces a balance-quality of:

$$U \text{ (lower)} = \frac{9.5.3000g. \ 1.0 \text{ mmm/sec}}{25,000}$$

.˙. U (lower) = 1.14 g-mm

This follows that the balance is between 1.14 and 2.85 g-mm which is towards the upper end for the maximum residual specific unbalance for the G 2.5 and approaching this level for G1.0 (see TABLE 2).

Even when tooling has been dynamically balanced in both planes, problems still exist particuarly in the fit of the spindle taper connection [19]. This is a result of the taper rate accuracy requirements between both the shank and taper socket (see FIG 3). In fact, the picture is very confusing, due to the relative Cone Angle Tolerance Grades AT-1 to AT-6 utilising the standard 7:24 taper. Not only do countries have their own standards, but individual machine tool manufacturers within each country adopt different standards. The proposed DIN 69892/3 standards would seem to offer a solution and may become the norm in due course? With cone angle tolerance for the shank and taper socket not a perfect fit, as the speed increases beyond 10,000 RPM there is every expectation that the balanced tooling will lose contact and register in its seating and become unbalanced (FIG 3). Moreover, as the speed increases the taper shank size — by proportion — must decrease, meaning that when UHSM very little in the way of taper fit register occurs, which can compound the problem still further.

In order to evacuate swarf from the workpiece correct chip control is necessary through judicious use of tool insert geometry. It is crucial to have effective chip handling/control [20] and not only must the volumes of swarf be accommodated efficiently, but pocket clearances require chips to be exhausted immediately. Work-hardened and highly abrasive swarf during machining can cause considerable damage to the machined pocket, which implies that large chip gussets and high shear cutting inserts — where applicable — can exhaust chips efficiently. However, these chip gussets are always a compromise between improved chip evacuation and reduced cutter rigidity although the former is possibly more important as if chips cannot be satisfactory exhausted then "chip jamming" and eventually cutter/workpiece damage will result [2]. Inevitably a compromise is reached between deployment of inserts around the cutter body and the relative size of the chip gusset. When cutters are "ramped-down" into a pocket — at a diagonal feedrate — cutters can produce some "workpiece abuse," particularly on aerospace aluminium alloy parts where large weight-saving pockets are machined. These components exhibit little in the way of surface damage, but can have a problem of surface integrity — in terms of the substrate damage — which can influence the part's fatigue life. Often the only way of establishing whether this "abusive" machining damage is present is by NDT assessment and speculation has been made that by providing a negative edge preparation to the insert's geometry this reduced the "recutting effect" which has been attributed to this "abusive" condition.

AXIS DRIVE CONTROL CAPABILITY

Of late, most CNC machine tools use "proportional servo-systems", where the velocity of an axis is proportional to the difference between the actual position and the command position [12]. An "error signal" is used by the system to determine any acceleration/deceleration necessary together with steady-state velocities. The term "servo-lag" is used to describe the distance between the actual and command position which is influenced by "gain" — which is a measure of the servo's responsiveness. The higher the "gain" the lower the lag. In order to appreciate the affect of "gain", consider what happens to the axes as one attempts to mill a right-angled corner. When moving along an axis — for example, the X-axis — whilst UHSM, causes the servo to develop a steady and progressive lag, until sufficient command signals have been generated to reach the "theoretical" corner. It is at this point that the controller begins to generate commands in the Y-axis, although due to "servo-lag" the actual slideway has not reached the corner.

Therefore, the X-axis begins decelerating while the Y-axis accelerates, that is the velocity is proportional to the distance between the command signal and the actual position. It is not until some distance has been travelled along the Y-axis that the X-axis slide actually ceases moving. This "servo -lag" problem causes an exponential curve, with the amount of variance from the sharp right-angled corner being dependent on the "servo-lag" magnitude – which itself depends upon the influence of feedrate and "gain" [21]. Obviously, these "servo-lag" inaccuracies become a major problem when machining complex contours, or even simple rectangular features at UHSM rates and are a function of several factors not least of which is block processing speed, as the next section shows.

CONTROLLER PROCESSING SPEED

Probably the main factor limiting contouring speed when UHSM is the processing speed of the CNC [12], with each "stroke" generated for the active axes requiring; reading, interpretation and activation. Such activities are referred to as the "block processing time", with the maximum allocated time for block processing the information being dependent on the length of stroke and the feedrate. The maximum block processing time "Tb" can be calculated in the following manner:

$$Tb = \frac{\text{maximum stroke length}}{\text{feedrate}}.$$

By way of illustration, supposing one requires a chord length – i.e. stroke length – of 0.50 mm in order to obtain a realistic contouring accuracy whilst milling at 3000 mm/min, or 50 mm/sec, then the maximum block processing time should be less than:

$$Tb = \frac{0.50}{3000/60} = \frac{0.50}{50} = 0.01/\text{sec, or 10 millisec}$$

Many current CNC's only have block processing times ranging between 60 to 80 millisec – with only a few 32-bit processors less than 10 millisec – this means that an "active" program suffers from "data starvation", whilst the CNC catches up on its data processing. This "starvation" causes hesitation in the slides, which will inevitably leave dwell marks on the workpiece and slow down the cutting time. Obviously this is unacceptable and a lower feedrate of necessity must be introduced, but at the expense of a longer cycle time. To press this point still further, if the block processing time of the CNC was 60 millisec, the contour machining of the part would take six times longer than one with a 10 millisec time.

By way of illustrating the problem of UHSM and the influence of CNC processing speed, the following two diverse examples will be addressed. When contour milling a die for the manufacture of intricate metal parts, having fine detailed work needing reproduction with radii as small as 0.25 mm – requiring a milling cutter tool tip radius of 0.025 mm – the spindle speeds must reach 40,000 RPM having a feed per revolution of 0.008 mm, relating to a feedrate of 320m/min. Assuming the controller has a "servo-gain" of 4 this results in a "servo-lag" of 0.075 mm/min – which is consistent with the production of radii of 0.25 mm. Whereas, if the "gain" was one, this causes a "servo-lag" of 0.320 mm/min and in this case it could not hope to machine the die. In such circumstances it is necessary to reduce the feedrate to 75 mm/min to generate the contour and this means that the cutting time increases four-fold. If consideration is given now to the requirements of block processing time–under such conditions – cutting a radius of 0.25 requires linear stroke lengths of 0.075 mm to reproduce acceptable detail. This means a block processing speed of 15 millisec, but if the CNC only has a 60 millisec available, then the feedrate must be constrained to 75 mm/run – increasing the milling time by a factor of four.

In the second example, consider the problem of machining a turbine fan casting pattern for an ECM electrode, made from aluminium having very gentle curvatures. In this case, the spindle might have a capability of 250,000 mm/min, with adequate power to machine at feeds of 0.25 mm/rev. This would incidate a feedrate of 62,500 mm/min would be possible. For accuracy, a chordal deviation of 0.005 mm indicates a stroke of 0.75 mm if the minimum radius of curvature was 25mm. Assuming a "servo-gain" of one occurred, then errors as large as 0.125mm would not produce an acceptable part. Further, at 62,500 mm/min feedrate a block processing time of 60 millisec would require stroke lengths of 2.5 mm, instead of the 0.75mm needed for the required accuracy. In order to eliminate the affects of "low gain", or slow processing time, it is necessary to decrease the feedrate — cutting time is increased by up to 400%.

Finally, by considering these two examples metaphorically [21], the first is similar to racing a go-cart on a small tightly curved track, whereas the second, is similar to a fast sports car racing on a longer smoother track. The go-cart might only reach 30km/hr, with the sports car touching 200 km/hr, however the corner forces and reaction times are similar even though the speeds differ markedly. Looked at from a different viewpoint, one can say that the frequency of response of driver and car — "servo-gain" and block processing time — are similar in both examples, even though their speeds — feedrates — are radically different.

CONCLUDING REMARKS

The advantages to be gained by adopting a UHSM strategy can be readily appreciated (FIG 1), but must be tempered by the fact that problems occur due to these high rotational speeds, not least of which are: spindle design and its life; differential centrifugal growth; momentum effects on linear axes; balancing tooling' axis drive capability — "servo-lag"; controller processing speed and lately a subject not discussed in this paper — the influence of coolant delivery systems. If these points are addressed by machine tool manufacturers and users alike, then "real" production and quality benefits will inevitably occur and companies embracing such technologies will obtain a significant competitive edge over their main rivals.

REFERENCES

1. GOUGH, J., 1991. "Going hell for leather", Manufacturing Engineer, June, pp 13 – 15.
2. SMITH, G.T., 1991. "Problems associated with high speed cutting on machining & turning centres", Seventh National Conference of Production Research, pp 115–120.
3. SALOMON, C., 1931. "Process for Machining of Metals or Similar Acting Materials When Being Worked by Cutting Tools", German Patent No 523594, April.
4. KING, I.R., 1985. "Historical Background", Handbook of High–speed Machining Technology, Chapman & Hall, pp 3–26.
5. VAUGHN,M R.L., 1958 "A theoretical approach to the solution of machining problems".
6. ARNDT, G., 1971. "On the Study of Metal–cutting & Deformation at Ultra– high Speeds," Proc. Harold Armstrong Conf.Prod. Sci. Industry, V01 30, Monash University.
7. RECHT, R.F., 1964. "Catastrophic Thermoplastic Shear," Trans. ASME, J. Appl.Mech., Vol 31, June, pp 189–193.

8. FENTON, R.G. & OXLEY, P.L.B., 1967. "Predicting Cutting Forces at Super -High Cutting Speeds From Work Material Properties & Cutting Conditions", Proc. 8th MTDR. Conf., Oxford, Pergamon Press, pp 247 - 258.
9. OKUSHIMA, K., et al., 1965. "A Fundamental Study of Super-High-Speed Machining", Bull. Japan Soc. Mech.Egnrs, Vol 8, 32, p 702.
10. TANAKA, Y., TSUWA, H & KITANO, M., 1967. "Cutting Mechanism in Ultra- High-Speed Machining," ASME Paper No. 67-Prod-14.
11. KING, R.I., 1979. "Phase IIA Summary Technical Report of the Feasibility Study for High-Speed Machining of Ships Propellers, "Contract No. 00140-79-C-0326, Lockheed Missiles & Space Company Inc., Sunnyvale, CA, Nov.
12. SMITH, G.T., 1992 "CNC Machining Technology-The Handbook on Turning & Machining Centres", Springer Verlag.
13. TLUSTY, J & MacNEIL, P., 1975. "Dynamics of Cutting Forces in End Milling", Annals of CIRP, Vol 24, 1, pp 21-25.
14. SMITH, G.T., 1989. "Advanced Machining - The Handbook of Cutting Technology", IFS/Springer Verlag.
15. SMITH, G.T., 1991. "Condition Monitoring for Sub-Micron Machining on Turning Centres", COMADEM 91, Southampton Institute, I O Publishing, pp 146-150.
16. KOMADURI, R., 1985. "High-Speed Machining," Mechanical Engineering, Dec, pp 64-76.
17. SPROW, E., 1991. "High-speed spindles for milling?", Tooling & Production, March, pp 32-35.
18. LAYNE, M., 1991. "On Balance," Cutting Tool Engineering, Aug, pp 36-38.
19. LEWIS, D.L., 1991. "Factors for successful rotating tool operation at Hi-speeds", SME High-Speed Machining Clinic, April, Raleigh, N.C.
20. MITCHELL, W.A., 1991. "When and Where to Think High-Speed Milling", Modern Machine Shop, October, pp 55-63
21. LINN, T., 1989 "Fundaments of precision in high-speed milling", Precision Toolmaker, October, pp 306-310.v

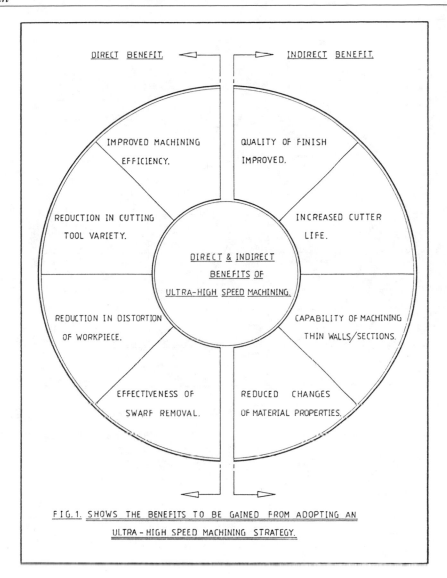

FIG.1. SHOWS THE BENEFITS TO BE GAINED FROM ADOPTING AN ULTRA-HIGH SPEED MACHINING STRATEGY.

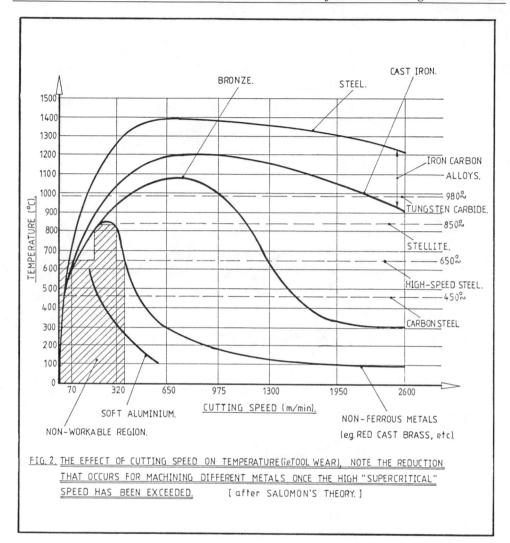

FIG. 2. THE EFFECT OF CUTTING SPEED ON TEMPERATURE (i.e. TOOL WEAR), NOTE THE REDUCTION THAT OCCURS FOR MACHINING DIFFERENT METALS ONCE THE HIGH "SUPERCRITICAL" SPEED HAS BEEN EXCEEDED.   [after SALOMON'S THEORY.]

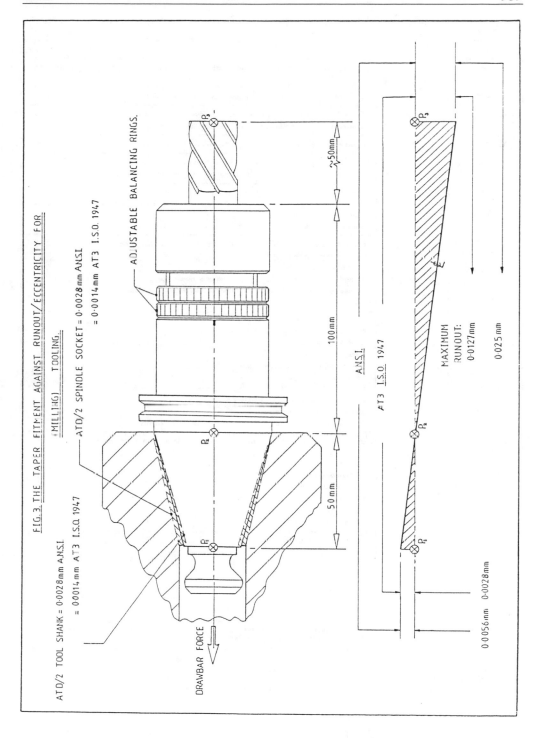

FIG. 3 THE TAPER FITMENT AGAINST RUNOUT/ECCENTRICITY FOR (MILLING) TOOLING.

ATD/2 TOOL SHANK = 0·0028mm A.N.S.I.
= 0·0014mm AT3 I.S.O. 1947

ATD/2 SPINDLE SOCKET = 0·0028mm ANSI
= 0·0014mm AT3 I.S.O. 1947

ADJUSTABLE BALANCING RINGS.

DRAWBAR FORCE

A.N.S.I.

AT3 I.S.O 1947

MAXIMUM RUNOUT:
0·0127mm
0·025 mm

0·0056mm 0·0028mm

Balance quality grades G	$e\omega$ [a,b] (mm/sec)	Rotor types—General examples
G 4 000	4 000	Crankshaft-drives[c] of rigidly mounted slow marine diesel engines with uneven number of cylinders.[d]
G 1 600	1 600	Crankshaft-drives of rigidly mounted large two-cycle engines.
G 630	630	Crankshaft-drives of rigidly mounted large four-cycle engines. Crankshaft-drives of elastically mounted marine diesel engines.
G 250	250	Crankshaft-drives of rigidly mounted fast four-cylinder diesel engines.[d]
G 100	100	Crankshaft-drives of fast diesel engines with six or more cylinders.[d] Complete engines (gasoline or diesel) for cars, trucks, and locomotives.[e]
G 40	40	Car wheels, wheel rims, wheel sets, drive shafts Crankshaft-drives of elastically mounted fast four-cycle engines (gasoline or diesel) with six or more cylinders.[d] Crankshaft-drives for engines of cars, trucks, and locomotives.
G 16	16	Drive shafts (propeller shafts, cardan shafts) with special requirements. Parts of crushing machinery. Parts of agricultural machinery. Individual components of engines (gasoline or diesel) for cars, trucks, and locomotives. Crankshaft-drives of engines with six or more cylinders under special requirements. Slurry or dredge pump impeller.
G 6.3	6.3	Parts or process plant machines. Marine main turbine gears (merchant service). Centrifuge drums. Fans. Assembled aircraft gas turbine rotors. Fly wheels. Pump impellers. Machine-tool and general machinery parts. Normal electrical armatures. Individual components of engines under special requirements.
G 2.5	2.5	Gas and steam turbines, including marine main turbines (merchant service). Rigid turbo-generator rotors. Rotors. Turbo-compressors. Machine-tool drives. Medium and large electrical armatures with special requirements. Small electrical armatures. Turbine-driven pumps.
G 1	1	Tape recorder and phonograph (gramophone) drives. Grinding-machine drives. Small electrical armatures with special requirements.
G 0.4	0.4	Spindles, disks, and armatures of precision grinders. Gyroscopes.

[a] $\omega = 2\pi n/60 \approx n/10$, if n is measured in revolutions per minute and ω in radians per second.

[b] In general, for rigid rotors with two correction planes, one-half of the recommended residual unbalance is to be taken for each plane; these values apply usually for any two arbitrarily chosen planes, but the state of unbalance may be improved upon at the bearings. For disk-shaped rotors the full recommended value holds for one plane.

[c] A crankshaft-drive is an assembly which includes the crankshaft, a flywheel, clutch, pulley, vibration damper, rotating portion of connecting rod, etc.

[d] For purposes of this Standard, slow diesel engines are those with a piston velocity of less than 9 m/sec; fast diesel engines are those with a piston velocity of greater than 9 m/sec.

[e] In complete engines, the rotor mass comprises the sum of all masses belonging to the crankshaft-drive described in Note c above.

TABLE 1. THE BALANCE QUALITY GRADES FOR VARIOUS GROUPS OF REPRESENTATIVE RIGID ROTORS.

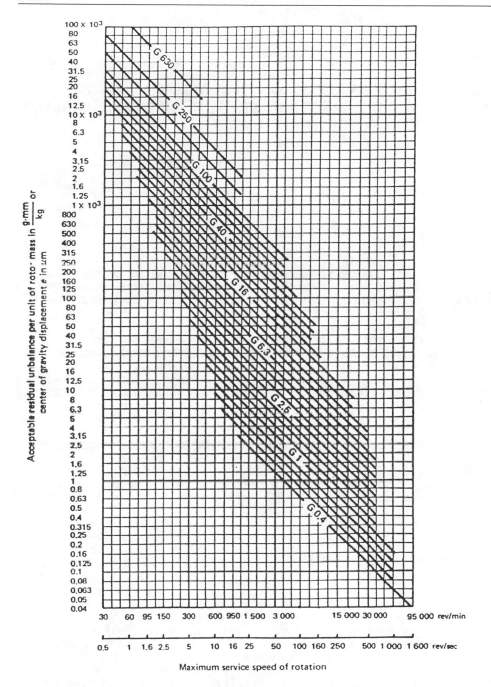

TABLE 2. THE MAXIMUM RESIDUAL SPECIFIC UNBALANCE CORRESPONDING TO VARIOUS BALANCE QUALITY GRADES, "G".

PERFORMANCE EVALUATION OF A MACHINING CENTRE USING LASER INTERFEROMETRY AND ARTIFACT-BASED TECHNIQUES

Paul R. Painter, Graham T. Smith, and Anthony D. Hope

Engineering Division, Southampton Institute, Southampton, U.K.

ABSTRACT

At present the Laser calibration of a machining centre takes approximately two days. However, using artifacts such as an LVDT Ballbar it is possible to evaluate the machine performance in a fraction of the time.

Nominally the Laser calibration involves a series of individual assessments for such factors as: linear displacement, straightness (ie horizontal/vertical), squareness, angular (ie pitch & yaw) and backlash errors. However, using a newly developed Ballbar system it is possible to generate data for dimensional/geometrical characteristics in a dynamic, or static form.

The paper discusses the merits of dynamic Ballbar testing and correlates the results to both the Laser interferometer and roundness polar plots obtained from machined disks at various feedrates. In the future such a test should give an accurate and speedy indication of machine tool performance, as a complement to a full Laser calibration, on a regular basis.

INTRODUCTION

Machine tools are available in a diverse range of axes configurations that can vary from the normal three linear cartesian types, to multiple linear/rotational relationships. However, the positioning accuracy of the tool slides are dependent upon the axes monitoring system [1] and rotary encoders are less accurate than the linear-scaled varieties. In the past, machine tools were calibrated optically by such instruments as autocollimators/alignment telescopes [2], however with the advent of the Laser and suitably user-friendly software the calibration task has been made simpler to perform, with the old-style tabulated techniques for assessing machine tool alignments being superceded by readily configured screen displays. If axis compensation is necessary, then error compensation values can be down-loaded to the machine's CNC controller and compensation tables are over-written with the desired corrections. No matter how laudable such calibration techniques are, it is still a time-dependant process, in that setting up the optics and capturing data requires considerable down-time on the machine tool. With the advent of the magnetic Ballbar [3,4,et al] and its deriatives, a quick method of establishing a machine tool's errors, in particular the circularity, repeatabilty, local squareness and linear scale errors can be achieved. Although by no means will such a system replace those tests carried out by Lasers, but complement them, as the periodic calibration of a machine tool is necessary in order to maintain consistent and accurate part quality. To those unfamiliar with the Ballbar concept, a brief description of a typical system is in order.

Most Ballbar gauges as their name implies, are ball-ended bar length transducers which are held kinematically between the machine spindle and a base, which is positioned and held fast at a convenient place within the volumetric envelope of the machine, typically on the table. Such LVDT transducer – based instruments have a resolution of 0.1 μm and an accuracy of $\pm$ 1 μm and can be used in either a dynamic, or static mode of motion control. In the dynamic method, the operation of calibration is achieved by programming the controller – using circular interpolation – to move the axes in a circle. The Ballbar being attached to the machine as illustrated in Fig 1, will through its transducer, detect deviations – inaccuracies – in the machine's circular path. Software has been developed which uses this captured deviation data to produce a speedy graphical polar plot representing the machine tool's contouring ability. A part program can be stored in the controller's memory for quick access whenever a test needs to be performed. Another test utilising static methods of data capture is possible, again, through suitable software, but is outside the remit of this paper.

DYNAMIC METHODS FOR CALIBRATING A VERTICAL MACHINING CENTRE

Using the Ballbar alluded to above, it is possible to achieve a degree of calibration which will assure the end-user that their machine is either within the specified performance limits, or more importantly, outside. With the latest type of Ballbar, carbon fibre extension bars can be fitted between the ball ends increasing the volumetric sweep for larger machine tools. However in this work a 150mm Ballbar was utilised throughout the tests and a vertical machining centre was assessed in the normal contouring plane – "X–Y". The machining centre in these tests is part of a flexible manufacturing system (FMS) and as such, has dedicated fixturing situated on the machine's table and this fact will mean that work pieces are machined within the constraints of the vertical chuck. Therefore any tests need to take account of this fact and indeed the Ballbar was positioned within the volumetric constraints of the chuck at two heights – just above the jaws and at a plane 100 mm higher (see Fig 1).

The tests were performed on a Φ 300mm circle in the "X-Y" plane at three different feedrates. The feedrates [6] were in accordance with normal machining practice for a solid carbide end mill milling aluminium and represented the low, middle and high ranges typical within the chosen cutting tool/workpiece material combination. This produced twelve profile error plots:

o   three at the lower position-clockwise and further three anti-clockwise.

o   three at 100 mm above this position - clockwise and a further three anti-clockwise.

In a similar fashion and with a 5mm thick aluminium disk held on a mandrel, twelve Φ 300mm disks were milled using a solid carbide end mill with only the smallest of finishing cuts (FIG 2). After machining, the disks were inspected on a roundness testing machine for their "least squares centre" and polar plots were produced as shown in Fig 3 (top). Once all of the disk machining and Ballbar plots had been completed, the machine tool was calibrated using a Laser interferometer (Fig 3-bottom). The data gathered from the three separate and distinct tests, allowed most of the dynamic characteristics of the machine to be assessed both: independently and simultaneously, to see if correlated data and trends could be established. In the following section each testing technique will be discussed further and finally some conclusions will be drawn about the correlated data on the machine's condition by such assessment methods.

DYNAMIC MACHINE CALIBRATION
THE BALLBAR - profile error plots

The profile error plots produced by the Ballbar software (Figs 4, 5, & 6) can be characterised into a series of trace patterns of motion error origins [7], with such traces having either one, or many of the errors superimposed onto them. These errors can be sub-divided into systematic and random errors [4].

Three Ballbar traces can be seen in Fig 4, taken at three different feedrates. These profile error plots show that as the feedrate increases the circularity error increases, with the general profile remaining similar but an increase in the "spikes" at the axes transitions - where reversal of an axis is necessary to generate the circular motion. Such "reversal spikes" can be a major contributor to a larger circularity error. However on some CNC controllers this "spike" can be eliminated by entering specific axis parameter compensations. In order to achieve this level of compensation, accurate measurements are essential and CNC controller builders nowadays provide a means of minimising these "reversal spikes" and prefer either Laser, Ballbar, or Circular masterpiece [8] techniques to be used, although the Circular masterpiece can cost up to ten times the cost of the relatively inexpensive Ballbar System. It is possible to measure the magnitude of each "spike" at a given feedrate this being the principal parameter for the compensation within the CNC controller. This is followed by further analysis to calculate the approximate time-base over which the "reversal spike" compensation is required. Estimation of the secondary machine parameter – "spike attenuation" per servo-loop cycle - can then be established. Such first/second order compensations can significantly reduce these "reversal spikes".

As previously mentioned (see Fig 1), the Ballbar was positioned in the same plane but at two heights with a 100 mm displacement in the "Z" axis – and these plots are shown in Fig 5. It can be seen by both the circularity values and the polar traces, that the results are similar. The change in the circularity values shown in Fig 5 (a,b) is due to the 100 mm height displacement of the "Z" axis.

A simple calculation will determine the magnitude of the squareness errors (see Fig 6a), when machining in one direction (anti-clockwise) with a "mirror-image" occurring if the motions are reversed (i.e. clockwise). From this profile error plot, it can be seen that the trace is made up of individual errors. These include: linear positioning errors on the "X-Y" axis, backlash, "reversal spikes" (i.e. attenuation per servo-loop cycle), and squareness in the "X-Y" plane.

As a general note of caution when undertaking the calibration of the machine tool using artifacts: Ballbars, Circular Masterpiece, NASA tests, etc. The test diameter – should be at least one third of the total travel of the shortest test machine axis [8]. This is to ensure that the relative linear motions are such that they will be influenced by the geometric errors on the machine and can be captured as realistic data within the confines of the Ballbar's rotational sweep. However, if a small "masterpiece" is used in the testing procedure, it will highlight the performance of the drives – hysteresis/linearity – and give an indication of the general CNC performance in terms of: interpolation, together with the influence of speed on servo-lag and processing speed. If the Ballbar is cycled through a circular move of less than the LVDT full-scale deflection, this will approximate the small "masterpiece" criteria and allow a measure of hysteresis/linearity to be determined.

DISK MACHINING - for roundness assessment.

In order to obtain a measure of correlation between the profile error plots and the anticipated workpiece condition after machining, a series of Φ 300m disks were machined (Fig 2) at a range of feedrates and with only a light finishing cut. Under such conditions the workpiece/cutter combination was barely influenced by the cutting forces and this arrangement resulted in little, if any, workpiece/ cutter deflection. Furthermore, when inspected on the roundness testing machine for "least squares centre" (Fig 3), a close agreement with the Ballbar plots occurred (Fig 6b). The smoothness of the roundness test trace was the result of "filtering" on the data and the smaller trace merely indicates a "scaled factor" so that a comparison can be made to the Ballbar plot. By comparing the Ballbar to the roundness trace, it is clear that the Ballbar profile error plots not only represent the unloaded performance characteristics of the machine tool, but show high correlation to the roundness profile of a machined part. The harmonics are in close agreement with those established elsewhere [7]. This means that the Ballbar can offer a speedier and cheaper alternative to the machining of a NASA testpiece.

LASER CALIBRATION – for "absolute" machine tool assessment.

Calibration by the Laser interferometer offers considerable scope when assessing a machine tool's ability to cut a high quality part with consistency. If one considers the information that can be gathered from the measurement on a single machine axis using a Laser, then it is possible to obtain:

o   linear positioning
o   angular measurement – pitch error
o   angular measurement – yaw error
o   straightness measurement – vertical plane
o   straightness measurement – horizontal plane

From these readings it is possible to derive further characteristics for the machine:

o   velocity
o   flatness
o   parallelism – linear and rotational
o   squareness

However, the Laser setup and data acquisition is time consuming (See FIG 3 bottom), but is important for the periodic detailed checks on the machine tool's performance. The Laser calibration of the vertical machining centre confirmed the information gained from the Ballbar data, that this machine tool has small squareness errors in the "X-Y" plane, together with variable backlash.

It should be noted that all of these assessment tests were undertaken with the vertical chuck in-situ. The chuck weighed over 100 kg and this meant that whenever the slideways were displaced to machine the disks, or generate profile error plots, the weight may exacerbate any angular errors (pitch and yaw), particularly for the machine's dynamic calibration characteristics.

CONCLUDING REMARKS

The Laser, Ballbar and Roundness tests – for the machined disks – showed excellent correlation of results, as illustrated by the values given in FIG 6b and the values given in Fig 7 over the whole range of assessment in terms of diverse feed rates and contouring directions – clockwise/anti-clockwise. This fact confirms that using artifacts – Ballbars etc., – as a complement to the Laser gives meaningful results. Particularly as the

Ballbar CNC programme can be stored in the memory allowing for speedy and precise execution for performance evaluation of the machine tool on either a daily/weekly interval, to reaffirm that no "drift" has occurred in any of the major machine tool parameters since the machine was last calibrated.

Finally, these tests were undertaken to gain an appreciation of the relative dynamic contouring capabilities of the machine tool. The future work concentrates on the static assessment tests using the Laser and Ballbar calibration methods on the same machine in an "unloaded" condition and on turning centres, but will be reported later. However, the Ballbar has shown itself to be a major step forward in the evaluation of machine tool performance and complements the periodic Laser calibration tests necessary in order to maintain the true cutting potential in terms of linear/ contouring capabilities.

REFERENCES

1. SMITH G.T., 1992 "CNC Machining Technolgy - The Handbook on Turning & Machining Centres", Springer Verlag.

2. GALYER J.F.W. & SHOTBOLT C.R., 1990 "Metrology for Engineers", Cassell.

3. BRYAN, J.B. 1982. "A simple method for testing measuring machines & machine tools":
    Part 1: Principles & applications, Precision Engineering, April, Vol 4, 2.
    Part 2: Construction details, Precision Engineering, July, Vol 4, 3.

4. KUNZMANN, H,. WALDELE, F & SALJE, E., 1983. "On Testing Coordinate Measuring Machines (CMM) with Kinematic Reference Standards (KRS), Annals of the CIRP, Vol 32, 1, pp 465 - 468.

5. SMITH G.T., 1991 "Determining & controlling the quality of prismatic parts machined in a flexible manufacturing cell environment", Production Research Approaching the 21st Century, Taylor & Francis, pp 513 -524.

6. SMITH, G.T., 1989. "Advanced Machining - The Handbook of Cutting Technology," IFS/Springer Verlag.

7. KAKINO, Y., IHARA, Y & E OKAMURA, K., 1987. "The Measurement of Motion Errors of NC Machine Tools & Diagnosis of their Origins by Using Telescoping Ball Bar Method", Annals of the CIRP Vol 36, 1, pp 377-380

8. KNAPP, W & E MATHIAS, E., 1983. "Test of the Three Dimensional Uncertainty of Machine Tools & Measuring Machines & its Relation to the Machine Errors". Annals of the CIRP, Vol 32, 1, pp 459 -464.

FIG.1. A "QUICK-CHECK" BALLBAR(LVDT) USED TO ASSESS A MACHINING CENTRE DYNAMICALLY AT TWO POSITIONS IN THE "X-Y" PLANE.

FIG.2. A 300mm ALUMINIUM DISK BEING MACHINED BY CIRCULAR INTERPOLATION : TWO POSITIONS; TWO DIRECTIONS (CLOCKWISE/ANTI-CLOCKWISE); THREE FEEDRATES.

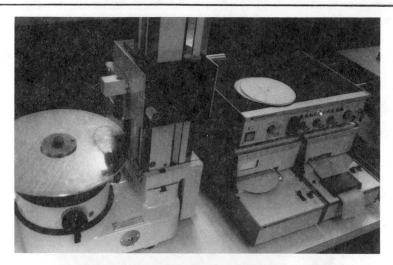

FIG.3. MACHINED DISKS BEING INSPECTED IN THE
       METROLOGY LABORATORY FOR "LEAST SQUARES" (TOP).
       LASER CALIBRATION OF THE MACHINING CENTRE (BOT.)

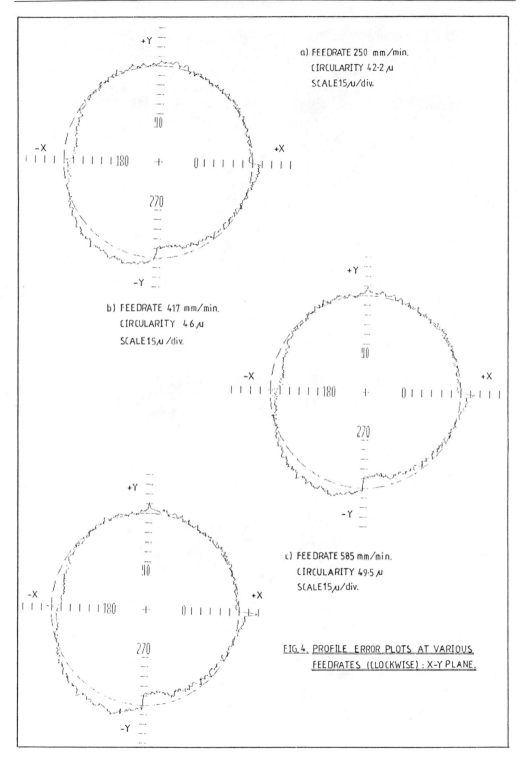

a) FEEDRATE 250 mm/min.
CIRCULARITY 42·2 μ
SCALE 15 μ/div.

b) FEEDRATE 417 mm/min.
CIRCULARITY 46 μ
SCALE 15 μ/div.

c) FEEDRATE 585 mm/min.
CIRCULARITY 49·5 μ
SCALE 15 μ/div.

FIG. 4. PROFILE ERROR PLOTS AT VARIOUS FEEDRATES (CLOCKWISE) : X-Y PLANE.

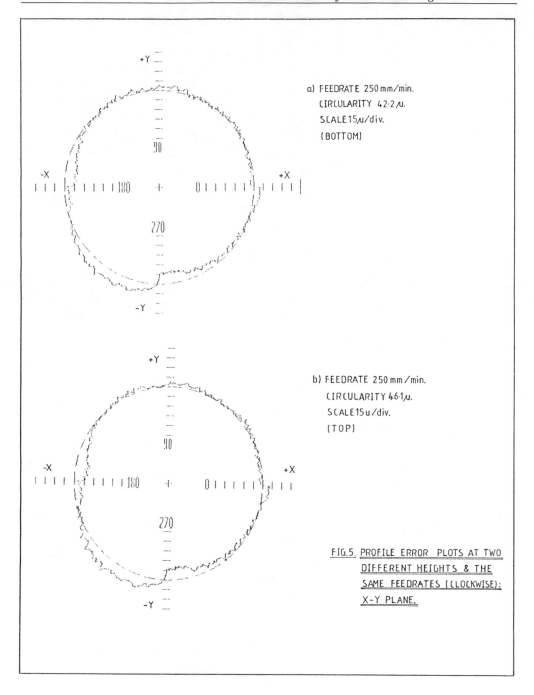

a) FEEDRATE 250 mm/min.
   CIRCULARITY 42·2 /u.
   SCALE 15 /u/div.
   (BOTTOM)

b) FEEDRATE 250 mm/min.
   CIRCULARITY 46·1 /u.
   SCALE 15 u/div.
   (TOP)

FIG.5. PROFILE ERROR PLOTS AT TWO
DIFFERENT HEIGHTS & THE
SAME FEEDRATES (CLOCKWISE):
X-Y PLANE.

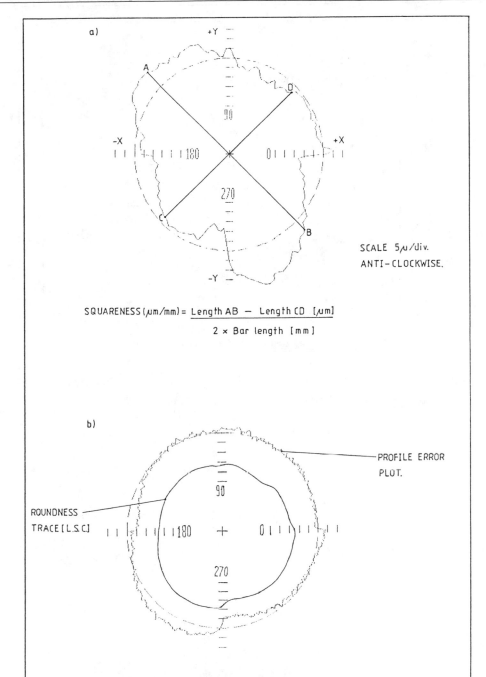

SQUARENESS $(\mu m/mm) = \dfrac{\text{Length AB} - \text{Length CD} \ [\mu m]}{2 \times \text{Bar length} \ [mm]}$

FIG.6. THE DETERMINATION OF SQUARENESS (TOP) & THE SUPERIMPOSED BALLBAR & ROUNDNESS TRACES (BOTTOM).

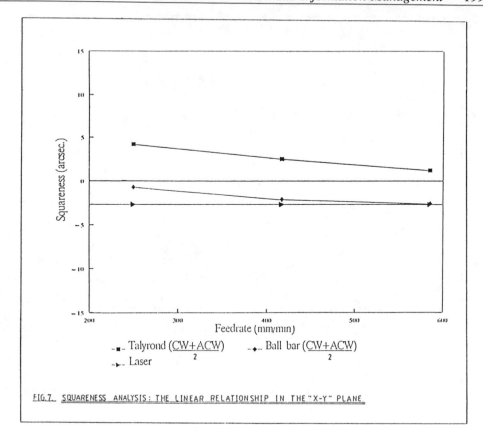

FIG.7. SQUARENESS ANALYSIS: THE LINEAR RELATIONSHIP IN THE "X-Y" PLANE

SHIELDED METAL ARC WELDING TRAINING — AN EXPERT SYSTEM APPROACH

Vivek Goel, T. Warren Liao, and Kwan S. Lee

Louisiana State University, Baton Rouge, Louisiana, U.S.A.

ABSTRACT

This paper describes a knowledge based approach for training in Shielded Metal Arc Welding (SMAW) domain. Although SMAW is a manual method, still it is used extensively in industry because of its versatility, portability and flexibility. SMAW is governed by a large variety of factors and so it is very difficult to gain expertise in SMAW procedures. The ignorance of the optimal method may result in the application of inappropriate technology.

A powerful training and learning tool is thus developed, using expert system approach, to help instructor effectively conduct the training courses and to shorten the training period.

The expert system has been implemented in Prolog. It accumulates all the available information on the SMAW process including edge preparation, electrode selection, economic evaluation, analysis of weld defects and trouble shooting. The expert system also includes an explanation module, which will show the reasoning process to reach a conclusion.

INTRODUCTION

Shielded metal arc welding (SMAW) has been the welding industry's mainstay and will remain a factor in many areas of welding in years to come. SMAW is used very heavily in construction because of its quality, portability, and versatility. SMAW can weld most of the metals such as stainless steels and carbon steels, and with special electrodes can weld high carbon steels, copper, brass, and even aluminum[1,2,3]. The training of SMAW operators involves a lot of concepts like selection of proper electrode, adoption of proper welding techniques and knowledge of weld defects and the means to avoid them. The Welders have also to be trained in setting the machine correctly and trouble shooting. The expert system developed takes care of all these aspects of training SMAW operators.

Some heuristics have been developed, for checking the applicability of shielded metal arc welding for a certain job, which can be codified in an expert system [2]. Thus using the expert system, the welding operators may know about the applicability of SMAW for a certain job.

The selection of welding parameters, consumables and procedures in SMAW, requires experience and the knowledge of certain thumb rules and heuristics. If all the knowledge for selection of parameters, consumables and procedures is codified in form of an expert system, then training inexperienced welders can become very easy.

Design of edge preparation is governed by certain rules[4] which are difficult to memorize and the process planner has to refer to manuals in order to design the joint. Cost and time calculation for SMAW requires either time study or the use of standard data. The latter involves tedious calculations and the planner has to refer to thick data tables to search for relevant information. A suitable search and inference strategy can be designed which can do these calculations for the SMAW Welding Operator.

Analysis of weld defects and trouble shooting requires considerable experience, and if all the knowledge regarding weld defects and problems encountered during welding are also codified in form of knowledge base then the task of planner/ welder will become very easy[5]. The Welder can be trained using this expert system, to use proper procedures and techniques of welding.

SHIELDED METAL ARC WELDING

The heat for SMAW comes from an arc which develops across an air gap between the end of an electrode and the base metal. The air gap produces a high resistance to the flow of current and this resistance generates an intense arc heat. The filler wire (electrode) forms molten droplets that deposit into the weld. The flux forms a gas that shields the molten weld pool. The arc force provides the digging action into the base metal for penetration. This process continues as the weld widens and the electrode continues across the joint. Transformers, motor generators and rectifiers are used as the power sources for SMAW. There are many kinds and sizes of electrodes and unless the correct one is selected it is difficult to do a good welding job.

THE EXPERT SYSTEM

The expert system developed has six modules which can aid the welder in the selection of the welding electrodes, parameters, and procedures for SMAW of mild steels only however the system is extendable for any type of metal with suitable addition of knowledge.

The expert system for training was implemented in TURBO PROLOG 2.0, supplied by Borland International. The hardware used was an IBM Compatible PC/AT with 1 MB RAM,

1.2 MB floppy disc and 20 MB hard disc.The specifications of the system are given in appendix I.

The Expert System for planning and training calculates the following variables for the job on hand:
1) Electrode specifications.
2) Cost of the job.
3) Time required for the job.
4) Welding parameters for the job.
5) Joint design for the job.

The determination of the above variables is dependent on a number of factors such as the nature of the job, parent metal composition, current range for the welding set, and many such factors. The Expert System also trains the Welding Operators in knowing about the various defects and the means to avoid them. A welder can also be trained in trouble shooting by this expert system.

Development Tool and Strategy: An approach utilizing an algorithmic language such as PASCAL or FORTRAN, to search a tabular database, would seem to be feasible at first and also easier to implement. But this approach could have the following problems :
1) **No Flexibility:** The conventional program is usually difficult to modify or update by any other person other than the original programmer, and hence difficult for researchers or Industrial collaborators to extend. There is no mechanism for discussing the knowledge with the user or express the knowledge in a natural way.
2) **No Fuzzy Logic:** They cannot be applied to problems that require decisions to be made with incomplete data.
3) **No Explanation Facility:** They cannot explain their line of reasoning for decisions etc.

Further Expert System development in TURBO PROLOG proved to be advantageous because of the following factors.
i) The number and complexity of the factors which influence the decision making process.
ii) The flexibility required from the system.
iii) The reasoning capabilities required for the selection approach.
iv) The opportunity of capturing, representing and preserving the knowledge and expertise for SMAW process in a declarative manner and using it for different purposes such as training inexperienced process planners.
v) The variables are "typed" or "declared" which provides more secure development control and better debugging, and minimizes memory requirement.
vi) Modular programming can be done.
vi) Predicates for random file access are provided.

Architecture of the Expert System: The system was based on the conventional concepts of the Expert System in which the User interface, Knowledge Base (KB), Working Data Base (WDB) and Inference Engine (IE) are developed. The structure of the expert system is as per figure 1.

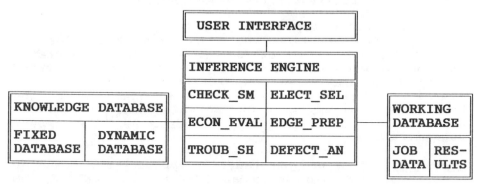

FIGURE 1: Architecture of the Expert System

User Interface: The system can be used in the menu mode or the sequential mode. In the menu mode all the modules are displayed in a pop-up menu format and the user can use any of the modules by selecting that particular choice. In the sequential format the user is asked about the job parameters and the system performs all the operations interactively with the user having the overall control.

Knowledge Base: The fixed Knowledge Base comprised of three forms of information - Data Files which hold different material properties, electrode properties, welding times for different electrodes, welding costs, procedures and other data held as facts or in rule based forms. The third form of information was contained in Installation Data file that is fed at the time of installing the expert system in any company. For e.g. Power Tariffs, Labor Costs, Overheads, Electrode Inventory etc.

Working Data Base: This holds two forms of information. First contains data about the type of job to be welded or which will be used for selection purposes eg. Job Material, Shape, Environmental conditions for the job, Service Requirements etc. which is different for each problem. The second is the information which is inferred and computed by the system.

Inference Engine: The working of the inference engine is controlled by the rules developed for the various modules of the expert system. The purpose of the inference engine is to match the relevant information on the knowledge base with that of the job and draw inferences regarding the electrode to be used, the welding parameters, cost, time and edge design.

Rules for the Expert System

Since the scope of the Expert System is large, a modular method of programming was used. Modularity is also desirable with regard to editing, readability, program testing and future extension. Few examples of the rules for the different modules are as under :

CHECK_SMAW : This module checks whether shielded metal arc welding is applicable for the job under consideration.

e.g. 1) IF   Production_Volume < 8
        AND Seam Length < 10 meters
        AND Many ranges of weld sizes exist
        AND Initial Equipment Cost is required to be
            less
        AND Arc Visibility is required for the job
        AND Skilled Operator is available
    THEN SMAW process is suitable for the job.

   2) IF   Production_Volume > 8
        AND Seam Length > 10 meters
        AND Many ranges of weld sizes donot exist
        AND Initial Equipment Cost is required to
            be medium
        AND Arc Visibility is not required for the
            job
        AND Skilled Operator is available
    THEN Submerged Arc Welding is suitable for the
        job.

ELECTRODE_SE: This module selects a suitable electrode based on the job parameters.

e.g. 1) IF   Strength desired > High
        AND   Bead Shape desired = Convex
        AND   Positions of Welding = All
        AND   Deposition Rate = High
        AND   Penetration Desired = Moderate
        AND   Base Metal Composition = High Sulfur
        AND   Hydrogen Content desired = Low
    THEN   Use E7015 electrode.

   2) IF   Strength desired > Medium
        AND   Bead Shape desired = Flat
        AND   Positions of Welding = All
        AND   Deposition Rate = Moderate
        AND   Penetration Desired = Moderate
    THEN   Use E6010 electrode.

ECON_EVAL: This module calculates the cost and time taken for completion of the job. The calculation is based on the standard data stored in the program. This module takes the electrode specification, diameter, type of weld(butt, fillet etc.), leg length / plate thickness (in butt weld) and position of welding as the input data and matches this information with that contained in the knowledge base and performs the necessary computations for time and cost calculation using the following equations:

Total Time = Length of bead (Arc Time + Auxiliary Time)

*Total Cost = Total Time * (Overhead Cost + Labor Cost) +*
*Number of Electrodes/meter * Length **
Electrode Cost.

The data for arc time, auxiliary time, number of electrodes/meter is available from standard data tables[] and has been stored in the knowledge base of the expert system.

TROUBLE_SHOOTER : This module is useful for solving problems that may be encountered during the welding process. The problems may be related to the welding set, spattery arc etc. The expert system diagnoses the causes for the trouble and suggests some remedies for them.

```
      e.g. IF      There is Loud crackling from the arc
           AND     The flux melts too rapidly
           AND     The bead is wide and thin
           AND     Spatter is taking place in large
                   beads
           THEN    The cause may be moisture contamination
                   of electrode
           AND     The remedy is the use of Dry Electrodes
```

DEFECT_ANALYZER : This module analyzes the various defects in the weldments and finds the causes and remedies of those defects.

```
e.g.       IF      Porosity is observed in the weldment
           AND     Surface holes are observed
           THEN    The cause may be dirt on the workpiece
           AND     The remedy may be to clean the work piece
```

Sample Run of the Expert System

The expert system was run for a variety of jobs and some of the sample sessions for various modules have been given in figures 2, 3, 4, 5, 6, 7, and 8. The expert system was run for the a set of data in the Elect_sel module and produced the output data as given in figure 3. The expert system was run in the econ_eval module and produced the output data as given in figure 5. These results were verified by performing the actual welding as per the job variables and were found to be quite matching.

```
================================ EX-SMAW ================================
 ┌─Choose──────────┐
 │ Job_data        │   CHECK_SMAW
 │ Firm_data       │       Production Volume: 8
 │ Check_SMAW      │       Seam Length: 10
 │ Elect_Sel       │       Many ranges of weld sizes: Yes
 │ Edge_Prep       │       Initial equipment cost should be
 │ Econ_Eval       │       less: Yes
 │ Def_Anal        │       Arc visibility required: Yes
 │ Troub_Sh        │       Is the operator skilled: Yes
 │ Quit            │       Chipping hammer/ deslagging
 └─────────────────┘       equipment is available: Yes
 ════════════════════CONSULTATION════════════════════
      Seam Length is not suitable for SMAW
      You may use some other process profitably
         Do you still want to continue? Yes
      Press F1 key for explanation...........

         ┌───────────────────────────────┐
         │ SMAW IS APPLICABLE FOR        │
         │ ALL CONDITIONS EXCEPT         │
         │ HIGHER SEAM LENGTH            │
         └───────────────────────────────┘
```

Figure 2: Sample Dialogue in the Check_SMAW subsystem.

```
================================ EX-SMAW ================================
 ┌─Choose──────────┐
 │ Job_data        │   ENTER THE JOB DATA:
 │ Firm_data       │       Type of Weld: FILLET
 │ Check_SMAW      │       Position of Welding: FLAT
 │ Elect_sel       │       Length of Weld (Meters): 10
 │ Edge_Prep       │       Strength: Moderate
 │ Econ_Eval       │       Bead Shape desired: Flat
 │ Defect_An       │       Deposition Rate: Fast
 │ Troub_Sh        │       Penetration Desired: Deep
 │ Quit            │
 └─────────────────┘
 ════════════════════CONSULTATION════════════════════
      Use E6010 electrode of diameter 3.25 mm
      Current: 80-120 A
      Polarity: DCRP
      Technique: Stringer Bead, Whip-and-pause motion
      Visual Inspection: Beads should be smooth and
      clear of any visible porosity, slag, or cracks.
      Light spatter is acceptable but large spatter
      should be removed with chipping hammer.
```

FIGURE 3: Sample Dialogue in the Elect_sel subsystem

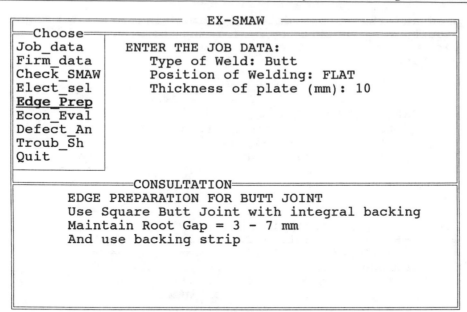

FIGURE 4: Sample Dialogue in the Edge_prep subsystem

```
╔══════════════════════ EX-SMAW ══════════════════════╗
║ ┌─Choose─┐                                           ║
║ │Job_data│   Input data:                             ║
║ │Firm_data│    Labor Cost: $20                       ║
║ │Check_SMAW│   Overhead Cost: $10                    ║
║ │Elect_Sel│    Electrode Cost: $2.10                 ║
║ │Edge_Prep│    AWS Specification selected by system: ║
║ │Econ_Eval│                 E6010                    ║
║ │Def_Anal │    Diameter of electrode: 4 mm           ║
║ │Troub_Sh │    Type of Weld: Fillet                  ║
║ │Quit     │    Position of Welding: Flat             ║
║ └─────────┘    Length of Weld Bead: 10 meters        ║
║ ═══════════════════CONSULTATION══════════════════════║
║               Output data:                           ║
║                 Total Time for Welding: 83.1 min     ║
║                 Recommended Current for Welding: 120 A║
║                 Weight of metal deposited: 1.53 kg    ║
║                 Total Cost of Welding: $129.75        ║
║                 No.of runs recommended: 1             ║
║                 Electrodes consumed: 42.0             ║
╚══════════════════════════════════════════════════════╝
```

Figure 5: Sample Dialogue in the Econ_Eval subsystem.

```
══════════════════ EX-SMAW ══════════════
┌─Choose──────────────────────────────────────────────┐
│ Job_data      DEFECT ANALYZER:                       │
│ Firm_data        1- UNDERCUTS                        │
│ Check_SMAW       2- WELD SPATTER                     │
│ Elect_sel        3- ROUGH WELDING                    │
│ Edge_Prep        4- ARC BLOW                         │
│ Econ_Eval        5- POROSITY                         │
│ Defect_An        6- SURFACE HOLES                    │
│ Troub_Sh         7- POOR FUSION                      │
│ Quit             8- SHALLOW PENETRATION              │
│                  9- QUIT                             │
│                                                     │
│              ═══════════CONSULTATION═══════════      │
│   Causes: 1- Work piece may be dirty                 │
│   Remedy: 1- Remove scale, dirt, rust, paint, and    │
│              moisture from the joint                 │
│   Cause:  2- Steel may have low carbon content or    │
│              manganese content or a high sulfur or   │
│              phosphorus content                      │
│   Remedy : 2-Use low Hydrogen electrodes. Minimize   │
│              admixture of base metal                 │
│   Cause  : 3- Too long an arc length                 │
│   Remedy: 3- Try using a shorter arc length          │
│              especially for low hydrogen electrodes  │
└─────────────────────────────────────────────────────┘
```

FIGURE 6: Sample Dialogue in the Defect An subsystem

```
══════════════════ EX-SMAW ══════════════
┌─Choose──────────────────────────────────────────────┐
│ Job_data      TROUBLE SHOOTING:                      │
│ Firm_data        1- Welding Set does not start       │
│ Check_SMAW       2- Welding Set starts but blows fuse │
│ Elect_sel        3- Welder welds but soon stops welding │
│ Edge_Prep        4- Variable or sluggish welding arc │
│ Econ_Eval        5- Welding arc is loud and spattery │
│ Defect_An        6- Polarity switch does not turn    │
│ Troub_Sh         7- Welder won't shut off            │
│ Quit             8- Arcing at ground clamp           │
│                                                     │
│              ═══════════CONSULTATION═══════════      │
│   Causes: 1- Current Setting is high                 │
│   Remedy: 1- Check setting and output with Ammeter   │
│   Causes: 2- Polarity may be wrong                   │
│   Remedy: 2- Change the polarity                     │
└─────────────────────────────────────────────────────┘
```

FIGURE 7: Sample Dialogue in the Troub_sh subsystem

```
╔══════════════════════ EX-SMAW ══════════════════════╗
║ ═══Choose═══                                         ║
║ Job_data     ┌─────────────────────────────────────┐║
║ Firm_data    │ CHECK_SMAW                          │║
║ Check_SMAW   │      Production Volume: 8           │║
║ Elect_Sel    │      Seam Length: 10                │║
║ Edge_Prep    │      Many ranges of weld sizes: Yes │║
║ Econ_Eval    │      Initial equipment cost should be│║
║ Def_Anal     │      less: Yes                      │║
║ Troub_Sh     │      Arc visibility required: Yes   │║
║ Quit         │      Is the operator skilled: Yes   │║
║              │      Chipping hammer/ deslagging     │║
║              │      equipment is available: Yes    │║
║              └─────────────────────────────────────┘║
║ ══════════════ EXPLANATION ══════════════           ║
║    SMAW is not applicable because                    ║
║       Seam Length > 10                               ║
║                                                      ║
║    Press any key to return to main menu......        ║
║              ┌──────────────────────────┐            ║
║              │ SMAW IS APPLICABLE FOR   │            ║
║              │ ALL CONDITIONS EXCEPT    │            ║
║              │ HIGHER SEAM LENGTH       │            ║
║              └──────────────────────────┘            ║
╚══════════════════════════════════════════════════════╝
```

Figure 8: Sample Dialogue in the Explanation subsystem.

Salient Features of the Expert System

The Expert System has the following salient features:

1. The system has been developed in a modular mode. The modular method provided higher efficiency and ease of development, updating and maintenance was also facilitated.

2. The system permits the user to define a core file consisting of electrodes suitable for the welding applications within the company. Initial electrode selection is done from the electrodes available in the core file. This approach maintains the principle of minimum inventory by avoiding the introduction of unnecessary electrodes into the system.

3. Given a set of jobs the system selects the minimum number of electrodes necessary for welding the given set of jobs, ie. If the core file is not defined by the user then the system develops it.

4. The system permits the user to override the decisions made by the system and carry on the consultation even if the recommendations made are not accepted.

5. Electrode selection for a particular welding operation is based on the suitability of the electrodes as determined by their manufacturers.

6. The system selects the electrode, parameters and procedures according to the criteria of minimum cost per piece and capable of trouble shooting (eg. It can advise on remedial action in case of excessive spatter or other welding defects or problems.

8. The system can be used for providing training to SMAW Operators as it has been designed in a user friendly manner.

CONCLUSIONS

Thus this paper provides a scientific procedure to overcome the problems encountered in training SMAW operators. Training on this Expert System may help the Welder in avoiding the non-optimal selection of parameters. The time and cost estimation also becomes less tedious as search for relevant standard data from tables is eliminated.

The time required for planning for a variety of jobs is also shortened as manual planning may require searching through huge standard data tables and also SMAW manuals. The customer can be informed about the estimated time and costs for the completion of the job within a short span of time.

This model can assist a planner to try to achieve optimality of the welding process for a welding job. This Model is one of the very few systems under development at the present time. The development of this Expert System also highlights the need for standardization of methods for implementing Expert Systems [6,7]. Expert System as outlined above can help the welding industry a lot if implemented properly.

REFERENCES

1. Stewart, J.P.: *The Welder's Handbook*, Reston Publishing Company Inc., 1981.
2. *The Procedure Handbook of Arc Welding*, The Lincoln Electric Company, 1973.
3. *Standard data for Arc Welding*, Welding Institute, U.K.
4. *Joint preparations for fusion welding of steels*, Welding Institute, U.K.
5. Goel, Vivek: *A knowledge based approach for the selection of parameters and economic evaluation of shielded metal arc welding*, M.E. Dissertation, 1991, University of Roorkee, Roorkee, India.
6. Andrew Taylor: An investigation of Knowledge based models for automated procedure generation in arc welding domain, *International Journal of Production Research*, Volume 27, No 11, 1855-1862.
7. Goel, Vivek, Singh, C.K.: "EX-SMAW: A knowledge based tool for shielded metal arc welding",*Proceedings of the international conference on systems and signals*, New Delhi, Dec 12-14, 1991.

APPENDIX I

TABLE 1: SYSTEM SPECIFICATIONS OF THE EXPERT SYSTEM FOR PLANNING TRAINING IN SMAW DOMAIN

SCOPE	Welding Process	Shielded Metal Arc Welding
	Steel Types	Mild Steel
	Plate Thickness	3 mm - 70 mm
	Joint Types	Butt, Fillet
	Electrode Diameter	Upto 15 mm
	Consumables	All types of MS electrodes.
MAIN OUTPUTS	Type of Electrode	All types
	Welding Current	40 A - 300 A
	Travel Speed	2.5 - 25 mm/s
	Edge Preparation	Single /double Square edge
	Complete Welding procedure Specifications	
	Trouble Shooting	Weld Defects & general problems.

Author Index

AUTHOR INDEX